Modeling and Application of Electromagnetic and Thermal Field in Electrical Engineering

Zhiguang Cheng · Norio Takahashi ·
Behzad Forghani
Editors

Modeling and Application of Electromagnetic and Thermal Field in Electrical Engineering

Science Press
Beijing

Springer

Editors
Zhiguang Cheng
Institute of Power Transmission
and Transformation Technology
Baobian Electric Co., Ltd.
Baoding, Hebei, China

Norio Takahashi (deceased)
Okayama, Japan

Behzad Forghani
Mentor Infolytica, a Siemens Business
Montreal, QC, Canada

ISBN 978-981-15-0172-2 ISBN 978-981-15-0173-9 (eBook)
https://doi.org/10.1007/978-981-15-0173-9

Jointly published with Science Press
The print edition is not for sale in China Mainland. Customers from China Mainland please order the print book from: Science Press.

This Springer imprint is published by the registered company Springer Nature Singapore Pte Ltd.
The registered company address is: 152 Beach Road, #21-01/04 Gateway East, Singapore 189721, Singapore

Foreword

The distribution and use of electrical energy is fundamental to the functioning of modern society. From the discovery of electromagnetic energy around 200 years ago to the present, devices based on converting energy between electromagnetic, mechanical and thermal forms have become so prevalent that they are hardly given a second thought and yet every one of those devices from large industrial machinery; through land, air and sea transportation to domestic devices ranging from washing machines to stoves, has to be designed, manufactured and tested. In addition, a generation and distribution system for electrical energy which is reliable, robust and efficient, has to be constructed. In 2017, about 26,000 TWh of electrical energy was generated, distributed and used globally. To minimize the losses in the transmission and distribution system and reduce the costs of the infrastructure, electrical energy is usually transmitted at a very high voltage, while it is generated and used at significantly lower levels. This implies the need for devices capable of changing voltages, i.e. transformers. It is interesting to consider that every Wh of electrical energy delivered through the distribution system and subsequently used has passed through at least two and probably nearer to ten, transformers.

From the very beginning of the electromagnetic era, the need for design tools has been paramount. Building physical prototypes is prohibitively expensive both in the cost of each prototype and in the time taken to realize a final device. Simple design tools based more on experience than theory evolved relatively quickly in the nineteenth century and the development of electromagnetic field theory provided the explanation of the physics underlying the operation of such devices. In effect, designers from the start have been using whatever tools and representations they can to create a virtual model of the device to determine the probable performance and explore the design space. With the advent of digital computers, the possibility of solving the field equations to simulate the actual performance of a device moved from a concept to reality. Over the past half century, both the computing hardware and the numerical methods necessary for solving the partial differential equations together with advanced representations of material properties, etc., have developed to a point where the simulations may now be considered accurate "digital twins" of the physical device allowing, in many cases, more detailed explorations of device performance than is possible on the physical system. These twins not only enable the total elapsed time from specification to full design to be decreased dramatically (along with a substantial reduction in costs) but also allow for manufacturing questions to be answered during the construction process and for performance monitoring during the operation of the physical device to identify developing faults before they become critical. This is a fundamental component of the concepts involved in moving to an "Industry 4.0" based world.

However, there are many requirements placed on the digital twin. First, it must represent the performance of the real device to the level of accuracy needed by the designer. This can vary through the design process and, typically, follows the well known "V-cycle", i.e. in the initial phases of a design, a system level representation of the device is needed—sometimes referred to as a Reduced Order Model—which incorporates as much of the multi-physics operation of the device as possible while allowing a fast exploration of the design space. This is sometimes referred to as "Front-Loading" the design process. As the design progresses, the simulation needs to become more detailed to answer questions such as the distribution of local losses in the device, the temperature rise in various components due to the losses, the forces on various components, etc. However, while it is tempting to just build extremely large models involving millions or tens of millions of degrees of freedom, the time taken to generate the performance of the twin and to explore the design space is critical. To be competitive, it is important that the overall design time is reduced as much as possible.

From the above discussion, the digital twin of an electromagnetic device should involve an appropriate numerical representation of the electromagnetic field. Since the behaviour of the field is controlled by the magnetic, electric, thermal and structural performance of the materials used to construct the device, it is crucial that any simulation system models the properties effectively. In addition, because all the areas of physics—magnetics, thermal, structural—are linked through the materials, a valid simulation must include a full multi-physics representation and, because the losses impact the thermal performance, the most important is an effective magnetic-thermal representation of the device. However, the behavior of the field also impacts the construction. For example, reducing losses in ferromagnetic components leads to a need to laminate the cores carrying the magnetic fluxes. These laminations are usually sub-millimeter in thickness while the dimensions of the entire device may often be measured in meters. The issues of scale can lead to huge numerical systems if all the details of the device are modeled accurately. This, in turn, can lead to extremely long simulation times. However, by representing some of the smaller components of the device with compact models, the problem sizes can be reduced significantly with no real loss in accuracy but with a massive gain in simulation speed allowing the digital twin to run on significantly smaller hardware systems.

This book provides an overview of the state-of-the-art for many of the issues described earlier. It has been created by authors who have significant experience in each of the areas critical to constructing and verifying the validity of a digital twin. They are recognized international authorities in each of their areas and several have been involved in organizations such as the International Compumag Society, the IEEE and standards organizations. They have made fundamental contributions to the representation and solution of electromagnetic field problems, the accurate modeling of materials, the measurement of material properties under the actual operating conditions experienced within a device, the construction of simulation systems, the development of verification and validation models for software and the development of optimization processes for an effective search of the design space. The book has been edited by three internationally recognized experts in the field: Dr. Zhiguang Cheng who has decades of experience in electromagnetic analysis, the validation of modeling and simulation tools, the measurement and prediction of material properties and, together with a large research and development team, has been involved in developing some of the world's largest transformers (with Baobian Electric, China); Prof. Norio Takahashi (from Okayama University, Japan), who received the Nikola Tesla award from the IEEE in 2013 for his work in modeling and design of electrical machines and was one of the leading developers of numerical formulations of electromagnetic field problems as well as having considerable expertise in material modeling; and Mr. Behzad Forghani who has been involved in the development of industrial software tools for electromagnetics design since the early 1980's (with Infolytica, Canada—now Mentor-Infolytica, A Siemens Business) and has been a member of the International Compumag Society Board for more than two decades. The resulting text represents man-centuries of experience in efficient modeling, numerical simulation and experimental verification of the complex engineering problems encountered in real electromagnetic devices and explains and identifies the issues that are crucial to anyone developing or using digital twin representations of such systems.

Although the contents have been created with the transformer designer in mind, much applies to almost any low frequency electromagnetic device. The initial chapters in the first part discuss the most often used approaches taken to developing a numerical representation of the magnetic field. The issues and advantages of each approach are discussed, and the reader is provided both with the theoretical development and with computational experiments which demonstrate the effectiveness of the approaches in terms of the problem sizes and typical solution times. While the representations deal with the basic field equations, constructing an effective system requires the introduction of knowledge and understanding to minimize the problem sizes without sacrificing accuracy. Thus, the next sections deal with issues which are of engineering importance, including optimization processes for exploring the design space. The theme here is very much "how can we develop an effective design tool?". This highlights the needs of the practicing designer of both speed of simulation and accuracy of solution. However, the importance of linking the thermal and magnetic field calculations is stressed and, in many devices, it is the thermal impact of the magnetic field that causes some of the most severe design issues and leads to many of the engineering problems which must be solved. The culmination of these discussions is demonstrated by the simulation system, "Simcenter™ MAGNET™". This is an example of a commercial tool that implements many of the concepts discussed previously. However, no tool can provide for every possibility that a designer wants so the ability to develop customizations, or shells, shows how the power of the digital twin can be leveraged for specific performance requirements. Finally, recognizing that the transient performance of devices is becoming ever more important, an approach for accelerating these computations is discussed.

Possibly the most interesting component of this book is the detailed review of material properties and their modeling. Material properties dominate in the solution of field problems and an understanding of the issues involved, from the behaviour under non-sinusoidal conditions, in the presence of dc bias and with rotational fluxes, to the impact of temperature on the behavior is critical in developing an effective and accurate simulation. An understanding of the properties and their variances can also help an engineer to understand what levels of accuracy it makes sense to request from the system. This section of the book draws on expertise in material property and measurement which is second to none in the world. The work of Prof. Norio Takahashi (Okayama, Japan) and Prof. Johannes Sievert (Braunshweig, Germany) is internationally recognized. In addition, the practical information on making measurements, the effect of core structures on properties and the design of experimental facilities, based on industrial experience, is extremely valuable in understanding what can realistically be done. This work is very timely—it deals with issues that are arising because of geomagnetically induced currents (a problem that all transformers must now be designed to survive), renewable energy systems, such as wind generation, that create huge time varying effects and high voltage dc (HVDC) transmission systems. If for no other reason, this book stands out in the way it discusses the issues with material performance in a real device.

Notwithstanding the above, no digital twin is acceptable unless its performance has been validated and verified. The authors of this book have been involved, for about two decades, with the development of a series of variations of an internationally accepted test model for software performance validation. The model, TEAM problem 21, includes many of the basic features found in a large power transformer and the experimental version of the problem has been built, and its performance modeled and measured, as co-research projects, jointly organized by Zhiguang Cheng, Norio Takahashi and Behzad Forghani.

While much of the information provided in the book is of general use to anyone working on the design of low frequency electromagnetic devices, the last part deals specifically with issues encountered in large modern power transformers expected to operate within the new grid architectures that are being proposed. The experience and knowledge embedded here is likely to be immensely valuable to anyone involved in transformer design to meet current operational requirements and international standards.

Overall, this work is an extremely comprehensive review of issues encountered in the design process for electromagnetic devices. It is a book which is targeted at both the research engineer and the practicing designer who want to understand the basis and capabilities of modern simulation systems for electromagnetics. The book contains knowledge and information from experts in the field developed over decades of both research work and practical experience. With over 450 references, the book contains one of the most comprehensive lists available of the key publications in the area of electromagnetic-thermal modeling and provides the reader with the opportunity to dig deeper into each of the areas covered.

David A. Lowther
Ph.D., A.K.C., C.Eng.(UK), P.Eng. (Ont), F.I.E.T., F.C.A.E., F.I.E.E.E.
Professor of Electrical Engineering
Department of Electrical and Computer Engineering
McGill University
Montreal, QC, Canada
e-mail: david.lowther@mcgill.ca

Preface

The co-research of the authors of this book, mainly involving 3-D electromagnetic and thermal modeling and simulation, measurement and prediction of material properties under standard and non-standard operating conditions, engineering-oriented benchmarking, based on well-established and collaborative research platforms, and transformer-related industrial applications, goes back to 30 years ago. This co-publication is based on its former version published in 2009, but is considerably extended, including the authors' major recent co-research works.

Motivation

The unprecedented high voltage and high capacity of today's electrical equipment, the economic pressures, as well as considerations, such as, the environmental protection, and high reliability within the life cycle, increasingly impose new and stringent requirements for the efficient and accurate analysis and design techniques, in particular with regards to the simulation of electromagnetic and thermal behavior, in large electromagnetic devices.

Modeling and prediction of the electromagnetic and thermal field behavior of large electrical equipment, especially in the UHV transmission and transformation engineering, lay the foundation for the in-depth study of topics, such as, vibration and noise, heating and cooling effects, under actual operating conditions. It involves material property modeling, large-scale multi-physics, multi-scale numerical analysis under complex conditions, and validation based on benchmark models, product-level models, and/or experiments with actual products.

This book aims to report the research works related to the above key projects, including many valuable measurement and simulation results, to motivate research teams to promote and participate in cooperation and exchanges in these fields, and to stimulate the exploration and discussion on future challenging topics.

Outline of This Book

This book focuses on the engineering electromagnetic and thermal field modeling and the related applications, taking large power transformers as its industrial background, and consists of the following five main parts:

(1) An overview of the electromagnetic and thermal field problems in electromagnetic devices, the new challenges posed by the UHV transformer engineering, the key research projects, and the foundation of the finite element method;
(2) The key technologies in solving the electromagnetic and thermal field problems, the effective solution of coupled electromagnetic and thermal fields, the API-based customized scripts, and the efficient harmonic balance finite element method;
(3) The foundation of the material property modeling, the improvement of classical magnetic measurement methods, the experimental study on magnetic anisotropy of grain-oriented silicon steel, the electromagnetic properties modeling based on product-level core models, the measurement and modeling of rotating magnetic properties, and the magnetic measurements of materials and components under complex conditions;
(4) The establishment and development of the engineering-oriented Benchmark Family (P21), the research on the additional core loss, caused by 3-D leakage magnetic flux, and the validation of engineering effectiveness of its analysis method and software;
(5) The engineering-oriented application research, including the modeling and simulation of product-based magnetic-thermal coupling, the transformer DC bias, and the heating and cooling behavior.

Co-authorship and Edition

This book is the result of long-term collaboration of an international R&D team composed of members from the Institute of Power Transmission and Transformation Technology, Baobian Electric, China; State Key Laboratory of Reliability and Intelligence of Electrical Equipment, Hebei University of Technology, China; Department of Electrical Engineering, North China Electric Power University, China; Department of Electrical and Electronic Engineering, Okayama University, Japan; Mentor-Infolytica, a Siemens Business, Montreal, Canada; and Physikalisch-Technische Bundesanstalt, IEC Technical Committee, Magnetic Alloys and Steels, Germany.

The Co-authorship

Herein, the co-authors of all 16 Chapters are: Chap. 1: Zhiguang Cheng; Chap. 2: Zhiguang Cheng and Norio Takahashi; Chap. 3: Norio Takahashi; Chap. 4:

Behzad Forghani; Chap. 5: Junjie Zhang; Chap. 6: Xiaojun Zhao; Chap. 7: Norio Takahashi; Chap. 8: Zhiguang Cheng, Lianbin Shi, and Johannes Sievert; Chap. 9: Tao Liu; Chap. 10: Yongjian Li; Chap. 11: Zhenbin Du, Meilin Lu, and Fulai Che; Chap. 12: Zhiguang Cheng, Norio Takahashi, Behzad Forghani, and Lanrong Liu; Chap. 13: Zhiguang Cheng, Chen Chang, and Dongjie Wang; Chap. 14: Lanrong Liu, Jie Li, and Fulai Che; Chap. 15: Mansheng Guo; Chap. 16: Weige Wu and Gang Liu.

Edition and Review

A number of illustrated figures and tables of the book were compiled and updated by Chen Chang and Lianbin Shi. Dongjie Wang organized the English edition. The English text for the illustrations in Chaps. 3 and 7 was edited by Meilin Lu.

Sajid Hussain and Dexin Xie were invited to review the manuscripts. All the authors contributed to the review process and the editorial work.

The final compilation, review, and edits of all the manuscripts of the book were performed by Zhiguang Cheng, Behzad Forghani, Yongjian Li, and Xiaojun Zhao.

The Authors' Expectations

This book is intended to be helpful to engineers, researchers, and postgraduate students majoring in electrical engineering, with an emphasis on efficient electromagnetic-thermal analysis methods, accurate material properties modeling, rigorous validation of the effectiveness of numerical modeling and simulation, and thorough considerations and discussions on the future research projects, e.g., the modeling and simulation of coupled electromagnetic-thermal-fluid fields, and coupled electromagnetic-vibration-noise fields. The authors are grateful for all the comments, suggestions, and discussions on this book, which will be very helpful for further co-research.

Baoding, China
Montreal, Canada
June 2019

Zhiguang Cheng
Behzad Forghani

Acknowledgements

The publication of this book was supported by all the related leaderships of both the Baobian Electric and Institute of Power Transmission and Transformation Technology, by the Hebei Key Laboratory of Electromagnetic & Structural Performance of Power Transmission and Transformation Equipment, and by the State Key Laboratory of Reliability and Intelligence of Electrical Equipment, Hebei University of Technology, China.

The related research works of this book were funded in part by the Natural Science Foundation of China under grants no. 59277296, no. 59924035, no. 51237005, and no. 50777042, by the joint project of the Chinese Academy of Sciences and National Natural Science Foundation of China under grant XK2018JSC001, by the National Key R&D Program of China under grant 2016YFB0300300, by the State Grid Science and Technology Projects under grant SGRI-WD-71-13-002, and by the Hebei Provincial Government Special Talents Training Funds.

In this book, a number of electromagnetic and thermal field simulations were performed, based on the jointly established industrial application center, using Simcenter™ MAGNET™ software. Simcenter, MAGNET and Simcenter™ FLOEFD™ are trademarks or registered trademarks of Siemens Product Lifecycle Management Software Inc., or its subsidiaries or affiliates, in the United States and in other countries. Restricted © Siemens 2019 is the Siemens copyright notice.

Here we are very grateful to all our respected predecessors, mentors, leaders, and experts in the field for their long-term countless support and help in so many ways.

At this moment, we would like to thank all our colleagues, postdoctoral fellows and postgraduate students for their contributions to the relevant experimental studies and numerical calculations in this book, over the years.

We are grateful to Ms. Hongmei Liu of Science Press (Beijing) and Ms. Jing Dou of Springer Beijing Office for their support and cooperation.

In particular, we wish to dedicate this book to Norio Takahashi, our co-editor and author. We are very sad that he deceased young, and will always remember his great cooperation and outstanding contribution to the field.

Finally, we are deeply grateful to all supporters and collaborators in our co-research throughout the decades.

Contents

Part III Measurement and Modeling of Magnetic Material and Component Properties

Part IV Validation Based on a Well-Established Benchmarking System

About the Editors

Zhiguang Cheng was born in Hebei, China, in 1942. He graduated from Peking University in 1967 and received a Ph.D. degree from Tsinghua University in 1995. He was a vice chief engineer of the R&D Center, and is currently a research advisor, Baobian Electric. He has received the National and Ministerial Science and Technology awards for his contributions to engineering science research and application. He is a senior member, IEEE, and a founding member of the International Compumag Society (ICS). His major interests are engineering electromagnetic field analysis, benchmarking, magnetic material modeling and industrial applications. Over the past decades, he has cooperated with his group to establish the joint industrial application center, the measurement and prediction system of magnetic material and component properties, and the international benchmark family (P21) approved by the ICS.

Norio Takahashi was born in Hyogo, Japan, in 1951. He received a B.E. degree from Okayama University in 1974 and M.E. and Ph.D. degrees from Kyoto University in 1976 and 1982, respectively. He was a professor of Department of Electrical and Electronic Engineering, Chair of Electromagnetic Device Laboratory of Okayama University, Vice President of Power and Energy Society, IEE of Japan, Vice President of International Compumag Society, and IEEE Fellow. His major interests were the development of numerical methods for calculating magnetic fields and optimal design methods for magnetic devices. Professor Norio Takahashi was deceased, and received the 2013 Nikola Tesla Award for his contributions to finite element modeling, analysis and optimal design tools of electrical machines, sponsored by the Grainger foundation and IEEE Power and Energy Society.

Behzad Forghani was born in Tehran, Iran, in 1957. He received a B.Eng. Degree, followed by a M.Eng. degree in 1981, both in Electrical Engineering, from McGill University in Montreal, Canada. Since 1981, he has been working at Infolytica Corporation, and then Mentor Infolytica, a Siemens Business, in the field of Computational Electromagnetics and is currently a Product Line Director. He is a senior member, IEEE, and a founding and Board member of the International Compumag Socienty. He is a member of OIQ (Order of Engineers in Quebec). He regularly serves on the Editorial Boards of Compumag and CEFC (two conferences with focus on the electromagnetic field computation). He is a co-author on several peer reviewed papers in the areas of solver formulation and material modeling, as applied to computational electromagnetics. In recent years, he has been involved in statistical analysis and the use of machine learning in cancer research, as an affilate member of the Departments of Diagnostic Radiology and Oncology at McGill University.

Part I
Engineering Electromagnetic and Thermal Field Problems and FEM Fundamentals

Chapter 1
General Survey of Engineering Electromagnetic and Thermal Field Problems

Zhiguang Cheng

Abstract A number of key problems in the modeling and application of engineering electromagnetic and thermal fields, involving the advanced material property modeling under complex working conditions, the efficient analysis method and simulation software, and the rigorous examination of the effectiveness and usefulness of large-scale modeling and simulation, are briefly outlined. Based on the industrial background in large power transformers, considering the rapid development of modern transmission and transformation technology and equipment, the major and very challenging research projects, mainly concerned with the modeling and prediction of transformer core loss and stray-field loss, and the multi-physics simulation requirements, for addressing the heating and cooling issues, are highlighted. This chapter provides a short overview of the evolution of modeling and simulation worldwide and stresses that today the simulation of the electromagnetic and thermal fields can be performed with considerable accuracy, even though there are still some important studies that need to continue. Finally, the overall composition of this book is introduced.

Keywords Engineering electromagnetic and thermal field · Numerical modeling and simulation · Digital twin · Material property · Measurement and prediction · Magnetic loss · Stray-field loss · Heating and cooling · Electrical equipment · Power transmission and transformation · Industrial application

---

Z. Cheng (✉)
Institute of Power Transmission and Transformation Technology, Baobian Electric, Baoding, China
e-mail: emlabzcheng@yahoo.com

Z. Cheng et al. (eds.), *Modeling and Application of Electromagnetic and Thermal Field in Electrical Engineering*, https://doi.org/10.1007/978-981-15-0173-9_1

1.1 Overview of Engineering Electromagnetic and Thermal Field Modeling

The analysis, design, manufacture, and operation of electromagnetic devices involve a large number of coupled multi-physics simulations. Engineering electromagnetic field analysis is the basis for studying the problems of loss, heating and cooling, electromagnetic force, vibration and noise, and so on. These performance parameters are closely related to reliable, safe, economical and environmentally friendly operation of electromagnetic devices throughout their life cycle.

It is difficult to analyze the large-scale engineering electromagnetic and thermal fields precisely for reasons including but not limited to the complex mathematical description and the numerical implementation of the coupled multi-physics problems; the modeling and simulation of very large electromagnetic devices, which become quite challenging, because of the existence of many types of multi-scale (or space multi-scale) problems brought about by factors such as the coexistence of large and complex geometric entities (e.g., more than 10 m level) and lamination structures (e.g., silicon steel sheet thickness usually not greater than 0.3 mm), and shallow field penetration depth (e.g., less than 1 mm); the lack of geometric or physical symmetry in the strict sense in the structure of large electromagnetic devices; the measurement and prediction of the characteristics of various materials and components in the solved region varying with external conditions such as excitation (e.g., multiple harmonics, DC bias), stress and temperature.

Looking back on the history of engineering electromagnetic and thermal field research, and its industrial application, in the past it was necessary to make significant simplifications, due to the limitations in the computing resources, when solving these problems. In the early stages of the industrial applications, simulations were performed using the 2D static field [1] solver, and later, the simulations gradually developed into 3D transient nonlinear field solutions. The problems more concerned about in electromagnetic design can be solved by decoupling the actual coupled fields or by implementing the so-called "weak coupling".

Much of this book is devoted to the problems related to the computation of low-frequency engineering electromagnetic and thermal fields. Based on large power transformers [2, 3], the engineering science problems and the key and generic technologies related to the modeling and application of engineering electromagnetic and thermal problems are deeply investigated herein and mainly include:

(1) The study of engineering-oriented material property modeling, involving the vector electromagnetic properties of magnetic materials [4, 5] and their standardization, and the working properties of materials accounting for the practical operating conditions; and then the establishment of a platform and database for material property measurement and prediction under "standard" and various "non-standard" conditions.

(2) Compelling requirements for research and development of effective analysis methods, with stable convergence, and software for large-scale electromagnetic and thermal field problems that supports deep saturation, strong nonlinearity, anisotropy and time asymmetry (including time multi-scale), the development of efficient and parallel solutions for super-large algebraic equations [6–9], and the improvement of the efficiency of large-scale product-level modeling and simulation [10].
(3) Verification of the effectiveness and usefulness of the electromagnetic and thermal analysis and corresponding computation software, i.e., these methods and software are supposed to provide sufficient, stable, and acceptable accuracy for the solutions to various complex problems and provide stable convergence in large-scale calculations, which needs to be validated with benchmark models, rigorously investigated and tested based on product-level models and actual products, and incorporated into industrial processes [11].
(4) Systematic study of the solutions to the problems related to engineering electromagnetic and thermal fields, in combination with the knowledge and experience of the experts in design, application, and manufacturing, is needed to establish a mature expert system by combining the material modeling, high-performance computation, and effective validation [10, 11].

As well known, it is not easy to make all problems very clear before product design, manufacturing, and even later product renewal. In other words, there will always be some problems to be further completely solved, whether or not the product is already made. Understandably, designers may rely more on the long-term accumulation of design, manufacture, test, and operation aspects when the problems they face are still at the research stage or without adequate technical support; that is, they may rely on "experience" to decide some new schemes and/or conduct costly "destructive" tests of products. In any case, we must always try to overcome those "bottlenecks," although it needs time and great perseverance.

1.2 New Challenges Posed by UHV Transformer Engineering

The rapidly growing demand of the global power industry and the pressure to save energy, protect the environment, improve the quality of power supply, among others, have challenged the traditional transformer design and manufacture. In the face of the emerging problems, it is essential to strengthen applied basic research and not stick to convention. In the field of modern transformer engineering, a number of new problems related to engineering science and technology need to be solved by promoting in-depth exchanges of ideas in the field, through the academia.

In this respect, J. Turowski took the lead in initiating the advanced research workshop on transformers and held international workshops (in Spain) in 2004 and 2007. The purpose was to provide a forum for scholars from industrial engineering and scientific research communities, all over the world, to exchange ideas on the development of modern transformers, the difficulties and the problems to be solved, and to further discuss issues of common interest, including transformer technology, high-voltage insulation materials, heating and cooling, coupled multi-physics analysis, fault diagnosis, stray loss control [12, 13], and energy saving and reliability of large transformers. J. Turowski's systematic research on the stray-field loss problems and the simplified and fast calculation of the stray losses over the years has attracted the attention of the researchers and designers. Based on years of research findings and accumulation of industrial application experience, he led the development of a 3D reluctance network method (RNM-3D), which has been verified by applying it to industrial problems, as a fast and effective expert approach. A series of improvements have been made through fundamental research and development to the electromagnetic computation program, which has been applied by transformer companies in several countries [14–19].

Since 2009, a series of International Colloquiums "Transformer Research and Asset Management" have been organized and held in Croatia by Z. Stih et al. So far, four (in 2009, 2012, 2014, and 2017, respectively) colloquiums have been held. Representatives from well-known transformer manufacturers, scientific research institutions, and universities have attended the conferences. Three aspects, materials and components, numerical modeling, and transformer life management, were the focus as main topics.

All the international workshops on transformers held in Europe have been actively supported by the CIGRE (International Council on Large Electric Systems) and have also attracted close attention from transformer manufacturers and experts in the field of transformer design, manufacturing technology, and advanced research. Z. Cheng of Baobian Electric (the author of this chapter) participated in the international workshops on transformer research held in Croatia in 2009 and 2012 [20, 21].

The continuous progress of power transformation technology and industry has been witnessed and driven by all of this. It has been a long course since the world's first closed core-type transformer came into being (in 1884, as shown in Fig. 1.1) [3], and so many milestones have been achieved in the voltage classes and single unit's capacity of the power transformers. With the tremendous progress of the global transformer manufacturing industry, the transformer manufacturing industry in China has shown sound development momentum. Having been tested through technological development and system operation over the time, all the key technologies of power transformers with voltage levels of 500 kV and above, including the design, manufacture, test and installation, operation, and maintenance, have matured and have accumulated much experience. Moreover, the domestic transformer design and manufacturing experience have been enriched through cooperation with overseas partners. A series of flagship products created by transformer manufacturers in China have broken one record after another.

The ODFPS-1000MVA/1000 kV UHV (ultra-high voltage) AC transformer with the world's highest voltage level and maximum capacity of a single unit was developed in Baobian Electric., China (in July 2008), and successfully passed all

Fig. 1.1 First closed core transformer (1884, Ganz)

specified tests. It was also developed for the world's first commercial UHV project, the 1000 kV Jindongnan-Nanyang-Jingmen UHV AC Test and Demonstration Project, as shown in Fig. 1.2a. The UHV transformer mainly includes an auto-transformer body, a voltage regulator, and a low-voltage compensator, with advanced technology and rational structure. The transformer has excellent electrical, mechanical and thermal performances, and its key indicators such as insulation tolerance, partial discharge, temperature rise and noise, all reach the level of excellence at the international level for transformer products, making it the flagship product of UHV transformers. Compared with 500 kV transmission lines, UHV lines have the advantages of high transmission power, low power loss, small floor area, and low comprehensive cost. With the development of UHV grid in China, problems such as insufficient transmission capacity of existing 500 kV grid can be solved, the transmission efficiency can be improved, and the investment cost and environmental pressure can be reduced.

The major transformer manufacturers in China have all kinds of excellent production and test facilities at the international level, as well as the design, manufacture, and test technologies for the AC transformers with a voltage of up to 1000 kV and a single capacity of 1000 MVA, the converter transformers and reactors with DC $\pm$ 800 kV (as shown in Fig. 1.2b) and above, and technologies for nuclear power transformers, large shell-type transformers, various special transformers, dry-type transformers, amorphous alloy transformers and distribution transformers, among others. This is recognized worldwide.

It should be noted that even if the milestone products such as 1000 kV/1000 MVA AC and $\pm$ 800 kV DC (even up to $\pm$ 1100 kV DC) transformers have been produced and put into operation, some deep-seated engineering science and key technical issues still remain and need to be further studied and face a series of new challenges from the UHV transformer engineering, including stricter requirements for energy saving and consumption reduction, economic, and

(a) 1000 MVA/1000 kV UHVAC transformer (2008)

(b) ±800 kV UHVDC Converter Transformer (2009)

Fig. 1.2 UHVAC/DC transformers (Baobian Electric., Baoding) environmentally friendly operation, and improvement in reliability, among other important projects. A forward-looking transformer manufacturer would see itself clearly, paying close attention to the progress of transformer research in the world, actively invest in manpower and R & D resources, and focus on core technology issues and work on a down-to-earth basis.

1.3 Some Key Research Projects

It is believed that under the new situation where the power system is highly developed, the key projects related to electromagnetic and thermal modeling of power transformers that need to be further studied mainly include, but are not limited to, the followings.

1.3.1 Accurate Analysis of Total Core Loss

The analysis of the electromagnetic and thermal fields in the transformer core is very complicated, in which the nonlinearity, anisotropy, 3D non-uniform distribution of alternating and rotating flux, over-excitation, harmonic and DC bias effects may be involved.

(1) The alternating flux in the transformer core limb and yoke, and the losses caused thereby, wherein problems related to the 3D large multi-scale and non-uniform electromagnetic and thermal fields are involved;
(2) The rotating flux at the T-joint of the iron core and the resulting loss density concentration caused thereby. It is reported that in large transformers, huge economic losses have been brought about by overheating caused by the excessive loss density at the joints of the core, which leads to the return of large products to the factory for processing. The complex flux distribution at the joint of the iron core, coupled with the 3D distribution of flux passing through the air gap in the joint, has brought difficulties to material modeling and accurate analysis of the electromagnetic field.
In addition, the determination of the exciting power at the joint of different core materials and joint types of lamination, and at the limb and yoke, as well as the separation of the total exciting power at different positions is also complex and worthy of attention.
(3) Additional core loss caused by the leakage magnetic flux of transformer windings, entering the core (particularly the component of leakage flux perpendicular to the core lamination). The iron core of the power transformer is formed by a stack of laminated sheets of different widths in a given inscribed

circle. The lamination with the smallest width (in multi-stage laminated core) is called the "outermost core lamination" in transformer engineering. The eddy current losses induced in the outermost core lamination, due to the perpendicular component of the leakage flux, cannot be ignored. Moreover, such loss is different from the "standard" iron loss measured by the standard method and can be called additional iron loss, which probably results in excessive concentration of loss density. Notice that this loss is also called "surprising loss" in reference [12]. As an effective response in transformer design, such core lamination is divided into "narrow strips" in the longitudinal direction to reduce eddy current loss. See Chap. 13 of this book for more detail.

A custom software tool for the design and calculation of the outermost core lamination (other layers closely adjacent thereto may also be considered) and the core tie-plate adjacent thereto needs to be developed to calculate the loss and the temperature rise of the components [22, 23]. The precise analysis of the rotational power loss caused by the rotating flux at the T-joint is a tough job and requires extensive experience in product design and testing. Designers are required to be experienced in dealing with rotational power losses for different capacities, core structures and joint types (e.g., multi-step lap joint). The core loss is often multiplied by a "loss factor" (or building factor) in design and calculation to give an overall consideration.

Experience in design and manufacture indicates that the advantage of high-quality silicon steel sheet is actually weakened when the ratio of the zone affected by rotating flux of core to the total volume of the total magnetic circuit is relatively large.

(4) In the construction process of UHV AC and DC power transmission project, multi-harmonic, DC bias, no-load and load over-excitation are found in large transformers. Therefore, manufacturers are required, by the power system, to promise that their products could withstand the specified level of over-excitation and DC bias. Therefore, the accurate measurement and prediction of electromagnetic properties of iron core under complex operating conditions (e.g., harmonic and DC bias excitation), including core loss, exciting power, magnetization, magnetostriction, etc., are required to further investigate the electromagnetic, thermal, mechanical and acoustic (vibration and noise) behaviors of the device under the above conditions, so as to formulate the corresponding standards for tolerance. Chapter 15 of this book shows the DC-biased transformer modeling in some detail.

On the other hand, it should be noted that significant progress has been made in silicon steel manufacturing technology and very low loss, very high magnetic induction, and low noise products. This also puts forward higher requirements for the design, modeling, and simulation of major equipment and the property modeling and selection of new materials.

1.3.2 *Efficient Solution of Transformer Winding Loss*

The total loss of transformer windings, as usually considered in electromagnetic design, includes several loss components: (1) the resistive loss, and it is the major part of the total winding loss and even the load loss; (2) the eddy current loss in conducting wires of the windings, caused by 3D leakage flux, and it is dependent on both the wire's size and the electromagnetic field distribution; (3) the circulating current loss generated in parallel wires of transformer winding, due to the non-uniform leakage flux linked with closed parallel wires of the windings, although the transposition technology for such parallel wire structure has been widely applied in transformer winding design and manufacture.

In order to model and predict the winding's losses, probably taking account of the effect of the heavy current leads, some specialized calculation method and design-oriented programs have been developed by designers and application engineers based on long-term design and application experience. Refer to Chap. 5 of this book for a script used for winding's performance analysis.

However, it is still challenging to accurately compute the total loss of transformer winding in the cases of very complex windings' structure and/or non-sinusoidal excitation conditions (e.g., including harmonic and/or DC bias). Moreover, in the related modeling and simulation of coupled electromagnetic and thermal fields, the effect of the non-uniform temperature on electromagnetic properties (e.g., conductivity or permeability) of all the related material and components should be taken into account.

1.3.3 *Modeling and Control of Stray-Field Loss in Structural Parts*

In addition to the losses in the windings, the stray-field losses in structural parts, mainly distributed in the transformer oil tank and various solid and laminated components (using magnetic or non-magnetic materials), including core tie-plates, clamps, and different kinds of shields, etc., are a problem that needs special attention and an in-depth study. The study of stray-field loss is of great significance to energy conservation and consumption reduction, avoidance of possible hazardous local overheating, and the improvement of operational reliability [11, 12].

To reduce the loss, first of all, the distribution of the loss, the key factors affecting the loss distribution and the effective measures to control the loss, should be ascertained. Some of the structural designs adopted in the project have gone through a lot of twists and turns from being conceived, tested, and improved to being optimized into a mature design and are not so easy to be quantitatively analyzed and evaluated, involving the large-scale complicated structures, the working properties of materials and components, the 3D multi-scale modeling and simulation, and the effective prediction of electromagnetic and thermal behavior.

In order to reduce the stray-field loss in structural components, for instance, the designers' efforts include as follows: 1. Non-magnetic steel is locally used in the oil tank of ordinary magnetic steel, i.e., to form a hybrid steel structure, to reduce loss, and to avoid hazardous local overheating; a natural question would be under what circumstances does non-magnetic steel need to be used and how can it be determined, and how to correctly evaluate the general effect of the hybrid structure composed of magnetic and non-magnetic steels; 2. the principle of material (magnetic steel and non-magnetic steel) selection and the structural design of the transformer core components (including core tie-plates and clamps) is determined; 3. the optimized design of magnetic and electromagnetic shields, the stray-field loss control, and the related application research on various effective measures for reducing the total loss and the local loss density must be carried out.

Note that the testing electromagnetic analysis methods (TEAM) benchmark problem 21 (3D stray-field loss model: Benchmark Family, available at www.compumag.org/team) has been well-established by the authors to study and validate the modeling and simulation available of the stray-field losses in transformer components. See the Chap. 12 of this book for more detail.

1.3.4 Numerical Prediction and Measurement of Electromagnetic and Thermal Fields

The heating and cooling in transformer engineering is a complex coupled multi-physics problem, involving 3D electromagnetic, thermal and oil flow fields, even considering the forced oil flow in large oil-immersed type transformers. The temperature distribution and local overheating, particularly the hot-point temperature rise, are still common concerns for manufacturers and the power system. It is considered a knotty issue in both experimental study and design calculation, as so many factors could affect the accuracy of the result, such as the deviation of 3D multi-field coupled modeling, the inaccuracy and incompleteness of the integrated performance parameters of materials, and the empirical formulae adopted, which often cannot meet the design requirements.

As mentioned above, the solution to the above problems is closely related to material property modeling. Establishing a property database for materials (and components consisting of such materials) operating under standard or possible working conditions to meet the needs of industrial applications is, therefore, a key topic.

In addition to these key projects that must be studied in-depth, the following practical problems or needs that may arise should be fully taken into account, such as: The post-processing of the general commercial software cannot meet the special but important needs of users; the software is not adapted to complete property data support provided by advanced material modeling; computational efficiency is not satisfactory, or application engineers cannot develop "customized scripts" according to certain necessary requirements based on the field calculation results of some commercial software (see Chap. 5 of this book for the development of script in some detail). The dilemma of high computational cost needs to be solved, and both the convergence and stability in large-scale calculation should be improved.

Note that Chap. 14 of this book demonstrates the electromagnetic and thermal modeling based on large power transformers, and furthermore, a typical heating and cooling model used for transformer engineering is well-established by the authors in Chap. 16 of this book, to facilitate the study of effective modeling and simulation for transformer heating and cooling.

1.4 Realization of Accurate Modeling and Simulation of Electromagnetic and Thermal Performance

The comprehensive performance analysis of the electromagnetic device has come a long way since the very beginning. Looking back, in the early days of computer application (more than 40 years ago), the author of this chapter and his colleagues did not have their own computers and rented computers from other units. Moreover, limited to the IO technology back then, the program input required paper tape; i.e., the program and the data had to be "punched" on the paper tape in advance, and the source program tape had to be manually modified. Figure 1.3 shows the punch machine used by the author in the 1970s and the manual tool used to modify the

(a) Puncher for making paper-tape program

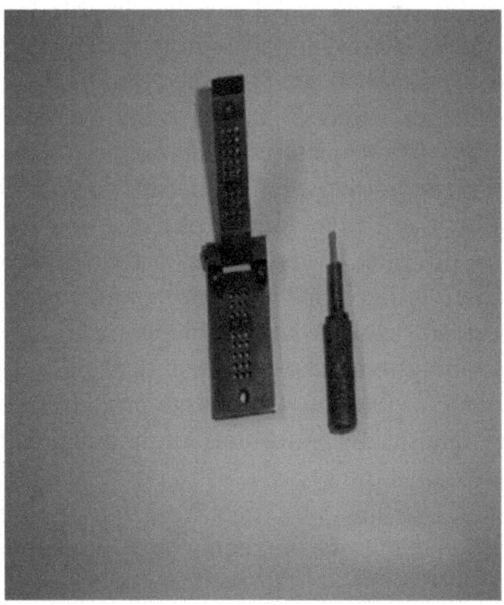

(b) Hand tool for modifying paper-tape program

Fig. 1.3 Old tools used for making and modifying paper tape program (the 1970s)

Fig. 1.4 3D finite element calculation model of typical transformer (by Simcenter MAGNET)

program (can be called as punch board). It must be hard to imagine the difficulty of manually modifying a program statement. Over time, the round holes in the steel punch board had been polished into oval shape. It was a witness of hard work of the application engineers in the "primitive days" of the computational electromagnetics (CEM). These pictures, like the first closed core transformer in 1884 to today's UHV transformers, hopefully, will inspire future generations to stay forward in the research and development process.

Fortunately, with the rapid development of power transmission and transformation systems and related manufacturing industries, it has become increasingly necessary to solve the problems using large-scale 3D coupled multi-physics simulation systems. As the theoretical basis for solving the thermal, mechanical and acoustic (vibration and noise) problems of electromagnetic devices, the research and industrial applications of international computational electromagnetics have made great progress [24]. Today, the 3D finite element model established in Fig. 1.4 (using Simcenter MAGNET) along with the corresponding electromagnetic and thermal modeling and simulation has become an ordinary application example.

As the two top international conferences held alternately, Compumag and IEEE CEFC have played an irreplaceable leading role in the sustainable development of computational electromagnetics. It is noted that the plenary speech (title:

Some Key Developments in Computational Electromagnetics and their Attribution) given by C. W. Trowbridge and J. K. Sykulski at the Compumag 2005, China, can be regarded as a historical overview on international computational electromagnetics. In the speech, the milestones in the field up to the turn of the century were reviewed, and the remarkable contributions of many pioneers were cited [25]; those outstanding achievements include:

- Delaunay meshing
- Kelvin transformation
- Automatic "cutting" of multiply connected regions
- Incomplete Cholesky conjugate gradient method (ICCG)
- 'Edge elements' and differential forms
- Dual energy methods
- Material modeling
- Forces
- Motion
- Fast multipole
- Transmission-line matrix (TLM) method
- Finite-difference time-domain method (FD-TD)
- Finite integration.

As well known, the analysis and design of electromagnetic devices are the two key links. The purpose of analysis is to model and predict the comprehensive performance of electromagnetic devices and key structures. Only the hot-spot temperature rise prediction [11, 17] in large transformers, which endangers the product safety and reliability service, is taken as an example here. It is still a problem to be solved for both numerical modeling and experimental study and involves the engineering effectiveness and feasibility of correctly handling the coupling of multiple electromagnetic fields, temperature fields, and fluid fields. It also involves reasonable simplification of practical engineering electromagnetic and thermal field problems, material property modeling under operating conditions, 3D finite element mesh generation, efficient solutions [26], improving the calculation efficiency and accuracy of large and complex engineering electromagnetic and thermal field problems [10], and scripting for special engineering needs.

In order to effectively reduce the computation cost of large and complex electromagnetic and thermal field problems, a series of efficient algorithms have been proposed and implemented, such as the homogenization models of the laminated core [6, 7, 26–28], the sub-region perturbation finite element method [29], the domain decomposition method [30], and the element by element parallel finite element method [31–34], which significantly improve the efficiency in solving large multi-scale complex problems.

By now, the accurate modeling and prediction of the comprehensive performance of large electromagnetic devices are no longer a dream, thanks to the rapid improvement of computer technology and high-performance computing capability, the construction of virtual numerical laboratory, the development of cloud computing, the multi-physical field coupling technology, and the maturing commercial software. As David Lowther pointed out that, in his plenary speech entitled "The Design of Electromagnetic Devices: From Simulation to Reality" at Compumag 2017, Korea, the simulation of the physics system can now be performed with considerable accuracy, the performance predicted often deviates from that measured on the actual physical device. This is due to the uncertainties involved in the input data to the system such as the material properties, the physical dimensions, the operating conditions.

The accurate modeling and prediction of the performance of electromagnetic devices, as "digital twins" of the physical devices, can be fully realized, so as to finally reach the design target. Certainly, it is needed to further study and solve the key problems that lead to the deviations encountered in modeling and simulation, in order to confidently construct and verify the validity of the digital twin.

1.5 Overall Composition of the Book

This book is primarily written to study the engineering electromagnetic and thermal modeling and related issues in the industrial application. Based on the large power transformers, the book brings together the major engineering scientific research achievements made through the long-term cooperation among the R&D teams of researchers from China, Japan, Canada, and Germany (Hebei Key Laboratory of Electromagnetic and Structural Performance of Power Transmission and Transformation Equipment, Baobian Electric; State Key Laboratory of Reliability and Intelligence of Electrical Equipment, Hebei University of Technology; North China Electric Power University, China; Okayama University, Japan; Mentor Infolytica, a Siemens Business, Canada; IEC Tech Committee (Magnetic Alloys and Steels), Germany), including the recent important contributions created by the joint team in material modeling [35–38], electromagnetic calculation methods [39–43], benchmarking-based validation [44–47], and industrial application [48–51].

It consists of five parts. The main contents of each part are as follows: (1) An overview of engineering electromagnetic and thermal field modeling, the new challenges arising from UHV transformer engineering with the rapid development of manufacturing industry, the modeling and simulation under complex operating conditions, and the foundation of finite element method; (2) a number of key problems in modeling and simulation of engineering electromagnetic and thermal fields, including the engineering-oriented coupled electromagnetic and thermal field solutions, customized API-based script development, and the harmonic balance finite element method (HBFEM) and its application; (3) the foundation of the magnetic materials' property modeling, the experimental study on the anisotropy of silicon steel sheet and the performance of square and ring iron cores of product level for engineering-oriented requirements, the measurement and modeling of rotational magnetic properties, and the magnetic measurement and prediction of materials and components under complex operating conditions; (4) the establishment and development of the Benchmark Family (P21), the determination of the additional core loss caused by 3D leakage flux in the GO silicon steel laminations, the validation of engineering effectiveness of analysis method; (5) industrial application, including electromagnetic and thermal modeling based on large power transformers, the engineering-oriented investigation of DC-biased transformer, and modeling and validation of thermal fluid field of transformer winding based on a well-designed heating and cooling model.

In this book, the close combination of advanced and efficient numerical analysis methods with reliable and accurate material's working property modeling and the rigorous validation of effectiveness of modeling and simulation is strongly emphasized in response to the growing demand of engineering science research and industry development.

In addition, the basic formulation and numerical implementation based on different potential sets, the well-established magnetic measurement and benchmarking system, the application research based on product-level model or large power transformer, and the valuable experimental and computational results are presented in detail and in a way that is easy to understand.

References

1. O. W. Andersen, "Transformer leakage flux program based on finite element method," *IEEE PAS*-92 (2), 1973, pp. 682-689.
2. Y. Xie et al, *Power Transformer Manual* (2nd edition, in Chinese). ISBN 978-7-111-46903-2. Machinery Industry Press, 2016.
3. K. Karsi, D. Kerenyi and L. Kiss, "*Large Power Transformers*" (book), Elsevier Science Publisher, 1987.
4. M. Enokizono and N. Soda, "Direct magnetic loss analysis by FEM considering vector magnetic properties," *IEEE Trans. Magn.* vol. 34, 1998, pp. 188–195.
5. K. Fujiwara, T. Adachi, and N. Takahashi, "A proposal of finite-element analysis considering two-dimensional magnetic properties," *IEEE Trans. Magn.* vol. 38, no. 2, 2002, pp. 889–892.
6. K. Preis, O. Bíró, and I. Ticar, "FEM analysis of eddy current losses in nonlinear laminated iron cores," *IEEE Trans. Magn.* vol. 41, no. 5, 2005, pp. 1412–1415.
7. H. Igarashi, K. Watanabe, and A. Kost, "A reduced model for finite element analysis of steel laminations," *IEEE Trans. Magn.* vol. 42, no. 4, 2006, pp. 739–742.
8. S. Yamada and K. Bessho, "Harmonic field calculation by the combination of finite element analysis and harmonic balance method," *IEEE Trans. Magn.* Vol. 24, no 6, pp. 2588–2590, Nov. 1988.
9. J. Lu, S. Yamada and H. B. Harrison, "Application of HB-FEM in the Design of Switching Power Supplies," *IEEE Trans. on Power Electronics*, Vol. 11, No 2, pp 347–355, March 1996.
10. D. Xie, Z. Zhu, D. Wu, and J. Wang, "Finding better solutions to reduce computational effort of large-scale engineering eddy current fields," *International Journal of Energy and Power Engineering*. Special Issue: Numerical Analysis, Material Modeling and Validation for Magnetic Losses in Electromagnetic Devices. Vol. 5, No. 1-1, 2016, pp. 12–20. https://doi.org/10.11648/j.ijepe.s.2016050101.12.

11. Bogdan Cranganu-Cretu,and Baden-Daettwil, "Coupled electromagnetic-thermal analysis for ABB power transformers," presented at International Colloquium Transformer Research and Asset Management Cavtat, Croatia, November 12–14, 2009.
12. J. Turowski: "Stray losses, screening, and local excessive heating hazard in large power transformers". Proceedings of ARWtr'04 and chapter in CD book "Transformers in practice", Vigo 2006. Publisher and Editor X. M. Lopez-Fernandez, Co-Editors: J. Turowski and M. Kazmierski, E. Lesniewska and B. Ertan.
13. J. Turowski, X. M. Lopez-Fernandez, A. Soto, and D. Souto, "Stray Losses Control in Core- and Shell Type Transformers," Presented at ARWtr-07, Baiona, Spain.
14. D. A. Koppikar, S. V. Kulkarni, and J. Turowski, "Fast 3-dimensional interactive computation of stray field and losses in asymmetric transformers," IEE Proc. of Generation, Transmission, Distribution. vol. 147, no 4, July 2000, pp. 197–201.
15. J. Turowski, M. Turowski, and M. Kopec, "Method of three-dimensional network solution of leakage field of three-phase transformers," *IEEE Trans. Magn*, Vol. 26, No 5, pp. 2911–2919, September 1990.
16. K. Komeza, G. Krusz, and J. Turowski, "Comparison of network and finite element approach to the solution of stray problems," Proc. ICEM, Part I, Lausanne, 1984, pp. 17–19.
17. M. Kazmierski, M. Kozlowski, J. Lasocinski, I. Pinkiewicz, and J. Turowski, "Hot spot identification and overheating hazard preventing when designing a large transformer," *CIGRE 1984 Plenary Session*. 29.08-6.09.1984. Report 12-12, pp. 1–6.
18. A. Demenko and J. Sykulski, "Network equivalents of nodal and edge elements in electromagnetics," *IEEE Trans. Magn.* vol. 38, no. 2, 2002, pp. 1305–1308.
19. A. Demenko, "Three-dimensional eddy current calculation using reluctance-conductance network formed by means of FE method," *IEEE Trans. Magn.* vol. 36, no. 4, 2000, pp. 741–745.
20. Z. Cheng, Q. Hu, N. Takahashi, and B. Foghani, "Stray-field loss modelling in transformers," presented at International Colloquium Transformer Research and Asset Management, Cavtat, Croatia, Nov. 12–14, 2009.
21. Z. Cheng, N. Takahashi, B. Forghani, X. Wang, L. Liu, Y. Fan, T. Liu, et al, "Engineering-oriented benchmarking and application-based magnetic material modeling in transformer research," (invited) presented at International Colloquium Transformer Research and Asset Management, Dubrovnik, Croatia, May 16–18, 2012.
22. Z. Cheng, S. Gao, J. Wang, H. He, Z. Liu, M. Wu, H. Li and Q. Hu, "Loss evaluation of non-magnetic tie-plates in transformers," COMPEL, vol. 17, no. 1/2/3, 1998, pp. 347–351.
23. D. A. Koppikar, S. V. Kulkarni, P. N. Srinivas, S. A. Khaparde, and R. Jain, "Evaluation of flitch plate losses in power transformers," *IEEE Trans. on Power Delivery*, vol. 14, No. 3, July 1999, pp. 996–1001.
24. D. Lowther, "Computational electromagnetics, research issues, challenges and commercial software," Lecture presented at Baobian Electric. Baoding, 2008-5-2.
25. C. W. Trowbridge and J. K. Sykulski, "Some Key Developments in Computational Electromagnetics and their Attribution," *IEEE Trans. Magn.* vol. 42, no. 4, pp. 503–507, 2006.
26. H. Kaimori, A. Kameari, and K. Fujiwara, "FEM computation of magnetic field and iron loss in laminated iron core using homogenization method," *IEEE Trans. Magn*, vol. 43, no. 4, 2007, pp. 1405–1408.
27. Z. Zhao, Z. Cheng, B. Forghani, F. Liu, Y. Li, and L. Liu, "Analytical study and corresponding experiments for iron loss inside laminated core under ac-dc hybrid excitation," *International Journal of Applied Electromagnetics and Mechanics*, 55(2017), 159–167.
28. D. Patrick, and G. Johan. "A 3-D magnetic vector potential formulation taking eddy currents in lamination stacks into account," *IEEE Trans. Magn.* Vol. 39, No. 3, pp. 1424–1427, 2003.
29. Z. Badics, Y. Matsumoto, K. Aoki, F. Nakayasu, M. Uesaka, and K. Miya. "An affective 3-D finite element scheme for computing electromagnetic field distortions due to defects in eddy-current nondestructive evaluation," *IEEE Trans. Magn.* Vol. 33, No. 2, pp. 1012–1020, 1997.

30. T. Lv, J. Shi, and Z. Lin. Domain Decomposition Algorithms—New Technology of Numerical Solution of Partial Differential Equation, Beijing: Science Press, 1997. (in Chinese).
31. T. Mifune, T. Iwashita, and M. Shimasaki. "A fast solver for FEM analysis using the parallelized algebraic multi-grid method," *IEEE Trans. Magn.* Vol. 38, No. 2, pp. 369–372, 2002.
32. S. Mcfee, Q. Wu, M. Dorica, et al. "Parallel and distributed processing for h-p adaptive finite-element analysis: a comparison of simulated and empirical studies," *IEEE Trans. Magn.* Vol. 20, No. 2, pp. 928–933, 2004.
33. T. J. R. Hughes, I. Levit, and J. Winget,"An element-by-element solution algorithm for problems of structural and solid mechanics," Computer Methods in Applied Mechanics and Engineering, Vol. 36, pp. 241–254, 1983.
34. Y. Liu, W. Zhou, and Q. Yang. "A distributed memory parallel element by element scheme based on Jacobi-conditioned conjugate gradient for 3D finite element analysis," Finite Elements in Analysis And Design, Vol. 43, pp. 494–503, 2007.
35. Z. Cheng, B. Forghani, X. Wang, L. Liu, T. Liu, Y. Fan, J. Zhang, X. Zhao, and Y. Liu, "Engineering-oriented investigation of magnetic property modeling and application," invited speech at the 1&2DM2016, *International Journal of Applied Electromagnetics and Mechanics*, 55(2017), 147–158.
36. Z. Cheng, N. Takahashi, B. Forghani, A. Moses, P. Anderson, Y. Fan, T. Liu, X. Wang, Z. Zhao, and L. Liu, "Modeling of magnetic properties of GO electrical steel based on Epstein combination and loss data weighted processing," *IEEE Trans. Magn*, vol. 50, no. 1, 6300209, 2014.
37. Q. Kong, X. Wang, Z. Cheng, Y. Fan, L. Liu, T. Liu, and J. Li, "Determination of the weighted mean path length of Epstein frame," COMPEL, 33, 1/2, pp. 224–233, 2014.
38. Z. Zhao, F. Liu, Z. Cheng, W. Yan, et al, "Measurement and calculation of iron loss inside silicon steel lamination under DC biasing, "*IEEE Trans. on Applied Superconductivity*,vol. 20, no. 3, pp. 1131–1134, 2010.
39. Z. Cheng, S. Gao, and L. Li, "Eddy Current Loss Analysis and Validation in Electrical Engineering," (supported by National Natural Science Foundation of China), ISBN 7-04-009888-1, Higher Education Press, 2001.
40. W. Zheng, and Z. Cheng, "Efficient finite element simulation for GO silicon steel laminationsusing inner-constrained laminar separation," *IEEE Trans. on Magetics*, vol. 48, no. 8, pp. 2277–2283, 2012.
41. X. Zhao, L. Li, Z. Cheng, Y. Zhong, and G. Liu, "Harmonic analysis of nonlinear magnetic field under sinusoidal and DC-biased magnetizations by the fixed-point method," *IEEE Trans. Magn.* 51(2015), 1–5, https://doi.org/10.1109/tmag.2014.2354234.
42. X. Zhao, L. Li, J. Lu, Z. Cheng, and T. Lu, "Characteristics analysis of the square laminated core under dc-biased magnetization by the fixed-point harmonic-balanced FEM," *IEEE Trans. Magn.* 48(2): 747– 750. 2012.
43. X. Zhao, J. Lu, L. Li, H. Li, Z. Cheng, and T. Lu, "Fixed-point harmonic-balanced method for dc-biasing hysteresis analysis using the neural network and consuming function," *IEEE Trans. Magn.* 48(11): 3356–3359, 2012.
44. Z. Cheng, N. Takahashi, B. Forghani, X. Wang, et al, "Extended progress in TEAM Problem 21 family," COMPEL, 33, 1/2, pp. 234–244, 2014.
45. Z. Cheng, N. Takahashi, B. Forghani, L. Liu, Y. Fan, T. Liu, J. Zhang, and X. Wang, "3-D finite element modeling and validation of power frequency multi-shielding effect," *IEEE Trans. Magn.* vol. 48, no. 2, pp. 243–246, 2012.
46. Z. Cheng, N. Takahashi, B. Forghani, Y. Du, Y. Fan, L. Liu, and H. Wang, "Effect of variation of B-H properties on both iron loss and flux in silicon steel lamination," *IEEE Trans. Magn.* vol. 47, no. 5, pp. 1346–1349, 2011.
47. Z. Cheng, N. Takahashi, B. Forghani, et al, "Effect of excitation patterns on both iron loss and flux in solid and laminated steel configurations," *IEEE Trans. Magn.* vol. 46, no. 8, pp. 3185–3188, 2010.

48. X. Zhao, F. Meng, Z. Cheng, L. Liu, J. Zhang, and C. Fan, "Stray-field loss and flux distribution inside magnetic steel plate under harmonic excitation," COMPEL, 36, 6, pp. 1715–1728, 2017.
49. Y. Du, Z. Cheng, et al, "Magnetic flux and iron loss modeling at laminated core joints in power transformers," *IEEE Trans. on Applied Superconductivity*, vol. 20, no. 3, pp. 1878–1882, 2010.
50. X. Zhao, L. Li, J. Lu, Z. Cheng and T. Lu, "Analysis of the saturated electromagnetic devices under DC bias condition by the decomposed harmonic balance finite element method," COMPEL, vol. 31, no. 2, 498–512, 2012.
51. X. Wang, Z. Cheng, L. Li, and J. Wang, "Calculation and validation of iron loss in laminated core of power and distribution transformers," COMPEL, 33, 1/2, pp. 137–146, 2014.

Chapter 2
Low-Frequency Electromagnetic Fields and Finite Element Method

Zhiguang Cheng and Norio Takahashi

Abstract Electromagnetic field analysis is the basis for solving the engineering coupled electromagnetic and thermal field problems. Based on the low-frequency Maxwell's equations, some key problems concerning the formulations and numerical implementations of typical 3-D eddy current analysis methods, using different potential sets, such as $\boldsymbol{A}$-V-$\boldsymbol{A}$ (or employing a reduced vector magnetic potential $\boldsymbol{A}_r$ to convert to $\boldsymbol{A}_r$-V-$\boldsymbol{A}_r$) and $\boldsymbol{T}$-Ψ-Ψ, are briefly explained. Furthermore, the numerical solvers based on different potential sets have been developed by the author's group and verified in the Testing Electromagnetic Analysis Methods (TEAM) benchmarking practices. In this chapter, the Galerkin weighted residual method, a key technique in numerical implementation, is elaborated, and the effectiveness of edge element, for example, in effectively reducing computational cost in industrial applications is discussed. Strengthening the theoretical basis of finite element analysis of electromagnetic fields and correctly understanding the significance of the combination of advanced numerical computation with accurate material property modeling will be more helpful in improving the effectiveness of modeling and simulation and further promoting the use of simulation in industrial applications.

Keywords Low-frequency electromagnetic field · Finite element method · Potential set · Nodal element · Edge element · Galerkin weighted residual technique · Formulation and implementation

Z. Cheng (✉)
Institute of Power Transmission and Transformation Technology,
Baobian Electric, Baoding, China
e-mail: emlabzcheng@yahoo.com

N. Takahashi (deceased)
Okayama University, Okayama, Japan

Z. Cheng et al. (eds.), *Modeling and Application of Electromagnetic and Thermal Field in Electrical Engineering*, https://doi.org/10.1007/978-981-15-0173-9_2

2.1 Introduction

The related problems involved in engineering electromagnetic and thermal fields are often highly complex. For example, in power transmission and transformation engineering, the critical heating and cooling issues of large power transformers are closely linked to the coupling of electromagnetic field, temperature field and the oil flow field of forced movement. In addition, the influence of temperature, and even stress, must also be considered in electromagnetic property modeling of materials and components, such as nonlinearity, anisotropy and hysteresis of various materials that are inherently complex. However, people do not leave themselves helpless, always trying to solve complex problems, step by step. Moreover, in order to solve the thermal field problem, the power loss, that is the heating source, must be accurately determined first, so electromagnetic field analysis is the basis for solving the problems related to coupled electromagnetic and thermal fields.

There are many classical literatures and monographs on the numerical computation of engineering electromagnetic fields [1–27]. The governing equations of the commonly used eddy current calculations, which can all be derived from Maxwell's equations, and appropriate boundary conditions, gauge and information of field source configuration are introduced to form a definite solution problem. The state variables used may be the field quantities to be solved, such as electric field intensity $\boldsymbol{E}$ and magnetic field intensity $\boldsymbol{H}$, or the vector potential and scalar potential, as well as potential sets thereof.

The finite element method (FEM) is a mature numerical method in industrial applications and is also the main numerical calculation method used in this book. Finite element, as a specialized term, first appeared in 1960 [3]. The usability of finite element was first demonstrated by O. C. Zienkiewicz in the late 1960s and was quickly extended to many application fields in the 1970s.

As far as electrical engineering is concerned, although A. M. Winslow adopted the complete concept of finite element as early as 1967 [6], the finite element method in its present form was applied and popularized by P. P. Silvester and his colleagues two years later [7]. At that time, the element type was mainly nodal element (or node-based element), the edge element (or edge-based element) was first proposed by Fraeijs de Veubeke [4], with its variable defined differently from nodal element, and the constraint of normal continuity of variables along the edge of the element was removed. Raviart and Thomas solved the two-dimensional problem by using the edge element method in 1977, and J. C. Nedelec established a 3-D mixed element model including edge elements and nodal elements by using tetrahedron and hexahedron elements in 1980 [8]. Since the early 1980s, the rapid development of edge element has attracted extensive attention from the international computational electromagnetics community, particularly in theory, numerical implementation and application [26–53]. Moreover, in some applications, the edge element shows its irreplaceable advantages. At the Compumag-1997, in Brazil, a specially arranged discussion meeting on edge element was held. A. Bossavit, Z. Cendes, T. V. Yioultsis, J. P. Webb and G. Mur shared their views in the discussions, which basically reflected the international computational electromagnetics community's understanding of all aspects of edge element. The theory and practice of edge element have been quite mature, promising a broad prospect of its academic and industrial application.

Edge element and nodal element are two main members of the finite element family and have achieved success in applications. As pointed out by the researchers, they do not replace each other and tend to develop in parallel. It is also shown in research and application that the combination of the two may be more promising, i.e., vector potentials are represented by edge elements, and scalar potentials are represented by nodal elements.

After a very brief review of the FEM as the foundation of electromagnetic and thermal field modeling, this chapter follows the low-frequency Maxwell's equations, introduces several eddy current analysis methods based on various potential sets [24, 25], deduces the basic formulation and numerical implementation and explores the internal relations among various algorithms. The edge element is also briefly discussed and compared with the nodal element, based on the discrete data for the same problem (e.g., total degrees of freedom, number of nonzero entries in the coefficient matrix, etc.). Moreover, based on a typical ***A***-*V*-***A*** method, the main derivation process is demonstrated.

Finally, it should be emphasized that the solid theoretical basis for the finite element analysis of electromagnetic fields, the efficient methods of computation, the accurate modeling of material properties and the tight link between them are most important in improving the effectiveness of modeling and simulation and further promoting the use of simulation in industrial applications.

2.2 Maxwell's Equations

The Maxwell's equations created by J. C. Maxwell (1831–1879) consist of the following four basic equations:

$$\nabla \times \boldsymbol{H} = \boldsymbol{J} + \frac{\partial \boldsymbol{D}}{\partial t} \tag{2.1}$$

$$\nabla \cdot \boldsymbol{B} = \boldsymbol{0} \tag{2.2}$$

$$\nabla \times \boldsymbol{E} = -\frac{\partial \boldsymbol{B}}{\partial t} \tag{2.3}$$

$$\nabla \cdot \boldsymbol{D} = \rho \text{ (when ignoring the effect of accumulated charge in the conductor } \rho = 0) \tag{2.4}$$

The fifth equation can be obtained from the above equations, that is, taking divergence to the first equation and using the result of the fourth equation, it is derived that

$$\nabla \cdot \boldsymbol{J} = -\frac{\partial \rho}{\partial t} \tag{2.5}$$

which satisfies the following constitutive equations:

$$\boldsymbol{B} = \mu \boldsymbol{H} \tag{2.6}$$

$$\boldsymbol{D} = \varepsilon \boldsymbol{E} \tag{2.7}$$

$$\boldsymbol{J} = \sigma \boldsymbol{E} \tag{2.8}$$

The low-frequency electromagnetic and thermal field are the main research project in this book. $\frac{\partial \boldsymbol{D}}{\partial t}$ in (2.1) can be omitted under the condition of low frequency, considering constitutive relation (2.8), and (2.1) can be rewritten as

$$\nabla \times \boldsymbol{H} = \sigma \boldsymbol{E} \tag{2.9}$$

In general, the phasor form of Eq. (2.1) is

$$\nabla \times \dot{H} = (\sigma + j\varepsilon\omega)\dot{E} \tag{2.10}$$

In the case of low frequency, the second right-hand term of (2.10) can be ignored.

The basic equations related to the complete low-frequency electromagnetic field include the field differential equation and the constitutive equations. Nowadays, with the development of international computational electromagnetics and its application, a series of electromagnetic field analysis methods, based on various "potential sets", have been highly developed and very mature. However, the electromagnetic properties of all the materials involved in the constitutive equations of the basic equations have become the focus of attention.

The material property parameters, σ, ε and μ, relate the basic field quantities $\boldsymbol{J}$, $\boldsymbol{D}$, $\boldsymbol{E}$, $\boldsymbol{B}$ and $\boldsymbol{H}$. Because of the urgent need of scientific research and industrial application, modern material modeling technology endows "σ-ε-μ" with more connotation, such as nonlinearity, time asymmetry, anisotropy and so on. The modeling of electrical, magnetic and thermal properties of electrical engineering materials under standard and non-standard conditions is the basic guarantee for performing effective electromagnetic and thermal analysis. Therefore, both, advanced methods of computation and advanced material models are required.

The material modeling under complex conditions, the implementation and validation of efficient algorithms are interdependent. It is very complicated to model the electromagnetic and other properties of materials under the actual conditions.

The real working conditions of component materials are usually different from the artificially imposed "standard" conditions for testing the properties of materials. In addition, the properties of materials often depend on external conditions such as temperature, frequency and stress, and the real excitation conditions are much more complicated than the imposed "standard" excitation conditions. The natural question is how to properly evaluate the engineering effectiveness of the standard property data and how to correctly handle the calculation results based on the standard property data. The author does not believe that the more complex the problem being studied, the better; however, the measurement and prediction of the "working characteristics" of materials and components under complex conditions is critical to ensuring the effectiveness of modeling and simulation. See Chaps. 7–11 of this book for further details on material modeling.

2.3 Governing Equations for Analysis of Low-Frequency Eddy Current Problems

In the 1980s, the methods of calculating 3-D eddy currents based on various potential sets were fully discussed. The $\boldsymbol{A}$-V-$\boldsymbol{A}$ and $\boldsymbol{T}$-ψ-ψ are two basic eddy current analysis methods, in which $\boldsymbol{A}$ and $\boldsymbol{T}$ are magnetic vector potential and current vector potential, and V and Ψ are electric scalar potential and magnetic scalar potential, respectively [24, 25].

$\boldsymbol{A}$-V-$\boldsymbol{A}$ can easily solve general complex problems, such as nonlinearity, multiply connected regions, multi-subdomains and non-uniform conductivity in conducting regions, which are often encountered in electrical engineering. In the implementation of Galerkin, the interface conditions are naturally met. However, it still has the following disadvantages: It uses vector potentials in the eddy current-free regions, requiring a large amount of memory and a long CPU time. Further research suggests that while $\boldsymbol{A}$-V-$\boldsymbol{A}$ based on nodal element uses magnetic vector potential $\boldsymbol{A}$ in the entire domain, when the Coulomb gauge condition is implemented by inserting penalty function term in the governing equation and there is a great difference in the magnetic permeability between the conducting region and the non-conducting region, the continuity of $\boldsymbol{A}$ and the discontinuity of permeability μ at the interface will force $\nabla \cdot \boldsymbol{A}$ at the interface to jump, which will lead to the decrease of accuracy at the interface. For this reason, O. Biro and others proposed to make the normal component A_n of $\boldsymbol{A}$ discontinuous at the interface to solve this difficulty [27].

$\boldsymbol{T}$-ψ-ψ is another effective method for eddy current analysis. However, despite the many advantages the traditional $\boldsymbol{T}$-ψ-ψ has, such as fewer unknowns and simple numerical implementation, it generally cannot solve the problems of multiply connected regions. There have been some remedies, for example, filling holes in the multiply connected domain with extremely low-conductivity materials, artificially transforming the multiply connected problems into simply connected ones.

Table 2.1 Dual relations of equations of $\boldsymbol{A}$-V-$\boldsymbol{A}$ and $\boldsymbol{T}$-ψ-ψ in conducting regions

Methods	$\boldsymbol{A}$-V-$\boldsymbol{A}$	$\boldsymbol{T}$-Ψ-Ψ
Definition of vector potential	$\boldsymbol{B} = \nabla \times \boldsymbol{A}$	$\boldsymbol{J} = \nabla \times \boldsymbol{T}$
Governing equations in conducting region	$\boldsymbol{H} = \frac{1}{\mu} \nabla \times \boldsymbol{A}$ $\nabla \times \boldsymbol{E} = -\frac{\partial}{\partial t}(\nabla \times \boldsymbol{A})$ $\nabla \times \left(\boldsymbol{E} + \frac{\partial \boldsymbol{A}}{\partial t}\right) = 0$ $\boldsymbol{E} + \frac{\partial \boldsymbol{A}}{\partial t} = -\nabla V'$ $\boldsymbol{E} = -\frac{\partial \boldsymbol{A}}{\partial t} - \nabla V'$	$\boldsymbol{E} = \frac{1}{\sigma} \nabla \times \boldsymbol{T}$ $\nabla \times \boldsymbol{H} = \nabla \times \boldsymbol{T}$ $\nabla \times (\boldsymbol{H} - \boldsymbol{T}) = 0$ $\boldsymbol{H} - \boldsymbol{T} = -\nabla \varphi$ $\boldsymbol{H} = \boldsymbol{T} - \nabla \psi$
	$\nabla \times \frac{1}{\mu} \nabla \times \boldsymbol{A} + \sigma\left(\frac{\partial \boldsymbol{A}}{\partial t} + \nabla V'\right) = 0$	$\nabla \times \frac{1}{\sigma} \nabla \times \boldsymbol{T} + \frac{\partial}{\partial t}(\mu(\boldsymbol{T} - \nabla \psi)) = 0$
	For nodal elements, Coulomb gauge is adopted and penalty function term is introduced $-\nabla \frac{1}{\mu} \nabla \cdot \boldsymbol{A}$ $\nabla \times \frac{1}{\mu} \nabla \times \boldsymbol{A} - \left(\nabla \frac{1}{\mu}(\nabla \cdot \boldsymbol{A})\right)$ $+\sigma\left(\frac{\partial \boldsymbol{A}}{\partial t} + \nabla V'\right) = 0$ it is derived that from $\nabla \cdot \boldsymbol{J} = 0$ $\nabla \cdot \left(-\sigma\left(\frac{\partial \boldsymbol{A}}{\partial t} + \nabla V'\right)\right) = 0$	For nodal elements, Coulomb gauge is adopted and penalty function term is introduced $-\nabla \frac{1}{\sigma} \nabla \cdot \boldsymbol{T}$ $\nabla \times \frac{1}{\sigma} \nabla \times \boldsymbol{T} - \left(\nabla \frac{1}{\sigma}(\nabla \cdot \boldsymbol{T})\right)$ $+\frac{\partial}{\partial t}(\mu(\boldsymbol{T} - \nabla \psi)) = 0$ it is derived that from $\nabla \cdot \boldsymbol{B} = 0$ $\nabla \cdot (\mu(\boldsymbol{T} - \nabla \psi)) = 0$

There are also handling means such as so-called cutting surface or auxiliary coils arranged around the holes (see Sect. 4.4 of Chap. 4 of this book) to solve the problem of "multiply connected" in the traditional $\boldsymbol{T}$-Ψ-Ψ. The systematic evaluation, comparison and discussion of various eddy current analysis methods and element types have been found in many references, which are of great significance for in-depth understanding of the advantages and disadvantages of various methods and correct selection of analysis methods.

The derivation of governing equations for $\boldsymbol{A}$-V-$\boldsymbol{A}$ and $\boldsymbol{T}$-Ψ-Ψ is of typical significance, and the derivation of governing equations for the corresponding conducting regions is shown in Table 2.1.

The dual relationship between $\boldsymbol{A}$-V-$\boldsymbol{A}$ and $\boldsymbol{T}$-ψ-ψ can be clearly seen from Table 2.1, and even the derivation steps are completely coherent. The derivation process of $\boldsymbol{A}$-V-$\boldsymbol{A}$ based on nodal element using Galerkin weighted residual technique is shown in Appendix.

It should be pointed out that in order to simplify the finite element pre-processing of complex excitation sources and improve the calculation accuracy, relevant reduced potentials are often used in the development of eddy current analysis software based on $\boldsymbol{T}$-Ψ-Ψ and $\boldsymbol{A}$-V-$\boldsymbol{A}$ potential sets. For instance, in $\boldsymbol{A}$-V-$\boldsymbol{A}$, a reduced magnetic vector potential, $\boldsymbol{A}_r$, is adopted, i.e., converting it to $\boldsymbol{A}_r$-V-$\boldsymbol{A}_r$; in the implementation of $\boldsymbol{T}$-Ψ-Ψ, a reduced magnetic scalar potential φ is adopted, i.e., $\boldsymbol{T}$-φ-φ [24, 25].

2.4 A_r-V-A_r-Based Method

$\boldsymbol{A}_r$-V-$\boldsymbol{A}_r$ [25] can simplify the 3-D grid meshing of complex excitation source structure and eliminate the error caused by inaccurate excitation conditions due to the difference between the grid and the source structure entity, reducing the amount of grid and the computing cost. The numerical solver based on $\boldsymbol{A}_r$-V-$\boldsymbol{A}_r$ potential set has been developed by the author and applied to the calculation of loss, magnetic flux density at designated position and interlinkage flux in the conducting component of the Testing Electromagnetic Analysis Methods (TEAM) Problem 21 benchmark models (the updated version of Problem 21 Family is posted at www.compumag.org/team), which has been validated through repeated comparisons of calculation and measurement results. Refer to the results of Chap. 12 of this book.

There is no substantial difference in the numerical implementation between $\boldsymbol{A}_r$-V-$\boldsymbol{A}_r$ and $\boldsymbol{A}$-V-$\boldsymbol{A}$. Where the total magnetic vector potential $\boldsymbol{A}$ is divided into two parts, i.e., the total magnetic vector potential $\boldsymbol{A}$ at any point in the field is synthesized by the contribution $\boldsymbol{A}_s$ of the excitation source and all contributions $\boldsymbol{A}_r$ other than the contribution of source

$$\boldsymbol{A} = \boldsymbol{A}_r + \boldsymbol{A}_s \tag{2.11}$$

where $\boldsymbol{A}_s$ is defined as being generated only by the excitation source in free space, which can be derived by Biot Savart Law, while $\boldsymbol{A}_r$ is a contribution other than the contribution of excitation source, which is an unknown variable to be solved and is called the reduced magnetic vector potential. $\boldsymbol{A}_s$ can be derived from (2.12)

$$\boldsymbol{A}_s = \frac{\mu_0}{4\pi}\int \frac{i\boldsymbol{dl}}{r} \tag{2.12}$$

Based on the magnetic field intensity $\boldsymbol{H}_s$ generated by the excitation source, it can be calculated by Eq. (2.13) that is

$$\boldsymbol{H}_s = \frac{1}{4\pi}\int \frac{i\boldsymbol{dl} \times \boldsymbol{r}}{r^3} \tag{2.13}$$

Problems needing attention in the numerical implementation and application of $\boldsymbol{A}_r$-V-$\boldsymbol{A}_r$ are as follows:

(1) When establishing the tangential continuity condition $\boldsymbol{H} \times \boldsymbol{n}$ of magnetic field intensity based on nodal elements, attention should be paid to the relationship

of $\boldsymbol{A} = \boldsymbol{A}_r + \boldsymbol{A}_s$, that is, the total magnetic vector potential is composed of two components;

(2) Similarly, based on the far field boundary (or magnetic symmetry plane) of nodal element, $\boldsymbol{B}_n = 0$ is expressed as

$$(\boldsymbol{A}_r + \boldsymbol{A}_s) \times \boldsymbol{n} = 0 \tag{2.14}$$

Far field boundary $\boldsymbol{H}_t = 0$ is expressed as

$$(\boldsymbol{A}_r + \boldsymbol{A}_s) \cdot \boldsymbol{n} = 0 \tag{2.15}$$

At infinitely far points

$$\boldsymbol{A}_r + \boldsymbol{A}_s = 0 \tag{2.16}$$

(3) The application practice shows that it is very important to calculate $\boldsymbol{A}_s$ and $\boldsymbol{H}_s$ correctly. In order to confirm the accuracy of the calculation, $\boldsymbol{A}_s$ and $\boldsymbol{H}_s$ can be calculated according to Eqs. (2.12) and (2.13), respectively, and $\boldsymbol{B}_s$ can be calculated based on $\boldsymbol{A}_s$ and $\boldsymbol{H}_s$, respectively, i.e.,

$$\boldsymbol{B}_{s1} = \mu_0 \boldsymbol{H}_s \tag{2.17}$$

$$\boldsymbol{B}_{s2} = \nabla \times \boldsymbol{A}_s \tag{2.18}$$

Then compare the results of $\boldsymbol{B}_s$ obtained according to Eqs. (2.17) and (2.18).

2.5 Scalar and Vector Galerkin Weight Function

Galerkin weighted residual method is one of the weighted residual techniques, and it is one of the key techniques in the numerical implementation of 3-D eddy current analysis method based on various potential sets (combination of vector potentials and scalar potentials).

After the Galerkin method was put forward in 1915, the weighted residual technique has gone through more than 40 years of development. The Galerkin method, integral method, subdomain method, least square method, moment method and other processing techniques were successively put forward and were collectively summarized as the weighted residual method by S. H. Crandall in 1956 [15].

As well known, the finite element method is an approximation of the actual continuous physical field. Taking the nodal element as an example, let the number of nodes be N_n, and the state variable φ can be expressed by the variable value φ_i and the shape function N_i (or experimental function) of the node as

$$\phi = \sum_{i=1}^{N_n} N_i \phi_i \tag{2.19}$$

As an approximate solution, it is impossible to accurately satisfy the governing equation, and the resulting error is called the residual $R(\varphi)$. It is necessary to find a weight function, for example, W_i, to force the inner product of the residual and the weight function to be zero, i.e.,

$$\langle R(\varphi), W_i \rangle = 0, \quad i = 1, 2, \ldots, N_n \tag{2.20}$$

This means that the residual is equal to 0 in the sense of weighted average. When the weight function W_i is consistent with the shape function N_i used, it is known as the Galerkin weighted residual method.

In the references on 3-D eddy current analysis, the weighted residual technique is widely used in numerical implementations. Galerkin method is used to deal with the vector eddy current equation, sometimes the scalar weight function W_i is used, and more often the vector weight function $\boldsymbol{N_i}$ is used. The formula derivation of nodal element based on scalar weight function and vector weight function shows that they present only differences in forms because the results obtained are completely consistent. The vector weight function makes the expression simple, and the scalar weight function has the advantages of clear hierarchy and easy deduction.

For the detailed deduction of Galerkin process based on edge element, it is suggested to refer to the related references [17]. Section 2.7 of this chapter only gives a comparison between the nodal element and the edge element with respect to Galerkin's residual.

2.6 Discussion on Edge Elements

A. Bossavit has made the following penetrating elaboration: In short, the finite element shall not be viewed in isolation; the finite element is expected to be used to approximately all the potentials and fields involved in Maxwell's equations, such as edge elements, nodal elements, facet elements and volume elements, which can be regarded as a consistent expression system, called the family of cell elements.

The edge element is closely related to the corresponding nodal elements. The difference between edge element and nodal element in discretization is a shape function. It can be said that if the analysis software based on edge element (nodal element) has been developed separately, it is not difficult to develop the software of new nodal element (edge element).

The application of edge element in electromagnetic field was pioneered by the French research group represented by A. Bossavit. Since the 1980s, edge element has been widely used in the field of electromagnetic fields. When it comes to the starting point of the development of edge elements, it is necessary to trace back the history of the development of $\boldsymbol{H}$-ψ. In $\boldsymbol{H}$-ψ, the magnetic field intensity $\boldsymbol{H}$ is directly used as the state variable in the eddy current region, instead of the traditional use of a certain potential, e.g., magnetic vector potential $\boldsymbol{A}$, and magnetic scalar potential ψ is used in the non-eddy current region. Since the magnetic field intensity $\boldsymbol{H}$ is used as the state variable, the required field quantity can be directly obtained, and it also has the advantages of easy handling of nonlinear problems (the permeability reflecting the nonlinearity of the material is not embedded in the differential operator). If the magnetic vector potential $\boldsymbol{A}$ is used as the state variable, the problem is that the coupling of $\boldsymbol{A}$-ψ generates a coefficient matrix that is either asymmetric or indefinite. However, if $\boldsymbol{H}$-ψ of nodal element is adopted, it will be difficult to deal with the tangential continuity of the coupling between eddy current region ($\boldsymbol{H}$) and non-eddy current region (ψ). It is in such a background that the edge element enters the field of computational electromagnetics.

The appearance of the edge element has attracted great attention in the field of computational electromagnetics and a great deal of research in theory, numerical implementation and gauge condition. Based on typical examples and very specific cases, a comprehensive and in-depth comparison is made between the edge element and the nodal element, and the advantages and disadvantages of edge element are discussed [28, 48]. It should be noted that Z. Ren and K. Shao et al. proposed hybrid FEM–BIM formulation and edge–nodal coupled model in 1990s, as well as related application [37, 51–53].

This chapter has made a systematic comparison of the two kinds of finite elements in another monograph [54] and pointed out that the role of nodal elements and edge elements should be correctly evaluated. Although the edge element has advantages over the nodal element in dealing with the problem of uniform or non-uniform regions (especially the latter), however, it is exaggerated to think that the edge element can surpass the nodal element in solving any difficulties encountered in electromagnetic field analysis, and that the accurate solution can be obtained only by using edge elements. It depends on what problems are being solved and what state variables are being used. We should consider the advantages of the edge element, as well as its disadvantages in some occasions or limitations in application. As G. Mur said, the point here is that when the right solution is derived, and it should be attributed to the applied finite element method, not to the edge element [28].

In edge element, when the magnetic vector potential $\boldsymbol{A}$ is a state variable, the normal continuity of magnetic flux density and current density ($\boldsymbol{B}$ and $\boldsymbol{J}$) is strictly guaranteed because the tangential component of $\boldsymbol{A}$ at the element interface is continuous, while the tangential continuity of magnetic field intensity and electric field intensity ($\boldsymbol{H}$ and $\boldsymbol{E}$) can only be weakly satisfied. When the magnetic field intensity $\boldsymbol{H}$ is a state variable, the continuity condition of the magnetic field intensity is strictly guaranteed because the tangential component of $\boldsymbol{H}$ at the element interface is continuous, while the normal continuity of the magnetic flux density ($\boldsymbol{B}$) also can only be weakly satisfied.

Besides, further research shows that if the continuity condition of the excitation current density cannot be strictly guaranteed, the convergence of the solution cannot be achieved through general ICCG methods when the edge element is used [30, 44].

It is hoped that these discussions will contribute to a better understanding of edge elements.

2.7 Comparison of Basic Equations and Galerkin Residuals of Nodal Elements and Edge Elements

When solving a problem related to low-frequency electromagnetic field, the basic equations in the conducting region may be different due to the different types of elements used. For example, the $\boldsymbol{A}$-V based on nodal elements uses magnetic vector potential $\boldsymbol{A}$ and electric scalar potential V in the conducting region, wherein each node contains four unknowns which requires four equations, and the basic equation consists of a vector equation and a scalar equation. However, based on the basic equation of the edge element, as shown in Table 2.2, when no electric scalar potential V is introduced, only the magnetic vector potential $\boldsymbol{A}$ is used. Therefore, there is only one vector equation and no scalar equation corresponding to the nodal element.

In addition, the gradient term of the electric scalar potential V in the expression G_e of Galerkin residual of the corresponding edge element will not appear, which is different from the expression G_n of Galerkin residual of the nodal element. Naturally, for the edge element, there is no expression G_{ns} of Galerkin residual corresponding to the nodal element, as shown in Table 2.2.

It should be pointed out that (1) For nodal elements, using scalar weight function N_i and vector weight function $\boldsymbol{N}_i$ will get the same results in Galerkin weighted residual processing [54]; (2) Although the weight functions $\boldsymbol{N}_i$ in the nodal element residual G_n (when vector weight function is used) and the edge element residual G_e are written using the same vector symbol, as shown in Table 2.2, their contents are completely different. Appendix shows the formulation of $\boldsymbol{A}$-V-$\boldsymbol{A}$ and Galerkin weighted residual processing.

2.8 Comparison of Nonzero Entries and Total Unknowns in Coefficient Matrix

The total unknowns and the number of nonzero entries in the coefficient matrix vary with the calculation method, the type of finite element (e.g., nodal element or edge element) and the ratio α of the number of elements in the conducting region to the number of elements in the entire solved domain, which directly affects the computer memory requirement and CPU time.

Table 2.2 Comparison of basic equations and Galerkin residuals of nodal elements and edge elements ($\boldsymbol{A}$-V-$\boldsymbol{A}$)

Element type	Basic equation (in conducting region)	Galerkin residual
Nodal element	$\nabla \times (\nu\nabla \times \boldsymbol{A}) = \boldsymbol{J}_0 - \sigma\left(\frac{\partial \boldsymbol{A}}{\partial t} + \nabla V\right)$	Using vector weight function $\boldsymbol{N}_i$: $G_n = \int_V \nabla \times \boldsymbol{N}_i \cdot (\nu\nabla \times \boldsymbol{A})\mathrm{d}v - \int_{vc} \boldsymbol{N}_i \cdot \boldsymbol{J}_0 \mathrm{d}v + \int_{ve}\left[\boldsymbol{N}_i \cdot \left(\frac{\partial \boldsymbol{A}}{\partial t} + \nabla V\right)\right]\mathrm{d}v - \int_s \boldsymbol{N}_i \cdot [(\nu\nabla \times \boldsymbol{A}) \times \boldsymbol{n}]\mathrm{d}s$ Using scalar weight function N_i: $G_n = -\int_v \nabla N_i \times (\nu\nabla \times \boldsymbol{A})\mathrm{d}v - \int_{vc} N_i \boldsymbol{J}_0 \mathrm{d}v + \int_{ve} N_i\left[\sigma\left(\frac{\partial \boldsymbol{A}}{\partial t} + \nabla V\right)\right]\mathrm{d}v + \int_s N_i \boldsymbol{n} \times (\nu\nabla \times \boldsymbol{A})\mathrm{d}s$
	$\nabla \cdot \left[-\sigma\left(\frac{\partial \boldsymbol{A}}{\partial t} + \nabla V\right)\right] = 0$	$G_{ns} = \int_{ve} \nabla N_i \cdot \left[\sigma\left(\frac{\partial \boldsymbol{A}}{\partial t} + \nabla V\right)\right]\mathrm{d}v + \int_{se} N_i\left[-\sigma\left(\frac{\partial \boldsymbol{A}}{\partial t} + \nabla V\right)\right] \cdot \boldsymbol{n}\mathrm{d}s$
Edge element	$\nabla \times (\nu\nabla \times \boldsymbol{A}) = \boldsymbol{J}_0 - \sigma\frac{\partial \boldsymbol{A}}{\partial t}$	$G_e = \int_V \nabla \times \boldsymbol{N}_i \cdot (\nu\nabla \times \boldsymbol{A})\mathrm{d}v - \int_{vc} \boldsymbol{N}_i \cdot \boldsymbol{J}_0 \mathrm{d}v + \int_{ve} \boldsymbol{N}_i \cdot \sigma\frac{\partial \boldsymbol{A}}{\partial t}\mathrm{d}v - \int_s \boldsymbol{N}_i \cdot [(\nu\nabla \times \boldsymbol{A}) \times \boldsymbol{n}]\mathrm{d}s$

This section proposes a method to estimate the total number of both the unknowns and the nonzero entries in the coefficient matrix for the same problem but using either edge elements or nodal elements and examines the effectiveness of edge elements by comparing the estimated results [49].

Only the ***A***-*V*-***A*** (also known as ***A***-*ϕ*) based on the brick element is taken as an example in the following analysis and comparison. It is assumed that the number of elements in the solved domain is so large that the reduction of unknowns due to the processing of boundary conditions can be ignored, and the use of gauge condition is not considered.

2.8.1 Unknowns and Number of Nonzero Entries in Matrix

2.8.1.1 Nodal Element

For a regular hexahedron (brick), the node number of an element is 8, and a node belongs to 8 elements, as shown in Fig. 2.1a. When the number of elements is very large, the average node number per element is 1 (=8/8). Therefore, the total number of elements *ne* and the total number of nodes *nt* have the following relationship:

$$nt = ne \tag{2.21}$$

In the nodal element-based ***A***-*V*-***A***, each node in the conducting region Rj has four unknowns, i.e., the electric scalar potential *V* and three components of magnetic vector potential ***A***, while in the air region Ro, each node has three unknowns, namely three components of ***A***. Therefore, the total unknowns *nu* can be expressed as

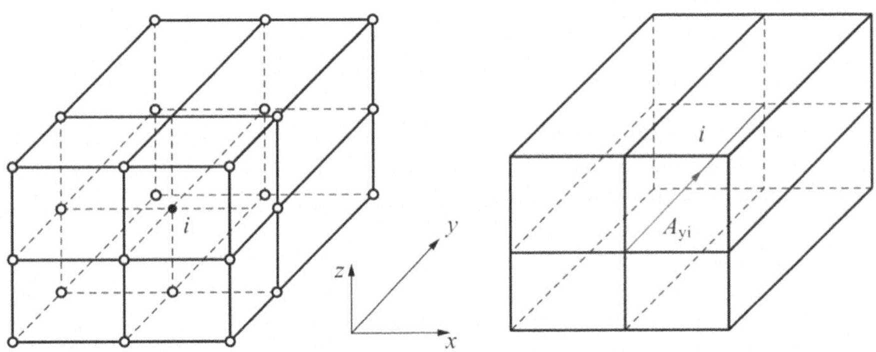

(a) Nodal element (b) Edge element

Fig. 2.1 Nodal element and edge element

$$\begin{aligned} nu(\text{nodal}, \boldsymbol{A}-V-\boldsymbol{A}) &= 4\alpha nt + 3\,(1-\alpha)nt \\ &= (\alpha+3)ne \end{aligned} \tag{2.22}$$

where α represents the percentage of the number of elements in the conducting region Rj to the total number of elements in the solution region (Rj + Ro). It is noted that the relation of $nt = ne$ under the condition that a large number of brick elements are introduced in Eq. (2.22). Since the node number associated with a node i is 27, as shown in Fig. 2.1a, the unknowns associated with node i in the conducting region are 27×4, while the unknowns associated with a node i in the air region are 27×3. In the case of nodal element, the number nz (nodal, $\boldsymbol{A}$-V-$\boldsymbol{A}$) of nonzero entries in the coefficient matrix can be expressed as

$$\begin{aligned} nz(\text{nodal}, \boldsymbol{A}-V-\boldsymbol{A}) &= 4\alpha nt \times 27 \times 4 + 3(1-\alpha)nt \times 27 \times 3 \\ &= 27(7\alpha+9)ne \end{aligned} \tag{2.23}$$

2.8.1.2 Edge Element

The number of edges in a brick element is 12, and the number of elements sharing one edge is 4, as shown in Fig. 2.1b. When the number of elements is very large, the average number of edges per element is 3 (=12/4). Therefore, the total number of elements ne and the total number of edges nh have the following relationship:

$$nh = 3ne \tag{2.24}$$

In $\boldsymbol{A}$-V-$\boldsymbol{A}$ using edge element, V can be set to zero [29]. Only one component of the magnetic vector potential $\boldsymbol{A}$ is defined on one edge. For example, only the y component Ayi is defined on the i-th edge, as shown in Fig. 2.1b. Therefore, the total unknowns nu (edge, $\boldsymbol{A}$-V-$\boldsymbol{A}$) can be expressed as

$$\begin{aligned} nu(\text{edge}, \boldsymbol{A}-V-\boldsymbol{A}) &= nh + \alpha nt = 3ne + \alpha nt \\ &= (\alpha+3)ne \end{aligned} \tag{2.25}$$

Here, the total number of edges associated with the i-th edge is 33, while the number of nodes associated with this edge is 18, as shown in Fig. 2.1b. Therefore, the total number of nonzero entries nz (edge, $\boldsymbol{A}$-V-$\boldsymbol{A}$) is

$$nz(\text{edge}, \boldsymbol{A}-V-\boldsymbol{A}) = 3ne \times 33 + \alpha nt \times 18 = (99+18\alpha)ne \tag{2.26}$$

where the meanings of ne, α and nt are the same as before.

According to the same approach, the number of unknowns in $\boldsymbol{T}$-Ψ-Ψ (also known as $\boldsymbol{T}$-Ω) and the number of nonzero entries in the matrix can be estimated.

2.8.2 Comparison of Nonzero Entries and Total Unknowns in Matrix

Equations (2.22) and (2.25) show that the total unknowns of the regular-hexahedron nodal element and the regular-hexahedron edge element are equal under the condition that the influence of the treatment of boundary conditions (possibly leading to a decrease in the unknowns) is not taken into account.

However, Eqs. (2.23) and (2.26) indicate that the numbers of nonzero entries in the element matrices are far from being equal.

The estimation method proposed in this section is verified by a 3-D eddy current verification model [50] proposed by IEEJ (IEE of Japan). IEEJ eddy current model is a section of square core surrounded and excited by a coil with a race track-shaped cross section. The core and the exciting coil are equal in height, and a square aluminum plate is symmetrically placed above and below the core and the exciting coil; the aluminum plate can be either with hole or without hole. See Fig. 2.2 for the structure and size, material property parameters and excitation ampere turns of the model of the aluminum plate with hole.

The model is calculated by ***A***-*V*-***A*** and ***T***-Ψ-Ψ and two types of elements (nodal element and edge element), respectively. See Table 2.3 for numerical computation data and CPU time.

The comparison of computation data of the same problem, two types of elements and two analysis methods shown in Table 2.3 can be summarized as follows:

(1) The total unknowns of ***A***-*V*-***A*** and ***T***-Ψ-Ψ are nearly equal, the number of unknowns of the nodal elements is slightly larger than that of the edge elements, and the increment is about 2–5%.

(2) The difference in the number of nonzero entries in the coefficient matrix is larger. The number of nonzero entries in the matrix of nodal elements in ***A***-*V*-***A*** is about 2.7 times that of edge elements, while the number of nonzero entries in the matrix of nodal elements in ***T***-Ψ-Ψ is about 1.5 times that of edge elements.

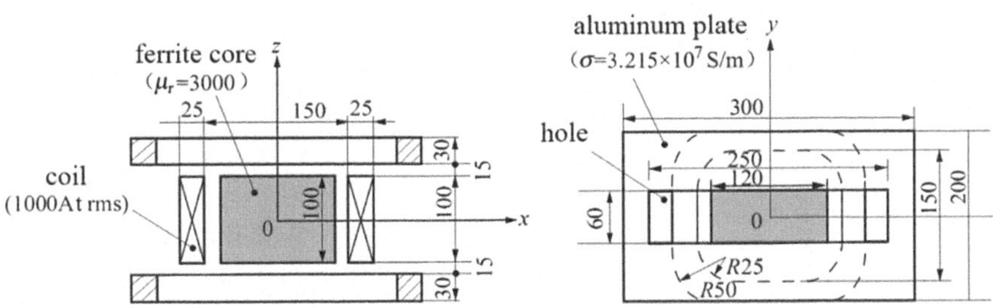

Fig. 2.2 Verification model (IEEJ)

Table 2.3 Numerical computation data and CPU time

Discrete and calculating data	Without hole				With hole			
	A*-*V*-*A		***T*-*Ψ*-*Ψ***		***A*-*V*-*A***		***T*-*Ψ*-*Ψ***	
	Nodal element	Edge element	Nodal element	Edge element	Nodal element	Edge element	Nodal element	Edge element
Total number of elements	14,400							
Total number of nodes	16,275							
Total number of unknowns	43,417	41,060	22,844	22,412	42,885	41,060	22,844	22,412
Number of matrix nonzero entries	1,781,644	653,718	632,859	423,056	1,734,684	653,718	632,859	423,056
Memory Requirements (MB)	72.2	28.4	30.7	19.4	70.5	28.4	30.7	19.4
Number of ICCG iterations	1306	513	172	192	1264	582	1141	327
CPU time (s)	6242	947	533	290	5870	1069	2001	442

Notes
(1) Computer used: NEC supercomputer SX-1E (285MFLOPS)
(2) ICCG convergence criterion: 10^{-7}
(3) The ratio of the number of elements in the conducting region to the total number of elements in solved region α
$\alpha = 0.14$ (with hole); $\alpha = 0.18$ (without hole)

(3) The difference in the memory requirement of the computer is larger. The memory requirement of the nodal element in ***A***-*V*-***A*** is about 2.5 times that of the edge elements, while the memory requirement ratio in ***T***-Ψ-Ψ is more than 1.5 times.
(4) The CPU time of the nodal element in ***A***-*V*-***A*** is 5.5–6.5 times that of the edge element, while the ratio of the CPU times in ***T***-Ψ-Ψ is 1.8–4.5 times or above.

Memory requirements, depending on the total unknowns of the problem to be solved and the number of nonzero entries in the finite element coefficient matrix, may become a "bottleneck" when using simulation in industrial applications. The above discussion is based on brick element. It is easy to understand that the number of nonzero entries in the coefficient matrix is related to the number of elements associated with nodes and edges. One of the hexahedron nodal elements is associated with eight brick elements, while one edge of the hexahedron edge element is associated with only four brick elements. Therefore, the "density" of nonzero entries in the coefficient matrix of edge elements is smaller than that of nodal elements.

2.9 Concluding Remarks

In this chapter, based on the fundamental equations of low-frequency electromagnetic fields, the dual relation between ***A***-*V*-***A*** and ***T***-ψ-ψ, as the basic eddy current analysis methods, is discussed, and the numerical implementation process of the Galerkin weighted residual method is deduced. The two most important members of the finite element family, nodal element and edge element are compared in multi-aspects, revealing their differences and internal relations. All the deductions of the formulation and the numerical implementation based on typical potential sets, in an easy to understand way, and the related discussions on the development and progress of the finite element method are of help for finite element investigation and application.

Taking the brick element used in large-scale finite element analysis as an example, when the edge element and the nodal element have the same mesh, the number of unknowns is similar between the two; however, the number of nonzero entries in the coefficient matrix of the edge element is significantly smaller than that of the nodal element. Therefore, the CPU time required in general is less, indicating that the edge element can effectively reduce the computing cost.

Furthermore, the combination of advanced numerical computation and accurate material property modeling is emphasized in order to improve the effectiveness of modeling and simulation and to promote the large-scale industrial application.

Acknowledgements This work was supported in part by the Natural Science Foundation of China (no. 59277296 and no. 59924035). In particular, the author appreciates the support of the leaders concerned and thanks all the colleagues for joint development of 3-D eddy current field solvers for years.

Appendix: Formulation of *A*-*V*-*A* and Galerkin Weighted Residual Processing

The Galerkin's weighted residual processing is demonstrated based on the potential set of ***A***-*V*-***A*** using nodal element. Although the model shown in Fig. 2.3 looks simple, but it does not lose generality. For the more general case, there is no substantial difference or difficulty in the derivation process here. This is helpful for knowing basic derivation process.

Basic Model of *A*-*V*-*A*

In the ***A***-*V*-***A*** model, ***A*** (magnetic vector potential) and *V* (electric scalar potential) are used in the conductor—eddy current region, and only the magnetic vector potential ***A*** is used outside the conducting region.

Where σ and μ are the conductivity and permeability of the conducting material, respectively, and the magnetic anisotropy and nonlinearity needed to be considered. In the non-conducting region, the conductivity σ is zero, and the relative permeability μ_r is equal to 1.

The symbols in Fig. 2.3 are given the following meanings:

Fig. 2.3 *A*-*V*-*A* model

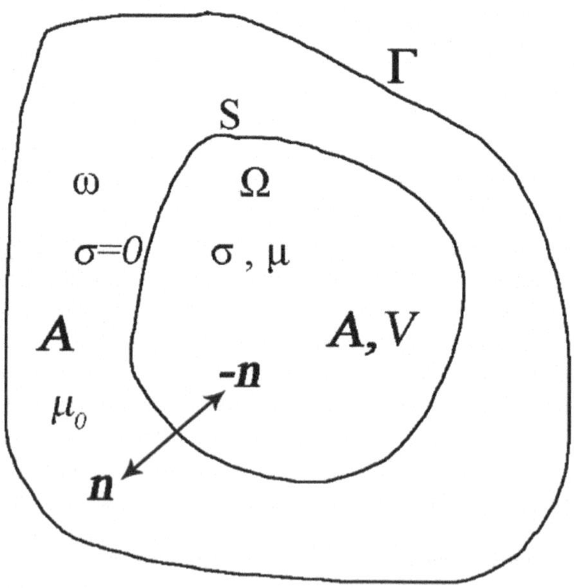

ω non-eddy current region, Ω: eddy current region
S interface between eddy current region and non-eddy current region
$\boldsymbol{n}$ exterior normal unit vector of conductor surface in eddy current region
$\boldsymbol{n}'$ internal normal unit vector of conductor surface in eddy current region, where

$$\boldsymbol{n}' = -\boldsymbol{n} \tag{2.27}$$

Γ: The outer boundary of the whole solution domain, including generally the following two types:

$$\Gamma_h: \boldsymbol{H} \times \boldsymbol{n} = 0 \tag{2.28}$$

$$\Gamma_b: \boldsymbol{B} \cdot \boldsymbol{n} = 0 \tag{2.29}$$

$\boldsymbol{A}$ magnetic vector potential (Wb/m)
V integral quantity of electric scalar potential V' to time (Vs)

$$V = \int V' \mathrm{d}t \tag{2.30}$$

μ permeability of conductor (H/m)
σ Conductor conductivity (S/m), where the material is set to electric linear and σ is constant

Considering that the permeability of the material in the eddy current region of conductor is treated as anisotropic in any direction, its general tensor form is

$$\bar{\mu} = \begin{bmatrix} \mu_{xx} & \mu_{xy} & \mu_{xz} \\ \mu_{yx} & \mu_{yy} & \mu_{yz} \\ \mu_{zx} & \mu_{zy} & \mu_{zz} \end{bmatrix} \tag{2.31}$$

The reluctivity of the material should, therefore, be expressed as follows:

$$\bar{\nu} = [\bar{\mu}]^{-1}$$

It should be pointed out that only for the sake of simplicity of deduction, the anisotropic permeability in the following formula is simply written as μ, and the anisotropic reluctivity is simply written as $\frac{1}{\mu}$.

Definition: $\boldsymbol{B} = \nabla \times \boldsymbol{A}$.

Governing Equation of *A*-*V*-*A*

The governing equations of ***A***-*V*-***A*** are as follows:

Eddy Current Region

$$\nabla \times \frac{1}{\mu} \nabla \times \boldsymbol{A} - [\nabla \frac{1}{\mu_c} \nabla \cdot \boldsymbol{A}] + \sigma \left(\frac{\partial \boldsymbol{A}}{\partial t} + \nabla \frac{\partial V}{\partial t} \right) = 0 \qquad (2.32)$$

In square brackets of Eq. (2.32) is a penalty function term to enforce the zero divergence condition, the same below, where μ_c needs to be isotropic to ensure the symmetry of the coefficient matrix. μ_c can be determined by the permeability of neighboring elements and updated during iteration.

$$\nabla \cdot \sigma \left(-\frac{\partial \boldsymbol{A}}{\partial t} - \nabla \frac{\partial V}{\partial t} \right) = 0 \qquad (2.33)$$

Non-Eddy Current Region

$$\nabla \times \frac{1}{\mu_0} \nabla \times \boldsymbol{A} - \nabla \frac{1}{\mu_0} \nabla \cdot \boldsymbol{A} = 0 \qquad (2.34)$$

where μ_0: permeability in air.

Galerkin Weighted Residual Processing

Galerkin Residuals

The Galerkin weighted residual technique is applied to (2.32)–(2.34), and the continuity condition of the boundary field quantity (***B***, ***H***) is considered.

$$(\boldsymbol{B}_1 - \boldsymbol{B}_2) \cdot \boldsymbol{n} = 0 \qquad (2.35)$$

$$(\boldsymbol{H}_1 - \boldsymbol{H}_2) \times \boldsymbol{n} = 0 \qquad (2.36)$$

The discretization equation is derived. The subscripts 1 and 2 indicate both sides of the interface.

The Galerkin residuals corresponding to (2.32) and (2.33), using scalar weight function, are, respectively, as follows:

$$\boldsymbol{R}_1 = \int\limits_{\Omega} N_i \left(\nabla \times \frac{1}{\mu} \nabla \times \boldsymbol{A} - \nabla \frac{1}{\mu_c} \nabla \cdot \boldsymbol{A} + \sigma \left(\frac{\partial \boldsymbol{A}}{\partial t} + \nabla \frac{\partial V}{\partial t} \right) \right) \mathrm{d}v \tag{2.37}$$

$$R_0 = \int\limits_{\Omega} N_i \nabla \cdot \sigma \left(-\frac{\partial \boldsymbol{A}}{\partial t} - \nabla \frac{\partial V}{\partial t} \right) \mathrm{d}v \tag{2.38}$$

where N_i is a scalar weight function. For ease of derivation, $\boldsymbol{R}_1$ is rewritten separately as three terms

$$R_{11} = \int\limits_{\Omega} N_i \nabla \times \frac{1}{\mu} \nabla \times \boldsymbol{A} \, \mathrm{d}v \tag{2.39}$$

$$R_{12} = -\int\limits_{\Omega} N_i \nabla \frac{1}{\mu_c} \nabla \cdot \boldsymbol{A} \, \mathrm{d}v \tag{2.40}$$

$$R_{13} = \int\limits_{\Omega} N_i \sigma \left(\frac{\partial \boldsymbol{A}}{\partial t} + \nabla \frac{\partial V}{\partial t} \right) \mathrm{d}v \tag{2.41}$$

Residuals Processing

Eddy Current Region

For (2.39), the vector formulation is

$$\nabla \times \left(N_i \frac{1}{\mu} \nabla \times \boldsymbol{A} \right) = \nabla N_i \times \frac{1}{\mu} \nabla \times \boldsymbol{A} + N_i \nabla \times \frac{1}{\mu} \nabla \times \boldsymbol{A} \tag{2.42}$$

(2.42) is sorted out to:

$$N_i \nabla \times \frac{1}{\mu} \nabla \times \boldsymbol{A} = \nabla \times \left(N_i \frac{1}{\mu} \nabla \times \boldsymbol{A} \right) - \nabla N_i \times \frac{1}{\mu} \nabla \times \boldsymbol{A} \tag{2.43}$$

Substituting (2.43) into (2.39)

$$R_{11} = \int\limits_{\Omega} \nabla \times \left(N_i \frac{1}{\mu} \nabla \times \boldsymbol{A} \right) \mathrm{d}v - \int\limits_{\Omega} \nabla N_i \times \frac{1}{\mu} \nabla \times \boldsymbol{A} \, \mathrm{d}v \tag{2.44}$$

The vector formulation is

$$\int_{\Omega} \nabla \times \boldsymbol{a}\, \mathrm{d}v = \oint_{s} \boldsymbol{n} \times \boldsymbol{a}\, \mathrm{d}s$$

(2.44) can be rewritten as follows:

$$R_{11} = -\oint_{s} N_i \boldsymbol{n} \times \frac{1}{\mu} \nabla \times \boldsymbol{A}\, \mathrm{d}s - \int_{\Omega} \nabla N_i \times \frac{1}{\mu} \nabla \times \boldsymbol{A}\, \mathrm{d}v \tag{2.45}$$

Apparently, $\boldsymbol{n} \times \frac{1}{\mu} \nabla \times \boldsymbol{A} = \boldsymbol{n} \times \boldsymbol{H}$ is the tangential component of the magnetic field intensity at the interface. According to the continuity condition of the tangential component of the magnetic field intensity at the interface, the surface integral in (2.45) will be canceled from each other with the corresponding surface integral in the non-eddy current region ω derived from (2.34), and then (2.45) contains only the volume integral term, i.e.,

$$R_{11} = -\int_{\Omega} \nabla N_i \times \frac{1}{\mu} \nabla \times \boldsymbol{A}\, \mathrm{d}v \tag{2.46}$$

The vector formulation is

$$\nabla N_i = \frac{\partial N_i}{\partial x}\boldsymbol{i} + \frac{\partial N_i}{\partial y}\boldsymbol{j} + \frac{\partial N_i}{\partial z}\boldsymbol{k} \tag{2.47}$$

$$\frac{1}{\mu}\nabla \times \boldsymbol{A} = \frac{1}{\mu_x}\left(\frac{\partial A_z}{\partial y} - \frac{\partial A_y}{\partial z}\right)\boldsymbol{i} + \frac{1}{\mu_y}\left(\frac{\partial A_x}{\partial z} - \frac{\partial A_z}{\partial x}\right)\boldsymbol{j} + \frac{1}{\mu_z}\left(\frac{\partial A_y}{\partial x} - \frac{\partial A_x}{\partial y}\right)\boldsymbol{k} \tag{2.48}$$

With the result of (2.47) and (2.48),

$$\begin{aligned} R_{11} &= -\int_{\Omega} \nabla N_i \times \frac{1}{\mu} \nabla \times \boldsymbol{A} \mathrm{d}v \\ &= -\left[\int_{\Omega} \left(\frac{1}{\mu_z}\frac{\partial N_i}{\partial y}\left(\frac{\partial A_y}{\partial x} - \frac{\partial A_x}{\partial y}\right) - \frac{1}{\mu_y}\frac{\partial N_i}{\partial z}\left(\frac{\partial A_x}{\partial z} - \frac{\partial A_z}{\partial x}\right)\right)\mathrm{d}v\right]\boldsymbol{i} \\ &\quad - \left[\int_{\Omega} \left(\frac{1}{\mu_x}\frac{\partial N_i}{\partial z}\left(\frac{\partial A_z}{\partial y} - \frac{\partial A_y}{\partial z}\right) - \frac{1}{\mu_z}\frac{\partial N_i}{\partial x}\left(\frac{\partial A_y}{\partial x} - \frac{\partial A_x}{\partial y}\right)\right)\mathrm{d}v\right]\boldsymbol{j} \\ &\quad - \left[\int_{\Omega} \left(\frac{1}{\mu_y}\frac{\partial N_i}{\partial x}\left(\frac{\partial A_x}{\partial z} - \frac{\partial A_z}{\partial x}\right) - \frac{1}{\mu_x}\frac{\partial N_i}{\partial y}\left(\frac{\partial A_z}{\partial y} - \frac{\partial A_y}{\partial z}\right)\right)\mathrm{d}v\right]\boldsymbol{k} \end{aligned} \tag{2.49}$$

For (2.40), the vector formulation is

$$\nabla(cd) = d\nabla c + c\nabla d$$

which leads to

$$\nabla\left(N_i \frac{1}{\mu_c} \nabla \cdot \boldsymbol{A}\right) = \nabla N_i \frac{1}{\mu_c} \nabla \cdot \boldsymbol{A} + N_i \nabla\left(\frac{1}{\mu_c} \nabla \cdot \boldsymbol{A}\right) \tag{2.50}$$

(2.50) is sorted out to:

$$N_i \nabla\left(\frac{1}{\mu_c} \nabla \cdot \boldsymbol{A}\right) = \nabla\left(N_i \frac{1}{\mu_c} \nabla \cdot \boldsymbol{A}\right) - \nabla N_i \frac{1}{\mu_c} \nabla \cdot \boldsymbol{A} \tag{2.51}$$

Substituting (2.51) into (2.40):

$$R_{12} = -\int_\Omega \nabla\left(N_i \frac{1}{\mu_c} \nabla \cdot \boldsymbol{A}\right) \mathrm{d}v + \int_\Omega \nabla N_i \frac{1}{\mu_c} \nabla \cdot \boldsymbol{A}\, \mathrm{d}v \tag{2.52}$$

The vector formulation is

$$\int_\Omega \nabla c\, \mathrm{d}v = \oint_s c\boldsymbol{n}\, \mathrm{d}s$$

The following form can be derived from (2.40):

$$R_{12} = -\oint_s N_i \left(\frac{1}{\mu_c} \nabla \cdot \boldsymbol{A}\right) \boldsymbol{n}\, \mathrm{d}s + \int_\Omega \nabla N_i \cdot \frac{1}{\mu_c} \nabla \cdot \boldsymbol{A}\, \mathrm{d}v \tag{2.53}$$

The continuity at the interface $(\frac{1}{\mu} \nabla \cdot \boldsymbol{A})$ becomes a natural interface condition. The surface integral of (2.53) will be canceled with the surface integral of (2.34), so only the volume integral term of

$$R_{12} = \int_\Omega \nabla N_i \frac{1}{\mu_c} \nabla \cdot \boldsymbol{A}\, \mathrm{d}v \tag{2.54}$$

The vector formulation is

$$\frac{1}{\mu_c} \nabla \cdot \boldsymbol{A} = \frac{1}{\mu_c}\left(\frac{\partial A_x}{\partial x} + \frac{\partial A_y}{\partial y} + \frac{\partial A_z}{\partial z}\right) \tag{2.55}$$

$$\nabla N_i = \frac{\partial N_i}{\partial x}\boldsymbol{i} + \frac{\partial N_i}{\partial y}\boldsymbol{j} + \frac{\partial N_i}{\partial z}\boldsymbol{k} \tag{2.56}$$

Substituting (2.55) and (2.56) into (2.54):

$$\begin{aligned}
R_{12} &= \int_{\Omega} \nabla N_i \frac{1}{\mu_c} \nabla \cdot \boldsymbol{A}\, \mathrm{d}v \\
&= \int_{\Omega} \frac{1}{\mu_c} \frac{\partial N_i}{\partial x} \left(\frac{\partial A_x}{\partial x} + \frac{\partial A_y}{\partial y} + \frac{\partial A_z}{\partial z} \right) \mathrm{d}v\, \boldsymbol{i} \\
&\quad + \int_{\Omega} \frac{1}{\mu_c} \frac{\partial N_i}{\partial y} \left(\frac{\partial A_x}{\partial x} + \frac{\partial A_y}{\partial y} + \frac{\partial A_z}{\partial z} \right) \mathrm{d}v\, \boldsymbol{j} \\
&\quad + \int_{\Omega} \frac{1}{\mu_c} \frac{\partial N_i}{\partial z} \left(\frac{\partial A_x}{\partial x} + \frac{\partial A_y}{\partial y} + \frac{\partial A_z}{\partial z} \right) \mathrm{d}v\, \boldsymbol{k}
\end{aligned} \tag{2.57}$$

For (2.41), we get

$$\begin{aligned}
R_{13} &= \int_{\Omega} \sigma N_i \left(\frac{\partial A_x}{\partial t} + \frac{\partial}{\partial x} \left(\frac{\partial V}{\partial t} \right) \right) \mathrm{d}v\, \boldsymbol{i} \\
&\quad + \int_{\Omega} \sigma N_i \left(\frac{\partial A_y}{\partial t} + \frac{\partial}{\partial y} \left(\frac{\partial V}{\partial t} \right) \right) \mathrm{d}v\, \boldsymbol{j} \\
&\quad + \int_{\Omega} \sigma N_i \left(\frac{\partial A_z}{\partial t} + \frac{\partial}{\partial z} \left(\frac{\partial V}{\partial t} \right) \right) \mathrm{d}v\, \boldsymbol{k}
\end{aligned} \tag{2.58}$$

Three components of the Galerkin residual of (2.32) can be obtained by synthesizing (2.49), (2.57) and (2.58):

$$\begin{aligned}
R_x &= -\int_{\Omega} \left(\frac{1}{\mu_z} \frac{\partial N_i}{\partial y} \left(\frac{\partial A_y}{\partial x} - \frac{\partial A_x}{\partial y} \right) - \frac{1}{\mu_y} \frac{\partial N_i}{\partial z} \left(\frac{\partial A_x}{\partial z} - \frac{\partial A_z}{\partial x} \right) \right) \mathrm{d}v \\
&\quad + \int_{\Omega} \frac{1}{\mu_c} \frac{\partial N_i}{\partial x} \left(\frac{\partial A_x}{\partial x} + \frac{\partial A_y}{\partial y} + \frac{\partial A_z}{\partial z} \right) \mathrm{d}v \\
&\quad + \int_{\Omega} \sigma N_i \left(\frac{\partial A_x}{\partial t} + \frac{\partial}{\partial x} \left(\frac{\partial V}{\partial t} \right) \right) \mathrm{d}v
\end{aligned} \tag{2.59}$$

$$R_y = -\int_\Omega \left(\frac{1}{\mu_x}\frac{\partial N_i}{\partial z}\left(\frac{\partial A_z}{\partial y} - \frac{\partial A_y}{\partial z}\right) - \frac{1}{\mu_z}\frac{\partial N_i}{\partial x}\left(\frac{\partial A_y}{\partial x} - \frac{\partial A_x}{\partial y}\right)\right)\mathrm{d}v + \int_\Omega \frac{1}{\mu_c}\frac{\partial N_i}{\partial y}\left(\frac{\partial A_x}{\partial x} + \frac{\partial A_y}{\partial y} + \frac{\partial A_z}{\partial z}\right)\mathrm{d}v + \int_\Omega \sigma N_i\left(\frac{\partial A_y}{\partial t} + \frac{\partial}{\partial y}\left(\frac{\partial V}{\partial t}\right)\right)\mathrm{d}v \tag{2.60}$$

$$R_z = -\int_\Omega \left(\frac{1}{\mu_y}\frac{\partial N_i}{\partial x}\left(\frac{\partial A_x}{\partial z} - \frac{\partial A_z}{\partial x}\right) - \frac{1}{\mu_x}\frac{\partial N_i}{\partial y}\left(\frac{\partial A_z}{\partial y} - \frac{\partial A_y}{\partial z}\right)\right)\mathrm{d}v + \int_\Omega \frac{1}{\mu_c}\frac{\partial N_i}{\partial z}\left(\frac{\partial A_x}{\partial x} + \frac{\partial A_y}{\partial y} + \frac{\partial A_z}{\partial z}\right)\mathrm{d}v + \int_\Omega \sigma N_i\left(\frac{\partial A_z}{\partial t} + \frac{\partial}{\partial z}\left(\frac{\partial V}{\partial t}\right)\right)\mathrm{d}v \tag{2.61}$$

To deduce R_0, the vector formulation is

$$\nabla\cdot\left(N_i\sigma\left(-\frac{\partial \boldsymbol{A}}{\partial t} - \nabla\left(\frac{\partial V}{\partial t}\right)\right)\right) = \nabla N_i\cdot\sigma\left(-\frac{\partial \boldsymbol{A}}{\partial t} - \nabla\left(\frac{\partial V}{\partial t}\right)\right) + N_i\nabla\cdot\left(\sigma\left(-\frac{\partial \boldsymbol{A}}{\partial t} - \nabla\left(\frac{\partial V}{\partial t}\right)\right)\right) \tag{2.62}$$

(2.62) is sorted out to:

$$N_i\nabla\cdot\left(\sigma\left(-\frac{\partial \boldsymbol{A}}{\partial t} - \nabla\left(\frac{\partial V}{\partial t}\right)\right)\right) = \nabla\cdot\left(N_i\sigma\left(-\frac{\partial \boldsymbol{A}}{\partial t} - \nabla\left(\frac{\partial V}{\partial t}\right)\right)\right) - \nabla N_i\cdot\sigma\left(-\frac{\partial \boldsymbol{A}}{\partial t} - \nabla\left(\frac{\partial V}{\partial t}\right)\right) \tag{2.63}$$

Substituting (2.63) into (2.38)

$$\begin{aligned} R_0 &= \int_\Omega \nabla\cdot\left(N_i\sigma\left(-\frac{\partial \boldsymbol{A}}{\partial t} - \nabla\left(\frac{\partial V}{\partial t}\right)\right)\right)\mathrm{d}v - \int_\Omega \nabla N_i\cdot\sigma\left(-\frac{\partial \boldsymbol{A}}{\partial t} - \nabla\left(\frac{\partial V}{\partial t}\right)\right)\mathrm{d}v \\ &= \oint_s N_i\sigma\left(-\frac{\partial \boldsymbol{A}}{\partial t} - \nabla\left(\frac{\partial V}{\partial t}\right)\right)\cdot\boldsymbol{n}\mathrm{d}s - \int_\Omega \nabla N_i\cdot\sigma\left(-\frac{\partial \boldsymbol{A}}{\partial t} - \nabla\left(\frac{\partial V}{\partial t}\right)\right)\mathrm{d}v \end{aligned} \tag{2.64}$$

According to the conductor surface condition of $J_n = 0$, the surface integral term of (2.64) is zero, i.e.,

$$R_0 = -\int_{\Omega} \nabla N_i \cdot \sigma\left(-\frac{\partial \mathbf{A}}{\partial t} - \nabla\left(\frac{\partial V}{\partial t}\right)\right)\mathrm{d}v \tag{2.65}$$

Upon a simple vector operation, (2.65) can be rewritten as

$$\begin{aligned} R_0 = &\int_{\Omega} \sigma\left(\frac{\partial N_i}{\partial x}\left(\frac{\partial A_x}{\partial t} + \frac{\partial}{\partial x}\left(\frac{\partial V}{\partial t}\right)\right)\right)\mathrm{d}v + \int_{\Omega} \sigma\left(\frac{\partial N_i}{\partial y}\left(\frac{\partial A_y}{\partial t} + \frac{\partial}{\partial y}\left(\frac{\partial V}{\partial t}\right)\right)\right)\mathrm{d}v \\ &+ \int_{\Omega} \sigma\left(\frac{\partial N_i}{\partial z}\left(\frac{\partial A_z}{\partial t} + \frac{\partial}{\partial z}\left(\frac{\partial V}{\partial t}\right)\right)\right)\mathrm{d}v \end{aligned} \tag{2.66}$$

Non-Eddy Current Region

In non-eddy current region ω, conductivity $\sigma = 0$; the third term in (2.59), (2.60) and (2.61) will disappear, and $\mu = \mu_0$, based on the above deduction, the corresponding Galerkin residuals of (2.34) are as

$$\begin{aligned} R'_x = &-\int_{\omega} \left(\frac{1}{\mu_0}\frac{\partial N_i}{\partial y}\left(\frac{\partial A_y}{\partial x} - \frac{\partial A_x}{\partial y}\right) - \frac{1}{\mu_0}\frac{\partial N_i}{\partial z}\left(\frac{\partial A_x}{\partial z} - \frac{\partial A_z}{\partial x}\right)\right)\mathrm{d}v \\ &+ \int_{\omega} \frac{1}{\mu_0}\frac{\partial N_i}{\partial x}\left(\frac{\partial A_x}{\partial x} + \frac{\partial A_y}{\partial y} + \frac{\partial A_z}{\partial z}\right)\mathrm{d}v \end{aligned} \tag{2.67}$$

$$\begin{aligned} R'_y = &-\int_{\omega} \left(\frac{1}{\mu_0}\frac{\partial N_i}{\partial z}\left(\frac{\partial A_z}{\partial y} - \frac{\partial A_y}{\partial z}\right) - \frac{1}{\mu_0}\frac{\partial N_i}{\partial x}\left(\frac{\partial A_y}{\partial x} - \frac{\partial A_x}{\partial y}\right)\right)\mathrm{d}v \\ &+ \int_{\omega} \frac{1}{\mu_0}\frac{\partial N_i}{\partial y}\left(\frac{\partial A_x}{\partial x} + \frac{\partial A_y}{\partial y} + \frac{\partial A_z}{\partial z}\right)\mathrm{d}v \end{aligned} \tag{2.68}$$

$$\begin{aligned} R'_z = &-\int_{\omega} \left(\frac{1}{\mu_0}\frac{\partial N_i}{\partial x}\left(\frac{\partial A_x}{\partial z} - \frac{\partial A_z}{\partial x}\right) - \frac{1}{\mu_0}\frac{\partial N_i}{\partial y}\left(\frac{\partial A_z}{\partial y} - \frac{\partial A_y}{\partial z}\right)\right)\mathrm{d}v \\ &+ \int_{\omega} \frac{1}{\mu_0}\frac{\partial N_i}{\partial z}\left(\frac{\partial A_x}{\partial x} + \frac{\partial A_y}{\partial y} + \frac{\partial A_z}{\partial z}\right)\mathrm{d}v \end{aligned} \tag{2.69}$$

On the treatment of complicated material properties in the formulation and numerical implementation based on the potential sets, and the measurement and prediction of the corresponding material properties, can be found in the related chapters of this book or other related literatures.

References

1. J. C. Maxwell, "A treatise on electricity and magnetism," Vols. I and II, Clarendon Press, England, 1904.
2. J. A. Stratton. *Electromagnetic Theory*. McGraw-Hill Book Company, Inc., New York & London, 1941.
3. R. W. Clough, "The finite element method in plane stress analysis," Proc. of 2nd ASCE Conf. Electronic Computation, Pittsburgh, PA., pp. 345–378, 1960.
4. Fraeijs B.X.de Veubeke, "Displacement and equilibrium models in the finite element method, Stress Analysis (ed. by O. C. Zienkiewicz & G. Holister), New York: Wiley. 1965.
5. O. C. Zienkiewicz. *The finite element method*. Fourth edition. McGraw-Hill. 1994.
6. A. M. Winslow, "Numerical solution of the quasi-linear Poisson equation in a non-uniform triangle mesh," Journal of Computational Physics, vol. 2, 1967, pp. 149–172.
7. P. P. Silvester, "Finite element solution of homogeneous waveguide problems," Alta Frequenza, 38, 1969, pp. 313–317.
8. J. C. Nedelec, "Mixed finite elements in R^3", Numerische Mathematik, vol. 35, pp. 315–341, 1980.
9. C. J. Carpenter, "Comparison of alternative formulations of 3-dimensional magnetic-field and eddy-current problems at power frequencies," Proc.IEE, Vol. 124, No. 11 Nov., 1977, pp. 1026–1034.
10. C. W. Trowbridge, "Electromagnetic computing: the way ahead!" *IEEE Trans. on Magn.*, vol. 24, 1988, pp. 3159–3161.
11. A. Bossavit. *Computational electromagnetism (Variational formulations, complementarity, edge elements)*. Academic Press. 1998.
12. C. Feng (editor), *Electromagnetic Field* (2nd Edition, in Chinese). Higher Education Press, 1983.
13. J. Sheng, et al., *Numerical Analysis for Electromagnetic Fields (in Chinese)*. Science Press, 1984.
14. X. Wang, *Theory and Application of Electromagnetic Fields* (in Chinese). Science Press, 1986.
15. K. Zhou, *On Engineering Electromagnetic Field (in Chinese)*. Huazhong Engineering College Press, 1986.
16. Y. Tang. *Electromagnetic Field in Motor* (in Chinese). Science Press, 1982.
17. N. Takahashi, *Three Dimensional Finite Element Methods*. IEEJ, Japan, 2006.
18. P. Chen, L. Yan, and R. Yao, *Theory and Calculation of Motor Electromagnetic Field* (in Chinese). Science Press, 1986.
19. X. Ma, J. Zhang, and P. Wang, *Fundamentals of Electromagnetic Fields* (in Chinese). Tsinghua University Press, 1995.
20. M. Fan, and W. Yan, *Electromagnetic Field Integral Equation* (in Chinese). China Machine Press, 1988.
21. G. Ni (editor), *Principle of Engineering Electromagnetic Field*(in Chinese). Higher Education Press, 2002.
22. W. Yan, Q. Yang, and Y. Wang, *Numerical Analysis of Electromagnetic Fields in Electrical Engineering* (in Chinese). China Machine Press, Aug. 2005.

23. D. Xie, et al., *Finite Element Analysis of 3D Eddy Current Field*. China Machine Press (2nd Edition, in Chinese), Mar. 2008.
24. O. Biro and K. Preis, "Finite element analysis of 3-D eddy currents," *IEEE Trans. on Magn.*, vol. 26, 1990, pp. 418–423.
25. O. Biro, K. Preis, W. Renhart, K. R. Richter, and G. Vrisk, "Performance of different vector potential formulations in solving multiply connected 3-D eddy current problems," *IEEE Trans. on Magn.*, vol. 26, 1990, pp. 438–411.
26. N. Ida and J. P. A. Bastos. *Electromagnetics and calculation of fields* (second edition). Springer, 1997.
27. O. Biro, "Edge element in eddy current computations," Lecture of seminar on advanced computational electromagnetics, May 26, 1999.
28. G. Mur, Edge elements, their advantages and their disadvantages, *IEEE Trans. on Magn.*, vol. 30, No. 5, 1994, pp. 3552–3557.
29. A. Kameari, "Three dimensional eddy current calculation using edge elements for magnetic vector potential," Applied electromagnetics in materials, 225(1989), Pergamon Press.
30. K. Fujiwara, "3D magnetic field computation using edge elements," Proc. of the 5th IGTE symposium on numerical field computation in electrical engineering, Graz, Austria, September 18–22, 1992, pp. 185–212.
31. A. Ahagon and K. Fujiwara, "Some important properties of edge shape functions," *IEEE Trans. on Magn.,* vol. 34, no. 5, Sep.1998, pp. 3311–3314.
32. M. Gyimesi and D. Ostergaard, "Non-conforming hexahedral edge elements for magnetic analysis," *IEEE Trans. on Magn.*, vol. 34, no. 5, 1998, pp. 2481–2484.
33. J. S. van Welij, "Calculation of eddy currents in terms of H on hexahedra," *IEEE Trans. on Magn.*, vol. 21, no. 6, 1985, pp. 2239–2241.
34. A. Kameari, "Calculation of transient 3D eddy current using edge elements," *IEEE Trans. on Magn.*, vol. 26, no. 2, 1990, pp. 466–469.
35. Z. J. Cendes, "Vector finite elements for electromagnetic field computation," *IEEE Trans. on Magn.*, vol. 27, 1990, pp. 3958–3966.
36. A. Bossavit, "Solving Maxwell equations in a closed cavity and the question of 'spurious modes'," *IEEE Trans. on Magn.*, vol. 26, 1990, pp. 702–705.
37. Z. Ren, C. Li and A. Razek, "Hybrid FEM-BIM formulation using electric and magnetic variables," *IEEE Trans. on Magn.*, vol. 28,1992, pp. 1647–1650.
38. A. Kameari and K. Koganezawa, "Convergence of ICCG method in FEM using edge elements without gauge condition," *IEEE Trans. on Magn.*, vol. 33, No. 2, March 1997, pp. 1223–1226.
39. G. Mur, "The finite element modeling of three-dimensional electromagnetic fields using edge and nodal elements," *IEEE Trans. on Antennas & Propagation*, vol. 41, No. 7, 1993, pp. 948–953.
40. K. Sakiyama, H. Kotera, and A. Ahagon, "3D electromagnetic field mode analysis using finite element method by edge element," *IEEE Trans. on Magn.*, vol. 26, no. 2, pp. 1759–1761, 1990.
41. M. L. Barton and Z. Cendes, "New vector finite elements for three-dimensional magnetic field computation," Journal of Applied Physics, vol. 61, No. 8, 1987, P. 3919–3921.
42. J. Wang and N. Ida, "Curvilinear and higher order 'edge' finite elements in electromagnetic field computation," *IEEE Trans. on Magn.*, vol. 29, No. 2, 1993, pp. 1491–1494.
43. K. Fujiwara, T. Nakata, and H. Ohashi, "Improvement of convergence characteristics of ICCG method for **A**-ϕ method using edge elements," *IEEE Trans. on Magn.*, vol. 32, no. 5, 1996, pp. 804–807.
44. O. Biro, K. Preis and K. R. Richter, "On the use of magnetic vector potential in nodal and edge finite element analysis of 3D magnetostatic problem," *IEEE Trans. on Magn.*, vol. 32, no. 5, 1996, pp. 651–654.
45. J. B. Manges and Z. J. Cendes, "A generalized tree-cotree gauge for magnetic field computation," *IEEE Trans on Magn.,* Vol. 31, no. 3, 1995, pp. 1342–1347.

46. J. P. Webb, "Edge elements and what they can do for you?" *IEEE Trans. on Magn.*, vol. 29, No. 2, 1993, pp. 1460–1465.
47. C. Bedrosian, M. V. K. Chari, and J. Joseph, "Comparison of full and reduced potential formulations for low-frequency applications," *IEEE Trans. on Magn.*, vol. 29, no. 2, 1995, pp. 1321–1324.
48. P. Dular, "Local and global constraints in finite element modeling and the benefits of nodal and edge elements coupling," *ICS Newsletter*, Vol. 7, no. 2, 2000, pp. 4–7.
49. T. Nakata, N. Takahashi, K. Fujiwara, and T. Imai, "Investigation of effectiveness of edge elements," *IEEE Trans. on Magn.*, vol. 28, no. 2, 1992, pp. 1619–1622.
50. T. Nakata, N. Takahashi, K. Fujiwara, and P. Olszewski, "Verification of software for 3-D eddy current analysis using IEEJ model," Advances in electrical engineering software, 349 (1990), Springer-Verlag.
51. H. Yu, K. Shao, L. Li, and C. Gu, "Edge-nodal coupled method for computing 3D eddy current problems," *IEEE Trans. on Magn.*, vol. 33, no. 2,1997, pp. 1378–1381.
52. J. Lu, L. Li, and K. Shao, *Calculation of Three-dimensional Transient Eddy Current by Using Finite Edge Element*, Proceedings of the CSEE, vol. 13, no. 5, 1993:34–41.
53. J. Lu, K. Shao, L. Li, and K. Zhou, "A new method to solve 3D eddy current using edge element in terms of A in the whole region, Proc. of ICEF, 1992, "Electromagnetic Field Problems and Application," (ed. by Jian Baiton), pp. 294–297.
54. Z. Cheng, S. Gao, and L. Li, "Eddy Current Loss Analysis and Validation in Electrical Engineering," (supported by National Natural Science Foundation of China), ISBN 7-04-009888-1, Higher Education Press, 2001.

Part II
Engineering Electromagnetic and Thermal Field Modeling

Chapter 3
Some Key Techniques in Electromagnetic and Thermal Field Modeling

Norio Takahashi

Abstract Some key techniques in electromagnetic and thermal field modeling, such as special elements, voltage-driven analysis, optimal design methods, and magneto-thermal coupled analysis, are discussed. Key points are as follows:

(1) For example, if a gap is narrow, the CPU time and the memory requirements can be reduced by using a special element.
(2) If an electrical machine is excited by voltage sources and the current is unknown, the "FEM" considering voltage sources should be used in order to accurately analyze the performance.
(3) If one uses the conventional optimization methods where the design variables are specified, only results which are foreseen in advance can be obtained. On the other hand, there is a possibility that a completely new result may be obtained by using the topology optimization method.
(4) For the accurate estimation of the temperature rise in an electromagnetic device, the precise magneto-thermal analysis is necessary considering not only the temperature dependence of material's thermal properties, such as thermal conductivity, but also the *B–H* curves at various temperatures.

Keywords Electromagnetic and thermal coupled modeling · Temperature effect · Special element · Voltage-driven analysis · Optimal design

3.1 Introduction

In the analysis of actual magnetic devices, such as transformers and motors, the following various kinds of techniques of the finite element method are necessary for the practical and accurate calculation:

N. Takahashi (✉) (deceased)
Okayama University, Okayama, Japan

Z. Cheng et al. (eds.), *Modeling and Application of Electromagnetic and Thermal Field in Electrical Engineering*, https://doi.org/10.1007/978-981-15-0173-9_3

(1) Narrow gaps (very short gap length) in a magnetic circuit of transformers, motors, etc., should not be sometimes negligible. If such a narrow gap is subdivided into finite elements, the number of elements near the gap is considerably increased. When the shielding plate is very thin, a similar problem occurs.
(2) Magnetic devices, such as motors, are usually operated by a constant voltage source. In this case, the amplitude of voltage is known, but the current is unknown. Then, the ordinary finite element method cannot be applicable, because the current is unknown.
(3) Various kinds of optimization methods combined with electromagnetic field analysis are proposed. The practical optimization methods applicable to actual electromagnetic devices should be examined.
(4) For the accurate estimation of temperature rise in an electromagnetic device, the precise magneto-thermal analysis is necessary, considering not only the temperature dependence of material constant, such as thermal conductivity, but also the *B–H* curves at various temperatures.

In this section, several techniques, such as, "special element" called as "gap element" and "shielding element" treating narrow gaps, and thin shielding plate, "method of analysis considering voltage source", which can analyze magnetic devices connected to a voltage source, various optimal design methods combined with electromagnetic field analysis, and magneto-thermal coupled analysis, are explained.

3.2 Special Elements

3.2.1 What Is Special Element?

In this section, 3-D special elements called the "gap element" [1] and "shielding element" [2] for discretizing narrow gaps in an iron core and thin shielding plates are discussed.

When the gap length D in a magnetic core is very short, it can be assumed that the flux is perpendicular to the gap between high-permeability materials as shown in Fig. 3.1a. Such a region V_s can be subdivided by, for example, 2-D quadrilateral elements as shown in Fig. 3.1c instead of conventional 3-D elements such as hexahedral elements. The conducting plate with a small thickness D (Fig. 3.1b) can also be represented by quadrilateral elements (Fig. 3.1c). Such additional 2-D elements are referred to as the "gap element," and the "shielding element," respectively. The "special element" is a general name for these elements. The special element has no volume, but has nearly the same energy as the gap and conducting plate shown in Fig. 3.1a, b.

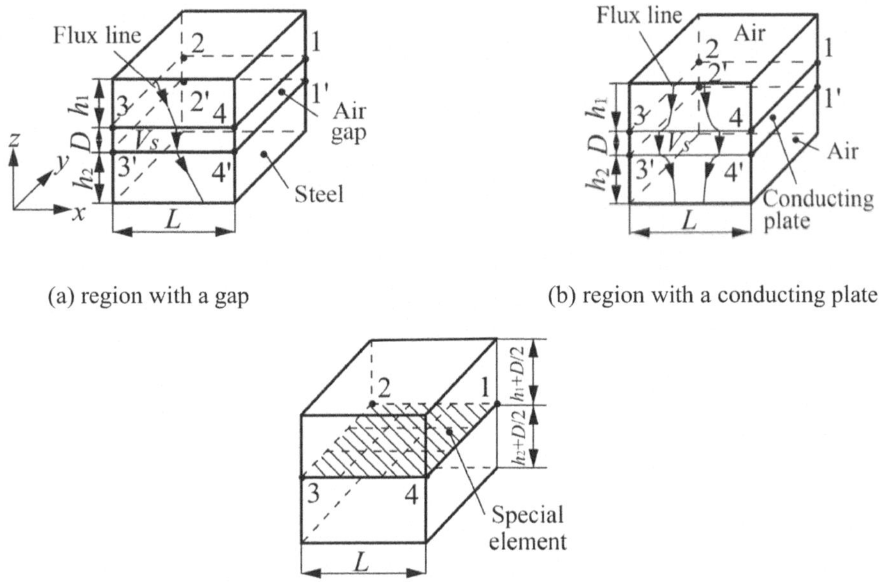

(a) region with a gap

(b) region with a conducting plate

(c) 2-D quadrilateral elements (special element)

Fig. 3.1 Special element

The special element has the following advantages:

(1) As the special element has no volume, it is easy to place new gaps or conducting plates in desired position in the mesh or to remove them.
(2) As the modification of the length D is also easy, the influence of D on the magnetic characteristics can be examined easily by using the special element.
(3) When the special element is used, the ill-conditioning of the coefficient matrix due to narrow gaps and thin conducting plates can be avoided. Therefore, the CPU time using the special element can be reduced when compared with that using the flat conventional element, because the number of iterations of the ICCG method [3] is decreased.

3.2.2 Distribution of Potentials in Special Elements

The x- and y-components of the flux density $\boldsymbol{B}$ are given by the following equations [4]:

$$B_x = \partial A_z/\partial y - \partial A_y/\partial z \tag{3.1}$$

$$B_y = \partial A_x/\partial z - \partial A_z/\partial x \tag{3.2}$$

where A_x, A_y, and A_z are the *x*-component, *y*-component, and *z*-component of the magnetic vector potential $\boldsymbol{A}$.

As the vector potential $\boldsymbol{A}$ in the regions V_s shown in Fig. 3.1a, b is constant in the *z*-direction, the following equation can be obtained:

$$\partial \mathbf{A}/\partial z = 0 \tag{3.3}$$

As the *x*- and *y*-components B_x and B_y are both equal to zero, the following equation can be obtained from Eqs. (3.1) to (3.3):

$$\partial A_z/\partial x = \partial A_z/\partial y = 0 \tag{3.4}$$

From Eqs. (3.3) and (3.4), A_z is constant in the region V_S. Thus, A_z is assumed to be zero as follows:

$$A_z = 0 \tag{3.5}$$

A_s the shielding element is to be used in the eddy current analysis, the distribution of the electric scalar potential ϕ as well as the vector potential $\boldsymbol{A}$ should be investigated. The *z*-component J_{ez} of the eddy current density J_e can be given by [4]

$$J_{ez} = -\sigma(\partial A_z/\partial t + \partial \phi/\partial z) \tag{3.6}$$

where σ is the conductivity. As the shielding plate is very thin, the scalar potential ϕ can be assumed to be constant in the *z*-direction as follows:

$$\partial \phi/\partial z = 0 \tag{3.7}$$

From Eqs. (3.3) and (3.7), the following relationships between each component of the vector potential A and the scalar potential ϕ at the nodes 1–4 and 1′–4′ in Fig. 3.1a, b can be assumed:

$$A_{ij} = A_{ij'}, \quad \phi_j = \phi_{j'} \tag{3.8}$$

($i = x, y, z, j = 1, \ldots, 4, j' = 1', \ldots, 4'$).

Equations (3.5) and (3.8) eliminate the vector potentials A_{zj}, $A_{xj'}$, $A_{yj'}$, $A_{zj'}$ and the scalar potential $\phi_{j'}$ from the unknowns.

3.2.3 Finite Element Formulation

Assuming that the potentials are constant along the flux line as shown in Eq. (3.8), the weighted residuals Goi* and Gei* of Galerkin's method for the special elements are given by

$$\begin{aligned} \mathrm{Goi}* = & -D\iint_{S_{\mathrm{s}}} \mathrm{rot}N_k \cdot (v\,\mathrm{rot}\boldsymbol{A})\mathrm{d}S \\ & + D\iint_{S_{\mathrm{se}}} N_k \cdot \sigma\left(\frac{\partial \boldsymbol{A}}{\partial t} + \mathrm{grad}\phi\right)\mathrm{d}S \end{aligned} \tag{3.9}$$

$$\mathrm{Gei}^* = D\iint_{S_{\mathrm{se}}} \mathrm{grad}N_i \cdot \left(\frac{\partial \boldsymbol{A}}{\partial t} + \mathrm{grad}\phi\right)\mathrm{d}S \tag{3.10}$$

where S_{s} is the region where the gap elements are defined. In the case of the gap element, the reluctivity v is changed to v_0 (the reluctivity of air). S_{se} is for the shielding element. As the magnetizing current does not flow in the special element, J_0 is removed from Eq. (3.9). Equations (3.9) and (3.10) show that a 3-D problem can be reduced to a 2-D problem by applying the special element. The gaps and shielding plates can easily be discretized by adding Eqs. (3.9) and (3.10) to the usual linear equations for the 3-D finite element method.

When the gap element is applied, both sets of nodes 1, 2, 3, 4 and 1′, 2′, 3′, 4′ are moved symmetrically to the centerline, between them, as shown in Fig. 3.1c. Therefore, the length of the iron part has to be increased by D. The energy of the increased iron part, however, can be neglected because the permeability of the iron part is very high compared with that of the gap part and the gap length D is short.

3.2.4 Some Examples

In order to verify the usefulness of the special elements, a few analyses are carried out. In the region where the special element is not applied, the ordinary first-order brick element is used.

1. Gap Element

The accuracy of the gap element is investigated by using the model with a gap as shown in Fig. 3.2 [2]. The current in the winding is DC. The 3-D magnetic field is analyzed neglecting the saturation. As the model is symmetric, only 1/4 region ($-40 < x < 40$, $0 < y < 60$, $0 < z < 120$ mm) need to be analyzed. The numbers of elements, etc., is shown in Table 3.1.

Figure 3.3 shows the effect of the gap length D on the error ε_{B} in the flux density. ε_{B} is defined by

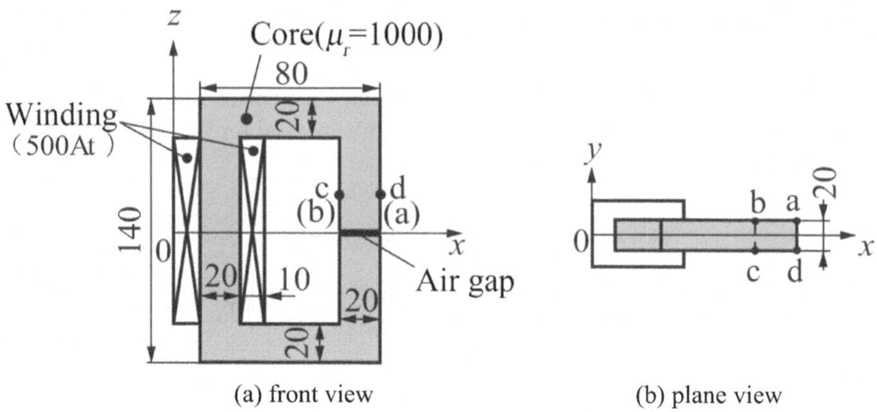

Fig. 3.2 Analyzed model with gap element

Table 3.1 Discrete data and CPU time

Item		Gap element	Conventional element
Number of elements	Out of conductor region	4928	5150
	Conductor region	150	450
Number of nodes		6072	6864
CPU time (s) (D = 0.3 mm)	ICCG	299	1445
	Total time	332	1482

Computer: NEC supercomputer SX-IE (maximum computing speed: 285MFLOPS)

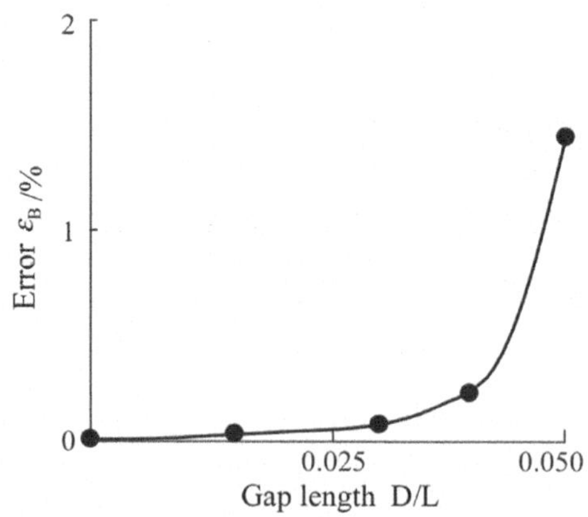

Fig. 3.3 Effect of the gap length on the error in the flux density

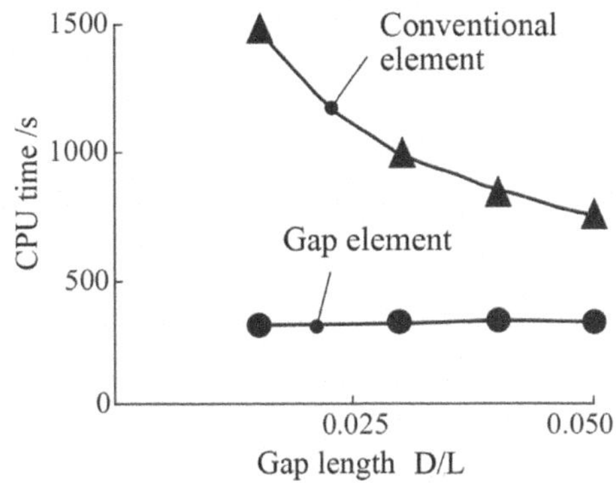

Fig. 3.4 Comparison of the CPU time

$$\varepsilon_{\mathrm{B}} = \frac{B_{\mathrm{g}} - B_{\mathrm{o}}}{B_{\mathrm{o}}} \times 100\ (\%) \tag{3.11}$$

where B_g is the average flux density on the plane *a–b–c–d–a* in Fig. 3.2, calculated using the gap element. B_0 is calculated using a conventional mesh subdivided into brick elements. L is the width of the gap (=20 mm). If the permissible error is less than 1%, the gap element can be used in practice under the condition that the ratio *D/L* is smaller than about 0.04, as shown in Fig. 3.3.

Figure 3.4 shows the comparison of the CPU time. The CPU time is not affected by the gap length D when the gap element is used. The CPU time using only conventional elements is significantly increased as D is decreased, because the coefficient matrix becomes ill-conditioned as D decreases.

2. Shielding Element

The effects of an aluminum shield on flux and eddy current distribution are investigated using the model shown in Fig. 3.5 [2]. The fields are produced by the exciting current at 50 and 200 Hz. The ampere-turns of the coil are set to 3000 AT (peak value). The conductivities of aluminum and steel are 3.56×10^7 and 7.51×10^6 S/m, respectively, and the relative permeability of the steel is assumed to be 1000. A quarter model ($0 < x < 500$, $0 < y < 500$, $-150 < z < 500$ mm) is analyzed. The numbers of elements, etc., are shown in Table 3.2. The CPU time using the shielding elements is reduced compared with that using only conventional brick elements. This tendency is similar to the case of the gap element.

Figure 3.6 shows the comparison of the z-component B_z of the flux density along the line *a–b* ($y = 12.5$, $z = 10$ mm) shown in Fig. 3.5. The thickness D of the plate is equal to 1 mm. The flux density is measured using a small search coil (diameter: 3 mm, height: 0.6 mm, 20 turns). The effect of the shielding plate on the flux density at 200 Hz is more remarkable than at 50 Hz.

Figure 3.7 shows the effect of the thickness D of the shielding plate on the error ε_J of the eddy current density. δ is the skin depth. The definition of the error ε_J is the

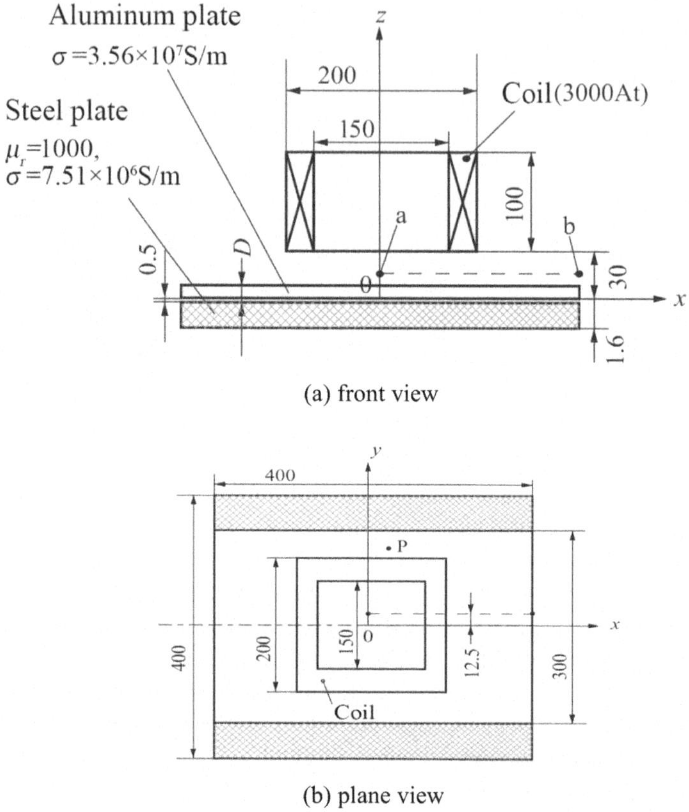

Fig. 3.5 Analyzed model with shielding plate

Table 3.2 Discrete data and CPU time

Item		Shielding element	Conventional element
Number of elements	Out of conductor region	2057	2115
	Conductor region	63	63
Number of nodes		2592	2736
CPU time (*s*)	ICCG	805	1142
(D = 1 mm, 200 Hz)	Total time	821	1161

Computer: NEC supercomputer SX-IE (maximum computing speed: 285MFLOPS)

same as Eq. (3.11). The eddy current density is examined at the point p (x = 12.5, y = 110, z = 1 mm) in the shielding plate. If the permissible error is less than 1%, the shielding element can be used when $D/\delta < 0.11$.

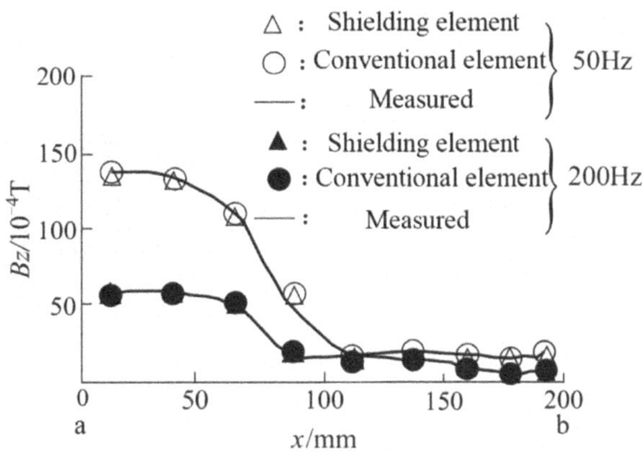

Fig. 3.6 Distributions of B_z (D = 1 mm)

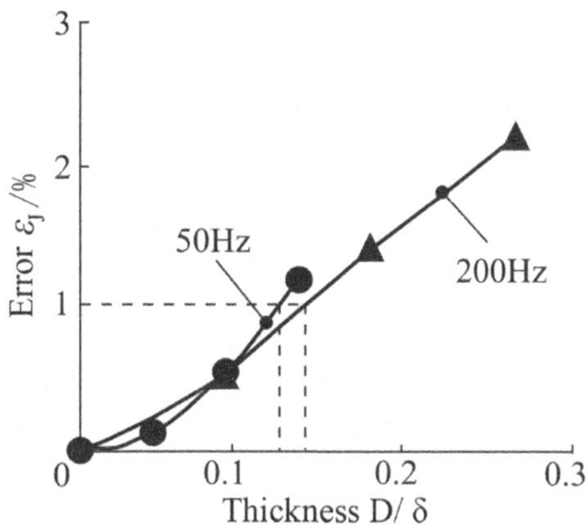

Fig. 3.7 Effect of the thickness of the shielding plate on the error of the eddy current density

3.3 Voltage-Driven Analysis

3.3.1 FEM Considering Voltage Source

When magnetic circuits are analyzed using Poisson's equation, the magnetizing current densities must be given. Electrical machines are usually excited from voltage sources as shown in Fig. 3.8, and the magnetizing current densities are unknown. Therefore, it is difficult to analyze magnetic fields in such machines using the conventional finite element method. If the loop equations obtained from Kirchhoff's second law are combined with Maxwell's equations for the magnetic field analysis, both magnetic fields and currents can be directly calculated. This method is called "FEM considering voltage source" [5, 6].

Although the magnetizing current density $\boldsymbol{J_0}$ has only one component in 2-D analysis, $\boldsymbol{J_0}$ has three components J_{0x}, J_{0y}, and J_{0z} in 3-D analysis, and these components change with the position in the exciting winding. If these three

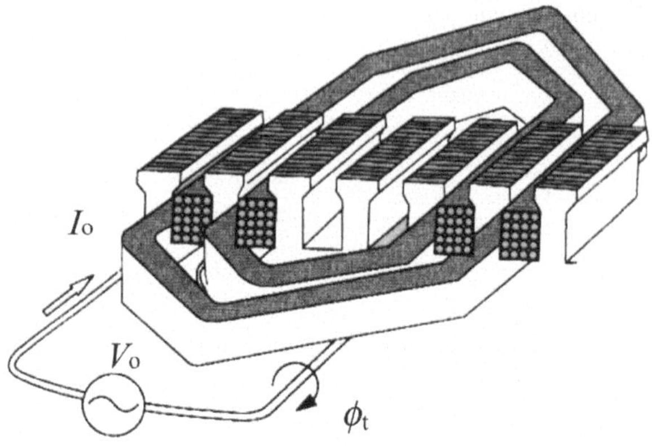

Fig. 3.8 FEM considering voltage source

components are treated as unknown variables, the number of unknown variables is increased above that of equations [1] which are obtained from loop equations derived from Kichhoff's second law and Maxwell's equations. Because each element in the winding has different values of J_{0x}, J_{0y} and J_{0z}, this is the cause for the difficulty in 3-D analysis. In many cases, however, the magnetizing current in the winding flows uniformly; therefore, if the magnetizing current densities, of which the amplitudes can be assumed to be all the same everywhere in a winding, are denoted by one unknown variable, the analysis becomes possible, because the number of unknown variables is equal to that of equations.

For easy understanding, using a simple model of the winding shown in Fig. 3.9, the method for reducing the number of unknown variables of $\boldsymbol{J_0}$ is explained in more detail. It is assumed that the magnetizing current flows in the x–y plane and the z-component of the current can be neglected. If the sign of current flowing

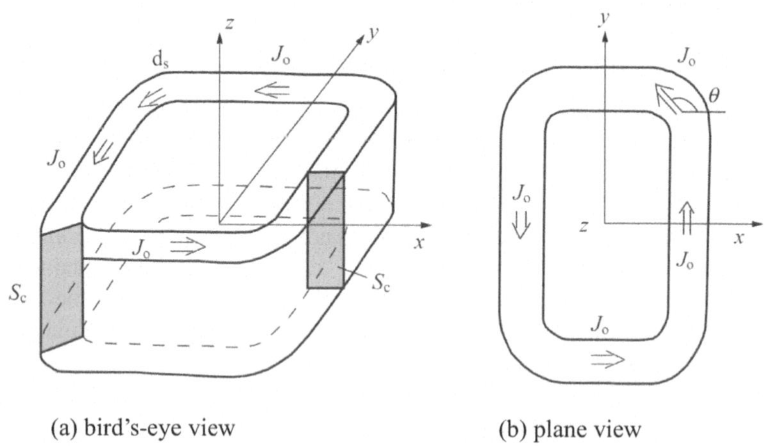

Fig. 3.9 Winding

anti-clockwise around the z-axis is positive, the magnetizing current density $\boldsymbol{J_0}$ is represented as follows:

$$\boldsymbol{J}_0 = J_0(\boldsymbol{i}\cos\theta + \boldsymbol{j}\sin\theta) \tag{3.12}$$

where θ is the angle from the x-axis. $\boldsymbol{i}$ and $\boldsymbol{j}$ are the x- and y-directional unit vectors, respectively. The amplitude I_0 of the magnetizing current is represented as follows:

$$I_0 = \frac{S_c J_0}{n_c} \tag{3.13}$$

where S_c and n_c are the cross-sectional area and the number of turns of the winding shown in Fig. 3.9, respectively.

From Eqs. (3.12) and (3.13), the third term Gvi becomes as follows:

$$\mathrm{Gvi} = \frac{-n_c}{S_c} \iiint_{\Omega_c} N_i I_0(\boldsymbol{i}\ \cos\theta + \boldsymbol{j}\ \sin\theta)\mathrm{d}V \tag{3.14}$$

In Eq. (3.14), the unknown variable is only the amplitude I_0 of the magnetizing current.

Figure 3.10 shows an equivalent circuit of the machine excited from a voltage source. The finite element region which is enclosed by the broken line in Fig. 3.10 corresponds to the winding shown in Fig. 3.9. R_c is the DC resistance of the winding. R_0 and L_0 are the resistance and the inductance which cannot be included in the finite element region. The following equation can be obtained from Kirchhoff's second law:

$$\eta = V_0 - \frac{\mathrm{d}\Psi}{\mathrm{d}t} - (R_0 + R_c)I_0 - L_0\frac{\mathrm{d}I_0}{\mathrm{d}t} = 0 \tag{3.15}$$

where Ψ is the interlinkage flux of the winding. By using the magnetic vector potential $\boldsymbol{A}$, Ψ can be rewritten as follows:

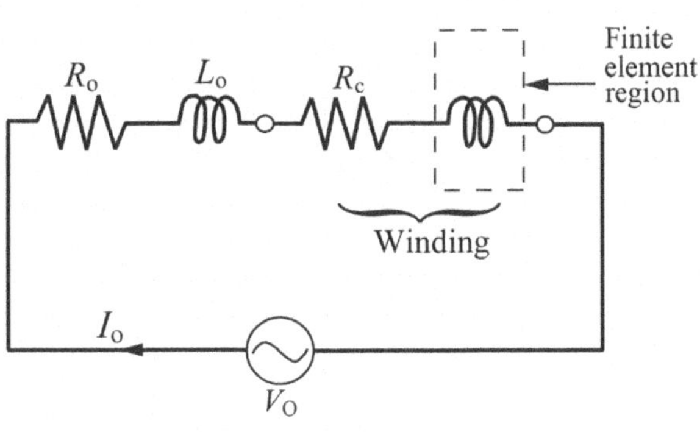

Fig. 3.10 Equivalent circuit

$$\Psi = \frac{n_c}{S_c}\iint_S\left(\int \boldsymbol{A}\cdot \mathrm{d}\boldsymbol{s}\right)\mathrm{d}S \tag{3.16}$$

where $\boldsymbol{s}$ is the unit tangent vector along the winding shown in Fig. 3.9. S is the cross section of the winding.

Ψ can be determined from the x-component and y-component, A_x and A_y, of $\boldsymbol{A}$ in the winding as follows:

$$\Psi = \frac{n_c}{S_c}\iiint_{\Omega_c}(A_x\cos\theta + A_y\sin\theta)\mathrm{d}V \tag{3.17}$$

In the nonlinear analysis, using the Newton–Raphson iteration technique, the increments of the unknown variables δA_{xj}, δA_{yj}, δA_{zj}, $\delta \Psi_j$, and δI_0 at the instant t are obtained from the following equation [1]:

$$\begin{bmatrix} \left[\dfrac{\partial \boldsymbol{G}_i}{\partial \boldsymbol{u}_j}\right] & \left\{\dfrac{\partial \boldsymbol{G}_i}{\partial \boldsymbol{I}_0}\right\} \\ \left\{\dfrac{\partial \eta}{\partial \boldsymbol{u}_j}\right\} & \dfrac{\partial \eta}{\partial I_0} \end{bmatrix} \begin{Bmatrix} \{\delta \boldsymbol{u}_i\} \\ \\ \delta I_0 \end{Bmatrix} = \begin{Bmatrix} -\boldsymbol{G}_i \\ \\ -\eta \end{Bmatrix} \tag{3.18}$$
$$(i,\ j = 1,\ 2,\ \ldots,\ n_u)$$

where n_u is the number of unknown nodes at which the potentials are unknown. $\{G_i\}$ and $\{u_j\}$ are denoted as follows:

$$\begin{aligned} \{G_i\} &= \{G_{oxi}, G_{oyi}, G_{ozi}, G_{di}\} \\ \{u_j\} &= \{A_{xj}, A_{yj}, A_{zj}, \Psi_j\} \end{aligned} \tag{3.19}$$

3.3.2 An Example

The currents in the primary and the secondary windings of a loaded transformer shown in Fig. 3.11 are analyzed [6]. The chained lines denote the analyzed region. The effective voltage and the frequency of the power source are 100 V and 50 Hz, respectively. The numbers of turns of the primary and the secondary windings are both equal to 30. The DC resistances R_c of the primary and the secondary windings are both equal to 1 Ω. The load is a pure resistance, and its value R_0 is equal to 0 Ω or 9 Ω. The magnetic characteristic of the core is assumed to be linear, and its relative permeability is 5000. The eddy current in the core is not taken into account.

Figure 3.12 shows the current waveforms obtained by transient analysis using the step-by-step method of which the time interval Δt is 1 ms. The solid and the

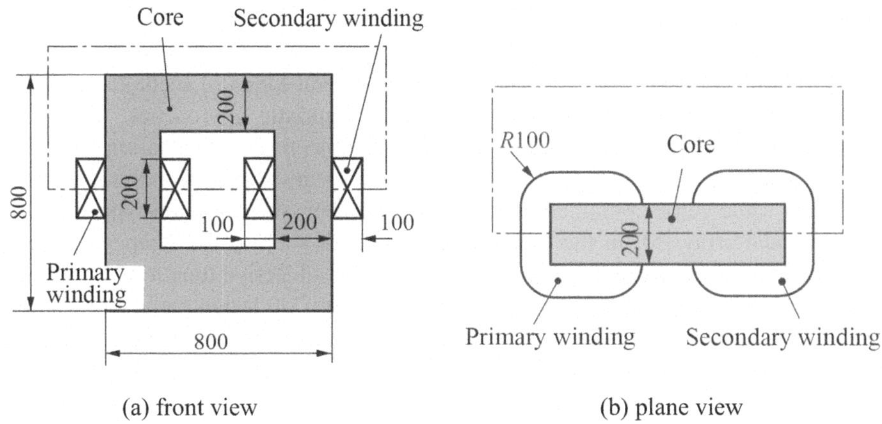

Fig. 3.11 Transformer model

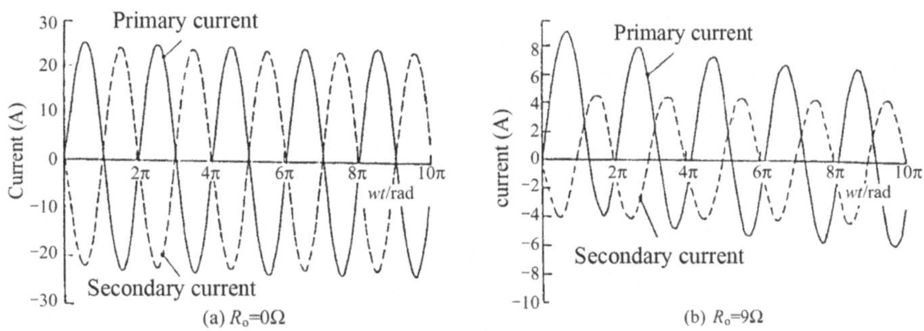

Fig. 3.12 Current waveforms

broken lines show the currents in the primary and the secondary windings, respectively. Figure 3.12 suggests that if the resistance R_0 is decreased, the current reaches steady state rapidly.

3.4 Optimal Design Method

3.4.1 Various Optimization Methods

The magnetic field analysis is widely used in the design of magnetic devices. If the design is carried out by changing shapes and dimensions of cores until the desired performance of the machine is obtained, a large number of calculations would be necessary. As a result, the CPU time would increase significantly, and such a parameter examination or cut-and-try procedure would not be practical. Hence, the optimal design method which combines the magnetic field analysis and an optimization method is used.

The optimization methods are classified into several kinds of methods, such as deterministic, stochastic, and hybrid techniques, heuristic approaches, design of experiment, topology optimization, and multi-objective optimization. In the experimental design method (EDM) which utilizes a statistical data analysis, a trend of optimal value can be obtained using a small number of calculations specified by an orthogonal array [5]. In the deterministic method, such as the steepest descent method and Newton method, the direction for smaller objective function is searched by using the derivative of an objective function in terms of design parameter. In this case, usually only a local minimum result can be obtained. In the stochastic method, such as the genetic algorithm (GA), simulated annealing method (SAM) [4, 7], and evolution strategy (ES) [8, 9], the probability is used for searching the global optimum. Rosenbrock's method [10] is one of the stochastic methods. In the topology optimization, such as the ON/OFF method [11, 12], the shape of core can be automatically determined judging the existence of magnetic material (ON) or not (OFF). As the imagination of optimal shape of the core is not necessary by the designer, in the case of ON/OFF method, we may obtain a new product which the designer could not imagine beforehand.

In this section, the experimental design method, Rosenbrock's method, the evolution strategy, and ON/OFF method are discussed.

3.4.2 *Experimental Design Method (EDM)*

The experimental design method is often used in the quality improvement process in the production and can give guidelines of the design of experiments that avoid unnecessarily complicated, inefficient, and sometimes ineffective ones by introducing the statistical data analysis. This method is called "Taguchi's method" [3].

Let us explain the method by using an example of increasing a torque of braking motor (retarder), of which the design variables are lengths L_1, L_2 and angles θ_1, θ_2 shown in Fig. 3.13. In this method, the torque which has nine patterns is calculated from an orthogonal array shown in Table 3.3.

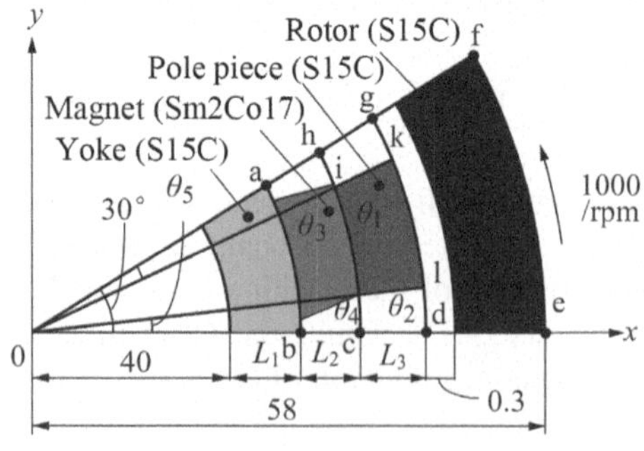

Fig. 3.13 Permanent magnet type of retarder (thickness: 16 mm)

Table 3.3 Orthogonal array

Item	Design variables			
	L_1	L_2	θ_1	θ_2
1	1	1	1	1
2	1	2	2	2
3	1	3	3	3
4	2	1	3	2
5	2	2	1	3
6	2	3	2	1
7	3	1	2	3
8	3	2	3	1
9	3	3	1	2

This is called as L_9 orthogonal array—a table of integers whose column elements (1, 2, and 3) represent the low (1), medium (2), and high (3) levels of the column factors. Each row of the orthogonal array represents a run, that is, a specific set of factor levels to be tested. The L_9 orthogonal array will accommodate four factors at three levels each in nine runs.

As the models No. 1–3 have a level 1 of L_1, the effective value of level 1 is the average value of models No. 1–3. The effective values of levels 2 and 3 are also the average values of models No. 4–6 and 7–9, respectively. The better value of design variable can be determined by considering the effect of each design variable on the torque shown in Fig. 3.14. These are the marginal means of each factor. The better values are determined as follows:

Search area can be reduced by comparing the torque of each level. If the level 1 is the largest and level 2 is smaller and level 3 is the smallest as shown in case 2 in Fig. 3.14, it is regarded that better value is less than level 2. In the same way, the better value for case 3 is regarded as larger than level 1 and less than level 3. If design variables are with less sensitivity, such as case 1, search area of the optimal value cannot be reduced.

The level of design variable where the braking torque becomes a maximum can be adopted as a candidate of optimal value. Or the combination of levels of design variables for maximum torque can be chosen.

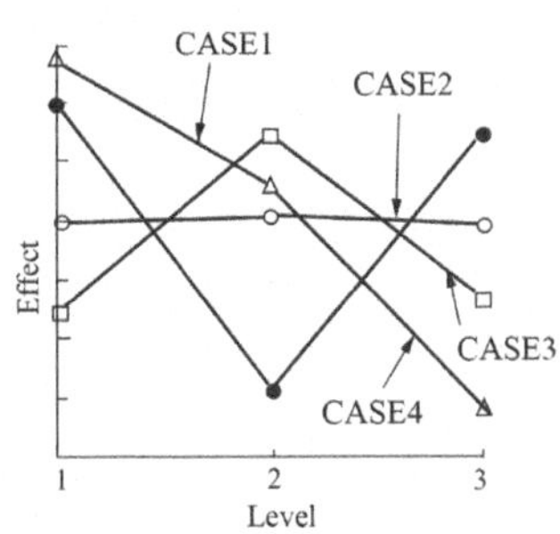

Fig. 3.14 Effect of design variables

3.4.3 Rosenbrock's Method (RBM)

If the optimal design method which is the combination of the finite element method and the direct search method, such as the Rosenbrock's method [10], is used, the nonlinearity of core can be easily considered and the CPU time required is not increased so much compared with other optimization methods, such as the simulated annealing method [4, 7].

Rosenbrock's method is explained by an example shown in Fig. 3.15. This is a 2-D model for determining an optimal dimensions of L_1 and L_2 of the core which produces the maximum y-directional electromagnetic force F_y of the conductor of which the current is equal to I. The optimal dimensions of L_1 and L_2 can be obtained by maximizing the objective function $W = F_y$. Figure 3.16 shows contour lines of equi-objective function W.

Rosenbrock's method is an iterative method for obtaining the optimal values of dimensions (L_1 and L_2) so that the objective function W will be either maximized or minimized. The method can easily be combined with the finite element method,

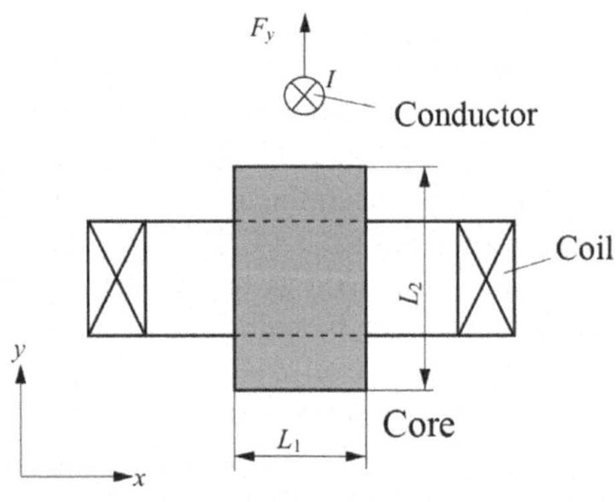

Fig. 3.15 Model with unknown dimension

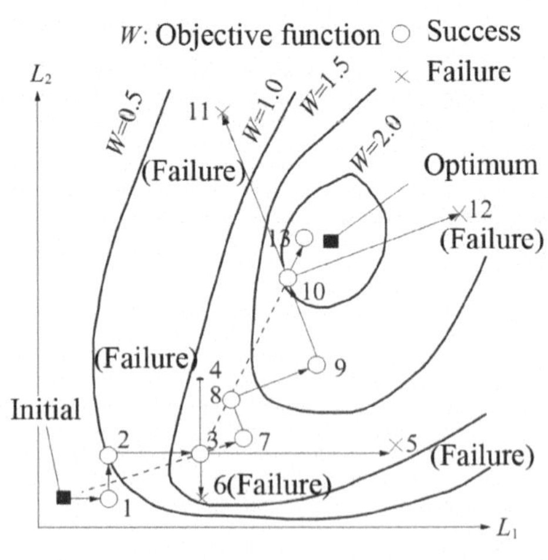

Fig. 3.16 Contour lines of equi-objective function

because the gradient of the objective function W is not necessary [1]. In executing the method, the optimal dimensions which maximize W are searched changing L_1 and L_2 with different step lengths ΔL_1 and ΔL_2, respectively. The initial values $\Delta L_1^{(0)}$ and $\Delta L_2^{(0)}$ of respective step lengths are chosen as 1/10 of the respective dimensions of the initial shape. When the search succeeds (W becomes larger), the (k + 1) th length $\Delta L_j^{(k+1)}$ (j = 1, 2) is changed as follows:

$$\Delta L_j^{(k+1)} = a \cdot \Delta L_j^{(k)} \tag{3.20}$$

When the search is not successful (W becomes smaller), $\Delta L_j^{(k+1)}$ is changed as follows:

$$\Delta L_j^{(k+1)} = -\beta \cdot \Delta L_j^{(k)} \tag{3.21}$$

where α and β are coefficients for the adjustment of ΔL_j and are chosen as 2 and 0.5, respectively. If ΔL_j becomes sufficiently small, the search is terminated, yielding the optimal shape.

Figure 3.17 shows the flowchart. The optimization process using Rosenbrock's method and the finite element method is as follows:

(1) Calculation of magnetic field for initial dimensions using the finite element method,
(2) Calculation of objective function W (electromagnetic force),
(3) Judgment whether W is increased or not,

Fig. 3.17 Flowchart

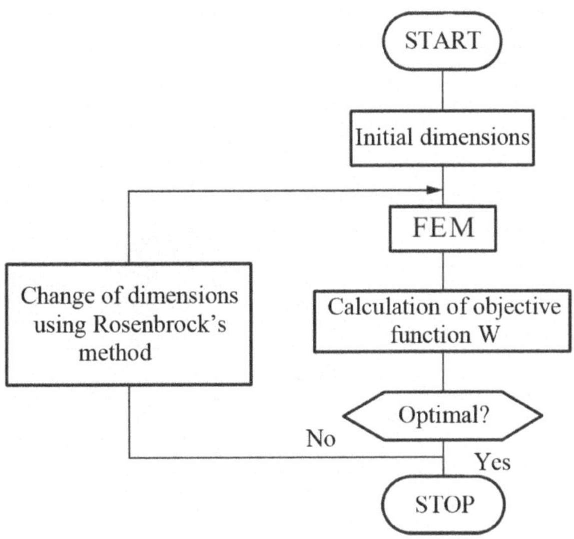

(4) Change of dimensions using Rosenbrock's method taking into account constraints.

The process of calculation of (3.2)–(3.4) is repeated until the optimal dimensions are obtained.

3.4.4 Evolution Strategy (ES)

(1 + 1)–evolution strategy (ES) is the method that generates one child vector from one parent vector comparing two objective functions of each vector as shown in Fig. 3.18 [8, 9, 13]. The vector with a good objective function is treated as a parent vector of next generation.

The child vector is defined by the following equation (mutation):

$$x_{\mathrm{c}}^{(k)} = x_{\mathrm{p}}^{(k)} + N_{\mathrm{n}}\left(0,\ \sigma^2(j)\right) \tag{3.22}$$

where $x_{\mathrm{p}}^{(k)}$ is a parent vector of the kth generation and $x_{\mathrm{c}}^{(k)}$ is a child vector. N_{n} (0, $\sigma^2(j)$) is a normal random vector ($\sigma^2(j)$): standard deviation, j: number of times that Rechenberg's 1/5 success law is carried out, and n: number of design variables.

The parent vector of next generation is determined as follows by comparing the objective function W_{p} of the parent vector and W_{c} of the child vector (selection):

$$x_{\mathrm{p}}^{(k+1)} = \begin{cases} x_{\mathrm{c}}^{(k)}\left(W_{\mathrm{c}} \le W_{\mathrm{p}}\right) \\ x_{\mathrm{p}}^{(k)}\left(W_{\mathrm{c}} > W_{\mathrm{p}}\right) \end{cases} \tag{3.23}$$

Equation (3.23) denotes that if the objective function W_{c} of the child vector $x_{\mathrm{c}}^{(k)}$ is smaller than W_{p} of the parent vector $x_{\mathrm{p}}^{(k)}$, the child vector $x_{\mathrm{c}}^{(k)}$ is adopted as parent vector $x_{\mathrm{p}}^{(k+1)}$ at the (k + 1)th generation. If W_{c} of the child vector $x_{\mathrm{c}}^{(k)}$ is larger than

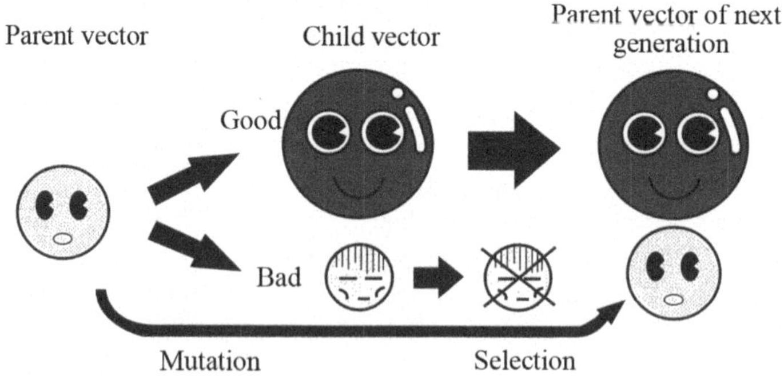

Fig. 3.18 (1 + 1)-Evolution strategy (ES)

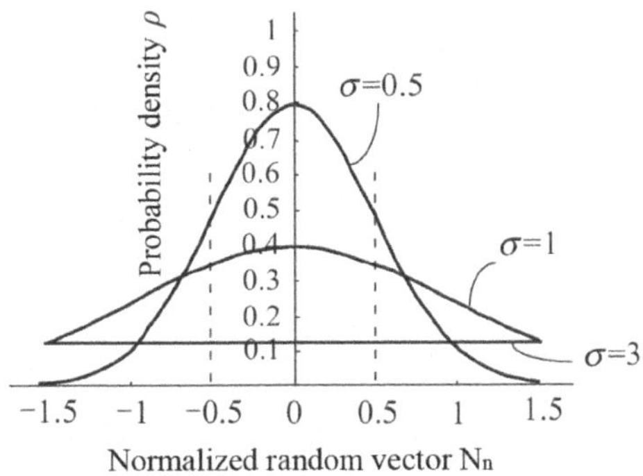

Fig. 3.19 Normalized distribution of probability density

W_p of the parent vector $x_p^{(k)}$, the child vector $x_c^{(k)}$ is not chosen as the parent vector $x_p^{(k+1)}$ at the $(k + 1)$th generation.

But the evolution is not good if optimized using only the mutation and selection operations. Then, Rechenberg's 1/5 success law, which changes the standard deviation at every several iterations, is used. The standard deviation $\sigma(j + 1)$ is decided by the following equation, according to the value of the probability P_s that the objective function gets smaller in, for example, 20(=10 × (number of design variables)) generations:

$$\sigma(j+1) = \begin{cases} \sigma(j) \cdot c_d & \left(P_s < \frac{1}{5}\right) \\ \sigma(j) & \left(P_s = \frac{1}{5}\right) \\ \sigma(j)/c_d & \left(P_s > \frac{1}{5}\right) \end{cases} \tag{3.24}$$

where c_d is a constant less than unity ($0.817 \leq c_d < 1.0$).

Figure 3.19 shows the normalized distribution of probability density at $\sigma = 0.5$, 1.0, and 3.0. If an initial value of a design variable is chosen as a middle value of its constraint (range of the amplitude of design variables), the normalized constraint N_n is within ±0.5 from the initial value. If the initial value σ_0 is equal to 0.5, the probability of searching a solution is higher (probability density p is nearly equal to 0.8) near the parent vector and lower (p is nearly equal to 0.5) at the boundary of the constraint (normalized coordinate $N_n = \pm 0.5$). In the case of $\sigma_0 = 3.0$, p is almost uniform within the constraint. Therefore, the solution can be searched in the entire range of constraint, when $\sigma_0 = 3.0$.

3.4.5 *ON/OFF Method*

1. Method of Analysis

In the usual optimization method, the dimensions, etc., of magnetic circuit are chosen as design variables. This means that the designer who wants to use the

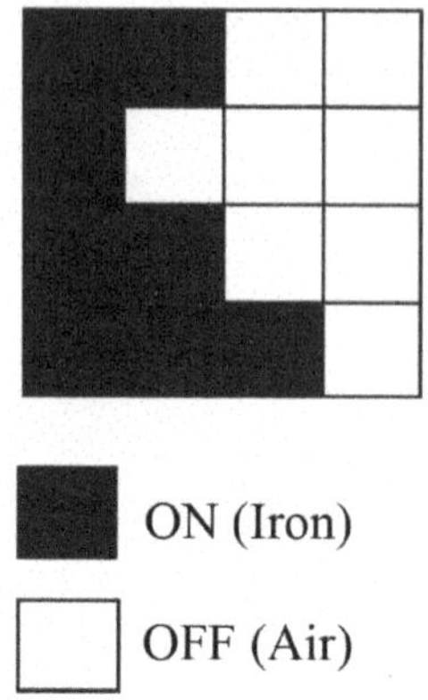

Fig. 3.20 ON/OFF method

optimization method must imagine the shape of the optimal design and its dimensions, etc. As long as we follow such a process, we cannot get a new (epoch-making) product which the designer could not imagine beforehand.

In the ON/OFF method, the design domain is discretized by the finite elements, then the existence of material (e.g., magnetic material) in each element is determined according to the principle that the objective function tends toward the optimal one or not [12]. For example, ON means that there exists a magnetic material and OFF means the air as shown in Fig. 3.20. By using such an ON/OFF method, we may get a new construction of magnetic circuits of electrical machines, etc.

The sensitivity is accurately calculated by using the adjoint variable method [8, 9]. The equation for FEM is given as

$$\boldsymbol{HA} = \boldsymbol{G} \tag{3.25}$$

where $\boldsymbol{H}$ is the coefficient matrix, $\boldsymbol{A}$ is the column vector of magnetic vector potential, and $\boldsymbol{G}$ is the right-hand side vector. Taking the derivative of Eq. (3.25) with design variable p_k in an element k,

$$\frac{\partial \boldsymbol{H}}{\partial p_k}\boldsymbol{A} + \boldsymbol{H}\frac{\partial \boldsymbol{A}}{\partial p_k} = \frac{\partial \boldsymbol{G}}{\partial p_k} \tag{3.26}$$

Equation (3.26) can be rewritten as follows:

$$\frac{\partial \boldsymbol{A}}{\partial p_k} = \boldsymbol{H}^{-1}\left(\frac{\partial \boldsymbol{G}}{\partial p_k} - \frac{\partial \mathbf{H}}{\partial p_k}\tilde{\boldsymbol{A}}\right) \tag{3.27}$$

where $\tilde{\boldsymbol{A}}$ is the value obtained by solving Eq. (3.25). If the objective function is expressed as the function $\boldsymbol{W}(p_k, \boldsymbol{A})$ of the permeability in the design domain and the magnetic vector potential, the sensitivity is given by:

$$\frac{d\mathbf{W}}{dp_k} = \frac{\partial \mathbf{W}}{\partial p_k} + \frac{\partial \mathbf{W}^{\mathrm{T}}}{\partial \mathbf{A}} \frac{\partial \mathbf{A}}{\partial p_k} \tag{3.28}$$

Substituting Eq. (3.27) into Eq. (3.28), Eq. (3.29) can be obtained.

$$\frac{d\boldsymbol{W}}{dp_k} = \frac{\partial \boldsymbol{W}}{\partial p_k} + \frac{\partial \boldsymbol{W}^{\mathrm{T}}}{\partial \boldsymbol{A}} \boldsymbol{H}^{-1} \left(\frac{\partial G}{\partial p_k} - \frac{\partial \boldsymbol{H}}{\partial p_k} \tilde{\boldsymbol{A}} \right) \tag{3.29}$$

In order to avoid the calculation of the inverse of H, an adjoint vector $\boldsymbol{\lambda}$ is introduced [8]. The adjoint equation is given by

$$\boldsymbol{H}^{\mathrm{T}} \boldsymbol{\lambda} = \frac{\partial \boldsymbol{W}}{\partial \boldsymbol{A}} \tag{3.30}$$

$\boldsymbol{\lambda}$ is obtained by solving Eq. (3.30), and $d\boldsymbol{W}/dp_k$ is calculated by substituting $\boldsymbol{\lambda}$ into Eq. (3.31).

$$\frac{d\boldsymbol{W}}{dp_k} = \frac{\partial \boldsymbol{W}}{\partial p_k} + \boldsymbol{\lambda}^{\mathrm{T}} \left(\frac{\partial \boldsymbol{G}}{\partial p_k} - \frac{\partial \boldsymbol{H}}{\partial p_k} \tilde{\boldsymbol{A}} \right) \tag{3.31}$$

Equation (3.31) suggests that only one extra solution for the adjoint vector is sufficient to determine the sensitivity to all parameters, rather than obtaining each value per parameter, providing a computationally fast method for deriving the gradients.

2. Process of Calculation

The flowchart of the ON/OFF method is shown in Fig. 3.21.

Step 1: decision of initial topology
The initial topology in the design domain is decided by the material arrangement in each element.
Step 2 and Step 7: FEM → **W**
The objective function **W** of the design domain is obtained from the calculated value using FEM.
Step 3: adjoint variable method
Solving the adjoint equation (Eq. (3.30)), the sensitivity is calculated by using Eq. (3.31).
Step 4: sorting of elements in order of sensitivity
Each element is ranked in order of the absolute value of sensitivity.
Step 5: set up of changeable elements
If the sensitivity $d\boldsymbol{W}/d\mu_{rk}$ with respect to permeability μ_{rk} in element k is negative, the permeability in the element k should be increased. Then, the magnetic material is located in the element k. On the other hand, if the sensitivity $d\boldsymbol{W}/d\mu_{rk}$ is positive, the permeability in the element k should be decreased. Then, the air is allocated in the element k. In this step, the element, of which the state is changeable, is selected.

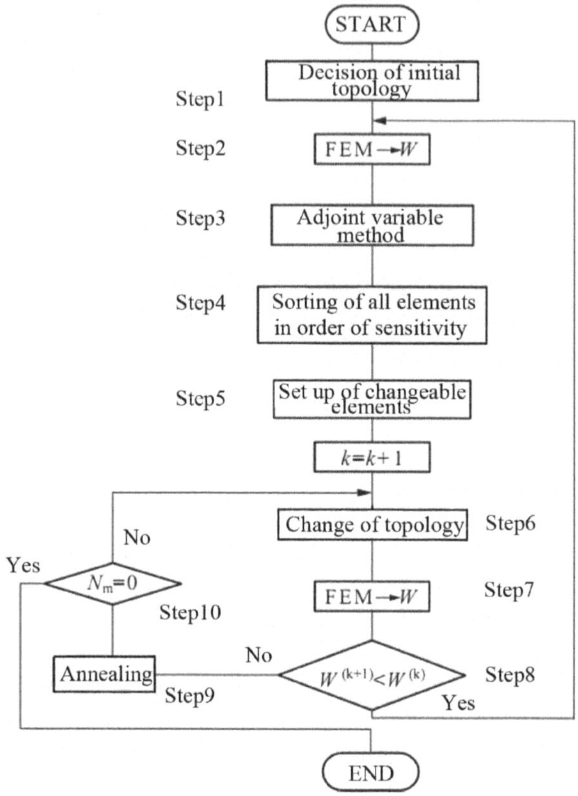

Fig. 3.21 Flowchart of ON/OFF method

Step 6: change of topology
The topology is modified following the information obtained in Step 5.
Step 8: $\boldsymbol{W}^{(k+1)} < \boldsymbol{W}^{(k)}$
If the objective function $\boldsymbol{W}^{(k+1)}$ is smaller than $\boldsymbol{W}^{(k)}$, return to Step 3, otherwise go to Step 9.
Step 9: annealing
If the objective function is not improved at all, the changeable elements N_m is decreased using the following equation:

$$N_{\mathrm{m}} = \gamma \cdot N_{\mathrm{m}} \tag{3.32}$$

where γ is the annealing factor, which is chosen as 0.9. Repeat Steps 6–9 until some improvement of the objective function is detected.
Step 10: $N_{\mathrm{m}} = 0$
If the number of changeable elements N_{m} becomes zero, the computation is terminated. Otherwise, return to Step 6.

Using this algorithm mentioned above, a fast convergence characteristic and a good topology can be obtained.

3.4.6 Example of Application

1. Optimal Design of Transformer Tank Shield Model using EDM and RBM

Figure 3.22 shows the transformer tank shield model [14]. The iron plate is made of steel, and eddy currents are taken into account (conductivity: 0.75×10^7 S/m). The shielding plate is made of grain-oriented silicon steel (grade (JIS): 30G140) in which it is assumed that eddy current flows. The rolling direction of steel is the y-direction. The ampere-turns of the coil are 5484AT (max) (12A (max), 457 turns, 60 Hz). As the steel is not saturated, the phasor method (so-called $j\omega$ method) is applied in 3-D eddy current analysis by assuming that the magnetic characteristic of magnetic material is linear. The relative permeability of iron plate is assumed as 1000, and those in rolling and transverse directions of silicon steel are assumed as 3000 and 30, respectively. The dimension of analyzed region is 1000 mm × 1000 mm × 1000 mm.

The following model having design variables in x-direction, y-direction, and z-direction are optimized by using Rosenbrock's method (RBM). No local heating occurs on the transformer tank at the same time:

L_1–L_9: unknown variables; V: shielding volume

$$V = (L_1L_6 + L_2L_7 + L_3L_8 + L_4L_9)L_5 \ (\mathrm{m}^3) \tag{3.33}$$

This model is optimized so that the objective function W shown in Eq. (3.34) becomes minimum, and the maximum value of eddy current density J_{em} in the iron plate should be less than the specified value J_{emo} (= 0.25×10^6 A/m^2) in order to avoid the local heating:

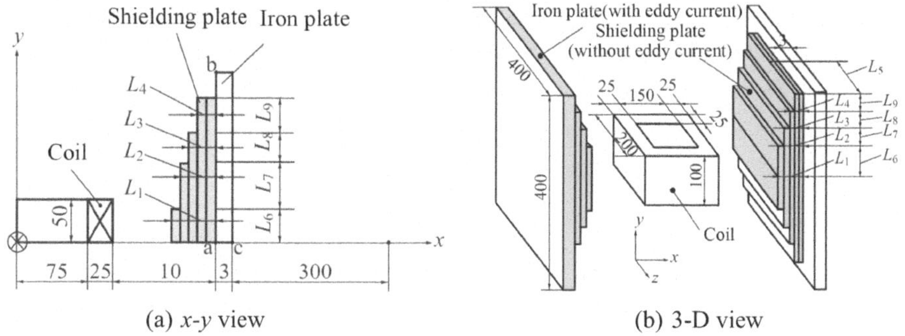

Fig. 3.22 Transformer tank shield model

$$W = V + P \tag{3.34}$$

where P is the penalty function which is defined by

$$P = \begin{Bmatrix} 0 & (J_{\mathrm{em}} < J_{\mathrm{emo}}) \\ J_{\mathrm{em}} & (J_{\mathrm{em}} \geq J_{\mathrm{emo}}) \end{Bmatrix} \tag{3.35}$$

J_{em} is given by

$$J_{\mathrm{em}} = \sqrt{|J_{ex}|^2 + |J_{ey}|^2 + |J_{ez}|^2} \tag{3.36}$$

where $J_{ex,}$ etc., are the complex numbers.

As J_{em} (A/m^2) is of the order of 10^5 and V (m^3) is of the order of 10^{-4}, Eq. (3.34) can be approximated as follows:

$$W = \begin{Bmatrix} V\ (\mathrm{m}^3) & (J_{\mathrm{em}} < J_{\mathrm{emo}}) \\ J_{\mathrm{em}}\ (\mathrm{A/m}^2) & (J_{\mathrm{em}} \geq J_{\mathrm{emo}}) \end{Bmatrix} \tag{3.37}$$

The constraint of L_1–L_9 is given by

$$\begin{aligned} &0 < L_1, L_2, L_3, L_4 < 0.01\ (\mathrm{m}) \\ &0.1 < L_5 < 0.2\ (\mathrm{m}) \\ &0 < L_6, L_7, L_8, L_9 < 0.05\ (\mathrm{m}) \end{aligned} \tag{3.38}$$

The constraints are divided into three levels as shown in Table 3.4. The objective function W which has 27 patterns is calculated from the orthogonal array (Nos. 1–27) as shown in Table 3.5. As J_{em} in all cases is larger than J_{emo} (=0.25 × 10^6 A/m^2), the objective function W is equal to J_{em}. J_{em} of model No. 3 is the smallest in Table 3.5. Then, the design variables of model No. 3 are chosen as initial values.

Table 3.6 shows the obtained result using the combined method of EDM and RBM. The results of 4 and 6 design variables are also denoted [14]. The initial values in Table 3.6 are obtained using EDM. If the change of design variables L_1–L_9 becomes less than 0.1 mm in the process of direct search of RBM, it is judged

Table 3.4 Three-level constraints

Design variables (10^{-3} m)	Levels		
	1	2	3
L_1, L_2, L_3, L_4	2.5	5	7.5
L_5	125	150	175
L_6, L_7, L_8, L_9	12.5	25	37.5

Table 3.5 Orthogonal array

Item	Design variables									V (10^{-4} m^3)	J_{cm} (10^6 A/m^2)
	L_1	L_2	L_3	L_4	L_5	L_6	L_7	L_8	L_9		
1	1	1	1	1	1	1	1	1	1	0.156	2.333
2	1	1	1	1	2	2	2	2	2	0.375	0.814
3	1	1	1	1	3	3	3	3	3	0.656	0.365
4	1	2	2	2	1	1	1	2	2	0.430	1.337
5	1	2	3	1	2	2	2	3	3	0.844	0.455
6	1	3	1	2	3	3	3	1	1	0.820	0.730
7	1	1	3	2	1	1	1	3	3	0.664	0.728
8	1	2	1	3	2	2	2	1	1	0.469	1.204
9	1	3	2	1	3	3	3	2	2	0.984	0.394
10	2	1	2	3	1	2	3	1	2	0.586	0.732
11	2	1	2	3	2	3	1	2	3	0.938	0.536
12	2	1	2	3	3	1	2	3	1	0.711	0.910
13	2	2	3	1	1	2	3	2	3	0.742	0.540
14	2	2	3	1	2	3	1	3	1	0.844	0.815
15	2	2	3	1	3	1	2	1	2	0.602	1.396
16	2	3	1	2	1	2	3	3	1	0.703	0.637
17	2	3	1	2	2	3	1	1	1	0.656	1.024
18	2	3	1	2	3	1	2	2	3	0.875	0.758
19	3	1	3	2	1	3	2	1	3	0.781	0.611
20	3	1	3	2	2	1	3	2	1	0.656	0.949
21	3	1	3	2	3	2	1	3	2	1.094	0.731
22	3	2	1	3	1	3	2	2	1	0.703	0.784
23	3	2	1	3	2	1	3	3	2	0.844	0.555
24	3	2	1	3	3	2	1	1	3	0.984	0.924
25	3	3	2	1	1	3	2	3	2	0.898	0.546
26	3	3	2	1	2	1	3	1	3	0.797	0.859
27	3	3	2	1	3	2	1	2	1	0.766	1.399

that the final (optimal) result is obtained. The convergence criterion of ICCG method is 5×10^{-4}.

Figures 3.23, 3.24, and 3.25 show the shapes, flux and eddy current distributions, and contour lines of eddy current density at the initial and final shapes. Figure 3.24 is the distribution on the surface of the iron plate which is observed from the coil side. Figures 3.23 and 3.25 and Table 3.6 show that the volume V of shielding plate is reduced by using the optimization method, and the eddy current density can be limited within the specified value by the shielding plate. The volume V can be reduced by approximately 35%.

Table 3.7 shows the obtained results using only RBM. The initial values are chosen as the middle value of each constraint of design variable. The volumes V of

Table 3.6 Result using the combined method

		Initial		Optimal		
Number of variables		6	9	4	6	9
Design variables (10^{-3} m)	L_1	2.5		4.25	4.01	4.38
	L_2	2.5		2.00	2.38	2.13
	L_3	2.5		1.00	2.38	1.00
	L_4	2.5		0.50	2.50	1.50
	L_5	175		–	124.4	130.0
	L_6	37.5		–	44.7	46.9
	L_7	37.5		–	44.7	47.5
	L_8	37.5		–	44.7	31.9
	L_9	37.5		–	44.7	37.4
V (10^{-4} m^3)		0.66		0.78	0.63	0.51
J (10^6 A/m^2)		0.365		0.238	0.249	0.247
Number of iterations		–		26	75	90
CPU time (h)		–		8.6	25.4	27.8

Computer application: VT-Alpha533 (SPECfp95: 22.5)
Convergence criterion of RBM: 0.3 mm
Convergence criterion of ICCG method: 5×10^{-4}

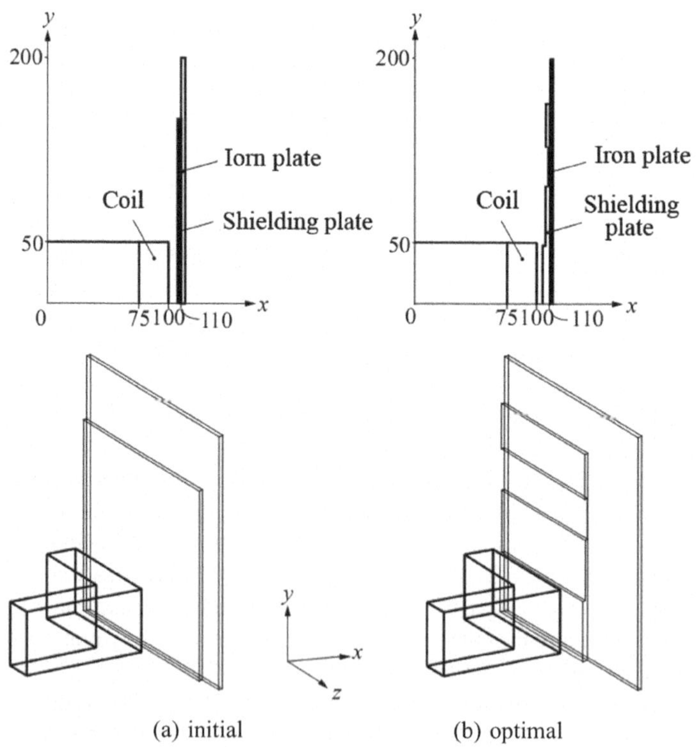

Fig. 3.23 Initial and optimal shapes (ωt = 0°, 9 design variables)

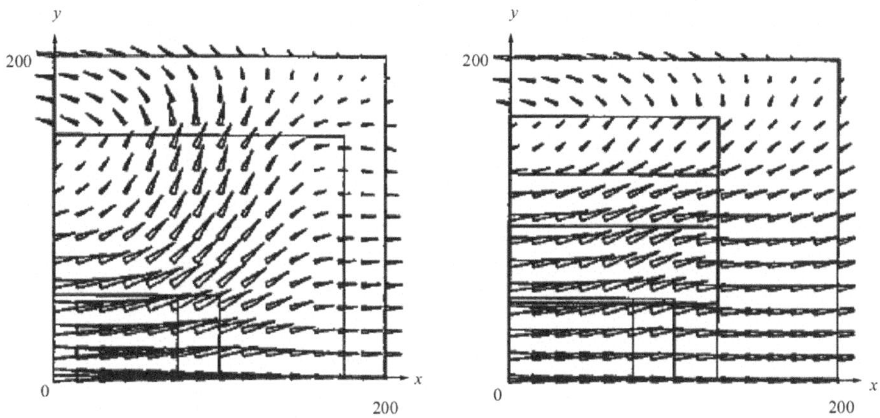

Fig. 3.24 Eddy current on surface of iron plate (ωt = 0°, 9 design variables)

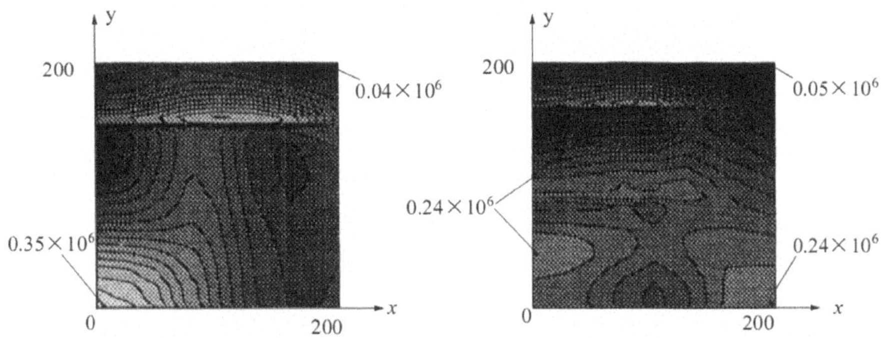

Fig. 3.25 Contour lines of eddy current density on surface of iron plate (ωt = 0°, 9 design variables)

obtained (optimal) results are increased compared with those at initial values. This suggests that the obtained result without EDM has reached to a local minimum. But J_{em} can be within the specified value J_{emo} (=0.25 × 10^6 A/m^2). The number of iterations of optimization was also increased considerably. This suggests that the combination of EDM and RBM is especially effective for such a 3-D optimization.

2. Optimal Design of Induction Heating Model using ES

Figure 3.26 shows the induction heating model [15]. The current of the main coil flows in the x–y plane, and its ampere-turns are 200 kAT(1.2 kHz). The current of the auxiliary coil flows in the z-direction. The steel is heated by the Joule loss due to the eddy current. The relative permeability of steel is assumed as unity. By using

Table 3.7 Results using only RBM

		Initial		Optimal		
Number of variables		6	9	4	6	9
Design variables (10^{-3} m)	L_1	5		4.23	4.81	6.03
	L_2	5		2	7.84	5.09
	L_3	5		0.83	6.22	7.88
	L_4	5		1.18	7.18	6.34
	L_5	150		–	142.7	176.8
	L_6	25		–	40.5	36.5
	L_7	25		–	40.5	35.7
	L_8	25		–	40.5	40.4
	L_9	25		–	40.5	38.1
V (10^{-4} m^3)		0.75		0.83	1.5	1.7
J (10^6 A/m^2)		0.805		0.241	0.247	0.249
Number of iterations		–		29	130	129
CPU time (h)		–		8.8	51.7	49.2

Computer application: VT-Alpha533 (SPECfp95: 22.5)
Convergence criterion of RBM: 0.3 mm
Convergence criterion of ICCG method: 5×10^{-4}

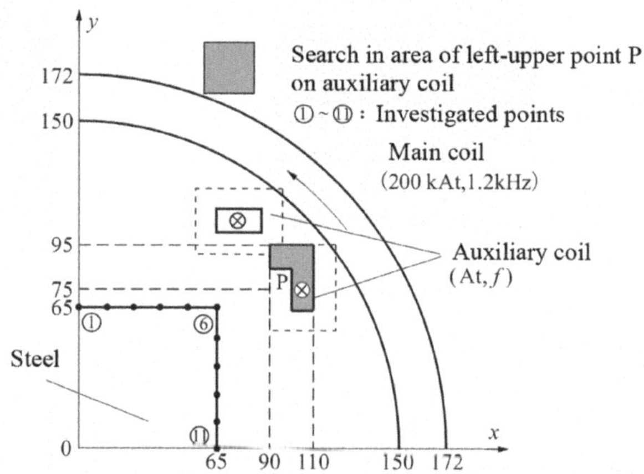

Fig. 3.26 Induction heating model

the evolution strategy (ES), x-coordinate and y-coordinate of the left-upper point P of the auxiliary coil, the maximum ampere-turns AT and the frequency f of the auxiliary coil are optimized so that the distribution of eddy current density at eleven points on the surface of steel shown in Fig. 3.26 becomes uniform. The objective function is defined as follows:

Table 3.8 Initial values and constraint of design variables

Design variables	Constraint	Initial values	
		Case 1	Case 2
x (mm)	$90 \leq x \leq 110$	100	109.9
y (mm)	$75 \leq y \leq 110$	85	75.1
AT (kAt)	$3 \leq \text{AT} \leq 9$	6	8.99
f (kHz)	$1.2 \leq f \leq 3.6$	2.4	1.21

$$\mathbf{W} = \max\{J_{ei}\} - \min\{J_{ei}\} \tag{3.39}$$

where J_{ei} is the maximum value of eddy current density at the investigated point i. Table 3.8 shows the initial values and the constraint of each design variable. Two kinds of initial values of design variables, the middle value of constraint (case 1, $x = 100$ mm, $y = 85$ mm, AT = 6 kAt, $f = 2.4$ kHz) and the value near the boundary (case 2, $x = 109.9$ mm, $y = 75.1$ mm, AT = 8.99 kAt, $f = 1.21$ kHz) are chosen.

The hexahedral edge elements having only one layer in the z-direction are used in the analysis. The number of elements is about 1400. The mesh is automatically produced at each iteration.As this is linear model, first, the flux distribution is calculated when only main coil $\left(J_{0x}^{\text{main}}, J_{0y}^{\text{main}}\right)$ is excited, and that only auxiliary coil $\left(J_{0z}^{\text{aux}}\right)$ is excited. J_{0x}^{main} and J_{0y}^{main} denote the x-component and y-component of the exciting current of main coil. J_{0z}^{aux} is the z-component of the exciting current of auxiliary coil. Second, both flux distributions are superposed. By superposing 2-D results, the CPU time can be considerably reduced compared with 3-D analysis.

Even if the frequency of the x-component and y-component $J_{ex}^{\text{main}}, J_{ey}^{\text{main}}$ of eddy current induced by the main coil is different from the z-component J_{ez}^{aux} of eddy current induced by the auxiliary coil, the maximum effective value $J_{e\max}$ of eddy current can be obtained by the following equation:

$$J_{e\max} = |J_e| = \sqrt{\left|J_{ex}^{\text{main}}\right|^2 + \left|J_{ey}^{\text{main}}\right|^2 + \left|J_{ez}^{\text{aux}}\right|^2} \tag{3.40}$$

where J_{ex} main, etc., are the complex numbers.

Figure 3.27a, b shows the eddy current distributions when only main coil or auxiliary coil is excited. These figures clearly illustrate the role of each coil.

The effects of the initial value of standard deviation and initial value of design variable on the convergence are examined. σ_0 is chosen as 0.01, 0.05, 0.1, 0.5, 1.0, and 3.0. Tables 3.9 and 3.10 show the effects of initial values of standard deviation on obtained value and CPU time, and so on. When the child vector is out of the constraint of design variables, it is not counted in the number of iterations, because the finite element analysis is skipped in such a case.

Tables 3.9 and 3.10 show that the smaller the initial value σ_0 of standard deviation, the faster the convergence. However, the solutions of smaller standard

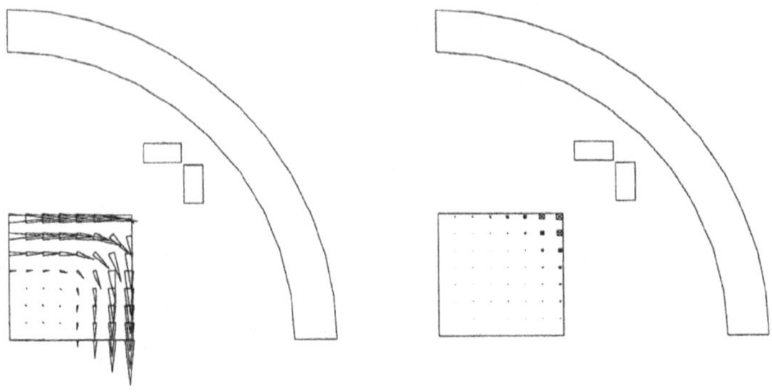

(a) main coil (b) auxiliary coil

Fig. 3.27 Eddy current distribution

Table 3.9 Optimal value (case 1)

Item	Initial values	0.01	0.05	0.1	0.5	1.0	3.0
x (mm)	100	100.2	99.3	101.8	99.1	98.0	97.5
y (mm)	85	89.7	94.4	95.0	94.5	92.2	93.9
AT (kAt)	6	5.91	5.80	6.06	5.78	5.60	5.59
f (kHz)	2.4	3.46	3.59	3.60	3.58	3.59	3.59
W (10^6 A/m$^{2)}$	4.59	1.08	1.05	1.06	1.05	1.04	1.04
Number of iterations	–	424	487	362	633	587	623
CPU time (s)	–	1988	2286	1742	2913	2704	2842

Computer: HP735 (45MFLOPS)

Table 3.10 Optimal value (case 2)

Item	Initial values	0.01	0.05	0.1	0.5	1.0	3.0
x (mm)	109.9	109.7	108.7	96.5	103.4	97.1	98.4
y (mm)	75.1	80.7	88.4	87.4	89.7	92.4	92.7
AT (kAt)	8.99	7.99	7.39	5.33	6.14	5.49	5.71
f (kHz)	1.21	2.94	3.15	3.60	3.60	3.60	3.54
W (10^6 A/m^2)	8.23	1.27	1.17	1.05	1.08	1.04	1.05
Number of iterations	–	490	546	561	583	636	717
CPU time (s)	–	2234	2552	2614	2676	2859	3252

deviations are not good. The solutions at σ_0= 0.01, 0.05, 0.1, and 0.5 fall into local minima. This suggests that the solution using ES converges to the global optimum in spite of the selection of initial values of design variables (case 2) if an initial value of the standard deviation is selected as larger than 1.0. Here, σ_0 is chosen as

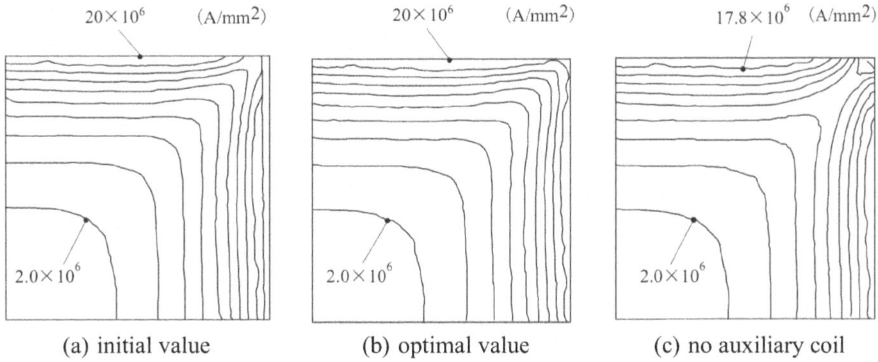

Fig. 3.28 Contour lines of eddy current density in steel (case 1)

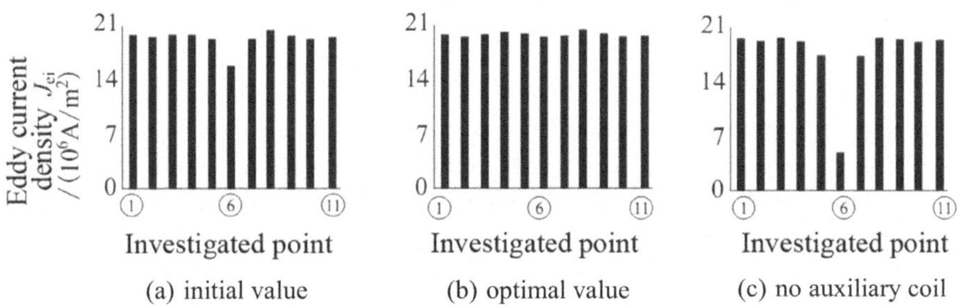

Fig. 3.29 Eddy current density of investigated point (case 1)

3.0, because the solution can be searched in the whole range and the number of iterations is not so much increased compared with the case of less than σ_0= 0.5.

Figure 3.28 shows the contour lines of eddy current density in the steel at the initial and optimal shapes (case 1) at σ_0 = 3.0 and that without auxiliary coil. Figure 3.29 shows the effective eddy current densities at the investigated points. Fairly uniform eddy current is induced in the case of optimized coil.

3. Topology Optimization of Recording Head using ON/OFF Method

A practical 3-D topology optimization technique combined with the edge-based finite element method called the ON/OFF method is applied to improve the design of a recording head [12]. The optimization of cusp field (CF) SPT head having magnetic shield is performed by using the topology optimization technique, so that the recording field becomes maximum and the leakage flux in the adjacent bit and adjacent track can be reduced.

Figure 3.30 shows the analyzed model of CF-SPT head. The ampere-turn of one coil is 0.1 AT. FeCoAlO (saturated magnetization: 2.4 T) is adopted as the magnetic material of the return yoke, the underlayer, a part of main pole, and the design domain corresponding to the yoke.

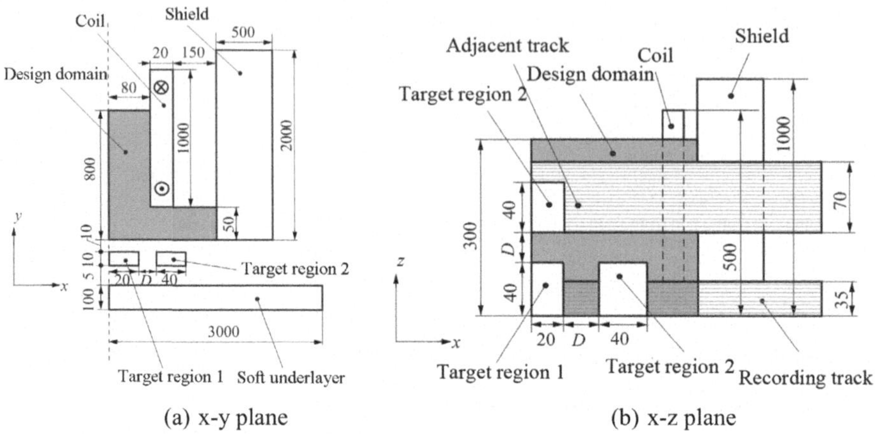

Fig. 3.30 Analyzed model of CF-SPT head

The medium is treated as air. The target region 1, on which a bit should be written, is set under the main pole. The target region 2, on which a bit should not be written is set on both the adjacent track and the adjacent bit in the recording track,

The design goal of SPT head is to maximize the flux density (recording flux) in the target region 1 and to minimize the flux density (leakage flux) in the target region 2, in the medium. The functions W_1 and W_2 in the target regions 1 and 2 to be minimized are given as:

$$W_1 = \iiint_{V_1} 1/B_y^2 \mathrm{d}V \tag{3.41}$$

$$W_2 = \iiint_{V_2} \left(B_x^2 + B_y^2 + B_z^2\right) \mathrm{d}V \tag{3.42}$$

where V_1 and V_2 are the volumes of target regions 1 and 2, and B_x, B_y, and B_z are the *x-component*, *y*-component, and *z*-component of flux density. The objective function W is the linear combination of W_1 and W_2 given as:

$$W = kW_1 + (1 - k)W_2 \tag{3.43}$$

where k ($0 \leqq k \leqq 1$) is the weighting coefficient.

Figure 3.31 shows the optimal topology obtained using the ON/OFF method. The obtained shape is changed by the distance D. The detailed dimension is shown in Fig. 3.32. Figure 3.33 shows the absolute value of the *y*-component of flux density in down-track direction. The figure denotes that the recording field is increased when the distance D between the target regions 1 and 2 is increased. In the case of D = 40 nm, the recording field of 1.75 T is obtained, which is more than the required values for 600 Gb/in^2 recording.

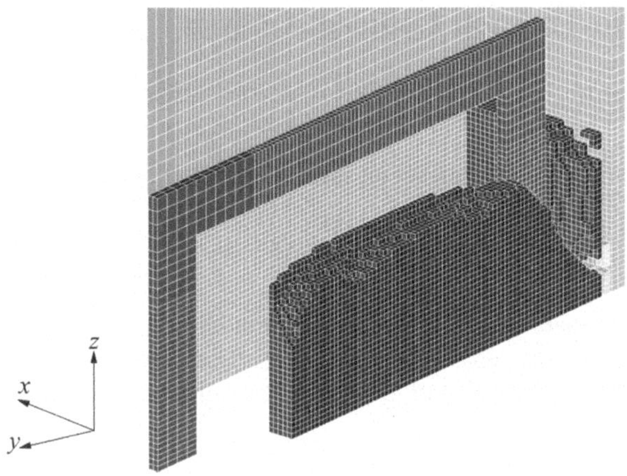

Fig. 3.31 Optimal topology

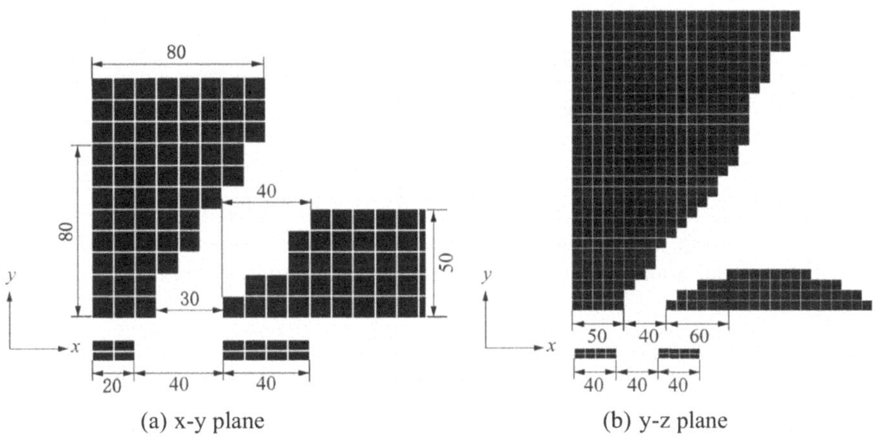

Fig. 3.32 Optimal dimension

Fig. 3.33 $|B_y|$ of target region 1 and target region 2

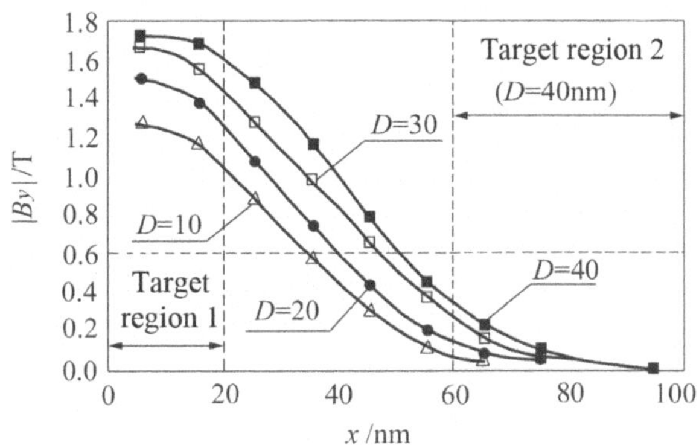

3.5 Magneto-Thermal Coupled Analysis

3.5.1 Thermal Analysis

In the thermal analysis, the following heat conduction equation is used [16]:

$$\mathrm{div}(\lambda\ \mathrm{grad}T) + Q = \rho c \frac{\partial T}{\partial t} \tag{3.44}$$

where T is the temperature, λ is the thermal conductivity, Q is the rate of internal heat generation, ρ is the material density, c is the specific heat, and t is the time.

Equation (3.44) can be rewritten in the x-coordinate, y-coordinate, and z-coordinate as follows:

$$\frac{\partial}{\partial x}\left(\lambda_{xx}\frac{\partial T}{\partial x}\right) + \frac{\partial}{\partial y}\left(\lambda_{yy}\frac{\partial T}{\partial y}\right) + \frac{\partial}{\partial z}\left(\lambda_{zz}\frac{\partial T}{\partial z}\right) + Q = \rho c\frac{\partial T}{\partial t} \tag{3.45}$$

The boundary condition of the heat conduction is given by the following equation:

$$q = -[\lambda]\frac{\partial T}{\partial \boldsymbol{n}} = -[\lambda](\boldsymbol{n}\cdot\nabla T) = -\begin{bmatrix}\lambda_{xx} & 0 & 0\\ 0 & \lambda_{yy} & 0\\ 0 & 0 & \lambda_{zz}\end{bmatrix}\begin{bmatrix}\frac{\partial T}{\partial x}n_x\\ \frac{\partial T}{\partial x}n_y\\ \frac{\partial T}{\partial x}n_z\end{bmatrix} \tag{3.46}$$

where q is the heat flow and $\boldsymbol{n}$ is the outgoing normal vector on the boundary.

The governing equation of heat conduction is discretized using three-dimensional nodal elements. The Galerkin equation is given by

$$\begin{aligned}
&\iint_S q\{N\}^{\mathrm{T}}\mathrm{d}S\\
&+\iiint_V\left\{\frac{\partial\{N\}}{\partial x}\left(\lambda_{xx}\frac{\partial\{N\}^{\mathrm{T}}}{\partial x}\right)+\frac{\partial\{N\}}{\partial y}\left(\lambda_{yy}\frac{\partial\{N\}^{\mathrm{T}}}{\partial y}\right)\right.\\
&\left.+\frac{\partial\{N\}}{\partial z}\left(\lambda_{zz}\frac{\partial\{N\}^{\mathrm{T}}}{\partial z}\right)\right\}\mathrm{d}V\{T\}_{\mathrm{e}}-\iiint_V Q\{N\}\mathrm{d}V\\
&+\iiint_V \rho c\{N\}\{N\}^{\mathrm{T}}\mathrm{d}V\frac{\partial\{T\}_{\mathrm{e}}}{\partial t}=0
\end{aligned} \tag{3.47}$$

where $\{N\}$ is the interpolation function.

The temperature distribution is affected by the heat transfer due to the convection on the boundary between the core or coil and the surrounding air shown in Fig. 3.34. The heat transfer is taken into account by introducing the heat flux q_c on the surface which is given by the following equation:

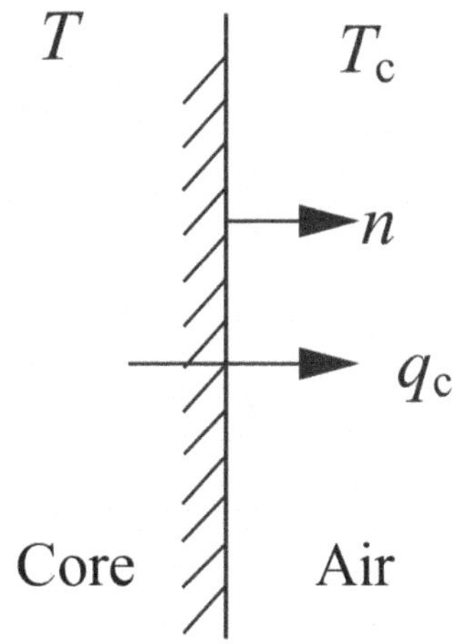

Fig. 3.34 Convection on the boundary

$$q_c = h(T - T_c) \tag{3.48}$$

where h is the heat transfer coefficient, T_c is the ambient temperature.

The equation of heat emission on the boundary is given by

$$q = h(T - T_c) \tag{3.49}$$

$$h = \varepsilon\sigma F^*(T + T_c)\left(T^2 + T_c^2\right) \tag{3.50}$$

where ε is the thermal emissivity, σ is the Stefan–Boltzmann coefficient, F^* is the radiation configuration factor, and h is the coefficient of heat emission.

The finite element (FE) equation of the heat conduction problem which includes the heat emission is given by

$$[K]\{T\} + \{C\}\left\{\frac{\partial T}{\partial t}\right\} = \{F\} \tag{3.51}$$

where $[K]$ is the heat conduction matrix, $[C]$ is the heat capacity matrix, and $\{F\}$ is the heat flux vector. These are given by

$$\begin{aligned}[K] = \iiint_V \bigg\{ & \frac{\partial\{N\}}{\partial x}\left(\lambda_{xx}\frac{\partial\{N\}^{\mathrm{T}}}{\partial x}\right) + \frac{\partial\{N\}}{\partial y}\left(\lambda_{yy}\frac{\partial\{N\}^{\mathrm{T}}}{\partial y}\right) \\ & + \frac{\partial\{N\}}{\partial z}\left(\lambda_{zz}\frac{\partial\{N\}^{\mathrm{T}}}{\partial z}\right)\bigg\}\mathrm{d}V + \iint_S h\{N\}\{N\}^{\mathrm{T}}\mathrm{d}S\end{aligned} \tag{3.52}$$

$$[C] = \iiint \rho c\{N\}\{N\}^{\mathrm{T}}\mathrm{d}V \tag{3.53}$$

$$\{F\} = \iiint_V Q\{N\}^{\mathrm{T}}\mathrm{d}V + \iint_S hT_{out}\{N\}^{\mathrm{T}}\mathrm{d}S \tag{3.54}$$

where T_{out} is the environment temperature.

3.5.2 Magneto-Thermal Analysis

1. Calculation Procedure

The magneto-thermal coupled analysis considering the heat emission, heat conduction, and temperature dependence of magnetic characteristics can be carried out by the following procedure:

As a first step, the magnetic field and eddy current are analyzed using 3-D FEM. The basic equation is given by

$$\mathrm{rot}(v\mathrm{rot}\boldsymbol{A}) = J_0 - \sigma\left(\frac{\partial \boldsymbol{A}}{\partial t} + \mathrm{grad}\phi\right) \tag{3.55}$$

where $\boldsymbol{A}$ is the magnetic vector potential, ϕ is the electric scalar potential, σ is the conductivity, v is the reluctivity, and J_0 is the current density of coil.

Next, the thermal analysis given by Eq. (3.51) is carried out. The eddy current loss obtained by the magnetic field analysis is used as the heat source.

The magnetic field analysis is carried out again by using the renewed material constants corresponding to the obtained temperature. Repeating these two kinds of analyses, the change of the temperature distribution with time is calculated.

2. Example of Analysis of Billet Heater

Figure 3.35 shows an example of billet heater [17]. The material of the billet is S45C (carbon steel). The insulating material is rolled out of the fire-resistant material. The Curie temperature of the billet is 760 °C.

3-D FEM using edge elements is used for the magnetic field analysis, and 3-D FEM using nodal elements is used for the thermal field analysis. Figure 3.36 shows the cross section of the examined model. The initial temperature of each material is 25 °C. The boundaries of analyzed region in Fig. 3.36 are assumed as the adiabatic boundaries.

The material constants of the billet, air, fire-resistant material, and insulating material at 25 °C are shown in Table 3.11. Figures 3.37, 3.38, 3.39, and 3.40 show the temperature dependence of material constants of the billet. The *B–H* curves are guessed from the *B–H* curve at 25 °C. The radiation configuration factors F^* of the billet and fire-resistant material of Eq. (3.50) are both assumed as unity.

Figure 3.41 shows the change of temperature at the center (r = 0 mm) and on the surface of billet (r = 27.5 mm) with time. The temperature is measured using thermocouples. The calculated results are in good agreement with the measured

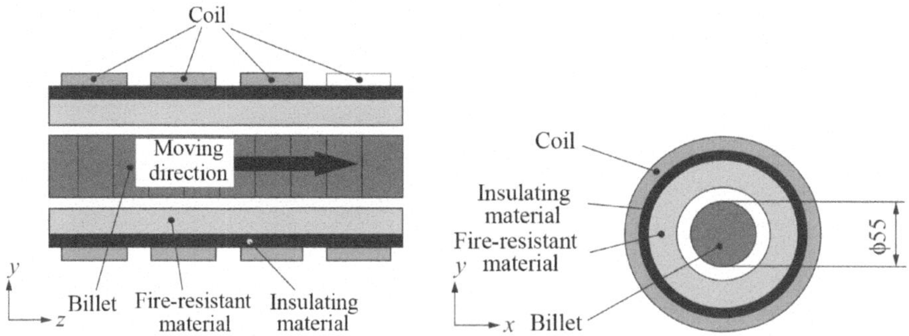

(a) z-y plane (b) x-y plane

Fig. 3.35 Billet heater

Fig. 3.36 Analyzed model of billet heater (1/4 model)

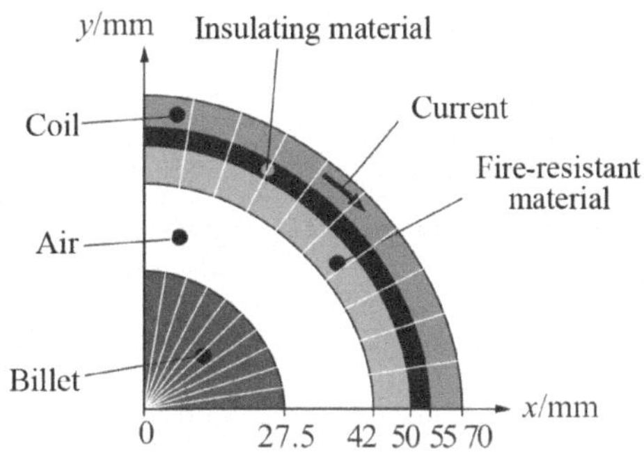

Table 3.11 Material constants at 25 °C

Material	σ (S/m)	C (J/(kg · K))	λ (W/(m · K))	ρ (kg/m^3)	ε
Billet (S45C)	4.17×10^6	444.1	51.9	7860	0.85
Fire-resistant material	–	1360	2.5	3160	0.7
Insulating material	–	1360	0.110	160	–
Air	–	1005	002	230	–

ones. The reason why the rate of increase of temperature is reduced at around 20 s is that the temperature of billet reaches near to the Curie temperature at this instant, and the specific heat rises rapidly.

Figure 3.42 shows the change of flux density in the billet with time. Figure 3.42 shows that the flux is invaded inside the billet and its amplitude is decreased with time. Then, the generation of power moves inside of the billet and its value is decreased.

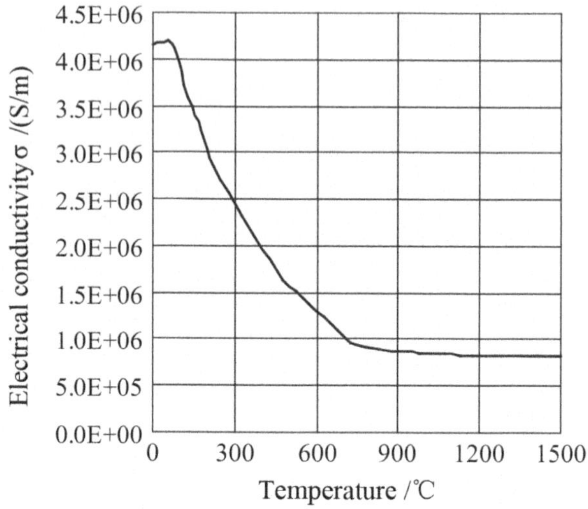

Fig. 3.37 Temperature dependence of electrical conductivity of billet

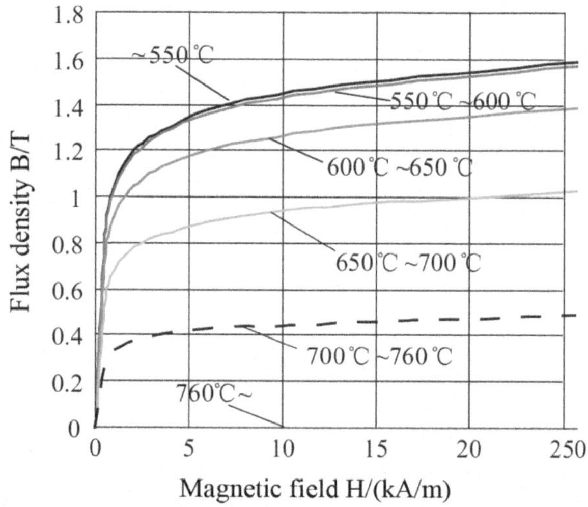

Fig. 3.38 Temperature dependence of *B–H* curves of billet

3.5.3 Magneto-Thermal-Fluid Analysis

1. Method of Analysis

In the magneto-thermal field analysis, in which the heat transfer coefficient is given on the surface of machine, the heat transfer coefficient is sometimes affected by the convection around the machine. In this case, the convection should also be taken into account in order to calculate the temperature rise accurately.

In this section, the magneto-thermal-fluid analysis method taking account of natural convection is discussed [18]. The semi-Lagrange (moving) coordinate

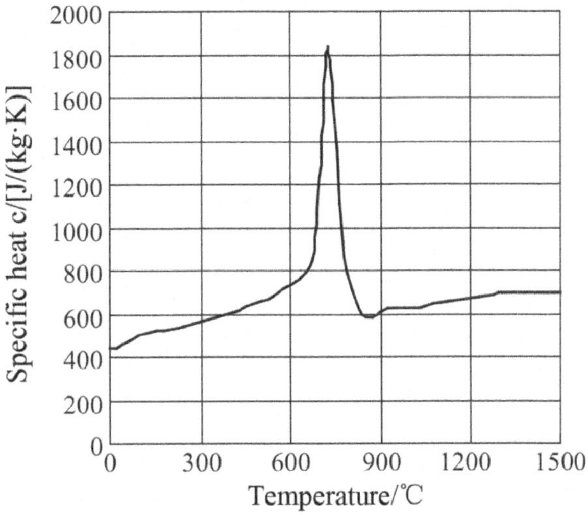

Fig. 3.39 Temperature dependence of specific heat of billet

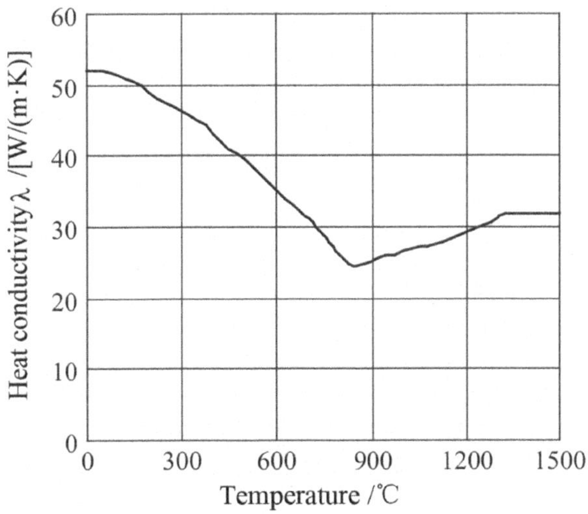

Fig. 3.40 Temperature dependence of heat conductivity of billet

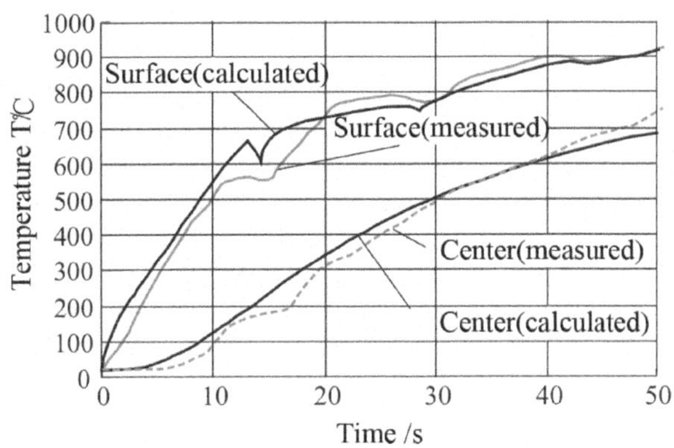

Fig. 3.41 Change of temperature at center and on the surface of billet with time

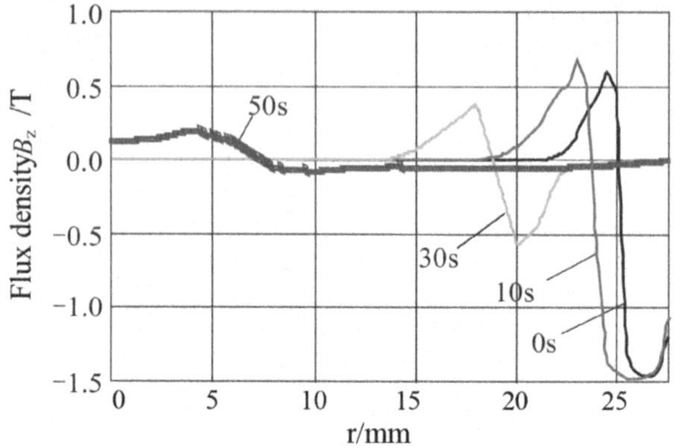

Fig. 3.42 Change of flux density in the billet with time

system, in which the velocity of fluid is defined by the Euler (fixed) coordinate system and the time derivative term is calculated using the Lagrange coordinate system. The Lagrange coordinate system has some advantages compared with the Euler coordinate system in the eddy current analysis in the moving conductor [19].

In the fluid analysis of natural convection, the following Navier–Stokes equation [20] with penalty function is used:

$$\rho_0 \frac{\mathrm{d}\boldsymbol{v}}{\mathrm{d}t} = \alpha \mathrm{grad}(\mathrm{div}\boldsymbol{v}) + \eta \nabla^2 \boldsymbol{v} + \rho_0 \boldsymbol{g} \beta (T - T_0) \tag{3.56}$$

where v is the velocity vector and $\boldsymbol{g}$ is the gravity acceleration vector (=9.8 m/s^2). η and β are the viscosity and the coefficient of volume expansion, respectively. α is the penalty number. d/dt denotes the time derivative using the Lagrange coordinate system. In the first term on the right-hand side (penalty term), the reduced integration technique is applied.

The Lagrange coordinate system is applied to the time differential term in Eq. (3.56). For example, the time derivative $\mathrm{d}T_p^{t+\Delta t}/\mathrm{d}T$ at the point p at the instant (t + Δt) can be discretized using the backward difference method as follows:

$$\frac{\mathrm{d}T_{\mathrm{p}}^{t+\Delta t}}{\mathrm{d}t} = \frac{T_{\mathrm{p}}^{t+\Delta t} - T_{\mathrm{q}}^{t}}{\Delta t} \tag{3.57}$$

where Δt is the time interval. q means the position where the point p at the instant (t + Δt) existed at the instant t. The position of point q_1 in the fluid shown in Fig. 3.43 (case 1) is obtained by deducting $u^t\Delta t$ from the position of point p_1. If the position of point q_2 is in the solid (case 2) or the position of point q_3 is out of the analyzed region (case 3) as shown in Fig. 3.43, the points q_2 and q_3 are moved to the positions of intersections q_2^* and q_3^*.

The flowchart for magneto-thermal-fluid coupled analysis is shown in Fig. 3.44, and it is explained as follows:

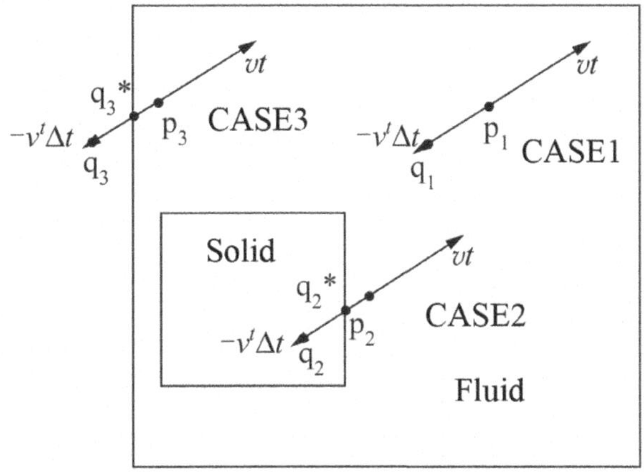

Fig. 3.43 Solid and fluid

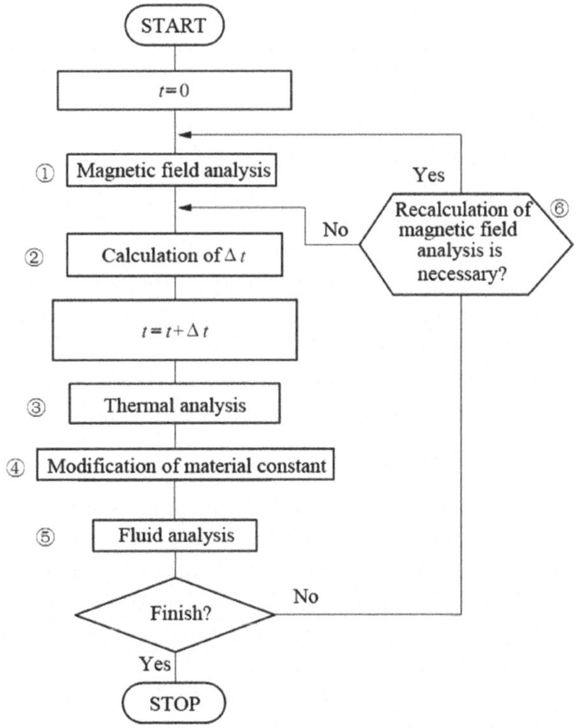

Fig. 3.44 Flowchart

- The distribution of heat source due to eddy current loss in a conducting body (solid) is obtained from the magnetic field analysis. The ac steady-state magnetic field analysis is carried out because the time constant is very small compared with that of thermal analysis.

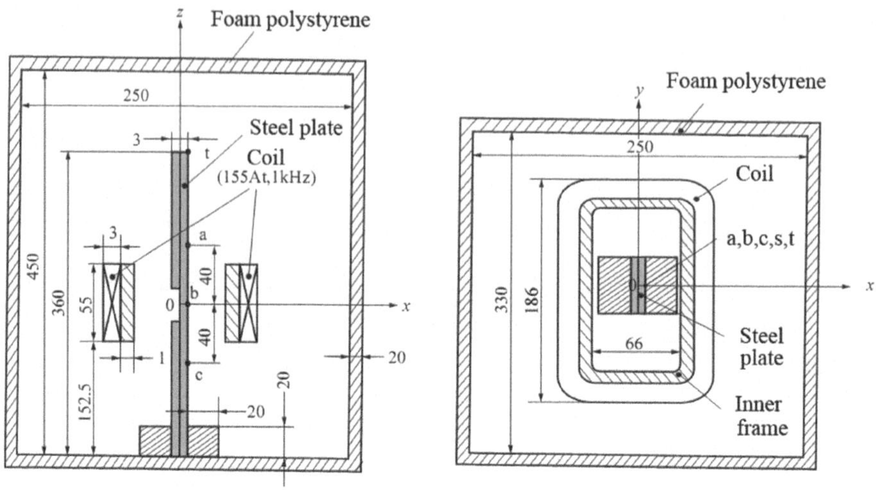

Fig. 3.45 Verification model

- The time interval Δt is calculated by the velocity of fluid considering the maximum movement during Δt.
- The temperature distribution in and around the conducting body is calculated by thermal analysis.
- The material constant for thermal and fluid analysis, such as η, modified by the temperature rise obtained from thermal analysis.
- The velocity of fluid due to natural convection around the conducting body is calculated by fluid analysis.
- The thermal and fluid analyses are iterated until the material constant for magnetic field analysis, such as σ, is modified due to the temperature rise.

2. Analysis of Verification Model

The developed code is verified by comparing the calculated result with measured one. A simple induction heating model composed of a coil and a steel plate (SS400) set in a box made of the heat insulating material (foam polystyrene) shown in Fig. 3.45 is chosen as a verification model [18]. The temperature in the steel plate rises due to the eddy current loss in the steel plate and joule loss in the coil when the coil is excited by ac current (max 155 AT, 1 kHz). The material constants at 20 °C are shown in Table 3.12. The relative permeability of the steel plate is set to 2000. This value is obtained from the measured average flux density (0.144 T) of the steel at $z = 0$. The transient analysis is carried out until $t = 600$ s from the initial temperature 20 °C.

3-D linear ac steady magnetic field analysis using the finite element is carried out. 1/8 of the entire region is subdivided into 15,912 first-order nodal brick elements. The flux and eddy current distributions and the contour lines of eddy current

Table 3.12 Material constants at 20 °C

Material	Steel plate (SS400)	Coil (copper)	Inner frame (paper)	Air
Relative permeability V_r	1/2000	Non magnetic material		
Electrical conductivity σ (S/m)	7.5×10^6	No eddy current		
Mass density ρ_0 (kg/m^3)	7850	8880	900	1.18
Heat capacity C (J/(kg · K))	465	386	1300	1007
Thermal conductivity λ (W/(m · K))	43	398	0.06	0.0261
Viscosity η (Pa · S)	Solid			18.62×10^{-6}
Volume expansion coefficient β				3.32×10^{-3}

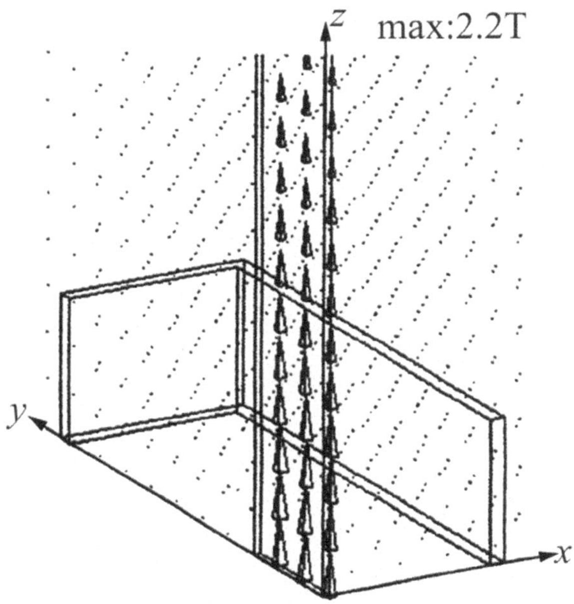

Fig. 3.46 Flux distribution (t = 0, current = maximum)

loss are shown in Figs. 3.46, 3.47, and 3.48. These figures denote that the skin effect in the steel plate is very large.

In the thermal and fluid analyses, 2-D analyses in x–z plane using the finite element method are carried out, because many time iterations are required. 1/2 of whole region is subdivided into 690 first-order nodal rectangular elements. The time interval Δt is decided so that $\upsilon_{max}\ \Delta t$ becomes 20 mm (υ_{max}: maximum velocity) and the maximum Δt is 1 s. The penalty number λ is chosen as 10^7.

Figures 3.49 and 3.50 show the temperature distribution and the natural convection at the instants t = 50, 300, and 600 s. It is shown that the temperature in the upper part is higher than that in the lower part, due to the natural convection. Figure 3.51 shows the effect of the natural convection on the temperature

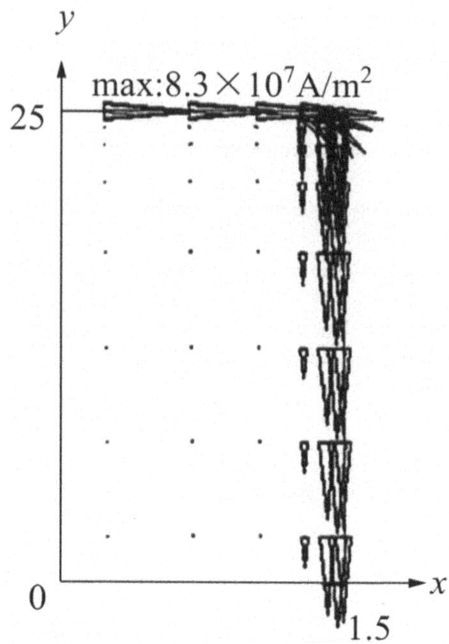

Fig. 3.47 Eddy current distribution ($z = 0$, current = 0)

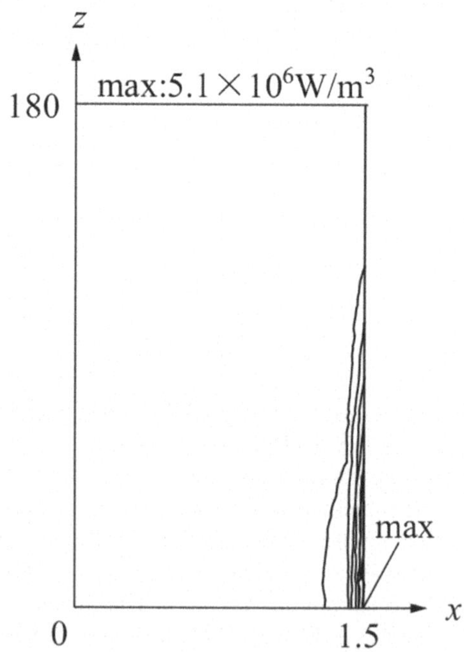

Fig. 3.48 Distribution of eddy current loss ($y = 0$)

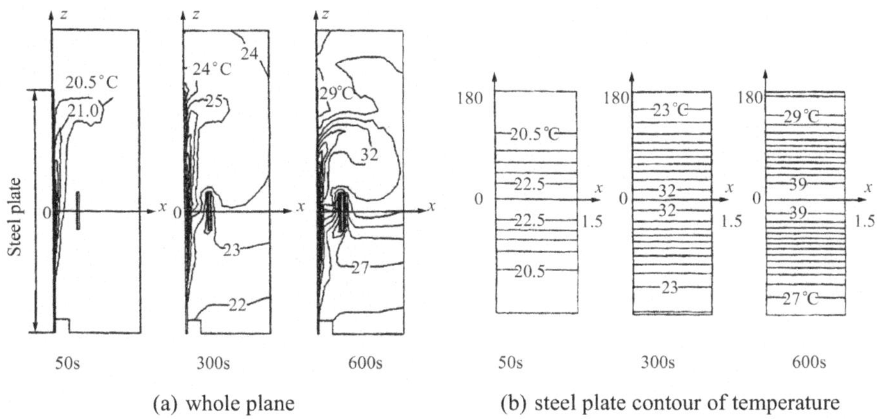

Fig. 3.49 Temperature distribution at the instants

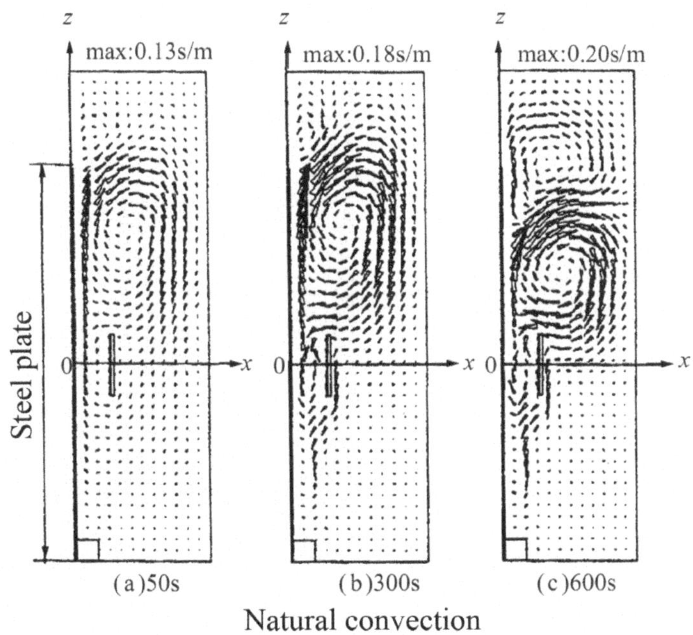

Fig. 3.50 Natural convection at the instants

distribution along the surface $s - t$ of steel plate. The figure denotes that the natural convection should be considered.

The experimental model is shown in Fig. 3.52. The temperature at the points *a, b,* and *c* shown in Fig. 3.53 is measured using the thermocouple (copper–constantan).

Fig. 3.51 Test model

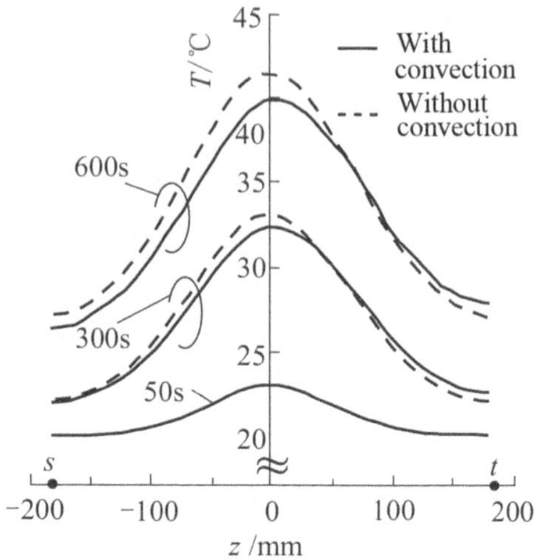

Effect of convection

Fig. 3.52 Test model

Figure 3.53 shows the comparison of temperature rise obtained by measurement and calculation taking account of natural convection. Although there is a difference between calculation and measurement, the tendency is almost the same. The reason for the discrepancy seems to be due to the linear analysis of magnetic field and 2-D thermal and fluid analyses.

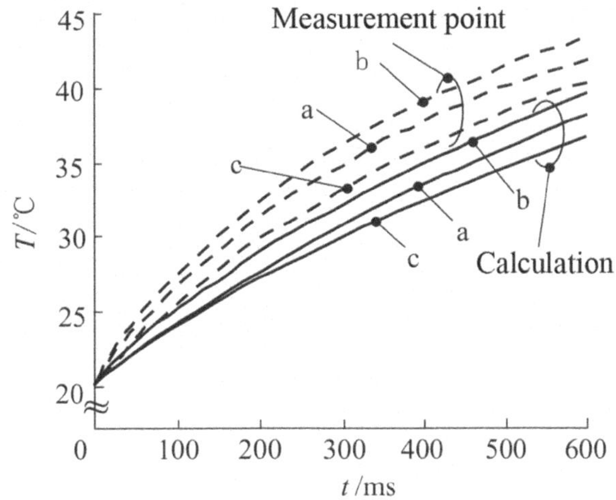

Fig. 3.53 Comparison of temperature rise obtained from measurement and calculation

3.6 Summary

In order to deal with the special and key modeling and simulation in electrical engineering, some practicable application technologies to be used in electromagnetic and thermal FE analysis have been discussed. It can be summarized as follows:

(1) The use of the special finite element in a special problem with narrow air gap can reduce the computing CPU time and the requirement of computer storage;
(2) In order to accurately analyze the overall characteristics of electrical device in the case when exciting currents are unknown, the voltage source excitation method should be used;
(3) Even though the usual optimization design method for the specified design parameters can get an expected result, the topology optimization method can obtain a new and much better result;
(4) The effects of the material temperature property on the electromagnetic and thermal analysis results are investigated and experimentally validated.

References

1. T. Nakata, Y. Ishihara and N. Takahashi: "Finite Element Analysis of Magnetic Fields by using Gap Element", Proceedings of Compumag Conference, No.5.7, 1978.
2. T. Nakata, N. Takahashi, K. Fujiwara, and Y. Shiraki: "3-D Magnetic Field Analysis using Special Element", IEEE Trans. Magn., Vol.26, No.5, pp. 2379–2381, 1990.
3. D.C. Montgomery: "Design and Analysis of Experiments (Third Edition)", John Wiley and Sons, 1991.

4. E. Aarts and J. Korst: "Simulated Annealing and Boltzmann Machines", John Wiley and Sons 1990.
5. T. Nakata and N. Takahashi: "Direct Finite Element Analysis of Flux and Current Distributions under Specified Conditions", IEEE Trans. Magn., Vol.18, No.2, pp. 325–330, 1982.
6. T. Nakata, N. Takahashi, K. Fujiwara, and A. Ahagon: "3-D Finite Element Method for Analyzing Magnetic Fields in Electrical Machines Excited from Voltage Sources": IEEE Trans. Magn., Vol.24, No.6, pp. 2582–2584, 1988.
7. J. Simkin and C.W. Trowbrige: "Optimizing electromagnetic devices combining direct search methods with simulated annealing", IEEE Trans. Magn., Vol. 28, No. 6, pp. 1545–1548, 1992.
8. T. Bäck: "Evolutionary Algorithm in Theory and Practice", Oxford University Press, 1996.
9. K. Preis, C. Magele, and O. Biro: "FEM and evolution strategies in the optimal design of electromagnetic devices", IEEE Trans. Magn., Vol. 26, No. 5, pp. 2181–2183, 1990.
10. D.M. Himmelblau: "Applied Nonlinear Programming", McGraw-Hill, 1972.
11. J.K. Byum, J.H. Lee, I.H. Park, H.B. Lee, K. Choi, and S. Y. Hahn: "Inverse problem application of topology optimization method with mutual energy concept and design sensitivity", IEEE Trans. Magn., vol.36, no.4, pp. 1144–1147, 2000.
12. Y. Okamoto, K. Akiyama, and N. Takahashi: "3-D Topology Optimization of Single-pole-type Head by using Design Sensitivity Analysis", IEEE Trans. Magn., Vol. 42, No. 4, pp. 1087–1090, 2006.
13. K. Hameyer and M. Kasper: "Shape optimization of a fractional horse-power dc-motor by stochastic methods", in Computer Aided Optimum Design Structures III, Optimization of Structural Systems and Applications, CMP: Elsevier Applied Science, pp. 15–30, 1993.
14. M. Horii, N. Takahashi, and J. Takehara: "3-D Optimization of Design Variables in *x*-, *y*- and *z*- Directions of Transformer Tank Shield Model", IEEE Trans. Magn., Vol. 37, No. 5, pp. 3631–3634, 2001.
15. M. Horii, N. Takahashi, and T. Narita: "Investigation of Evolution Strategy and Optimization of Induction Heating Model", IEEE Trans. Magn, Vol. 36, No. 4, PP. 1085–1088, 2000.
16. Y. Lee, H.B. Lee, S.Y. Hahn, and K.S. Lee: "Temperature Analysis of Induction Motor with Distributed Heat Sources by Finite Element Method", IEEE Trans. Magn., Vol. 33, No. 2, pp. 1718–1721, 1996.
17. H. Kurose, D. Miyagi, N. Takahashi, N. Uchida, and K. Kawanaka: "3D Eddy Current Analysis of Induction Heating Apparatus Considering Heat Emission, Heat Conduction and Temperature Dependence of Magnetic Characteristics", Proceedings of CEFC Conference, 2008.
18. K. Muramatsu, N. Takahashi, and T. Mimura: "Magneto-Thermal-Fluid Analysis Taking Account of Natural Convection using Lagrange Coordinate System", IEEE Trans. Magn., Vol. 35, No. 3, pp. 1670–1673, 1999.
19. K. Muramatsu, T. Nakata, N. Takahashi, and K. Fujiwara: "Comparison of Coordinate Systems for Eddy Current Analysis in Moving Conductors", IEEE Trans. Magn., Vol.28, No.2, pp. 1196–1189, 1992.
20. T.J. Chung: "Finite Element Analysis in Fluid Dynamics", McGraw-Hill, 1978.

Chapter 4
Solution of Coupled Electromagnetic and Thermal Fields

Behzad Forghani

Abstract The objective of this chapter is to discuss the electromagnetic and thermal simulation requirements when designing large power transformers; in particular, the focus will be on the study of overheat problems in the transformer tank due to the leakage flux and the induced eddy currents. There are a number of requirements for the model specification, the field solution, and the evaluation of the results, related to the electromagnetic performance, as there are a number of requirements for the model specification, the field solution and the evaluation of results, related to the thermal performance, of a power transformer. The model specification covers the geometric description, material properties of the components used in the device, current and voltage sources, as well as the numerical parameters, such as those related to the finite element method (FEM) (Silvester and Ferrari in finite elements for electrical engineers, Cambridge University Press, 1996 [1]). The coupled electromagnetic–thermal simulation requires the solution of two sets of equations. Since, in the overwhelming majority of cases, the time constants between the electromagnetic and thermal response are very different, the two sets of field equations can be solved separately; in other words, they are loosely coupled. The electromagnetic field equations may require a static, time-harmonic, or transient field solver, whereas the thermal field equations may require a static (steady-state) or transient field solver. The coupled electromagnetic–thermal simulation addresses both aspects of physics, i.e., electromagnetic and thermal, and the way the effect of one on the other is taken into account, by considering the temperature rise due to electromagnetic losses, and the effect on the material properties due to the change in temperature. Material property modeling plays a key role in the accurate simulation of the device. Since the magnetic properties of steel are non-linear and hysteretic, and anisotropic for the grain-oriented steel used in transformers, advanced material models are needed for an accurate representation of the material, under the different operating conditions of a transformer. Simcenter™ MAGNET™ software is a general-purpose simulation tool on which the content of this chapter is based. This tool can be used for the design and analysis of many

B. Forghani (✉)
Mentor Infolytica, a Siemens Business, Montreal, QC, Canada
e-mail: Behzad.Forghani@siemens.com

Z. Cheng et al. (eds.), *Modeling and Application of Electromagnetic and Thermal Field in Electrical Engineering*, https://doi.org/10.1007/978-981-15-0173-9_4

devices. More in-depth information on its general capabilities can be found at (Simcenter MAGNET knowledge base articles [2]).

Keywords Leakage flux · Overheat problems · Coupled simulation · Electromagnetic–thermal · Material properties

4.1 Simulation as a Design and Analysis Tool

The demand for electrical power has been increasing globally, and as a result, power transformers, which are one of the important components in a power generation and transformation network, are regularly re-designed in order to meet this demand. Building power transformers is quite costly, both in terms of the material and the labor cost. The time it takes to build one unit of 500 or 1000 kV transformer is of the order of 6–10 months (or more for UHV DC converter transformer, up to 1 year), because of the high standard manufacturing and testing procedures that should be followed, as usual. For this reason, it is not practical to build prototypes or make any mistakes in the design, making the simulation and the accurate computation of the performance parameters critical steps in the design process.

Additionally, there are safety issues which are of utmost importance to avoid and prevent potential failures. The reliability in life cycle of the operation of a transformer is key since the downtime of a transformer could have serious economic consequences. There are also regulations that are concerned with the efficiency of the operation of transformers and the reduction of losses, such as the minimum energy performance standard (MEPS), which pose challenging requirements on the design, and the accuracy and capability of the design software tools.

There are several types of simulations that a designer of a large power transformer would be interested in. The forces experienced on the windings of a transformer, the reactance and the impedance of the windings, the losses in the core, in the windings and in the tank, and a number of other design parameters could be the result of computations performed in a simulation tool. The study of the potential overheat in key transformer components is, however, a class on its own and deserves its own treatment. Simcenter MAGNET simulates the electromagnetic behavior of a device by solving the static, time-harmonic, or transient electromagnetic field due to either permanent magnets, current distribution sources, or both. Additionally, it simulates the temperature distribution of specified heat sources in the presence of thermally conducting materials and convective/radiative boundary losses. Full integration between the electromagnetic and thermal solvers provides accurate results for coupled electromagnetic–thermal analysis needs. Two-way coupling helps to simulate the heating effects due to eddy current and hysteretic losses in the magnetic system. For the majority of cases, a time-harmonic electromagnetic field solution which is coupled to a steady-state thermal field solution is adequate for the analysis of stray-field losses and the resulting overheat problem. In this type of simulation, it is required to determine the leakage flux distribution in a 3D model and the resulting loss density inside various magnetic and non-magnetic components. From the loss density, the thermal profile can be determined. The leakage flux and the eddy current losses are determined from a time-harmonic magnetic solver, and the temperature profile is determined from a steady-state thermal solver.

Simcenter MAGNET can be used to simulate the transformer in operation and to determine its behavior using a virtual laboratory, rather than a physical one. The ability to construct a model in a CAD system and simulate the design is the only way to make sure that all design failures are caught before an expensive unit is built. The design cost is reduced significantly since building a large transformer requires a significant investment in material and labor. Since it is possible to study what-if scenarios quickly in a design system, the design time can be significantly reduced. The accuracy that can be achieved in the fields and other quantities of interest in the design can be very high with a simulation system that is based on the finite element method (FEM). Using FEM, the electromagnetic and thermal field equations can be solved for quite accurately and very complex structures with complicated material characteristics can be modeled.

Using a simulation tool based on FEM, it is possible to determine the current distributions through the windings and the bus bars, no matter how complex the geometry and the arrangements. It is also possible to obtain the impedance/reactance of complex bus bar arrangements. Given the distribution of the fluxes and the currents throughout the tank and other areas in the model, the losses can be obtained. It is then possible to change the geometry and introduce shields to reduce such losses. From the losses, the temperature profile can be determined and examined throughout the model.

One of the requirements for a simulation is the ability to perform the same tests in the virtual laboratory that the designer would normally perform in the real laboratory. These tests are meant to validate the design under operating conditions. For this purpose, electrical circuits can be defined in which the windings or the bus bar terminals are connected to sources or loads. The circuit components can take complex topologies when connected in parallel and in series.

Simcenter MAGNET has an advanced geometric and material modeling capability, including the support for temperature-dependent material properties. The parameterization capabilities allow for "what-if" analysis. The programmability and customization support make it possible to use the system in an established design flow environment of an organization.

4.2 Modeling

The models for the magnetic and thermal solutions of large power transformers are truly 3D and require 3D field equation solvers. The magnetic model consists of several components, and each component may be built with much of its geometric detail. The components are typically the core, the low-voltage (LV) and high-voltage (HV) windings, the tank, and the bus bar structure. The tank itself is built with the tank details such as the walls, the cover, and flanges.

The thermal model has fewer components and consists of only those components where power is dissipated and where heat is transferred to, through the conduction mechanism. The surrounding air or the oil are not included in the thermal model, but the transfer of heat, through either convection or radiation, to the surrounding environment is modeled by specifying coefficients of heat transfer for convection and radiation on the surfaces of the objects.

In this type of application, the heat is normally generated due to the local power losses in a component, such as the tank, which then travels through that component and to other touching components, through the mechanism of conduction or is transferred to the environment through convection. The environment could be the surrounding air or the oil.

4.2.1 Geometry

The complex 3D geometry of the transformer, with all the details, can be created either in Simcenter MAGNET or imported from third-party CAD systems.

If built-in Simcenter MAGNET, the geometric model of a device is based on familiar component-by-component building concepts; i.e., a geometric model is formed from one or more components. A component is created by drawing the cross section and then sweeping that cross section in an arbitrary direction to form a 3D object. The 2D drawings can also be imported as a *DXF* file. When importing a 2D or 3D geometry, the system helps in detecting potential modeling errors that are inconsequential as a CAD drawing but cause errors when doing a numerical solution. The cause of these errors could be tolerance issues with the coordinates of vertices where edges do not meet exactly, ending up in either a gap or an overlap.

The imported 3D components from other CAD systems are usually in standard formats. In a model in Simcenter MAGNET, some components may be imported from other CAD packages and some may be built inside Simcenter MAGNET. There are several Boolean operations, such as *union, subtract,* and *intersect*, that let the user further modify the geometry. The operations of *scale, rotate, mirror,* and *shift* can be used on a component or a set of components in order to transform and possibly duplicate the components making the construction of a repetitive structure fast and easy.

Figure 4.1 shows a portion of the model with all the geometric detail necessary for performing an accurate electromagnetic and thermal simulation of a transformer.

Figure 4.2 shows several conducting parts that are connected to form the complex structure of the bus bars.

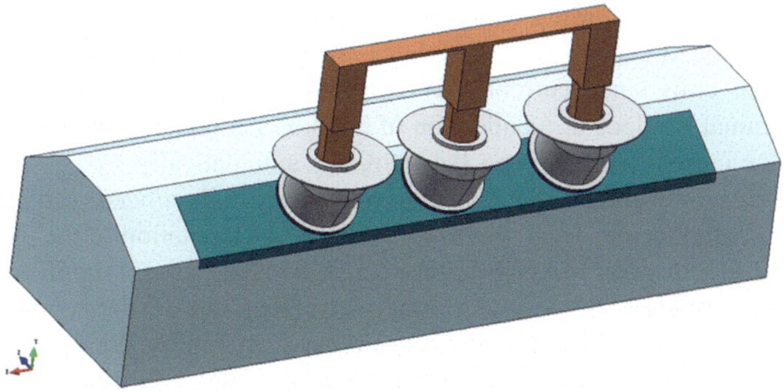

Fig. 4.1 Outside section of the tank (Figure, courtesy of Baobian Electric)

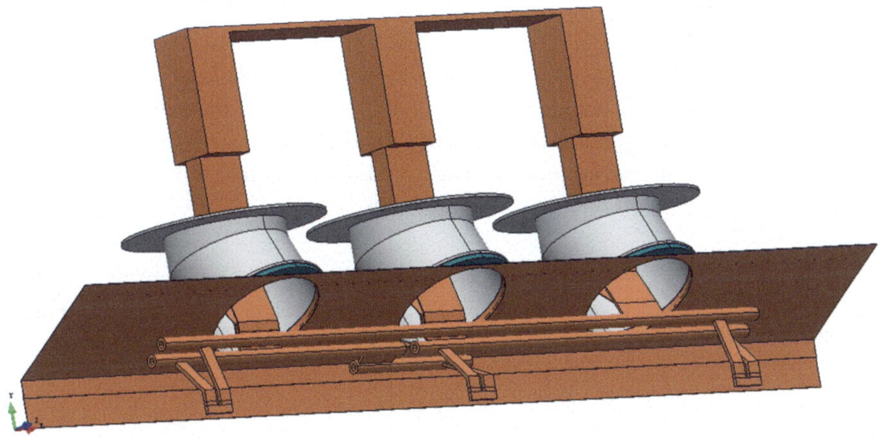

Fig. 4.2 Three bus bars inside of the tank (Figure, courtesy of Baobian Electric)

4.2.2 Coils and Sources

Coils are conducting components with attributes, such as stranded or solid, and the number of turns, when stranded. The windings around the limbs of a transformer are specified to be stranded coils. In this case, the individual turns are not modeled and it is assumed that the cross-sectional current density is uniform; this is an acceptable simplification and reduces the computational cost significantly. The correct copper cross-sectional area can be determined, if the strand area or wire gauge is known, which helps with the correct calculation of the winding resistance, affecting other computed parameters. The terminals of the coils are face or faces that become the connecting points to an electrical circuit.

The bus bar structure is specified to be a solid coil. When a coil is defined as solid, the system solves for the current density throughout the conductors and the current redistributes according to Maxwell's equations. The bus bars have a complex geometric structure. In Simcenter MAGNET, they are modeled as multi-terminal coils since current may enter and exit at a number of terminals, where a terminal is a face or a collection of faces of a conducting body. The system computes the current paths among the terminals automatically.

Sources are circuit components that are connected to coils either directly or through other circuit components. They are responsible for the excitation of the electromagnetic system. In Simcenter MAGNET, the sources could be current supplies or voltage supplies. They could be DC, AC, or arbitrary waveforms that are defined as a function of time, such as *sinusoidal, exponential,* or *pulse.* Arbitrary waveforms may be specified by providing a series of current or voltage values over time.

4.2.3 Circuits

In Simcenter MAGNET, there is a *circuit editor* for connecting electrical circuit components to the coils. There is no restriction with regard to the complexity of the circuit connections. The circuit components supported are *current source, voltage source, resistor, capacitor, inductor, switch,* and *diode*. Figure 4.3 shows a circuit with a number of parallel and series connections.

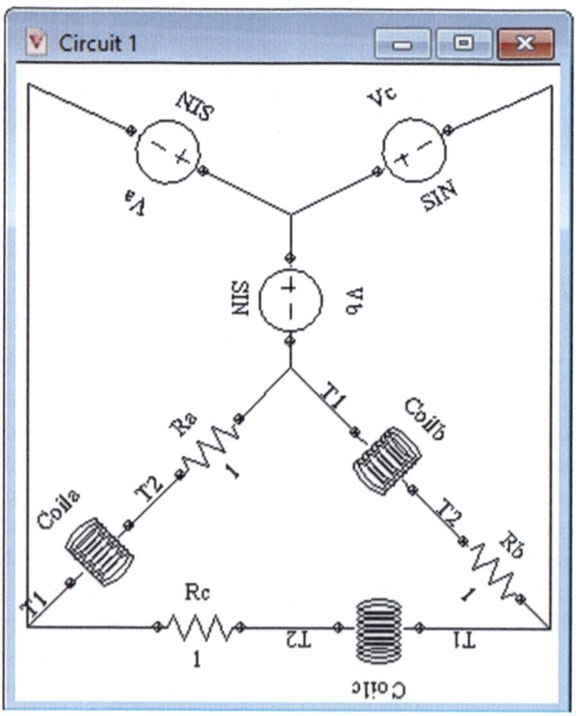

Fig. 4.3 An example of a circuit in Simcenter MAGNET *circuit editor*

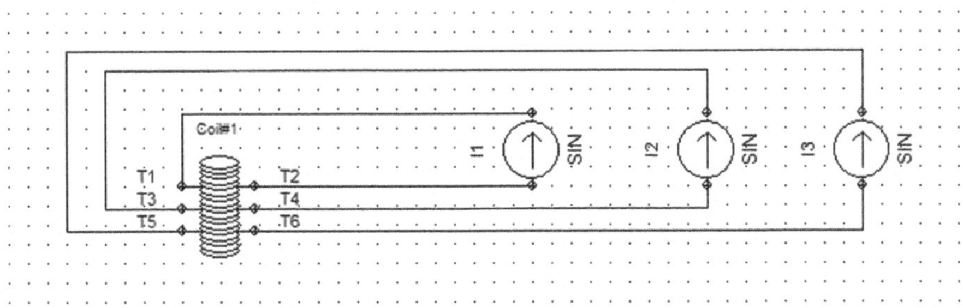

Fig. 4.4 Three-phase current source connected to a bus bar

Figure 4.4 shows the circuit that is used to set up a typical simulation where the three-phase current sources are connected to the six terminals of the bus bars in a transformer.

The solution of the electromagnetic field and electrical circuit equations can be obtained in a number of ways. One of which is a fully coupled system of equations where all the unknowns, i.e., the electromagnetic field unknowns over the discretized space, and the branch currents or the node voltages of the electrical circuit, are all solved for, at the same time. The case shown in the above figure is one such condition where it is possible to solve for the entire system of equations in one pass. In this case, the electromagnetic and the electrical circuit equations are said to be tightly coupled.

There may be other applications where there are much more complex electrical circuits involved and it may be more practical to decouple the two systems of equations and solve them separately and link them through either a *response surface model* [3] or by running two time-domain solution systems in parallel, otherwise known as *co-simulation*.

The ability to use the full circuit capability with the time-harmonic analysis, in Simcenter MAGNET, enables the user to have phase-shifting circuits, without having to use the transient analysis, saving significant solution time, since the AC analysis is a single-step solution while the transient analysis requires many steps.

4.2.4 Material Properties

The material properties are needed when solving for the electromagnetic and thermal fields. The electromagnetic equations require the permeability and the conductivity of all the materials in the problem domain. The materials used in the core are electrical grade laminated steel. The grain-oriented steel, normally used in the core, exhibits an anisotropy in its permeability such that the properties are not the same in the rolling and transverse directions. The material used in the tank is usually a mild steel. Some materials may be nonlinear; e.g., the permeability of the laminated steels is nonlinear and is provided by the manufacturer as a series of

B–H values. The thermal equations require the thermal conductivity, specific heat capacity, and the mass density for each material.

The material properties could change as a function of temperature. Therefore, each property can be specified at a number of temperatures and interpolation can be used to derive values at other temperatures in-between.

The material manufacturer makes measurements of the iron losses in the lamination steels, at different induction levels and source frequencies. The core loss can be calculated directly from these measurements, at the post-processing stage, or through advanced material models that simulate the magnetic hysteresis.

4.2.5 Material Modeling

The accuracy of the solution is only as good as the input data and the accuracy of the material properties play a significant role in the accuracy of the results. Simcenter MAGNET uses a *material editor*, for entering the material data. The following is a list of the primary material properties for the simulation of electromagnetic and thermal fields:

- Magnetic permeability
- Core loss data
- Electric conductivity
- Thermal conductivity
- Specific heat capacity
- Mass density.

All the above material properties are associated with a temperature. Therefore, the user can enter many values representing the change in the property as temperature changes. Depending on the type of analysis, one or more of the properties are used during the solution process.

4.2.5.1 Permeability

Figure 4.5 shows the dialog box for entering the properties of a nonlinear isotropic permeability for a material. For each temperature that the user has data, a set of *B–H* values are entered and a smooth nonlinear curve is automatically constructed.

The *material editor* can also be used to enter the nonlinear anisotropic *B–H* properties of laminated steel. Figure 4.6 shows the nonlinear *B–H* properties in different directions.

When assigning a nonlinear anisotropic material to a component, the user can orient the material direction to match the position and orientation of the component. For example, in the case of an anisotropic laminated sheet, the user would identify the rolling and transverse directions to the system by setting the orientations.

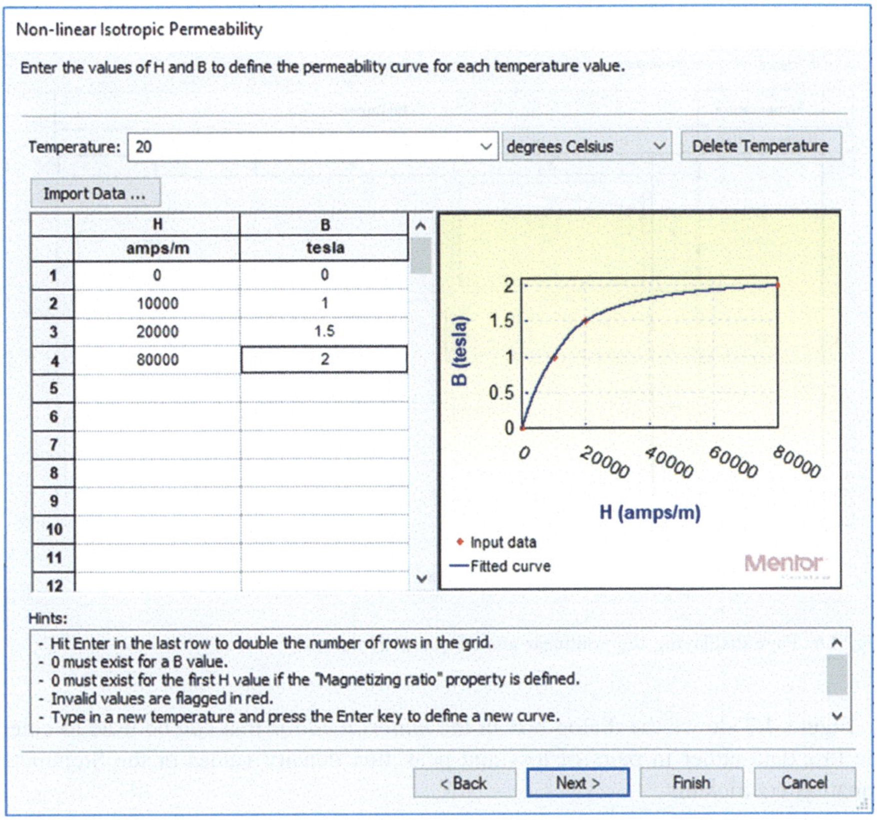

Fig. 4.5 Dialog box for entering the nonlinear isotropic permeability data

4.2.5.2 Core Loss

In general, the winding losses and the eddy current losses in the conductors are the dominant losses in a device. However, the core which is made of thin insulated steel laminations also experiences a loss, when subjected to time-varying fields. Since this loss is usually small, it is not included as part of the electromagnetic field solution, but rather, it is estimated at the post-processing stage, by looking up the measured loss data, based on the induction level and the frequency. The measurements are usually made for sinusoidally varying fields, at certain frequencies and at several induction levels, using the Epstein frame. This simple model is basically a lookup table, and the variation of the loss with the induction level can be represented by a simple curve which satisfies the Steinmetz equation [4], having a term that represents the lamination eddy current loss and a term that represents the lamination hysteresis loss. This is an empirical approach; it is quick to compute and is most often used in simulations.

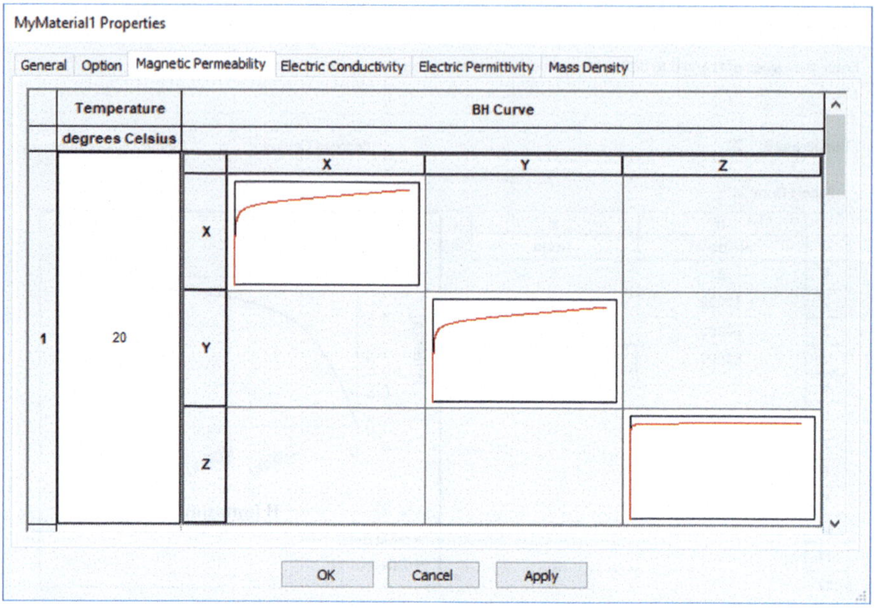

Fig. 4.6 Page displaying the nonlinear anisotropic *B–H* curves

Figure 4.7 shows the dialog box in the *material editor* that can be used to enter the loss data either in pairs of loss and peak flux density values or the Steinmetz equation coefficients.

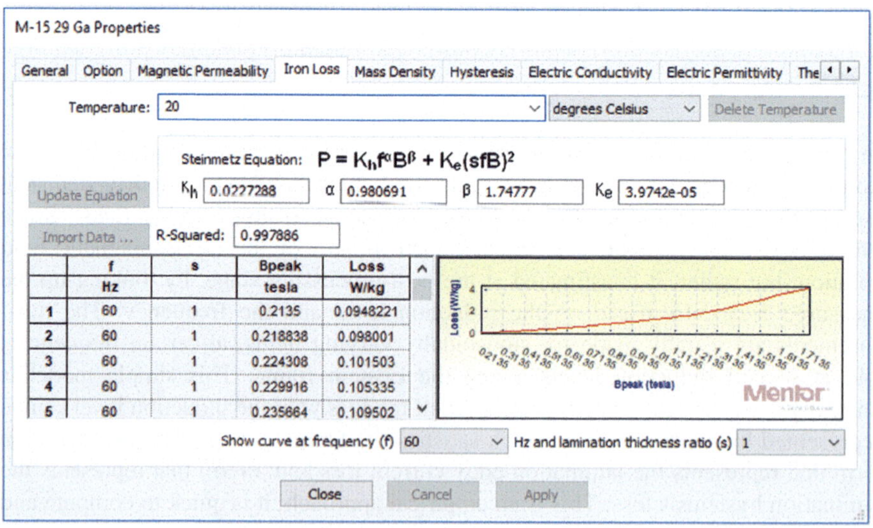

Fig. 4.7 Dialog box in *material editor* for entering core loss data

However, using empirical approaches, one cannot accurately compute the core loss if the external fields are non-sinusoidal. When the accuracy of the core losses becomes significant, more advanced material models that simulate the hysteretic behavior of the material are required [5]. These models carry a much higher computational cost; however, they are general purpose. The other option, for a better estimation of losses, would be through a series of very carefully designed benchmark problems that represent the operating conditions of a transformer, with a series of measurements on the benchmark models [6–16].

The ability to compute core losses accurately is quite significant since the losses affect the efficiency of the device and determine the thermal management required for a safe operation. There is a wide range of approaches that are used for predicting the core loss. Very often, single-valued *B–H* curves are adopted for describing the material's magnetic behavior, as described above. In doing so, the irreversible behavior of the material is neglected. Moreover, iron loss is calculated at the post-processing stage and is not included in the field solutions, so their effects on the global quantities such as current, voltage, forces, and flux linkage, are not directly taken into account [17]. The incorporation of the *B–H* loops (as experienced in ferromagnetic materials, under varying magnetic field) in computer simulations will overcome these issues.

There are a number of phenomenological models that describe the hysteretic behavior of a magnetic material. The Jiles–Atherton (JA) model proposed by Jiles and Atherton in 1986 [18] is one such model, which is based on the physics of ferromagnetism and explains the hysteresis phenomenon inside ferromagnetic materials. The data representing the JA model can be entered in Simcenter MAGNET, using a number of methods. See Fig. 4.8, showing the material editor and the data describing the JA model.

Using advanced material models, with access to material data, representing the magnetic hysteresis, and more efficient and advanced numerical techniques that avoid numerical stability issues, when handling the hysteretic nonlinearities in the finite element solver, it is possible to include hysteretic effects in the field solutions.

Benchmark Test for Hysteresis Models

Testing electromagnetic analysis methods (TEAM) benchmark problem 32 [19] is a benchmark test that was designed for testing the accuracy of numerical models, representing the hysteresis effects. The simulation model of the TEAM problem in Simcenter MAGNET software is shown in Fig. 4.9b which ran for 125 ms using both the single-valued (SV) and the vector JA hysteresis (Hys) models.

The shaded plots for computed $\boldsymbol{B}$ fields at $t = 75$ ms using both models are shown in Fig. 4.9c and d, respectively [19].

The voltages of both coils using both material models are shown in Fig. 4.10a. The phase difference in the Hys case is because of the phase lag between *B* and *H* fields. In Fig. 4.10b, the results for the computed and measured magnetic flux densities at point *P* of Fig. 4.9a are shown [19].

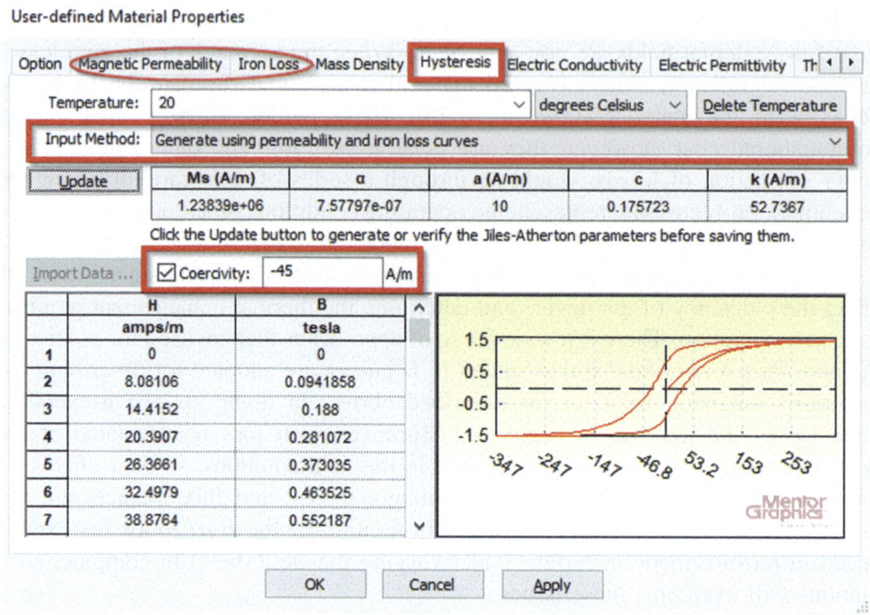

Fig. 4.8 Dialog box in *material editor* for entering JA model data

4.2.5.3 Electric Conductivity, Thermal Conductivity, Specific Heat Capacity, and Mass Density

Electric conductivity is usually isotropic and linear, i.e., the relationship between the electric field E and the current density J; however, it changes with temperature, and therefore, it can be specified at different temperature values. If the conductivity is anisotropic, then a tensor may be specified for each temperature. Although it is not common, the relationship between E and J could be nonlinear. There are similar options in the *material editor* for entering the data for other material properties.

4.2.6 Boundary Conditions

Wherever possible, boundary conditions should be used to take advantage of the symmetry and reduce the problem size. In the magnetic problem, there are two boundary conditions that are used routinely. One is the *field normal* and as suggested by the name, this boundary condition imposes the tangential component of the field to be zero. The other is the *field tangential* boundary condition which does the opposite; i.e., it imposes the normal component of the field to the boundary to be zero. In addition to these two boundary conditions, the tank wall can be modeled by applying the *surface impedance* boundary condition to its faces. This will help

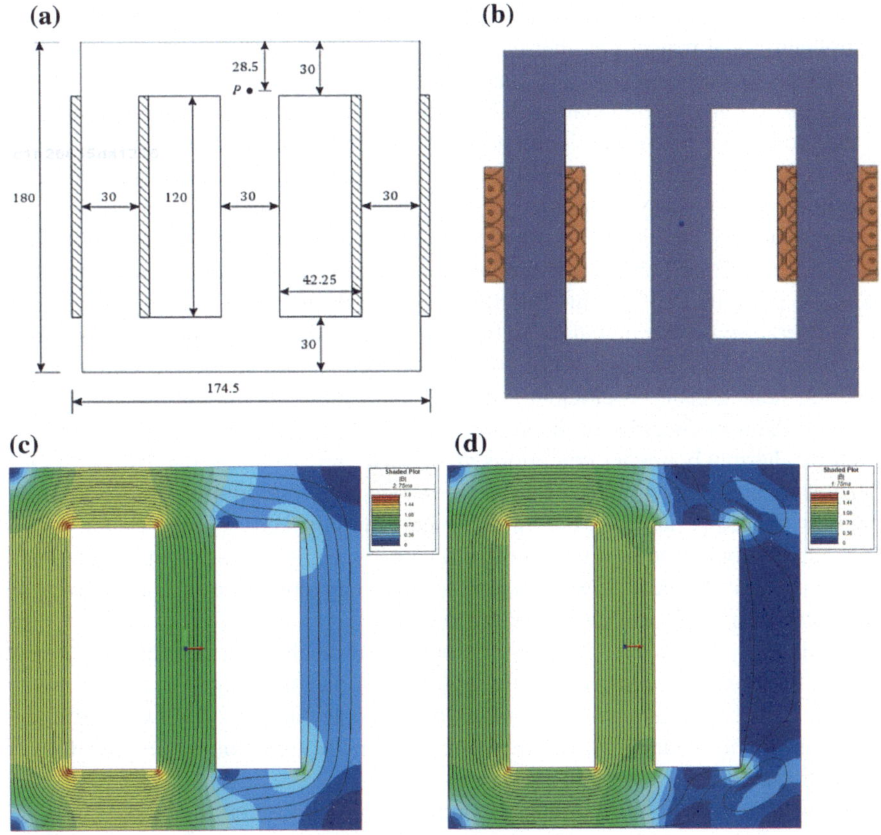

Fig. 4.9 **a** Geometry of the three-limb transformer [19] (dimension in mm) **b** simulation model. B field shaded plot at t = 75 ms computed using the **c** single-valued, and the **d** vector JA models. [5] (Figure, courtesy of Mentor Infolytica, a Siemens Business)

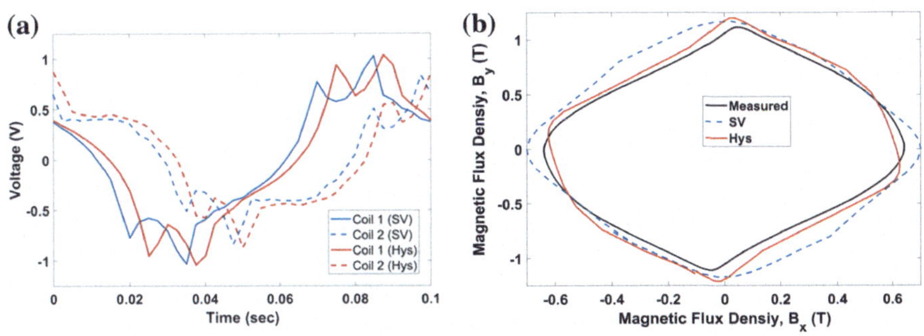

Fig. 4.10 **a** Voltages across two coils using the SV and the Hys models. Computed and measured **b** B_x and B_y flux densities at point P [5] (Figure, courtesy of Mentor Infolytica, a Siemens Business)

avoid modeling the tank wall volume and as a result reduces the problem size significantly, and hence, the time-harmonic simulation solution time is reduced by as much as a couple of orders of magnitude.

The *flux tangential* boundary condition is applied by default to all outer surfaces of the model that are not assigned any other boundary conditions. Thus, for each outer surface on which the current flow is normal (entering or exiting) and the fields tangential, no boundary condition needs to be explicitly specified.

The boundary conditions in the thermal problem are the *perfect insulator*, *prescribed* temperature, and the *environmental* condition. With the *perfect insulator*, no heat is permitted to flow through the boundary. With the *prescribed* temperature, the temperature on the boundary is fixed to a value defined by the user. With the *environmental* conditions, defined by the convection heat transfer coefficient, the radiation heat transfer coefficient, and the temperature of the surrounding environment, the user specifies to the system how the heat flows through the boundary.

Of the thermal boundary conditions mentioned, the one that is mainly used in the transformer application is the *environmental* boundary condition in which the user specifies the temperature of the environment, the convective heat transfer coefficient, and the radiation heat transfer coefficient. Since, in this application, the main mechanism for the heat flow through the boundary condition is convection, the radiation heat transfer coefficient is set to zero.

Additionally, Simcenter MAGNET supports a mechanism for defining a heat transfer between two components through some other medium in air. This mechanism is called the *convective link* which implements convection between different boundaries. This boundary condition is not normally used for power transformer applications.

As an example of the boundary conditions in Simcenter MAGNET that are normally used for the transformer application and the way they are applied, Figs. 4.11 and 4.12 illustrate the case of a three-phase transformer model.

There is a front–rear symmetry, and additionally, a top–bottom symmetry, depending on where the air gap is situated, in which case a *field normal* boundary condition must be explicitly specified on the symmetry plane, as shown in Fig. 4.12.

4.2.7 Accuracy Considerations

The accuracy of the field and the quantities that are derived from them depends, to a large extent, on the finite element mesh and the convergence tolerances used in the solution system. The following sections describe the tools in the simulation system that may be used in order to improve the accuracy of the results. These tools include the mesh controls, adaption, polynomial order, time step refinement, solution tolerances, and solver options.

Fig. 4.11 Three-phase transformer: full geometry

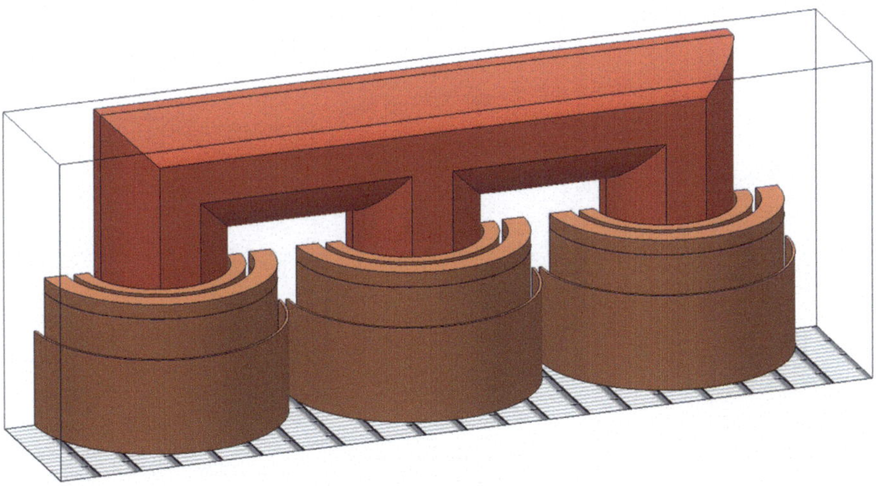

Fig. 4.12 Three-phase transformer: quarter model, exploiting rear–front, and top–bottom symmetries

4.2.7.1 Mesh Controls

Maximum Element Size

One of the commonly used mesh controls is the *maximum element size* which can be applied at different levels to specify the mesh density. By setting this parameter, the mesh density can be increased or decreased based on a criterion which relates to the element average edge length. This control can be applied at the component, surface, edge, and vertex levels.

Boundary Layer Mesh

There are situations, such as skin depth modeling, Figs. 4.13 and 4.14, when there is a requirement to model sharp field variations, taking place near a component surface. In this case, a *maximum element size* specified on the component surface itself might be adequate. However, this might produce a prohibitively large number of elements in directions parallel to the surface, which might not be optimal if the fields in these directions do not vary nearly as they do toward the interior of the component. One solution is to use larger elements, e.g., by increasing the *maximum element size*, and to specify *mesh layer* on the component surface, which will result in a smaller, more regular and anisotropic mesh. The anisotropic mesh technique can also be used for modeling air gap, across which the fields often vary rapidly.

The *mesh layer*, used in the block, in Fig. 4.14, shows four layers of a structured mesh, starting from the top face of the block. This mesh control is useful for

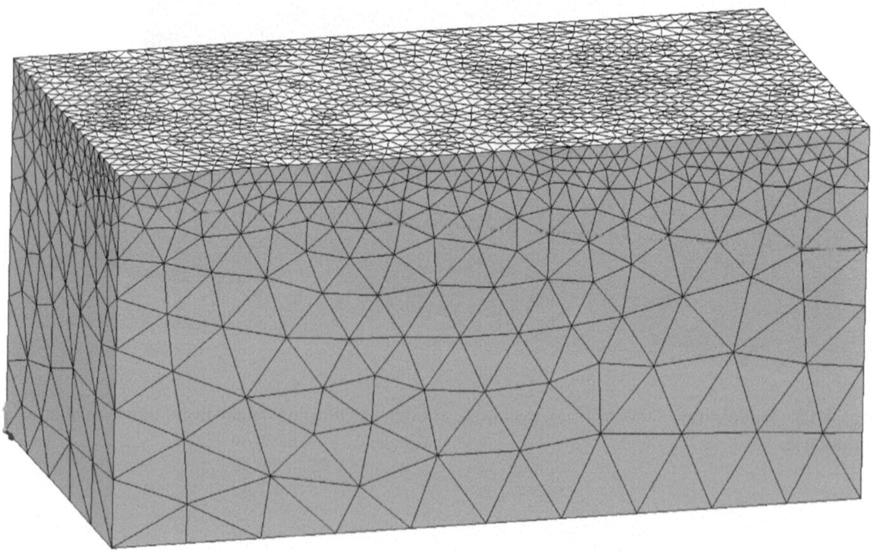

Fig. 4.13 Mesh resulting from the use of *maximum element size* on the top surface of the block

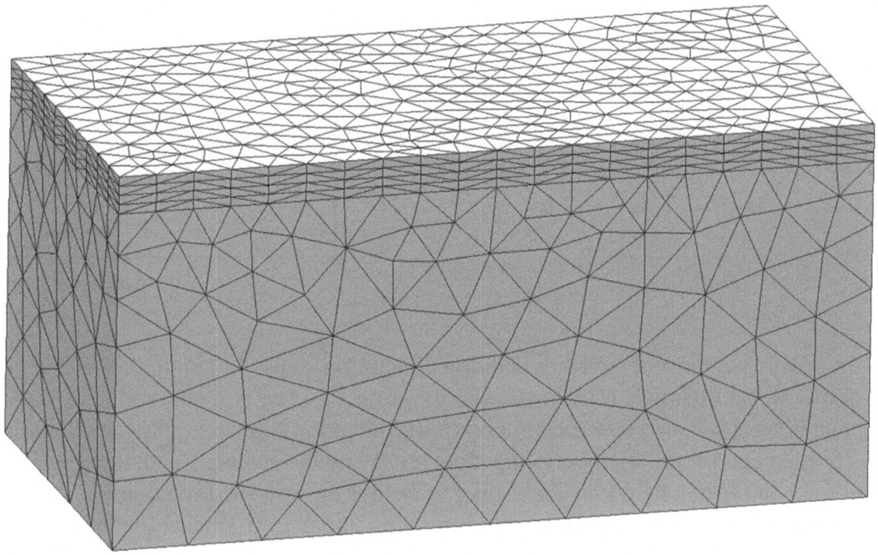

Fig. 4.14 Mesh resulting from the use of *mesh layers* starting from the top surface of the block

modeling skin depth or thin air gap regions where multiple layers of long, thin elements will suffice. With this control, the elements are used where they are needed the most, and the finer mesh does not propagate to other regions where they are not required. *Mesh layers* may be assigned to a component or a component face, and if assigned on component faces, they can propagate toward the inside, outside, or in both directions.

Mesh Edge Subdivisions

Mesh edge subdivisions can be specified on edges and have a very localized effect on a 3D mesh, as they affect the mesh only in close proximity of the edge.

4.2.7.2 Adaption

The mesh can be automatically modified during the solving stage through adaption. Adaption is an iterative process [20] that stops when the relative variation of global stored magnetic energy between solver runs is smaller than a prescribed tolerance, as shown in Fig. 4.15.

Adaption works based on the error in the local field values, usually determined according to the discontinuity of the field at the element boundaries. The error would be largest where the field is changing rapidly and there is not enough discretization in the region, to support the sharp variation of the field. During each adaption step, errors in the field are computed over all the elements, and those with the largest error are automatically refined.

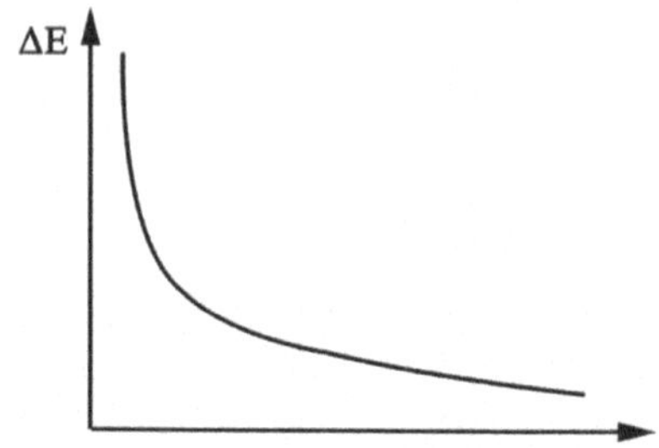

Fig. 4.15 Change in the stored energy during adaption

In order to determine whether additional adaption steps are required, the change in some global quantity, usually the energy in the system, is monitored, and if the change is greater than a specified tolerance, the adaption step takes place. Following the refinement of the elements, the field problem is resolved, and this process continues until the change in the energy is below the specified tolerance.

When elements are refined, by subdividing them into smaller elements, the process is called *h*-adaption. Following the refinement, the *Delaunay* algorithm is applied to the mesh, in order to produce optimum aspect ratio elements in the vicinity of the subdivided element.

When the solution is refined by raising the polynomial order in an element, the process is called *p*-adaption. Increasing the polynomial order results in greater accuracy, because additional degrees of freedom are inserted to support a greater variation of the field.

h-adaption and *p-adaption* can be combined in order to benefit from the advantages of the two schemes. In Simcenter MAGNET, *h*-adaption can be applied in 2D field solutions, whereas in 3D field solutions, it is possible to use both *h-adaption* and *p-adaption*.

4.2.7.3 Polynomial Order

In the finite element method, the polynomial order is a measure of the accuracy of the shape functions that are used to represent the field. A higher polynomial order is equivalent to a more accurate field representation. In Simcenter MAGNET, the user can use polynomial orders one to four for 2D field solutions and polynomial orders one to three for 3D field solutions.

In addition, Simcenter MAGNET supports mixed order elements in 3D. In a mixed order mesh, some elements may be first order, some second order, and some third. To take advantage of a mixed order mesh, the user sets the polynomial order for each component in the model.

It should be noted that higher polynomial orders give better accuracy, but the solutions times are also longer, since there are more degrees of freedom in a higher-order mesh. Roughly, the change from polynomial order one (1) to two (2) generates eight (8) times the number of degrees of freedom and a matrix that has fourteen (14) times the number of nonzeros. A change from polynomial order one (1) to three (3) generates twenty-seven (27) times the number of degrees of freedom and a matrix that has seventy-four (74) times the number of nonzeros.

Where solution time and memory usage are of concern, which is the case for a large 3D problem such as a transformer, the polynomial order should be increased only for individual components where the solution accuracy is critical, instead of setting it globally.

Hierarchal Elements

When elements of different polynomial order are used in the same mesh, the elements are said to be hierarchal [21]. Using hierarchal elements, the polynomial order can be raised where accuracy is needed the most. In areas, where the magnetic field is not changing rapidly or the field is weak and the accuracy is not of any concern, a polynomial order of one (1) can be used. However, if there are areas where the field is changing rapidly and the accuracy of the results depends on the accurate representation of the field in those areas, then polynomial orders of two (2) or three (3) can be used. Having this flexibility in 3D problems is very important since very accurate solutions can be obtained with reasonable number of degrees of freedom and matrix sizes. In problems where there are large volumes of conducting material and the eddy current effects are significant, the solution may not be practical without a hierarchal mesh. Elements in a conducting material require a much larger number of degrees of freedom. As will be described later, the conducting material of the tank will use a surface impedance boundary condition to represent the field inside, and hence, there are huge savings in the solution size. However, the 3D model of the transformer with the tank still includes a significant amount of detail and there is a large contrast in the dimensions of the components resulting in very large meshes.

There are situations where high-order field accuracy is needed, not only in a component, but also in the immediate elements surrounding that component. This situation arises in the conducting tank of large power transformers. The conducting tank is where the eddy currents circulate and where the hot spots are. In this case, it is found that the correct solution of the power loss in the tank depends on an accurate field computation in the surrounding elements as well as the tank itself.

The transformer tank itself is modeled with the surface impedance boundary condition, and this is made possible because of the very small depth of penetration of the fields inside of the tank. Nevertheless, a high polynomial order in the tank components, even with the surface impedance boundary condition, helps with the accuracy of the computed power losses. To achieve even better accuracy, it is found that the surrounding elements in the air or in the oil, whatever the case may be, would also require high field accuracy. In this case, it is recommended that the tank components use high polynomial order and the surrounding elements to use the same high polynomial order. It is found that with a reasonable mesh, the polynomial order of two (2) is usually adequate for the tank components and the surrounding elements. Polynomial order one (1) is adequate in the remainder of the elements, in the air, or in the oil.

The way hierarchal elements work is that the layer of elements between elements of differing orders is an order in-between. For example, if the tank is set to be polynomial order 2 and the surrounding oil is set to be polynomial order one (1), the layer of elements joining the mesh of the two different polynomial orders is neither a second-order element nor is it a first-order element. This layer of elements has a polynomial order with shape functions that change from order two (2) to order one (1) over the volume of the layer. Let us call this polynomial order 2–1, when polynomial order two (2) is used on one side and polynomial order one (1) is used on the other side. Similarly, one could have an element that could be order 3–1 or 3–2, depending on the polynomial orders on the two sides of the layer of elements. In this layer of elements, the shape functions have an order that matches both sides and is hence somewhere in-between. Now in the transformer tank example, one choice is for the system to set the polynomial order of the boundary elements of the tank to full second order and the first layer of elements in the oil or in the air to be the 2–1 order. The other option would be for the system to choose the boundary elements of the tank to be the 2–1 order. It is useful if one is able to control in which mesh layer the transition takes place, i.e., one layer away from the tank, two layers away from the tank, and so on. This level of control provides the user with a flexibility to achieve very high accuracy, at the lowest possible computational cost, when using hierarchal elements.

Benchmark Test to Demonstrate Accuracy Versus Computational Resources

The Bath Plate benchmark model (TEAM 3) [22] can be used to demonstrate the use of hierarchal elements and to examine the accuracy versus computational resources. Intentionally, the specified mesh discretization has been set to be very coarse, which is not adequate to capture the re-distribution of the current in the conducting ladder, and the rapid change in the field, in the surrounding air region (Fig. 4.16).

Here, the benchmark results, i.e., the flux density along a line over the ladder, at the source frequency of 50 Hz [22], are used to demonstrate the accuracy. The following are the cases that are compared:

- When all the components, i.e., the ladder, the coil, and the air, are set to polynomial order one (1), the size of the problem is 1577 degrees of freedom.
- When all the components, i.e., the ladder, the coil, and the air, are set to polynomial order three (3), the size of the problem is 35,682 degrees of freedom, i.e., nearly 22.6 times larger, in terms of problem size.
- With a combination of first- and third-order elements, i.e., third order in the ladder and the immediate surrounding air, and first order in the coil and the remaining air region, the size of the problem is 11,301 degrees of freedom, i.e., nearly 7 times larger, in terms of problem size. This is a much smaller problem size and achieves the same level of accuracy, as shown in Fig. 4.17.

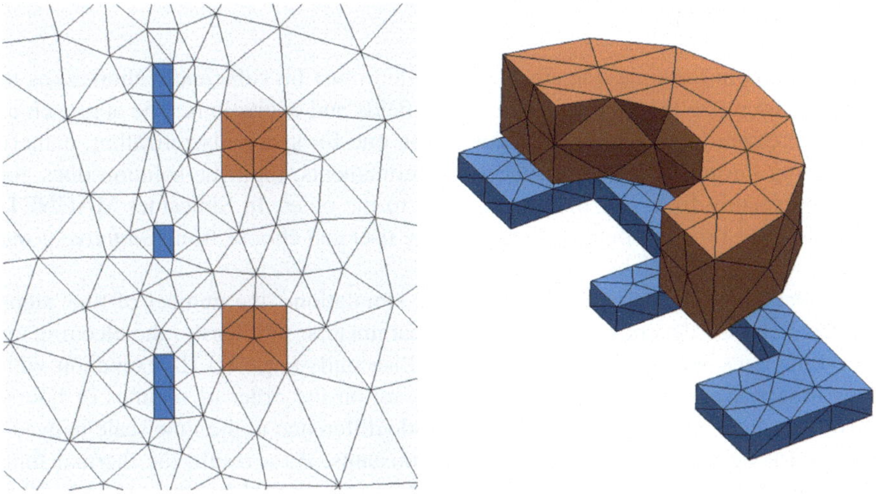

Fig. 4.16 TEAM model 3, with a very coarse discretization in Simcenter MAGNET (Figure, courtesy of Mentor Infolytica, a Siemens Business)

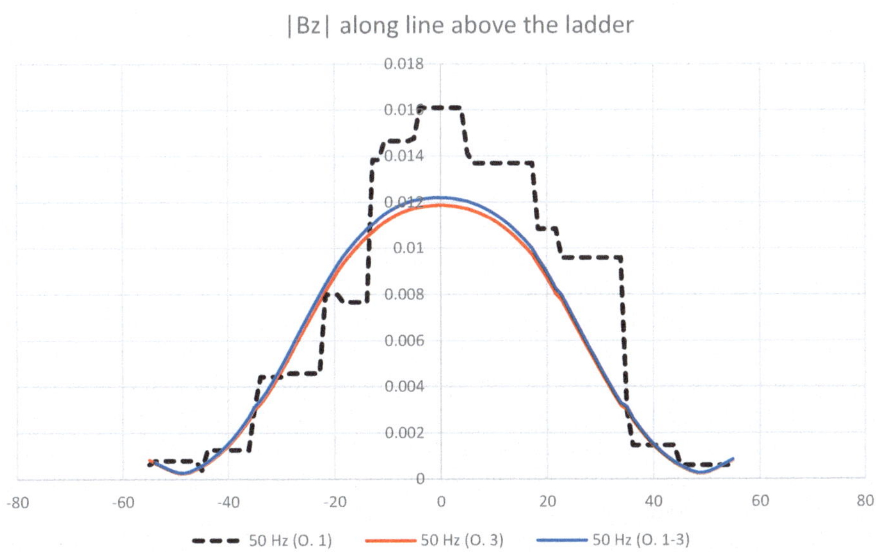

Fig. 4.17 Flux density along a line above the ladder [22], when using polynomial order one (1) for all components, polynomial order three (3) for all the components and a combination of polynomial orders' one (1) and three (3) (Figure, courtesy of Mentor Infolytica, a Siemens Business)

4.2.7.4 Time Steps

For solutions in the time domain, the time step must be sufficiently short so as to properly discretize the time variation of the fields and excitations. The approach to time discretization should be the same as the one for space discretization, namely proceeding through a series of successive refinements, until the solution does not vary appreciably from one time step length to the other. In Simcenter MAGNET, there are three time stepping methods that the user can choose from: *adaptive*, *fixed interval,* and *user defined.*

For the coupled electromagnetic–thermal simulations, the number of time steps and the step size depends on a number of parameters. Ordinarily, the electromagnetic time constants are on the order of milliseconds (e.g., 60 Hz excitation with 1 ms time step) and thermal time constants are on the order of minutes or hours. Therefore, there are many orders of magnitude difference in the time scale between the electromagnetic and the thermal time constants. As a result, the thermal time steps are set very differently and the magnetic field should be updated only after there has been sufficient changes in the material properties.

4.2.7.5 Solver Options

The solver options have an impact on the accuracy of the results. For example, when choosing the time-harmonic solver, the material nonlinearity is not treated as exactly as the transient solver. However, it is adequate for most transformer simulations with AC sources, in particular when the sources are current-driven, and the solution can be obtained much quicker than using the time-domain solver. To obtain a much better accuracy, e.g., for the flux density waveforms, the *B*–*H* properties of the nonlinear materials can be transformed [23], so that a better fit is achieved between the outcome of a time-harmonic simulation and one obtained using the time-domain simulation.

Figure 4.18 shows the outcome when a nonlinear material is saturated with a perfectly sinusoidal *H* waveform. The time-harmonic solver returns a sinusoidal *B* waveform, whereas the time-domain solver would return a truncated *B* waveform.

Simcenter MAGNET provides a number of transformation options in the way the *B*–*H* properties of the steel are to be interpreted, in order to obtain the best fit, depending on whether the sources that are producing the electromagnetic field are current-driven or voltage-driven.

In nonlinear models, an outer iteration loop is initiated during the solution process in order to linearize the system of equations to be solved. The *Newton* tolerance and *Max Newton* iterations options are used to specify when this process terminates. The smaller the *Newton tolerance*, the more accurate is the solution.

The *CG tolerance* is used to determine when the iterative matrix system solution should be considered to have converged to the solution of the linearized matrix system resulting from the discretization of the model. The lower the *CG* tolerance, the more accurate and uniform is the solution of the linearized matrix system.

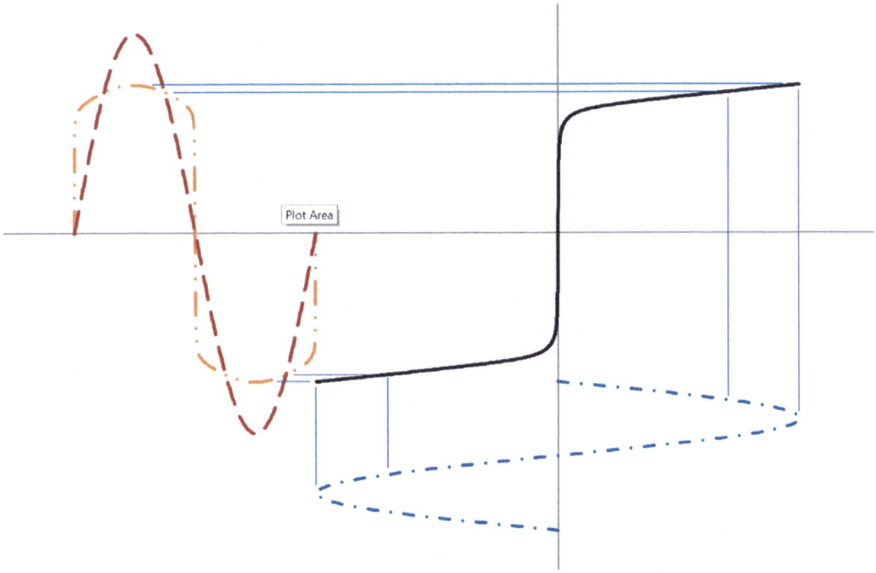

Fig. 4.18 Nonlinear *B*–*H* curve (black), sinusoidal *H* waveform (blue), sinusoidal *B* waveform (red), true nonlinear *B* waveform (orange) (Figure, courtesy of Mentor Infolytica, a Siemens Business)

4.3 Result Evaluation

Once the electromagnetic and thermal models are created and the field problem has been solved, it is possible to query the system for results through a set of post-processing tools. The goal of post-processing is to obtain results that the designer would ordinarily measure in the laboratory through a set of experiments. What comes out of a solution system is a set of raw data, usually in terms of potential values at the nodes of the finite elements. In a numerical simulation system like Simcenter MAGNET, there are a number of tools that let the user obtain useful results from the raw potential field obtained from the solver. These tools give the user the ability to mathematically manipulate the fields and display them in different ways.

It could be said that all quantities of interest can be derived from the potential field. In Simcenter MAGNET, the fields that the majority of users are interested in are readily available in the user interface. If a particular field is not available in the user interface, there are mathematical operations available that let the user derive that field from the other fields available.

Simcenter MAGNET provides the user with a number of global quantities. These are not fields, but are quantities that are derived from the fields, usually through integration. They are usually single numbers and correspond to what can be measured at the terminals of a device. Impedances, currents, voltages, and other similar quantities are examples of global quantities.

4.3.1 Fields

The following is a description of some of the fields that are readily available in the user interface. It should be noted that certain fields depend on the type of solution that the user has requested. In Simcenter MAGNET, there are three types of solvers for electromagnetic field simulation and these are static, time-harmonic, and transient. And, there are two types of solvers for the thermal field simulation and these are steady state and transient. The static solver produces a single potential value at each node (or for each degree of freedom). The time-harmonic solver produces a complex potential value, i.e., two numbers, at each node. The transient (or time-domain) solver produces a series of values for the potential at each node representing the change that takes place over time.

There are some fields that are useful for viewing; e.g., the *flux* function in 2D models, either with Cartesian or axisymmetric geometries, visually gives the user ample information about the way the device works since the user can examine the flux path through the magnetic circuit.

Some of the field quantities that are of immediate interest to a user when solving a magnetic problem are the magnetic flux density $\boldsymbol{B}$, the magnetic field $\boldsymbol{H}$, and the current density $\boldsymbol{J}$. Equivalently, the field quantities that are of immediate interest to a user when solving a thermal problem are the temperature field $\boldsymbol{T}$, the heat flow, and the loss density. The system obtains these fields from the potential by simple mathematical operations of *gradient* and *curl*. The potential field obtained from the thermal solver is the temperature, and hence, there is no need to use any mathematical operation to derive it.

The system computes other fields that may be of interest to the user. These are fields such as the ***force density***, ***ohmic loss density***, and the ***iron loss density***.

Error field: This field represents the errors in each element that the system computes when adaption is enabled. By displaying this field, the user can visually inspect where in the mesh, the system could use some refinement, in order to improve the accuracy of the results.

User-defined fields: Any other field that the user derives by performing mathematical operations on the existing fields is a user-defined field. The same way as the standard fields in the system that are listed above, the user-defined fields can be probed for values and displayed.

The form of the results can vary depending on the analysis type. As was mentioned before, the static and transient solvers solve for potentials that are expressed in real numbers, whereas the time-harmonic solvers solve for potentials that are expressed in complex numbers. For all the fields mentioned above, the user may ask for the magnitude, spatial components of *x, y, or z* and the time-averaged values. In the *time-harmonic* context, the magnitude $|\boldsymbol{A}| = (\boldsymbol{A} \cdot \boldsymbol{A})^{1/2}$ is replaced by the root-mean-squared magnitude $|\boldsymbol{A}|_{\mathrm{rms}}$ and phasors are used to represent the time dependence and to take into account the phase effects.

4.3.2 Global Quantities

Global quantities, such as energy, force, torque, flux linkage, are related to the type of analysis performed (e.g., static, time-harmonic, or transient).

The stored magnetic energy W_{m} is the integral over the entire volume of the solution domain of the magnetic energy density w_{m}:

$$W_{\mathrm{m}} = \int w_{\mathrm{m}} \mathrm{d}V \tag{4.1}$$

This is related to the area under the H versus B curve.

The (net) force vector $\boldsymbol{F}$ calculated for a solid body results from the integration of the electromagnetic force density vector field over that body. The effect of the force density vector field on the body is equivalent to $\boldsymbol{F}$ acting on the center of mass of the solid body, plus a torque vector with respect to the same point and which can be obtained provided the center of mass coordinates are known.

Three different methods can be used to obtain the electromagnetic net forces on bodies or components:

- **Lorentz**: when the materials are conducting and not permeable
- **Maxwell stress**: for all bodies, regardless of the material
- **Virtual work**: can be obtained from the difference in energy when a small displacement of the object takes place.

The flux linkage is automatically calculated and reported for each coil. For the stranded coil, this corresponds to the *total* flux; i.e., it includes the number of turns:

$$\Lambda = N\phi \tag{4.2}$$

Since the flux linkage is related to the voltage through $V = \mathrm{d}\phi/\mathrm{d}t$, the flux linkage for coils connected in series and parallel obeys the same rules as the voltage.

The *ohmic* loss P (in W or J/s) is related to the finite resistivity ρ of the material:

$$P = \int \rho \boldsymbol{J} \cdot \boldsymbol{J} \, \mathrm{d}v = \int \frac{\boldsymbol{J} \cdot \boldsymbol{J}}{\sigma} \mathrm{d}v \tag{4.3}$$

where $\sigma = 1/\rho$ is the conductivity. In the *time-harmonic* context, the time-averaged *ohmic* loss is calculated as

$$\overline{P} = \int \frac{\overline{\boldsymbol{J} \cdot \boldsymbol{J}}}{\sigma} \mathrm{d}v = \int \frac{\boldsymbol{J}_{0\mathrm{rms}} \cdot \boldsymbol{J}^{*}_{0\mathrm{rms}}}{\sigma} \mathrm{d}v \tag{4.4}$$

where $\mathbf{J}_0$ is the phasor associated with $\boldsymbol{J}$.

If a component is modeled with the *surface impedance* boundary approximation, the surface loss density p_S (in W/m^2) is computed from the surface impedance, using:

$$p_s = R_s |\boldsymbol{J}_s|^2 \tag{4.5}$$

where $\boldsymbol{J}_S$ is the surface current density.

The voltage across a coil is obtained by differentiating the flux through the coil with respect to time and then adding the resistive component.

4.3.3 Scripting

Scripting can be used to automate repetitive tasks, to execute a set of operations in the software as a batch process, and to customize the system, based on the user requirements.

Customizing the system would mean that the general interface is replaced with an interface which is more specific to the application in hand. For example, routine transformer calculations can be programmed up and made available to the user through the user interface menu or the toolbar system. Chapter 5 covers the scripting concepts and demonstrates the implementation of a customized interface for the calculation of transformer winding parameters.

Simcenter MAGNET can communicate with many other tools, and its application programming interface (API) can be called within programs written in several programming languages [24]. This *interoperability* allows Simcenter MAGNET to work within a design flow environment where it becomes one component of a much larger set of software components that are used within an organization to keep track, analyze, and design all or parts of a large power transformer.

4.4 Electromagnetic Field Computation

4.4.1 Solving the Electromagnetic Field Problem

The 3D formulation in Simcenter MAGNET is based on the *T*-Ω method [21], where the magnetic field is represented as the sum of two parts: the gradient of a scalar potential and, in conductors, an additional vector field represented with vector edge elements. As a result, the solution vector consists of the magnetic scalar potential at nodes plus edge degrees of freedom associated with current flow in solid conductors. This approach is memory efficient because the non-conducting regions can be solved with a scalar potential. The T-Ω method does not run into the convergence and instability issues that other formulations run into. In addition, the formulation uses a number of trial function spaces which are hierarchically constructed, therefore allowing different polynomial orders to be mixed in the same mesh. The formulation supports hierarchical elements based on polynomial orders' one (1) to three (3).

The following constitutive tensor relations are used:

$$\boldsymbol{B} = \boldsymbol{\mu} \cdot \boldsymbol{H} \tag{4.6}$$

$$\boldsymbol{E} = \boldsymbol{\rho} \cdot \boldsymbol{J} \tag{4.7}$$

where $\boldsymbol{\mu}$ is the permeability and $\boldsymbol{\rho}$ is the resistivity of the material.

4.4.1.1 Conducting Components

In conducting components, like the tank and the bus bar structure, the following equation is solved for in the time-domain solver:

$$\nabla \times [\boldsymbol{\rho} \cdot \nabla \times \boldsymbol{H}] + \boldsymbol{\mu} \cdot \frac{\partial \boldsymbol{H}}{\partial t} = 0 \tag{4.8}$$

4.4.1.2 Non-conducting Components and Stranded Coil Regions

In non-conducting components, like the core or the stranded coil regions, like the windings wound around the core limbs where current density is assumed to be uniform, the following equation holds.

$$\nabla \times \boldsymbol{H} = \boldsymbol{J}_s \tag{4.9}$$

where $\boldsymbol{J}_s$ is the current density in the stranded coil ($\boldsymbol{J}_s = 0$ everywhere else). To compute an accurate magnetic field, the current density $\boldsymbol{J}_s$ needs to be solenoidal [25].

The magnetic field $\boldsymbol{H}$ can be written as

$$\boldsymbol{H} = -\nabla\phi + \boldsymbol{H}_s \tag{4.10}$$

where $\boldsymbol{H}_s$ is any source field that satisfies

$$\nabla \times \boldsymbol{H}_s = \boldsymbol{J}_s \tag{4.11}$$

and ϕ is a scalar potential such that

$$\nabla \cdot \boldsymbol{B} = \nabla \cdot [\boldsymbol{\mu} \cdot (-\nabla\phi + \boldsymbol{H}_s)] = 0 \tag{4.12}$$

4.4.1.3 Conductors with Holes

Conducting objects with a hole in them require that the system impose a condition where there is current flowing around the hole [26]. In some formulations, the user is responsible for these topological holes and imposing the condition manually. In Simcenter MAGNET, holes in a conductor are found and a zero-volt voltage-driven coil is implicitly constructed around the hole, automatically, to indicate that there might be a nonzero net current flow around the hole.

4.4.2 Boundary Conditions

Field normal and *flux tangential* boundary conditions set a zero reference point for the scalar potential ϕ on the corresponding boundaries.

4.4.3 Problem Size

In the 3D formulation used, conducting regions are modeled using vector edge elements, which add significantly more degrees of freedom to the system matrix than when the same region is filled with a non-conducting material, causing in turn a substantial increase in problem size and solving time. In some situations, the *surface impedance* boundary condition can be used in order to reduce the problem size.

The size of the matrix representing the electromagnetic field problem could become very large, in particular, in applications where there are large volumes that are conducting and the eddy current modeling is of significance. In these applications, it is crucial to control the problem size efficiently, in order to end up with a matrix whose size is manageable. The typical sizes of the problems involved in simulating the power transformer operation require millions of degrees of freedom and the sparse matrices have tens of millions of nonzeros.

4.4.4 Surface Impedance Modeling

In three dimensions, solid conductors increase significantly the computer resources that are required to solve the field problem. Because of the additional degrees of freedom introduced by the shape functions representing the conducting domain, the problem size grows rapidly, and as a result, more memory is required during the solving process. The solution times could also become prohibitive if the problem size grows too large. For this reason, any way that a simplification can be made to reduce the problem size, as long as it is not at the sacrifice of accuracy, should be considered. Fortunately, there is a simplification that can be made with this regard in some classes of applications including the power loss calculation in a transformer tank.

In order to assess whether it is necessary to model the conductors, one should calculate the skin depth which is a function of the frequency of the sources and the material properties of permeability and conductivity of the conductor. The skin depth is the distance that it takes the electromagnetic field to drop to $1/e$ of its initial value when penetrating a conductor. It is defined as:

$$\delta = \frac{1}{\sqrt{\pi f \mu \sigma}} \tag{4.13}$$

where f is the frequency, μ is the permeability, and σ is the conductivity.

There are two limiting cases. One limiting case is when the skin depth is much larger than the conductor thickness. In this case, it is not necessary to model the volume as conducting, and while the problem size is significantly reduced by this assumption, the results are not affected. The other limiting case is when the skin depth is much smaller than the thickness of the solid conductor. In this case, there are no fields that penetrate the conductor and the component can be assumed to be a perfect electric conductor. In practical cases like the power loss calculation in a transformer tank, however, the skin depth is small, but the limiting case of the perfect electric conductor does not apply. In such a situation, the model can still be simplified and the problem size can still be greatly reduced if the impedance over the surfaces of the conductors is known and the surfaces of the conducting component are specified to be the boundary of the computational domain.

In Simcenter MAGNET, when the boundary surfaces of a conductor are assigned to be the surface impedance boundary condition, the volume of the conducting region is not meshed, since in this case, there is no need to compute the field inside of the conducting volume. Given that there would normally be a large number of degrees of freedom for the solution of the field inside a conductor, using the surface impedance boundary condition reduces the problem size significantly.

In the power transformer studies, the surface impedance boundary condition is mainly used for the tank components.

The surface impedance of a conductor is defined as

$$Z_s = \frac{E_t}{H_t} \tag{4.14}$$

It can be shown that

$$Z_s = (1 + j)\sqrt{\frac{\pi f \mu}{\sigma}} \tag{4.15}$$

This can be written in terms of the skin depth δ as

$$Z_s = \frac{1 + j}{\delta \sigma} \tag{4.16}$$

Given the equation above, one can see that the limiting cases discussed above are when Z_s approaches either zero or infinity. Any value for Z_s represents an impedance that is seen at the boundary of the conductor which is based on its material properties and the source frequency.

It can be shown that the condition above is exact for an infinite plane. For the transformer case, the tank walls are very large compared to the skin depth and this assumption holds very well. In cases where the infinite plane assumption is not valid or where there are effects near the corners that should be accounted for, Simcenter MAGNET allows the user to specify a value for the impedance so that the memory and time-saving benefits of the surface impedance boundary condition can still apply.

It should be noted that the *surface impedance* boundary condition is used in the context of a time-harmonic solution which is also what is used when dealing with power loss computation inside of transformer tanks.

Even though the fields inside of a component that uses the surface impedance boundary condition are not calculated, quantities such as the power loss are calculated. Therefore, in the case of the power loss inside of a transformer tank, the power loss in the tank is calculated and is used inside of the Simcenter MAGNET system to determine the temperature rise in the tank.

To investigate the usefulness of the surface impedance boundary condition, consider a typical transformer wall. Typically, the width and the height of the wall are a few meters long, but the thickness is around a couple of centimeters. Given the material properties of mild steel and a source frequency which is either 50 or 60 Hz, the skin depth is around a few tenths of a millimeter. Considering the contrast in the dimensions of width or height and the thickness, it is evident that any mesh that is created in this volume ends up with elements with very large aspect ratios unless the number of elements increases to a prohibitively large size. Given the typical thickness and the skin depth dimensions, it is evident that creating a mesh that is accurate enough to model the eddy currents also becomes too large to handle. Therefore, not only does the surface impedance boundary condition eliminate the requirement for this very difficult mesh generation, it makes it possible to obtain accurate results for the power loss in the tank.

A nonlinear model for the surface impedance is required, in order to obtain accurate, and yet very efficient, in terms of computing resources, stray losses in the mild steel used for the tank. Using a nonlinear surface impedance, the solution can be reduced from hours, even days, to minutes.

4.4.5 Skin Depth Modeling

In other large solid conductors, e.g., the copper bars where the skin depth may be too large to use the surface impedance boundary condition, the skin depth region must be modeled in such a way as to properly capture its field variations. This is accomplished by first estimating the skin depth based on the equations above and then using adequate mesh refinements over a thickness of about two to three skin depths.

Since the field decays rapidly with increasing depth into the conductor, the *mesh layer* feature can be used to construct a mesh that supports large variation of the field into the conductor and hardly any variation in the other directions. In such a case, one can construct two or three mesh layers of second polynomial order, or one mesh layer of third order, over a total height of two skin depths.

4.5 Temperature Field Computation

4.5.1 Solving the Thermal Field Problem

The thermal module in Simcenter MAGNET can be used to solve for the steady-state and time-varying temperature distributions caused by time-varying heat sources, in the presence of thermally conducting materials and convective/radiative boundary losses. It handles heat transfer through *conduction*, *convection,* and *radiation.*

The equations to be solved are as follows:

$$\nabla \cdot k\nabla T = -Q + \rho c \frac{\partial T}{\partial t} \tag{4.17}$$

where k is the thermal conductivity; Q is a given source density; ρ is the mass density; c is the specific heat. All the material properties are a function of temperature.

Boundary conditions are specified temperature T or specified heat flux:

$$-(k\nabla T)\cdot \boldsymbol{n} = q + q_{\mathrm{c}} + q_{\mathrm{r}} \tag{4.18}$$

where q is a prescribed outward heat flux (W m^{-2}), q_{c} is the heat flux due to convection, and q_{r} is the heat flux due to radiation:

$$q_{\mathrm{c}} = h_{\mathrm{c}}(T)(T - T_{\mathrm{e}}) \tag{4.19}$$

$$q_{\mathrm{r}} = h_{\mathrm{r}}(T)\left(T^4 - T_{\mathrm{e}}^4\right) \tag{4.20}$$

where h_{c} and h_{r} are user-specified convection and radiation heat transfer coefficients (functions of temperature) and T_{e} is the user-specified temperature of the surrounding environment.

4.5.2 Problem Sizes

When solving for the thermal fields, the problem sizes are usually much smaller. Not only, the problem domain in this case contains fewer components, but also the scalar field that is used to solve for the temperature distributions lends itself to a much smaller matrix.

4.6 Mechanism of Coupling Electromagnetic and Thermal Field Solutions

4.6.1 Sources of Heat Generation

The sources of heat generation in an electromagnetic–thermal problem are the ohmic and iron losses. Once, the electromagnetic field is solved for, the electromagnetic losses are computed and automatically fed to the thermal module in Simcenter MAGNET, as input, to solve for the temperature distribution. There is a possibility for the user to augment the electromagnetic losses by some value in each component; in case, there are other physical or chemical phenomena contributing to the loss.

Core losses are an area of growing interest in fields such as advanced electric machines and transformers. Traditionally, losses have been specified and calculated using empirical loss curves provided by manufacturers, which specify the power loss per unit mass at a given frequency as a function of the maximum magnetic flux density $\boldsymbol{B}$. However, in some cases, there often exist regions of high magnetic saturation that remain relatively constant with time. In this case, the core loss model has to take into account the minor loops. Simcenter MAGNET uses a method that takes into account minor loops as well as changes in the orientation of $\boldsymbol{B}$. Loss due to changing orientation occurs, e.g., if $\boldsymbol{B}$ is constant in magnitude but changing in direction. The more accurate core loss calculations are available for a time-domain magnetic field equation solution since the changes in the flux density over time can be tracked.

As discussed earlier, advanced material models can be used to model the full hysteretic effects, which results in a more accurate core loss computation.

4.6.2 Solving the Coupled Electromagnetic–Thermal Problem

Full integration of Simcenter MAGNET provides accurate results for coupled electromagnetic–thermal analysis needs when there are eddy current and hysteretic losses in the magnetic system. In addition to the ohmic and hysteretic losses, it is possible to specify another source for the amount of power that is dissipated in each component in the form of density. This additional power density may represent other physical or chemical phenomena that the user may be interested to incorporate in the simulation.

Electromagnetic losses affect the temperature distribution, and the temperature changes affect the material properties which in return change the power losses. To account for the coupling effect between the electromagnetic and thermal solutions, an iterative scheme can be used. An iterative coupling to magnetic static or time-harmonic solution will help to predict the temperature distribution caused by the core and the eddy current losses. The corresponding temperature effects on electrical conductivity and magnetic material properties can be taken into account for the next iteration. Due to the different time constants in the electromagnetic response and thermal response, it is desirable to control how frequently the magnetic analysis is repeated.

In a coupled simulation context, the static thermal solver simulates the steady-state temperature distribution due to conduction, convection, and radiation once the power losses are computed from an electromagnetic field solution. The temperature distribution in this case is the one that is achieved after the sources have been turned on and some time has elapsed for the temperature to rise to its final steady-state value.

In a coupled simulation context, the transient thermal solver simulates the time-varying temperature distribution due to conduction, convection, and radiation once the power losses are computed from an electromagnetic field solution. The temperature distribution in this case keeps on changing with time, and the user is interested in the way temperature rises in different parts of a device. The transient solvers begin by finding a static solution, that is, the fields that would exist in the device assuming the conditions at the start time had been held unchanged for all earlier times. The transient solution develops in time from this starting condition.

Coupling thermal with electromagnetic fields takes into account the effect of ohmic power and core losses over time. The static thermal solver or the transient thermal solver can be coupled to any of the electromagnetic solvers of magneto-static, time-harmonic magnetic or transient magnetic.

The bidirectional link and the temperature-dependent material properties involved in the coupled thermal–electromagnetic solvers ensure that the losses and temperature are updated at every step of the iterative solution process.

A key feature when using a model for a coupled analysis is the *enable* and *disable* options. Depending on the type of analysis, it may be sometimes necessary to ignore a feature, which is essential in the magnetic or thermal solution, but not essential in the other. In other words, modeling a feature in one aspect of physics may be important, while in some other aspects, unnecessary. The *enable/disable* options allow users to optionally enable or disable any component during the analysis. For example, in transformers, the modeling of the core, the windings, the tank and the bus bar structure and the air is necessary for the electromagnetic analysis, but only the heated components in the tank and perhaps the bus bar structure are required for the thermal analysis.

4.6.3 Coupled Solution Controls

When performing coupled simulations, a number of controls are required in order to guide the system as to the different solution that is of interest.

For the case of coupled static thermal solutions solver, the following controls are available:

- Specify which problem (magnetic or thermal) should be solved first
- Maximum number of iterations
- Convergence tolerance for the coupled solutions.

For the case of coupled transient thermal solutions, the solver controls allow for specifying the number of times to solve the electromagnetic problem:

- Solve at every step of the thermal transient
- Solve at the beginning
- Solve at every user-specified interval.

As an example, Fig. 4.19 shows the thermal simulation results for the temperature in the tank and the measured temperature at the specified points. The user can display the shaded plot of the loss density, temperature profile, and probe for the temperature values at different points in the tank.

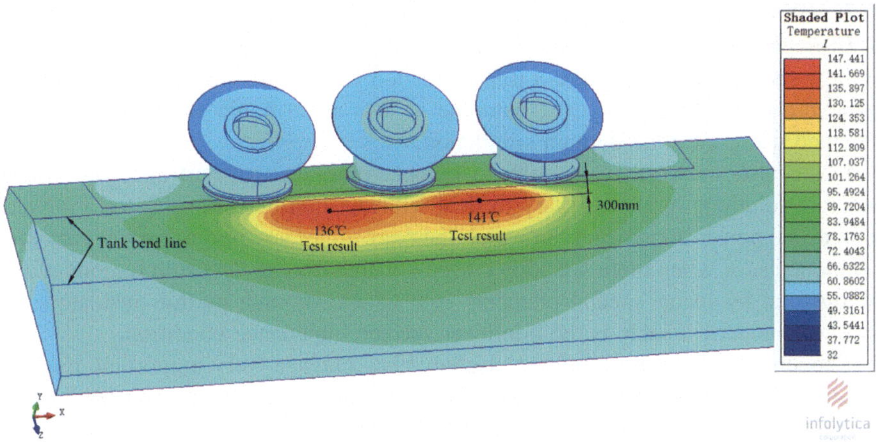

Fig. 4.19 Thermal simulation and test results for the temperature in the tank (Figure, courtesy of Baobian Electric) (ambient temperature: 32.5 °C top oil temperature 68.0 °C)

4.6.4 Coupled Electromagnetic–Thermal–Flow Simulation

The coupled electromagnetic–thermal simulation described above uses empirical values for the convective heat transfer coefficient in order to model the flow, e.g., in the case of oil which is used as the cooling medium, and as can be seen in Fig. 4.19, the results for the temperature are predicted with reasonable accuracy, when comparing the computed with measured values of the temperature in the tank. See Chap. 3 for cases when using the empirical values for the convective heat transfer coefficient may not be accurate. In such cases, the convective heat transfer coefficient values should be either re-calibrated, based on measurement, or a flow simulation, or a coupled electromagnetic–thermal–flow simulation should be performed by coupling with tools such as Simcenter™ FLOEFD™.

4.7 Concluding Remarks

In this chapter, the requirements for an efficient and accurate simulation of the electromagnetic and thermal behavior of a power transformer were outlined. It was shown that the material data and its modeling play a significant role for an accurate simulation. In cases, when the irreversible behavior of the material should be taken into account, in order to compute the correct flux and the resulting losses, e.g., in the case of a rotating flux at a joint, then a hysteresis model is required because a single-valued *B–H* curve is no longer adequate for solving for the magnetic field accurately and the empirical approach is no longer accurate for estimating the loss. For more detail, see Chap. 7.

It was demonstrated that using a time-harmonic solver, when there are nonlinear *B–H* properties, as is in the case of the transformer core, and in the presence of AC voltage sources, would require a transformation to the *B–H* properties, in order to achieve a better fit between the outcome of a time-harmonic simulation and one obtained using the time-domain simulation.

It was also shown that, in order to reduce computational cost and, yet, obtain accurate solutions, it is necessary to deal with the multi-scale nature of a device, such as the power transformer, in which there are either very large, but yet very thin components, or when there are eddy currents which flow in very thin regions of the conductors, by using special elements (see Chap. 3), such as thin insulators, or special boundaries, such as the nonlinear surface impedance condition.

The efficiency of the solution system also plays a very important role, in obtaining field solutions in a reasonable time. For example, in non-conducting regions, the potential can be scalar which reduces the size of the matrix, and therefore, a mixed formulation, such as the T-Ω method [21], in particular, with the use of hierarchal elements, can significantly reduce the size of the matrix and yet deliver very accurate field solutions. See Chap. 2 for more information on different formulations.

It is also very important to be able to simulate the real operating conditions of a device by driving it using an electrical circuit, even in time-harmonic simulations, or to simulate dynamic conditions, such as faults, in the time-domain simulations. See Chap. 3 for more information on the use of voltage sources.

References

1. P. P. Silvester, R. L. Ferrari, "Finite elements for electrical engineers", Cambridge University Press, Third Edition 1996.
2. Simcenter MAGNET Knowledge Base articles at https://support.sw.siemens.com/en-US/knowledge-base/MG601540.
3. D. N. Dyck, B. S. Murray, "Transient Analysis of an Electromagnetic Shaker Using Circuit Simulation With Response Surface Models", IEEE Trans. on Magnetics, Vol. 37, No. 5, September 2001.
4. C. P. Steinmetz, "On the law of hysteresis," *Proceedings of the IEEE*, vol. 72, no. 2, pp. 197–221, 1984.
5. S. Hussain, K. Chang, "Effects of Incorporating Hysteresis in the Simulation of Electromagnetic Devices Using MAGNET v7.9", White Paper, Mentor Graphics Corporation, 2018.
6. Z. Cheng, N. Takahashi, B. Forghani, G. Gilbert, J. Zhang, L. Liu, Y. Fan, X. Zhang Y. Du, J. Wang, and C. Jiao, "Analysis and measurements of iron loss and flux inside silicon steel laminations," *IEEE Trans. on Magnetics*, vol. 45, no. 3, pp. 1222–1225, 2009.
7. Z. Cheng, N. Takahashi, B. Forghani, Y .Du, J. Zhang, L. Liu, Y. Fan, Q. Hu, C. Jiao, and J. Wang, "Large power transformer-based stray-field loss modeling and validation", Electric Machines and Drives Conference, IEEE International, pp. 548–555 (Digital Object Identifier: https://doi.org/10.1109/iemdc.2009.5075260), 2009.

8. Z. Cheng, N. Takahashi, B. Forghani, G. Gilbert, Y. Du, Y. Fan, L. Liu, Z. Zhai, W. Wu, and J. Zhang, "Effect of Excitation Patterns on both Iron Loss and Flux in Solid and Laminated Steel Configurations", *IEEE Trans. on Magnetics*, vol. 46, no. 8, August 2010.
9. Z. Cheng, N. Takahashi, B. Forghani, Y. Du, Y. Fan, L. Liu, Z. Zhao, and H. Wang, "Effect of Variation of B-H Properties on Loss and Flux Inside Silicon Steel Lamination", *IEEE Trans. on Magnetics*, vol. 47, no. 5, May 2011.
10. J. Zhang, L. Li, L. Liu, Y. Fan, B. Forghani, and Z. Cheng, "Measurement and 3D FEM Analysis of Additional Loss in Laminated Silicon Sheets Caused by Leakage Flux", Transactions of China Electrotechnical Society, 2013, Vol. 28, Issue (5): 148–153.
11. Z. Cheng, N. Takahashi, B. Forghani, L. Liu, Y. Fan, T. Liu, Q. Hu, S. Gao, J. Zhang, and X. Wang, "Extended progress in TEAM Problem 21 Family", COMPEL 2014, Vol. 33 Iss: 1/2 2014, pp. 234–244.
12. Z. Cheng, N. Takahashi, B. Forghani, A.J. Moses, P.I. Anderson, Y. Fan, T. Liu, X. Wang, Z. Zhao, and L. Liu, "Modeling of Magnetic Properties of GO Electrical Steel Based on Epstein Combination and Loss Data Weighted Processing", IEEE Trans. on Magnetics, Vol. 50, No. 1, January 2014.
13. Z. Cheng, B. Forghani, Y. Liu, Y. Fan, T. Liu, and Z. Zhao, "Magnetic Loss Inside Solid and Laminated Components under Extreme Excitations", International Journal of Energy and Power Engineering, Vol. 5, Issue 1–1, pp. 21–30, September 2015.
14. Z. Cheng, B. Forghani, T. Liu, Y. Fan, and L. Liu, "Extended P21-Based Benchmarking", International Journal of Energy and Power Engineering, Issue 1–1, pp. 1–11, February 2016.
15. Z. Zhao, Z. Cheng, B. Forghani, and L. Liu, "Analytical study and corresponding experiments for iron loss inside laminated core under ac-dc hybrid excitation", International Journal of Applied Electromagnetics and Mechanics · August 2017.
16. Z. Cheng, B. Forghani, X. Wang, L. Liu, Y. Fan, X. Zhao, and Y. Liu, "Engineering-oriented investigation of magnetic property modeling and application", International Journal of Applied Electromagnetics and Mechanics, August 2017.
17. S. Hussain, "Development of advanced material models for the simulation of low-frequency electromagnetic devices," Ph.D. Thesis, McGill University, Montreal, Canada, Feb. 2017.
18. D. C. Jiles and D. L. Atherton, "Theory of ferromagnetic hysteresis," *Journal of Magnetism and Magnetic Materials*, vol. 61, no. 1–2, pp. 48–60, 1986.
19. Description of TEAM Problem 32: A Test-Case for Validation of Magnetic Field Analysis with Vector Hysteresis, O. Bottauscio, M. Chiampi, C. Ragusa, L. Rege and M. Repetto, https://www.compumag.org/wp/team/.
20. J. P. Webb, B. Forghani, "Adaptive Improvement of Magnetic Fields using Hierarchal Finite Elements", IEEE Trans. on Magnetics, Vol. 30, No. 5, pp. 3511–3514, November 1994.
21. J.P. Webb, B. Forghani, "A T-Omega method using hierarchal edge elements", IEE Proceedings, Sci. Meas. Technol., Vol. 142, no. 2, pp. 133–141, March 1995.
22. Description of TEAM Problem 3: Eddy Current Workshop (3), Bath Plate with 2 Holes, https://www.compumag.org/wp/wp-content/uploads/2018/06/problem3.pdf.
23. D. A. Lowther, B. Forghani, and E. Freeman, "On equivalent reluctances for time harmonic eddy current analysis", Elsevier Studies in Applied Electromagnetics in Materials, Vol. 6, 1995, Pages 647–650.
24. Simcenter MAGNET Knowledge Base article at https://support.sw.siemens.com/en-US/knowledge-base/MG601540#Scripting.
25. D.N. Dyck, J. P. Webb, "Solenoidal Current Flows for Filamentary Conductors", IEEE Trans. on Magnetics, Vol. 40, No. 2, March 2004.
26. J. P. Webb, B. Forghani, and D.A. Lowther, "An Approach to the Solution of Three-Dimensional Voltage Driven and Multiply Connected Eddy Current Problems", IEEE Trans. on Magnetics, Vol. 28, No. 2, pp. 1193–1196, March 1992.

Chapter 5
Development of Customized Scripts

Junjie Zhang

Abstract Well-established and fully validated computer-aided engineering (CAE) software is essential for the analysis and design of electrical equipment. However, general-purpose commercial software may not meet specific requirements from designers and application engineers, and it may be necessary to customize the software by developing an application-specific script which uses the application programming interface (API) technology. This chapter introduces the basic preparation and the implementation of a script mainly based on the Simcenter MAGNET™ API. An example of the script which is used to calculate the parameters of the transformer winding is presented.

Keywords Script · Syntax · Application programming interface (API) · Power transformer · Magnetic field · Winding

5.1 Introduction

For large-scale commercial CAE software, its universality should be considered, the main objective being the accurate computation of the key parameters of electrical equipment. For users in different fields, however, special analysis and design may be required for a certain type of product, and in many cases, such analysis is beyond the capability of general-purpose software. Here are a few cases often encountered in engineering:

- The human–machine interactive operation of the software is suitable, intuitive, and convenient for the initial analysis of a product. However, during the routine design and optimization phase, calculations for products with similar structures require a large number of repetitive tasks, such as in modeling, setting of properties and mesh parameters, extracting calculation results, and so on.

J. Zhang (✉)
Institute of Power Transmission and Transformation Technology,
Baobian Electric, Baoding, China
e-mail: zhangjunjie@btw.cn

Z. Cheng et al. (eds.), *Modeling and Application of Electromagnetic and Thermal Field in Electrical Engineering*, https://doi.org/10.1007/978-981-15-0173-9_5

In particular, for the modeling and calculation of electric machines with complex cogging structures, it will take more time for designers if experienced manual intervention is required for each step of the operation. Users want the complex, repetitive, and dynamic calculations to be performed automatically.
- For problems, such as the transient electromagnetic simulation of a short-circuit fault of a single phase to ground, in a three-phase transformer, or the transient simulation of a short circuit in a three-phase electric machine, the electromagnetic field simulation software is required to exchange data with other commercial software tools, such as MATLAB®, SimuLink® and PSIMLink®, or Microsoft Word or Microsoft Excel. Users are interested in customizing the analysis flow between different simulation tools, in order to meet user requirements independently and be able to control the entire process, during which, the electromagnetic simulation data and the data calculated in other commercial software can flow between two or more software tools.
- For certain equipment, some specialized calculation and analysis may be needed to extract post-processing information based on the field results, such as obtaining the magnitude or waveform of magnetic flux density, so as to calculate the loss using different algorithms. Another example is the calculation of the short-circuit force and the eddy current loss in a winding caused by the leakage magnetic field [1]. Since, in this case, the cross-sectional dimensions of the wires are relatively small, it is difficult to obtain the eddy current loss directly, using the field solution, whereas an approximate solution may be obtained based on the leakage flux distribution, using the theoretical equations (to be described in detail in Sect. 5.4 under this Chapter).
- Not only does the script provide the user with the opportunity of secondary development, it also successfully builds a bridge between the commercial CAE software, the user, and other third-party applications, surmounting the obstacles among the three. The batch-driven, complex, and repetitive tasks, such as parameter-driven modeling, automatic finite element solution, subsequent secondary calculations, and interactive simulation can be realized through executable script command files, forms containing scripts, and event handling scripts.

Some of the large commercial CAE software tools, such as the Simcenter MAGNET suite provide a set of APIs that allow scripts to run in-process. In this chapter, starting from the basic knowledge of script programming and taking the typical scripts in the development environment of Simcenter MAGNET, and an actual development process as an example, the basic-to-advanced implementation of scripts is described.

5.2 Basics of the Script

5.2.1 Definition and Role of the Script

A script is a set of instructions composed of executable statements, written in a programming language that can be interpreted and run by the scripting engine in an application program [2].

Scripts are industry standard tools for implementing extensibility. By using script:

(1) Complex and repetitive tasks, including preprocessing and post-processing, can be completed automatically, e.g., parametric modeling, parametric solution, subsequent calculation of the macroscopic quantity with the post-processing information (such as calculation of impedance with energy and calculation of loss with magnetic flux density waveforms), auto screenshot for figures, and analysis of data.
(2) A customized application menu may be provided to meet the user requirements.
(3) Direct calls to and from other *ActiveX* interface compliant software (interoperability).

The extension and customization of the software and the co-simulation and invocation with other software can be implemented through the powerful, yet convenient scripting, such as the establishment of batch processing calculation, automatic preprocessing, secondary development using various results, and the solution of some physical quantities using certain specialized equations. Users may customize and expand CAE processes according to the needs of their respective products and integrate CAE as a technical product with self-relied intellectual property.

5.2.2 Classification of the Script

1. **Internal Script**

The internal script is implemented by *OLE Automation*, a branch of *ActiveX*, as a bridge connecting various software components.

Some of the large commercial CAE applications are provided with API interfaces that allow scripts to run in-process, among which Simcenter MAGNET and other series can run script command files, forms containing scripts, and event handling scripts. The form provides a basic graphical user interface (*GUI*) and is programmed with a window suitable for users, while the event handling script can respond to run a specific script program according to an event, such as "opening a file" or "during the solution phase of a model."

Generally, internal scripts are not used to call one application inside another, although that can be achieved, but internal scripts are normally used for unidirectional functions such as exporting data to *Excel*®.

Internal scripting languages include *Visual Basic Script* (also called *VBScript* or *VBS*), *Java Script* (also called *JScript*), etc. Many application software products commonly use the VBScript.

2. **External Script**

Scripts can also be run using scripting engines other than the one in the proprietary application, such as *Windows Script Host*, *Excel*®, and *MatLab*®. External scripts are usually used to invoke and establish communication between different applications.

The external scripting languages can be any of the programming languages. They can be internal script languages (*VBS*, *JScript, Perl Script*), *Visual Basic for Applications* (also called *VBA*, which can be found in *Excel*®, *Word*®, etc.) and *Visual Basic* (VB), etc. This chapter focuses on VBScript.

5.2.3 Concise Basic Syntax of VBScript

5.2.3.1 Basic Format

The VBScript language cannot be used on its own. It must be executed in conjunction with the application. The code of VBScript is plain text and can be written using any text editor, such as Notepad, WordPad, or Word®. Its basic syntax is similar to Visual Basic, both of which are object-based programming (OBP) languages, not object-oriented programming (OOP) languages, and can be programmed following object properties, methods, and events. If one knows VB, one is also able to program in VBScript very easily. For example:

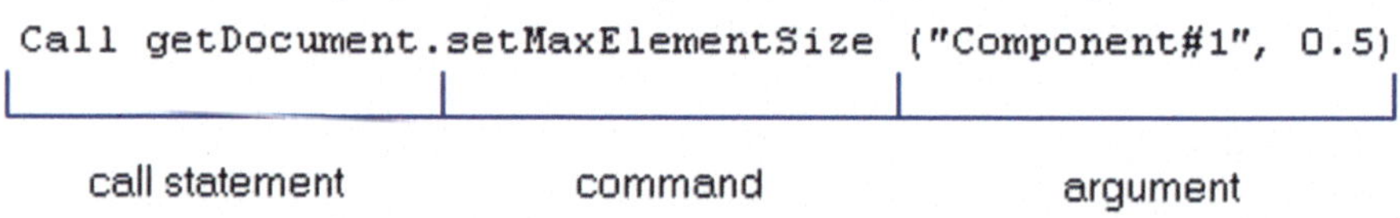

Except for strings, VBScript code is case-insensitive, but one should still try to keep it tidy in order to make the code legible. One should make sure to use English punctuation marks in the code; otherwise, the code will not run.

5.2.3.2 Special Symbols

When writing VBScript programs, it should be noted that some symbols have special uses in statements.

1. **“.”: Dependent symbol**

The objects can have their own properties, methods, or events. To read or change the property of an object, to use a method, or to respond to an event in a program, one can simply insert a dependent symbol “.” between the object name and its method name or the property name or the event name to indicate a subordinate relationship. To set the name of a button to “Start,” one can write:

```
Command1. value = “Start”
```

2. **“()”: Parentheses**

Generally, the parameters can be enclosed within the parentheses behind the function, method, or event handler names to input and output. For example:

```
CStr (123456) ‘The CStr function converts the expression enclosed within the
parentheses into a string and returns.
```

3. **“_”: Multiple Line Symbol**

When a long statement is to be written using multiple lines, an underscore “_” can be added at the end of each line to implement continuation. The following two lines of statements will be treated as one line when the program is executed. The continuation symbol can be used to improve readability by not making a line of statement to exceed the window width or print span. For example:

```
a = (1 + 2 + 3) * _
(1 + 2 + 3)
```

4. **“:”: Line Combination Symbol**

If multiple lines of statements in a program are very short, the statements can be written on one line and separated by colon “:”, which not only increases readability but also shortens the length of the program. The statements will be executed one by one when the program runs. Following is an example of two short statements:

```
n1 = 3
n2 = 12
```

These two lines can be combined into:

```
n1 = 3:n2 = 12
```

5. **“'”: Comment Symbol**

Any comment may be added following a single quote “'”, generally as a brief explanation of the code. When the program is executed, the text following the comment symbol is ignored and the following line of statement will be executed directly. Comments are often placed before the statement that is to be described or on the same line following the statement, as shown below:

```
'This program calculates the addition of two numbers.
Sub add()
N1=3:n2=12              'Two lines of statements are combined into one.
Write n1+n2
End Sub
```

6. **“"”: String Symbol**

The string value must be enclosed in double quotes. Such as the following three strings:

```
"Welcome to Beijing!", "abc," "123"
```

7. **“#”: Date Symbol**

The value for a variable of data type should be enclosed within hash symbol “#”, such as:

```
Writeday = #22/9/2008 08:28:23 AM#
```

8. **“=”: Assignment Symbol**

5.2.3.3 Data Type of VBScript

There is only one data type in VBScript, called Variant. It is also the data type returned by all functions in VBScript. The data type stored in the variables of variant is called the subtype of data.

```
Variable=2008        'treated as an integer by VBScript
Variable="2008"      'treated as a string by VBScript
Variable="Beijing"   'treated as a string by VBScript
Variable=2008.01     'treated as a decimal by VBScript
```

The common subtypes are string, number, date and logical type, as shown in Table 5.1.

5.2.3.4 VBScript Constants

A constant contains a numeric value with a certain name. Constants can represent strings, numbers, dates, or other constants. Once a constant is declared, its value can no longer be changed. Many predefined intrinsic constants are provided in VBScript. It is not necessary to explicitly declare such symbolic constants before use, and they can be used anywhere in the code. For example, vbCrLf represents a combination of carriage return and line feed, and vbGreen represents a green value.

Table 5.1 Data subtypes of variant

Subtype		Description
Character	String	Contains a variable-length string, and the maximum length can be 2 billion characters
Numeric	Byte	Contains an integer in the range 0–255
	Integer	Contains an integer in the range −32,768 to 32,767
	Long	Contains an integer in the range −2,147,483,648 to 2,147,483,647
	Single	Contains a single-precision, floating-point numbers in the range −3.402823E38 to −1.401298E−45 for negative values; 1.401298E−45 to 3.402823E38 for positive values
	Double	Contains a double-precision, floating-point number in the range −1.79769313486232E308 to −4.94065645841247E−324 for negative values; 4.94065645841247E−324 to 1.79769313486232E308 for positive values
Date	Date (time)	Contains a number that represents a date between January 1, 100 A.D. to December 31, 9999 A.D.
Logic	Boolean	Contains either True or False. When other data types are converted to logical data, nonzero is converted to True and 0 is converted to False
Currency	Currency	−922,337,203,685,477.5808 to 922,337,203,685,477.5808
Object	Object	Objects are included, such as obtaining a mesh: set mesh = get Document().get Solution().get Mesh(prob_id)
Error	Error	Contains an error number

Users can declare constants, such as:

```
Const PI=3.1415926                       'It represents a numeric constant.
Const MyString = "This is a string. "    'It represents a string constant.
Const ConstString1="100"                 'It represents a string constant.
Const ConstDate=#2008-08-08#             'It represents a date constant.
```

5.2.3.5 VBScript Variables

Variables refer to the name of the address storing the data in memory.

1. **Naming Rules**

To name a variable, one must follow the naming rules of VBScript:

The first character of the name must be an alphabet; inserted period "." cannot be contained; the name cannot exceed 255 characters and must be unique in the declared context; it cannot be the same as the VBScript's keywords or intrinsic keywords in the application.

2. **Declaring and Scope of Variables**

Dim, Public, and Private statements are generally used to explicitly declare variables and allocate storage space in VBScript. The syntax format is:

```
{Dim|Private|Public} <variable name1> [,<variable name2>] [,<variable name3>]...[,<variable name>]
```

Description:

(1) The variables declared in the procedure with the Dim statement are called procedure-level variables, and those declared outside the procedure are called script-level variables; the former can only be applied within a procedure, while the latter can be applied to all procedures in the script;
(2) The variables declared with the Public statement can be used for all procedures in the associated script;
(3) The variables declared with the Private statement have scope only within that script in which they are declared;
(4) Multiple variables to be declared should be separated by commas, e.g.,

```
Dim a, b, c, d.
```

The variable name can be used directly without declaring the variable in VBScript, which is a simple implicit declaration. However, this is not a good practice generally, as it can sometimes lead to unexpected results when the script is run due to misspelled variable name. Therefore, it is best to use the Option Explicit statement to require explicit declaration of all variables, and it shall be the first statement in the script:

```
Option Explicit
```

3. **Lifetime of Variables**

The time during which a variable exists is called the lifetime of the variable. The lifetime of a script-level variable extends from the time it is declared until the time the script has finished running, while the lifetime of procedure-level variable is only the running duration of the procedure, i.e., the variable extinguishes as soon as the procedure is exited. Local variables are ideal as temporary storage space when a procedure is executing. Local variables with the same name can be used in different procedures because each local variable is only recognized in the procedure in which it is declared.

5.2.3.6 VBScript Array Variables

In most cases, only one variable value needs to be assigned to the variable to be declared. Variables that contain only one value are called scalar variables. In some cases, it is more convenient to assign a plurality of correlative values to a variable, whereby a variable containing a series of values can be created, called an array variable. The declaration of an array variable and a scalar variable is basically the same, except that the name of the array variable is followed by parentheses enclosing numbers that specify the number of elements of the array. For example:

```
Dim myArray(10) ‘An array variable of 11 elements is declared.
```

All arrays are zero-based, so myArray(10) actually contains 11 elements.

An array is not limited to one dimension, and its dimension can be as large as 60. When a multidimensional array is to be declared, commas should be used to separate each number in parentheses that represents the size of the array.

```
Dim MyTable(5, 9) 'A 2D array consisting of 6 rows and 10 columns is declared.
```

A dynamic array, i.e., an array whose size changes when the script is running, can also be declared. The Dim or ReDim statement is used for the initial declaration of the dynamic array, and numbers cannot be included in parentheses. For example:

```
Dim MyArray()
ReDim AnotherArray()
```

To use a dynamic array, the number of dimensions and size of each dimension must subsequently be determined using ReDim. In the following example, ReDim is used to set the initial size of the dynamic array to 25, while the following ReDim statement resizes the array to 30, and the Preserve keyword is used to preserve the contents of the array when resizing.

```
ReDim MyArray(25)
...
ReDim Preserve MyArray(30)
```

Although there is no limit to the number of times a dynamic array can be resized, it should be noted that data in the eliminated elements will be lost when the array is reduced in size.

5.2.3.7 Operators of VBScript

VBScript inherits all kinds of operators from Visual Basic, including arithmetic operators, comparison operators, logical operators, and concatenation operators.

1. **Arithmetic Operator**

The names and expressions of arithmetic operators are shown in Table 5.2.

2. **Comparison Operator**

The comparison operator is used to compare two data, which must be of the same type, and the result returned is usually a Boolean value. The comparison operators or the expression formed by them are often applied to conditional statements, as shown in Table 5.3.

Table 5.2 Arithmetic operator

Operator	Name	Expression	Operator	Name	Expression
^	Exponentiation	a^b	*	Multiplication	a*b
/	Floating-point division	a/b	\	Integer division	a\b
+	Addition	a + b	–	Subtraction	a−b
Mod	Modulus arithmetic	a Mod b	–	Negation	−c

3. **Logical Operator**

The names, examples, and descriptions of logical operators are shown in Table 5.4.

4. **Concatenation Operator**

To concatenate two strings or values into one string, simply insert a string concatenation operator "&" between the two strings or values:

```
"Welcome to"&" Beijing!"   ' = "Welcome to Beijing!"
3&5                        ' = "35"
"No." & "3"                ' = "No. 3"
```

Sometimes "+" can also be used to concatenate strings, but if the two expressions are numbers, it will be the addition operator, for instance:

```
3+5                        ' =8
```

5. **Precedence of Operators**

When a plurality of operators are included in an expression, the precedence of operators should be noted. It is executed from the inside to the outside of the parentheses, and the standard precedence of operator should still be followed inside the parentheses; operators with the same precedence should be calculated from left to right; during calculation, the order of arithmetic operator—concatenation operator—comparison operator—logical operator should be followed, as shown in Table 5.5.

5.2.3.8 Control Statements

There are two types of control statements in VBScript: conditional (decisional) control statements and looping control statements. The conditional statements are usually used to control the commutation under conditionals and selection problems of program process, including selection statements (If…Then…Else) and switch

Table 5.3 Comparison operator

Operator	Description	True	False	Null
<	Less than	Expression1< Expression2	Expression1> = Expression2	Expression1 or Expression2 = Null
<=	Less than or equal to	Expression1 <= Expression2	Expression1 > Expression 2	Expression1 or Expression2 = Null
>	Greater than	Expression1 > Expression2	Expression1 <= Expression2	Expression1 or Expression2 = Null
>=	Greater than or equal to	Expression1 > = Expression2	Expression1 < Expression2	Expression1 or Expression2 = Null
=	Equal to	Expression1 = Expression2	Expression1 <> Expression2	Expression1 or Expression2 = Null
<>	Unequal to	Expression1 <> Expression2	Expression1 = Expression2	Expression1 or Expression2 = Null
Is	Object equivalence	Same object	Different object	Object1 or Object2 = Null

Table 5.4 Logical operator

Operator	Name	Example	Description
And	Logical conjunction	(4 > 5) and (3 < 4)	The result is True only if the values of both expressions are True
Or	Logical disjunction	(4 > 5) or (3 < 4)	As long as one of the two expressions is True, the result is True. Only if the values of both expressions are False, the result is False
Not	Logical negation	not (1 > 0)	The negation is carried out from True to False or from False to True
Xor	Logical exclusion	10 > 8 xor 8 > 6	If the values of both expressions are True or False, the entire expression is True; otherwise, it is False
Eqv	Logical equivalence	(3 − 1) eqv (4 − 2) (3 > 1) eqv (4 > 2)	The value is true only if the two expressions have the same value. A bit-by-bit comparison can also be performed on the same bits in two numeric expressions
Imp	Logical implication	(3 > 1) imp (2 < 4)	The value is False only if the first expression is True and the second expression is False. A bit-by-bit comparison can also be performed on the same bits in two numeric expressions

Table 5.5 Precedence of operator

Precedence	Operator type	Operator
1	Arithmetic operator	^ (exponentiation)
2		− (negation)
3		*, /(multiplication and division)
4		\ (integer division)
5		Mod (modulus)
6		+, −(addition and subtraction)
7	Concatenation operator	& (concatenation)
8	Comparison operator	=, <>, < , > , <=, >=, Is
9	Logical operator	Not
10		And
11		Or
12		Xor
13		Eqv
14		Imp

statements (Select…Case). The looping control statements are used to write loop flow under specific conditions in the program, including For (For…Next), Do (Do…Loop) and While (While…Wend) looping statements, as shown in Table 5.6.

Table 5.6 Control statement

Name of statement	Syntax format	Functional description

If… Then… Else (conditional statement)	If <Conditional Expression 1> Then
	<Statement Body 1>
	[Else Statement 2]
	[Else If <Conditional Expression 2> Then
	…
	End If
		A set of statements is conditionally executed based on the value of the expression

Select Case… (switch statement)	Select Case test expression [Case Expression list 1 [Statement Body 1]] [Case Expression list 2 [Statement Body 2]] … [Case Else [Statement Body n]] End Select	One of several sets of statements is executed according to the value of the expression
For…Next For Each…Next (For looping statement)	For counter = start To end [Step step]For [Statement Body 1] [Exit For] [Statement Body 2] Next [counter]	A set of statements that is executed specified times
	For Each Element In group [Statement body1] [Exit For] [Statement Body2] Next [element]	A set of statements is repeatedly executed for each element in an array or group
Do…Loop (Do looping statement)	Do [{While | Until} condition] [Statement body1] [Exit Do] [Statement Body 2] Loop [{While | Until} condition]	While: Repeat execution of a statement body when it is True Until: Repeat execution of a statement body until it becomes True
While…Wend (While looping statement)	While (loop condition) [Statement body] Wend	A series of statements are executed when the specified condition is True. This is an old syntax structure and a Do Loop statement is recommended
Set	Set objectvar = {object expression | New class name | Nothing} Set object.event name = Get Ref(procname)	Assign an object reference to a variable or attribute, or associate a procedure reference with an event

5.2.3.9 VBScript Procedures

When a script is written where the same set of statements is to be executed in multiple places, the procedures are often used to make the program reusable and to keep it simple. VBScript is provided with two kinds of procedures, one is Sub procedure and the other is Function procedure.

1. **Sub Procedure**

The Sub procedure is a procedure without a return value. Sub procedures can take arguments such as constants, variables, or expressions. A Sub procedure without parameters must contain a pair of empty parentheses. The general format is as follows:

```
[Private | Public]Sub Procedure Name ([Parameter 1, Parameter 2, ...])
 [Statement Body]

 [Exit Sub]
[Statement Body]

End Sub
```

2. **Function Procedure**

Function is similar to Sub, except that a function can return a value. Function procedures can take arguments such as constants, variables, or expressions. For a Function procedure without argument, its Function statement must also contain a pair of empty parentheses. The function returns a value by assigning a value to the function name in one or more statements of the procedure. The general format is as follows:

```
[Private | Public] Function Procedure Name ([Parameter 1, Parameter 2, ...])

[Statement Body]

Function Procedure Name = Expression

[Exit Function]
[Statement Body]

End Function
```

3. **Using Sub and Function Procedures**

When a Sub or Function procedure is called from another procedure, the procedure can be called with a Call statement, and any argument must be enclosed in parentheses, even if the procedure takes no argument. The syntax is as follows:

```
Call Sub | Function Procedure Name ([Parameter 1, Parameter 2,...])
```

In addition to calling a procedure with a Call statement, a Sub procedure may be provided as a single statement by simply typing the procedure name and any arguments required, or just the procedure name if the procedure takes no argument.

```
Sub Procedure Name [Parameter 1, Parameter 2,...]
```

Where the return value of the Function procedure is not taken into account, the Function procedure can be called like the Sub procedure, as shown in the following example.

```
MsgBox "Task Completed!", 0, "Task Box"
```

Including parenthesis in the example above will give rise to a syntax error.

To use the value returned form a Function procedure, the Function procedure name must be assigned to the variable, and the parameters should be enclosed in parentheses, as shown in the following example:

```
Answer3 = MsgBox ("Are you happy with your salary?", 4, "Question 3")
```

Functions can be involved in operations like variables, just as internal functions are used. For example:

```
sum = sum + int (Myinput ("Please enter a transaction amount"))
```

5.3 Script Development in Simcenter MAGNET

5.3.1 Automatic Modeling

5.3.1.1 Script File

The Simcenter MAGNET scripting toolbar is shown in Fig. 5.1.

The script can be run by clicking on the order of *Tools menu/Scripting/Run Script*…or by clicking  (*Run Script*) button on the scripting toolbar [3].

The following is an example of the automatic building of a cubic model, illustrating the implementation of automatic modeling by the script file.

(1) Create a new plain text file (.txt) with notepad, change its file name to *SquareComponent.vbs*, open it with notepad and enter the following contents and save it.

```
Call newDocument()                    'Create a new document.
Call getDocument().getView().showGrid(True)       'Display construction grid.
Call getDocument().beginUndoGroup("Set Default Units", true)
Call getDocument().setDefaultLengthUnit("Millimeters") 'Change the default length unit to mm.
Call getDocument().endUndoGroup()
Call getDocument().getView().newLine(-5, -5, -5, 5) 'Draw a construction line.
Call getDocument().getView().newLine(-5, 5, 5, 5)
Call getDocument().getView().newLine(5, 5, 5, -5)
Call getDocument().getView().newLine(5, -5, -5, -5)
Call getDocument().getView().selectAt(-0.0402161553502083, -0.160793229937553, infoSetSelection, Array(infoSliceSurface))                    'Select the construction surface.
Call getDocument().getView().selectAt(-0.0402161553502083, -0.160793229937553, infoSetSelection, Array(infoSliceSurface))
REDIM ArrayOfValues(0)
ArrayOfValues(0)= "Component#1"
Call getDocument().getView().makeComponentInALine(1, ArrayOfValues, "Name=CR10: Cold rolled 1010 steel", True)                    'Stretch the construction surface to generate a cube.
Call getDocument().save("C:\Windows\Temp\SquareComponent.mn") 'Save file.
```

In the above script commands, a lot of API functions which are internally integrated in Simcenter MAGNET are widely adopted.

(2) Open the Simcenter MAGNET module and create a new document.

(3) Click on the order of *Tools/Scripting/Run Script*…or click  (*Run Script*) button on the scripting toolbar.

Fig. 5.1 Scripting toolbar

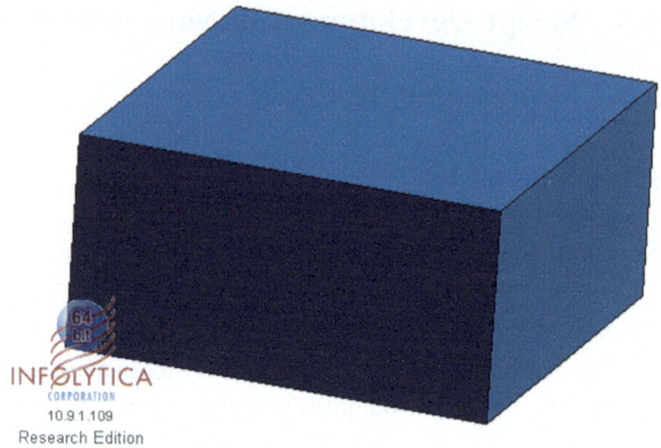

Fig. 5.2 A cube model built automatically by script file

(4) Select *SquareComponent.vbs* file, and click Open to execute the script, then a cube can be generated automatically, as shown in Fig. 5.2.

5.3.1.2 Script Form

A script form can also be created by the user to accept user's input and to control the process of the program in stages.

The following example demonstrates how to use the Simcenter MAGNET script form to automatically create a cube model whose side length is specified by the user.

(1) Start the Simcenter MAGNET module.
(2) Click on the order of *Tools/Scripting/New Script Form*…or click (*New Script Form*) button on the scripting toolbar to open a new form. Click *File/Save as* …to save it as a *SquareComponent.frm* file.
(3) Click (*Command*) button to add a command button on the opened form, and then click **A** (*Label*) button and **ab|** (*TextBox*) button to add a label and text input, as shown in Fig. 5.3.
(4) Right-click on the Command button to select the "Properties," change the "ID" value to "GoButton" and the "Caption" value to "Make Square Component," as shown in Fig. 5.4.
(5) Also, right-click on the Label and select "Properties," and change the value of "Caption" to "Side." Change the "Text" value of TextBox to "20" and the "ID" value to "Inputbox." In addition, the user can customize other properties on the property page, such as the text font. The adjusted form is shown in Fig. 5.5.

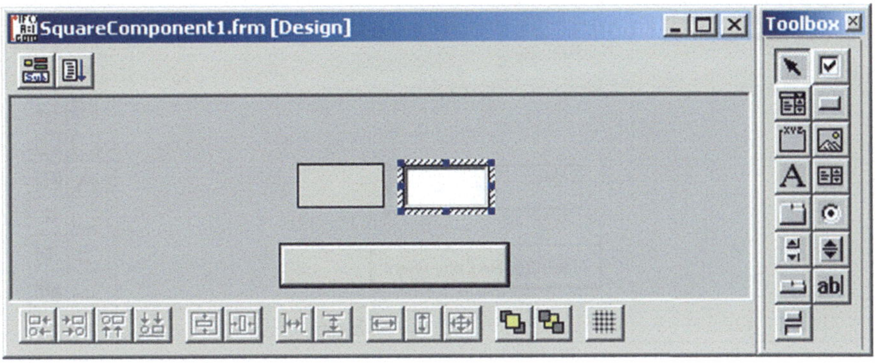

Fig. 5.3 A label and text input are added to the form

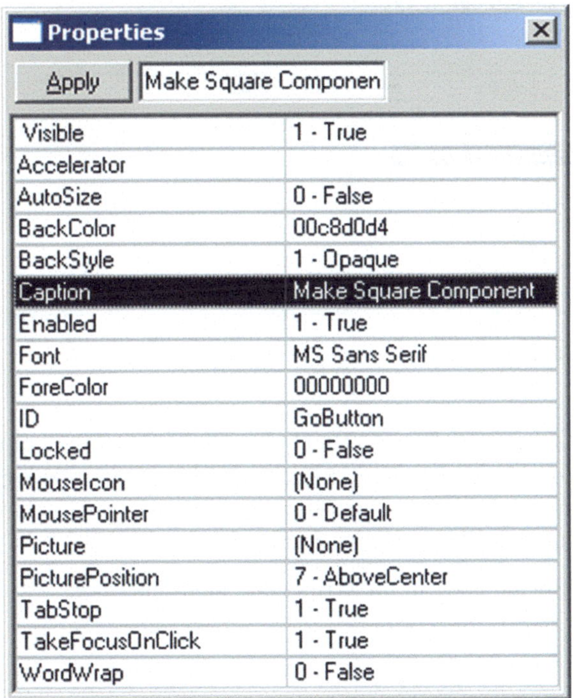

Fig. 5.4 "Properties" page for command buttons in the script form

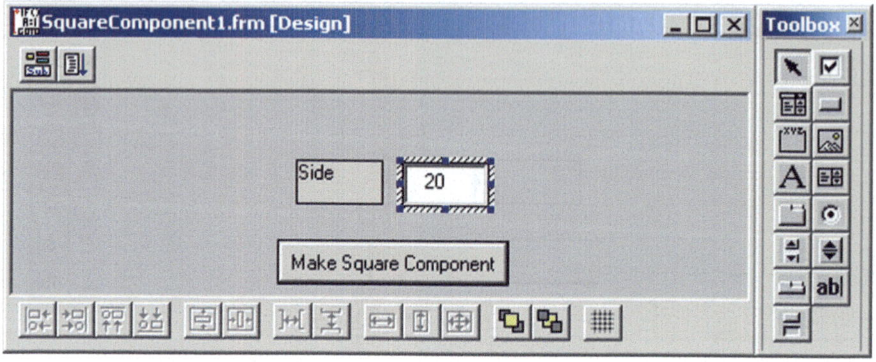

Fig. 5.5 User script form interface

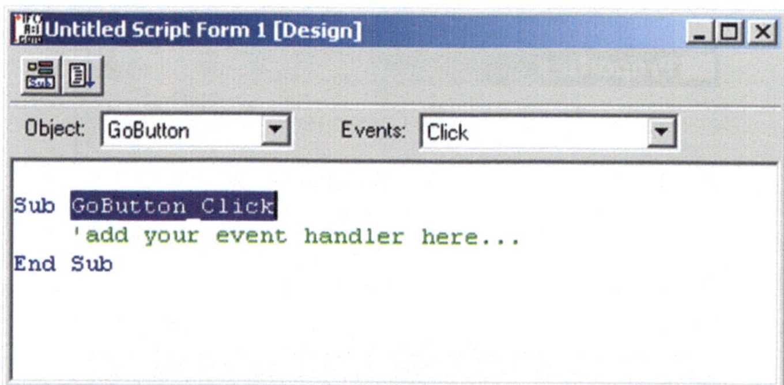

Fig. 5.6 Script code input interface

(6) Click (*Object View/Script View*) button at the top of the script form to open the script code input interface. Select "GoButton" from the Object list and "Click" from the Events list, and then the code will appear in the code input interface automatically, as shown in Fig. 5.6.

Enter the following code in the script code input interface:

```
Sub GoButton_Click

    Call newDocument()
    DIM Side
    Side = CDbl(Inputbox.text)    ' obtain user data from the TextBox.
    Call getDocument().getView().showGrid(True)
    Call getDocument().beginUndoGroup("Set Default Units", true)
    Call getDocument().setDefaultLengthUnit("Millimeters")
    Call getDocument().endUndoGroup()
    Call getDocument().getView().newLine(-Side/2, -Side/2, -Side/2, Side/2)
    Call getDocument().getView().newLine(-Side/2, Side/2, Side/2, Side/2)
    Call getDocument().getView().newLine(Side/2, Side/2, Side/2, -Side/2)
    Call getDocument().getView().newLine(Side/2, -Side/2, -Side/2, -Side/2)
    Call getDocument().getView().selectAt(0, 0, infoSetSelection, Array(infoSliceSurface))
    REDIM ArrayOfValues(0)
    ArrayOfValues(0)= "Component#1"
    Call getDocument().getView().makeComponentInALine(Side, ArrayOfValues, "Name=CR10: Cold
rolled 1010 steel", True)
    Call getDocument().save("C:\Windows\Temp\SquareComponent.mn")

    End Sub
```

(7) Click (*Run Script*) button at the top of the script form to start the customized script form that was just created. The side length of the cube can be modified in the TextBox, and then click "Make Square Component" button, to view the change of the cube model.

5.3.2 *Recording Script File*

In software such as the Simcenter MAGNET suite, the user is allowed to record any operation in the software as a script file, and such an operation can be reproduced by running the recorded script file. This can be achieved by using (*Record User Script*) button, (*Pause*) button, (*Resume Recording*) button and (*Stop Recording*) button.

5.3.3 *Interoperability*

Direct interoperability between electromagnetic simulation software and other ActiveX-compliant software can be achieved using a script, which is very important for users. For example, the following can be realized:

- Data can be exchanged between electromagnetic simulation software and Excel spreadsheet. The data can be imported and exported from one to the other, and some calculations can be made by VBA program in spreadsheet [4].

- The resulting data from an electromagnetic simulation being queried or exported can be used in a third-party simulation program to deal with multi-physics problems or in a self-compiled program for subsequent calculations.
- The user is allowed to connect the electromagnetic simulation software with the commercial optimization software to complete the product optimization analysis.
- The electromagnetic simulation software can be connected to Word to generate automatic reports.
- Users may use Simcenter MAGNET as a client through an internal script, or call Simcenter MAGNET as a server through an external script.

Next, the interoperability between Microsoft Excel 2010 and Simcenter MAGNET can be demonstrated by taking the input data in Excel and calling the MAGNET's module to create a cube model automatically. In this example, the MAGNET's module acts as a server, and Excel is a client, using an external script.

(1) Open Excel and save the file as *SquareComponent.xls*.
(2) Select "File/Options/Customize Ribbon" menu and check "Developer" in Main Tabs list to make Developer Ribbon visible.
(3) Click  (*Button*) on the Form Controls and drag a button onto a Sheet, and then press "OK," as shown in Fig. 5.7.
(4) Right-click the button and select "Edit Text," and change the value of "Button 1" to "*SquareComponent.*"

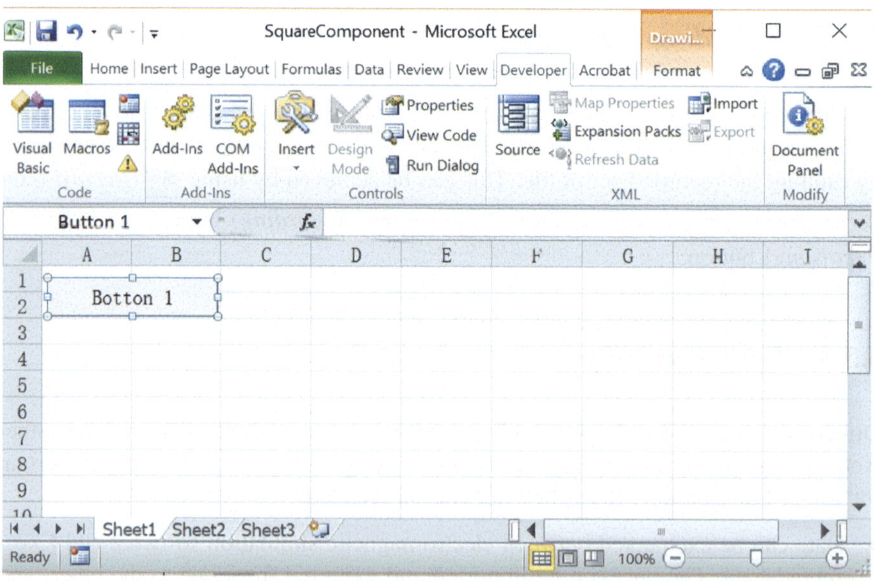

Fig. 5.7 A button is created on a sheet

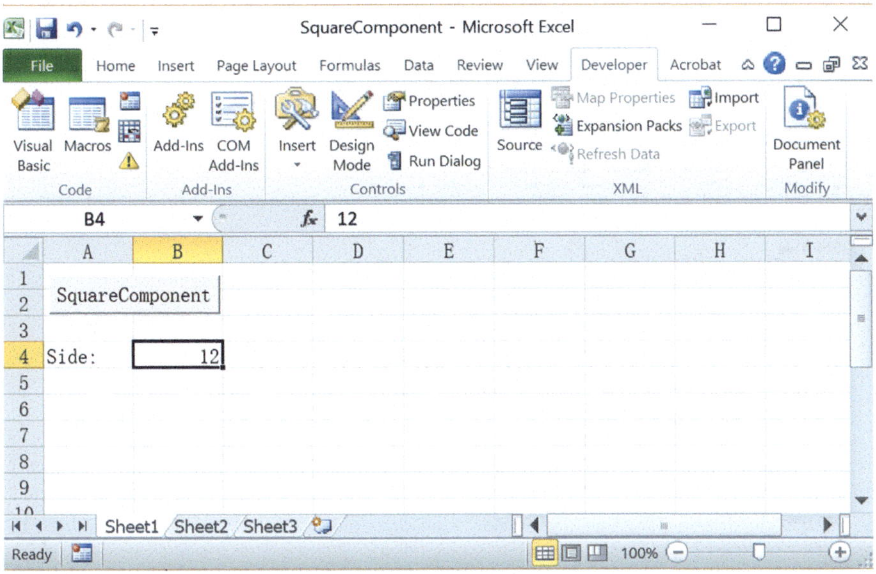

Fig. 5.8 Drive parameters are entered in the sheet

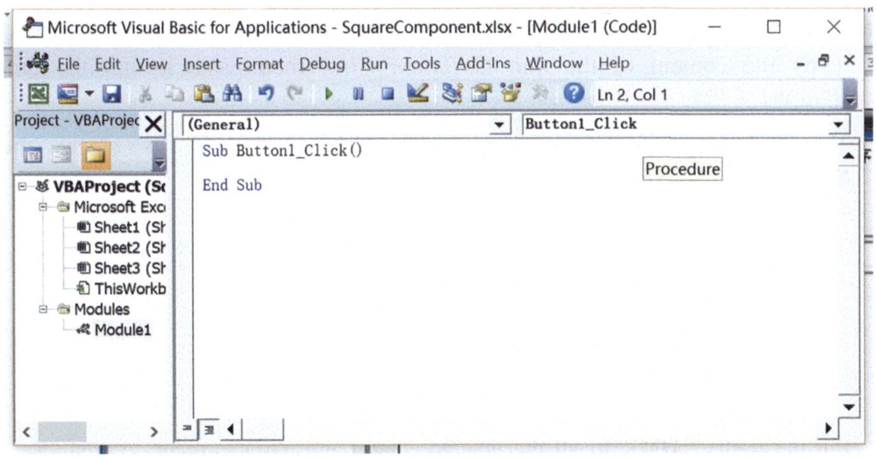

Fig. 5.9 VBA programming mode in excel

(5) Enter, respectively, "Side" and "12" in cells A4 and B4 in the Sheet, and the results are shown in Fig. 5.8.
(6) Right-click the button and select "Assign Macro…" and press "Edit" button to enter the VBA programming mode. The statement for Sub of button click event has been automatically created, as shown in Fig. 5.9.

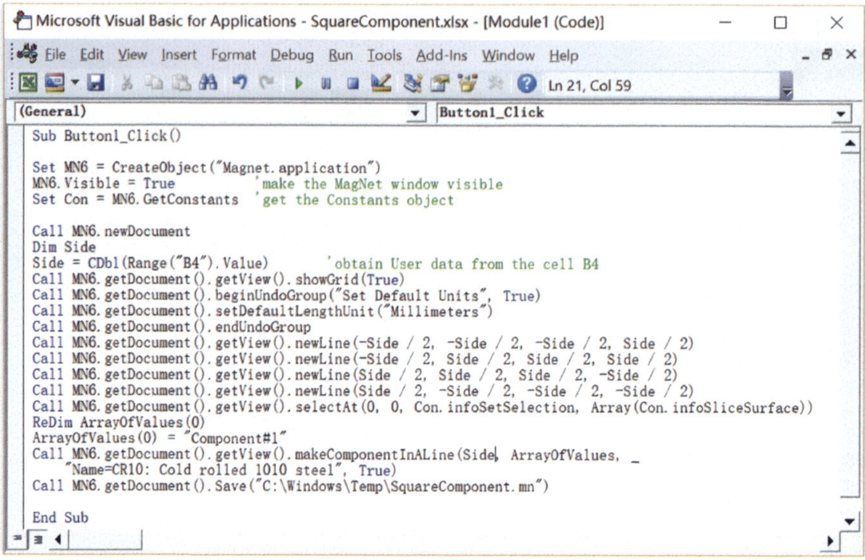

Fig. 5.10 Code editing window

(7) Open the *SquareComponent.frm* (script file) established in Sect. 5.3.1.2 and copy the content of the *Sub GoButton_Click* then paste it into the *Sub Button1_Click*.

- The following three commands are added in front to create the Simcenter MAGNET program and its constants:

```
Set MN6 = CreateObject("Magnet.application")
MN6.Visible = True
Set Con = MN6.GetConstants
```

- Put the prefix "*MN6*" to all the *newDocument* and *getDocument* commands.
- Put the prefix "*Con*" to all the *infoSetSelection* and *infoSliceSurface* commands.

The above codes can be applied as a reference format to call a Simcenter MAGNET program. It is very convenient.

- Replace the "Inputbox.text" by "Range("B4").Value," then the value of cell B4 will be passed to the "Side" variable.

The final results are shown in Fig. 5.10.

(8) Close the VBA code editing window and save this Excel document. Then, if one clicks the *SquareComponent* button, the Simcenter MAGNET program will start and create a cube with a side length of 12 units automatically.

5.3.4 Export the Field Data

One of the most important uses of scripts is to export the solved field data. It is not easy to extract the field data on the specified component and its corresponding element node coordinate value from the huge result data without using scripting language, and using the scripting functions that are encapsulated in the Simcenter MAGNET module make it so much easier.

The following is an example of exporting field data:

```
Dim Doc, Sol, Field, Mesh
Dim SolutionId(1)
Set Doc = getDocument()
Set Sol=Doc.getSolution()
SolutionId(0)= 1
SolutionId(1)= 1
Set Mesh=Sol.getMesh(SolutionID, "Component#2")   'To obtain the mesh of Component#2 in a solution
problem
Set Field=Sol.getSystemField(Mesh,"|B|")  'To obtain the target field on the specified component, in this
case, the target field is "|B|"
Call Field.getFieldData(Data)          'To obtain the data of the target field and pass it to the Data array
Dim text                    'Next, output the raw data to a text file.
text=""
NumOfFieldNodes= Field.getNumberOfFieldNodes() 'Number of nodes in the target field
For i=0 to NumOfFieldNodes-1
text=text & Data(i) & vbNewLine
Next
Call writeTextFile("D:\datafile.txt", text)    'Write data to a file in the hard disk.
```

The above codes can be applied as a reference format to export the Simcenter MAGNET field data. It is very convenient to use those commonly used variables, such as “Doc,” “Sol,” and so on.

The user may define the format of the output data or save the results in different document forms such as Txt, Word, or Excel. Users can also save the specified data as needed, e.g., include the node coordinate values of mesh, the magnetic flux density, eddy current, loss and stress values, and so on. The data can be local or integrated global. All of them can be called with corresponding functions in the Simcenter MAGNET module.

Figure 5.11 shows the results exported and stored in an Excel document using scripting commands, which include the mesh node information and magnetic flux densities $|B|_{smoothed}$ on specified component in the Simcenter MAGNET model.

L28 fx 0

	A	B	C	D	E	F	G	H	I	J	K	L
1	Component Name: hv5											
2		|B| smoothed:		Node1 No.	Node1 x	Node1 y	Node1 z		Node2 No.	Node2 x	Node2 y	Node2 z
3		0.157472		30	536	554.58	0		31	546.2728	554.58	0
4		0.142504		25	649	554.58	0		51	649	564.6062	0
5		0.144438		5	649	715	0		7	638.7272	715	0
6		0.01923		4	536	715	0		21	536	704.9738	0
7		0.021045		3	561.6818	715	0		2	556.5454	715	0
8		0.017102		58	566.8182	715	0		3	561.6818	715	0
9		0.050238		59	560.7691	710.6779	0		2	556.5454	715	0
10		0.057723		61	563.7234	704.5953	0		58	566.8182	715	0
11		0.0528		63	599.7438	561.4597	0		35	607.9091	554.58	0
12		0.098925		36	618.1818	554.58	0		65	619.3937	563.9919	0
13		0.10858		66	574.9461	707.0559	0		58	566.8182	715	0
14		0.105267		67	629.2145	701.8398	0		68	619.4606	700.9565	0
15		0.089826		62	649	684.9213	0		70	636.113	691.1032	0
16		0.091766		12	536	584.6587	0		72	547.9975	582.3442	0
17		0.090327		68	619.4606	700.9565	0		67	629.2145	701.8398	0
18		0.087916		74	640.1556	696.8008	0		67	629.2145	701.8398	0
19		0.089826		0	618.1818	715	0		75	607.9091	715	0
20		0.090327		77	637.0469	572.3915	0		78	632.2493	578.1314	0
21		0.090327		36	618.1818	554.58	0		80	624.8467	560.083	0
22		0.091766		80	624.8467	560.083	0		36	618.1818	554.58	0
23		0.094369		82	544.7624	695.6766	0		83	548.4	688.5278	0
24		0.089201		60	555.1689	704.4704	0		61	563.7234	704.5953	0

Sheet1 Sheet2 Sheet3

Fig. 5.11 Exported field data and node information on specified component

The results exported and stored in an Excel document using scripting commands include the mesh node information and magnetic flux densities $|B|_{\text{smoothed}}$ on a specified component in the Simcenter MAGNET model.

5.4 Development of a Script for Transformer Winding Parameters Calculation

5.4.1 Requirements for Script to Calculate Transformer Winding Parameters

Winding parameters such as the eddy current loss and the short-circuit force, caused by the leakage electromagnetic field in the transformer, are important technological indexes in the electrical design of a transformer. The small cross section of wire strands should be taken into account when these winding parameters are being calculated. To model every single wire strand and to calculate the field directly is a massive computational expense and not practical. The leakage electromagnetic field can be solved for first, and then an approximate solution can be obtained using an analytical equation, based on the leakage flux distribution. Also, since the transformer winding parameters are the basic technological indexes at the electrical design stage, and each transformer product requires this calculation, there is a requirement to avoid repetitive and tedious interactive manipulations and to obtain the results quickly and easily. All these requirements can be met using scripting. The calculation script for transformer winding parameters:

- allows seamless connection with the Excel electrical calculation sheet and the user does not have to manually re-input the data in this sheet.
- is driven by the input parameters, and therefore, the electromagnetic field calculation does not require highly skilled human–computer interaction and maintenance, and the data is input once and then the solution is completed automatically. The advantage is that it is simple to operate, saves labor, there are fewer calculation, and the results are accurate.
- uses the flexible preprocessing functions in the Simcenter MAGNET module and is not limited by various modeling and calculation restrictions.
- makes full use of the resources in the Simcenter MAGNET module, including the flexible meshing method and the stable solution kernel.
- calculates and returns results in Excel automatically and generates reports in the required format.
- uses the rich post-processing functions of the Simcenter MAGNET module and obtains the color-shaded plot, showing the eddy current loss, the short-circuit force, the winding temperature and even customized field quantities intuitively and aesthetically.
- has greater extensibility. It can be expanded from 2D to 3D, from the problems of steady-state short-circuit leakage electromagnetic field to the problems of transient short-circuit transient field, DC bias, over excitation, and others.
- is relatively stable, compared with other specialized software.

5.4.2 Goal of the Script Used to Calculate Transformer Winding Parameters

Through the script, used to calculate the short-circuit leakage magnetic field and winding parameters of the transformer, the following can be realized:

- the data import from Excel electrical calculation sheet, automatic parametric modeling, parametric solution, and automatic finite element method (FEM) solution for the short-circuit leakage magnetic field of the transformer;
- according to the stored magnetic energy obtained from the solved field, calculating the short-circuit impedance of the transformer by the energy method;
- obtaining the contour line of flux and the color-shaded plot of magnetic flux density, as well as the field values at the specified lines which are users concerned;
- to get the total eddy current loss of windings and its distribution, obtaining color-shaded plot of eddy current loss, including horizontal and vertical components, and extracting numerical curves at specified lines;
- calculating the total electromagnetic force of the windings and its distribution, obtaining color-shaded plot of the electromagnetic force, including the

horizontal and vertical components, and extracting numerical curves at specified lines;
- to get the temperature of the windings and the oil flow distribution, obtaining the color-shaded plot of temperature on the windings and the temperature rise of the hottest spot and its location;
- further expansion of the functionality, for example, to calculate lost-stability of windings.

5.4.3 FEM Method to Calculate the Eddy Loss of Windings

When the leakage magnetic field of the transformer passes through the winding, electromotive force (EMF) is induced in the wires. Under the EMF, eddy currents are generated in the section of the wire, and loop currents are also formed between the parallel wires of the winding. The loss caused by the eddy current is called the eddy current loss, and its value mainly depends on the size of the wire, the value and the distribution of the leakage magnetic field, and varies with the position of the windings. The eddy current loss inside the volume of one element can be calculated by:

For eddy current loss in the transverse direction:

$$P_{\mathrm{er}i} = \frac{1}{24\rho}\omega^2 b^2 B_{\mathrm{r}i}^2 V_i \tag{5.1}$$

For eddy current loss in the longitudinal direction:

$$P_{\mathrm{eh}i} = \frac{1}{24\rho}\omega^2 d^2 B_{\mathrm{h}i}^2 V_i \tag{5.2}$$

where

$B_{\mathrm{r}i}$ the transverse magnetic flux density in the ith element, T;
$B_{\mathrm{h}i}$ the longitudinal magnetic flux density in the ith element, T;
b, d the size of the wire, m.

See Fig. 5.12 for definitions of the above symbols.

ρ the resistivity of the material, Ω m;
ω the angular frequency, rad/s;
V_i the volume of ith element, m^3.

Replace V_i with S_i in 2D Calculation, S_i—area of ith element, m^2.
Now, the total eddy current loss P_{e} in the winding is:

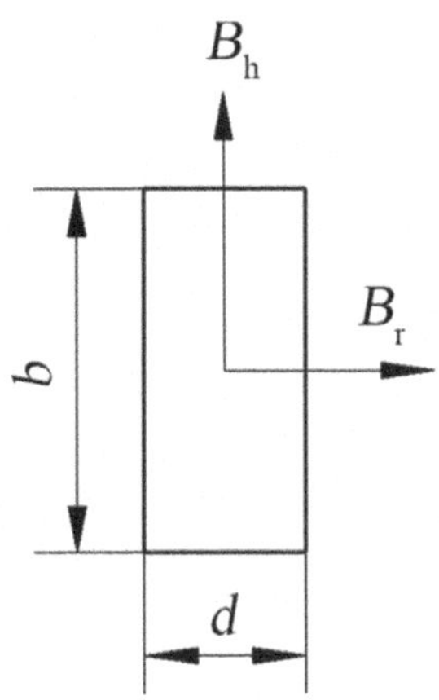

Fig. 5.12 Sectional dimensions of one wire

$$P_e = \sum_{i=1}^{N} P_{ei} = \sum_{i=1}^{N} (P_{eri} + P_{ehi}) \tag{5.3}$$

where

N Total number of the winding elements.

5.4.4 Implementation Process

1. **Creating a Basic GUI Interface Form**

At first, the method to implement the tool using scripting should be determined, i.e., whether to implement using script files or script forms, or be in third-party software and set up the mode of principal and subordinate interaction. In general,

- the programs which require no or little input, or require less manual intervention, are usually implemented using script files;
- the programs which require more input and control are implemented using script forms;
- when the program needs to exchange data with another application, one of the applications can serve as the principal interface and call the other, as the subordinate application.

In this example, the program uses a script for calculating the winding parameters of large power transformers and adopts a GUI written using Microsoft Visual Basic [5, 6] and can be called interactively with the Simcenter MAGNET module, and the form can be started independently without Simcenter MAGNET. In the VB environment, elements such as labels, text boxes, buttons, and pictures should be placed

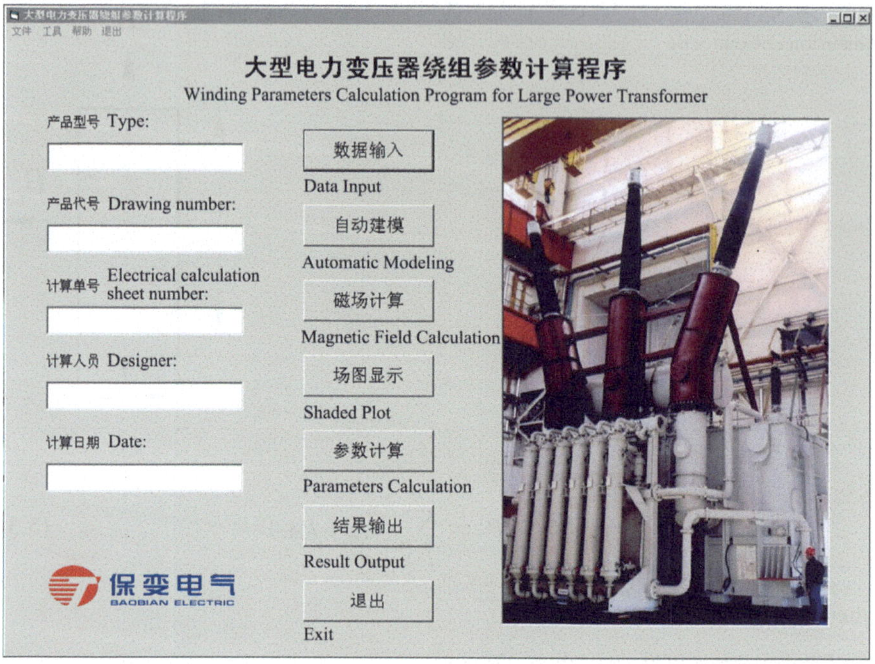

Fig. 5.13 Main interface of winding parameters calculation program for large power transformer

in the form, and their respective properties should be set according to demand. The implementation approach is the same as the VBScript mentioned before, therefore no more restatement. The main interface can be obtained after arranging the elements in the form based on the aesthetics and the practicality concerns. The result is shown in Fig. 5.13.

2. **Data Import**

The more complex a model, the more electrical parameters there are. To save time on entering data, Excel documents can be established to import data, and then the following command can be used to import data in batches:

```
Side = CDbl (Range ("B4").Value)
```

Here, "B4" is the cell number of Excel, and CDbl command is used to convert the data into double-precision data.

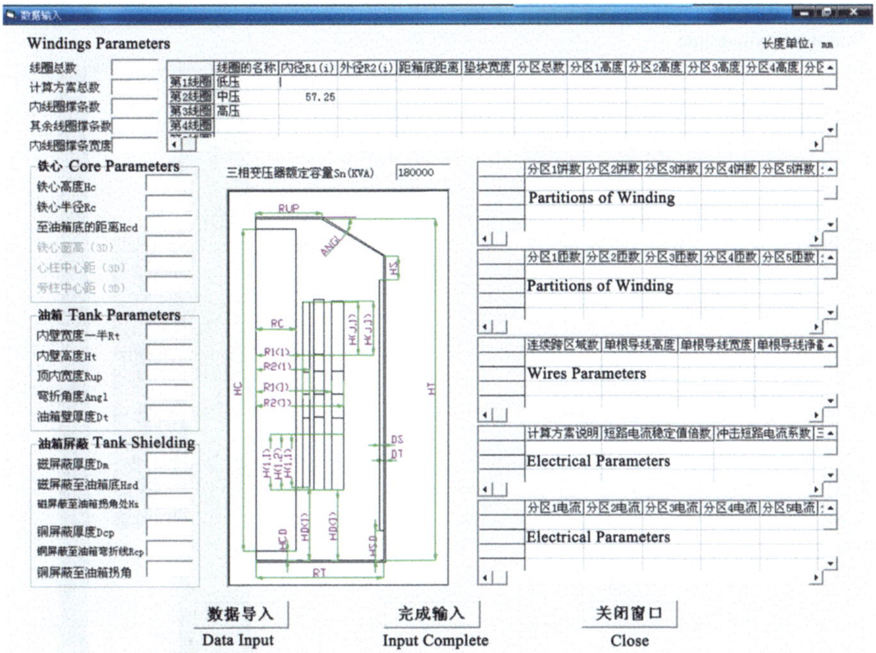

Fig. 5.14 Data input interface

The winding parameters calculation program for large power transformers imports data using an Excel document, which are hyperlinked with the electrical calculation sheet, so the user does not need to input extra for the electrical parameters.

The imported data are displayed in the data input interface, as shown in Fig. 5.14. The user is allowed to modify data in the interface, or enter data directly without using import.

Each textbox on the data input interface needs to be numbered. After the user finishes proofreading, the data is assigned to the variables in memory for subsequent calculation. The format of assign command is as follows:

```
Ttype = InputForm.Text1.text
```

3. **Automatic Modeling**

After all data are proofread, click the "Auto Modeling" button to call the Simcenter MAGNET script function, and then the transformer will be modeled automatically. Automatic modeling includes several parts of FEM preprocessing:

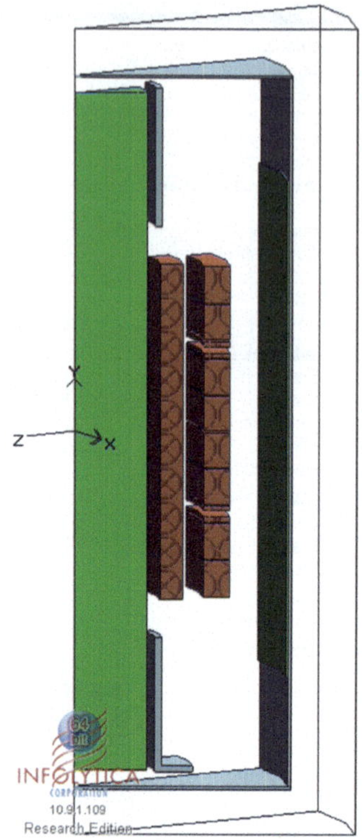

Fig. 5.15 Model generated by automatic modeling

- geometric model can be established according to the input parameters;
- the corresponding material property can be assigned for each component according to the material code (serial number or character pre-assigned);
- script can be used to adjust the mesh generation method, mesh size, or order of components, etc., to obtain more accurate results, as described in Chap. 4;
- the exciting source can be established and the number of turns and the current in each partition of the winding can be entered;
- the corresponding boundary conditions can be specified;
- other settings in the solver need to be adjusted;
- after automatic modeling, users can partially modify the model manually in Simcenter MAGNET in particular cases. The model generated by automatic modeling is shown in Fig. 5.15.

4. **Magnetic Field Calculation**

Once the model is created, the preparations for the solver are complete. When the "Magnetic Field Calculation" button is clicked, Simcenter MAGNET is called by the script and the model is automatically meshed, and then, the finite element simulation is performed using the time-harmonic solver. Figure 5.16 shows the mesh of the Model.

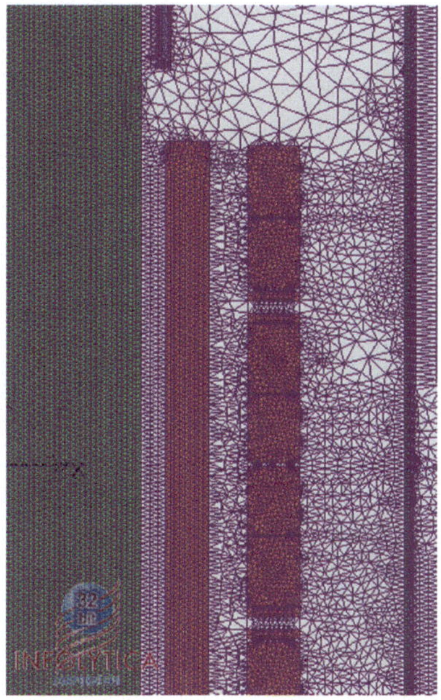

Fig. 5.16 Mesh of the model

5. **Field Plot**

Once the solution is completed, if the "Shaded Plot" button is clicked, the program extracts typical field screenshots, and field value curves over the coils or the magnetic shielding. The distribution and variety of field quantities can be directly observed through the field shaded plot and curves, referenced to designer. Figure 5.17 shows the contour lines of flux and shaded plot of magnetic flux density extracted from the solution.

6. **Parameter Calculation**

Click the "Parameter Calculation" button after the leakage magnetic field result are obtained, and the script will export the field data. The eddy current loss of the windings, the short-circuit force on the windings and the winding temperature are calculated using the method outlined in Sect. 5.4.3, and the temperature rise of the hottest spot and its location on the winding is obtained.

For non-standard field data in Simcenter MAGNET such as eddy current loss and temperature rise of the windings, script can also be used to display these results by using Simcenter MAGNET's post-processing function. Such results can be displayed in color-shaded plot, and horizontal and vertical components can also be displayed separately. And, then, the numerical curve can also be extracted. Figure 5.18 shows the shaded plot of the short-circuit force.

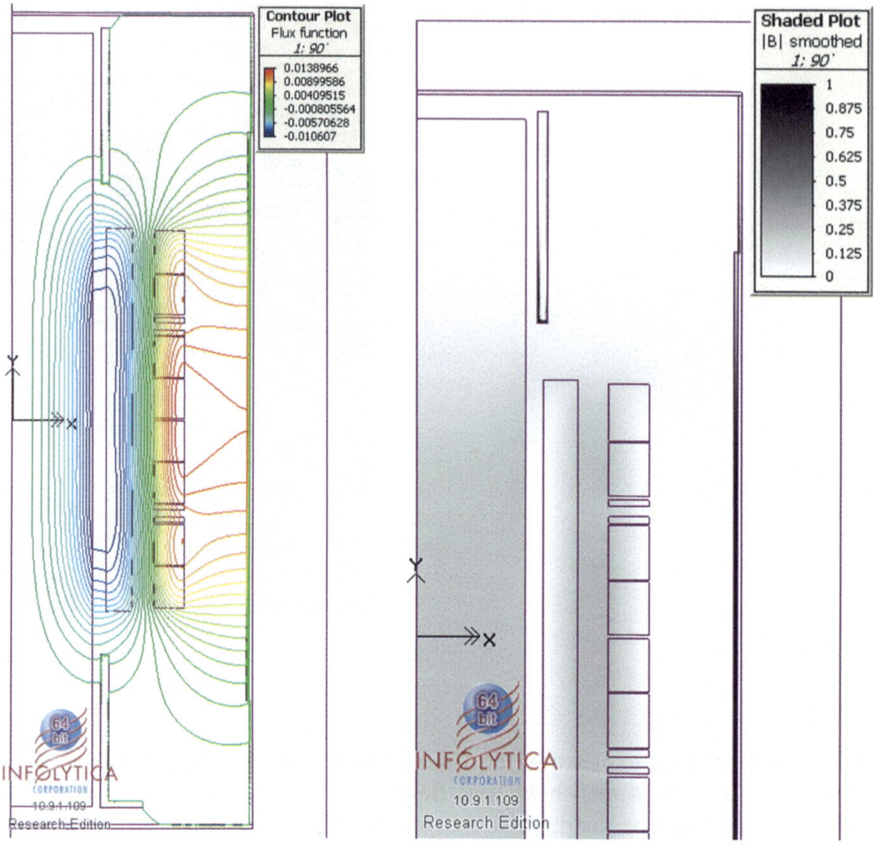

Fig. 5.17 Contour line of flux and shaded plot of magnetic flux density

7. **Output**

The shaded plot, field value curves, and the aforementioned data, including magnetic field, loss and electromagnetic force, are collected together and output in the report. The report can be saved in the form of Excel documents, and thus, users can conveniently draw some reports in the Excel for analysis with the resulting data.

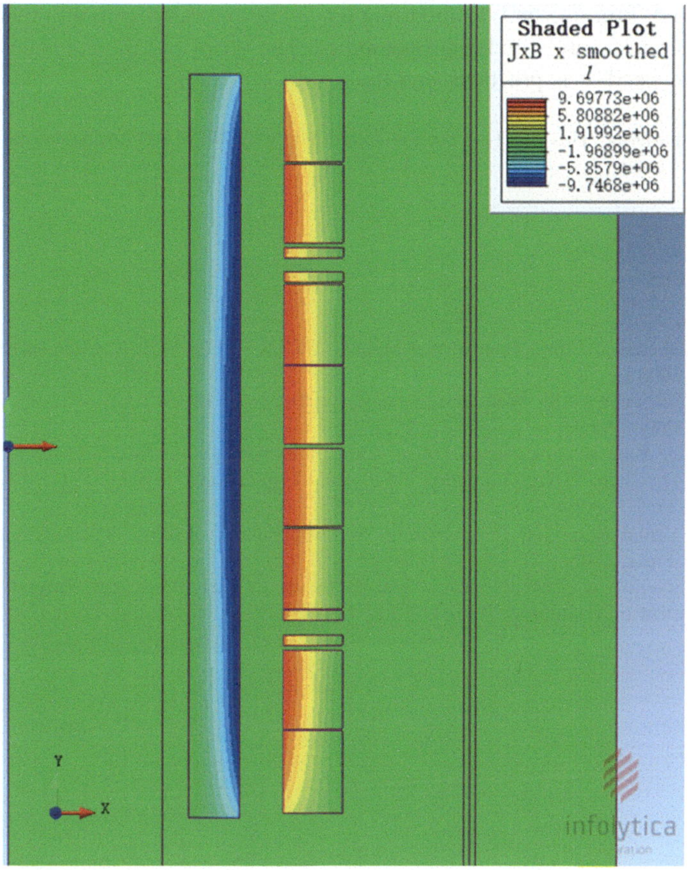

Fig. 5.18 Color-shaded plot of the short-circuit axial force

5.5 Summary

With a universal and flexible programming language, script has built a bridge between a commercial electromagnetic CAE software, users, and third-party applications, making secondary development for users' specific needs feasible and simple. Batch-driven, complex and repetitive tasks, parameter-driven modeling, automatic finite element solution, subsequent secondary calculations, and interactive simulations can be realized through executable script files, forms containing scripts, and event handling scripts, and the human–computer interaction can also be greatly improved and broadened through the GUI interface when using forms.

The program for calculating winding parameters of large power transformers is based on the VBScript technology and VB programming. It uses GUI and makes full use of electromagnetic CAE software resources. The parametric solution has been realized. Intuitive and detailed reports can be provided without repetitive and tedious data preparation and interoperability. The script has the advantages of fast calculation speed, it is practical, and the results are stable.

Acknowledgements This R&D project was supported by the BTW-RD/Infolytica Joint Industrial Application Center.

References

1. Y.Xie et al (eds.). "Power Transformer Manual," ISBN 7-111-11082-X, China Machine Press, Beijing, 2003.
2. P. Yu, "VBScript / ASP Web Design Language Tutorial," ISBN 7-5053-9829-6, Electronics Industry Press, 2004.
3. Knowledge Base at https://support.mentor.com/en/knowledge-base/MG601540#Scripting.
4. X. Liu, D. Chen, "Excel in Office - Efficient Application," ISBN 7-111-18580-3, China Machine Press, 2006.
5. Y. Li, "Visual Basic 6.0 Programming and Application," ISBN 7-115-14269-6, Posts & Telecom Press, 2006.
6. Qiushi Science & Technologies. "The Book of Visual Basic 6.0 Programming and Development of Technology," ISBN 7-115-12514-7, Posts & Telecom Press, 2004.

Chapter 6
Harmonic-Balanced Finite Element Method and Its Application

Xiaojun Zhao

Abstract This chapter mainly introduces the harmonic-balanced finite element method (HBFEM) and its application in electrical engineering. Different from the traditional frequency domain method, the HBFEM is able to compute the nonlinear magnetic field with electrical equipments with significant saturation. Compared with the time-stepping method, transient process can be avoided in HBFEM to reduce the computational time. Furthermore, the hybrid with the fixed-point technique realizes the decomposition of harmonic solutions, which greatly improves the efficiency of numerical computation in the frequency domain. The HBFEM is employed to compute the nonlinear electromagnetic field under DC bias condition and to investigate the force characteristic in a gapped reactor core model under harmonic magnetization. Actually the principle of harmonic-balanced method is applicable to steady-state thermal problems, and it is predicted to have further contribution on loss modeling and thermal analysis of power transformers.

Keywords Harmonic-balanced · DC bias condition · Harmonic magnetization · Eddy current · Power transformer

6.1 Development of HBFEM

The time-varying electromagnetic field described by Maxwell's equations can be solved in time domain or frequency domain [1, 2] based on different potential sets. In electrical engineering, almost all electrical equipment works under steady-state excitation in the normal operation except for breakdown; therefore, the numerical method to solve time-periodic magnetic field problems has drawn much attention from many international scholars.

X. Zhao (✉)
Department of Electrical Engineering, North China Electric Power University, Box 10, Yonghua North Street No. 619, Baoding, China
e-mail: zxjncepu@ncepu.edu.cn

Z. Cheng et al. (eds.), *Modeling and Application of Electromagnetic and Thermal Field in Electrical Engineering*, https://doi.org/10.1007/978-981-15-0173-9_6

The solutions in time domain generally require several periods to reach steady state by using time-stepping technique, starting from arbitrary initial values. Variables in electromagnetic field under steady-state excitations can be approximated by triangular series. Harmonic-balanced theory is used in the finite element method and the integral equation method to calculate nonlinear magnetic field [3–6]. The electromagnetic field can be solved directly in harmonic frequency domain, without long computational time. In recent years, the harmonic-balanced finite element method (HBFEM) has been further developed to investigate the DC bias phenomena in power transformers [7, 8].

The fixed-point technique can be used in time-stepping finite element method to analyze the hysteretic characteristics and eddy current problems. Different strategies are presented to determine the fixed-point reluctivity in each iterative step [9, 10]. A smoothing algorithm is proposed to guarantee the stability of solution in the iterative procedure when hysteresis modeling is involved in the numerical computation of magnetic field and iron loss [11].

In order to solve time-periodic nonlinear magnetic field and compute the iron loss more accurately, a novel numerical method in frequency is proposed, combining the fixed-point technique, the FEM and the harmonic-balanced method. Therefore, the so-called fixed-point HBFEM [12, 13] is presented to calculate the nonlinear magnetic field in the harmonic frequency domain and to obtain the iron loss of laminated core under sinusoidal and DC-biased excitations.

It can be expected that the HBFEM, as a frequency domain method, will play an important role in solving the practical engineering problems, including efficient computation of electromagnetic fields, loss determination of iron core and even solution of multi-physical field.

6.1.1 Basic Theory of HBFEM

The nonlinear magnetic field can be described by the following equation:

$$\nabla \times \nu \nabla \times A - J = 0, \tag{6.1}$$

where $\boldsymbol{A}$ is the magnetic vector potential, ν is the reluctivity, and $\boldsymbol{J}$ is the current density, which includes exciting current density and eddy current density.

In the two-dimensional case, the nonlinear magnetic field equation can be written as follows:

$$\frac{\partial}{\partial x}\left(\nu \frac{\partial A}{\partial x}\right) + \frac{\partial}{\partial y}\left(\nu \frac{\partial A}{\partial y}\right) + \sigma \frac{\partial A}{\partial t} - J_s = 0. \tag{6.2}$$

where A and J_s are the magnetic vector potential and impressed exciting current density in the z-direction, respectively.

In light of Galerkin's method, the weighted residual can be obtained from (6.2) as follows:

$$
\begin{aligned}
G = & \iint\limits_{\Omega_e} \left(\frac{\partial N_i^e}{\partial x} \nu \frac{\partial A}{\partial x} + \frac{\partial N_i^e}{\partial y} \nu \frac{\partial A}{\partial y} \right) \mathrm{d}x\mathrm{d}y \\
& + \iint\limits_{\Omega_e} \left(\sigma \frac{\partial A}{\partial t} N_i^e \right) \mathrm{d}x\mathrm{d}y - \iint\limits_{\Omega_e} \left(J_s N_i^e \right) \mathrm{d}x\mathrm{d}y,
\end{aligned}
\tag{6.3}
$$

where N_i is the interpolation function for a linear triangular element, J_s is the exciting current density, and σ is the conductivity.

The time-periodic solutions are focused on for the DC bias phenomenon, since it is a harmonic problem with alternating and direct excitations. The magnetic flux density and magnetic vector potential are both periodic functions in the time domain. According to the Fourier transformation theory, all variables such as vector potential A, flux density B and exciting current density J_s can be approximated by a triangular series. Therefore, a HBFEM matrix equation for a single element can be obtained based on the harmonic-balanced method as follows:

$$
\begin{aligned}
\boldsymbol{G}^e & = \boldsymbol{S}^e \boldsymbol{A}^e + \boldsymbol{M}^e \boldsymbol{A}^e - \boldsymbol{K}^e \\
& = \frac{1}{4\Delta^e} \begin{bmatrix} (b_1 b_1 + c_1 c_1)\boldsymbol{D} & (b_1 b_2 + c_1 c_2)\boldsymbol{D} & (b_1 b_3 + c_1 c_3)\boldsymbol{D} \\ (b_2 b_1 + c_2 c_1)\boldsymbol{D} & (b_2 b_2 + c_2 c_2)\boldsymbol{D} & (b_2 b_3 + c_2 c_3)\boldsymbol{D} \\ (b_3 b_1 + c_3 c_1)\boldsymbol{D} & (b_3 b_2 + c_3 c_2)\boldsymbol{D} & (b_3 b_3 + c_3 c_3)\boldsymbol{D} \end{bmatrix} \\
& \quad \cdot \begin{Bmatrix} \boldsymbol{A}_1^e \\ \boldsymbol{A}_2^e \\ \boldsymbol{A}_3^e \end{Bmatrix} + \frac{\sigma\omega\Delta^e}{12} \begin{bmatrix} 2N & N & N \\ N & 2N & N \\ N & N & 2N \end{bmatrix} \begin{Bmatrix} \boldsymbol{A}_1^e \\ \boldsymbol{A}_2^e \\ \boldsymbol{A}_3^e \end{Bmatrix} - \begin{Bmatrix} \boldsymbol{K}_1^e \\ \boldsymbol{K}_2^e \\ \boldsymbol{K}_3^e \end{Bmatrix} \\
& = \boldsymbol{0},
\end{aligned}
\tag{6.4}
$$

in which $b_i = y_j - y_k, c_i = x_k - x_j$. The harmonic forms of the magnetic vector potential $\boldsymbol{A}_e^i$ and the exciting term $\boldsymbol{K}_e^i$ are expressed in (6.5) and (6.6), respectively,

$$
\boldsymbol{A}_i^e = \{ A_{i0}^e \quad A_{i1s}^e \quad A_{i1c}^e \quad A_{i2s}^e \quad A_{i2c}^e \quad \ldots \} \tag{6.5}
$$

$$
\boldsymbol{K}_i^e = \{ K_{i0}^e \quad K_{i1s}^e \quad K_{i1c}^e \quad K_{i2s}^e \quad K_{i2c}^e \quad \ldots \}. \tag{6.6}
$$

6.1.2 Coupling Between Electric Circuits and the Magnetic Field

When electromagnetic devices, such as transformers under DC bias, are excited by voltage, the exciting currents are unknown. Therefore, Eq. (6.2) is no longer applicable to solve the coupled problem. In that case, the coupling between the electric circuit and the magnetic field should be taken into account [14–16].

According to Kirchhoff's Law, the applied voltage on the external port of the electric circuit can be defined as follows:

$$V_{\mathrm{in}k} = V_k + R_k I_k + L_k(\mathrm{d}I_k/\mathrm{d}t) + (1/C_k)\int I_k \mathrm{d}t, \tag{6.7}$$

where $V_{\mathrm{in}k}$ is the input voltage of circuit k, and V_k is the corresponding induced electromotive force. C_k and L_k are the capacitance and inductance in circuit k, respectively.

Combined with the finite element method, the equations of electric circuits coupled to the magnetic field can be expressed in the form of matrix as follows:

$$\boldsymbol{V}_{\mathrm{in}k} = \boldsymbol{C}_k\boldsymbol{A} + S_{ck}\boldsymbol{Z}_k\boldsymbol{J}_k, \tag{6.8}$$

where $\boldsymbol{V}_{\mathrm{in}k}$ is the harmonic vector of the input voltage, $\boldsymbol{C}_k$ is the coupling matrix, and $\boldsymbol{Z}_k$ is the corresponding impedance matrix.

Finally, the new system matrix equation, considering the applied voltage, can be obtained by combining (6.4) with (6.8) as follows:

$$\begin{bmatrix} \boldsymbol{H} & -\boldsymbol{G}_1 & -\boldsymbol{G}_2 & \cdots & -\boldsymbol{G}_k & \cdots \\ \boldsymbol{C}_1 & S_{c1}\boldsymbol{Z}_1 & \boldsymbol{0} & \boldsymbol{0} & \boldsymbol{0} & \cdots \\ \boldsymbol{C}_2 & \boldsymbol{0} & S_{c2}\boldsymbol{Z}_2 & 0 & \boldsymbol{0} & \cdots \\ \vdots & \boldsymbol{0} & \boldsymbol{0} & \ddots & \boldsymbol{0} & \cdots \\ \boldsymbol{C}_k & \boldsymbol{0} & \boldsymbol{0} & \boldsymbol{0} & S_{ck}\boldsymbol{Z}_k & \cdots \\ \vdots & \vdots & \vdots & \vdots & \vdots & \ddots \end{bmatrix} \begin{Bmatrix} \boldsymbol{A} \\ \boldsymbol{J}_1 \\ \boldsymbol{J}_2 \\ \vdots \\ \boldsymbol{J}_k \\ \vdots \end{Bmatrix} = \begin{Bmatrix} \boldsymbol{0} \\ \boldsymbol{V}_{\mathrm{in}1} \\ \boldsymbol{V}_{\mathrm{in}2} \\ \vdots \\ \boldsymbol{V}_{\mathrm{in}k} \\ \vdots \end{Bmatrix}. \tag{6.9}$$

where the DC voltage component is also included in the following triangular series as follows:

$$V_{\mathrm{in}k} = V_{k0} + \sum_{n=1}^{\infty}\{V_{kns}\sin(n\omega t) + V_{knc}\cos(n\omega t)\}. \tag{6.10}$$

The vector potentials and current densities can be solved simultaneously by the above equation.

6.1.3 *Epstein Frame-like Core Model Under DC-Biased Magnetization*

An Epstein frame-like core model made by the Tianwei Group, Baoding, China, has been tested under different DC bias conditions. Harmonic analysis of the

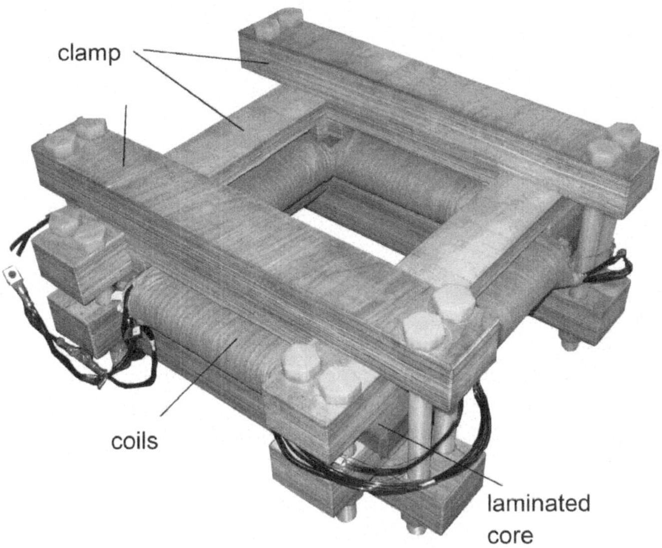

Fig. 6.1 Epstein frame-like core model

magnetizing current and magnetic field is carried out based on the consistency between the computation and the measurement.

The Epstein frame-like core model for the DC-biased test is shown in Fig. 6.1. The iron core is made up of silicon steel lamination (model number 30Q140). Figure 6.2 shows the fitted magnetizing curve of the silicon steel. There are two windings on the ferromagnetic core: the exciting coil (fed by an alternating voltage) and the measuring coil.

Fig. 6.2 Magnetizing curve of the silicon steel sheet

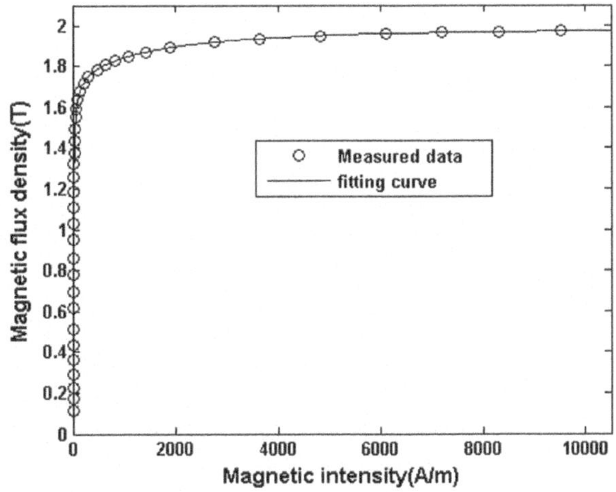

Table 6.1 Different DC bias conditions specified by quantity in the magnetic field

Cases (i/j)	DC bias			AC excitation	
	P_i (% I_0)	$I_{dc,i}$ (A)	$H_{dc,i}$ (A/m)	$U_{m,j}$ (V)	$B_{m,j}$ (T)
1	25	0.4256	105.68	26	0.09
2	50	0.847	213.12	133	0.49
3	75	1.273	320.30	240	0.88
4	100	1.697	425.23	370	1.37
5	150	2.530	636.58	420	1.57
6				495	1.82

The peak value of the excitation current without a DC bias is selected as a reference. This reference current causes the flux density in the silicon steel to reach the rated value (1.7 T) in the transformer's no-load operation. The DC bias in the form of direct current is then applied, in proportion to the reference current, to the exciting coil.

The DC bias current should be integrated with the input voltage of the corresponding circuit for the implementation of computation in (6.9). The input voltage $V_{\mathrm{in}k}$ in (6.10) is actually a harmonic vector that includes DC and AC components. The DC component of $V_{\mathrm{in}k}$ is connected with the DC bias current through the resistance of the transformer's winding.

The value of reference current I_0 measured on the square ferromagnetic core model is 1.68 A. The DC bias is applied in incremental proportions of the reference current, which are represented by $P_i(i = 1, 2, 3, 4)$ in Table 6.1. The AC excitation is also applied in four different cases indicated by the subscript $j(j = 1, 2, 3, 4)$. I_{dc} represents the DC bias current that corresponds to different proportions of the reference current I_0, while H_{dc} is the subsequently generated magnetic intensity. The peak value of alternating flux density B_{m} in the magnetic core varies with the step increase of alternating voltage U_{m} (peak value) [17], which is also shown in the same table.

6.1.4 *Simulation and Analysis*

A. ***Computed and Measured Exciting Current***

The calculated results are compared with the experimental data in different excitation cases, as shown in Figs. 6.3 and 6.4. It is observed that there is consistency in the computational and measured results obtained from the magnetizing current waveforms.

A quantitative comparison between the calculated and measured results is necessary to estimate the calculated errors by HBFEM. There are two main causes of inaccuracy in computation. The first is the hysteresis effect of the magnetic core in the model, which is neglected in this paper for the simplicity of computation. The second cause is a truncation error that plays a key role in the calculation of exciting

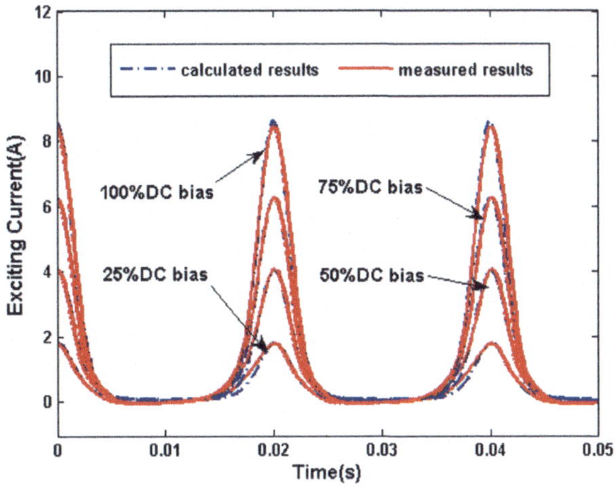

Fig. 6.3 Exciting current under different DC bias ($U_m = U_{m,2} = 133$ V; $B_m = B_{m,2} = 0.49$ T)

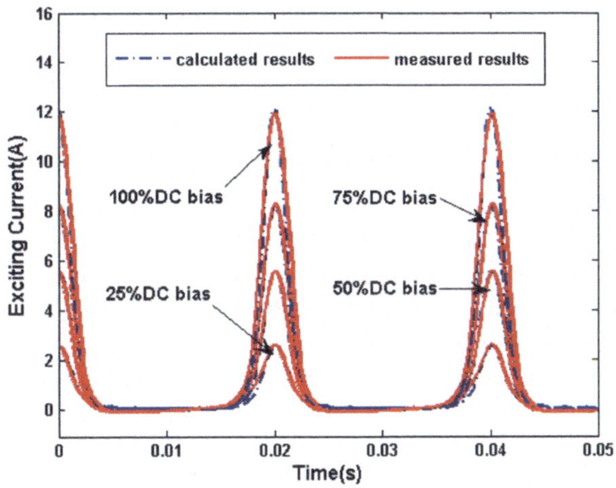

Fig. 6.4 Exciting current under different DC bias ($U_m = U_{m,3} = 240$ V; $B_m = B_{m,3} = 0.88$ T)

currents and magnetic fields under DC bias. The higher the harmonic order considered, the more accurate the results are. The truncated harmonic order depends on the DC and AC excitation. A deficiency of considered harmonic numbers in computation will result in ripples in waveforms of exciting currents.

Because the waveform of the exciting current is non-sinusoidal under DC bias conditions, the root mean square value is selected to carry out the quantitative comparison between the calculated and measured results [17]. $I_{c,rms}$ and $I_{m,rms}$ represent the root mean square values of the exciting current obtained from calculation and measurement, respectively. The error data reflects the inaccuracy resulting mainly from neglect of the hysteresis effects and truncation in computation. The error analysis, as shown in Table 6.2, is performed in different cases related to DC and AC excitations, which can be combined with the waveform

Table 6.2 Errors between calculated and measured results in exciting current

H_{dc} (A/m)	B_m (T)	$I_{c,rms}$ (A)	$I_{m,rms}$ (A)	Error (%)
105.68	0.49	0.8677	0.9088	4.5258
105.68	0.88	1.0861	1.1449	5.1419
105.68	1.37	1.2943	1.3736	5.7773
105.68	1.82	2.9859	3.0984	3.630
213.12	0.49	1.9141	1.9312	0.8869
213.12	0.88	2.3463	2.3695	0.9794
213.12	1.37	2.7315	2.7613	1.0789
213.12	1.82	4.9705	5.3355	6.8412
320.30	0.49	2.9749	2.9878	4.3294
320.30	0.88	3.4717	3.5846	3.1513
320.30	1.37	4.2376	4.5009	5.8500
320.30	1.82	10.042	10.513	4.4700
425.23	0.49	4.0284	4.0519	0.5790
425.23	0.88	4.9335	5.2631	6.2600

comparison in Figs. 6.3, 6.4, 6.5 and 6.6 to evaluate the validity of the computation by HBFEM.

B. ***Harmonic Analysis of the Magnetizing Current***

There are only odd harmonics in the magnetizing current when the transformer is fed by AC excitation. However, additional harmonics appear when the direct current invades the transformer windings. The generation of large harmonics results in significant saturation of the magnetic core and half-cycle saturation of the magnetizing current. Therefore, the relationship between the DC bias and harmonic components should be considered by using harmonic analysis.

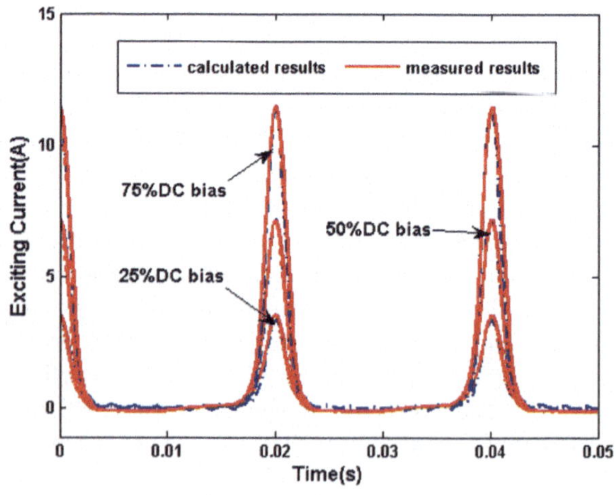

Fig. 6.5 Exciting current under different DC bias ($U_m = U_{m,4} = 370$ V; $B_m = B_{m,4} = 1.37$ T)

Fig. 6.6 Exciting current under different DC bias ($U_m = U_{m,6} = 495$ V; $B_m = B_{m,6} = 1.82$ T)

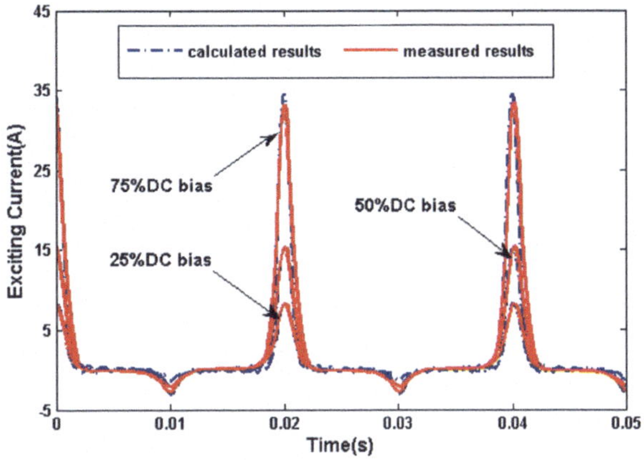

Unlike the time-domain iterations and the Fourier transforming process of the solution in the time-stepping finite element method, all harmonic components in the magnetizing current can be obtained directly from the harmonic solution using the HBFEM. The histograms in Figs. 6.7 and 6.9 show the contribution of different harmonic components to the magnetizing current under different DC biases.

Figure 6.7 shows that while the size of all harmonic components increases when additional DC bias is applied, the growth rate varies in different components. The growth tendency of each harmonic is shown in Fig. 6.8. The numbers 1, 2, 3, 4 in the horizontal coordinate represent different proportions (25%, 50%, 75%, 100%) of the DC bias reference current, respectively. It is obvious that the fundamental and second harmonic components increase near linearly, while higher-order harmonics (the third and fourth) grow faster than linearly.

The contribution of each harmonic component is different when the peak value of alternating voltage is increased up to 495 V, which is given in Fig. 6.9.

Fig. 6.7 Each harmonic component of exciting current under different DC bias ($U_m = U_{m,3} = 240$ V; $B_m = B_{m,3} = 0.88$ T)

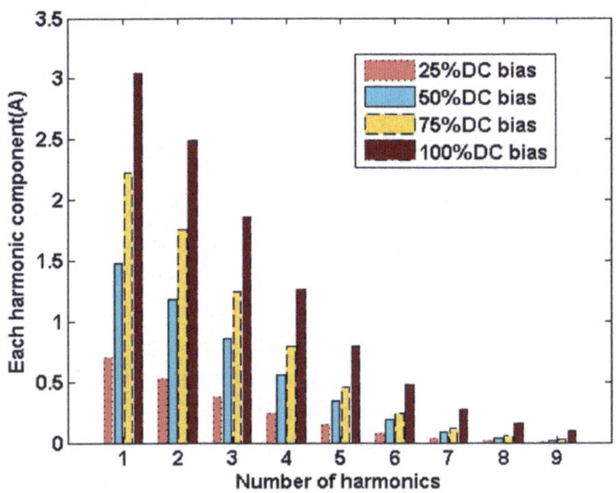

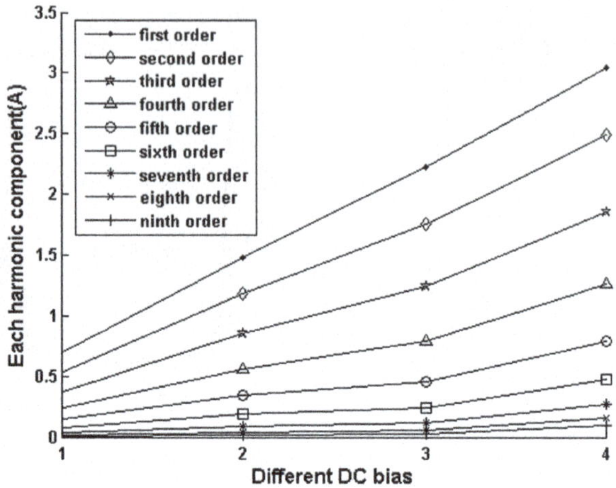

Fig. 6.8 DC bias effect on different harmonics ($U_m = U_{m,3} = 240$ V; $B_m = B_{m,3} = 0.88$ T)

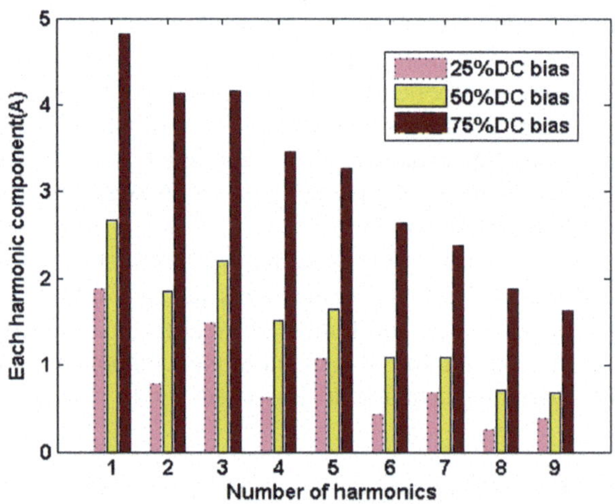

Fig. 6.9 Each harmonic component of exciting current under different DC bias ($U_m = U_{m,6} = 495$ V; $B_m = B_{m,6} = 1.82$ T)

Odd harmonics are greater than even order components under 25% and 50% DC bias, respectively. It is implied that the growth of odd harmonic components is related to the increased AC excitation.

Curves in Fig. 6.10 display a relationship between odd harmonics and AC excitation. With the increased alternating voltage, odd harmonics grow faster (and are greater in size) than the even harmonics. On the other hand, the negative influence of DC bias on each harmonic is analyzed in Fig. 6.11 when the ferromagnetic core is significantly saturated as a consequence of high alternating voltage. Even harmonics increase faster than the odd harmonics with the increased DC bias and constant AC excitation.

It can be concluded that the appearance of DC bias in exciting current leads to the generation of even harmonics in the DC-biased problem, and each harmonic

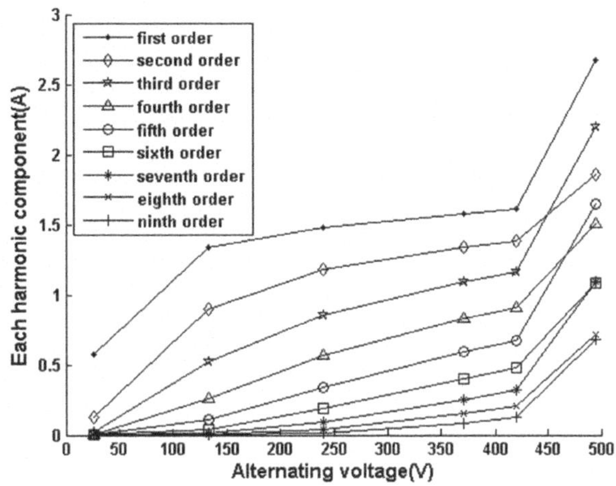

Fig. 6.10 AC voltage (peak value) effect on each harmonic component under 50% DC bias ($I_{dc} = I_{dc,2} = 0.847$ A; $H_{dc} = H_{dc,2} = 213.12$ A/m)

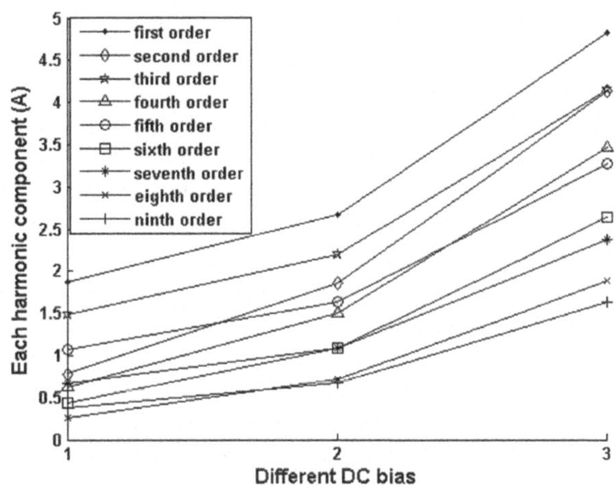

Fig. 6.11 DC bias effect on different harmonics ($U_m = U_{m,6} = 495$ V; $B_m = B_{m,6} = 1.82$ T)

component in the exciting current is affected by DC and AC excitation simultaneously. The applied alternating voltage makes the main contribution to the growth of odd harmonics while the DC bias plays a more important role in the variation of even harmonics, especially when the ferromagnetic core is significantly saturated.

C. ***Harmonic Analysis of the Magnetic Field***

DC flux exists in the ferromagnetic core of power transformers when the DC bias is applied to the external port of electric circuits. Combined with AC flux, DC flux creates high-order harmonics in exciting current and flux density. This results in severe saturation of the magnetic core and reduces the operational efficiency of transformers.

The relationship between DC bias and DC flux requires further study because the DC flux is not affected linearly by the DC bias. DC and AC harmonic components

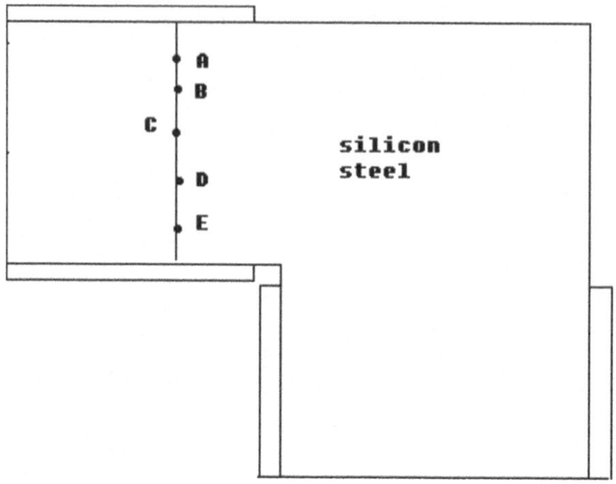

Fig. 6.12 One quarter of the computational model

of the magnetic flux density can be computed directly by the HBFEM. Analysis of the DC component of the magnetic flux density is carried out through the calculated harmonic solutions.

A quarter of the Epstein frame-like core model in Fig. 6.12 is computed considering its structural symmetry. One point on the cross section of the core, such as point C, can be selected to observe the effect of DC bias on the DC flux density. The magnetic flux density in point C has two components: B_x and B_y in the x-direction and y-direction, respectively. However, the harmonic components of flux density B_y are too small to analyze the variation under different excitations. Therefore, in the following part of the paper, the DC component $B_{x,0}$ and the AC component $B_{x,1}$ (in the flux density B_x) are mainly focused on the harmonic analysis under DC bias conditions.

The horizontal coordinate in Fig. 6.13 has the same meaning as that in Fig. 6.8. The peak value of alternating voltage is increased gradually to calculate the DC flux density at point C. If the DC bias is kept constant, the DC flux density decreases with the increment of AC excitation. On the contrary, when the alternating voltage is constant, the DC flux density increases with the growth of the DC bias. It is clear that the DC bias and AC excitation affect the DC flux density at the same time. In fact, the DC component of flux density is due to the balanced effect of DC and AC excitation.

The waveforms of flux density at point C are analyzed under different DC bias conditions. In Fig. 6.14, there are four waveforms of magnetic flux density corresponding to 25%, 50%, 75% and 100% DC bias, respectively (under the condition that the alternating voltage is 240 V). The waveform is raised by the increased DC bias, and the negative peak value of the magnetic flux density approaches −2.0 T. This leads to the rapid saturation of the ferromagnetic core.

As shown in Fig. 6.15, 50% DC bias is selected to observe the effect of alternating voltage on the waveform of the magnetic flux density under DC bias conditions. When the applied AC voltage is low, the magnetic core is saturated only at

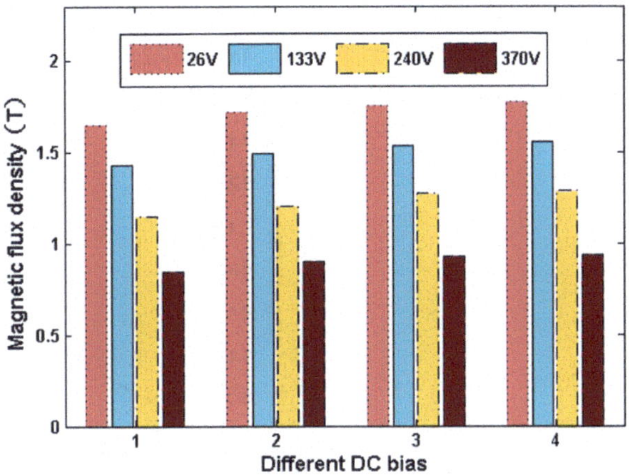

Fig. 6.13 DC component of flux density ($B_{x,0}$) under different DC bias conditions

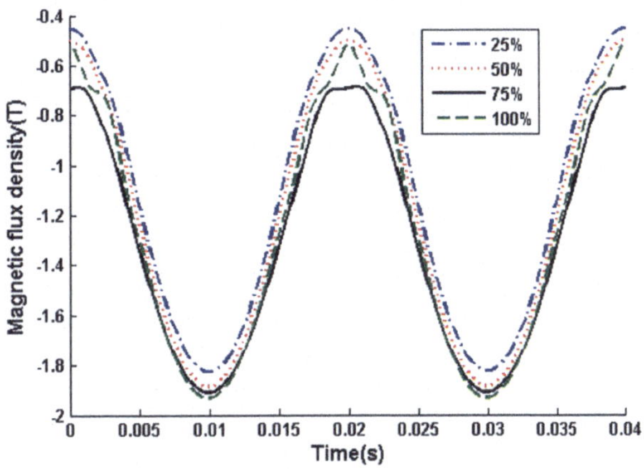

Fig. 6.14 Magnetic flux density (B_x) under different DC bias when the AC excitation is constant ($U_m = U_{m,3} = 240$ V)

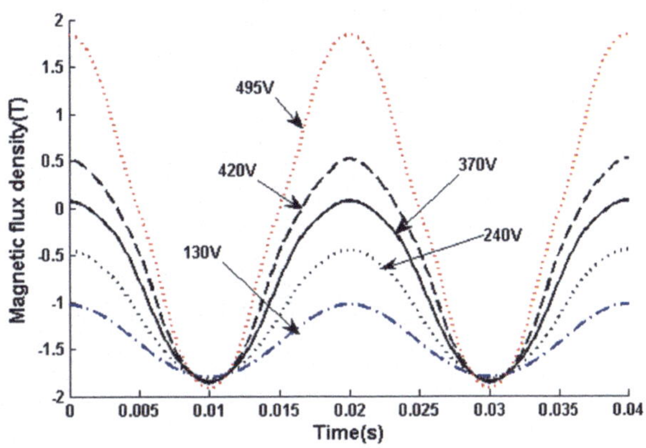

Fig. 6.15 AC voltage effect on waveforms of magnetic flux density (B_x) under 50% DC bias ($I_{dc} = I_{dc,2} = 0.847$ A)

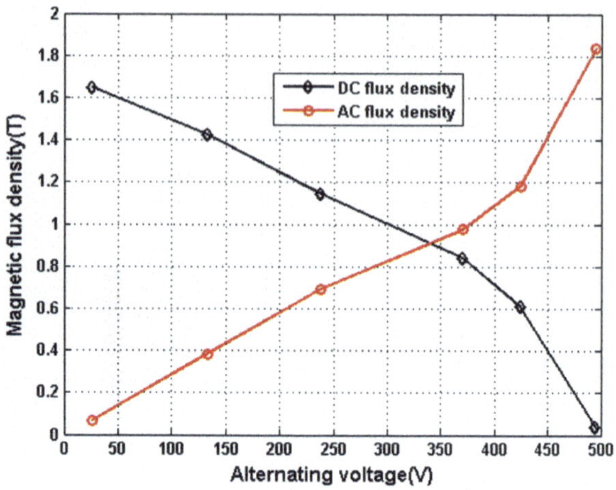

Fig. 6.16 AC voltage effect on DC and AC components of the magnetic flux density ($B_{x,0}$ and $B_{x,1}$) under 25% DC bias

the negative peak value of the magnetic flux density, which corresponds to the half-cycle saturation of the magnetizing current in Figs. 6.3, 6.4 and 6.5. An increase in alternating voltage causes the positive peak value of the flux density to increase so rapidly that the magnetic core is also saturated in the other half-cycle. This is consistent with the appearance of a negative peak of magnetizing current shown in Fig. 6.6.

The detailed harmonic analysis of DC and AC components in the magnetic flux density is given in Figs. 6.16, 6.17, 6.18, 6.19, 6.20 and 6.21. Of the AC components of flux density, the fundamental component is dominant and much larger than other high-order components (see Table 6.3). Therefore, in Figs. 6.16, 6.17, 6.18, 6.19, 6.20 and 6.21, the AC flux density refers to the first harmonic component in the magnetic flux density, neglecting high-order components in the analysis.

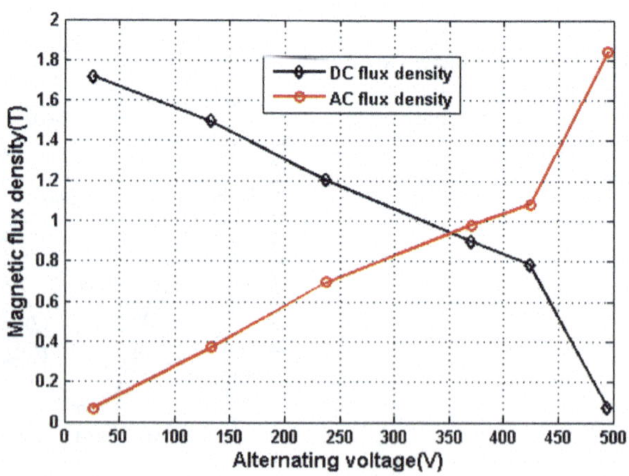

Fig. 6.17 AC voltage effect on DC and AC components of the magnetic flux density ($B_{x,0}$ and $B_{x,1}$) under 50% DC bias

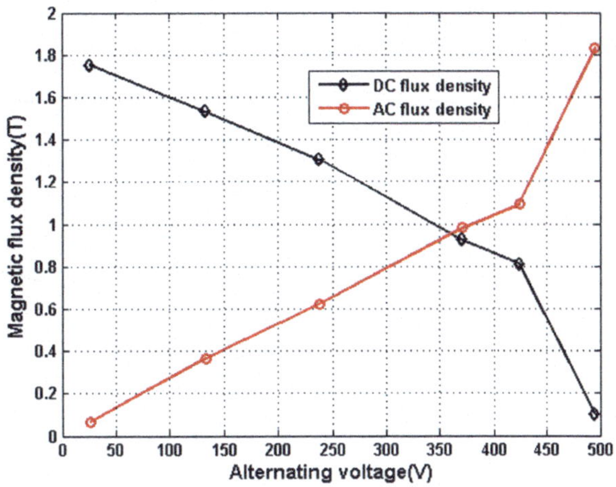

Fig. 6.18 AC voltage effect on DC and AC components of the magnetic flux density ($B_{x,0}$ and $B_{x,1}$) under 75% DC bias

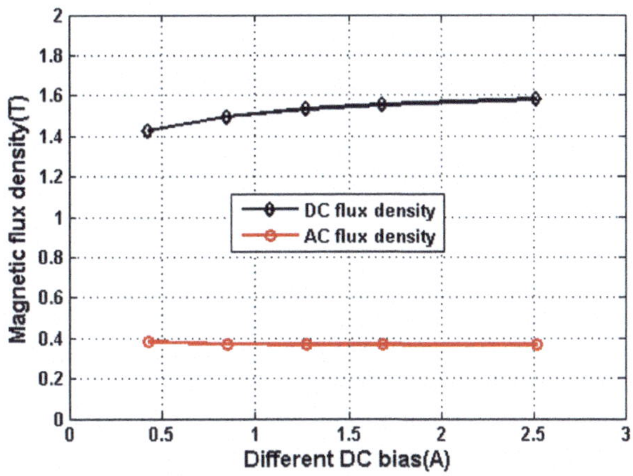

Fig. 6.19 DC bias effect on DC and AC components of the magnetic flux density ($B_{x,0}$ and $B_{x,1}$) with alternating voltage of 133 V

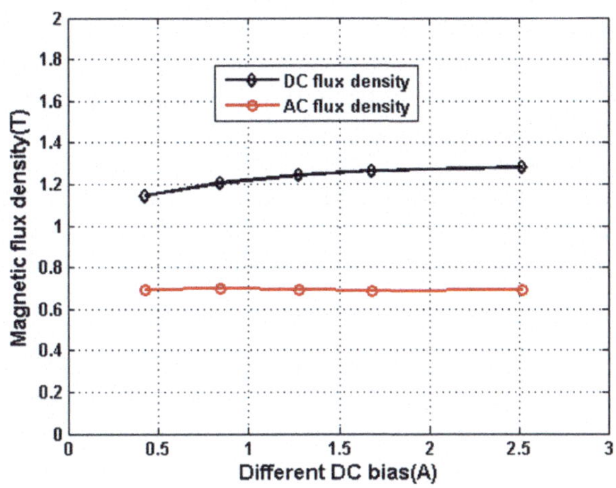

Fig. 6.20 DC bias effect on DC and AC components of the magnetic flux density ($B_{x,0}$ and $B_{x,1}$) with alternating voltage of 240 V

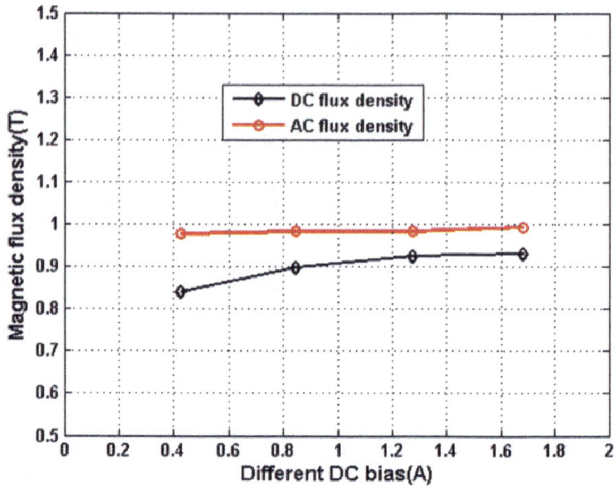

Fig. 6.21 DC bias effect on DC and AC components of the magnetic flux density ($B_{x,0}$ and $B_{x,1}$) with alternating voltage of 370 V

Table 6.3 Comparison of the computational cost

N_h		7	9	13
Traditional	M_c/Mb	13.64	20.90	39.99
	T_c/s	208.93	339.31	707.07
Proposed	M_c/Mb	4.32	5.48	7.82
	T_c/s	277.35	394.60	711.38

(U_{ac} = 1.4 V/m, U_{dc} = 0 V/m, e_{mean} = 0.1% and e_{max} = 1%)

Figures 6.16, 6.17 and 6.18 show that the DC flux density decreases with higher AC voltage while the AC flux density increases at the same time. This variation is even more apparent when the AC voltage is very high because the core has been significantly saturated.

Compared with Figs. 6.16, 6.17, 6.18, Figs. 6.19, 6.20, 6.21 reflect the influence of different DC bias on DC and AC flux density. The DC flux density increases slowly and tends to remain constant (the value of the DC bias is 2.5 A). In that case, the peak value of the magnetic flux density approaches 2.0 T, which means significant saturation of the ferromagnetic core. The AC flux density varies little with the dramatic increase in the DC bias.

It can be predicted that a nonlinear relationship exists between the DC bias and the DC flux and between the AC voltage and the AC flux. The DC flux is a necessary result of the DC bias but it is actually affected by the simultaneous excitation of the DC and AC. The AC flux depends mainly on the alternating excitation.

D. ***Harmonic Analysis of Flux Distribution***

The magnetic flux density in each element is represented in the form of a harmonic component and can be calculated directly from (6.9). Therefore, the harmonic flux distribution in the steel region can be given directly through the solution of the magnetic field.

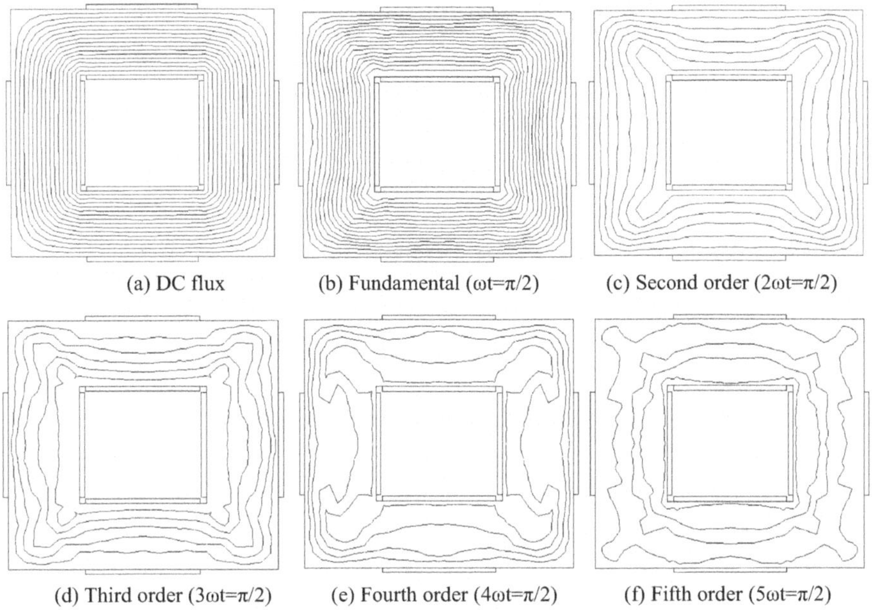

Fig. 6.22 Flux distribution of harmonic components (I_{dc} = 1.27 A, U_{ac} = 370 V)

Harmonic flux distributions under DC bias conditions are presented in Fig. 6.22. The harmonic flux distributions vary with the time (related to phase angle) and excitations (related to DC bias and alternating voltage). The high-order harmonic distribution is apparently different from the fundamental and DC component under DC bias conditions. That may affect the total flux, which is the superposition of all harmonic components.

6.2 The Fixed-Point Harmonic-Balanced Method

Nonlinear eddy current problems can be solved by using the time-stepping method [18] in the time domain or the harmonic-balanced method [19] in the frequency domain. The harmonic-balanced theory has been widely used in the finite element method [20] and integral method [21] to compute the nonlinear magnetic field. The time-stepping method requires many periods to attain the accurate steady-state solution; whereas, the harmonic-balanced method computes the nonlinear magnetic field directly in the frequency domain by introducing the fixed-point reluctivity ν_{FP}.

6.2.1 The Fixed-Point Technique

A relationship between magnetic field intensity $\boldsymbol{H}$ and magnetic flux density $\boldsymbol{B}$ is represented by introducing the fixed-point reluctivity ν_{FP} [22] as follows:

$$\boldsymbol{H}(\boldsymbol{B}) = \nu_{\mathrm{FP}}\boldsymbol{B} - \boldsymbol{M}_{FP}(\boldsymbol{B}) \tag{6.11}$$

where ν_{FP} is a constant value. $\boldsymbol{M}_{\mathrm{FP}}$ is a magnetization-like quantity which varies nonlinearly with $\boldsymbol{B}$. Therefore, the magnetic field intensity $\boldsymbol{H}$ is divided into two parts, the linear part related to ν_{FP} and the nonlinear part related to $\boldsymbol{M}_{\mathrm{FP}}$.

6.2.2 Fixed-Point Harmonic-Balanced Equation

The vector potential equation is used to describe the two-dimensional nonlinear magnetic field by substituting (6.11) into Maxwell's equations as follows:

$$\nabla \times \nu_{\mathrm{FP}}(\nabla \times A) + \sigma\frac{\partial A}{\partial t} = J - \nabla \times \boldsymbol{M}_{\mathrm{FP}} \tag{6.12}$$

where $\boldsymbol{A}$ is the magnetic vector potential, and σ is the electric conductivity.

The fixed-point reluctivity determines the convergence of solutions. If ν_{FP} is a constant, the harmonic solution converges slowly and even unstably, especially when the value of ν_{FP} is not selected properly. In fact, ν_{FP} can be updated in each iterative step, and a fast convergence can be achieved based on the locally convergent method. The local fixed-point reluctivity can be determined by

$$\nu_{\mathrm{FP}} = \left(\partial H_x/\partial B_x + \partial H_y/\partial B_y\right)/2 \tag{6.13}$$

The periodic variables in the electromagnetic field under DC-biased excitation can be approximated by the Fourier series with a finite number of harmonics [23] as follows:

$$W(t) = W_0 + \sum_{n=1}^{\infty}\left(W_{2n-1}\sin n\omega t + W_{2n}\cos n\omega t\right) \tag{6.14}$$

where $W(t)$ can be replaced by current density $\boldsymbol{J}$, vector potential $\boldsymbol{A}$, magnetic flux density $\boldsymbol{B}$, magnetic field intensity $\boldsymbol{H}$ and the magnetization-like quantity $\boldsymbol{M}_{\mathrm{FP}}$. Equation (6.11) can be rewritten for the isotropic material by means of a harmonic vector in the harmonic domain as follows:

$$\begin{cases} \boldsymbol{H}_x = \nu_{\text{FP}} \boldsymbol{B}_x - \boldsymbol{M}_{\text{FP}x} \\ \boldsymbol{H}_y = \nu_{\text{FP}} \boldsymbol{B}_y - \boldsymbol{M}_{\text{FP}y} \end{cases} \tag{6.15}$$

in which

$$\begin{cases} \boldsymbol{H}_x = \begin{bmatrix} H_{x,0} & H_{x,1} & H_{x,2} & H_{x,3} & H_{x,4} & \cdots \end{bmatrix}^{\text{T}} \\ \boldsymbol{H}_y = \begin{bmatrix} H_{y,0} & H_{y,1} & H_{y,2} & H_{y,3} & H_{y,4} & \cdots \end{bmatrix}^{\text{T}} \end{cases} \tag{6.16}$$

Each of the harmonic vectors $\boldsymbol{B}_x$, $\boldsymbol{B}_y$, $\boldsymbol{M}_{\text{FP}x}$ and $\boldsymbol{M}_{\text{FP}y}$ has a similar expression with (6.16). The fixed-point reluctivity can also be expressed in the harmonic domain as follows:

$$\nu_{\text{FP}}(t) = \text{d}H(B)/\text{d}B = \nu_0 + \sum_{n=1}^{\infty} \{\nu_{2n-1} \sin(n\omega t) + \nu_{2n} \cos(n\omega t)\} \tag{6.17}$$

where ν_{2n-1} and ν_{2n} are the harmonic coefficients of the fixed-point reluctivity, while ν_0 is the DC component. The fixed-point harmonic-balanced equation can be obtained by substituting (6.14) and (6.17) into the weak form of (6.12) based on Galerkin's method as follows:

$$\begin{aligned} \boldsymbol{S}_e\boldsymbol{A}_e + \boldsymbol{M}_e\boldsymbol{A}_e - \boldsymbol{K}_e - \boldsymbol{P}_e = & \begin{bmatrix} S_{11}\boldsymbol{D}_{\text{FP}} & S_{12}\boldsymbol{D}_{\text{FP}} & S_{13}\boldsymbol{D}_{\text{FP}} \\ S_{21}\boldsymbol{D}_{\text{FP}} & S_{22}\boldsymbol{D}_{\text{FP}} & S_{23}\boldsymbol{D}_{\text{FP}} \\ S_{31}\boldsymbol{D}_{\text{FP}} & S_{32}\boldsymbol{D}_{\text{FP}} & S_{33}\boldsymbol{D}_{\text{FP}} \end{bmatrix} \cdot \begin{bmatrix} \boldsymbol{A}_{e1} \\ \boldsymbol{A}_{e2} \\ \boldsymbol{A}_{e3} \end{bmatrix} \\ & + \begin{bmatrix} M_{11}\boldsymbol{N} & M_{12}\boldsymbol{N} & M_{13}\boldsymbol{N} \\ M_{21}\boldsymbol{N} & M_{22}\boldsymbol{N} & M_{23}\boldsymbol{N} \\ M_{31}\boldsymbol{N} & M_{32}\boldsymbol{N} & M_{33}\boldsymbol{N} \end{bmatrix} \begin{bmatrix} \boldsymbol{A}_{e1} \\ \boldsymbol{A}_{e2} \\ \boldsymbol{A}_{e3} \end{bmatrix} \\ & - \begin{bmatrix} \boldsymbol{K}_{e1} \\ \boldsymbol{K}_{e2} \\ \boldsymbol{K}_{e3} \end{bmatrix} - \begin{bmatrix} \boldsymbol{P}_{e1} \\ \boldsymbol{P}_{e2} \\ \boldsymbol{P}_{e3} \end{bmatrix} = \boldsymbol{0} \end{aligned} \tag{6.18}$$

where $S_{i,j} = \int\limits_{\Omega_e} \nabla N_i \cdot \nabla N_j \text{d}\Omega$, $T_{i,j} = \int\limits_{\Omega_e} \sigma N_i \cdot N_j \text{d}\Omega$, and the subscripts i and j represent the node number in finite elements. N_i is the shape function on node i in the finite element region. $\boldsymbol{A}$ is the harmonic form of magnetic vector potential, and $\boldsymbol{N}$ is related to the harmonic number (and called the harmonic matrix) [24]. $\boldsymbol{D}_{\text{FP}}$ is related to the reluctivity and can be called the fixed-point reluctivity matrix, which has a similar expression with the reluctivity matrix in [25]. $\boldsymbol{P}$ is derived from the magnetization-like quantity $\boldsymbol{M}_{\text{FP}}$ as follows:

$$\boldsymbol{P}_{ei} = \begin{bmatrix} P_{ei,0} & P_{ei,1} & P_{ei,2} & P_{ei,3} & P_{ei,4} & \cdots \end{bmatrix}^{\text{T}} \tag{6.19}$$

$$P_{ei,m} = \iint\limits_{\Omega_e} \left(M_{y,m} \frac{\partial N_{ei}}{\partial x} - M_{x,m} \frac{\partial N_{ei}}{\partial y} \right) \mathrm{d}x\mathrm{d}y \tag{6.20}$$

where $\boldsymbol{X}$ represents all harmonic solutions, including $\boldsymbol{A}$ and $\boldsymbol{J}$.

6.2.3 *Electromagnetic Coupling*

Nonlinear eddy current problems can be solved directly with the prescribed impressed current density in two-dimensional problems. However, the impressed current density is unknown when the solid conductor or strand coil is connected to the voltage source. Therefore, the coupling between the magnetic field and electric circuits should be investigated if the nonlinear eddy current problem is solved in the harmonic domain.

A. **Solid Conductor Connected to the Voltage Source**

When the solid conductor is fed by the voltage source, the eddy current exists in the solid conductor and the other conducting materials. The magnetic vector potential $\boldsymbol{A}$ and the scalar potential V can be linked to the current density J by the equation as follows:

$$J = -\sigma \frac{\partial A}{\partial t} - \sigma \nabla V \tag{6.21}$$

The nonlinear equation, including the magnetic and electric fields, can be presented as follows [26]:

$$\nabla \times \nu_{\mathrm{FP}}(\nabla \times A) + \sigma \frac{\partial A}{\partial t} + \sigma(\nabla V) = -\nabla \times \boldsymbol{M}_{\mathrm{FP}} \tag{6.22}$$

The fixed-point harmonic-balanced equation can be established by applying the finite element method on the entire problem domain as follows:

$$\sum_{\Omega_{ew}} \boldsymbol{S}^e \boldsymbol{A}^e + \sum_{\Omega_{ed}} \boldsymbol{T}^e \boldsymbol{A}^e = \sum_{\Omega_{ec}} \boldsymbol{K}^e + \sum_{\Omega_{en}} \boldsymbol{P}^e \tag{6.23}$$

By integrating (6.21) on the solid conductor, we can obtain

$$\sum_{\Omega_{ec}} \boldsymbol{C}^e \boldsymbol{A}^e + \boldsymbol{Z}\boldsymbol{I} = \boldsymbol{U} \tag{6.24}$$

since

$$\boldsymbol{K}_i^e = \int\limits_{\Omega_{ec}} \sigma N_i \boldsymbol{U} \mathrm{d}\Omega, \quad \boldsymbol{C}_i^e = \boldsymbol{N} \frac{\omega}{S_{cd}} \int\limits_{\Omega_{ec}} N_i \mathrm{d}\Omega,$$

$$\boldsymbol{Z} = \begin{bmatrix} R & 0 & \cdots \\ 0 & R & \cdots \\ \vdots & \vdots & \ddots \end{bmatrix},$$

where Ω_{ec} represents the finite element in the conducting region, R is the conductor's resistance per unit length, S_{cd} is the cross-sectional area of the solid conductor, and $\boldsymbol{U}$ is the harmonic vector of the voltage per unit length.

B. **Strand Coil Connected to the Voltage Source**

The strand coil consists of fine wires where the eddy current is generally too small to be considered for computation.

The supplied voltage U and the exciting current I in the coil can be linked by Kirchhoff's Law and Faraday's Law [27] as follows:

$$N_{\text{coil}} \int\limits_{\Omega_c} \frac{\partial A}{\partial t} \mathrm{d}\Omega + RI = U \tag{6.25}$$

where N_{coil} is the number of turns of the strand coil.

The nonlinear magnetic field is governed by the following equation:

$$\nabla \times \nu_{\text{FP}}(\nabla \times A) + \sigma \frac{\partial A}{\partial t} = \frac{N_{\text{coil}}}{S_{\text{coil}}} I - \nabla \times \boldsymbol{M}_{\text{FP}} \tag{6.26}$$

where S_{coil} is the cross-sectional area of the strand coil.

The frequency domain system equation considering electromagnetic coupling can be obtained according to the harmonic-balanced theory as follows:

$$\sum_{\Omega_{ew}} \boldsymbol{S}^e \boldsymbol{A}^e + \sum_{\Omega_{ed}} \boldsymbol{T}^e \boldsymbol{A}^e + \sum_{\Omega_{ec}} \boldsymbol{G}^e \boldsymbol{I} = \sum_{\Omega_{en}} \boldsymbol{P}^e \tag{6.27}$$

$$\sum_{\Omega_{ec}} \boldsymbol{C}^e \boldsymbol{A}^e + \boldsymbol{Z}\boldsymbol{I} = \boldsymbol{U} \tag{6.28}$$

since

$$\boldsymbol{G}_i^e = -\boldsymbol{I}_u \frac{N_{\text{coil}}}{S_{\text{coil}}} \int\limits_{\Omega_{ec}} N_i \mathrm{d}\Omega,$$

where $\boldsymbol{I}_u$ is a unit matrix of the same size with $\boldsymbol{D}_{\text{FP}}$ and $\boldsymbol{N}$.

Consequently, the harmonic solutions of the magnetic field and magnetizing current can be computed simultaneously by solving (6.29) when the solid conductor and strand coil are both connected to the voltage source as follows:

$$\begin{bmatrix} \boldsymbol{S}+\boldsymbol{T} & \boldsymbol{G} \\ \boldsymbol{C} & \boldsymbol{Z} \end{bmatrix} \begin{Bmatrix} \boldsymbol{A} \\ \boldsymbol{I} \end{Bmatrix} = \begin{Bmatrix} \boldsymbol{K}+\boldsymbol{P} \\ \boldsymbol{U} \end{Bmatrix} \tag{6.29}$$

where $\boldsymbol{G}$ is related to the spatial distribution of the magnetizing current when the strand coil is fed by the voltage source, while $\boldsymbol{K}$ appears on the right side of the equation when the solid conductor is connected to the voltage source.

C. **Convergent Strategy for Harmonic Computation**

Two different methods to determine fixed-point reluctivity ν_{FP} are presented and compared in the time-stepping finite element analysis of nonlinear eddy current problems [10]. The locally convergent method [28] is superior to the globally convergent algorithm, since the local fixed-point reluctivity in each time step is determined by the differential reluctivity (which can speed up the convergence of solutions). At the same time, the convergence factor is required to ensure the efficiency and excellence of the locally convergent method [10, 28]. However, the determination of the convergence factor [10] is empirical, and it is not feasible to use harmonic computation. The optimal strategies to determine the fixed-point reluctivity or permeability [29, 30] have been investigated to achieve the stable convergence of solutions.

The fixed-point reluctivity ν_{FP} can be regarded as a periodic variable when it is determined in each time step. Consequently, the harmonic coefficients of ν_{FP} can be used to calculate $\boldsymbol{D}_{\text{FP}}$ in the harmonic-balanced method [7]. All elements in the square matrix $\boldsymbol{D}_{\text{FP}}$ are nonzero, which indicates harmonic solutions are coupled with each other. In that case, the memory demand will increase significantly in the large-scale computation, although fast convergence is achieved. In fact, ν_{FP} can be a constant in the harmonic domain and is determined as follows:

$$\nu_{\text{FP}} = \partial H(B_{\max})/\partial B_{\max} \tag{6.30}$$

where $B_{\max}$ represents the maximum value of the magnetic induction in one period.

As a result, (6.29) can be solved iteratively according to the following procedure (p represents the iterative step): $[\boldsymbol{A}(p),\ \boldsymbol{I}(p)] \rightarrow [\boldsymbol{B}_x(p),\ \boldsymbol{B}_y(p)] \rightarrow [\boldsymbol{H}_x(p),\ \boldsymbol{H}_y(p)] \rightarrow [\boldsymbol{P}(p), \boldsymbol{D}_{\text{FP}}(p)] \rightarrow [\boldsymbol{A}(p+1), \boldsymbol{I}_k(p+1)]$. The mean ($e_{\text{mean}}$) and maximum ($e_{\max}$) variation of the reluctivity defined by $\nu = H/B$ can be observed to check the convergence of the harmonic solutions.

6.2.4 Validation and Discussion

An example (the conducting copper surrounded by the ferromagnetic wall) is calculated by the proposed method and analyzed to investigate eddy current problems under DC-biased magnetization.

The two-dimensional problem consists of a solid copper conductor and an iron screen with an air gap. As shown in Fig. 6.23, the iron screen surrounds the conductor. The eddy current exists in both the copper conductor and iron screen. The conductivities of the copper and iron are $\sigma = 5.7 \times 10^7$ S/m and $\sigma = 1.0 \times 10^6$ S/m, respectively. The copper conductor is connected to a voltage source of 50 Hz. The *B–H* curve is detailed in [31]. Nearly four thousand (3950) first-order elements with 2040 nodes are used in the numerical computation. Computational costs of the proposed method and the traditional method [7] are compared in Table 6.3. M_c and T_c represent the memory demand and computational time, respectively. N_h is the truncated harmonic number. Compared to the traditional method, the proposed method significantly reduces memory requirements with a slight increase in computational time due to a few more nonlinear iterations.

The number of nonlinear iterations required to achieve a convergence of harmonic solutions is presented in Table 6.4. Figure 6.24a depicts the magnetizing current in the copper conductor connected to the voltage. The corresponding

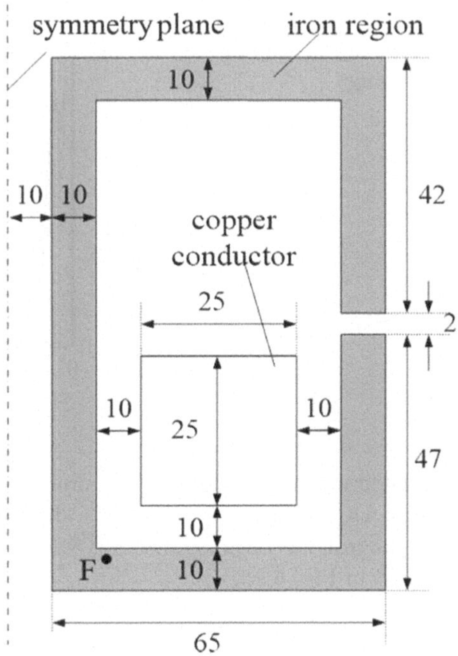

Fig. 6.23 Geometric structure of the two-dimensional model (unit: millimeter)

Table 6.4 Number of iterations under DC-biased magnetization

	U_{ac} = 1.4 V/m	U_{ac} = 2.0 V/m
U_{dc} = 0 V/m	40	57
U_{dc} = 0.01 V/m	51	63
U_{dc} = 0.02 V/m	67	88
U_{dc} = 0.03 V/m	77	91

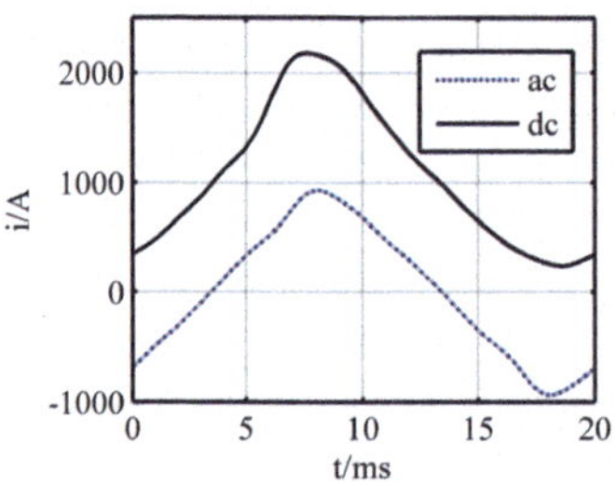

(a) Magnetizing current

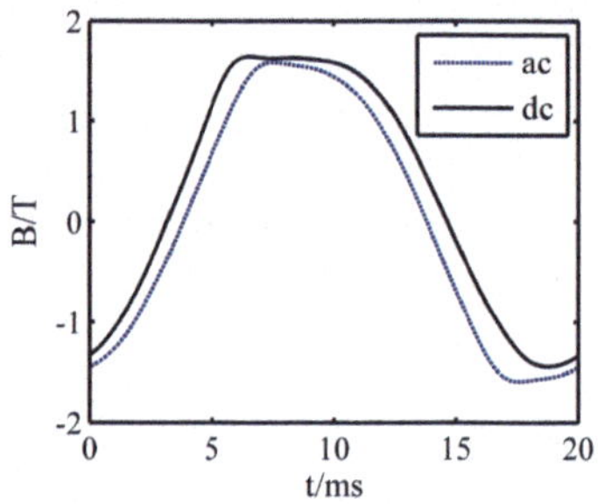

(b) Magnetic induction

Fig. 6.24 Calculated results under sinusoidal and DC-biased magnetization (ac: U_{ac} = 1.4 V/m, U_{dc} = 0 V/m; dc: U_{ac} = 1.4 V/m, U_{dc} = 0.03 V/m)

Fig. 6.25 Revolution of the relative error

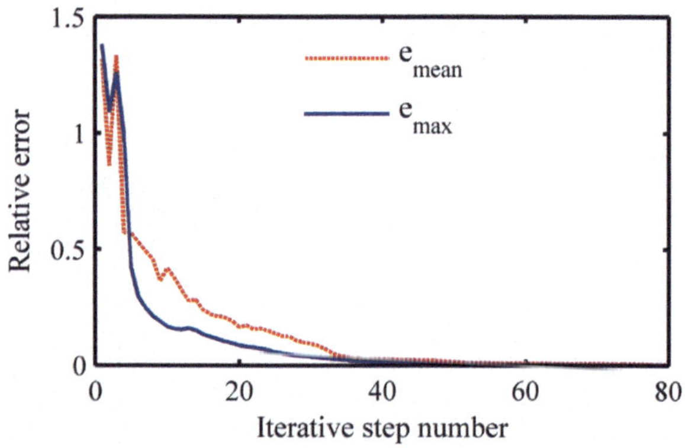

calculated magnetic induction on point F is shown in Fig. 6.24b. It is apparent that a slight DC voltage gives rise to a dramatic increase in the magnetizing current since the iron has been significantly saturated. The variation of the relative errors is depicted in Fig. 6.25.

6.3 Decomposed Harmonic-Balanced Method

6.3.1 The Fixed-Point Reluctivity

A fixed-point reluctivity ν_{FP} can be defined as follows [32]:

$$\nu = \nu_{\mathrm{FP}} + (\nu - \nu_{\mathrm{FP}}) \tag{6.31}$$

where ν_{FP} is time-independent and not necessarily independent of space coordinate. In contrast to ν_{FP}, the reluctivity ν is time-dependent and determined by the computed magnetic field.

Different from the fixed-point reluctivity in (6.11), ν_{FP} in (6.31) does not change the magnetic constitutive relation and allows for the decoupled computation of harmonic solutions in parallel computing.

6.3.2 Linearization and Decomposition

The vector potential equation is used to describe the two-dimensional nonlinear magnetic field by substituting (6.31) into Maxwell's equations as follows:

$$\nabla \times \nu_{\mathrm{FP}}(\nabla \times \boldsymbol{A}) + \sigma \frac{\partial \boldsymbol{A}}{\partial t} = \boldsymbol{J} - \nabla \times (\nu - \nu_{\mathrm{FP}})(\nabla \times \boldsymbol{A}) \tag{6.32}$$

Owing to the time periodicity of the electromagnetic field under harmonic excitations, periodic variables such as current density $\boldsymbol{J}$, magnetic vector potential $\boldsymbol{A}$ and reluctivity ν can be represented by a summation of trigonometric function as follows:

$$W(t) = \sum_{n=-N_h}^{N_h} W_n e^{jn\omega t} \tag{6.33}$$

$$\boldsymbol{W}_f = [\, W_0 \quad W_{-1} \quad W_1 \quad \cdots \quad W_{-N_h} \quad W_{N_h} \,] \tag{6.34}$$

where W can be replaced by $\boldsymbol{J}$, $\boldsymbol{A}$ and ν. N_{h} is the total number of harmonics truncated in computation, and ω is the angular frequency. W_n is the n-th component in frequency domain. $\boldsymbol{W}_f$ is the spectrum of the periodic variable W.

Essentially, a new system of equations in the frequency domain can be obtained, as follows, by using the harmonic-balanced theory [33] as well as applying Galerkin's method and the finite element method over the entire problem domain:

$$(\boldsymbol{S} * \boldsymbol{D}_{\mathrm{FP}} + \mathrm{M} * \boldsymbol{N})\boldsymbol{A}_{f,i+1} = \boldsymbol{K}_f + \boldsymbol{S} * (\boldsymbol{D} - \boldsymbol{D}_{\mathrm{FP}})\boldsymbol{A}_{f,i} \tag{6.35}$$

$$S_{p,q} = \int_{\Omega_e} \nabla N_p \cdot \nabla N_q \mathrm{d}\Omega \tag{6.36}$$

$$M_{p,q} = \int_{\Omega_e} \sigma N_p \cdot N_q \mathrm{d}\Omega \tag{6.37}$$

$$\boldsymbol{A}_f = \left[\boldsymbol{A}_{f,0}, \boldsymbol{A}_{f,-1}, \boldsymbol{A}_{f,1}, \ldots, \boldsymbol{A}_{f,-N_\mathrm{h}}, \boldsymbol{A}_{f,N_\mathrm{h}}\right]^\mathrm{T} \tag{6.38}$$

$$\boldsymbol{K}_f = \left[\boldsymbol{K}_{f,0}, \boldsymbol{K}_{f,-1}, \boldsymbol{K}_{f,1}, \ldots, \boldsymbol{K}_{f,-N_\mathrm{h}}, \boldsymbol{K}_{f,N_\mathrm{h}}\right]^\mathrm{T} \tag{6.39}$$

$$\boldsymbol{D}_{\mathrm{FP}} = \begin{bmatrix} \nu_{\mathrm{FP}} & 0 & 0 & 0 & 0 & \cdots \\ 0 & \nu_{\mathrm{FP}} & 0 & 0 & 0 & \cdots \\ 0 & 0 & \nu_{\mathrm{FP}} & 0 & 0 & \cdots \\ 0 & 0 & 0 & \nu_{\mathrm{FP}} & 0 & \cdots \\ 0 & 0 & 0 & 0 & \nu_{\mathrm{FP}} & \cdots \\ \vdots & \vdots & \vdots & \vdots & \vdots & \ddots \end{bmatrix} \tag{6.40}$$

$$\boldsymbol{D} = \begin{bmatrix} \nu_0 & \nu_1 & \nu_{-1} & \nu_2 & \nu_{-2} & \cdots \\ \nu_{-1} & \nu_0 & \nu_{-2} & \nu_1 & \nu_{-3} & \cdots \\ \nu_1 & \nu_2 & \nu_0 & \nu_3 & \nu_{-1} & \cdots \\ \nu_{-2} & \nu_{-1} & \nu_{-3} & \nu_0 & \nu_{-4} & \cdots \\ \nu_2 & \nu_3 & \nu_1 & \nu_4 & \nu_0 & \cdots \\ \vdots & \vdots & \vdots & \vdots & \vdots & \ddots \end{bmatrix} \tag{6.41}$$

$$\boldsymbol{N} = \omega \begin{bmatrix} 0 & 0 & 0 & 0 & 0 & \cdots \\ 0 & -j & 0 & 0 & 0 & \cdots \\ 0 & 0 & j & 0 & 0 & \cdots \\ 0 & 0 & 0 & -2j & 0 & \cdots \\ 0 & 0 & 0 & 0 & 2j & \cdots \\ \vdots & \vdots & \vdots & \vdots & \vdots & \ddots \end{bmatrix} \tag{6.42}$$

where $\boldsymbol{M}$ is the mass matrix and $\boldsymbol{S}$ the stiffness matrix. $\boldsymbol{D}$ and $\boldsymbol{N}$ are, respectively, the reluctivity and harmonic matrices. p, q denote the finite element node numbers. i represents the number of iterative steps, and $\boldsymbol{A}_f$ is the vector potential in frequency domain. $\boldsymbol{K}_f$ is related to spatial distribution of the impressed current density,

$$\boldsymbol{K}_{f,n} = \left[K_{1,n}, K_{2,n}, \ldots, K_{N_t,n}\right]^\mathrm{T} \quad (n = -N_\mathrm{h}, \ldots, N_\mathrm{h}) \tag{6.43}$$

$$K_{p,n} = \int_{\Omega_e} (J_n \cdot N_p) \mathrm{d}\Omega \tag{6.44}$$

where N_t is the total number of nodes, and N_p is the shape function on node p in the finite element region Ω. J_n is the n-th harmonic component of the impressed current density.

The operator $*$ in Eq. (6.35) can be defined as follows:

$$\boldsymbol{S} * \boldsymbol{D}_{\mathrm{FP}} + \mathbf{M} * \boldsymbol{N} = \begin{bmatrix} \boldsymbol{S}(\nu_{\mathrm{FP}}) & \mathbf{0} & \mathbf{0} & \cdots \\ \mathbf{0} & \boldsymbol{S}(\nu_{\mathrm{FP}}) & \mathbf{0} & \cdots \\ \mathbf{0} & \mathbf{0} & \boldsymbol{S}(\nu_{\mathrm{FP}}) & \cdots \\ \vdots & \vdots & \vdots & \ddots \end{bmatrix} + j\omega \begin{bmatrix} 0 & 0 & 0 & \cdots \\ 0 & -\mathbf{M} & 0 & \cdots \\ 0 & 0 & \mathbf{M} & \cdots \\ \vdots & \vdots & \vdots & \ddots \end{bmatrix} \tag{6.45}$$

As shown in (6.45), the block matrix including ν_{FP}, which is the fixed-point reluctivity, is time-independent. Therefore, a new decomposed harmonic-balanced system of equations can be obtained to decouple harmonic solutions as follows:

$$[jn\omega\mathbf{M} + \boldsymbol{S}(\nu_{\mathrm{FP}})]\boldsymbol{A}_{f,n} = \boldsymbol{P}_{f,n} + \boldsymbol{K}_{f,n}(n = -N_{\mathrm{h}}...N_{\mathrm{h}}) \tag{6.46}$$

$$\boldsymbol{P}_{f,n} = \sum_{m=-N_h}^{N_{\mathrm{h}}} [\boldsymbol{S}(\nu_{n-m}) - \boldsymbol{S}(\nu_{n-m})]\boldsymbol{A}_{f,m} \tag{6.47}$$

where $\boldsymbol{A}_{f,m}$ is the m-th harmonic solution of the magnetic vector potential. $\boldsymbol{P}_{f,n}$ is obtained from the convolution product of coefficient related to reluctivity and vector potential in harmonic domain.

When electromagnetic devices are excited by voltage sources, the magnetic field can be coupled with electric circuits in frequency domain as follows:

$$\boldsymbol{U}_{f,n} = \boldsymbol{Z}\boldsymbol{I}_{f,n} + \boldsymbol{C}_n\boldsymbol{A}_{f,n} \tag{6.48}$$

where $\boldsymbol{U}_f$ and $\boldsymbol{I}_f$ are, respectively, the voltage excitation and exciting current in the frequency domain, and $\boldsymbol{Z}$ and $\boldsymbol{C}_n$ denote the n-th impedance and coupling matrices, respectively.

The frequency domain finite element equation of two-dimensional nonlinear field under voltage excitation is obtained.

$$\begin{bmatrix} jn\omega\mathbf{M} + \boldsymbol{S}(\nu_{\mathrm{FP}}) & \boldsymbol{G}_n \\ \boldsymbol{C}_n & \boldsymbol{Z}_n \end{bmatrix} \begin{Bmatrix} \boldsymbol{A}_{f,n} \\ \boldsymbol{I}_{f,n} \end{Bmatrix} = \begin{Bmatrix} \boldsymbol{P}_{f,n} \\ \boldsymbol{U}_{f,n} \end{Bmatrix} \quad (n = -N_{\mathrm{h}}, \ldots, N_{\mathrm{h}}), \tag{6.49}$$

where $\boldsymbol{G}_n$ is related to the spatial distribution of the n-th harmonic component of the current density.

The decomposed system of equations in (6.49) can be solved separately and in parallel. Meanwhile, only $N_h + 1$ equations are required to be solved due to the conjugate symmetry of the harmonic solutions shown in (6.34). Compared with the method presented in [34], the proposed method can reduce the memory cost and computation time in the calculation of the nonlinear magnetic field.

The fixed-point reluctivity ν_{FP} can be determined optimally [32] in order to guarantee stable and fast convergence of harmonic solutions.

The magnetic quality of electrical equipment is usually made up of ferromagnetic material or permanent magnetic material, and the magnetic substance is subjected to force in the magnetic field. The Maxwell force $\boldsymbol{F}$ can be calculated by the surface integral of the tensor $\boldsymbol{T}$ as follows:

$$\boldsymbol{F} = \oint\limits_{\Omega_s} \boldsymbol{T} \cdot \mathrm{d}s = \oint\limits_{\Omega_s} \begin{bmatrix} B_xH_x - \frac{1}{2}BH & B_xH_y \\ B_yH_x & B_yH_y - \frac{1}{2}BH \end{bmatrix} \begin{bmatrix} n_x \\ n_y \end{bmatrix} \mathrm{d}\Omega \tag{6.50}$$

where $\boldsymbol{T}$ is a two-order tensor which contains four components; Ω_s is a closed surface that surrounds the whole magnetic field in the air. According to the energy principle and Eq. (6.50), the Maxwell force on the node n is as follows:

$$[F_n] = \begin{bmatrix} F_{nx} \\ F_{ny} \end{bmatrix} = -\oint\limits_{\Omega} \begin{bmatrix} T_{xx} & T_{xy} \\ T_{yx} & T_{yy} \end{bmatrix} \begin{bmatrix} \frac{\partial}{\partial x} \\ \frac{\partial}{\partial y} \end{bmatrix} N_n \mathrm{d}v \tag{6.51}$$

Through the finite element analysis, the magnetic induction and magnetic field intensity of each unit have been obtained. If the closed surface is properly selected, the Maxwell force acting on the node can be obtained by Eq. (6.51).

In the nonlinear iterative computation, v is chosen as the convergence variable. The convergence is determined by the mean relative error ε_{mean} and maximum relative error ε_{max} of the medium permeability. If the difference in the two iterative permeabilities meets the criteria ε_1 and ε_2, the iteration process ends. The computation process is illustrated in Fig. 6.26, and the nonlinear iterative strategy is shown in Fig. 6.27.

6.3.3 Force Computation of a Gapped Reactor Core Model

A. **Experimental Setup and Model for Test**

A gapped-core reactor model made of grain-oriented silicon steel sheet (B30P105) produced by Baosteel is used in the experiment for the computation and analysis of magnetic forces when the model is magnetized by sinusoidal and harmonic excitation. Figure 6.28 shows the gapped laminated core and the experimental setup. The magnetization curve used for calculation is shown in Fig. 6.29.

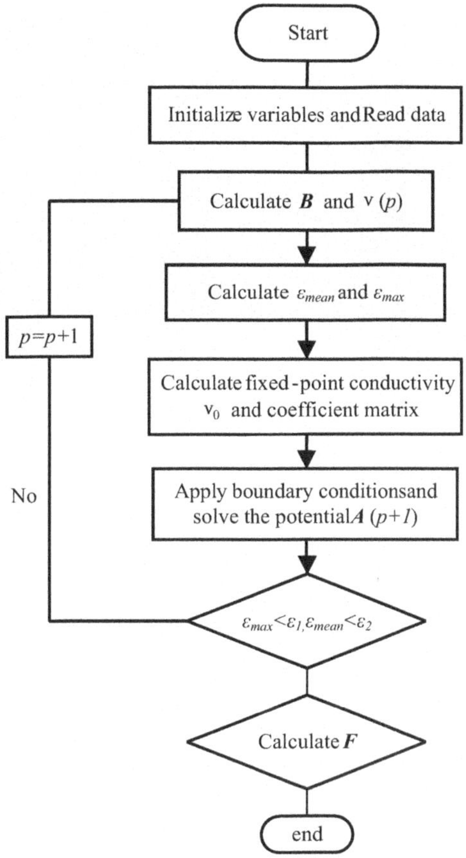

Fig. 6.26 Flowchart of computation process

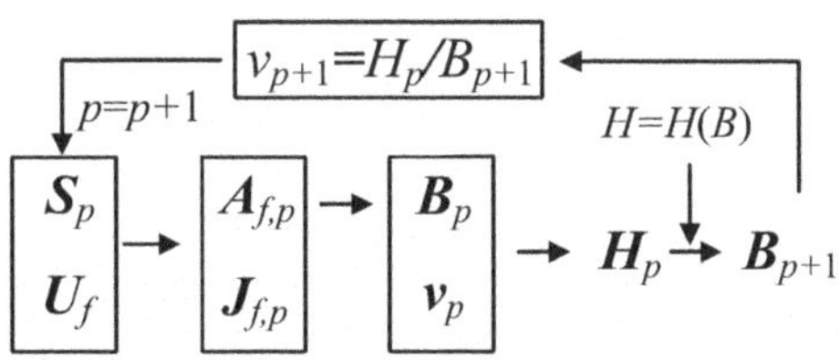

Fig. 6.27 Nonlinear iterative strategy

The thickness of the laminated core is 60 mm, and the width of the air gap in the middle limb is 1.8 mm. There are two exciting coils in parallel on the side yoke, and the number of turns of each coil is 115.

In the experimental setup, a harmonic voltage source generates the voltage signal at different frequencies. In the experiment, only one high-order harmonic is superposed on the fundamental harmonic excitation due to the capacity limitation of the harmonic voltage source. The fundamental frequency is 50 Hz in the experiment. The generated signal can be represented as follows:

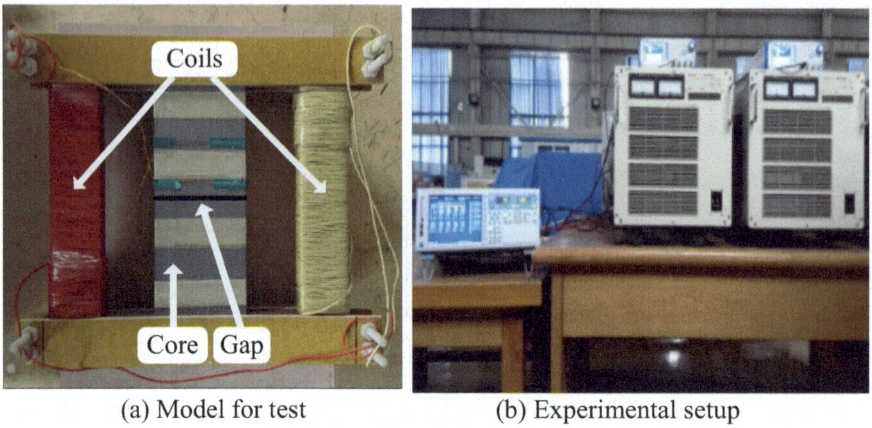

Fig. 6.28 Gapped-core reactor model and experimental setup of harmonic magnetization

Fig. 6.29 Magnetization curve

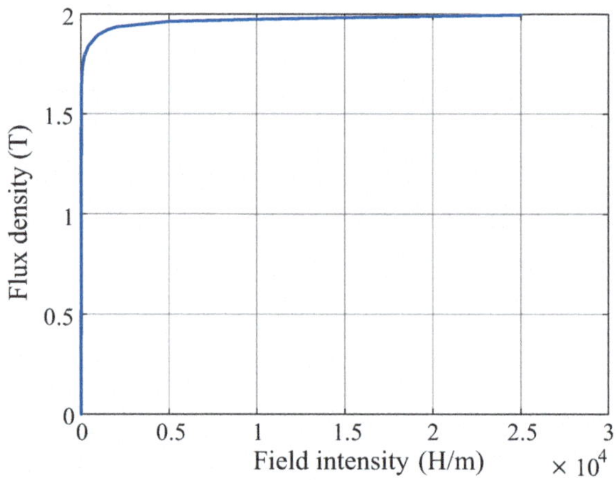

$$U(t) = U_1 \sin(\omega t) + U_n \sin(n\omega t) \tag{6.52}$$

where U_1 and U_n represent the fundamental and n-th harmonic component, respectively.

The hybrid harmonic excitations are generated and controlled by the multi-function generator (NFWF1974) and precision power amplifier (NF4520). The precision power analyzer (WT3000 YOKAGAWA) is used to measure the exciting current. One search coil is wound on the side yoke to measure the flux density in the iron core. The root mean square (RMS) value of the fundamental harmonic U_1 is 195 volts, and the high-order harmonic U_n is superposed in incremental proportions of the fundamental one in the experiment, so as to investigate the influence of frequency and amplitude of harmonic excitation on the flux

Table 6.5 Harmonic voltage excitation in experiment

$U_n = \alpha U_1$ (V)	α				
	0.1	0.2	0.3	0.4	0.5
$n = 3, 5, 7$	19.5	39.0	58.5	78.0	97.5

density and magnetic force. The amplitude of high-order harmonics is shown in Table 6.5, where α denotes an incremental proportion related to the n-th harmonic components.

B. **Convergent Performance of the Proposed Method**

The gapped-core reactor model in Fig. 6.28 is used for the computation of the nonlinear magnetic field and the corresponding magnetic nodal force [35]. In this paper, nonlinear iteration is stopped when the mean and the maximum relative variations of reluctivity become smaller than 0.0005 and 0.005, respectively.

Variations of reluctivity with iterative steps in the proposed method, shown in Fig. 6.30, demonstrate that the decomposition algorithm used in the harmonic-balanced method can guarantee the smooth convergence of harmonic solutions. Table 6.6 shows the convergent performance of the proposed method. N_{FP} denotes the number of iterations by fixed-point method.

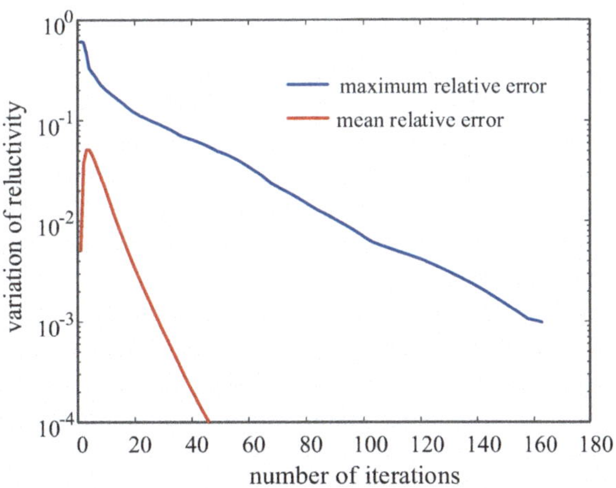

Fig. 6.30 Variation of reluctivity with nonlinear iterative steps ($U_1 = 195$ V, $U_5 = 97.5$ V, $n = 5$, $\alpha = 0.5$)

Table 6.6 Comparison of convergence performance between the fixed-point method and the proposed method

Number of iteration		$\alpha = 0.3$	$\alpha = 0.5$
n	3	116	118
	5	113	112
	7	115	114

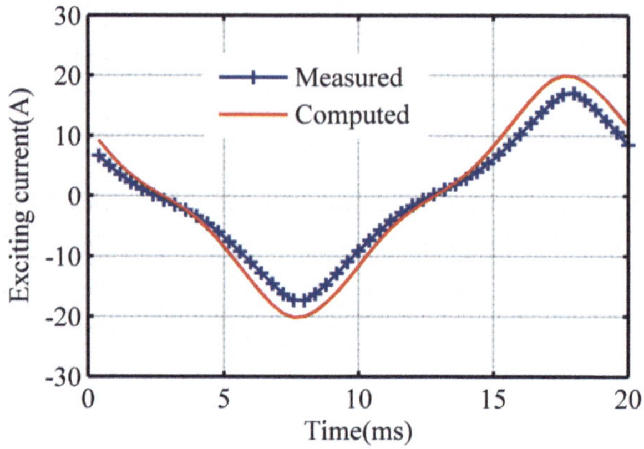

Fig. 6.31 Computed and measured exciting currents (U_1 = 195 V, U_3 = 97.5 V, n = 3, α = 0.5)

Table 6.7 Computed and measured flux density in the core

n	3		5		7	
α	0.3	0.5	0.3	0.5	0.3	0.5
B_c (T)	1.634	1.733	1.574	1.634	1.549	1.592
B_m (T)	1.722	1.825	1.658	1.720	1.630	1.675
E_r (%)	5.11	5.04	5.07	5.00	4.97	4.96

C. **Computed Flux Density and Magnetic Nodal Force**

Figure 6.31 shows that the computed exciting current agrees well with the measured one. The average flux density in the side yoke measured by the search coil is compared with that computed by the proposed method. Table 6.7 shows the agreement between the measurement and computation. B_c and B_m are, respectively, the computed and measured flux density, and E_r is the relative error. It is noticed that lower-order rather than higher-order harmonic excitation leads to more serious saturation of the core.

One node on the interface between the iron core and the air gap is selected to compare the calculated magnetic nodal forces (F_y) by the proposed method in this paper and the time-stepping method in Simcenter MAGNET, which is shown in Fig. 6.32. The comparison verifies the accuracy and validity of the decomposed harmonic-balanced method. The waveforms of magnetic nodal force in Fig. 6.32 show that the period of the force is 10 ms, which equals to half of the period of flux density.

In addition, it can be seen that the waveform of magnetic nodal force is distorted when the gapped-core reactor model works under an increased harmonic voltage excitation, which indicates that there are more harmonics in the spectrum of the magnetic force.

The magnetic force on the gapped core can be obtained based on the computed flux density [36]. The peak value of magnetic force under different harmonic magnetizations is depicted in Fig. 6.33. It can be concluded that harmonic order

Fig. 6.32 Comparison of computed magnetic nodal force

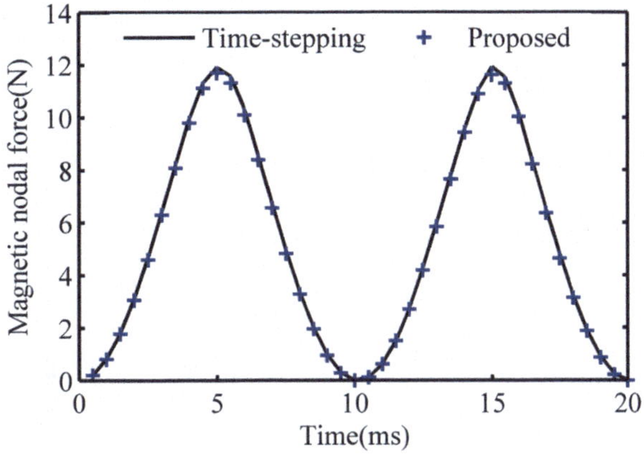

(a) U_1=195V, U_3=19.5V (n=3, α=0.1)

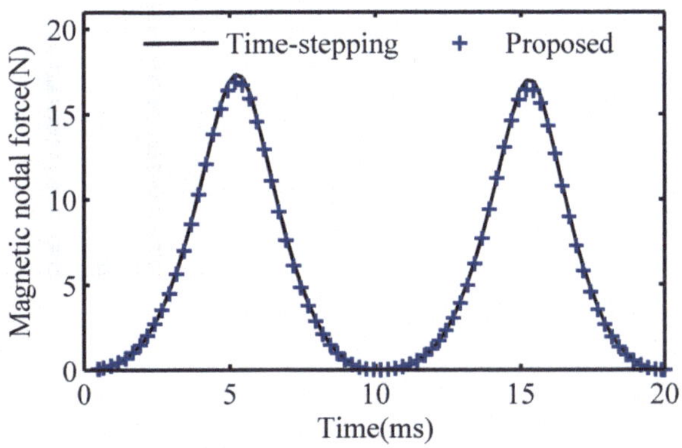

(b) U_1=195V, U_3=97.5V (n=3, α=0.5)

Fig. 6.33 Peak value of magnetic force under different harmonic magnetizations (U_1 = 195 V, $U_n = \alpha U_1$, α = 0.1–0.5)

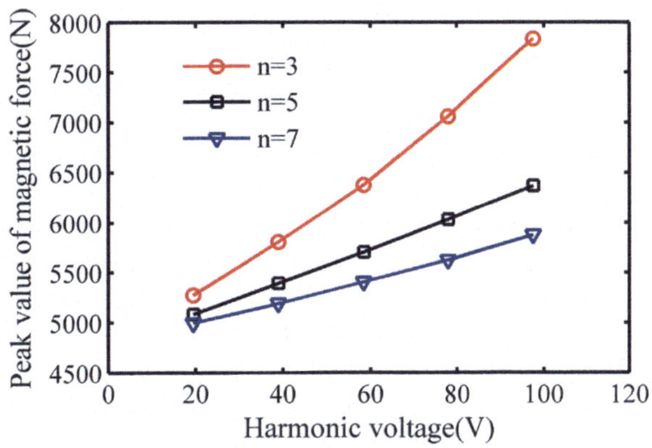

Table 6.8 Spectrum of magnetic force on the gapped core under different harmonic magnetizations (U_1 = 195 V, α = 0.5, $U_n = \alpha U_1$ = 97.5 V)

n	F_m (N) (m = 0–9)									
	F_0	F_1	F_2	F_3	F_4	F_5	F_6	F_7	F_8	F_9
3	2569.24	119.18	3480.67	39.99	1159.44	29.94	343.47	19.60	148.50	7.90
5	2372.67	84.96	2469.21	18.33	679.35	24.24	616.24	34.16	103.11	6.56
7	2319.08	81.02	2384.52	13.43	164.24	32.37	461.68	26.14	408.59	22.75

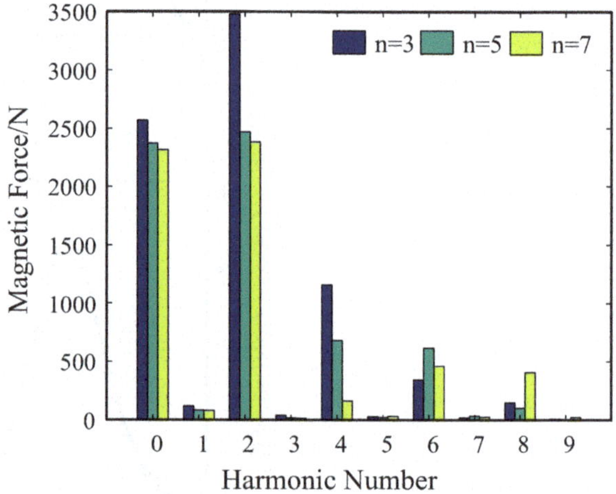

Fig. 6.34 Distribution of magnetic force on the gapped core under different harmonic magnetizations (U_1 = 195 V, α = 0.5, $U_n = \alpha U_1$ = 97.5 V)

plays an important role on the maximum value of magnetic force. The lower the harmonic order, the larger is the magnetic force. Furthermore, the peak value of magnetic force increases linearly with the RMS value of harmonic voltage when the fifth or seventh harmonic is superposed on the fundamental harmonic excitation; however, the peak value of magnetic force increases slightly nonlinearly with the RMS value of the third harmonic voltage.

The spectrum of magnetic force on the gapped core is shown in Table 6.8 and Fig. 6.34, where n represents n-th harmonic, m represents harmonic number, and F_m denotes the m-th harmonic components of magnetic force.

It is noticed that the DC component, corresponding to the static magnetic force, is considerable in the frequency spectrum of magnetic force. In addition, even harmonics are dominant in the spectrum of magnetic force, in spite of odd harmonic excitation. It can be seen that high-order harmonics account for a large proportion of the magnetic force especially when the seventh harmonic is superposed in the excitation. When the n-th harmonic is superimposed on the voltage excitation, according to (6.52), the DC component as well as the second, the $(n-1)$-th and $(n+1)$-th harmonic in the magnetic force are dominant in spectrum. With the increase of harmonic frequency, the content of higher even harmonic which is close to the n-th harmonic may increase significantly.

Acknowledgements This work is supported in part by the National Key Research and Development Program of China (Grant No. 2017YFB0902703), in part by the National Natural Science Foundation of China (Grant No. 51777073), in part by Hebei Province Natural Science Foundation (Grant No. E2017502061), in part by State Key Laboratory of Reliability and Intelligence of Electrical Equipment (Grant No. EERIKF2018011) and in part by the Fundamental Research Funds for the Central Universities (Grant No. 2019MS078).

References

1. Y. Yao, C. S. Koh, G. Ni and D. Xie, "3-D nonlinear transient eddy current calculation of online power transformer under DC bias," IEEE Trans. Magn., Vol. 41, no. 5, pp. 1840–1843, May. 2005.
2. S. Ausserhofer, O. Biro and K. Preis, "An efficient harmonic balance method for nonlinear eddy-current problems," IEEE Trans. on Magn., vol. 43, no. 4, pp. 1229–1232, 2007.
3. O. Deblecker and J. Lobry, "A new efficient technique for harmonic-balance finite-element analysis of saturated electromagnetic devices," IEEE Trans. on Magn., vol. 42, no. 4, pp. 535–538, 2006.
4. S. Yamada, K. Bessho, "Harmonic field calculation by the combination of finite element analysis and harmonic balance method," IEEE Trans. Magn., vol. 24, no, 6, pp. 2588–2590, Nov. 1988.
5. J. Lu, S. Yamada and K. Bessho, "Time-periodic magnetic field analysis with saturation and hysteresis characteristics by harmonic balance finite element method," IEEE Trans. Magn., vol. 2, no. 2, pp. 995–998, Mar. 1990.
6. I. Ciric, and F. Hantila, "An efficient harmonic method for solving nonlinear Time-periodic eddy-current problems," IEEE Trans. on Magn., vol. 43, no. 4, pp. 1185–1188, 2007.
7. X. Zhao, L. Li, J. Lu, Z. Cheng and T. Lu, "Characteristics analysis of the square laminated core under dc-biased magnetization by the fixed-point harmonic-balanced FEM," IEEE Trans. on Magn., vol. 48, no. 2, pp. 747–750, 2012.
8. X. Zhao, L. Li, J. Lu, Z. Cheng and T. Lu, "Analysis of saturated electromagnetic devices under DC bias condition by the decomposed harmonic balance finite element method," COMPEL., vol. 31, no. 2, pp. 498–513, 2012.
9. E. Dlala, A. Belahcen and A. Arkkio, "A fast fixed-point method for solving magnetic field problems in media of hsyteresis," IEEE Trans. on Magn., vol. 44, no. 6, pp. 1214–1217, 2008.
10. E. Dlala and A. Arkkio, "Analysis of the convergence of the fixed-point method used for solving nonlinear rotational magnetic field problems," IEEE Trans. on Magn., vol. 44, no. 4, pp. 473–478, 2008.
11. M. Mathekga, R. Mcmahon and A. Knight, "Application of the fixed-point method for solution in time stepping finite element analysis using the inverse vector Jiles-Atherton model," IEEE Trans. on Magn., vol. 47, no. 10, pp. 3048–3051, 2011.
12. Xiaojun Zhao*, Junwei Lu, Lin Li, Huiqi Li, Zhiguang Cheng, Tiebing Lu. Fixed-point harmonic-balanced method for dc-biasing hysteresis analysis using the neural network and consuming function. IEEE Transactions on Magnetics. 48(11): 3356–3359, 2012. (SCI).
13. Xiaojun Zhao, Lin Li, Zhiguang Cheng, Yuting Zhong and Gang Liu. Harmonic analysis of nonlinear magnetic field under sinusoidal and dc-biased magnetizations by the fixed-point method. IEEE Transactions on Magnetics. 51(3): 7000705. 2015. (SCI).
14. P. Zhou, W. N. Fu, D. Lin, and Z. J. Cendes, "Numerical modeling of magnetic devices," *IEEE Trans. Magn.*, vol. 40, no. 4, pp. 1803–1809, Mar. 2004.
15. K. Yamazaki, "An efficient procedure to calculate equivalent circuit parameter of induction motor using 3-D nonlinear time-stepping finite element method," *IEEE Trans. Magn.*, vol. 38, no. 2, pp. 1281–1283, Mar. 2002.

16. J. Lu, S. Yamada and K. Bessho, "Harmonic balance finite element method taking account of external circuit and motion," *IEEE Trans. Magn.*, vol. 27, no. 5, pp. 4024–4027, Sep. 1991.
17. Z. Cheng, N. Takahashi, and B. Forghani, Electromagnetic and Thermal Field Modeling and Application in Electrical Engineering. Beijing: Science Press, 2009, pp. 386–393.
18. O. Bíró, S. Außerhoher, G. Buchgraber, K. Preis, and W. Seitlinger, "Prediction of magnetizing current waveform in a single phase power transformer under DC bias," *IET Sci., Meas. Technol.*, vol. 1, no. 1, pp. 2–5, 2007.
19. P. R. Price, "Geomagnetically induced current effects on transformers," *IEEE Trans. Magn.*, vol. 17, no. 4, pp. 1002–1008, Oct. 2002.
20. Picher, P., Bolduc, L., Dutil, A. and Pham, V.Q. (1997), "Study of the acceptable DC current limit in core-form power transformers", IEEE Trans. on Power Delivery, Vol. 12, No. 1, pp. 257–65.
21. N. Takasu, T. Oshi, F. Miyawaki, and S. Saito, "An experimental analysis of DC excitation of transformers by geomagnetically induced currents," *IEEE Trans. Power Delivery*, vol. 9, no. 2, pp. 1173–1182, Apr. 1994.
22. F. I. Hantila, G. Preda and M. Vasiliu, "Polarization method for static field" *IEEE Trans. on Magn.*, vol. 36, no. 4, pp. 672–675, 2000.
23. V. D. Albertson, B. Bozoki, W. E. Feero, J. G. Kappenman, E. V. Larsen, D. E. Nordell, *et al*, "Geomagnetic disturbance effects on power systems," *IEEE Trans. Power Delivery*, vol. 8, no. 3, pp. 1206–1216, Jul. 1993.
24. S. Yamada, K. Bessho and J. Lu, "Harmonic balance finite element method applied to nonlinear AC magnetic analysis" *IEEE Trans. on Magn.*, vol. 25, no. 4, pp. 2971–2973, 1989.
25. Zhao, X., Lu, J., Li, L., Cheng, Z. and Lu, T. (2011), "Analysis of the DC bias phenomenon by the harmonic balance finite element method", IEEE Trans. Power Delivery, Vol. 26, No. 1, pp. 475–85.
26. W. Fu, P. Zhou, D. Lin, S. Stanton and Z. Cendes, "Modeling of solid conductors in two-dimensional transient finite element analysis and its application to electrical machines," IEEE Trans. Magn., 2004, 40(2): 426–434.
27. P. Zhou, W. Fu, D. Lin, S. Stanton and Z. Cendes, "Numerical modeling of magnetic devices," IEEE Trans. Magn., 2004, 40(4): 1803–1809.
28. E. Dlala, A. Belahcen, and A. Arkkio, "Locally convergent fixed-point method for solving time-stepping nonlinear field problems," IEEE Trans. Magn., 2007, 43(11): 3969–3975.
29. S. Ausserhofer, O. Biro, and K. Preis, "A strategy to improve the convergence of the fixed-point method for nonlinear eddy current problems," IEEE Trans. Magn., 2008, 44(6): 1282–1285.
30. G. Koczka, S. Auberhofer, O. Biro and K. Preis, "Optimal convergence of the fixed point method for nonlinear eddy current problems," IEEE Trans. Magn., 2009, 45(3): 948–951.
31. O. Biro and K. Preis, "An efficient time domain method for nonlinear periodic eddy current problems," IEEE Trans. Magn., 2006, 42(4): 695–698.
32. G. Koczka, S. Auberhofer, O. Biro and K. Preis, "Optimal convergence of the fixed-point method for nonlinear eddy current problems," IEEE Trans. Magn., vol. 45, no. 3, pp. 948–951, Mar. 2009.
33. J. Lu, X. Zhao and S. Yamada, Harmonic Balance Finite Element Method: Applications in Nonlinear Electromagnetics and Power Systems. Singapore, 2016.
34. O. Biro and K. Preis, "An efficient time domain method for nonlinear periodic eddy current problems," IEEE Trans. Magn., vol. 42, no. 4, pp. 695–698, Apr. 2006.
35. A. Kameari, "Local force calculation in 3D FEM with edge elements" Int. J. of Applied Electromagnetics in Materials, vol. 3, pp. 231–240, 1993.
36. D. Xie and S. Yang, Numerical Analysis and Synthesis of Engineering Electromagnetic Field. Beijing: China Machine Press, 2009, pp. 176–187.

Part III
Measurement and Modeling of Magnetic Material and Component Properties

Chapter 7
Fundamentals of Magnetic Material Modeling

Norio Takahashi

Abstract Various kinds of modeling methods of magnetic material properties, such as the *B–H* curve at high flux density, anisotropy, hysteresis, iron loss in laminated core, magnetic properties under stress, rotating flux and DC bias conditions, are discussed. Key points are as follows:

(1) One must be careful when modeling the *B–H* curve at nearly saturated operating condition. After extreme saturation, the gradient of the *B–H* curve is equal to the permeability of vacuum.
(2) Several *B–H* curves in arbitrary directions may be necessary in the modeling of an anisotropic material.
(3) Various kinds of hysteresis models, such as Preisach model and Jiles–Atherton model, are proposed. One must be careful of the accuracy of the modeling in practical complicated problems, such as inverter-fed motors.
(4) The iron loss is composed of hysteresis loss and eddy current loss. In order to accurately estimate the iron loss under distorted flux waveform and rotating flux, various kinds of measurement data are necessary.
(5) In order to accurately analyze the flux and the eddy current distribution in the laminated core in a practical machine, modeling methods, such as homogenization method and two-zone method, should be utilized.
(6) The magnetic properties of electrical steel are affected by the distortion due to the manufacturing processes and operating conditions, e.g., cutting, compression, DC bias and temperature. As an example, the permeability is reduced and the iron loss is increased by the compressive stress.

Keywords Magnetic property · Magnetic material · Magnetic measurement · Modeling and simulation

---

N. Takahashi (✉) (deceased)
Okayama University, Okayama, Japan

Z. Cheng et al. (eds.), *Modeling and Application of Electromagnetic and Thermal Field in Electrical Engineering*, https://doi.org/10.1007/978-981-15-0173-9_7

7.1 Introduction

In the analysis of magnetic field in an iron core, the nonlinearity of the permeability μ or the reluctivity ν should be considered. The permeability or reluctivity can be obtained from the relationship between $\boldsymbol{B}$ and $\boldsymbol{H}$ of iron steel. When the excitation of magnetic field is limited to one dimension and the variation of both magnetic field strength $\boldsymbol{H}$ and flux density $\boldsymbol{B}$ is in phase, the relationship between the flux density $\boldsymbol{B}$ and the magnetic field strength $\boldsymbol{H}$ which should be used in the magnetic field analysis is reduced to the well-known scalar B–H curve. When the variation of flux density $\boldsymbol{B}$ lags that of magnetic field strength $\boldsymbol{H}$ in either magnitude or direction, this phenomenon is known as the magnetic hysteresis.

In order to analyze magnetic fields taking account of the nonlinearity of permeability (B–H curve), magnetic anisotropy and magnetic hysteresis, the B–H curves in various directions and hysteresis loop should be modeled. In this chapter, practical modeling methods of magnetization characteristics are shown.

7.2 Modeling of *B–H* Curve

7.2.1 Relationship Between B *and* H

In the ordinary nonlinear analysis, the B–H curve shown in Fig. 7.1 is used. The B–H curve to be used in eddy current analysis should be a DC curve, because the eddy current term is already included in the equation to be solved. As the measurement of DC curve at the completely DC condition is not easy, the B–H curve can be measured under a very low frequency, for example, 0.01 Hz. If it should be measured at a frequency of the order of 10 Hz, the measured AC B_m–H_b curve can be regarded as a DC curve, where B_m is the maximum value of flux density on the hysteresis loop and H_b is the magnetic field intensity at the instant when the flux density becomes maximum (B_m). This is because the eddy current becomes almost zero at the instant when the change of flux density is nearly equal to zero.

The relationship between the flux density $\boldsymbol{B}$ and the magnetic field intensity $\boldsymbol{H}$ is written using the permeability μ as follows:

$$\boldsymbol{B} = \mu \boldsymbol{H} \tag{7.1}$$

Equation (7.1) can be rewritten as follows using the reluctivity ν:

$$\boldsymbol{H} = \nu \boldsymbol{B} \tag{7.2}$$

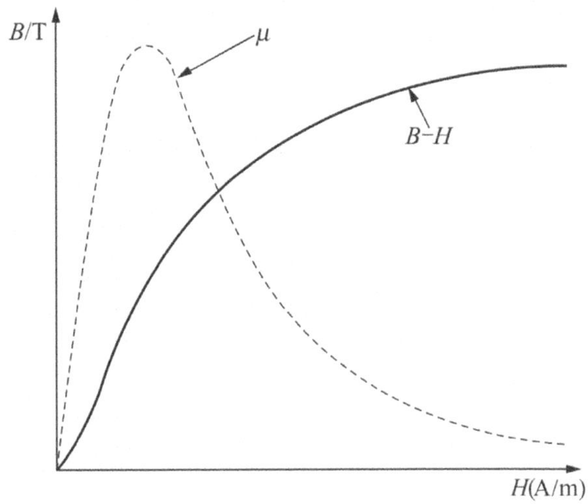

Fig. 7.1 *B–H* curve and *μ*–*H* curve

The reluctivity is a tensor. The relationship between the *x*-, *y*- and *z*-components B_x, B_y, B_z, H_x, H_y and H_z is written by [1]

$$\begin{Bmatrix} H_x \\ H_y \\ H_z \end{Bmatrix} = \begin{pmatrix} v_{xx} & v_{xy} & v_{xz} \\ v_{yx} & v_{yy} & v_{yz} \\ v_{zx} & v_{zy} & v_{zz} \end{pmatrix} \begin{Bmatrix} B_x \\ B_y \\ B_z \end{Bmatrix} \tag{7.3}$$

where v_{xx} and v_{xy} are components of the reluctivity tensor. If the non-diagonal components can be treated as zero for simplicity, Eq. (7.3) can be written as

$$\begin{Bmatrix} H_x \\ H_y \\ H_z \end{Bmatrix} = \begin{pmatrix} v_x & 0 & 0 \\ 0 & v_y & 0 \\ 0 & 0 & v_z \end{pmatrix} \begin{Bmatrix} B_x \\ B_y \\ B_z \end{Bmatrix} \tag{7.4}$$

where v_x, v_y and v_z are the *x*-, *y*- and *z*-components of reluctivity.

7.2.2 Sectional Polynomial Approximation

The initial magnetization curve is approximated by the piecewise linear lines, polynomials, etc. If the *B–H* curve is divided into many sections and approximated by a chain of third-order cubic splines, it sometimes waves and it is not easy to get a monotonic curve. On the other hand, a monotonic curve can tend to be obtained using Akima's method. In the case when the measured points are P_1–P_6 shown in Fig. 7.2, *B* can be denoted by Akima's method as follows:

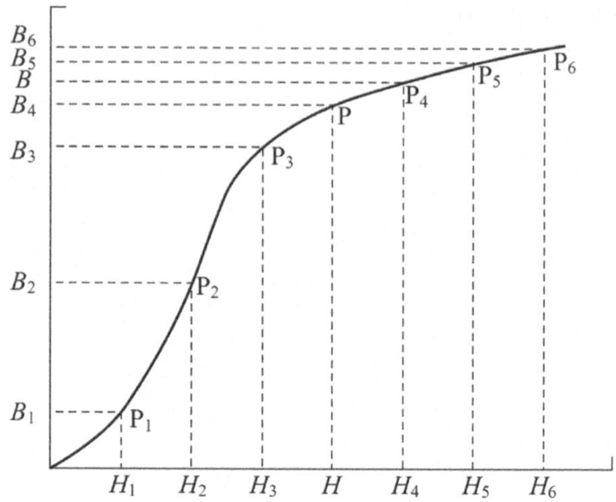

Fig. 7.2 *B–H* curve (Akima's method)

$$B = a_3(H - H_3)^3 + b_3(H - H_3)^2 + c_3(H - H_3) + d_3 \tag{7.5}$$

In the piecewise linear approximation, the flux density $\boldsymbol{B}$ should be continuous at the point on the boundary between sections and the derivative d$B(H)$/dH of an adjacent section should be the same at the sectioned point. If the values B_3, H_3, B_4, H_4 and d$B(H_3)$/dH, d$B(H_4)$/dH are given at the point between sections, the following equations are obtained by putting them to Eq. (7.5):

$$B_3 = d_3 \tag{7.6}$$

$$B_4 = a_3(H_4 - H_3)^3 + b_3(H_4 - H_3)^2 + c_3(H_4 - H_3) + d_3 \tag{7.7}$$

$$\mathrm{d}B(H_3)/\mathrm{d}H = c_3 \tag{7.8}$$

$$\mathrm{d}B(H_4)\Big/\mathrm{d}H = 3a_3(H_4 - H_3)^2 + 2b_3(H_4 - H_3) + c_3 \tag{7.9}$$

The derivatives d$B(H_3)$/dH and d$B(H_4)$/dH are approximated using the slope m_j of adjacent section as follows:

$$\frac{\mathrm{d}B(H_3)}{\mathrm{d}H} = \frac{|m_4 - m_3|m_2 + |m_2 - m_1|m_3}{|m_4 - m_3| + |m_2 - m_1|} \tag{7.10}$$

$$\frac{\mathrm{d}B(H_4)}{\mathrm{d}H} = \frac{|m_5 - m_4|m_3 + |m_3 - m_2|m_4}{|m_5 - m_4| + |m_3 - m_2|} \tag{7.11}$$

where the slope m_3 of the line P_3–P_4 can be given by

$$m_3 = \frac{B_4 - B_3}{H_4 - H_3} \tag{7.12}$$

By giving these values to Eqs. (7.8) and (7.9) and solving Eqs. (7.6)–(7.9) simultaneously, the coefficients a_3, b_3, c_3 and d_3 of Eq. (7.5) can be obtained. The undesired oscillation of the approximated *B–H* curve can be avoided sometimes using Akima's method, because the derivatives at the sectioned points are determined considering the adjacent slope of curve.

7.2.3 Approximation of B–H *Curve at High Flux Density*

In order to develop miniature machines, the iron core sometimes works at high flux density. Meanwhile, in the field analysis at high flux density, the *B–H* curve should be measured under high flux density. However, the measurement is difficult at very high flux density region, e.g., over 2 T. Thus, a technique of extrapolation is applied.

The permeability under over-saturation condition (e.g., saturation magnetization of iron: 2.158 T, silicon steel: 2.03 T) is approaching to the permeability of vacuum (μ_0). In this case, the measured *B–H* curve can be extrapolated in segments, as shown in Fig. 7.3, where the coefficients *a*, *b* and *c* can be determined by the curve fitting method.

When the flux density is low like the case of the magnetic field analysis of shielding room made of permalloy under the geomagnetism, μ_r is changed by the flux density considerably as shown in Fig. 7.4. Therefore, the nonlinear analysis is necessary even in such a case of low flux density in order to get an accurate result.

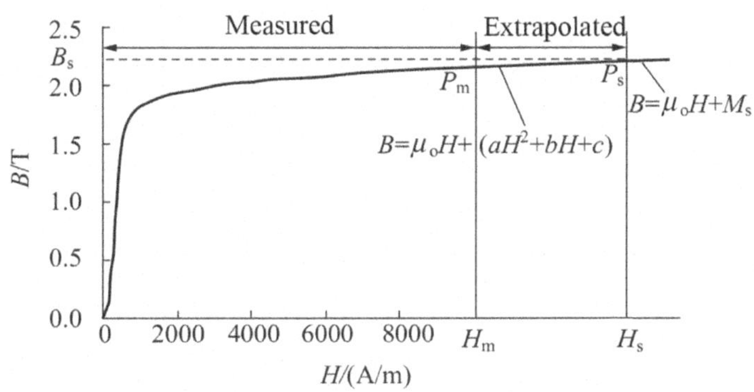

Fig. 7.3 Approximation of *B–H* curve at high flux density

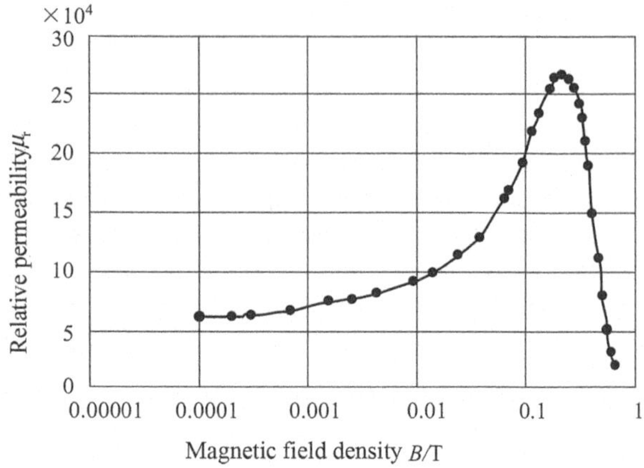

Fig. 7.4 μ_r–B curve of permalloy

7.3 Modeling of Magnetic Anisotropy

7.3.1 Problem of Two B–H *Curves Model*

Some magnetic materials exhibit magnetic anisotropy in *x*-, *y*- and *z*-directions. The measurement of magnetic characteristics in the thickness (*z*-)direction is very difficult; for example, in the case of grain-oriented silicon steel sheet, only the anisotropy along the sheet (*x*-*y* plane) is considered.

In the grain-oriented silicon steel sheet, the permeability in the rolling direction is larger than that of the transverse direction. Also, the magnetic field strength vector $\boldsymbol{H}$ is not always parallel to the flux density vector $\boldsymbol{B}$ in the anisotropic material as shown in Fig. 7.5. This magnetic property is called as "vector magnetic property" or "two-dimensional magnetic property" [2, 3].

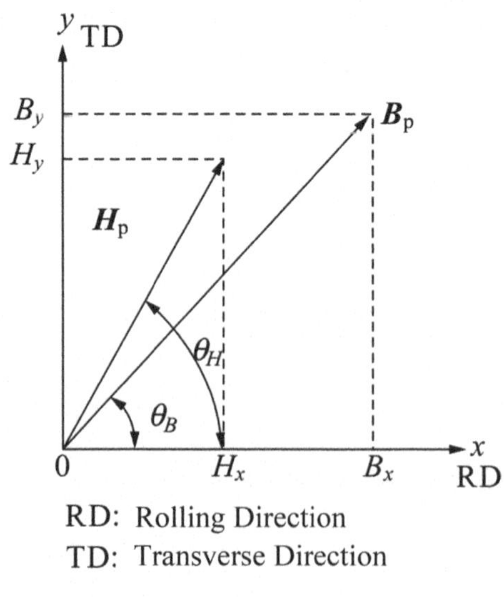

Fig. 7.5 Vector $\boldsymbol{B}$ and vector $\boldsymbol{H}$

The magnetic field strengths H_x and H_y in the x- and y-directions are functions of flux densities. B_x and B_y in respective directions are shown in Eq. (7.13) [2, 4, 5].

$$\left.\begin{aligned} H_x &= f_x\left(B_x, B_y\right) \\ H_y &= f_y\left(B_y, B_y\right) \end{aligned}\right\} \tag{7.13}$$

where f denotes a function. A simple way of treating the anisotropy is to use two kinds of B–H curves in rolling and transverse directions. Here, this is called as "two B–H curves model." In this case, for example, H_x is assumed to be a function of B_x, and then, Eq. (7.14) has been used instead of Eq. (7.13).

$$\left.\begin{aligned} H_x &= f_x^*\left(B_x\right) \\ H_y &= f_y^*\left(B_y\right) \end{aligned}\right\} \tag{7.14}$$

where f^* denotes a function. In this case, Eq. (7.14) can be written as

$$\begin{Bmatrix} H_x \\ H_y \end{Bmatrix} = \begin{bmatrix} \nu_x & 0 \\ 0 & \nu_y \end{bmatrix} \begin{Bmatrix} B_x \\ B_y \end{Bmatrix} \tag{7.15}$$

But, the magnetic characteristics along the magnetic hard axis cannot be represented exactly by this method [3]. Let us explain this phenomenon. We define that the x-direction is the rolling direction (RD), namely the magnetic easy axis, and the y-direction is the transverse direction (TD). Figure 7.6 shows an example of B–H curves of 0.3-mm thick highly grain-oriented silicon steel (JIS: 30P110) measured using the two-dimensional single-sheet tester [6]. In the conventional method [1], only two B–H curves, namely B_x–H_x curve for $B_y = 0$ and B_y–H_y curve for $B_x = 0$, are used.

Figure 7.7 shows the loci of $\boldsymbol{B}$ for constant $|\boldsymbol{H}|$. Figure 7.7a shows the measured curves, and Fig. 7.7b shows the curves calculated using Eq. (7.14). The distance between the origin and a point on the locus corresponds to $|\boldsymbol{B}|$. When the distance is long, the permeability is large, because $|\boldsymbol{H}|$ is constant on the locus. Therefore, the

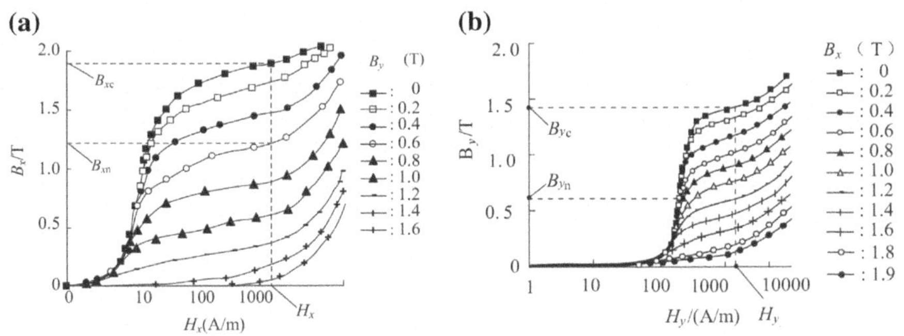

Fig. 7.6 *B–H* curve (JIS:30P110)

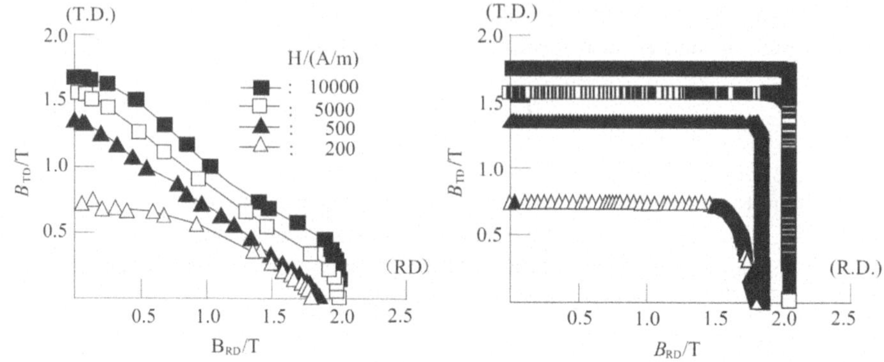

(a) Measured curves (b) Curves calculated using Eq.(7.14)

Fig. 7.7 Loci of ***B*** for constant |***H***|

magnetic easy axis estimated from the conventional modeling method shown in Fig. 7.7b is completely different from the actual easy axis, which coincides with the rolling direction (x-direction). Figure 7.7 denotes that the permeability estimated from Eq. (7.13) is higher than that for the magnetic easiest axis at high flux density region. This discrepancy is due to the assumption in Eq. (7.13).

A method of analysis taking account of *B–H* curves in various directions is used. Here, this is called as "multi-*B–H* curve model." The "E&SS model" which takes into account the vector magnetic characteristics is also shown in this section.

7.3.2 *Multi-***B–H** *Curve Model*

1. Equations and Technique of Interpolation

In order to realize the precise and fast analysis of magnetic fields in anisotropic material, the multi-*B–H* curve model is used [1]. In this technique, the anisotropic reluctivity is treated as a function of the amplitudes of ***B*** (flux density), ***H*** (magnetic field intensity), θ_B (direction of ***B***) and θ_H (direction of ***H***). The Bézier interpolation technique is applied for smoothing the measured data. It is shown that such extrapolation and smoothing techniques are considerably effective for the fast convergence of the *N-R* iteration using a simple magnetic circuit composed of grain-oriented silicon steel.

By using the relationships of $H_x = H\cos\theta_H$, $H_y = H\sin\theta_H$, the following expression of reluctivity can be obtained:

$$\nu = \begin{vmatrix} \frac{H\cos\theta_H}{B\cos\theta_B} & 0 \\ 0 & \frac{H\sin\theta_H}{B\sin\theta_B} \end{vmatrix} = \begin{bmatrix} \nu_x & 0 \\ 0 & \nu_y \end{bmatrix} \tag{7.16}$$

Equation (7.16) denotes that the reluctivity is represented as a function of $\boldsymbol{B}$, θ_B, $\boldsymbol{H}$ and θ_H. When a first-order triangular element is used in the 2-D finite element method, the derivative $\partial G_i^{(e)}/\partial A_j$ of weighted residual by the vector potential A_j is given by

$$\frac{\partial \boldsymbol{G}_i^{(e)}}{\partial \boldsymbol{A}_j} = \frac{1}{4\Delta}\left(v_x d_i d_j + v_y c_i c_j\right) + \frac{\partial v_x}{\partial \boldsymbol{A}_j}\frac{1}{4\Delta}\sum_{k=1}^{3} d_i d_{ke}\boldsymbol{A}_{ke} + \frac{\partial v_y}{\partial \boldsymbol{A}_j}\frac{1}{4\Delta}\sum_{k=1}^{3} c_i c_{ke}\boldsymbol{A}_{ke} \tag{7.17}$$

where c_i, d_i are the functions of coordinates of nodes, and Δ is the area of the first-order triangular finite element e. $\partial v_x/\partial \mathbf{A}_j$ and $\partial v_y/\partial \mathbf{A}_j$ are given by the following equations:

$$\begin{aligned}\frac{\partial v_x}{\partial \boldsymbol{A}_j} = {} & \frac{1}{\boldsymbol{B}\cos\theta_B}\left[\cos\theta_H\left(\frac{\partial \boldsymbol{H}}{\partial \boldsymbol{B}}\right) - \boldsymbol{H}\sin\theta_H\left(\frac{\partial \theta_H}{\partial \boldsymbol{B}}\right)\right. \\ & \left. + \left(\frac{\left\{\left(\frac{\partial \boldsymbol{H}}{\partial \boldsymbol{B}}\right)\cos\theta_H - \boldsymbol{H}\sin\theta_H\left(\frac{\partial \theta_H}{\partial \boldsymbol{B}}\right)\right\}\boldsymbol{B} - \boldsymbol{H}\cos\theta_H}{\boldsymbol{B}}\right)\right]\frac{\partial \boldsymbol{B}}{\partial \boldsymbol{A}_j} \\ & + \frac{1}{\boldsymbol{B}\cos\theta_B}\left[\cos\theta_H\left(\frac{\partial \boldsymbol{H}}{\partial \theta_B}\right) - \boldsymbol{H}\sin\theta_H\left(\frac{\partial \theta_H}{\partial \theta_B}\right)\right. \\ & \left. + \left(\frac{\left\{\left(\frac{\partial \boldsymbol{H}}{\partial \theta_B}\right)\cos\theta_H - \boldsymbol{H}\sin\theta_H\left(\frac{\partial \theta_H}{\partial \theta_B}\right)\right\}\cos\theta_B + \boldsymbol{H}\cos\theta_H\sin\theta_B}{\cos\theta_B}\right)\right]\frac{\partial \theta_B}{\partial \boldsymbol{A}_j}\end{aligned} \tag{7.18}$$

$$\begin{aligned}\frac{\partial v_y}{\partial \boldsymbol{A}_j} = {} & \frac{1}{\boldsymbol{B}\sin\theta_B}\left[\sin\theta_H\left(\frac{\partial \boldsymbol{H}}{\partial \boldsymbol{B}}\right) + \boldsymbol{H}\cos\theta_H\left(\frac{\partial \theta_H}{\partial \boldsymbol{B}}\right)\right. \\ & \left. + \left(\frac{\left\{\left(\frac{\partial \boldsymbol{H}}{\partial \boldsymbol{B}}\right)\sin\theta_H + \boldsymbol{H}\cos\theta_H\left(\frac{\partial \theta_H}{\partial \boldsymbol{B}}\right)\right\}\boldsymbol{B} - \boldsymbol{H}\sin\theta_H}{\boldsymbol{B}}\right)\right]\frac{\partial \boldsymbol{B}}{\partial \boldsymbol{A}_j} \\ & + \frac{1}{\boldsymbol{B}\sin\theta_B}\left[\sin\theta_H\left(\frac{\partial \boldsymbol{H}}{\partial \theta_B}\right) + \boldsymbol{H}\cos\theta_H\left(\frac{\partial \theta_H}{\partial \theta_B}\right)\right. \\ & \left. + \left(\frac{\left\{\left(\frac{\partial \boldsymbol{H}}{\partial \theta_B}\right)\sin\theta_H + \boldsymbol{H}\cos\theta_H\left(\frac{\partial \theta_H}{\partial \theta_B}\right)\right\}\sin\theta_B - \boldsymbol{H}\sin\theta_H\cos\theta_B}{sin\theta_B}\right)\right]\frac{\partial \theta_B}{\partial \boldsymbol{A}_j}\end{aligned} \tag{7.19}$$

When the Newton–Raphson method is applied to the nonlinear analysis, the calculation of the derivatives, $\partial \boldsymbol{H}/\partial \boldsymbol{B}$, $\partial \boldsymbol{H}/\partial \theta_B$, $\partial \theta_H/\partial \boldsymbol{B}$ *and* $\partial \theta_H/\partial \theta_B$, is required. In order to get a stable convergence for nonlinear iteration, the smooth approximation

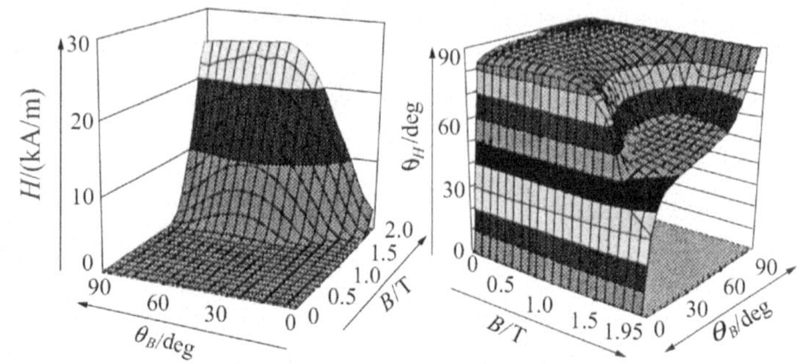

(a) ***H*- (*B*,θ_B)** function (b) **θ_H- (*B*,θ_B)** function

Fig. 7.8 Measured 2-D magnetic properties

Fig. 7.9 Bézier approximation

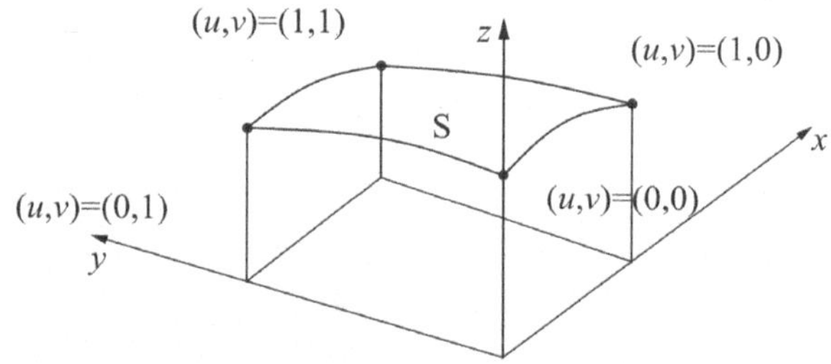

of measured 2-D magnetic properties shown in Fig. 7.8 is strictly necessary. Then, 3-D functions, $\boldsymbol{H}$-($\boldsymbol{B}$, θ_B) and θ_H-($\boldsymbol{B}$, θ_B), are smoothly interpolated by using the Bézier approximation shown in Fig. 7.9. In this case, x and y correspond to $\boldsymbol{B}$ and θ_B, respectively. z corresponds to $\boldsymbol{H}$ or θ_H. Then, the approximated values S_x, S_y and S_z in the x-, y- and z-components are given by

$$
\begin{aligned}
&S = \sum_{i=0}^{n}\sum_{j=0}^{m} k_{ij} B_i^n(u) B_j^m(v) \\
&S = \begin{bmatrix} S_x & S_y & S_z \end{bmatrix}^{\mathrm{T}} \quad k_{ij} = \begin{bmatrix} x_{ij} & y_{ij} & z_{ij} \end{bmatrix}^{\mathrm{T}} \\
&B_i^n(u) = {}_nC_i u^i (1-u)^{n-i} \quad B_j^m(v) = {}_mC_j v^j (1-v)^{m-j} \\
&{}_nC_i = \frac{n!}{i!(n-i)!}
\end{aligned}
\tag{7.20}
$$

where m and n denote measured points in u- and v-directions. For example, x_{ij} is the measured value of B at $u = i$ and $v = j$. The maximum values of them are normalized to unity, and the minimum values are set to zero. In order to utilize the Bézier interpolation, firstly, the values of $\boldsymbol{H}$ (or θ_H) are sampled or interpolated at lattice points of $\boldsymbol{B}$ and θ_B, and then the Bézier interpolation is carried out.

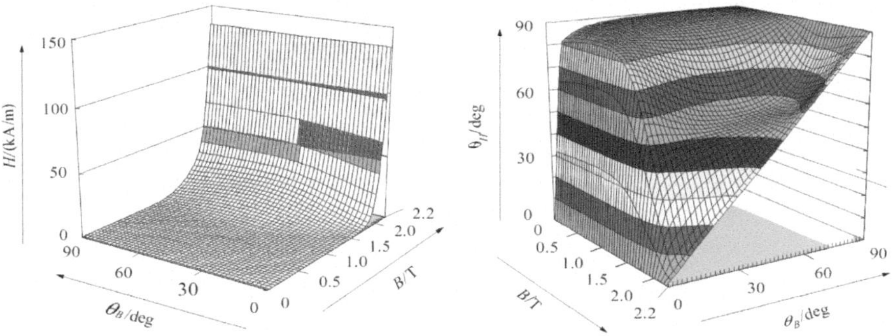

(a) ***H*-(*B*,θ_B)**function (b) **θ_H-(*B*,θ_B)**function

Fig. 7.10 Two-dimensional magnetization property approximated by the Bézier interpolation

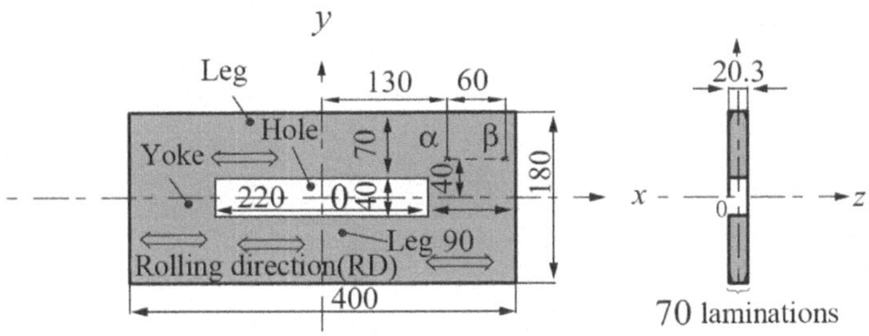

Fig. 7.11 Test model

Figure 7.10 shows the magnetization property approximated by the Bézier interpolation.

2. Analysis of Verification Model

Figure 7.11 shows a single-phase transformer core made of the same material as that in Fig. 7.6 [3]. The shape is chosen so that the result calculated using the conventional method is very much different from that using the improved method. The core is laminated by 70 sheets (sheet dimension: 0.3 mm × 180 mm × 400 mm). A rectangular window (40 mm × 220 mm) is bored in each sheet.

The magnetic vector potential ***A*** is used in the 2-D finite element analysis. Figure 7.12 shows the flux distributions at a practical flux density (B_{leg} = 1.7 T) and at low flux density (B_{leg} = 0.5 T). B_{leg} is the average flux density in the leg.

Figure 7.13 shows the distributions of the absolute value $|B|$ and the angle θB of the flux density deviated from the rolling direction along the line α-β in Fig. 7.11. The flux densities can be measured using a search coil which is wound through 0.6 mm holes drilled in the 70 laminations as shown in Fig. 7.13. Figure 7.13 shows that the results obtained by the multi-*B–H* curve method are nearer to the measured ones than those calculated by the two *B–H* curves method.

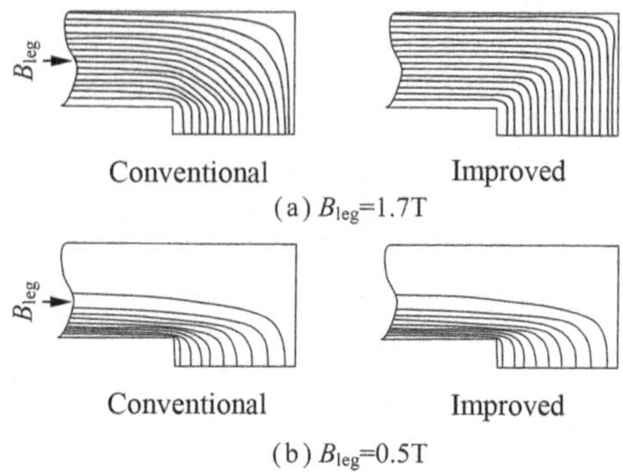

Fig. 7.12 Flux distributions (AISI:M-0H, 0.3 mm)

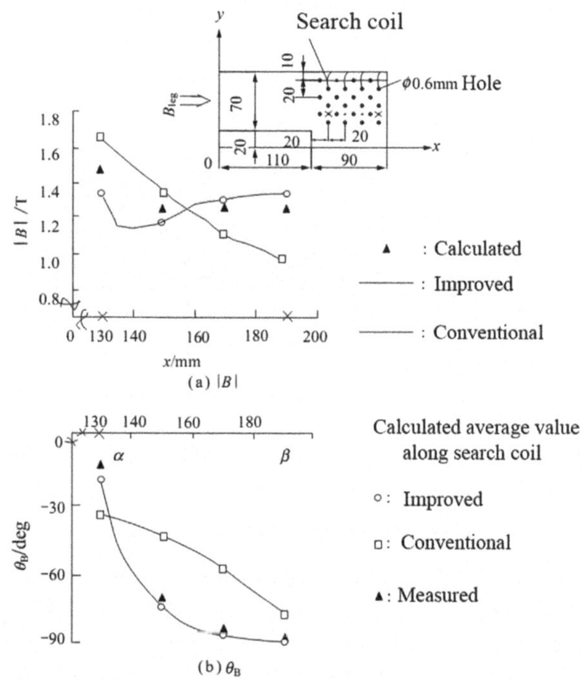

Fig. 7.13 Comparison between the calculated value and the measured value (B = 1.7 T)

3. Analysis of Reactor Made of Grain-Oriented Silicon Steel [4]

Figure 7.14 shows the analyzed model of a three-phase reactor. The frequency of power source is 50 Hz. The terminal voltage of the winding is 6072 V (rms). The current in the winding is assumed as sinusoidal, and the amplitude is determined so that the average flux density becomes nearly 1.4 T. The yoke width is 210 mm. In order to measure the flux density waveform in the yoke, search coils are set along the lines L_1, L_2 and L_3 in Fig. 7.15 by making holes of 1-mm diameter in 17 sheets of silicon steel.

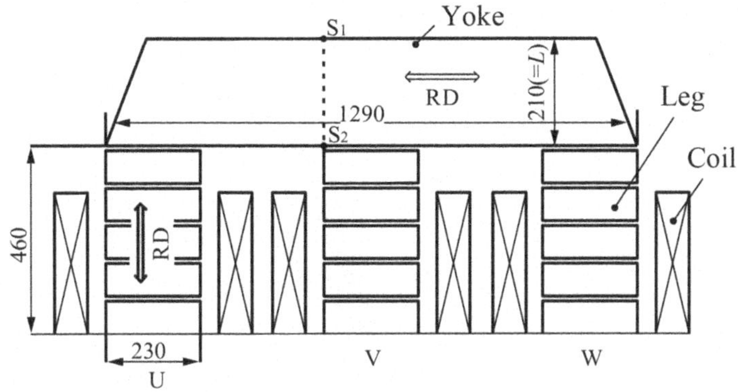

Fig. 7.14 Model of three-phase reactor

Fig. 7.15 Investigated point

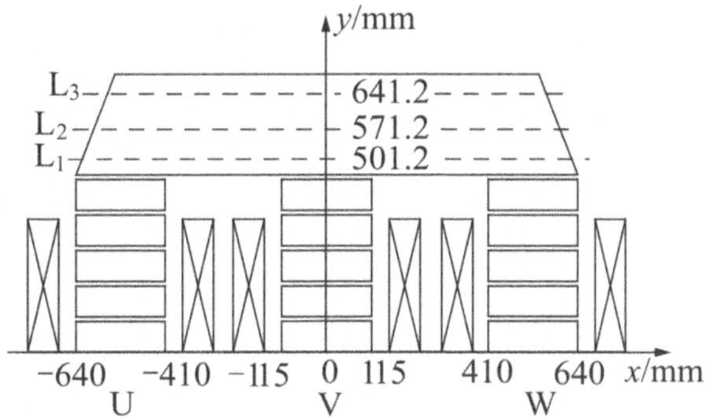

The core of reactor is made of grain-oriented silicon steel (JIS: 35G165). In order to measure the 2-D *B–H* curves and iron loss in arbitrary directions of grain-oriented silicon steel up to 2 T as shown in Figs. 7.16 and 7.17, an improved single-sheet tester, having *H*-coils in the *x*- and *y*-directions, is used [3]. The components of flux density (B_x, B_y) and magnetic field strength (H_x, H_y) in the rolling and transverse directions are measured using various rectangular specimens which are cut in the θ_B directions. θ_B is the direction of flux density vector from the rolling direction.

The flux distribution is analyzed using the 2-D *B–H* curves shown in Fig. 7.16. The coefficient, $\partial \mathbf{G}_i^{(k)}/\partial \mathbf{A}_i^{(k)}$, at the *k*th nonlinear iteration in Newton–Raphson method can be represented by the following function:

$$\frac{\partial \boldsymbol{G}_i^{(k)}}{\partial \boldsymbol{A}_j^{(k)}} = g\left(\frac{\partial H_x^{(k)}}{\partial B_x^{(k)}}, \frac{\partial H_x^{(k)}}{\partial B_y^{(k)}}, \frac{\partial H_y^{(k)}}{\partial B_x^{(k)}}, \frac{\partial H_y^{(k)}}{\partial B_y^{(k)}}\right) \tag{7.21}$$

where g denotes a function. $\boldsymbol{G}_i$ and $\boldsymbol{A}_j$ are the residual at a node i and the magnetic vector potential at a node j, respectively. The coefficient matrix is asymmetric due

Fig. 7.16 Two-dimensional *B–H* curves (35G165)

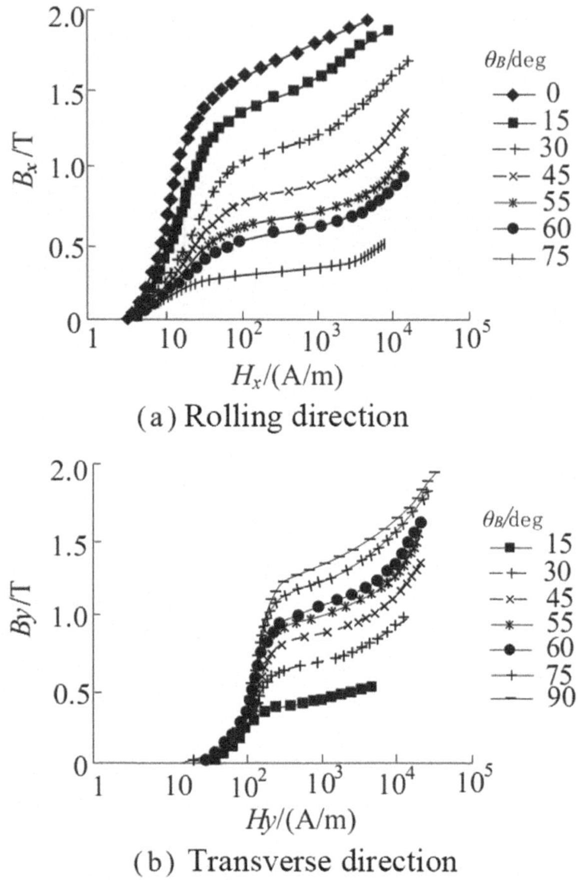

Fig. 7.17 Two-dimensional iron loss curves (35G165)

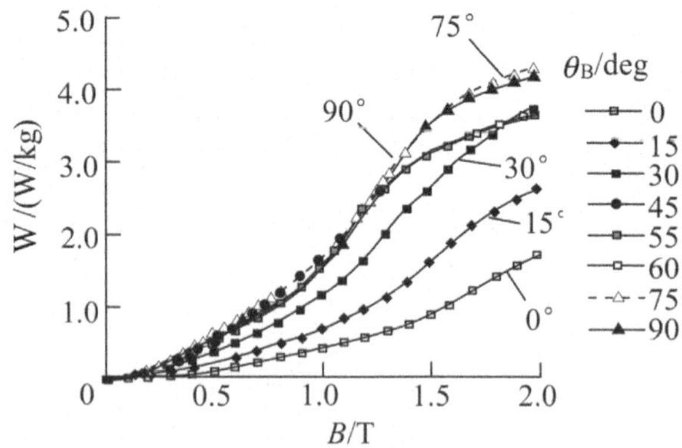

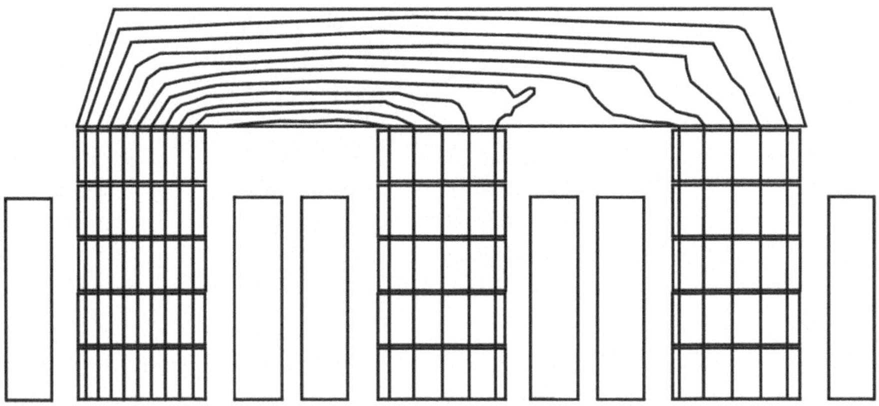

Fig. 7.18 Flux distribution (φ_u = max)

to $\partial H_x/\partial B_y$ and $\partial H_y/\partial B_x$. $\partial H_x/\partial B_y$ and $\partial H_y/\partial B_x$ in Eq. (7.21) are assumed to be zero.

Only one *B–H* curve is used in an element during the Newton–Raphson iteration. Then, the new *B–H* curve is chosen corresponding to the newly calculated θ_B. This process is iterated until the convergent results can be obtained [1]. The *B–H* curve is calculated by interpolating the measured *B–H* curves stored in a computer.

As the convergence of the nonlinear analysis of magnetic field in such anisotropic material is not easy or, in worst case, the solution does not converge, the modified Newton–Raphson method which introduces a relaxation factor [7] is used.

Figure 7.18 shows the flux distribution at the instant when the flux Φ_u in the *U*-leg is maximum. The flux flows along the yoke edge. This is the feature of anisotropic material [5]. Figure 7.19 shows the distribution of maximum value of density along the lines L_1, L_2 and L_3 shown in Fig. 7.15.

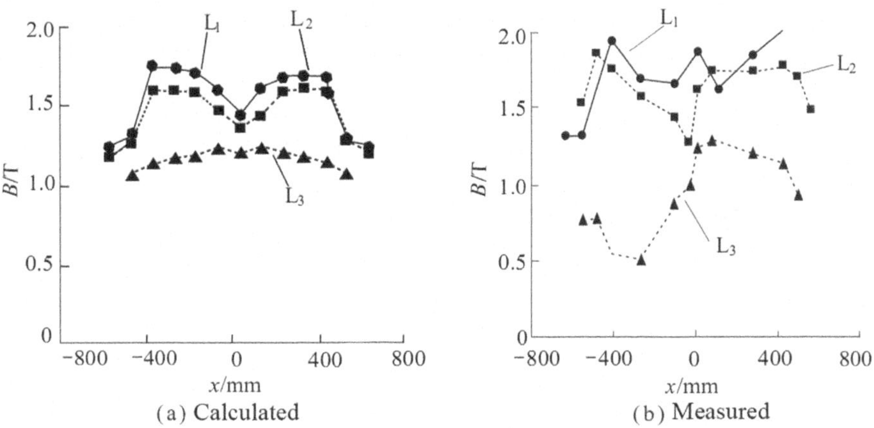

Fig. 7.19 Flux density distribution (L = 210 mm)

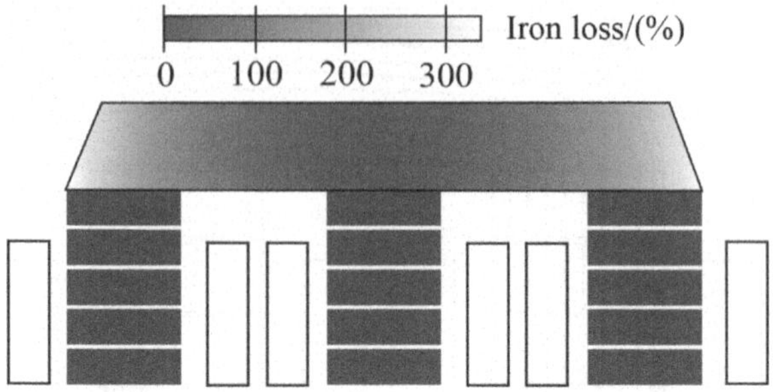

Fig. 7.20 Iron loss distribution (L = 210 m, calculated)

The flux density in the middle part of yoke between legs is larger than in the other part. The discrepancy between the calculation and the measurement along the line L_3 may be due to the insufficient number of measured B–H curves ($\theta_B = 0°$, 15°, 30°, 45°, 55°, 60° and 75°) shown in Fig. 7.17.

Figure 7.20 shows the iron loss distribution calculated using the 2-D iron loss curves shown in Fig. 7.17. The average iron loss of the core L = 210 mm is normalized to 100%. The iron loss is calculated by assuming that the iron loss is the function of the absolute value B of the maximum AC flux density and the direction θ_B of flux density. The fact that the flux is nearly alternating is confirmed by plotting the locus of the calculated flux density vectors. Figure 7.20 shows that the iron loss is increased near both sides of yoke. This is because the direction of flux density vector is considerably deviated from the rolling direction near the side of yoke.

7.3.3 E&SS Model

In order to express not only alternating magnetic property in arbitrary directions but also magnetic property under rotating field, the E&SS model is proposed by using the vector relationship between the flux density vector $\boldsymbol{B}$ and the magnetic field strength vector $\boldsymbol{H}$ [2, 4, 8].

The vector $\boldsymbol{H}$ is not always parallel to the vector $\boldsymbol{B}$ in the anisotropic material. By using E&SS model, various kinds of magnetic characteristics, such as hysteresis loop and iron loss, can be directly obtained.

In the E&SS model, H_x and H_y can be written by

$$\begin{aligned} H_x &= \nu_{xr}(\boldsymbol{B}, \theta_B, \alpha, t)B_x \\ &\quad + \nu_{xi}(\boldsymbol{B}, \theta_B, \alpha, t)\int B_x \mathrm{d}t \end{aligned} \tag{7.22}$$

$$H_y = \nu_{yr}(\boldsymbol{B}, \theta_B, \alpha, t)B_y + \nu_{yi}(\boldsymbol{B}, \theta_B, \alpha, t)\int B_y \mathrm{d}t \tag{7.23}$$

where $\boldsymbol{B}$ is the amplitude of flux density, α is the axis ratio of ellipse (Lissajous of $\boldsymbol{B}$ vector), ν_{xr} and ν_{yr} are the magnetic reluctivity coefficients, and ν_{xi} and ν_{yi} are the magnetic hysteresis coefficients, these are functions of $\boldsymbol{B}$, θ_B, α and t. ν_{xr}, ν_{yr}, and ν_{xi} and ν_{yi} are obtained from the measured results of vector magnetic property.

The iron loss can be obtained directly from $\boldsymbol{B}$ and $\boldsymbol{H}$ as follows without loss data:

$$W = \frac{1}{\rho T}\int_0^T \left(H_x\frac{\mathrm{d}B_x}{\mathrm{d}t} + H_y\frac{\mathrm{d}B_y}{\mathrm{d}t}\right)\mathrm{d}t \left[\frac{W}{kg}\right] \tag{7.24}$$

where ρ is the material density and T is the period.

7.4 Hysteresis Modeling

7.4.1 *Various Hysteresis Models*

When the magnetic field $\boldsymbol{H}$ is impressed to a magnetic material, the flux density $\boldsymbol{B}$ is increased along the initial B–H curve shown in Fig. 7.21. When $\boldsymbol{H}$ is reduced, $\boldsymbol{B}$ is reduced along with another curve as shown in Fig. 7.21. This phenomenon is called

Fig. 7.21 Hysteresis loop

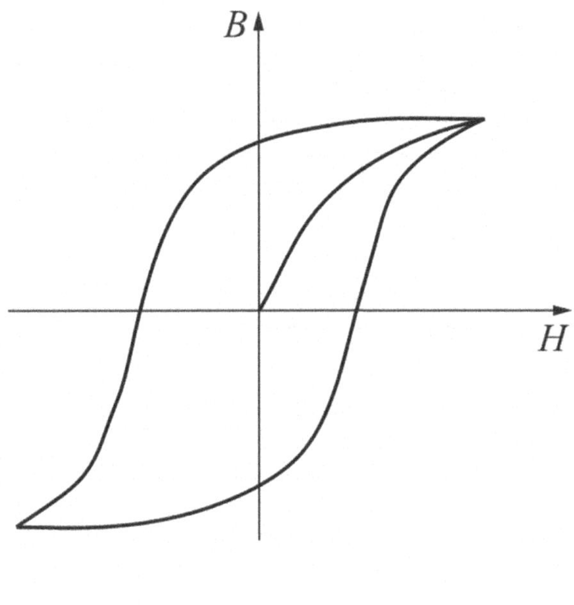

as "hysteresis." Many models are proposed to represent this phenomenon [9]. The modeling of the magnetic properties of ferromagnetic materials for describing components of electrical circuits has proved to be a difficult problem because of hysteresis in the variation of magnetic flux density $\boldsymbol{B}$ with the applied magnetic field $\boldsymbol{H}$. There are two main classes of models that have been developed to address this problem of describing hysteresis. One is the purely mathematical models, such as the Preisach model [10–13] and the Hodgdon model [14, 15]. The other is the physical models such as Stoner–Wohlfarth [16–19] and Jiles–Atherton [20–22], which follows the phenomenon in nature.

The Preisach model relies on the description of bulk hysteresis as the summation of a large number of single-domain hysteresis loops with different switching fields or localized coercivities. This has enabled the Preisach function to be described over the region of interest on the B, H plane, representing the density of domains in the material with a particular pair of switching fields (i.e., positive and negative coercivities).

The advantage of the Preisach model is that it can describe the shapes of arbitrary hysteresis loops to a high degree of accuracy. The disadvantage of the model is that, since it is not physically based, any changes in external conditions, such as stress or temperature, cannot be described by extension and instead require the whole of the Preisach function to be recalculated over the region of the B, H plane of interest.

The physically based models usually have fewer degrees of freedom and are based either on the concept of domain rotation (Stoner–Wohlfarth) or on the domain boundary displacement (Jiles–Atherton). The need to consider the physical basis for hysteresis in these models leads to some advantages and disadvantages over the purely mathematical models.

The advantages are the fact that the models can easily be adapted by extension to include effects of stress, temperature and external field frequency. In this way, the values of the model parameters once determined can be used for the material under a variety of external conditions. Another advantage of the physically based models is that the limitations imposed by the consideration of the physical mechanisms mean that the number of degrees of freedom is much less than the Preisach model.

In this section, the Preisach model, Jiles–Atherton model and Stoner–Wohlfarth model are discussed as well as the simple interpolation method using measured hysteresis loops.

7.4.2 Interpolation Method

The simple way of considering the hysteresis is the interpolation method using the measured hysteresis loops [23].

Typical DC hysteresis loops are stored in a computer to represent hysteresis phenomena as shown in Fig. 7.22. Hysteresis loop 1 with the maximum flux density B_m is interpolated from these typical hysteresis loops as shown in Fig. 7.23.

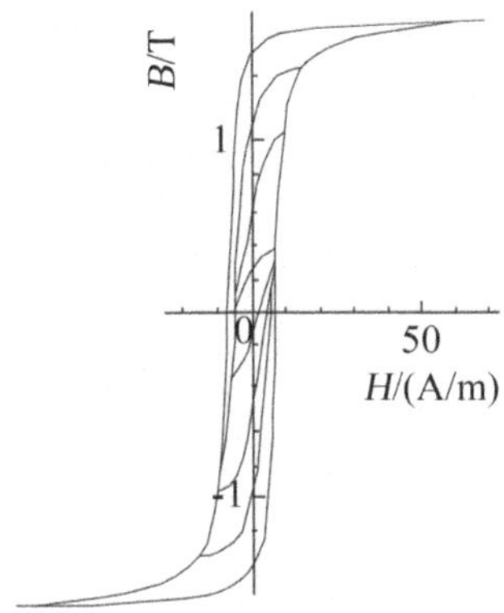

Fig. 7.22 Typical DC hysteresis loops (JIS:30P110)

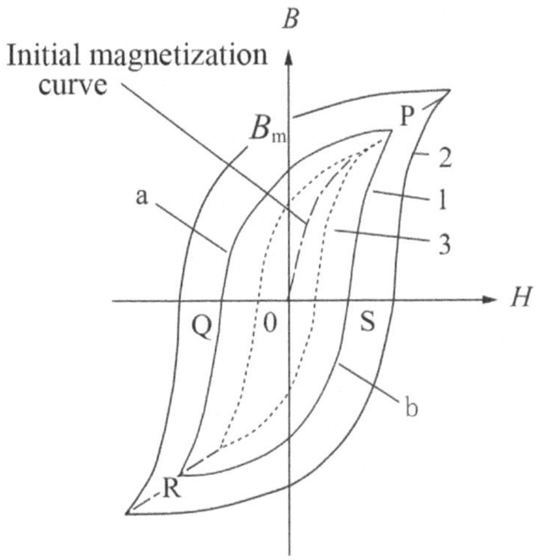

Fig. 7.23 Determination of hysteresis loop

The new loop *P–Q–R–S–P* is obtained from two nearest typical loops 2 and 3. Two forms of the *B–H* curves, in the rolling and the transverse directions, are necessary to analyze the magnetic circuit when anisotropic material such as grain-oriented silicon steel is used.

The process of calculation is as follows: In the first few time steps (ωt), the flux distribution is calculated using the initial magnetization curve denoted by the chain line in Fig. 7.23. If the flux density reaches the maximum value B_m at the point *P*, demagnetization progresses along the upper branch *a* of the hysteresis loop. When the demagnetization reaches a state of opposite magnetization at the point *R*, remagnetization takes along the lower branch b. The above-mentioned process is repeated until a steady state is reached.

7.4.3 Preisach Model

1. Scalar Preisach Model

In a scalar Preisach model [10, 11], it is assumed that the magnetic material consists of many elementary interacting particles and each of them can be represented by a rectangular elementary hysteresis loop having positive or negative magnetized state as shown in Fig. 7.24. By increasing H, the elementary particles of which the switching field H_u is lower or equal to H turn up-magnetized state of magnetization $+M_s$. With the decreasing H, the particles of which the magnetization is positive remain in their position until H is decreased to the switching field H_l. At $H \le H_l$, the particles reverse from one stable magnetization position to the other one providing magnetization of $-M_s$. The change of magnetization of magnetic material can be represented as the reversal of domains. If the number of particles, having switching fields (H_u, H_e), is $\kappa(H_u, H_l)$, the magnetization of particles having (H_u, H_l) is equal to $+M_s \cdot \kappa(H_u, H_l)$ or $-M_s \cdot \kappa(H_u, H_l)$.

According to the Preisach model, the magnetization M can be determined as the magnetization assembly of particles having the distribution function $\kappa(H_u, H_l)$.

$$M = \int_{H_u} \int_{H_l} \alpha(H_u, H_l) \cdot M_s \cdot \kappa(H_u, H_l) \mathrm{d}H_u \mathrm{d}H_l \tag{7.25}$$

The definition region of switch fields (H_u, H_l), the Preisach triangle, is shown in Fig. 7.25. The number of particles $\kappa(H_u, H_l)$ which are defined in the Preisach

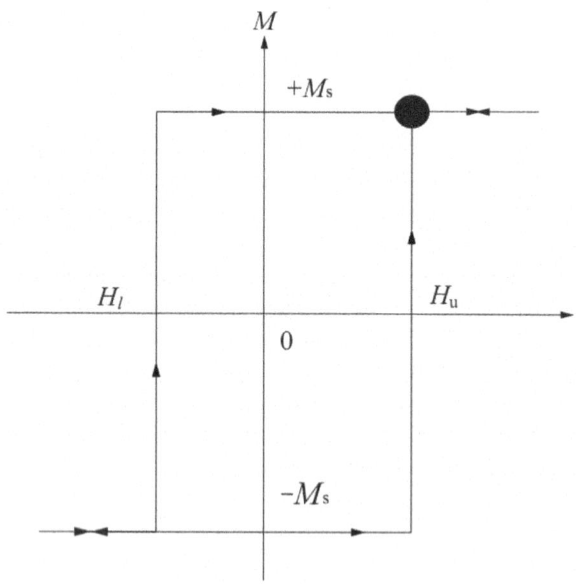

Fig. 7.24 Basic hysteresis loop

Fig. 7.25 Preisach triangle

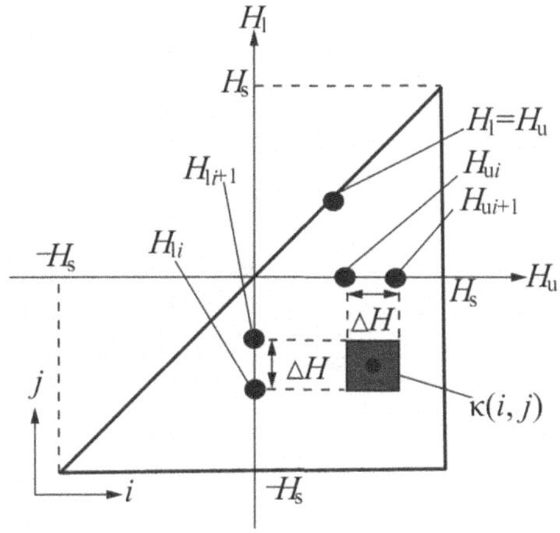

Fig. 7.26 Transition curves

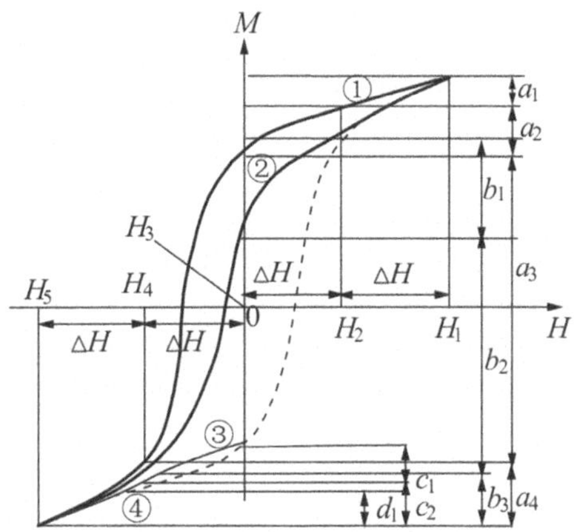

triangle is called as a distribution function. As M_s can be arbitrarily determined, it is set at unity for all elementary particles. $\alpha(H_u, H_l)$ is the elementary hysteresis operator and has values of +1/2 and −1/2 corresponding to up and down positions of the elementary hysteresis loop.

The distribution function can be obtained by using the transition curves shown in Fig. 7.26. Applying finite change in the descending field, it results finite change in the magnetization. If the magnetic field strength between $-H_s$ and H_s is discretized to n parts having the interval ΔH, it results n cells in both directions. Figure 7.26 shows an example of $n = 4$. When the applied magnetic field strength H is decreased from H_2 to H_3 along the curve ①, the integration K of the following equation along the curve ① corresponds to a_2 in Fig. 7.27a.

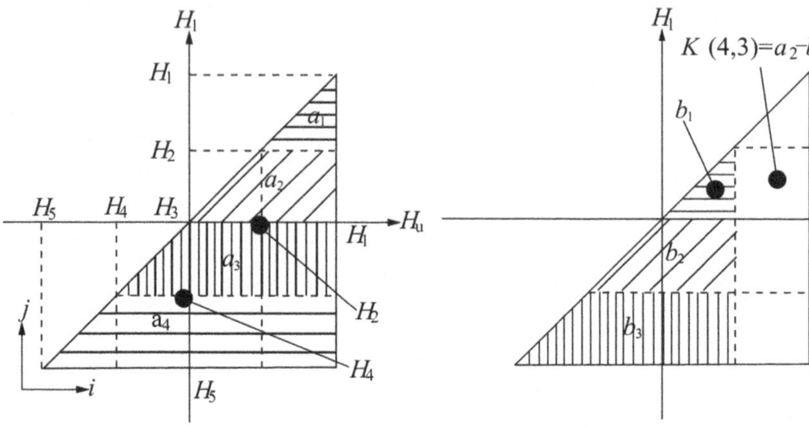

(a) Transition curve ① (b) Transition curve ②

Fig. 7.27 Relationship between transition curve and distribution function

$$K = \int_{H_u} \int_{H_l} \kappa(H_u, H_l) \mathrm{d}H_u \mathrm{d}H_l \tag{7.26}$$

The integration K along the curve ② in the range of $H_3 \leqq H \leqq H_2$ is equal to b_1 in Fig. 7.27b. Then, $K(4, 3)$, which is defined in the region of $H_2 \leq H_u \leq H_1$, $H_3 \leq H \leq H_2$ in Fig. 7.27b, is obtained as $(a_2 - b_1)$.

By using the above-mentioned $K(i, j)$, Eq. 7.25 can be rewritten as

$$M = \sum_{i}^{n} \sum_{j}^{n} \alpha(i,j) \cdot M_s \cdot K(i,j) \tag{7.27}$$

2. Inverse Distribution Function Method

The distribution function with $\boldsymbol{H}$ as variables is not suitable for analysis of hysteresis properties by using the usual $\boldsymbol{B}$-oriented finite element method of which the unknown variable is the magnetic vector potential, because the distribution function is a function of $\boldsymbol{H}$. Therefore, many iterations are necessary to obtain H from the calculated $\boldsymbol{B}$.

The inverse distribution function method for obtaining $\boldsymbol{H}$ directly from $\boldsymbol{B}$ is conceived to avoid such iterations [24]. In this method, the inverse transition curves shown in Fig. 7.28 are utilized to obtain the inverse distribution function. The inverse distribution function can be obtained in a similar way as the case of the distribution function. Figure 7.29 shows an example of cells of $n = 4$. The integration $K'(4, 3)$ of the inverse distribution function at $i = 4$ and $j = 3$ is obtained as $(a_2 - b_1)$ as shown in Fig. 7.29. The magnetization history is stored in α (H_u, H_l).

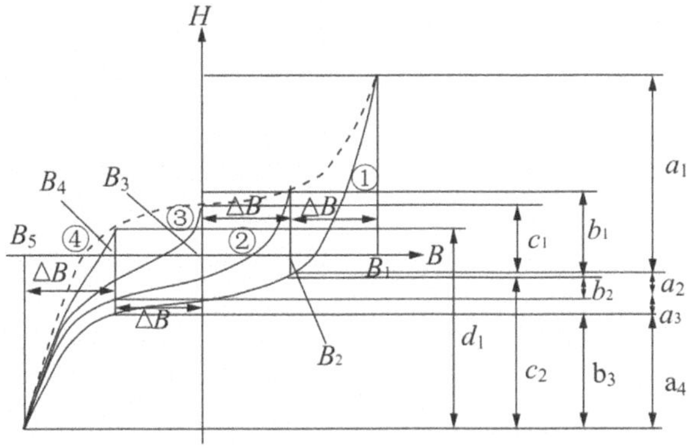

Fig. 7.28 Inverse transition curves

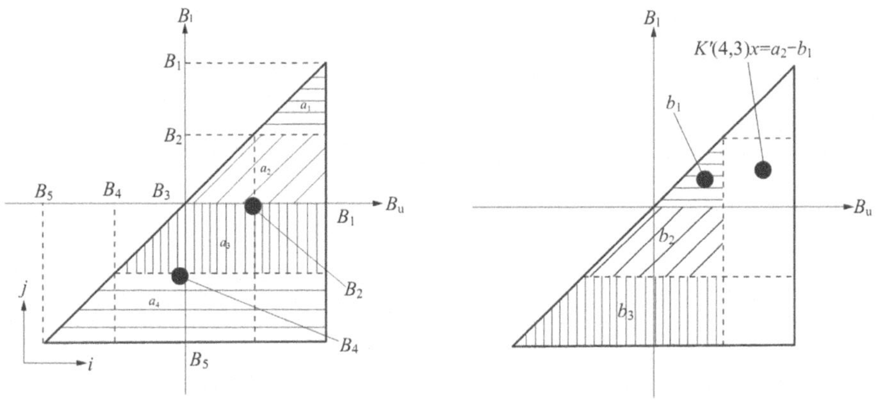

(a) Inverse transition curve①

(b) Inverse transition curve②

Fig. 7.29 Relationship between inverse transition curve and distribution function

In this case, huge memory ($=n^2/2$) is required in storing +1/2 and −1/2 regions in the Preisach triangle. Alternately, the magnetization history can be stored by memorizing the position of the borderline between +1/2 and −1/2 regions [10]. This reduced the memory requirement from $n^2/2$ to only n.

The inverse distribution function method is applied to a simple model as shown in Fig. 7.30 and compared with the ordinary distribution function method and the neural network (NN) method [24]. A coil with the current shown in Fig. 7.31 is located between two cores made of steel (SS400). The eddy current is ignored. Figure 7.32 shows the inverse transition curves of steel measured by using a very low-frequency excitation system to avoid the eddy current effect. Figure 7.33 shows the obtained inverse distribution function. Figure 7.34 shows the calculated magnetization process.

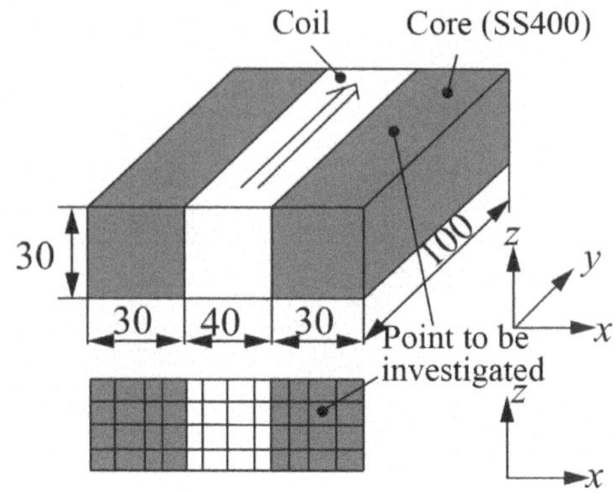

Fig. 7.30 Simple model

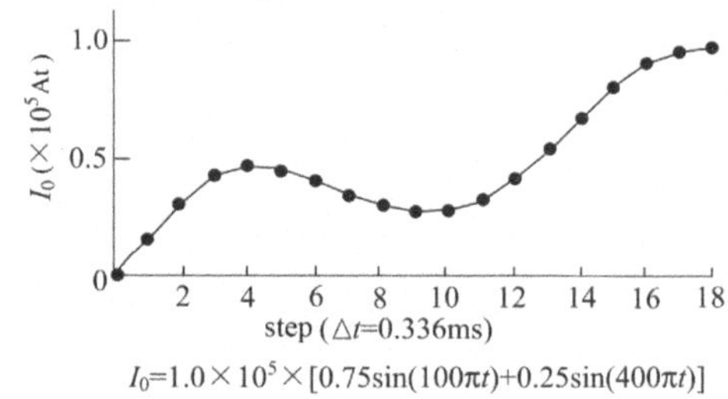

Fig. 7.31 Exciting current waveform

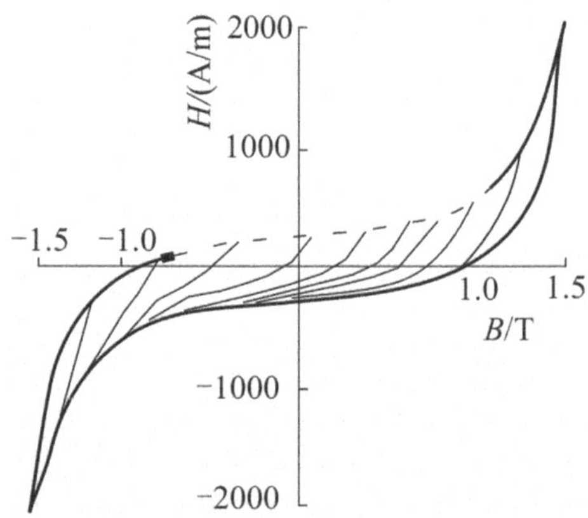

Fig. 7.32 Measured inverse transition function

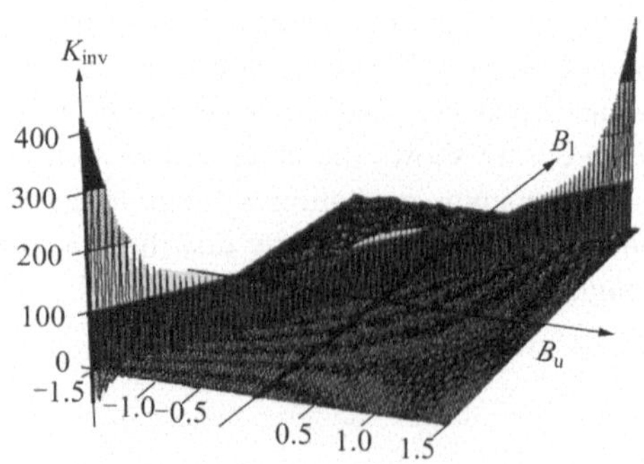

Fig. 7.33 Inverse transition function

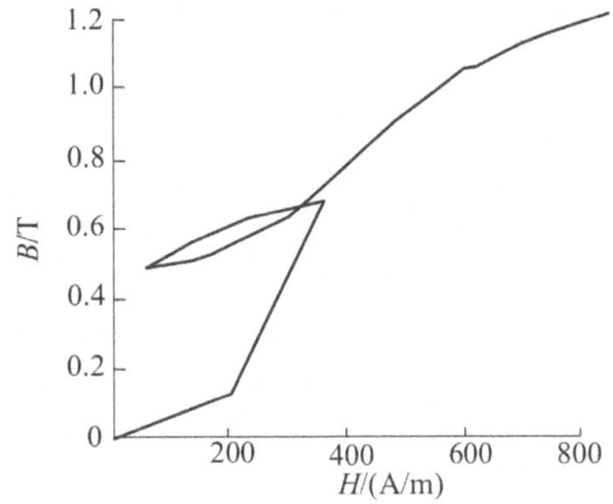

Fig. 7.34 Calculated magnetization process

Table 7.1 Memory requirements and CPU time

Method	Memory (MB)	CPU (s)
Distribution function	70.1	62,569
Neural network	8.8	2806[a]
Inverse distribution function	10.0	2579

Computer used: HP735 (45 MFLOPS)
Dimension n: 200
[a]CPU time for training of NN: 32,800 (s)

Table 7.1 shows the comparison of memory requirement and CPU time. The number n of cells (dimension) of the Preisach triangle is chosen as 200. The table denotes that the memory requirements for the neural network method and the inverse distribution function method are considerably reduced compared with the distribution function method. As the neural network method needs a long CPU time for the training of the network, it can be concluded that the inverse distribution function method is preferable from the viewpoint of memory requirement and CPU time.

The effect of the dimension n of inverse distribution function on the memory requirements and CPU time is investigated using the model shown in Fig. 7.35. The current waveform is denoted in Fig. 7.31. The eddy current is ignored. The analyzed region is subdivided into 4710 first-order brick nodal elements.

Figure 7.36 shows the effect of dimension n on memory requirement and CPU time. The CPU time increases nearly quadratically with n. The figure denotes that the CPU time for calculating hysteresis loop is negligible if n is nearly equal to 200.

7.4.4 Jiles–Atherton Model

The Jiles–Atherton model is a physically based model considering the domain wall motion [20, 21]. The hysteresis-free curve, or anhysteretic curve, is described by

Fig. 7.35 Investigated model

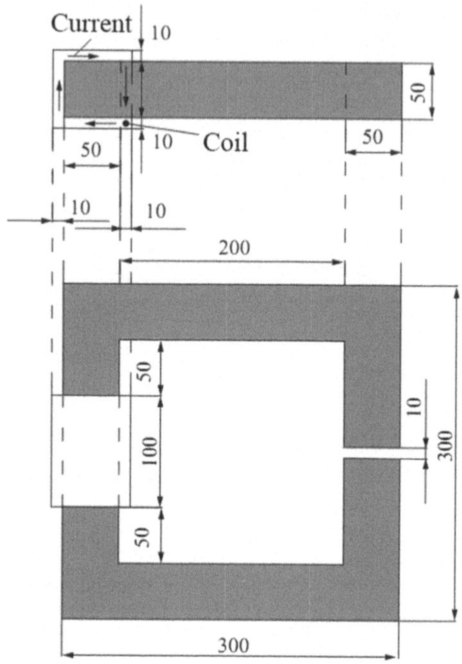

$$M_{an}(H) = M_s\left\{\coth\left(\frac{H+\alpha M}{a}\right) - \left(\frac{a}{H+\alpha M}\right)\right\} \tag{7.28}$$

where M_s is the saturation magnetization and α is the internal coupling together of the domain magnetizations, and $a = \frac{\kappa_B T}{\mu_0 \langle m \rangle}$ where k_B is the Boltzmann constant, T is the temperature, $\mu_0 = 4\pi \times 10^{-7}$ H m^{-1}, $\langle m \rangle$ is the mean effective domain size, M_{an} is the anhysteretic magnetization and H is the magnetic field.

The hysteresis curve is given by

$$\begin{aligned}\frac{\mathrm{d}M(H)}{\mathrm{d}H} = &\frac{M_{an}(H) - M_{irr}(H)}{k\delta - \alpha[M_{an}(H) - M_{irr}(H)]} \\ &+ c\left(\frac{\mathrm{d}M_{an}(H)}{\mathrm{d}H} - \frac{M_{an}(H) - M_{irr}(H)}{k\delta - \alpha[M_{an}(H) - M_{irr}(H)]}\right)\end{aligned} \tag{7.29}$$

where δ is a directional parameter having the value +1 for dH/dt > 0 and −1 for dH/dt < 0.

These equations can then be solved to give the hysteresis curves of the material. Here, k is the energy loss per unit change in magnetization and c gives a measure of the amount of reversible domain wall motion, which can be related directly to domain wall surface energy.

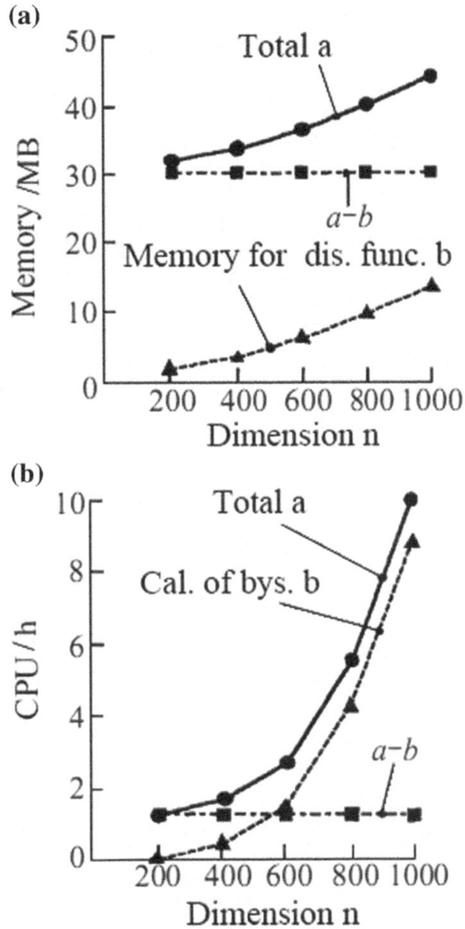

Fig. 7.36 Effect of dimension n on memory requirement and CPU time

Various kinds of hysteresis loops can be modeled by adequately choosing the parameters, M_s, α, a, c and k. In order to determine these parameters, various techniques are proposed [22].

7.4.5 Stoner–Wohlfarth Model

A Stoner–Wohlfarth (S-W) model consists of an ensemble of non-interacting S-W particles or single-domain uniaxial ones whose easy axes are distributed non-uniformly in 3-D space [25]. In this model, S-W particles are assumed to have the same magnetic moment and volume, but their anisotropy fields are assumed to follow a Gaussian distribution. The magnetization of the ensemble in an applied magnetic field is given by summing the magnetic moment of each S-W particle calculated by taking account of the history (binary bits are stored for two easy magnetic moment directions) and is fitted to experimental data of the magnetic material.

In the S-W model, it is assumed that a magnetic material consists of a collection of small particles, each with anisotropy due to stress, crystal structure or particle shape. Each particle is uniformly magnetized to saturation in the direction of the easy axis, giving a single magnetic domain with moment m_s which is free to rotate in any direction. Such a particle is called the S-W particle. The particle interaction, either due to quantum exchange forces or to magnetic dipole–dipole forces, is not considered.

When a magnetic field $\boldsymbol{H}$ is applied, the magnetic moment $\boldsymbol{m}_s$ of an S-W particle rotates to the orientation which results in a minimum energy, as shown in Fig. 7.37.

The total energy of a single domain with moment $\boldsymbol{m}_s$ can be expressed as

$$E(\theta, \boldsymbol{H}) = K \sin^2 \theta - \mu_0 \boldsymbol{m}_s \bullet \boldsymbol{H} \tag{7.30}$$

where K is the domain crystal anisotropy constant and θ is the angle between $\boldsymbol{m}_s$ and the easy axis. The positions of minimum energy can be found by solving the equation

$$\frac{\partial E(\theta, \boldsymbol{H})}{\partial \theta} = 0 \tag{7.31}$$

for $\partial^2 E(\theta, \boldsymbol{H})/\partial \theta^2 > 0$. It can be shown that there are two energy minima if a small field is applied. As the field strength increases, the positions of these minima change. Initially, these changes are reversible. When the applied field strength exceeds a certain critical value H_c, however, one of the energy minima becomes unstable, and the domain magnetization jumps to the other minimum, which is the global energy minimum. This critical point at which the irreversible domain rotation occurs is the point of minimum energy for which

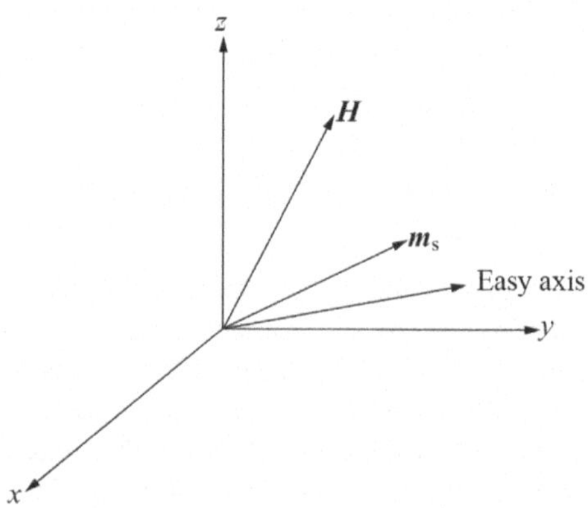

Fig. 7.37 Relationship between magnetic moment $\boldsymbol{m}_s$ and magnetic field strength $\boldsymbol{H}$

Fig. 7.38 Asteroid boundary

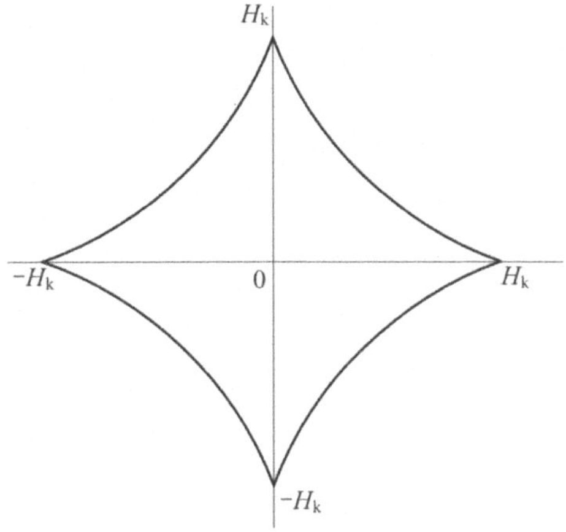

$$\frac{\partial^2 E(\theta, \boldsymbol{H})}{\partial \theta^2} = 0 \tag{7.32}$$

Figure 7.38 plots the rotation of a single S-W particle, where the asteroid boundary,

$$H_e^{2/3} + H_p^{2/3} = H_k^{2/3} \tag{7.33}$$

which separates the reversible and irreversible domain, is obtained by solving Eqs. (7.31) and (7.32) simultaneously. H_e and H_p are the components of $\boldsymbol{H}$ on the easy axis of the particle and on the axis perpendicular to the easy axis, respectively, and $H_k = 2K/(\mu_0 \boldsymbol{m}_s)$

It can be shown that the equilibrium position of $\boldsymbol{m}_s$ is on one of the lines tangent to the asteroid and passing through the tip of H. When H is outside the asteroid, two such tangent lines can be drawn, and the equilibrium magnetization m_1 is parallel to the one making a smaller angle θ_1 with the easy axis, as shown by m_1^+ in Fig. 7.39. When H is inside the asteroid, four tangent lines can be drawn, and two possible equilibrium magnetic moments are parallel to the two lines making smaller angles θ_2 and θ_3 with the easy axis, as shown by m_2 and m_3 in Fig. 7.40. Such a response of a single S-W particle can be viewed as a vector elemental hysteresis operator. Similarly, if the tip of the magnetic field H lies in any of the other quadrants, the same consequences should also be obtained.

An irreversible jump of the magnetization from one direction to another may occur, when H exceeds and crosses the asteroid boundary. However, the direction of the magnetic moment depends on the previous history of magnetization as well. This is known as the asteroidal rule.

The magnetization in a bulk material is the vector sum of the contributions of all of constituent domains, that is

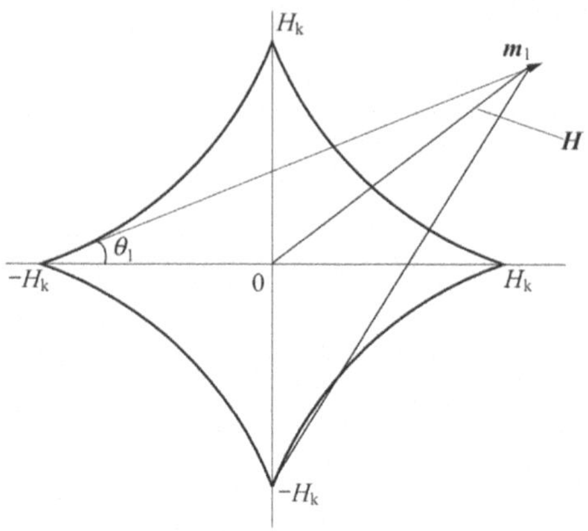

Fig. 7.39 Magnetization angle θ of $\boldsymbol{H}$ (outside the asteroid)

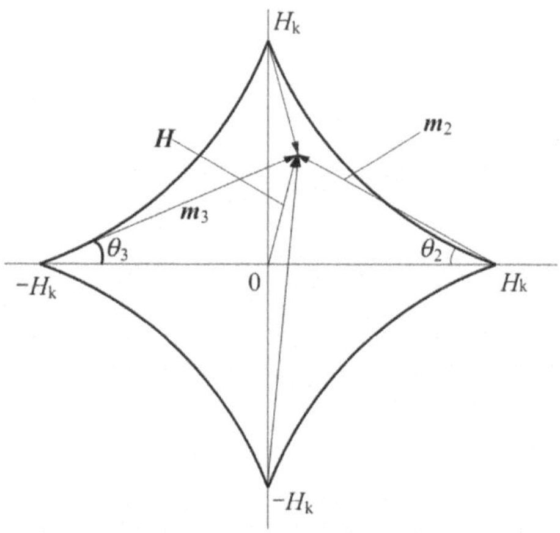

Fig. 7.40 Magnetization angle θ_1 and θ_2 of $\boldsymbol{H}$ (inside the asteroid)

$$\boldsymbol{M} = \frac{1}{V}\int_0^{2\pi}\int_0^{\pi} \boldsymbol{m}_\mathbf{s}(\xi, \psi, \boldsymbol{H})\rho(\xi, \psi)\sin(\psi)\mathrm{d}\psi\mathrm{d}\xi \tag{7.34}$$

where

$$\int_0^{2\pi}\int_0^{\pi} \rho(\xi, \psi)\sin(\psi)\mathrm{d}\psi\mathrm{d}\xi = 1 \tag{7.35}$$

V is a sample volume and $\rho(\xi, \psi)$ is the distribution of the S-W particles in the spherical coordinates (ξ, ψ).

In the assumptions of the S-W model, the interaction between S-W particles and the pinning effects of domain walls are ignored. In real magnetic materials, however, these effects are important and should not be neglected. To account for the interaction between domains, the S-W model is modified by adding a mean-field term [17], $\boldsymbol{H}_{en} = \boldsymbol{H} + k\boldsymbol{M}$, where k is a constant feedback coefficient. This modifies the energy of an arbitrary particle, and Eq. (7.34) becomes

$$\boldsymbol{M} = \frac{1}{V}\int_0^{2\pi}\int_0^{\pi} \boldsymbol{m}_s(\xi, \psi, \boldsymbol{H} + k\boldsymbol{M})\rho(\xi, \psi)\sin(\psi)\mathrm{d}\psi\mathrm{d}\xi \tag{7.36}$$

This macroscopic mean-field interaction is qualitatively correct, but requires further adjustments, in particular, to the easy axes distribution.

7.4.6 Effect of Hysteresis on Flux Distribution of Single-Phase Transformer

Magnetic characteristics of single-phase transformer core are analyzed. In this problem, hysteresis characteristics should be taken into account in order to analyze the flux waveforms accurately.

Figure 7.41 shows a quarter of the analyzed single-phase, two-limbed core [23]. The core is made of 0.3-mm thick highly oriented silicon steel (JIS: 30P110). The hysteresis loops are shown in Fig. 7.22. The core is excited by a sinusoidal voltage, and the overall flux density in the limb is 1.7 T. Figure 7.42 shows the calculated flux distributions. Zero time is taken to be the instant when the flux density in the limb is at a maximum. Figure 7.43 shows the waveforms of the localized flux densities at the center of the limb. *a-b* corresponds to the position *a-b* in Fig. 7.41. As the waveforms are not so much distorted, the calculated results of such two-limbed core are fairly in good agreement with the measured ones as shown in Fig. 7.43.

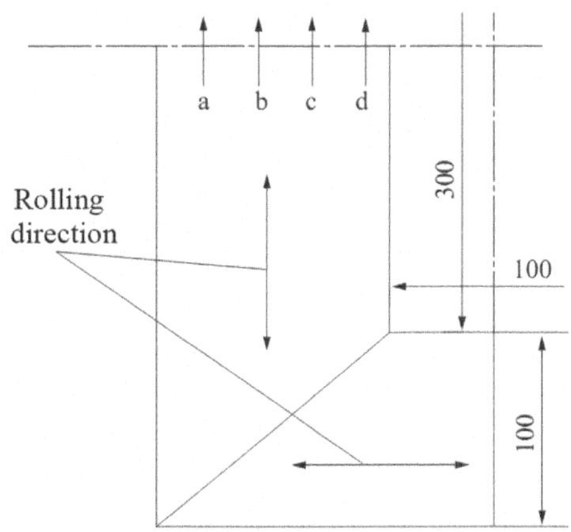

Fig. 7.41 1/4 single phase

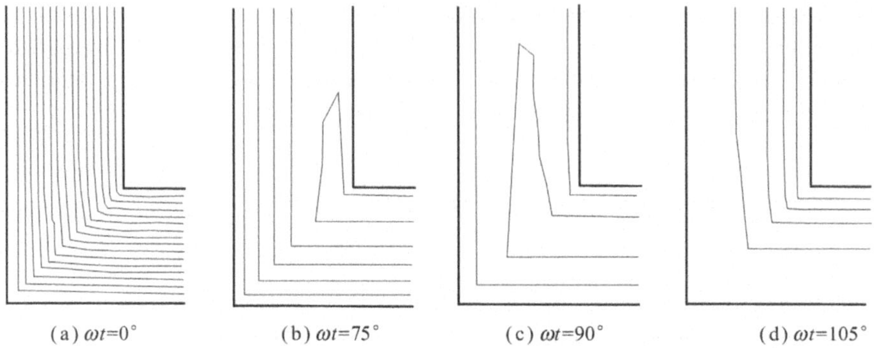

Fig. 7.42 Calculated flux distributions

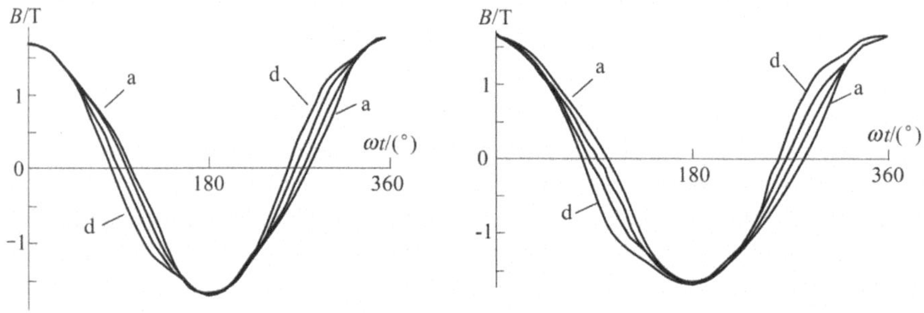

Fig. 7.43 Waveforms of the localized flux densities

7.5 Estimation of Iron Loss

7.5.1 Iron Loss Under Alternating Flux

1. Sinusoidal Flux

Figure 7.44 shows an example of iron loss curve (JIS: 50A1300). The iron loss per weight W (W/kg) under the sinusoidal alternating excitation can be expressed approximately as follows [26]:

$$W = K_e f^2 B_{max}^2 + K_h f B_{max}^2 \tag{7.37}$$

where f is the frequency, K_e, K_h are the eddy current and hysteresis loss coefficients, and B_{max} is the maximum flux density during the period of AC excitation. By approximating the W/f curve in Fig. 7.45 by a straight line, K_h is obtained as the intersection point of ordinate ($K_h = 6.012\times10^{-2}$), and K_e is obtained as the slope of the line ($K_e = 2.923 \times 10^{-4}$). The first and second terms in the right-hand side correspond to the eddy current loss and the hysteresis loss, respectively.

Fig. 7.44 Iron loss curve (JIS: 50A1300)

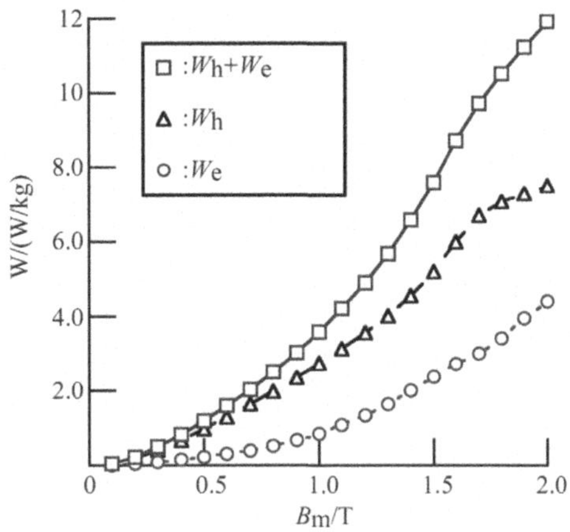

Fig. 7.45 *W/f-f* curve (B = 1 T, measured, 50A1300)

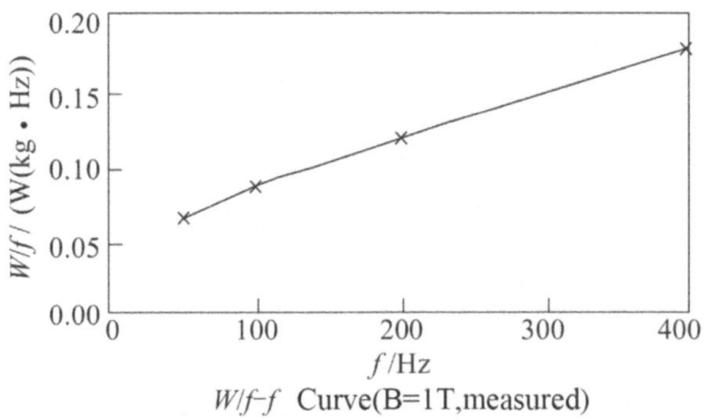

In Eq. (7.37), it is assumed that the eddy current loss and hysteresis loss are proportional to f^2 and f, respectively.

2. Distorted Flux

Recently, various kinds of magnetic devices, such as motors, actuators and transformers, are excited by inverters. The iron losses of such devices excited by a distorted voltage (PWM, etc.) are increased compared with the case of sinusoidal voltage excitation. Several estimation methods of iron loss under the distorted flux excitation are proposed. The method using the effective flux density B_{eff} is discussed here.

The iron loss W can be separated to the hysteresis loss W_h and the eddy current loss W_e as shown in Fig. 7.44. Figure 7.46 shows the examples of distorted waveforms and hysteresis loops. The amplitudes of the third harmonic component are the same. Figure 7.46a shows the case when the phase of the third harmonic component is opposite. There is no minor loop. In the case of Fig. 7.46b, the phase of the third harmonic component is the same, and there occur minor loops.

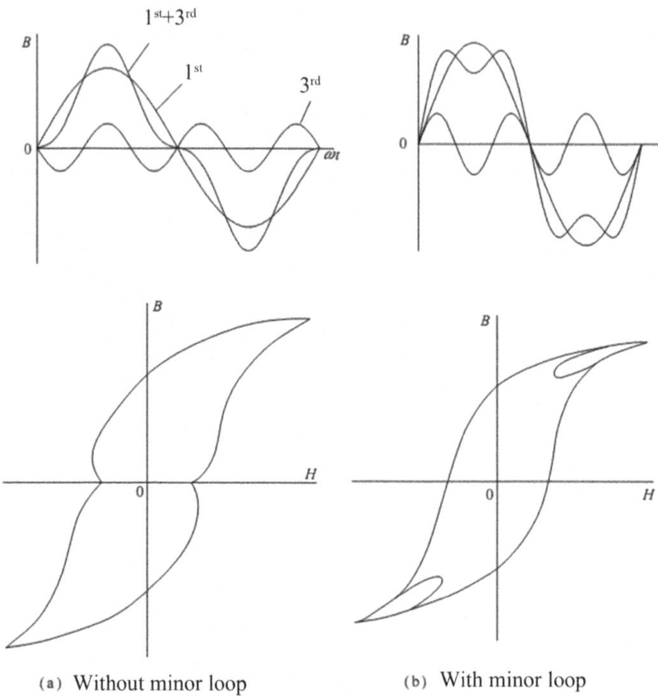

Fig. 7.46 Distorted waveforms and hysteresis loops

In the case of Fig. 7.46a without minor loops, the iron loss can be estimated by [27]

$$W = W_{\mathrm{h}}(B_{\mathrm{max}}) + W_{\mathrm{e}}(B_{\mathrm{eff}}) \tag{7.38}$$

where $W_{\mathrm{h}}(B_{\mathrm{max}})$ denotes that the hysteresis loss is the function of the maximum flux density B_{max} and this curve is shown in Fig. 7.44. B_{max} of Fig. 7.46a is different from that of Fig. 7.46b. $W_{\mathrm{e}}(B_{\mathrm{eff}})$ denotes that the eddy current loss is the function of the effective flux density B_{eff}. B_{eff} corresponds to the maximum flux density which has an equivalent effective voltage equal to the effective voltage of distorted flux waveform, and is given by

$$B_{\mathrm{eff}} = \sqrt{\Sigma(nB_n)^2} = \sqrt{B_1^2 + 4B_2^2 + 9B_3^2 + \cdots} \tag{7.39}$$

Equation (7.39) means that even if the amplitude of harmonic component is the same, the iron loss changes if the phase of harmonic component is different.

In the case of Fig. 7.46b with minor loop, the iron loss can be estimated by [27]

$$W = W_h(B_{max}) + W_e(B_{eff}) + 2\sum_k \eta W_h(B_k) \tag{7.40}$$

The third term of Eq. (7.40) corresponds to the minor loop hysteresis loss. As there are two minor loops when the distorted flux waveform is symmetric as shown in Fig. 7.46b, two is multiplied to this term. The minor loop hysteresis loss is the function of the maximum flux density B_k of minor loop k, and the loss is changed by the position of the minor loop. η is the displacement coefficient which represents the rate of increase of iron loss due to the minor loop which is the function of the position of it.

Equation (7.40) can be applicable when the distorted waveform contains harmonic components. For the estimation of iron loss under the distorted waveform with high harmonic components, the following formula is proposed [28]:

$$W = \sum_n \{nW_h(B_n) + n^2 W_e(B_n)\} \tag{7.41}$$

where B_n is the maximum flux density of the nth harmonics.

7.5.2 *Iron Loss Under Rotating Flux*

There are various kinds of estimation methods of iron loss. In this section, two kinds of methods are explained.

(1) Method I

In method I, the iron loss under rotating flux is assumed as the summation of iron losses under alternating fluxes in the x- and y-directions [29]. The iron loss under alternating flux is calculated by Eq. (7.37).

(2) Method II

In method II, the iron loss under distorted elliptical rotating flux is estimated by using the iron losses measured under elliptical rotating fluxes having various axis ratios [29].

The iron loss W under a distorted rotating flux is represented as a summation of the loss W_0 under an elliptical rotating flux of fundamental component shown in Fig. 7.47 and the iron loss W_n under alternating flux of harmonic components shown in Fig. 7.48 as follows:

$$\begin{aligned} W &= W_0 + W_n \\ &= W_0(\alpha, B_{max}) + \sum_{n=2,3\ldots} W_n(B_n) \end{aligned} \tag{7.42}$$

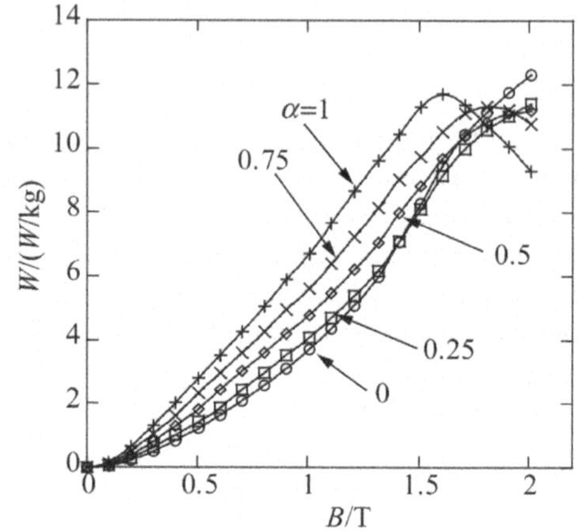

Fig. 7.47 Iron loss *W* under a rotating flux

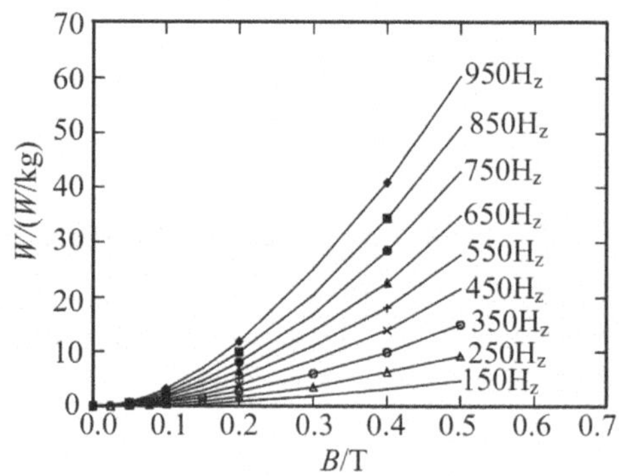

Fig. 7.48 *B*–*W* curve of alternating flux at different frequencies

where $W_0(\alpha, B_{max})$ denotes that the iron loss under rotating flux is the function of axis ratio α and the maximum flux density (major axis) B_{max}. $W_n(B_n)$ is the iron loss due to the alternating flux as a function of harmonic components B_n.

The calculation is performed by the following procedure: Firstly, the fundamental components in the *r*- and θ-components are obtained by a harmonic analysis. Then, the iron loss W_0 under the rotating flux of fundamental component at any axis ratios α and flux density B_{max} can be obtained by the interpolation of *B*–*W* curve shown in Fig. 7.47. The iron loss due to harmonic component is obtained as a summation of the iron loss at each harmonics using the *B*–*W* curves at various frequencies. Figure 7.49 shows the comparison of estimated and measured iron losses under the rotating flux composed of ellipse ($\alpha = 0.25$) and 10% third harmonic field [26]. The excitation field for 2-D-SST in Fig. 7.50 is shown in Fig. 7.49a. The figure suggests that the result of method II is superior to that of method I, when $B > 1.0$ T.

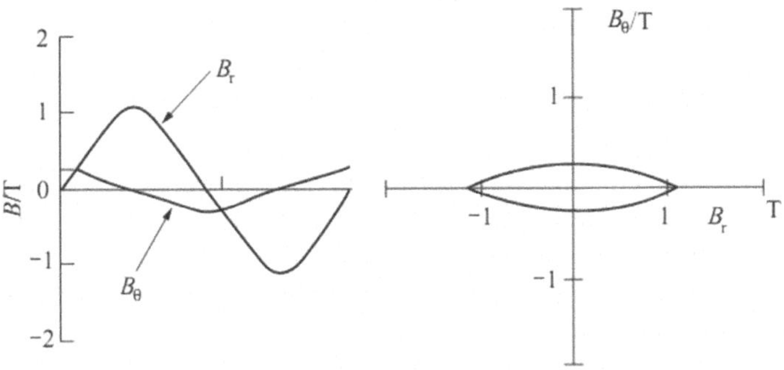

(a) Waveforms of excitation fields

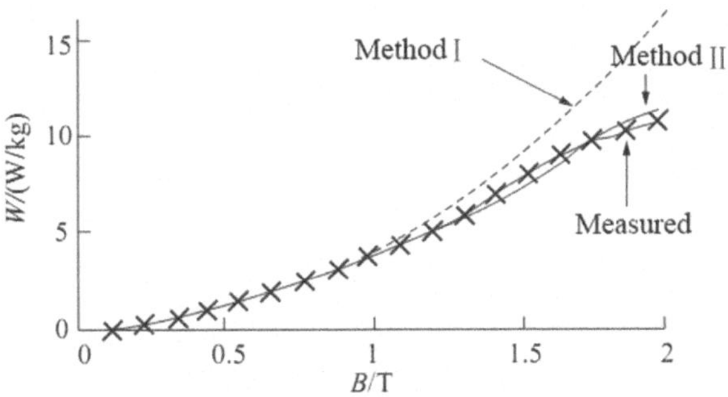

(b) Comparison of estimated and measured iron losses

Fig. 7.49 Comparison of estimated curve and measured curve

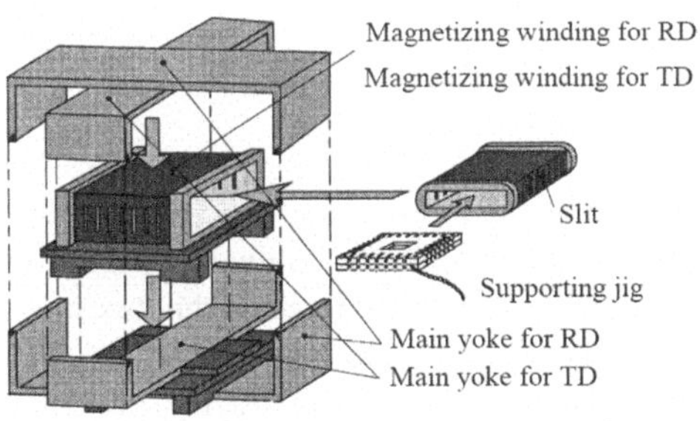

(a) Schematic structure of double-excitation type of 2-D SST(DET)

(b) Real product of 2-D SST

Fig. 7.50 Double-excitation type of 2-D-SST (DET)

7.6 Modeling of Laminated Core

7.6.1 Laminated Core and Various Modeling Methods

The laminated iron core is commonly used in transformers, motors, etc. For example, in an air-gapped iron core-type power reactor, there occurs some fringing flux at the air gap portion of the core block [30]. The eddy current is induced in the silicon steel sheets due to the fringing flux, and this causes a local overheating or burning insulating materials in some extreme cases. In order to avoid the local heating and to design a reactor with high efficiency, the exact and quick analysis of flux and eddy current distributions is necessary. If all silicon steel sheets in the core are subdivided into a fine mesh so that accurate results can be obtained, the number of elements becomes huge, and the calculation is impractical.

Several techniques for modeling laminated cores are proposed. Some of them are for linear case, and an analytical value of magnetic field is used [31, 32]. A technique for modeling the laminated core as a non-laminated bulk material with anisotropic conductivity and the method using an effective permeability are proposed in order to take into account the nonlinearity of *B*–*H* curve of laminated sheet, and only the total loss is examined [33, 34].

In this section, two kinds of modeling methods of laminated core are shown. One is the so-called homogenization method. In this method, the laminated core is treated as a bulk core having anisotropic conductivity, and the equivalent permeability and conductivity of the laminated core are utilized [35]. The other is the "two-zone method." In this method, the analyzed region is classified into two parts: One part near the surface of laminated core is subdivided into a fine mesh, and the inner part is modeled by a bulk core having anisotropic conductivity [36].

7.6.2 Homogenization Method

In this method, the laminated core is homogenized, and the equivalent values of reluctivity and conductivity are used. This method is applicable when the flux flows mainly in parallel to the lamination and the fringing flux is not remarkable. Figure 7.51a shows a periodic part of laminated core, and Fig. 7.51b shows the equivalent circuit. W_0 and W_s are the air gap length and the thickness of sheet, respectively. L is the length of sheet. B_y using the homogenized (equivalent) reluctivity $N_{\|}$ parallel to the lamination, the following equation is obtained from Fig. 7.51b [35]:

$$\frac{W_s + W}{N_{\|} L} = \frac{W_s}{\nu_s L} + \frac{W_0}{\nu_0 L} \tag{7.43}$$

where ν_0 and ν_s are reluctivities of air and lamination steel, respectively.

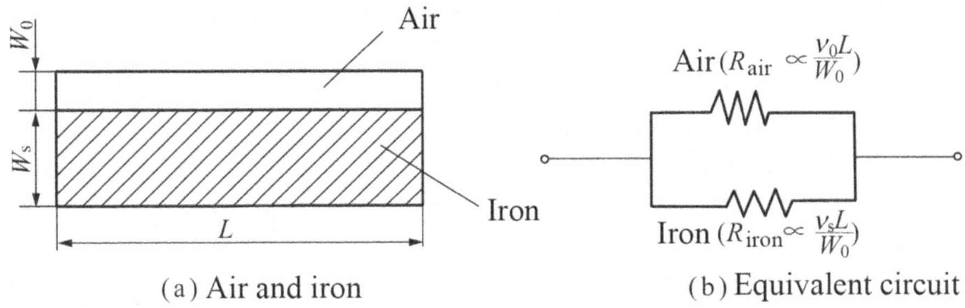

Fig. 7.51 Laminated core model

By introducing the space factor $\alpha(=W_s/L)$, Eq. (7.43) can be written by

$$\frac{1}{N_{\parallel}} = \frac{\alpha}{\nu_s} + \frac{1-\alpha}{\nu_0} \tag{7.44}$$

The homogenized reluctivity $N_{\perp}$ perpendicular to the lamination can be easily derived as follows:

$$N_{\perp} = \alpha\nu_s + (1-\alpha)\nu_0 \tag{7.45}$$

If the thickness direction of laminated core is parallel to the z-direction, the amplitude B_s of flux density in the iron sheet is given by

$$B_s = \sqrt{\left(\frac{N_{\parallel}}{\nu_s}\right)^2 \left(B_x^2 + B_y^2\right) + B_z^2} \tag{7.46}$$

As the eddy current does not flow across sheets, the homogenized conductivity $\Sigma_{\parallel}$ parallel to the lamination and that $\Sigma_{\perp}$ perpendicular to the lamination are assumed as follows:

$$\Sigma_{\parallel} = \alpha\sigma, \quad \Sigma_{\perp} = 0 \tag{7.47}$$

where σ is the conductivity of iron steel.

7.6.3 Two-Zone Method

1. Outline of the Method

In this method, only several sheets near the surface of laminated core are subdivided into a fine mesh, and the inner part is modeled by a bulk core [36]. This method is applicable when the eddy current flows on the surface of lamination due to the

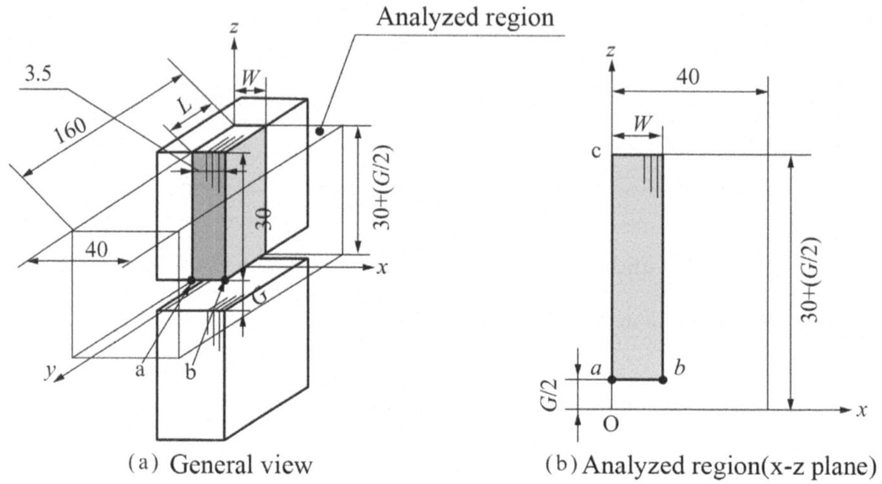

Fig. 7.52 Simple model

fringing flux. As it can be assumed that the fringing flux does not penetrate into the central part of the laminated core, it may not be necessary to make a fine mesh near the central part. Then, the central part is treated as a bulk of rough mesh with anisotropic conductivity shown in Eq. (7.47).

2. Examination of the Number of Sheets to be Divided into Fine Mesh in Two-Zone Method

Figure 7.52 shows a simple model of a core block. The core block is laminated in the x-direction. The lamination is composed of 20 sheets. The core is made of grain-oriented silicon steel 35G165 (thickness: 0.35 mm, coating: 0.005 mm, iron loss: 1.65 W/kg at 1.5 T, 50 Hz). The conductivity of silicon steel is assumed to be 2.08×10^6 S/m. The average flux density of core is 1.0 T, and the frequency f is 60 Hz. In order to impress the flux in the z-direction, the Dirichlet boundary condition is imposed on the outer surfaces ($x = 40$ mm and $y = 160$ mm) of analyzed region shown in Fig. 7.52b. One-eighth region (gray region in Fig. 7.52a) is analyzed using the 3-D edge-based hexahedral edge element ($\boldsymbol{A} - \phi$ method). Silicon steel is assumed to be isotropic, and only the B–H curve in the rolling direction is used.

The skin depth δ is equal to 0.202 mm (at relative permeability $\mu_s = 50{,}000$, $f = 60$ Hz). It is required to subdivide three-layer mesh within the skin depth region in order to obtain an accurate result. This means that each silicon steel sheet (0.35-mm thickness) near the surface of laminated core should be subdivided into six layers.

It is sufficient that only the surface sheets are divided into a fine mesh, and the central region is treated as a bulk core having anisotropic conductivity [6] in order to examine the overheating. In order to examine how to subdivide the laminated

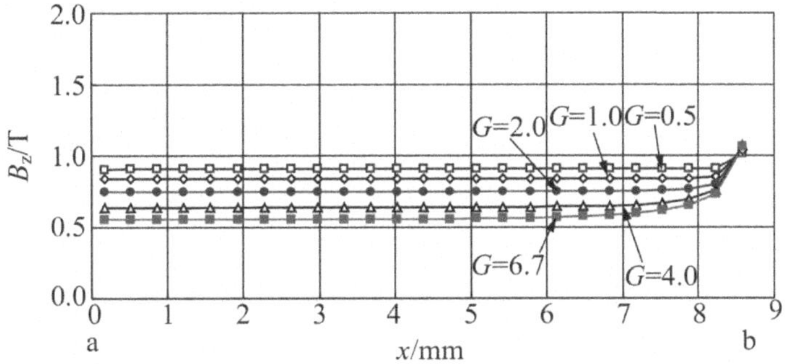

Fig. 7.53 Flux distributions along the x-axis

cores, the flux distribution in the center part (z = 30 + G/2 mm) of core and the part (z = G/2 mm) facing the gap is examined. Figure 7.53 shows the flux distributions along the x-axis. The ordinate is the average value B_z of the z-component of the flux density in each sheet. The figure suggests that the flux in the laminated core is concentrated in the surface of core near the center (z = 30 + G/2 mm) of core and near the gap (z = G/2 mm). The concentration of flux is remarkable, when the gap length G is large. The field gradient (dB_z/dx) is not so changed by the width W of core.

Figure 7.54 shows the distribution of eddy current loss. This is the average value of eddy current loss in each sheet. The eddy current loss in the laminated core is concentrated in several sheets from the surface of core due to the fringing flux.

Figure 7.55 shows the error ε of eddy current loss. The error ε is defined by

$$\varepsilon = \frac{W_n - W_o}{W_o} \times 100(\%) \tag{7.48}$$

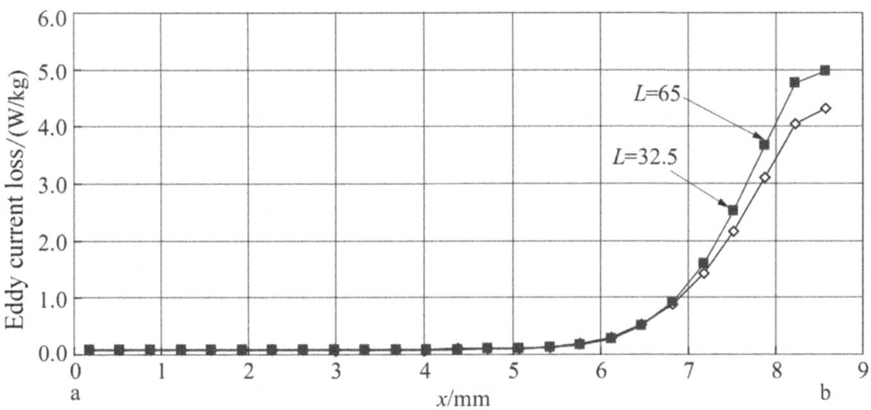

Fig. 7.54 Distribution of eddy current loss

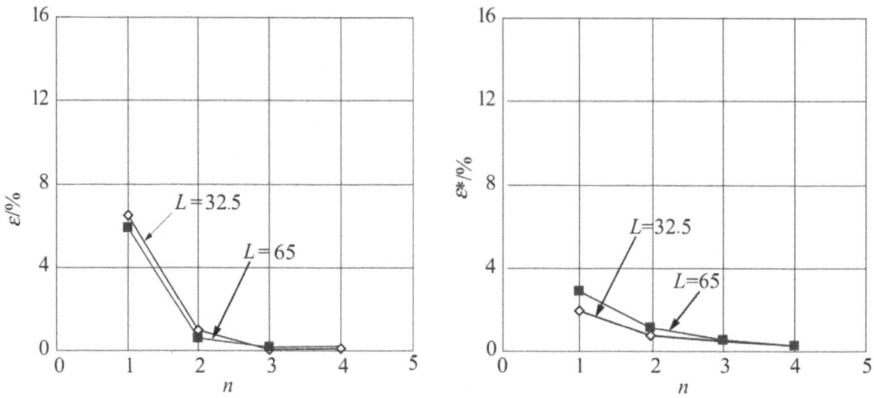

Fig. 7.55 Error ε of eddy current loss (*W*=8.75mm, *G*=4.0mm)

where W_o is the eddy current loss in a surface sheet when all sheets are subdivided into a fine mesh (six layers in each sheet). W_n is that when only n sheets in the surface side are subdivided into a fine mesh (each surface sheet is subdivided into six layers). For example, $n = 2$ means that only two sheets near the surface are subdivided into a fine mesh. The figure suggests that ε becomes almost zero at $n = 4$. Therefore, it can be concluded that it may be sufficient to subdivide only four sheets in the surface side into a fine mesh.

Figure 7.56 shows the effect of *n* on the CPU time and the number of unknowns. The computer used for calculations is CPU Pentium 4 (3.2 GHz) and RAM 2 GB (6.4 MB/s). When all sheets are subdivided into a fine mesh, the CPU time and the number of unknown increase greatly. When four sheets in the surface side are subdivided into a fine mesh, the increase in CPU time and the number of unknowns

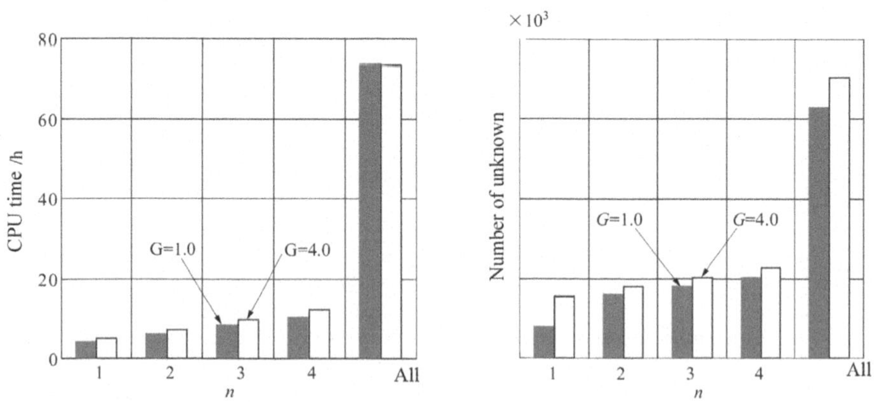

Fig. 7.56 Effect of *n* on the CPU time and the number of unknowns

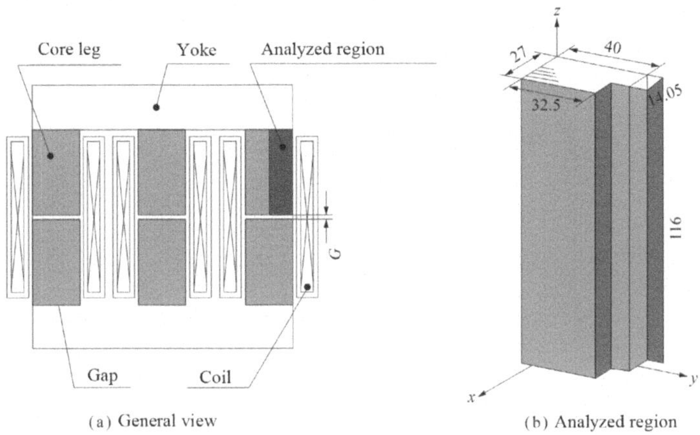

Fig. 7.57 Reactor model

is not remarkable. Therefore, the proposed modeling is very effective in an actual model composing many sheets in terms of CPU time and the number of unknowns.

3. Application to Practical Model

The validity of the proposed modeling method is examined by comparing the calculated results and measurements of local loss on the surface of core. Figure 7.57 shows the model reactor [36] used for experiment and analysis. Figure 7.57b shows the analyzed region (1/8 of the whole region). It is a serial reactor for capacitor facilities for three phases, 60 Hz and 100 kVAR. The core is made of grain-oriented silicon steel 35G165. Since the rating capacity is small, there is just one air gap at the center of the core leg. The air gap length G between core blocks is equal to 6.7 mm. The average flux density of the core block is assumed to be 1.10 T.

A thermocouple method is employed to measure the local loss on the surface of the core. Thermocouples are installed at five points at intervals of 1 mm along the z-axis from the gap (x = 27 mm, y = 0 mm).

Figure 7.58 shows the comparison of calculated and measured results of local loss on the surface sheet along the z-axis. In the result of the proposed modeling, four sheets in the surface side are subdivided into a fine mesh. In the large fine mesh modeling, the core block is classified into three areas with different mesh densities in terms of the amount of fringing flux penetration [30]. About 1/10 region from the surface of core block is subdivided with a fine mesh. Each sheet in this region is subdivided into six layers. Each sheet in the region between about 1/10 and about 1/2 from the surface is subdivided into two layers. The central area inside about 1/2

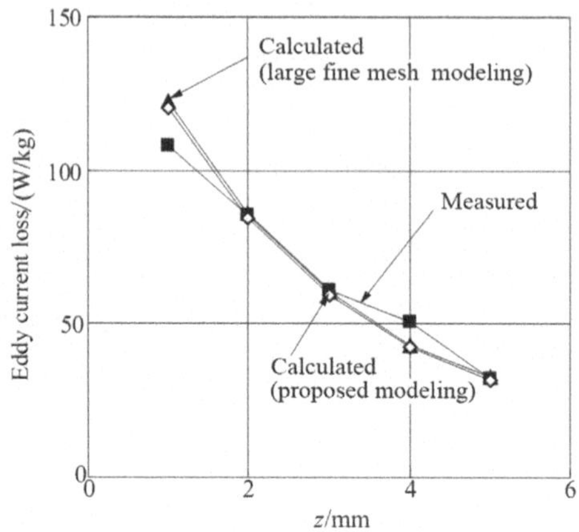

Fig. 7.58 Comparison of calculated and measured results of eddy current loss on the surface ($x = 27$ mm, $y = 0$ mm)

Table 7.2 Comparison between proposed modeling and large fine mesh modeling

	Proposed modeling	Large fine mesh modeling
Maximum loss on the surface (W/kg)	215	211
Maximum loss per one sheet of surface of laminated core (W/kg)	7.83	7.91
Number of elements	40,664	151,164
CPU time (h)	45.5	400.6

Computer used: Intel Pentium IV 3.2 GHz, PC3200 memory: 2.0 GB

region is treated as a bulk assuming that the x-component of conductivity is equal to zero. The calculated eddy currents are changed in the thickness direction (x-axis) of sheet. But, the average value of them is adopted, because the sheet is thin (=0.35 mm) and the heat produced by eddy currents is not negligible. Although the change of eddy current loss along the z-axis is remarkable, the calculated results of the proposed modeling agree with the measured ones as shown in Fig. 7.58. Table 7.2 shows the comparison between the proposed modeling and the large fine mesh modeling. The table denotes the effectiveness of the proposed modeling. The computation cost of the proposed modeling is much smaller than that of the large fine mesh modeling, and the accuracy of both the proposed modeling and the large fine mesh modeling is almost the same.

7.7 Factors Affecting Magnetic Properties of Electrical Steel

Magnetic properties of electrical machines, such as transformers and motors, are affected by the distortion due to the cutting and the compression caused by bolt bundle, shrink fitting, etc. [37–42]. The iron loss is increased under rotating flux condition [29, 43] and DC bias condition [44]. Also, the magnetic property is affected when the temperature of electrical steel is changed [45]. If the magnetic circuit design of electrical machine is performed by the numerical analysis using the catalog data of steelmaker, accuracy may suffer, because the actual magnetic properties of the core are changed due to cutting, compressive stress, DC bias, etc. Therefore, such property of magnetic characteristics should be taken into account in the design of electrical machine. In this section, the magnetic properties under residual stress by cutting and compressive stress due to shrink fitting are shown. The iron losses under rotating flux excitation and DC bias excitation are also illustrated.

7.7.1 Residual Stress by Cutting

The effect of cutting on the magnetic properties is examined [46]. If the specimen is set in parallel as shown in Fig. 7.59, the flux is deviated in the specimen, and the obtained result is different from the actual one. Then, the correlation between the cutting distortion and the magnetic properties is measured using only one piece of sheet in an SST [47].

Figures 7.60 and 7.61 show the measured *B–H* curves and *B–W* curves. Figure 7.62 shows the measured hysteresis loops of the specimens of 5- and 30-mm width. These figures denote that the necessary H_b of the deteriorated specimen to produce the desired B_m increases when the permeability is high (B_m = 1.4 T), and the iron loss increases when B_m is low (B_m = 1.4 T). The rate of increases of H_b and iron loss is significant, when the width of specimen is narrow. The rate of increase of H_b is larger than that of iron loss. The increase of iron loss is related to the area of hysteresis loop, as shown in Fig. 7.62.

In this case, *H* is increased about two times and iron loss is increased about 1.2 times by cutting. The rate of the deterioration is less than 10% around 2.0 T. As the deterioration of 50A1300, of which the iron loss is originally large, is not

Fig. 7.59 Magnetic flux offset

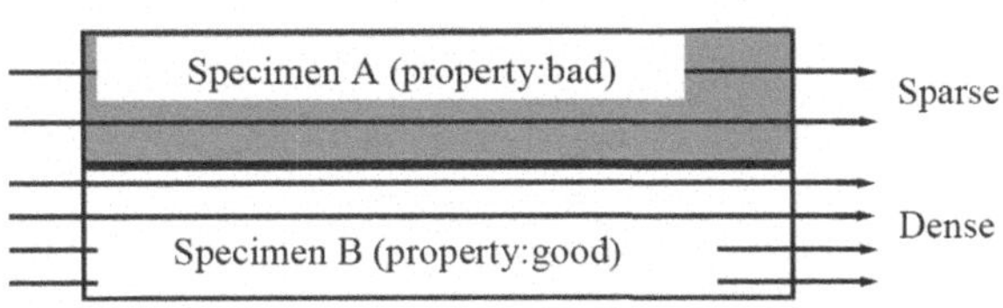

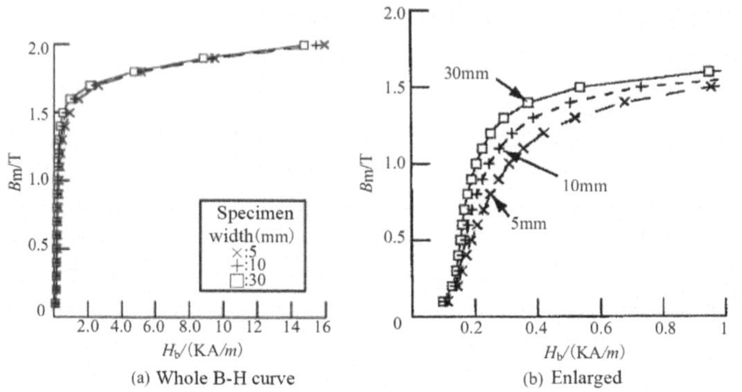

Fig. 7.60 Effect of cutting on *B–H* curve

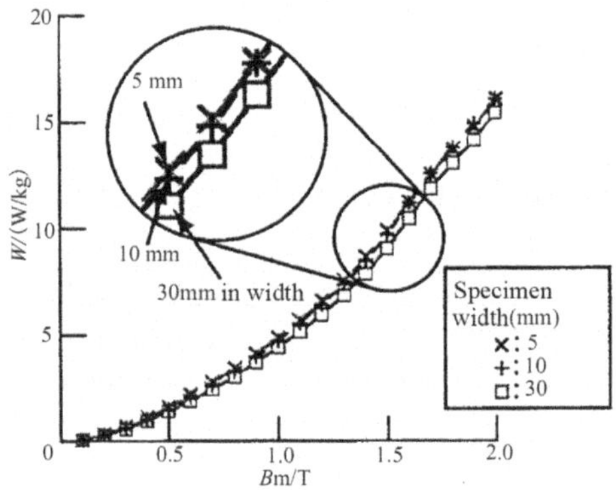

Fig. 7.61 Effect of cutting on B_m–*W* curve

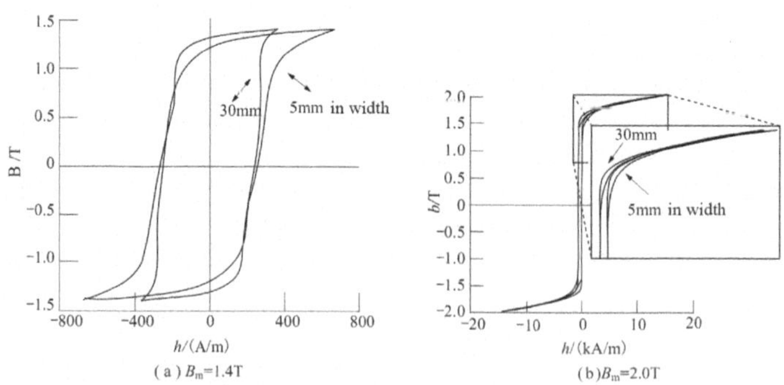

Fig. 7.62 Effect of cutting on hysteresis loop

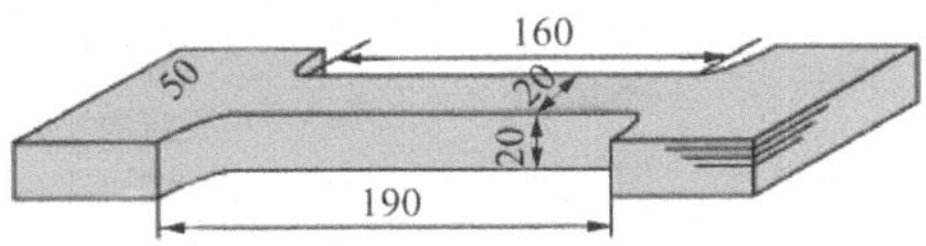

Fig. 7.63 Measured under compressive stress

remarkable around 2.0 T, the property of a real machine may not be different from that designed using a catalog data. On the other hand, in the case of the electrical steel sheet of which the iron loss is originally small, the increase of iron loss by cutting is significant.

7.7.2 *Compressive Stress*

The magnetic properties under compressive stress in the longitudinal direction are measured using the laminated specimen in order to avoid the buckling [46]. Figure 7.63 shows the specimen (thickness: 20 mm) of which 57 non-oriented electrical steel sheets (grade: JIS 35A360, 0.35 mm thick, 3.6 W/kg at 1.5 T, 50 Hz) are laminated.

Figure 7.64 shows the measurement apparatus. The yokes are set on both sides of laminated specimen. The B coil (search coil, three turns) is wound around the specimen. The one *H*-coil (area turns: 13.0924×10^{-3} m^2) is set on the surface of the laminated specimen. The compressive stress is applied by a hydraulic equipment (maximum pressure: 7 tons). In order to apply the stress to the specimen uniformly, the output of the strain gauges on the surface of the specimen is monitored. The frequency is 50 Hz. The magnetic measurements are carried out under the sinusoidal flux condition.

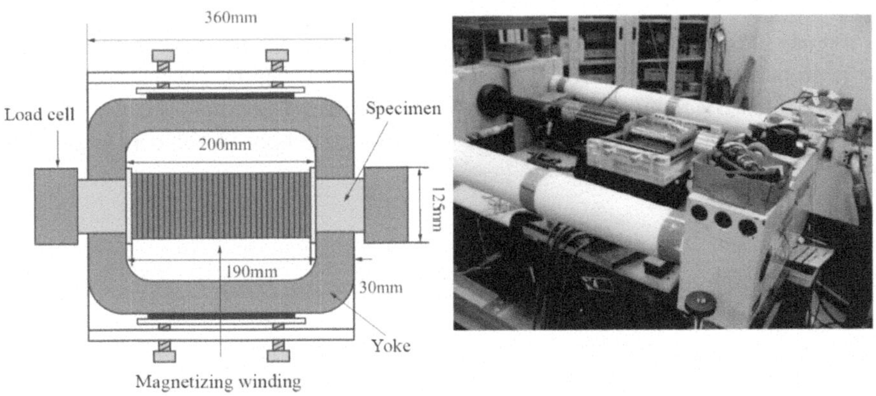

(a) Schematic (b) Real product

Fig. 7.64 Magnetic properties measuring system

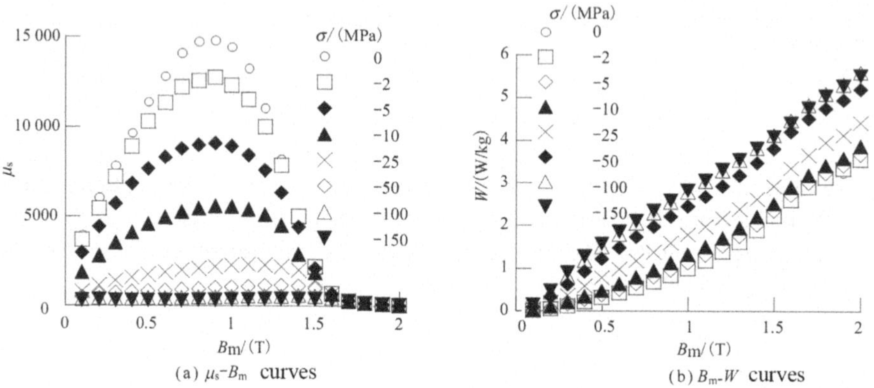

Fig. 7.65 Magnetic properties along longitudinal axis under compressive stress

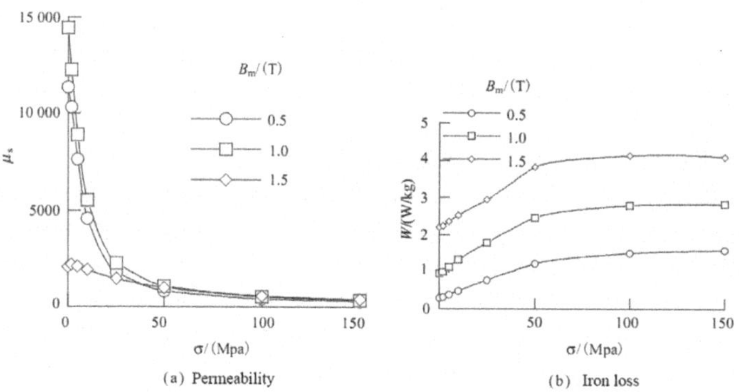

Fig. 7.66 Effect of the compressive stress on the magnetic properties along longitudinal axis

Figures 7.65 and 7.66 show the effect of the compressive stress σ on the magnetic properties. The permeability is rapidly decreased even by small stress, and the rate of decrease is reduced when the stress is increased. The iron loss is increased by the stress, but it almost does not increase when the amplitude of σ is larger than about 50 MPa.

7.7.3 Effect of Press and Shrink Fitting on Iron Loss of Motor Core

The effect of press and shrink fitting on iron loss of motor core is examined systematically [39, 46]. Figure 7.67 shows a six-pole surface permanent magnet (SPM) motor model. In this model, the dashed line position is also pressed, and

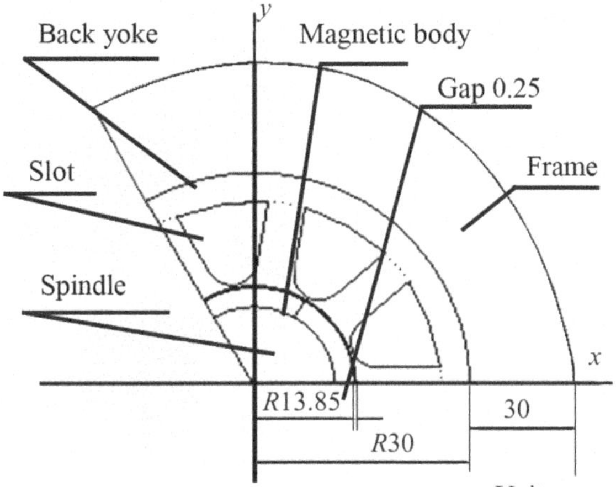

Fig. 7.67 Six-pole SPM motor model

then the stator back core and the teeth part can be separated, so that we can easily insert coils. Adjacent teeth are connected to each other at the tooth tip. The stator core is made of non-oriented electrical steel (grade: 35A360). The rotor is made of carbon steel (S45C). The residual magnetism of permanent magnet of rotor is 1.25 T (radial orientation). Three kinds of motor cores of different manufacturing processes (press and wire cut) are produced. Two kinds of shrink fittings (10.7 and 29.4 MPa) are investigated. Table 7.3 shows the investigated motor models.

The torque of the SPM motor model is measured by a torque meter when the permanent magnet rotor is driven by another motor as shown in Fig. 7.68. The iron loss of the stator is obtained by subtracting the torque (corresponding to the mechanical loss) measured by rotating a rotor having a non-magnetized permanent magnet from the torque when the permanent magnet rotor is driven by another motor.

Figure 7.69 shows the comparison of iron losses under various conditions with or without press and shrink fitting. The iron loss calculated by the finite element method (FEM) using the material properties (*B*–*H* and iron loss curves under no stress condition is also shown (model E)). The reason why the iron loss by FEM

Table 7.3 Motor models

Model	Anneal	Compressive force by shrink fitting (MPa)
A (wire electric discharge machine)	Yes	–
B (press)	No	–
C-1 (press)	No	10.7
C-2 (press)	No	29.4
D-1 (wire electric discharge machine)	Yes	10.7
D-2 (wire electric discharge machine)	Yes	29.4
E (numerical analysis using FEM)	–	–

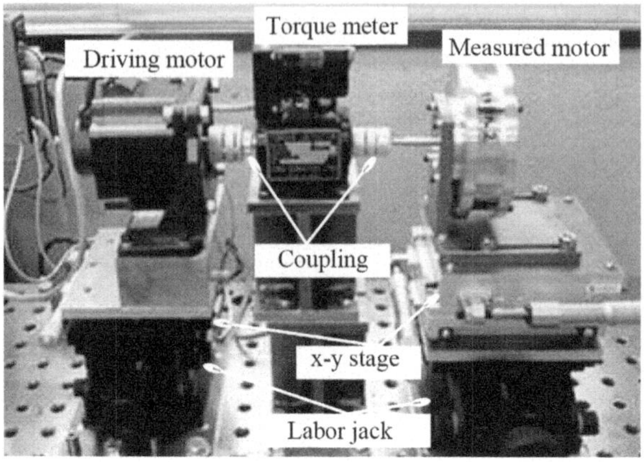

Fig. 7.68 Torque meter

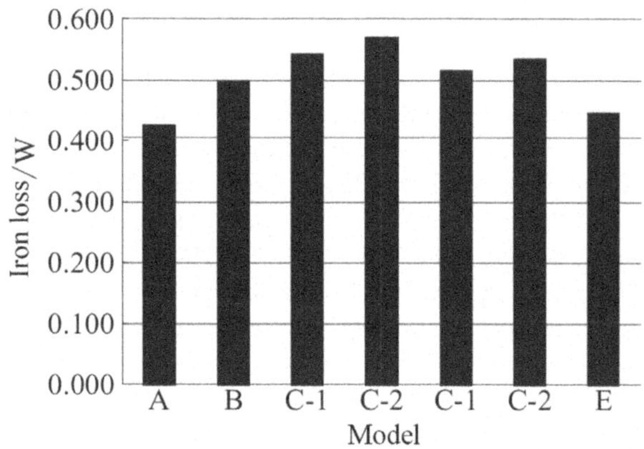

Fig. 7.69 Iron losses under various conditions (1000 min^{-1})

(model E) is larger than that of motor core with wire cut (model A) may be due to the fact that the actual magnetization of permanent magnet is different from 1.25 T which is used in the FEM analysis. The iron loss of core with press (model B) is about 16% larger than model A due to the residual stress by press.

The rate of increase (20%) due to the shrink fitting of wire cut model (model D-1) is larger than the increase (9%) of press model (model B). This is because the rate of increase of iron loss due to the compressive stress is decreased when the stress is large as shown in Fig. 7.66b.

7.7.4 Iron Loss Under Rotating Flux Excitation

Figure 7.70 shows the schematic structure of the SST having two diagonal-type magnetizing windings (2-D-SST) for the rolling direction (RD) and the transverse direction (TD) [48]. As a specimen, a square shingle sheet of 150 mm × 150 mm

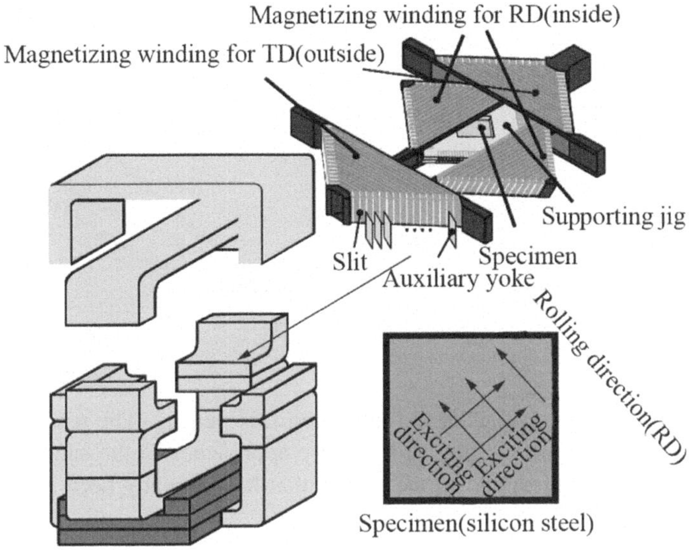

Fig. 7.70 Schematic structure of the SST having two diagonal-type magnetizing windings (2-D-SST)

is used. The diagonal direction of the specimen is RD. The rotating flux condition as well as alternating one in arbitrary directions can be satisfied, because the amplitude and the direction of flux density can be controlled by two windings. Remarkable structural features are the crosswise overlapped diagonal-type windings. By adopting this structure, the relation between the reluctance and the magnetomotive force becomes the same in all magnetic paths. The winding for RD is set inside that for TD to increase the maximum flux density.

The flux density $\boldsymbol{B}$ is detected in 20 mm $\times$ 20 mm region at the center of specimen using probes equivalent to single-turn search coils. The magnetic field strength $\boldsymbol{H}$ is detected in 20 mm $\times$ 16 mm region by the double H-coils [49]. The number of turns of each H-coil is 400.

The iron losses of non-oriented silicon steel JIS 50A290 (JIS: Japanese Industrial Standards, thickness: 0.5 mm, $W15/50 < 2.90$ W/kg) at 30 and 50 Hz are measured. The ratio of the axis of the rotating flux is assumed to be unity. The iron loss is separated to a hysteresis loss W_h and a eddy current loss W_e by the two-frequency method. The error of measurement of iron loss under the rotating flux is caused by the displacement of the probes and the double H-coils. Therefore, the measurement results in clockwise rotation and counterclockwise rotation are averaged in order to delete the error of measurement [50].

Figure 7.71 shows the iron loss W, the hysteresis loss W_h, the eddy current loss W_e and the classical eddy current loss W_e' calculated from the theoretical formula under a rotating flux at 50 Hz. W_h reaches the maximum value at almost 1.5 T and decreases under rotating flux at high flux density. The measurement result shows

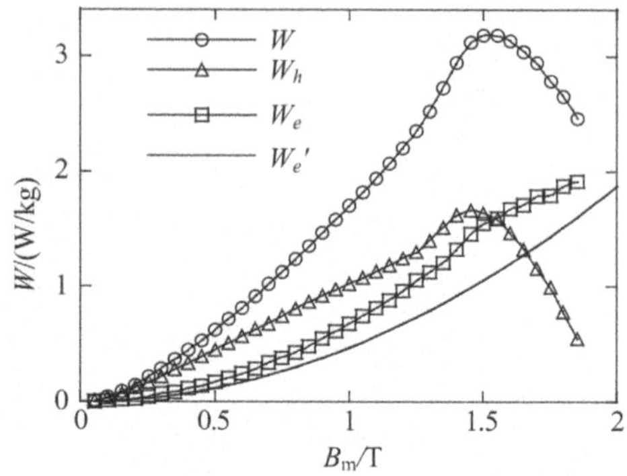

Fig. 7.71 Iron loss curves under rotating flux at 50 Hz

the tendency that W_h vanishes at the saturation flux density. The tendency that W_e approaches to W'_e as the rotating flux density approaches to the saturation. These loss characteristics correspond to the results that have already been reported [51].

7.7.5 Iron Loss Under DC Bias Excitation

The use of iron core under DC-biased magnetization generates a distorted asymmetrical hysteresis loop, and the iron loss under DC-biased magnetization is increased compared with the sinusoidal excitation.

In this section, a measuring system of magnetic properties of electrical steel sheet under DC-biased magnetization using an open-type SST [52] and a Helmholtz coil is explained, and DC-biased magnetic properties of silicon steels are shown [44]. The method has an advantage that the control is easy because there is no magnetic coupling between AC and DC exciting coils. The magnetic properties of two kinds of non-oriented silicon steel sheets, JIS 50A290 (thickness: 0.5 mm, iron loss $W15/50 \leq 2.9$ W/kg at 1.5 T and 50 Hz) and 6.5% Si–Fe sheet (thickness: 0.1 mm), under DC-biased magnetization are measured.

Figure 7.72a shows the hysteresis loop and the definitions of physical values under DC-biased magnetization. Figure 7.72b, c shows waveforms of flux density b and magnetic field strength h, respectively. ΔB is the DC-biased flux density, and H_{dc} is the DC-biased magnetic field strength. H_b is the magnetic field strength at the instant when flux density becomes the maximum when an average value of eddy current in an electrical steel sheet is zero. B_m is the amplitude of AC component of flux density, and the waveform of AC component of b is controlled as a sinusoidal wave.

Figure 7.73 shows the exciting coil arrangement composed of the SST of the open-type magnetic circuit and Helmholtz coil. The AC magnetic field and DC magnetic field are independently excited by separate coils. The AC field is impressed by an open-type SST, and DC field is impressed by a Helmholtz coil (162 turns per one coil). As shown in Fig. 7.73, the SST of open-type magnetic

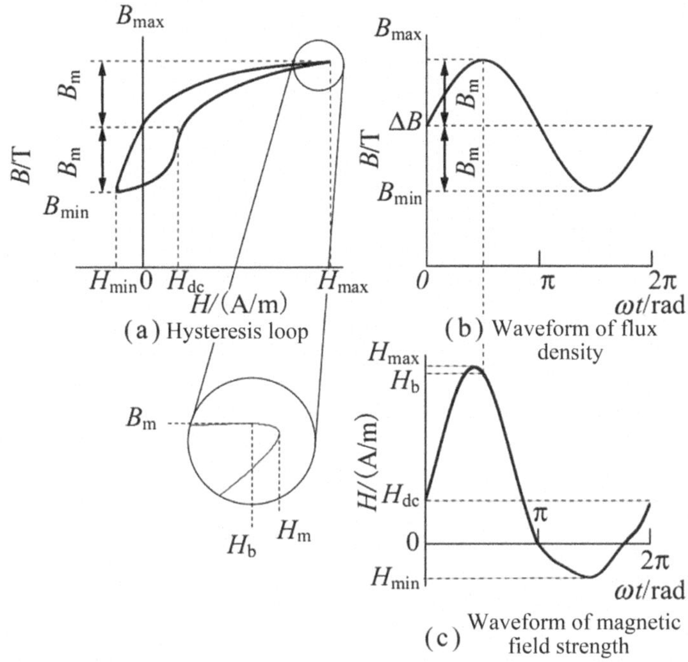

Fig. 7.72 Hysteresis loop under DC bias excitation

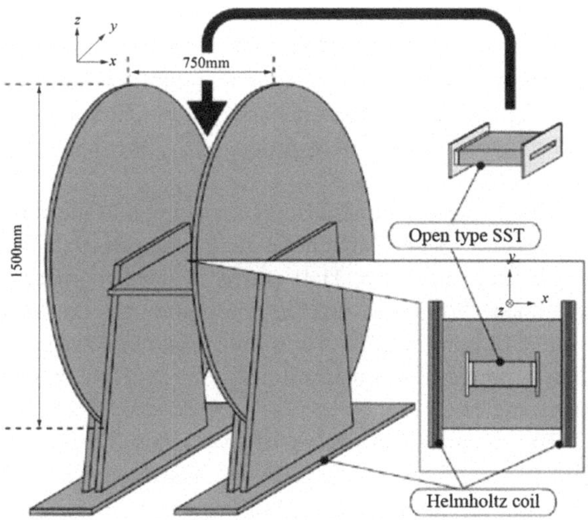

Fig. 7.73 Exciting coil arrangement composed of the SST of the open-type magnetic circuit and Helmholtz coil

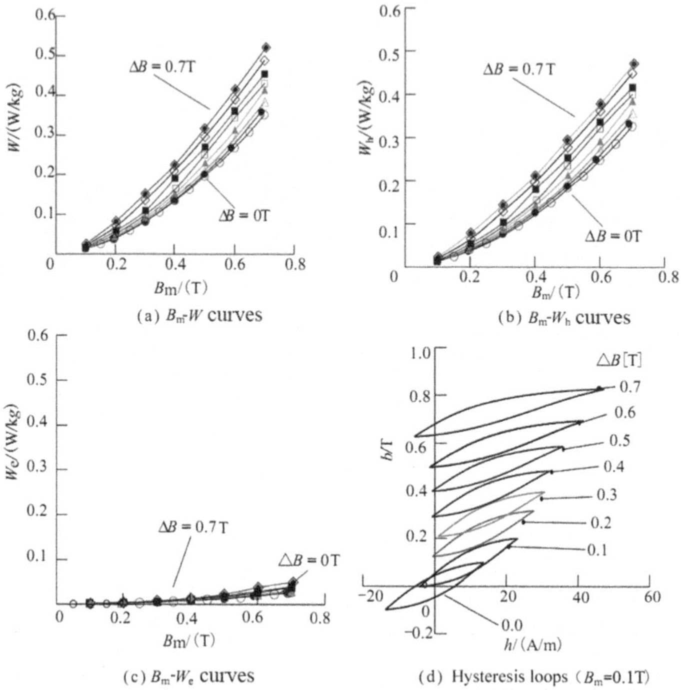

Fig. 7.74 Magnetic property under DC bias excitation (6.5% Si–Fe sheet, 50 Hz)

circuit is set inside the Helmholtz coil. The specimen is set inside the SST. In this measuring system, the magnetic coupling between AC exciting coil and DC exciting coil does not occur, because the Helmholtz coil is separated from the SST and the AC leakage field produced by the SST does not interlink the Helmholtz coil.

The B coil is wound around a specimen to measure the flux density. It is difficult to measure the DC magnetic field strength H_{dc} using an H-coil like the general single-sheet tester. A Hall probe is used to measure H_{dc} on the surface of the specimen. ΔB is measured by integrating the output signal of the **B** coil during the change of current of the Helmholtz coil from zero to a specified value. The AC flux density is measured using the B coil, and AC magnetic field strength is measured using the H-coil.

The magnetic properties of two kinds of non-oriented silicon steel sheets, JIS 50A290 and 6.5% Si–Fe sheet, are measured. The AC field of 50 Hz or 100 Hz and DC field are applied. These values can be observed, for example, in a reactor for inverter. The iron loss is separated to the hysteresis loss W_h and the classical eddy current loss W_e by the two-frequency method (using losses at 50 and 100 Hz).

Figure 7.74a shows the effect of DC bias on the iron loss W at 50 Hz of 6.5% Si–Fe sheet.

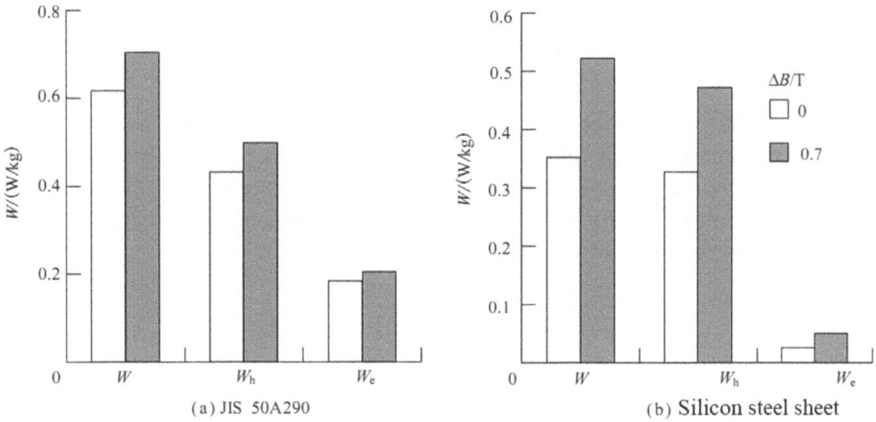

Fig. 7.75 Iron loss distribution under DC bias excitation (B_m = 0.7 T, 50 Hz)

Figure 7.74b, c shows the effect of DC bias on hysteresis loss W_h and classical eddy current loss W_e, respectively.

Figure 7.74d shows the effect of DC bias on minor loops at B_m = 0.1 T (50 Hz). Figure 7.74 illustrates that the area of minor loop is increased when B is large. As a result, the iron loss under large DC bias is increased as shown in Fig. 7.74a. Figure 7.75 shows the comparison of iron losses at ΔB = 0 T and ΔB = 0.7 T (B_m = 0.7 T, 50 Hz). The iron losses at ΔB = 0.7 T are increased about 14 and 50% than those at ΔB = 0 T for JIS 50A290 and 6.5% Si–Fe sheet, respectively.

7.8 Summary

This chapter discusses several magnetic property modeling methods and presents the measurement results, which can be summarized as follows:

(1) It is necessary to accurately measure the *B–H* curves, especially when the magnetic flux density is approaching to magnetic saturation, and for the anisotropic material (e.g., GO silicon steel) it is suggested to measure the *B–H* curves at different directions.
(2) The accurate modeling of hysteresis behavior under complex conditions is really required, even though some hysteresis models have been proposed.
(3) In order to estimate the iron loss under the distortion flux and rotating flux accurately, it is necessary to conduct an in-depth study on both experiment and analysis.
(4) In order to accurately analyze the flux and eddy current distribution in the laminated core of the actual motor equipment, the “homogenization method” and “zoning method” can be used to model the laminated core.

(5) The effects of shear, stress, DC bias and temperature on the magnetic properties of electrical steel sheets are studied, which is helpful to improve the engineering effectiveness of electromagnetic thermal field simulation.

It should be pointed out that advanced material simulation technology is the fundamental guarantee to provide correct material properties, and it is very important for industrial applications. It is more important and complex to develop experimental equipment for material property measurement and prediction, especially under nonstandard or extreme conditions.

References

1. K. Fujiwara, T. Adachi, and N. Takahashi, "A Proposal of Finite-Element Analysis Considering Two-Dimensional Magnetic Properties," IEEE Trans. Magn., Vol. 38, No. 2, pp. 889–892, 2002.
2. M. Enokizono, and N. Soda, "Magnetic Field Analysis by Finite Element Method using Effective Anisotropic Field," IEEE Trans. Magn., Vol. 31, No. 3, pp. 1793–1796, 1995.
3. T. Nakata, K. Fujiwara, N. Takahashi, M. Nakano, and N. Okamoto, "An Improved Numerical Analysis of Flux Distributions in Anisotropic Materials," IEEE Trans. Magn., Vol. 30, No. 5, pp. 3395–3398, 1994.
4. M. Enokizono, H. Shimoji, A. Ikariga, S. Urata, and M. Ohoto, "Vector Magnetic Characteristic Analysis of Electrical Machines," IEEE Trans. Magn., vol. 41, No. 5, pp. 2032–2035, 2005.
5. T. Kosaka, N. Takahashi, S. Nogawa, and M. Kuwata, "Analysis of Magnetic Characteristics of Three- Phase Reactor Made of Grain - Oriented Silicon Steel," IEEE Trans. Magn., Vol. 36, No. 4, pp. 1894–1897, 2000.
6. M. Nakano, H. Nishimoto, K. Fujiwara, and N. Takahashi, "Improvements of Single Sheet Testers for Measurement of 2-D Magnetic Properties up to High Flux Density," IEEE Trans. Magn., Vol. 35, No. 5, pp. 3965–3967, 1999.
7. K. Fujiwara. T. Nakata. N. Okamoto and K. Muramatsu, "Method for Determining Relaxation Factor for Modified Newton-Raphson Method," IEEE Trans. Magn., Vol. 29, No. 2, pp. 1962–1965. 1993.
8. M. Enokizono, and H. Shimoji, "Vector Magneto-hysteretic Engineering Model," Journal of Materials Processing Technology, Vol. 161, pp. 136–140, 2005.
9. F. Liorzou, B. Phelps, and D. L. Atherton, "Macroscopic Models of Magnetization," IEEE Trans. Magn., Vol. 36, No. 2, pp. 418–428, 2000.
10. F. Preisach, "Über die magnetische nachwirkung," Zeitschrift für Physik, Vol. 94, pp. 277–302, 1935.
11. I. D. Mayergoyz, "Mathematical Models of Hysteresis," Springer-Verlag, 1991.
12. A. Ivanyi, "Hysteresis Model in Electromagnetic Computation," Akadémiai Kiadó, 1997.
13. F. Vaja and E. Della Torre, "Efficient Numerical Implementation of Complete-moving Hysteresis Models," IEEE Trans. Magn., Vol. 29, No. 2, pp. 1532–1537, 1993.
14. B. D. Coleman, and M. L. Hodgdon, "A Constitutive Relation for Rate-independent Hysteresis in Ferromagnetical Soft Materials," Int. J. Engng. Sci., Vol. 24, No. 6, pp. 897–919, 1986.
15. M. L. Hodgdon, "Applications of a Theory of Ferromagnetic Hysteresis," IEEE Trans. Magn., Vol. 24, No. 1, pp. 218–221, 1988.
16. E. C. Stoner, and E. P. Wohlfarth, "A Mechanism of Magnetic Hysteresis in Heterogeneous Alloys," Phil. Trans. Roy. Soc., Vol. 240A, pp. 599–642, 1948.

17. D. L. Atherton, and J. R. Beattie, "A Mean Field Stoner-Wohlfarth Hysteresis Model," IEEE Trans. Magn., Vol. 26, No. 6, pp. 3059–3063, 1990.
18. I. A. Beardsley, "Modeling the Record Process," IEEE, Trans. Magn., Vol. 22, No. 5, pp. 454–459, 1986.
19. G. Friedman, and I. D. Mayergoyz, "Stoner-Wohlfarth Hysteresis Model with Stochstic Input as a Model of Viscosity in Magnetic Materials," IEEE Trans. Magn., Vol. 28, No. 5, pp. 2262–2264, 1992.
20. D. C. Jiles, and D. L. Atherton, "Theory of Ferromagnetic Hysteresis," J. Appl. Phys., Vol. 55, No. 6, pp. 2115–2120. 1984.
21. D. C. Jiles, and D. L. Atherton,"Theory of Ferromagnetic Hysteresis," Journal of Magnetism and Magnetic Materials, Vol. 61, pp. 48–60, 1986.
22. D. C. Jiles, J. B. Thoelke, and M.K. Devine, "Numerical Determination of Hysteresis Parameters for the Modeling of Magnetic Properties Using the Theory of Ferromagnetic Hysteresis," IEEE Trans. Magn., Vol. 28, No. 1, pp. 27–35, 1992.
23. T. Nakata, N. Takahashi, and Y. Kawase, "Finite Element Analysis of Magnetic Fields Taking into Account Hysteresis Characteristics," IEEE Trans. Magn., Vol. 21, No. 5, pp. 1856–1858, 1985.
24. N. Takahashi, S. Miyabara, and K. Fujiwara, "Problems in Practical Finite Element Analysis Using Preisach Hysteresis Model," IEEE Trans. Magn., Vol. 35, No. 3, pp. 1243–1246, 1999.
25. N. Fujiwara, K. Sinagawa, K. Ashiho, and K. Fujiwara, N. Takahashi, "Development of 3-D Read/Write Simulation System for Higher Areal Recording Density," IEEE Trans. Magn., Vol. 40, No. 3, pp. 2613–2615, 2004.
26. H. Domeki, Y. Ishihara, C. Kaido, Y. Kawase, S. Kitamura, T. Shimomura, N. Takahashi, T. Yamada, and K. Yamazaki, "Investigation of Benchmark Model for Estimating Iron Loss in Rotating Machine," IEEE Trans. Magn., Vol. 40, No. 2, pp. 794–797, 2004.
27. T. Nakata, Y. Ishihara, and M. Nakano, "Iron Losses of Silicon Steel Core Produced by Distorted Flux," Electrical Engineering in Japan, Vol. 90, No. 1, pp. 10–20, 1970.
28. Y. Ishihara, "Characteristics of Iron Core of High Frequency Motor Driven by Power Electronics," JMA Applied Magnetics Symposium, No. 5–2, 2008 (in Japanese).
29. N. Takahashi, A. Fukuma, and D. Miyagi, "Analysis of Iron Loss under Distorted Elliptical Rotating Flux of SPM Motor," COMPEL, Vol. 24, No. 2, pp. 385–395, 2005.
30. S. Nogawa, M. Kuwata, D. Miyagi, T. Hayashi, H. Tonai, T. Nakau, and N. Takahashi, "Study of Eddy Current Loss Reduction of Slit in Reactor Core," IEEE Trans. Magn., Vol. 41, No. 5, pp. 2024–2027, 2005.
31. K. Hollaus and O. Biro, "A FEM Formulation to Treat 3D Eddy Current in Laminations," IEEE Trans. Magn., Vol. 36, No. 5, pp. 1289–1292, 2000.
32. P. Dular, J. Gyselinck, C. Geuzaine, N. Sadowski, and J. P. A. Bastos, "A 3-D Magnetic Vector Potential Formulation Taking Eddy Currents in Lamination Stacks into Account," IEEE Trans. Magn., Vol. 39, No. 3, pp. 1424–1427, 2003.
33. I. Sebestyen, S. Gyimothy, J. Pavo, and O. Biro, "Calculation of Losses in Laminated Ferromagnetic Materials," IEEE Trans. Magn., Vol. 40, No. 2, pp. 924–927, 2004.
34. K. Muramatsu, T. Okitsu, H. Fujitsu, and F. Shimanoe, "Method of Nonlinear Magnetic Field Analysis Taking into Account Eddy Current in Laminated Core," IEEE Trans. Magn., Vol. 40, No. 2, pp. 896–899, 2004.
35. H. Kaimori, A. Kameari, and K. Fujiwara, "FEM Computation of Magnetic Field and Iron Loss in Laminated Iron Core Using Homogenization Method," IEEE Trans. Magn., Vol. 43, No. 4, pp. 1405–1408, 2007.
36. S. Nogawa, M. Kuwata, T. Nakau, D. Miyagi, and N. Takahashi, "Study of Modeling Method of Lamination of Reactor Core," IEEE Trans. Magn., Vol. 42, No. 4, pp. 1455–1458, 2006.
37. R. Rygal, A. J. Moses, N. Derebasi, J. Schneider, and A. Schoppa, "Influence of Cutting Stress on Magnetic Field and Flux Density Distribution in Non-oriented Electrical Steels," Journal of Magnetism and Magnetic Materials, no. 215–216, pp. 687–689, 2008.

38. A. Pulnikov, V. Permiakov, M. D. Wulf, and J. Melkebeek, "Measuring Setup for the Investigation of the Influence of Mechanical Stresses on Magnetic Properties of Electrical Steel," Journal of Magnetism and Magnetic Materials, no. 254–255, pp. 47–49, 2008.
39. N. Takahashi, D. Miyagi, R. Usui, M. Nakaoka, and M. Nakano, "Measurement of Deterioration of Magnetic Properties due to Shrink Fitting," Journal of The Japan Society of Applied Electromagnetics and Mechanics, vol. 15, no. 3, pp. 222–225, 2007.
40. N. Takahashi, H. Morimoto, Y. Yunoki, and D. Miyagi, "Effect of Shrink Fitting and Cutting on Iron Loss of Permanent Magnet Motor," Journal of Magnetism and Magnetic Materials, 2008.
41. K. Fujisaki, R. Hirayama, T. Kawachi, S. Satou, C. Kaido, M. Yabumoto, and T. Kubota, "Motor Core Iron Loss Analysis Evaluating Shrink Fitting and Stamping by Finite-Element Method," IEEE Trans. Magn., vol. 43, no. 5, pp. 1950–1954, 2007.
42. V. Permiakov, L. Dupre, A. Pulnikov, and J. Melkebeek, "Loss Separation and Parameters for Hysteresis Modeling under Compressive and Tensile Stresses," Journal of Magnetism and Magnetic Materials, no. 272–276, pp. e553–e554, 2004.
43. J. Sievert, "Recent Advances in the One-and Two-Dimensional Magnetic Measurement Technique for Electrical Sheet Steel," IEEE Trans. Magn., Vol. 26, No. 5, pp. 2553–2558, 1990.
44. D. Miyagi, T. Yoshida, M. Nakano, and N. Takahashi, "Development of Measuring Equipment of DC-Biased Magnetic Properties Using Open-Type Single-Sheet Tester," IEEE Trans. Magn., Vol. 42, No. 10, pp. 2846–2848, 2006.
45. M. Nakaoka, A. Fukuma, H. Nakaya, D. Miyagi, M. Nakano, and N. Takahashi, "Examination of Temperature Characteristics of Magnetic Properties Using a Single Sheet Tester," Trans. IEE of Japan, A, Vol. 125, No. 1, pp. 63–68, 2005.
46. N. Takahashi, and D. Miyagi, "Examination of Magnetic Properties of Electrical Steels under Stress Condition," Proc. Inter. Conf. Electrical Engineering, No. O -003, 2008.
47. Japanese Industrial Standards, JIS C 2556, 1996.
48. D. Miyagi, Y. Yunoki, M. Nakano, and N. Takahashi, "Study on Measurement Method of 2 Dimensional Magnetic Properties of Electrical Steel Using Diagnonal Exciting Coil," Proc. 10th Int. Workshop on 1&2 DM, 2008.
49. T. Nakata, Y. Kawase, and M. Nakano, "Improvement of Measuring Accuracy of Magnetic Field Strength in Single Sheet Testers by Using Two H Coils," IEEE Trans. Magn., vol. 23, no. 5, pp. 2596–2598, 1987.
50. Y. Maeda, H. Shimoji, T. Todaka, and M. Enokizono, "Study of the Counterclockwise/clockwise (CCW/CW) Rotational Losses Measured with a Two-dimensional Vector Magnetic Property Measurement System," IEEJ Trans. Electrical and Electronic Engineering, Vol. 3, No. 2, pp. 222–228, 2008.
51. T. Yamaguchi, and K. Narita, "Rotational Power Loss in Commercial Silicon-iron Laminations," IEE J Trans. Fundamentals and Materials, Vol. 51, No. 2, pp. 341–348,1976.
52. T. Nakase, M. Nakano, K. Fujiwara, and N. Takahashi, "Single Sheet Tester Having Open Magnetic Path for Measurement of Magnetostriction of Electrical Steel Sheet," IEEE Trans. on Magnetics, Vol. 35, No. 5, pp. 3956–3958, 1999.

Chapter 8
Magnetic Measurement Based on Epstein Combination and Multi-angle Sampling

Zhiguang Cheng, Lianbin Shi and Johannes Sievert

Abstract Up to now the Epstein frame, as a standard magnetic measurement method, is still widely used in magnetic measurement, even though its advantages and disadvantages have been well recognized. As an application-oriented improvement, magnetic measurements based on the combination of Epstein frames of different sizes and loss data weighted processing have been proposed by the authors and briefly demonstrated in this chapter. According to a frequent request from industrial users, the multi-directional electromagnetic properties of the grain-oriented silicon steel are modeled using the 25 cm Epstein frame, in which the specimens are cut at different angles to the rolling direction. The magnetization curves (*B*–*H*) and the specific total loss curves (B_m–W_t) are measured at different sampling angles, meanwhile, the effects of the stress relief annealing on the electromagnetic properties are also examined.

Keywords Magnetic measurement · GO silicon steel · Magnetic anisotropy · Combined Epstein method · Multi-angle sampling

8.1 Introduction

The research topic of material modeling proposed in the computational electromagnetics is not targeted to develop new materials, but mainly to study the macroscopic characteristics of materials. As is known, the fundamentals of material property measurements are mainly based on Faraday's law of electromagnetic induction, and on Ampere's law, i.e. the current law, as well as the continuity

---

Z. Cheng · L. Shi
Institute of Power Transmission and Transformation Technology,
Baobian Electric, Baoding, China

J. Sievert (✉)
IEC Techn. Committee "Magnetic Alloys and Steels", Physikalisch-Technische Bundesanstalt (retired), Braunschweig, Germany
e-mail: johannes.sievert@t-online.de

Z. Cheng et al. (eds.), *Modeling and Application of Electromagnetic and Thermal Field in Electrical Engineering*, https://doi.org/10.1007/978-981-15-0173-9_8

condition of magnetic flux density and the parallel component of the magnetic field intensity at the materials' surface. With the increasing requirements of computational electromagnetics and industrial applications, material modeling is becoming more important to improve the accuracy and effectiveness of calculations, which has become extremely complex due to the nonlinearity, magnetic anisotropy and hysteresis of soft magnetic materials and the dependence of the electromagnetic properties on temperature, frequency, stress and multi-dimensional impressed field (excitation) conditions. Numerous scholars have made historic contributions to the R&D of material modeling and related computational electromagnetics [1–12]. Despite this, up to now some electromagnetic properties of electrical engineering materials are still very difficult to measure and to predict accurately under complex working conditions. As a result, serious challenges remain on many aspects [13].

From the perspective of industrial applications, for the modeling and simulation of a complex electromagnetic device or system, the integral quantity of the physical field, the degree of the local concentration of the field quantity and the overall distribution of the field are generally important. The integral quantity is related to the general technical level and economic index of the device or products; high concentration of local field quantity may endanger the safety and reliability in operation, and the overall distribution of the field may provide valuable inspiration and suggestions for structural design and optimization. Various levels of material modeling may have a substantial or non-substantial impact on the above three kinds of data. The influence of the performance data given, based on the traditional and improved material modeling techniques, on the calculated results of field should be investigated.

It has been shown that the effectiveness of the numerical modeling and simulation are closely dependent on the accurate material properties. At the same time, it can be seen that the standard magnetic measurement methods, e.g., Epstein frame [14] and single sheet tester (SST) [15], are still commonly used, and, beyond that, some special needs from product design and industrial application, e.g., the modeling of magnetic property of GO silicon steel at different sampling angles, are often asked for.

For this reason, some related research and special magnetic measurements have been carried out by the authors and are briefly introduced in this chapter, as follows:

(1) As an application-oriented improvement of the standard magnetic measurement method, the Epstein combination (i.e., a combination of the different size Epstein frames, including standard and scaled-down Epstein frames) and loss data weighted processing has been proposed by the authors [16], and the methodology and typical results are briefly presented in the Sect. 8.3.2 of this chapter.
(2) In response to the requirement from the R&D and design of electrical products and electromagnetic analysis, the authors have cut Epstein specimens under different angles to the rolling direction for studying the effects of different sampling angles on the magnetic measurement results, and investigated the effects of stress relief annealing on the properties of the specimens. See Sect. 8.4 of this chapter.

It should be noted that this is certainly not the equivalent to the determination of the magnetic properties under two-dimensional (2-D) excitation [17], but if we look at it from the perspective of industrial application, it may be used as an approximate processing method to apply magnetic properties measured in this way in different directions to the numerical analysis of electromagnetic fields.

8.2 Magnetic Properties Under Rotating Flux Conditions

The magnetic anisotropy of silicon steel sheets has attracted the attention of scientists and engineers for a long time. Nowadays, the measurement and prediction of material properties are carried out based on 1-D and 2-D, even 3-D testers under standard and non-standard conditions [18].

It is shown that, when the *B*–*H* curves, measured under the two orthogonal directions, i.e., along and perpendicular to the rolling direction, are used for the calculation of the magnetic field, the resulting magnetic flux density will be considerably different from the actual values. So it is necessary to study the 2-D magnetic properties, that is, to consider the magnetic anisotropy inherent in the plane [11, 12, 17]. One of the key problems in anisotropic modeling is to consider the different directions of magnetic flux density ($\boldsymbol{B}$) and magnetic field intensity ($\boldsymbol{H}$) in space. The directions of $\boldsymbol{B}$ and $\boldsymbol{H}$ are different, resulting in non-diagonal elements of reluctivity tensor not being zero. Further experimental studies on $\boldsymbol{B}$ and $\boldsymbol{H}$ in non-oriented materials show that even the so-called non-oriented materials also exhibit a certain anisotropic behavior [13]. In addition, transformer design experts found that the higher the performance of the silicon steel sheet, the higher is the sensitivity with regard to directionality.

It is noteworthy that A. D. Napoli and R. Paggi published reluctivity tensors characterizing anisotropy in the early 1980s, including reluctivity models of orthogonal anisotropy and arbitrary anisotropy, and transformed them into the form of diagonal matrix by a reference coordinate system transformation [5, 6, 8]. They also calculated the transformer magnetic field, reactance and coil short-circuit force based on the reluctivity models of isotropy, orthogonal anisotropy and arbitrary anisotropy, and compared them with the results measured, and they pointed out that the so-called orthogonal anisotropy is a simplification and is correct only if the magnetic field is either along, or perpendicular to the rolling direction.

As M. Enokizono indicated, in the case of strong nonlinearity and rotating flux, some problems have been found in the traditional orthogonal anisotropy processing, the essence of which is that usually $\boldsymbol{B}$ and $\boldsymbol{H}$ have different directions [17]. The rotational power loss occurs at the T-joints of the transformer core. Two different forms of reluctivity tensors were derived and established by him, because the magnetic properties of materials are different under the conditions of alternating flux and rotating flux.

Based on the work of M. Enokizono, K. Fujiwara et al. defined the anisotropic reluctivity as a function of the magnitudes of $\boldsymbol{B}$ and $\boldsymbol{H}$ and their respective direction angles θ_B (e.g., the angle between $\boldsymbol{B}$ and X-axis to the rolling direction) and θ_H (the angle between $\boldsymbol{H}$ and X-axis), and derived the expression of the definition of effective anisotropic reluctivity, v, which has a similar form as the traditional orthogonal anisotropic permeability, except that the directions of $\boldsymbol{B}$ and $\boldsymbol{H}$ are different [19].

$$\upsilon = \begin{bmatrix} \upsilon_x & 0 \\ 0 & \upsilon_y \end{bmatrix} \tag{8.1}$$

where

$$\begin{aligned} \upsilon_x &= \frac{H\cos\theta_H}{B\cos\theta_B} \\ \upsilon_y &= \frac{H\sin\theta_H}{B\sin\theta_B} \end{aligned} \tag{8.2}$$

In fact, H cos θ_H and B cos θ_B are projections H_x and B_x of $\boldsymbol{H}$ and $\boldsymbol{B}$ in the x-direction, respectively, and $\frac{\boldsymbol{H}\cos\theta_H}{\boldsymbol{B}\cos\theta_B}$ in Eq. (8.2) represents the reluctivity in the x-direction; H sin θ_H and B sin θ_B are projections H_y and B_y of $\boldsymbol{H}$ and $\boldsymbol{B}$ in the y-direction, respectively, $\frac{H\sin\theta_H}{B\sin\theta_B}$ represents the reluctivity in the y-direction. Therefore, it can be further written as

$$\upsilon = \begin{bmatrix} \frac{H_x}{B_x} & 0 \\ 0 & \frac{H_y}{B_y} \end{bmatrix} \tag{8.3}$$

Equation (8.3) is formally the same as the reluctivity under the traditional 2-D orthogonal anisotropy, but has different connotations.

In addition, D. Lin et al. proposed a simplified model to deal with the nonlinearity and anisotropy of soft magnetic materials based on the anisotropy of energy density. With this model, only the *B–H* curves in the rolling and anti-rolling directions need to be measured, that is, only two *B–H* curves are needed for the 2-D problem and three for the 3-D problem [20]. The *B–H* curve in the principle direction can generally be provided directly by the silicon steel sheet manufacturer. The numerical experiments of the proposed method are carried out and the application example is given. There is no doubt that such thinking is understandable, only that more testing is needed.

8.3 Application and Improvement of Epstein Frame Measurement

8.3.1 Epstein Frame

The Epstein Frame method is the commonly used standardized method to measure the magnetic properties of electrical sheet steel [14]. The square-shaped frame comprises a primary winding, a secondary winding and the specimen to be tested as a laminated core, and forms a no-loaded transformer. Specifically, an Epstein frame consists of four coils (each coil is provided with a set of primary and secondary windings). A mutual inductor for air flux compensation is included with the Epstein frame. The winding formers supporting the coils are made of hard insulating material. The frame shall be fixed to an insulating and non-magnetic base in such a way as to form a square. Figure 8.1 shows the basic structure of a 25 cm Epstein frame for measuring the magnetic properties of electrical steel sheet and strip.

The total weight of the specimen is about 1 kg in conventional experiments (for 25 cm Epstein frame). The specimen in the frame is double-lapped at the corners, as shown in Fig. 8.2, and forms a square magnetic circuit immediately adjacent to the inner side of the frame. The magnetic material properties of silicon steel sheet specimens can be measured by applying AC or DC current to the primary winding of the frame to measure the electric signals from the primary and the secondary windings.

The following should also be noted when measuring the magnetic properties of electrical steel sheets with a standard Epstein frame:

(1) First, the upper-frequency limit of the Epstein frame for AC measurements (e.g., not more than 400 Hz).
(2) Since the specimen is provided with a small width, the stress produced in the sample's edges by the cutting cannot, in the case of grain-oriented material, be ignored, and stress relief annealing treatment [21] is required. The specifications of the specimen of the Epstein frame (25 cm) are shown in Fig. 8.1.

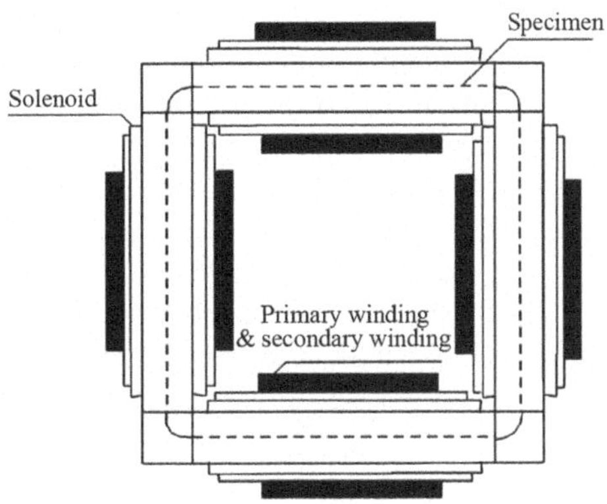

Fig. 8.1 Structure of 25 cm Epstein frame

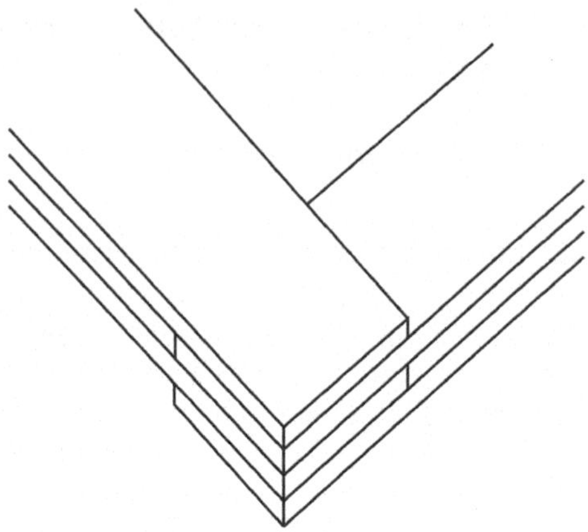

Fig. 8.2 Double-lapped specimen

$$\text{Width}: 30\,\text{mm} \pm 0.2\,\text{mm};$$
$$\text{Length range}: (280 \sim 320)\text{mm} \pm 0.5\,\text{mm}.$$

(3) The silicon steel strips are arranged in the double over-lapping form at the corners, and the length of effective magnetic path is defined as 0.94 m according to the standard 25 cm Epstein frame. However, further research suggests that this is just conventional. To this end, Wolfson Center for Magnetics, Cardiff University, UK, used the double Epstein frame method to study the average magnetic path length [22]. Moreover, it should be noted that this problem has been studied by many researchers [23].

(4) On the different directions of ***B*** and ***H*** in the specimen (strip/sheet):
N. J. Layland, A. J. Moses, N. Takahashi and T. Nakata collaborated in a valuable work: an experimental study on the effect of the specimen width on magnetic flux density (***B***) and magnetic field intensity (***H***) in anisotropic silicon steel sheets [10]. The widths of the two kinds of specimens of oriented silicon steel sheet (3.2% silicon iron) were 100 mm and 25 mm, respectively, and the lengths of both are 300 mm. The experimental results show that:

(a) In a specimen with a width of 100 mm, the magnetic flux density (***B***) is substantially along the rolling direction at a lower magnetic flux density, and ***B*** and ***H*** are no longer in the same direction at a higher magnetic flux density.
(b) However, this is not the case in the narrower specimen (width, 25 mm), in that the magnetic field intensity and magnetic flux density are always consistent with the length direction.
(c) It can also be seen that the angle at which the magnetic flux density and magnetic field intensity deviate from the rolling direction will increase with the increase of magnetization for the wider specimen.

(5) Further studies show that in the Epstein frame experiment with multi-direction sampling, attention should be paid to the order in which the specimens cut at different angles are placed in the frame. The wrong order will affect the measurement results of magnetic properties in varying degrees [24].

8.3.2 Epstein Combination and Loss Data-Based Weighted Processing Method

As mentioned above, the standard Epstein frame method for measuring magnetic material properties has been used for many years, having some recognized advantages, however, some problems are still worthy of further investigation, e.g., its mean magnetic path length, the non-uniformity of the electromagnetic field and magnetic loss inside the frame. Besides, the stress relief annealing process for all grain-oriented samples is needed. However, annealing is not allowed for samples of high permeability domain-refined GO silicon steel because their properties would be deteriorated by the annealing.

8.3.2.1 On the P_{loss}-Based Weighted Method

The extended modeling of the magnetic properties of GO electrical steels was proposed and implemented by the authors [16], based on a set of standard (25 cm) and scaled-down (17.5 and 20 cm) Epstein frames, referred to as E-25, E-17.5 and E-20, respectively, as shown in Fig. 8.3a. In fact, the goal of this co-research is to investigate or find a way to eliminate the effect of the non-uniformity of both magnetic field and loss in Epstein frame on the magnetic measured results.

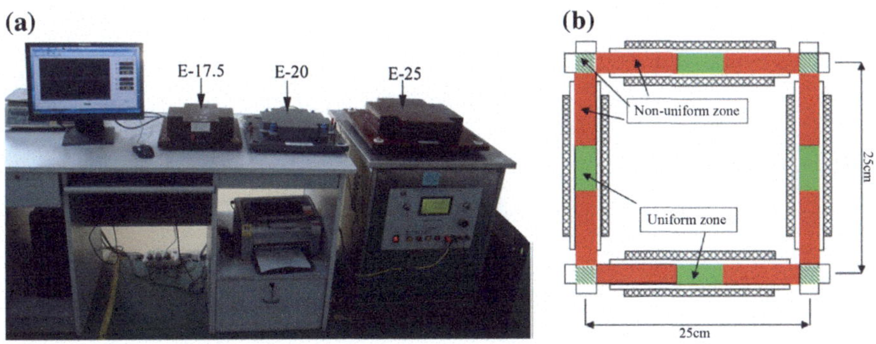

Fig. 8.3 Extended Epstein application: **a** Epstein combination (E-25/20/17.5); **b** non-uniform field in frame

Here, two assumptions were made:

(1) The non-uniform magnetic field and loss distribution over the corner regions of both the standard Epstein frame and the scaled-down Epstein frame are identical, despite the difference in their limb lengths.
(2) The magnetic field and loss is uniform over the middle section of each Epstein limb. See Fig. 8.3b.

The mean magnetic path length, l_m, of the standard Epstein frame is given by,

$$l_m = 4l \cdot \frac{P_n}{m_t P_{\text{loss}}} \tag{8.4}$$

where l [m]: the total length of each Epstein strip; P_n[W]: the absolute total magnetic loss of the standard Epstein frame (E-25); m_t[kg]: the total mass of all the laminations inside the frame; and P_{loss}[W/kg]: the specific magnetic loss.

Obviously, the mean magnetic path length, l_m, is dependent on the specific magnetic loss, as shown in (8.4). However, the specific magnetic losses in the uniform and non-uniform zones of the entire Epstein frame (denoted by P_{loss1} and P_{loss2}, respectively) differ due to the different field distributions. Therefore, two forms of mean magnetic path lengths, l_{m1} and l_{m2}, for the standard Epstein frame, can be determined based on the specific magnetic losses, P_{loss1} and P_{loss2}, of the uniform and non-uniform sub-regions [16].

In order to obtain a closer approximation to the mean magnetic path length of the standard Epstein frame, l_e, a weighted processing method, based on the already obtained l_{m1} and l_{m2}, is proposed, i.e., l_e becomes a weighted sum of l_{m1} and l_{m2}, incorporating the corresponding weight factors α and β, as given by

$$l_e = \alpha \cdot l_{m1} + \beta \cdot l_{m2} \tag{8.5}$$

where α and β represent the contribution of $\frac{1}{P_{\text{loss}}}$ to the weighted magnetic path length l_e, let

$$\begin{cases} P_x = \frac{1}{P_{\text{loss1}}} \\ P_y = \frac{1}{P_{\text{loss2}}} \end{cases} \tag{8.6}$$

Then, the weighted factors α and β are as

$$\begin{cases} \alpha = \frac{P_x}{P_x + P_y} \\ \beta = \frac{P_y}{P_x + P_y} \end{cases} \tag{8.7}$$

After a simple derivation, α and β can be expressed as,

$$\begin{cases} \alpha = \frac{P_{\text{loss2}}}{P_{\text{loss1}} + P_{\text{loss2}}} \\ \beta = \frac{P_{\text{loss1}}}{P_{\text{loss1}} + P_{\text{loss2}}} \end{cases} \tag{8.8}$$

A proposed weighted processing of the Epstein data covers the following themes:

(1) It shows the benefit of establishing an Epstein set, combining one standard frame (25 cm) and two scaled-down frames (17.5 and 20 cm), E-25, E-20 and E-17.5, two Epstein combinations can be alternately formed. i.e., 2E (25–17.5) and 2E (25–20), respectively.
(2) It demonstrates the use of a weighted processing method, proposed by the authors, which is based on the loss data and can be applied to reasonably determine the mean magnetic path length of the Epstein frame under various conditions.
(3) It examines the effect of the grade and texture of GO electrical steel, flux density, magnetizing frequency ambient temperature, and the angle at which the Epstein strips are cut to the rolling direction (RD), on the specific magnetization loss and exciting power (or specific apparent power).

Note that the determinations of the mean magnetic path lengths (l_{m1} and l_{m2}), the detailed description of Epstein combination method and the related results can be found in [16].

8.3.2.2 Typical Results of Path Length

The weighted mean magnetic path lengths of the standard Epstein frame (E-25) were determined for the 30P120 grade steel, using Epstein group 2E(25–17.5) at 50 Hz, as shown in Fig. 8.4.

The typical trends in the variation of the path length with magnetic flux density and strip angles can be summarized as follows:

(1) While l_{m1} and l_{m2} of the Epstein frame (25 cm) are different, the weighted mean magnetic path length, l_e, of the Epstein frame lies between them. It can be seen from Fig. 8.4a that it varies within a narrow range, i.e., from 0.940 to 0.945 m, and is not a constant value as the flux density increases, for strips cut parallel to the RD.

2) The mean magnetic path length of the standard Epstein frame (E-25) is not always 0.94 m, as specified in the IEC standard [14], e.g., shown in Fig. 8.4b, c. It is approximately 0.93 m for the strip angle 55° and 0.92 m for the strip angle 90°.

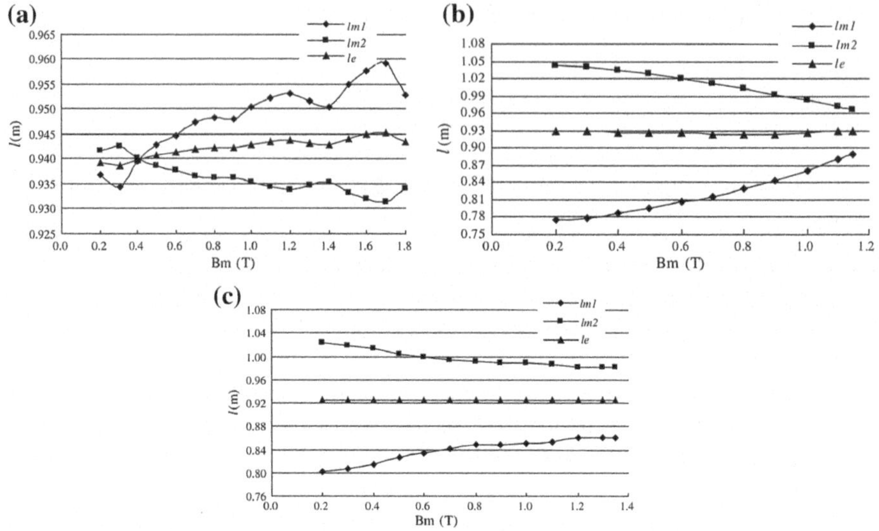

Fig. 8.4 Variation of mean magnetic path length of the Epstein frame (25 cm) with flux density measured using 2E(25–17.5), at 50 Hz, 30P120: **a** Strip angle 0°; **b** strip angle 55°; **c** strip angle 90°

8.3.2.3 Remarks on the Improved Epstein Measurement

The extended modeling of magnetic properties of GO electrical steel based on an Epstein combination and weighted processing, and a number of experimental results have been obtained by the authors, which can be summarized as:

(1) The double Epstein frame method, in fact based on an Epstein subtraction scheme (e.g., using 2E(25–17.5) or 2E(25–20)), eliminates the effect of the non-uniformity of the specific magnetization loss over the corner regions of the entire frame, and enables the accurate measurement of the specific magnetization loss and exciting power in the uniformly magnetized limb regions.
(2) The weighted processing methods based on the loss data, obtained by an Epstein set (E-25, E-20, and E-17.5), are proposed and implemented, offering two benefits: (i) the first-level weighted method takes the non-uniformity of the magnetic field and the power loss inside the entire Epstein frame into account, making it possible to accurately determine the mean magnetic path length of the Epstein frame; (ii) the second-level weighted method can be used to further examine the effect of different Epstein groups (2E(25–17.5) and 2E(25–20)) on the mean magnetic path length, specific power loss and exciting power.
(3) All the measurements covering the many Epstein test cases, adequately demonstrate the Epstein combination and weighted processing methods and can be safely carried out. The corresponding results also show the impacts of the related factors, such as magnetic flux density, non-RD (rolling direction) magnetization, excitation frequency and ambient temperature, on Epstein magnetic properties.

(4) Measurements using the standard Epstein frame (E-25) are subject to errors because the value of magnetic path length is fixed. As the demand for more accurate measurements under non-standard conditions increases, it is becoming more important to quantify, or even eliminate, these errors.

While intended for academic interest, it is strongly hoped that this work stimulates discussion and debate among the steel manufacturers, users and researchers to develop the possibly improved magnetic methodologies, or address the unsatisfactory nature of the existing magnetic measurements for modern and the future industrial needs.

8.4 Magnetic Measurement Based on Multi-angle Sampling

8.4.1 Multi-direction Magnetic Measurement

Following the frequently encountered requirement for multi-direction sampling measurement proposed in the design of electrical products and electromagnetic analysis, the magnetic property measurement under multi-angle sampling conditions using Epstein frame, including *B–H* curves and loss curves, is described in detail below, and the influence of stress relief annealing on the magnetic property of test specimen is investigated.

The magnetic properties of eight types of specimens with standard sizes at different angles to the rolling direction of silicon steel sheets, i.e., specimens with angles of 0°, 15°, 30°, 45°, 55°, 60°, 75° and 90°, respectively, are measured before and after stress relief annealing, respectively.

It should be noted that the method of sampling at different angles to rolling direction not only involves heavy workload, but also has certain limitations, making it difficult to accurately determine and predict the magnetic properties at arbitrary magnetization direction.

The weight data measured with a digital balance before and after annealing are shown in Table 8.1 when the silicon steel sheet 30P120 is at different angles (0°–90°) to the rolling direction.

8.4.2 Multi-angle Sampling

1. The specimen is cut at different angles, i.e., θ = 0°, 15°, 30°, 45°, 55°, 60°, 75° 90°, from the rolling direction of the silicon steel sheet as shown in Figs. 8.5, 8.6 and 8.7. The samples from different angles to the rolling direction can be divided into several groups (or called as columns) according to the size of the sampled grain oriented steel sheet, as shown in Fig. 8.7.

Table 8.1 Specimen weight measured before and after annealing

Sampling angle to rolling direction	Before annealing (kg)	After annealing (kg)
0°	0.948	0.948
15°	0.936	0.935
30°	0.933	0.933
45°	0.937	0.936
55°	0.940	0.940
60°	0.937	0.940
75°	0.939	0.938
90°	0.949	0.949

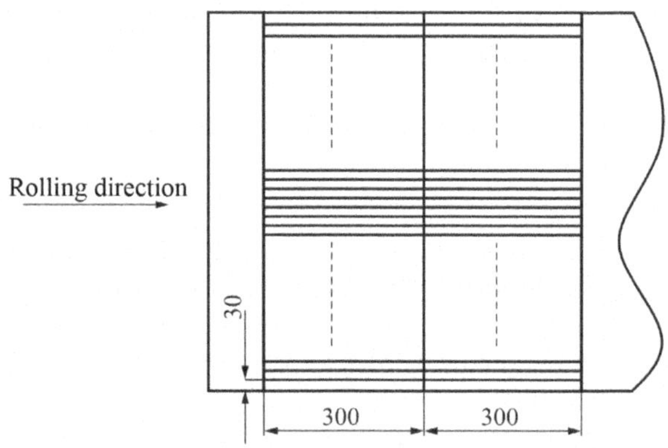

Fig. 8.5 0° to rolling direction during cutting

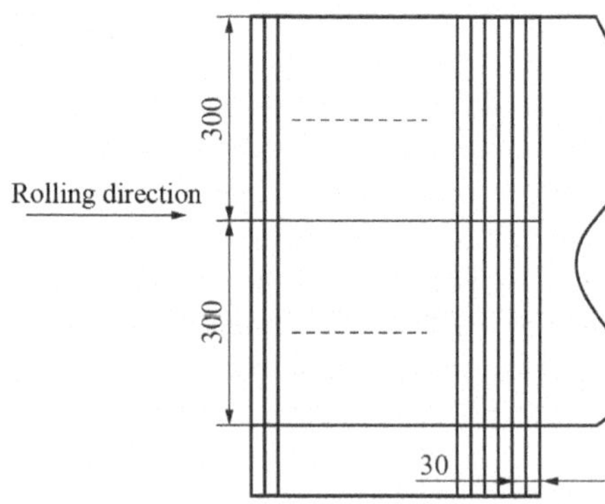

Fig. 8.6 90° to rolling direction during cutting

2. Specimen Requirements

(a) Dimensions of the specimen cut: 300 × 30 mm, each set of specimens weighs about 1 kg;

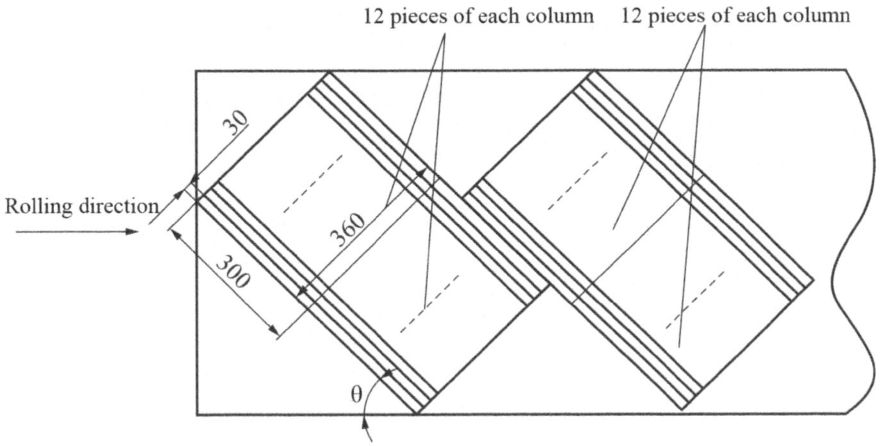

Fig. 8.7 Sampling angle θ to rolling direction (θ = 15°, 30°, 45°, 55°, 60°, 75°)

(b) if the difference between the thickness of silicon steel sheet and that of 0.3 mm (e.g., 30P120) is large, the number of sheets should be adjusted, but must be in multiples of 4;
(c) stress relief annealing of the specimen shall be carried out in accordance with the standard;
(d) the burr shall be <0.015 mm in specimen cutting.

Note that if a test specimen is cut from the edge of a large steel coil, it may not represent the loss or other magnetic properties of the entire coil.

3. Stress Relief Annealing

In order to investigate the influence of stress relief annealing on magnetic properties, the magnetization curves and loss curves measured before and after annealing of the specimen are compared.

8.5 Measurement Results and Discussions

8.5.1 $\mathrm{B_m}$–$\mathrm{H_m}$ *Curve Before Annealing (30P120)*

The magnetization curves (B_m–H_m) measured before annealing for 30P120 silicon steel sheet specimens with different angles are shown in Fig. 8.8. The definition of the magnetization curve B_m–H_m is given in [25].

The change of magnetization curve with sampling angle can be seen from the 30P120's B_m–H_m curves of 0°–90° before annealing in Fig. 8.8. The results show that the specimen with an angle of 90° to the rolling direction is not the worst in terms of permeability. In a certain range of magnetic field intensity (***H***), the

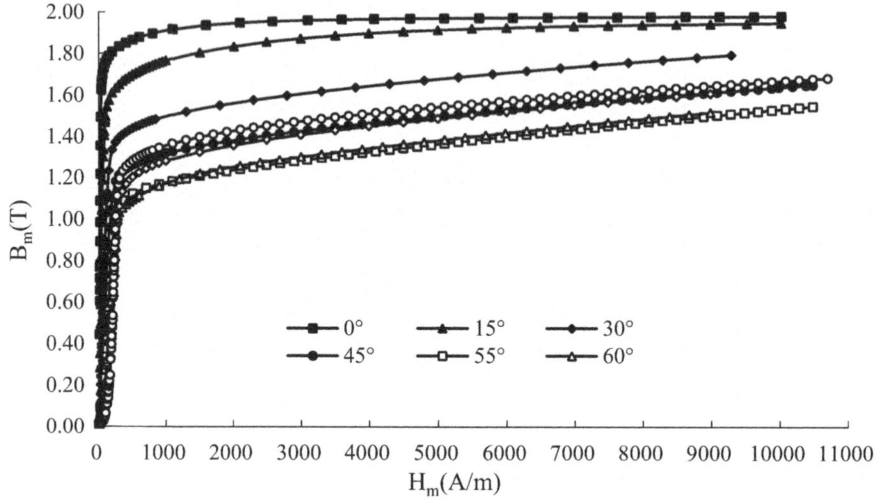

(a) B_m-H_m curves before annealing

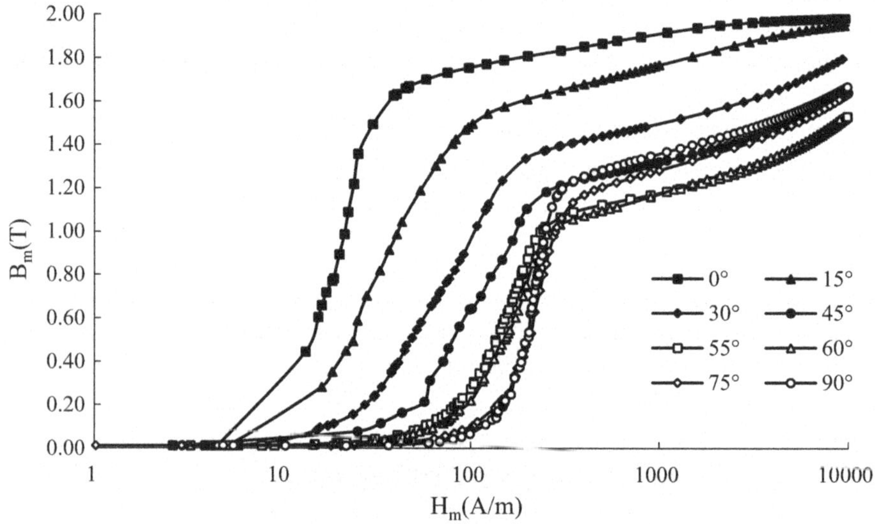

(b) B_m-H_m curves before annealing (logarithm)

Fig. 8.8 B_m–H_m curves of 30P120 before annealing

permeability of the specimen with a sampling angle of 55° is lower than that of the specimens with other angles (the angles to the rolling direction). However, since sampling from different angles is limited, it does not mean that the direction of 55° is the worst in permeability.

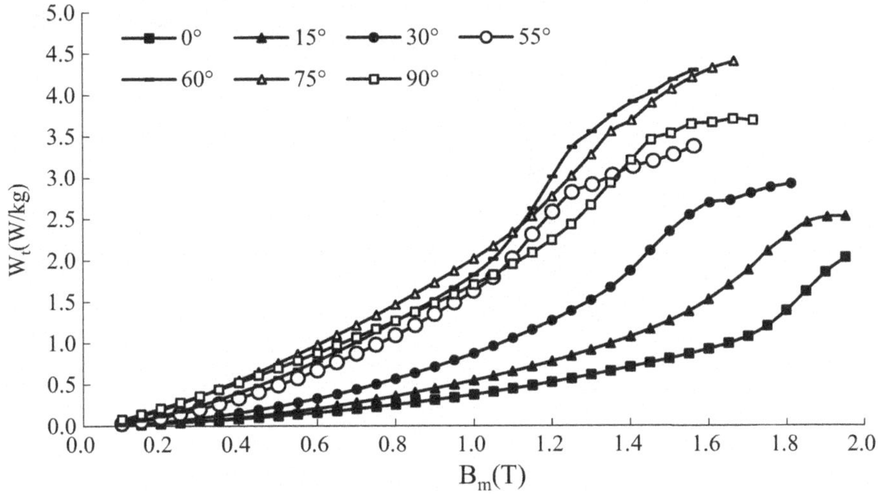

Fig. 8.9 B_m–W_t curves measured before annealing of 30P120

8.5.2 B_m–W_t *Curve Before Annealing (30P120)*

The B_m–W_t curves measured before annealing of specimens 30P120 with different angles (0°–90°) to the rolling direction are shown in Fig. 8.9.

The change of specific total loss with sampling angle can be seen from the 30P120's B_m–W_t curves measured at 0°–90° before annealing in Fig. 8.9. Similarly, the specimen with an angle of 90° to the rolling direction is not the worst in terms of loss characteristics. According to the results shown in Fig. 8.9, when the maximum magnetic induction is greater than 1.15 T, the specific loss of the specimen with the angle of 60° is higher than that of the specimens with other angles (the angles to the rolling direction). However, since sampling from different angles is limited, it does not mean that the direction of 60° is the worst in specific total loss.

8.5.3 Comparison of B_m–H_m *Curves Measured Before and After Annealing*

By comparing B_m–H_m curves measured before and after annealing of each specimen (30P120) with the angles of 0°–90° to the rolling direction, it is found that there is no obvious change in B_m–H_m curves measured before and after annealing of the specimens with different angles, however, improvement is visible for some sampling angles. See Figs. 8.10, 8.11, 8.12, 8.13, 8.14, 8.15, 8.16 and 8.17. Further analysis based on measurement data [26] also confirms this.

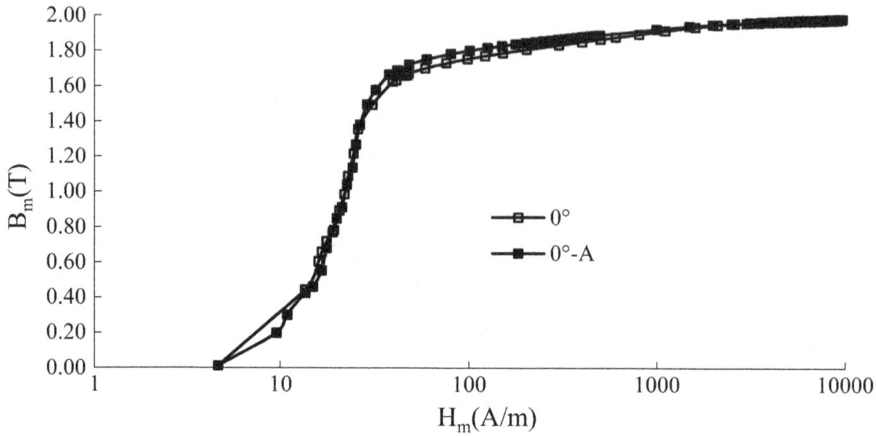

Fig. 8.10 Comparison of B_m–H_m curves measured before and after annealing of silicon steel sheet with an angle of 0° to the rolling direction. *Note* In the Figure the sign "−A" denotes after annealing. The same below

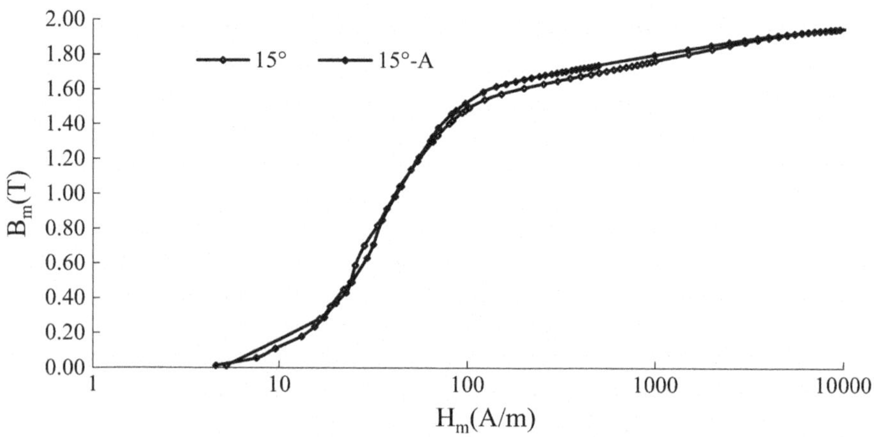

Fig. 8.11 Comparison of B_m–H_m curves measured before and after annealing of silicon steel sheet with an angle of 15° to the rolling direction

8.5.4 Comparison of B_m–W_t *Curves Measured Before and After Annealing*

The comparison of B_m–W_t curves measured before and after annealing of specimens (30P120) with different angles to the rolling direction is shown in Figs. 8.18, 8.19, 8.20, 8.21, 8.22, 8.23 and 8.24.

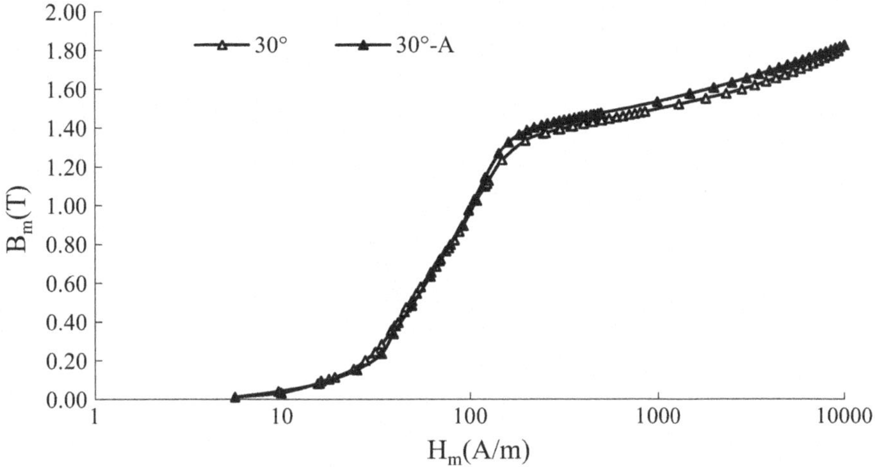

Fig. 8.12 Comparison of B_m–H_m curves measured before and after annealing of silicon steel sheet with an angle of 30° to the rolling direction

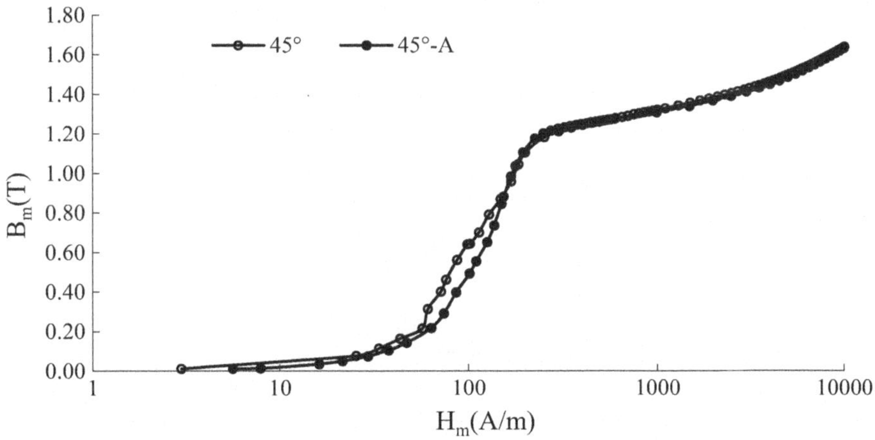

Fig. 8.13 Comparison of B_m–H_m curves measured before and after annealing of silicon steel sheet with an angle of 45° to the rolling direction

Through the comparison of B_m–W_t curves measured before and after annealing of the silicon steel sheet specimens with the angles of 0°–90° to the rolling direction, it is found that:

(a) According to the experimental results of all the test specimens, the specific total loss is decreased after annealing.

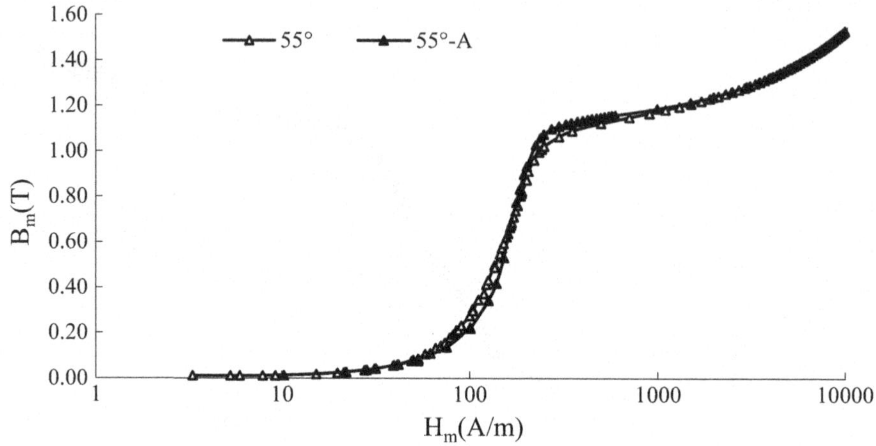

Fig. 8.14 Comparison of B_m–H_m curves measured before and after annealing of silicon steel sheet with an angle of 55° to the rolling direction

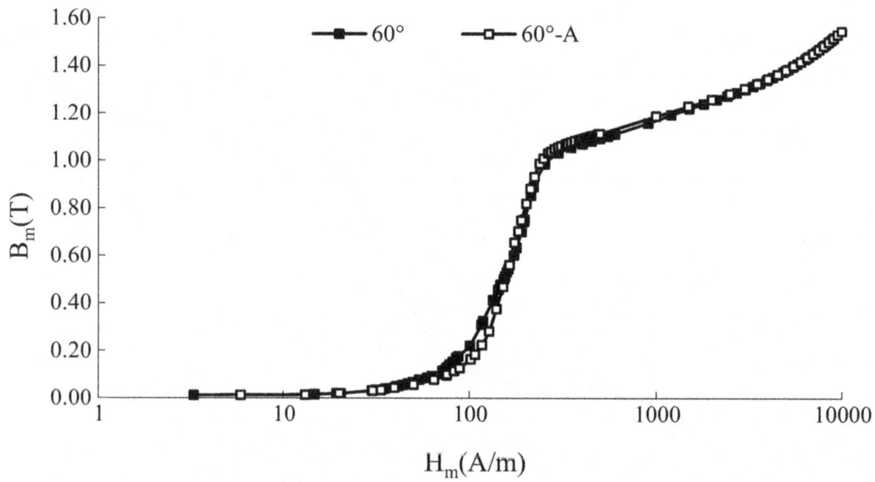

Fig. 8.15 Comparison of B_m–H_m curves measured before and after annealing of silicon steel sheet with an angle of 60° to the rolling direction

(b) the loss curves of specimens with the angles of 0°, 15° and 30° almost showed no change.

(c) the specimens with the angles of 55°, 60°, 75° and 90° show obvious differences in the loss measured before and after annealing as the magnetic flux density is gradually increased, and the specific total loss measured after annealing is reduced by up to 10%.

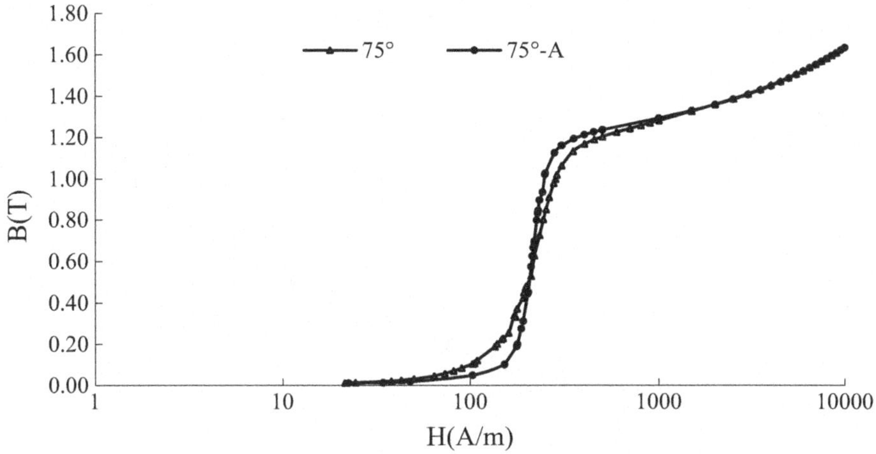

Fig. 8.16 Comparison of B_m–H_m curves measured before and after annealing of silicon steel sheet with an angle of 75° to the rolling direction

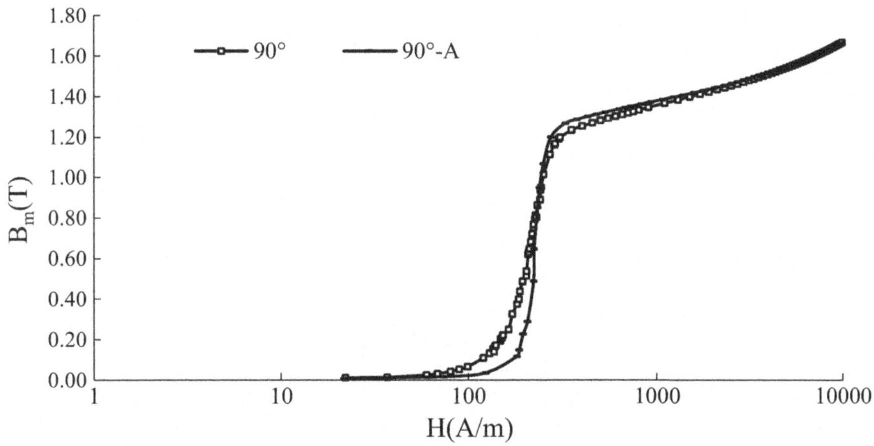

Fig. 8.17 Comparison of B_m–H_m curves measured before and after annealing of silicon steel sheet with an angle of 90° to the rolling direction

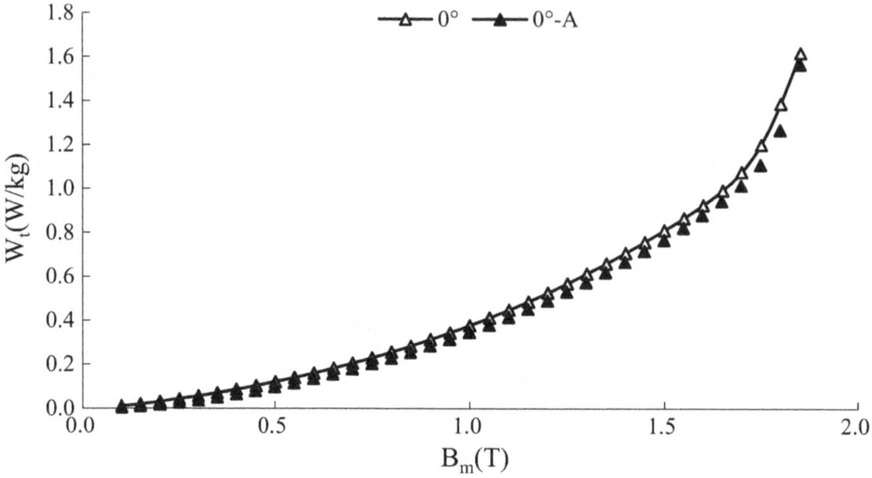

Fig. 8.18 Comparison of B_m–W_t curves measured before and after annealing of silicon steel sheet with an angle of 0° to the rolling direction

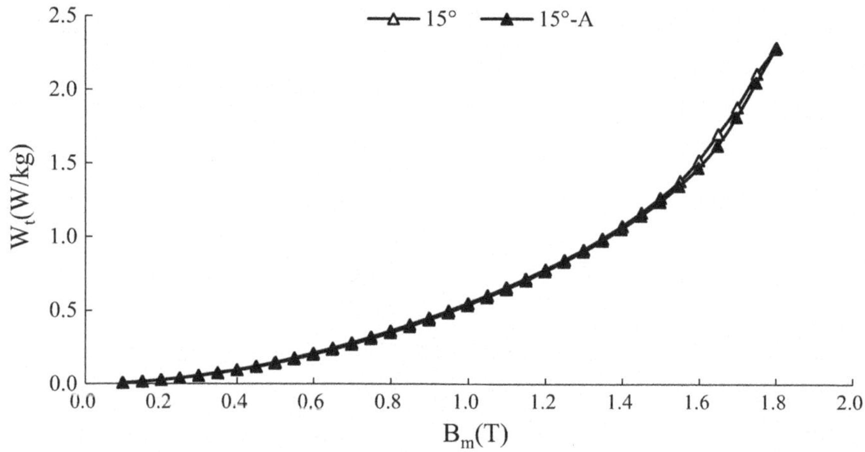

Fig. 8.19 Comparison of B_m–W_t curves measured before and after annealing of silicon steel sheet with an angle of 15° to the rolling direction

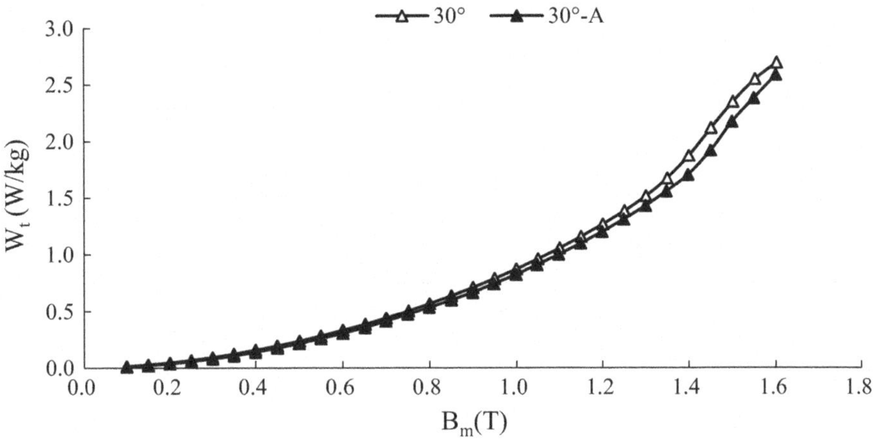

Fig. 8.20 Comparison of B_m–W_t curves measured before and after annealing of silicon steel sheet with an angle of 30° to the rolling direction

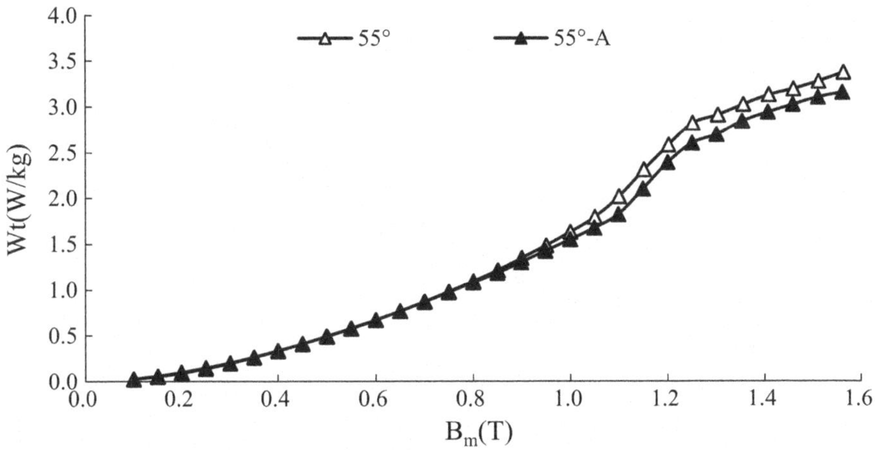

Fig. 8.21 Comparison of B_m–W_t curves measured before and after annealing of silicon steel sheet with an angle of 55° to the rolling direction

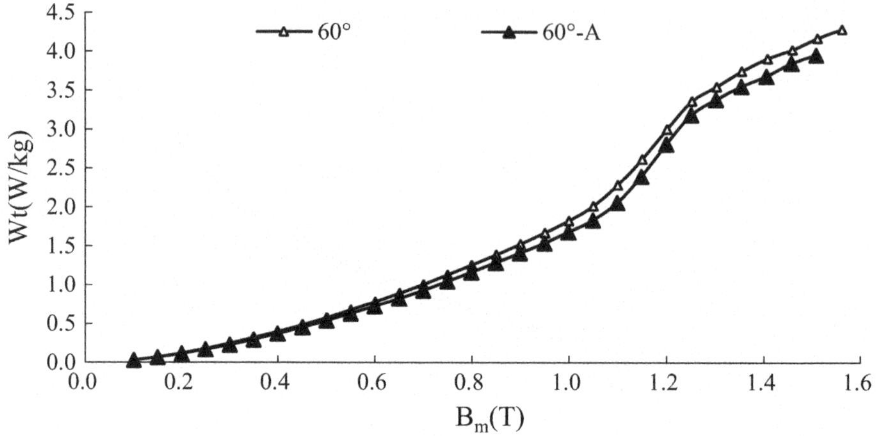

Fig. 8.22 Comparison of B_m–W_t Curves Measured before and after annealing of silicon steel sheet with an angle of 60° to the Rolling Direction

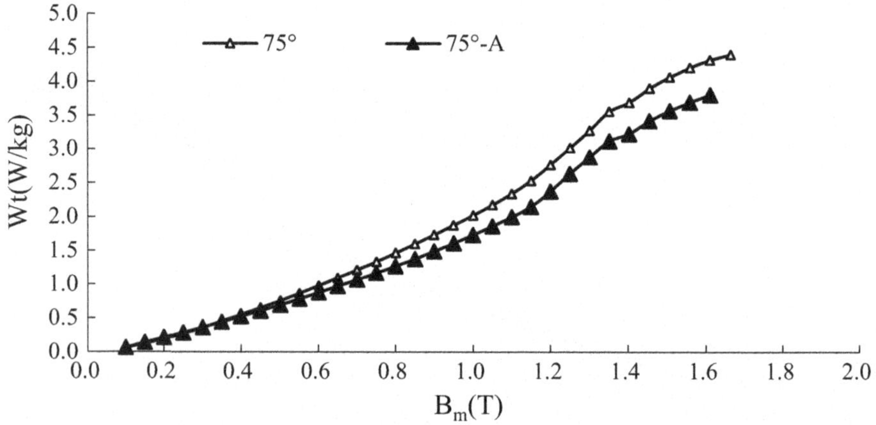

Fig. 8.23 Comparison of B_m–W_t curves measured before and after annealing of silicon steel sheet with an angle of 75° to the rolling direction

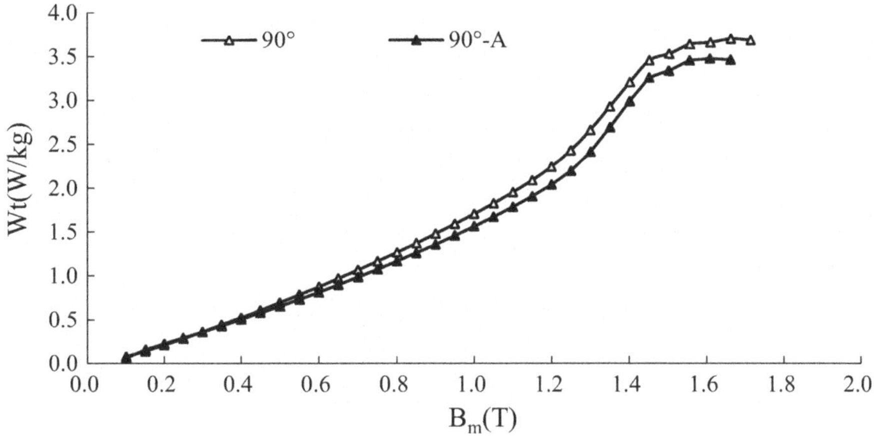

Fig. 8.24 Comparison of B_m–W_t curves measured before and after annealing of silicon steel sheet with an angle of 90° to the rolling direction

8.6 Measuring Record of Voltage Starting Distortion in Magnetic Measurement Before and After Sample-Annealing

The records of the B_m–H_m curves measured before and after annealing of test specimens (30P120), and the voltage, current, magnetic field intensity and magnetic induction measured when voltage waveform distortion is found during B_m–W_t curve experiment are shown in Tables 8.2 and 8.3.

At the same time, the corresponding voltage and current waveforms were obtained by the authors from the B_m–H_m curves and B_m–W_t curves measured before and after annealing of the specimens with different sampling angles. The influence of the sampling angle, stress relief annealing and the measured magnetic properties (e.g. magnetization curve and loss curve) on the waveforms of voltage and current was therefore investigated in detail [26].

8.7 Concluding Remarks

The solution of the Epstein-combination measurement and the loss data weighted processing is proposed based on the non-uniformity of the field quantity and magnetic properties in the Epstein frame. It helps to reasonably determine the Epstein equivalent magnetic path length and related magnetic properties.

According to the demands in industrial applications, the magnetic properties of silicon steel sheet specimens (30P120) with different angles to the rolling direction were measured before and after stress relief annealing by means of 25 cm Epstein

Table 8.2 Record of B_m–H_m curves measured before and after annealing when voltage waveform distortion is found during experiment

Angle between rolling and shearing directions	Before annealing				After annealing			
	Average voltage (V)	Average current (A)	Peak value of magnetic field (A/m)	Peak value of magnetic induction (T)	Average voltage (V)	Average current (A)	Peak value of magnetic field (A/m)	Peak value of magnetic induction (T)
0°	27.40	0.1908	802.4	1.897	27.69	0.1792	794.8	1.917
15°	23.58	0.1056	300.2	1.654	24.84	0.1462	494.6	1.741
30°	20.21	0.1540	401.4	1.421	21.81	0.2941	990.6	1.534
45°	18.63	0.2847	846.2	1.306	18.68	0.3147	998.5	1.309
55°	16.46	0.2762	700.2	1.149	16.60	0.2337	594.8	1.159
60°	15.61	0.2446	499.9	1.094	16.57	0.3083	793.8	1.160
75°	16.64	0.2753	398.5	1.164	16.60	0.2374	299.2	1.160
90°	12.14	0.2216	238.2	0.9083	19.04	0.2978	500.5	1.316

Table 8.3 Record of B_m–W_t curves measured before and after annealing when voltage waveform distortion is found during experiment

Angle between rolling and shearing directions	Before annealing				After annealing			
	Average voltage (V)	Average current (A)	Specific total loss (W/Kg)	Peak value of magnetic induction (T)	Average voltage (V)	Average current (A)	Specific total loss (W/Kg)	Peak value of magnetic induction (T)
0°	28.20	0.5312	2.038	1.954	27.72	0.1862	2.032	1.918
15°	24.23	0.1574	1.902	1.700	26.37	0.4884	2.476	1.850
30°	20.62	0.2010	2.115	1.450	22.73	0.5467	2.583	1.600
45°	19.27	0.4618	247.8	1.352	18.55	0.2792	2.291	1.300
55°	17.19	0.4978	2.461	1.201	17.18	0.3943	2.392	1.200
60°	16.41	0.3526	2.624	1.151	17.12	0.4171	2.818	1.199
75°	17.88	0.3700	3.018	1.251	18.61	0.4342	2.878	1.301
90°	19.53	0.4204	2.932	1.351	17.36	0.2627	2.041	1.200

frame. Such multi-directional electromagnetic properties are generally not provided by silicon steel sheet suppliers. By means of the Epstein frame, with test specimens sampled in different directions, the authors obtained the magnetic properties of grain-oriented silicon steel sheets in different directions, and investigated the effects of both sampling angle and stress relief annealing on the magnetic properties. As an approximation, the obtained magnetic properties can be considered for modeling and simulation of 3-D magnetic field and calculation of iron loss in silicon steel lamination, but its usefulness is yet to be proven in more detail.

Moreover, it should be noted that further research is needed to correctly evaluate the obtained multi-directional magnetic properties from the viewpoint of the multi-angle specimens' microscopic magnetic structure [27].

Acknowledgements The authors thank all colleagues for the co-research on the Epstein combination method. Under the guidance of former Chief Engineer of Baobian Electric., Qifan Hu, the magnetic measurement based on multi-angle sampling was completed by Guisheng Han, Lanrong Liu, Yana Fan, and Junjie Zhang from the Institute of Power Transmission and Transformation Technology, Baobian Electric., and Dr. Xian Zhang of Tianjin Polytechnic University. It was also supported by colleagues from other relevant departments of Baobian Electric.

References

1. P. Silvester and M. V. K. Chari, "Finite element solution of saturable magnetic field problems," *IEEE Power App. Syst.*, vol. PAS-89, pp; 1642–1651, Dec. 1970.
2. T. Yamamoto and Y. Ohya, "Single Sheet tester for measuring core losses and permeabilities in a silicon steel sheet," *IEEE Trans. on Magnetics*, 1974, 10(2): 157–159.
3. T. Nakata, "Analysis of flux distribution of three phase three-limbed transformer core," Electrical Engineering in Japan, vol. 95, no. 3, pp. 43–50,1975.
4. T. Nakata, Y. Ishihara, K. Yamada, and A. Sasano, "Nonlinear analysis of rotating flux in the T-joint of a three phase three phase transformer core," in Proc. EPS Soft Magnetic Materials Conf., Bratislava, Sep. 1977, pp. 57–62.
5. A. Di. Napoli and R. Paggi, "Sizing of magnetic shield in saturated condition for large transformers," in Proc. of IEEE PAS 1977, A77 663-8.
6. A. Di. Napoli and S. Vranco, "Anisotropic effect on magnetic field distribution in ferromagnetic region of transformer core," presented at INTERMAG IEEE Conf., Boston, MA, Apr. 1980.
7. M. Chiampi, A. Negro, M. Tartaglia, "A finite element method to compute three-dimensional magnetic field distribution in transformer cores," *IEEE Trans. on Magnetics,* vol. 16, pp. 1413–1419, Nov. 1980.
8. A. Di. Napoli and R. Paggi, "A Model of Anisotropic Grain-Oriented Steel," *IEEE Trans. on Magnetics*, vol. 19, no. 4, 1983, pp. 1557–1561.
9. P. P. Silvester, and R. P. Gupta, "Effective computational models for anisotropic soft B-H curves," *IEEE Trans. on Magnetics*, vol. 27, no. 5, 1991, pp. 3804–3807.
10. N. J. Layland, A. J. Moses, N. Takahashi, and T. Nakata, "Effects of shape on samples of silicon-iron on the directions of magnetic field and flux density," Nonlinear Electromagnetic Systems, A. J. Moses and A. Basak (Eds.), IOS Press, 1996, pp:800–803.
11. M. Enokizono, K. Yuki, and S. Kawano, "An improved magnetic field analysis in oriented steel sheet by finite element method considering tensor reluctivity," *IEEE Trans. on Magnetics*, vol. 31, no. 3, 1995, pp. 1797–1800.

12. M. Enokizono and N. Soda, "Finite-element analysis of transformer model core with measured reluctivity," *IEEE Trans. on Magn*etics, vol. 33, pp. 4110–4112, May 1997.
13. D. Xie and Z. Cheng, "Development of magnetic characteristic modeling of electrical steel sheet in computational electromagnetics," Electromagnetic field problems and applications-II, volume 22, TSI Press Series, USA, 2008, p. 351–355.
14. IEC 60404-2 Ed.3.1(2008): Magnetic Materials - Part 2: Methods of measurement of the magnetic properties of electrical steel sheet and strip by means of an Epstein frame; Amendment 1.
15. IEC 60404-3Ed.2.2(2010): Magnetic Materials–Part 3: Methods of measurement of the magnetic properties of electrical steel strip and sheet by means of a single sheet tester-Edition2.2;ConsolidatedReprint.
16. Z. Cheng, N. Takahashi, B. Forghani, A. Moses, P. Anderson, Y. Fan, T. Liu, X. Wang, Z. Zhao, and L. Liu, "Modeling of magnetic properties of GO electrical steel based on Epstein combination and loss data weighted processing," *IEEE Trans. on Magnetics*, vol. 50, no. 1, 6300209, 2014.
17. M. Enokizono and N. Soda, " Direct magnetic loss analysis by FEM considering vector magnetic properties," *IEEE Trans. on Magnetics*, vol. 34, 1998, pp. 188–195.
18. J. Zhu, J. Zhong, Z. Lin, and J. D. Sievert, "Measurement of magnetic properties under 3-D magnetic excitations," *IEEE Trans. on Magnetics*, vol. 39, no. 5, Sep. 2003, pp. 3429–3431.
19. K. Fujiwara, T. Adachi, and N. Takahashi, "A proposal of finite-element analysis considering two-dimensional magnetic properties," *IEEE Trans. on Magnetics*, vol. 38, no. 2, 2002, pp. 889–892.
20. D. Lin, P. Zhou, Z. Badics, W. Fu, Q. Chen, and Z. J. Cendes, "A new nonlinear anisotropic model for soft magnetic material," *IEEE Trans. on Magnetics*, vol. 42, no. 4, 2006:963–966.
21. A. J. Moses, N. Derebasi, G. Loisos and A. Schoppa, Aspects of the cut edge effect stress on the power loss and flux density distribution in Electrical steel sheets, J. Magnetism and Magnetic Materials, 215–216, 2000, 690-692.
22. P. Marketos, S. Zurek, and A. J. Moses, "Calculation of the mean path length of the Epstein frame under non-sinusoidal excitations using the double Epstein method," J. Magn. Magn. Mater., vol. 320, pp. 2542–2545, 2008.
23. J. D. Sievert, "Determination of AC magnetic power loss of electrical steel sheet: present status and trends," *IEEE Trans. on Magnetics*, vol. 20, no. 5, PP. 1702–1707, 1984.
24. A. J. Moses and P. S. Phillips, "Effect of stacking methods on Epstein-square power loss measurements," Proc. IEE, 124 (1977) 413–416.
25. Z. Cheng, N. Takahashi, B. Forghani, Y. Du, Y. Fan, L. Liu, and H. Wang, "Effect of variation of B-H properties on both iron loss and flux in silicon steel lamination," *IEEE Trans. on Magnetics*, vol. 47, no. 5, pp. 1346–1349, 2011.
26. Magnetic measurement Report using Epstein frame based on multi-angle samples, Baobian Electric., April 2008.
27. A. J. Moses, "Relevance of microstructure and texture to the accuracy and interpretation of 1 and 2 directional characterisation and testing of grain-oriented electrical steels," 55(2), pp. 3–13, IJAEM(International Journal of Applied Electromagnetics and Mechanics), Oct. 2017.

Chapter 9
Electromagnetic Property Modeling Based on Product-Level Core Models

Tao Liu

Abstract The magnetic properties of power transformer core are dependent on the joint type, the magnetic material, the machining, and assembly technology level. The "two-core method" is proposed by the authors to determine the magnetic properties in the joint area and the middle uniform area, for which two product-level core models of the same material (27ZH100) and the same joint type but different lengths of magnetic path are specially designed and manufactured. The total exciting power and loss of the two-core models are measured separately, and then the exciting power and the specific total loss of the joint area and the middle uniform area of the core can be calculated based on the measurement results and the design parameters of the core. Each of the models is assembled entirely in accordance with the standard manufacturing technology of the power transformer core and is actually one-staged lamination of the multi-stage laminated core, which can be used to study the magnetic properties of the actual transformer core, such as the *B*–*H* curve, the specific total loss curve, especially the magnetic property curve under oversaturation, as well as the building factor and the difference in magnetic properties between laminated core and single sheet of material. Considering the temperature rise occurred during the operation of a transformer, the experimental study on oriented silicon steel material and core model at different ambient temperatures is carried out, which provides a reference for power transformer simulation and further design optimization. In addition, the influence of stress-relief annealing on the electromagnetic properties of the ring core is also studied.

Keywords Magnetic properties · Core model · Magnetic measurement · Exciting power · Specific total loss · Temperature effect · Stress-relief annealing

T. Liu (✉)
Institute of Power Transmission and Transformation Technology,
Baobian Electric, Baoding, China
e-mail: liutaojszx@btw.cn

Z. Cheng et al. (eds.), *Modeling and Application of Electromagnetic and Thermal Field in Electrical Engineering*, https://doi.org/10.1007/978-981-15-0173-9_9

9.1 Introduction

The magnetic properties of transformer core, such as exciting power and no-load power loss, are dependent on the joint type, the magnetic material, the processing technology, and the assembly technology. The 3D magnetic field distribution in the joint area is complex and is difficult to analyze accurately due to insufficient data on the magnetic properties of the material. The electromagnetic property data, such as no-load voltage and current waveform, iron loss, hysteresis loop, and average magnetization curve of the iron core, under different conditions, are the basis for studying the problems of no-load over-excitation or DC bias of transformer. The "two-core method" is proposed by the authors to measure the magnetic properties in the joint area and the middle uniform area, for which two product-level core models of the same material (27ZH100), the same joint type but different lengths of magnetic path is specially designed and manufactured, and based on the two-core models, the magnetic properties are measured and studied. In addition, the influence of temperature on the magnetic properties of iron core is also investigated. The main experimental research contents in this chapter are as follows:

(1) The no-load magnetic properties of product-level core model under standard excitation condition are measured, including voltage, current, *B–H* curve, loss curve, exciting power, etc. [1, 2].
(2) The difference in magnetic properties measured under standard condition between the oriented silicon steel and the transformer core, based on actual conditions, is studied.
(3) Based on two-core models with the same joint type, material, and lamination section but different external structures, the exciting power and specific total loss of the core models are measured respectively by means of the two-core method, and the distribution of exciting power and specific total loss in the joint area and the middle uniform area of the iron core is calculated according to the experimental results [3, 4]. In reference [3], the equation for calculating the components of total exciting power is given without explaining the measurement method. The "two-core method" is proposed to implement the separation of the exciting power and specific total loss of the core in the joint area and the middle uniform area.
(4) Assuming that the magnetic field distribution in the joint areas of the two-core models is uniform, the total excitation power (or total loss) and total weight of two-core models are subtracted separately, then the total exciting power (or total loss) difference value is divided by the total weight difference value, and finally, the exciting power (or specific total loss) can be obtained, which eliminate the influence of joints.
(5) The building factor (BF) of transformer cores is studied based on the product-level core model manufactured according to the assembly technology level of transformer cores.

(6) The variation of magnetic properties of oriented silicon steel material and transformer core model under different ambient temperatures is studied based on the actual working conditions of the transformer.
(7) The influence of stress-relief annealing on the electromagnetic properties of the ring core is also investigated based on two ring cores with the same design specification. Two ring cores of the same specification are used as test specimens to further ensure the validity of the measurement results, which can be compared with the magnetic property data of the original oriented silicon steel material. The experimental results can be used as a reference for the R&D and design of mutual inductors.

9.2 Measurement of Magnetic Properties of Product-Level Core Model

9.2.1 Two-Laminated Core Models

The main design parameters of the two product-level core models (Model C1 and Model C2) in this experiment are shown in Table 9.1.

The specific design size of the two-core models is shown in Figs. 9.1 and 9.2.

The joint type of core: 45° mitered joint, two sheets as one stack in core stacking, and three-level step lap, is shown in Fig. 9.3. Among the parameters related to lamination thickness, the lamination coefficient of the core model is 0.97.

The exciting coils and search coils are designed and wound considering equal turns and interlinkage flux. The configurations of the search coils and exciting coils in the enlarged cross sections are shown in Figs. 9.1 and 9.2. The exciting coil and the search coil are wound clinging to the core, and the search coil is located between two exciting coils connected in parallel. The iron core is clamped by insulating non-magnetic components. The four local search coils of Model C1 and three local search coils of Model C2 are wound at their respective positions as shown in Figs. 9.4 and 9.5.

The actual lamination thickness of the two-core models assembled is 20.7 mm, and the manufactured core models are shown in Figs. 9.6 and 9.7.

9.2.2 Experimental Equipment

The harmonic power supply system is established, which is used to realize magnetic measurement under standard and non-standard excitations, as well as the instruments. See Table 9.2.

Table 9.1 Design parameters of the core models

Name	Model C1	Model C2
Type of silicon steel sheet	27ZH100 (Nippon Steel, unannealed)	27ZH100 (Nippon Steel, unannealed)
Density of silicon steel sheet (kg/m^3)	7.65×10^3	7.65×10^3
Design lamination section area of the core (mm^2)	2.0×10^3	2.0×10^3
Design weight of the core (kg)	41.55	29.68
Turns of exciting coil	144 turns (two-paralleled conductors, single layer)	144 turns (two-paralleled conductors, single layer)
Density of conductor (kg/m^3)	8.90×10^3	8.90×10^3
Wire gauge of exciting coil (mm)	Ø1.6	Ø1.6
Turns of search coil	144 turns (single layer)	144 turns (single layer)
Conductor conductivity at 20 °C (S/m)	5.71×10^7	5.71×10^7
Wire gauge of search coil (mm)	Ø1.6	Ø1.6
Wire gauge of local search coil (mm)	Ø1.6	Ø1.6
Turns of local search coil	10 turns	10 turns

9.2.3 Experimental Content and Circuit

The magnetic property measurement experiment based on two-core models (Model C1 and Model C2) is as follows:

(1) The induced voltage, exciting current, and other corresponding waveform data of the core models at different saturation levels are measured.
(2) The magnetic properties of core model, such as exciting power and magnetization curve, are measured and compared with the magnetic property data provided by silicon steel sheet manufacturers.

The experimental circuit for measuring the magnetic properties of the core model is shown in Fig. 9.8.

9.2.4 Measurement Procedure and Key Points

The measurement procedure and key points are as follows:

(1) Connect and check the experimental circuit according to Fig. 9.8.
(2) The following steps should be taken to check the measurement circuit: close K, open K1, start the power supply, increase the output voltage, and read the voltmeter. If the reading increases gradually, the power supply is normal. Reset

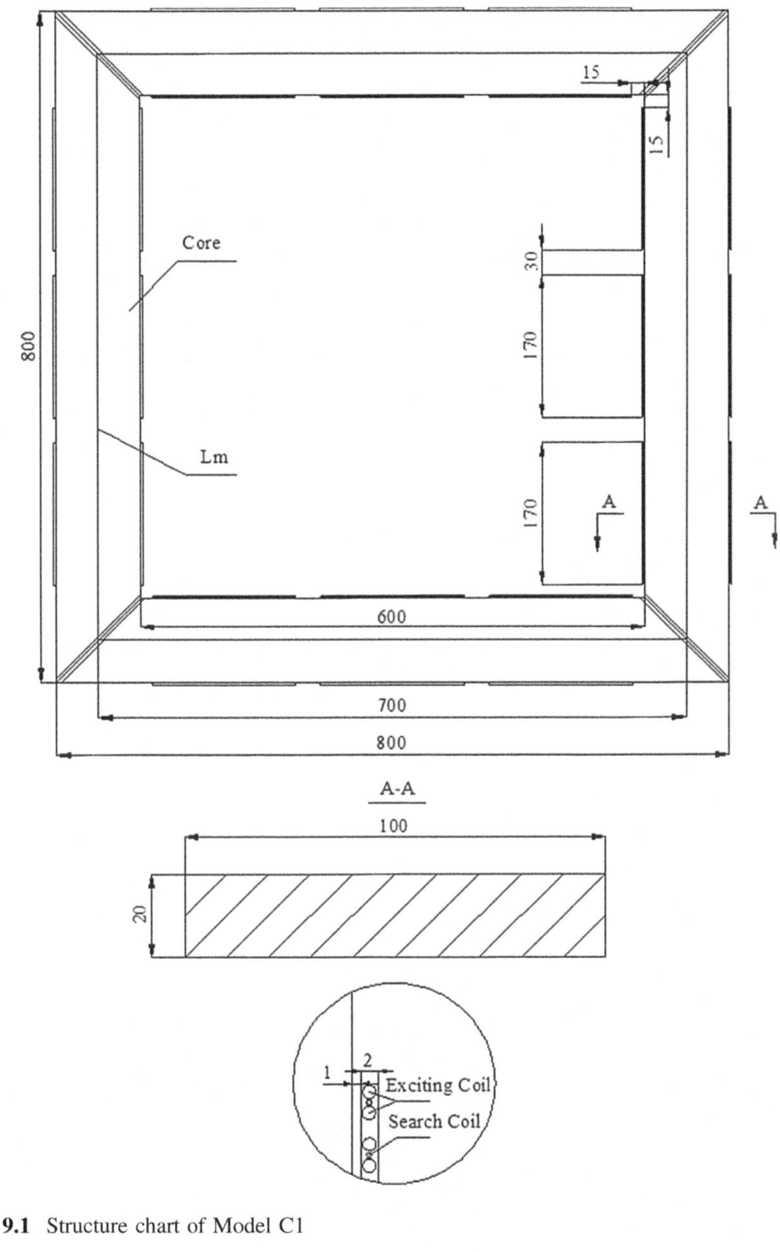

Fig. 9.1 Structure chart of Model C1

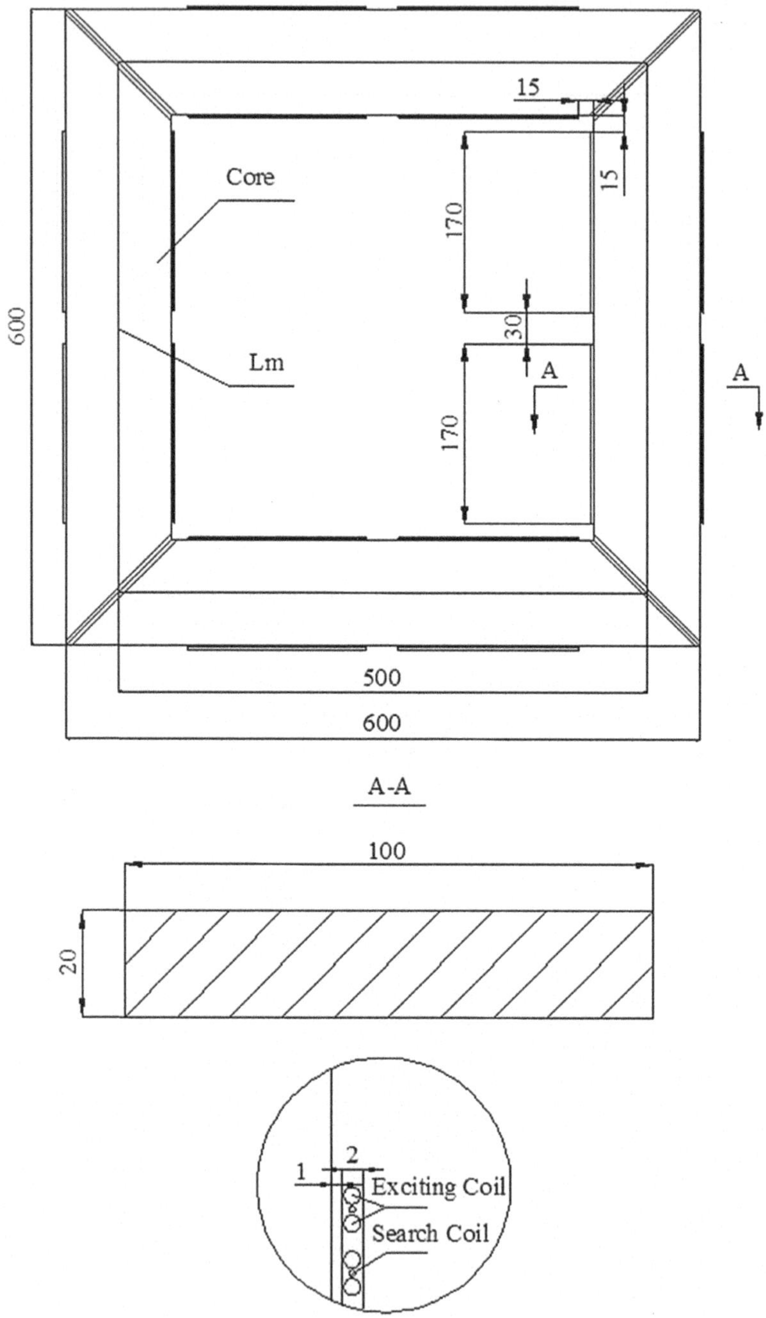

Fig. 9.2 Structure chart of Model C2

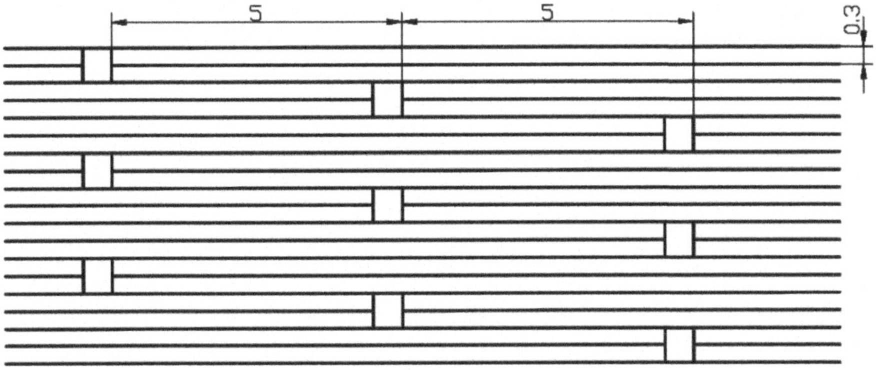

Fig. 9.3 Three-step lap joint (two sheets as one stack in core stacking, unit: mm)

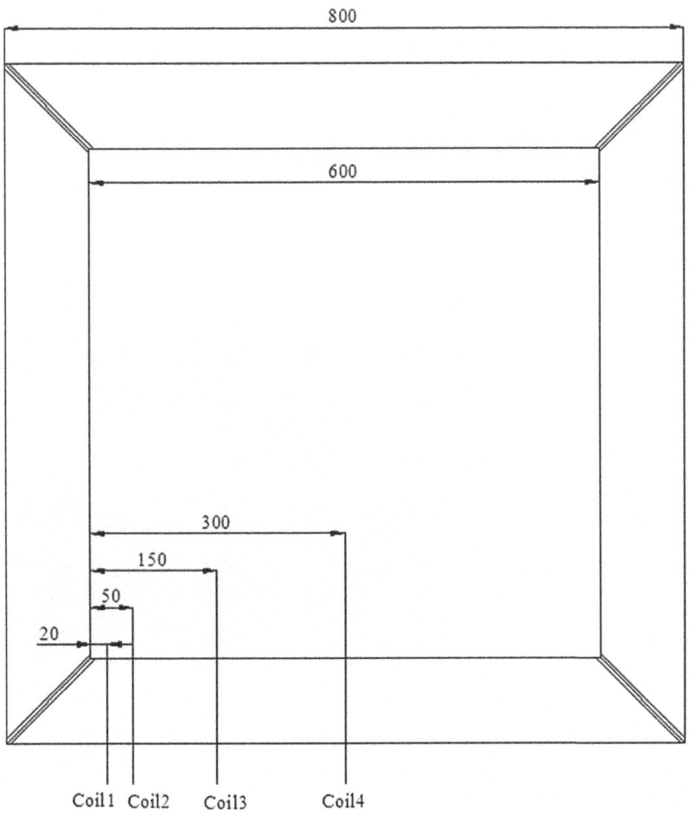

Fig. 9.4 Distribution of local search coils of Model C1

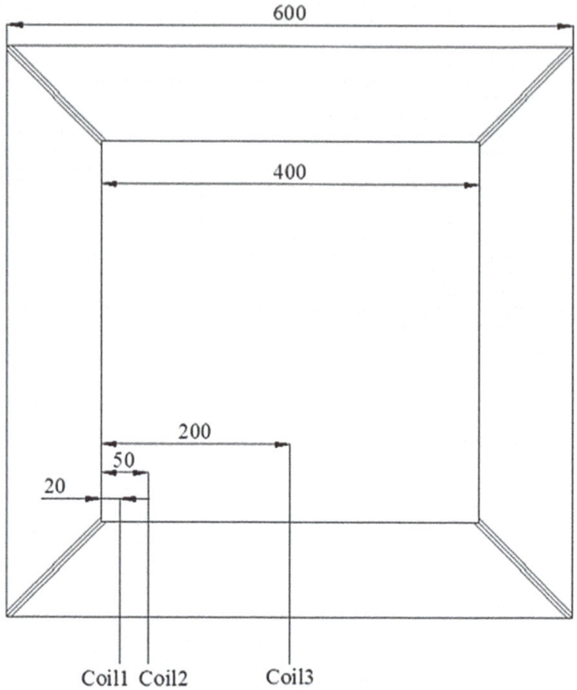

Fig. 9.5 Distribution of local search coils of Model C2

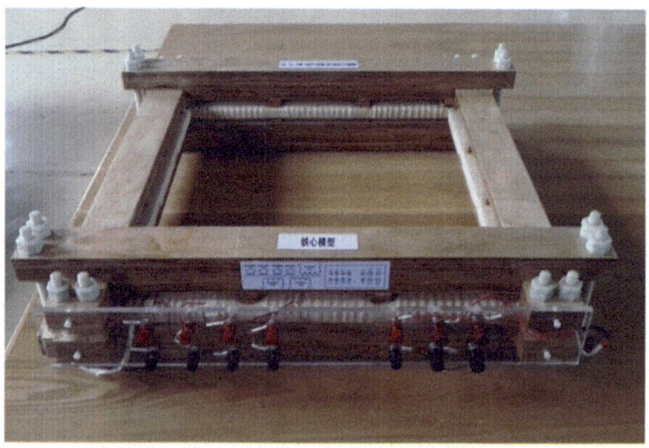

Fig. 9.6 Model C1

the voltage regulator to zero; close K and K1; and slightly boost the voltage regulator. If the voltage and current readings of the power analyzer increase, the circuit is normal, and measurement can be carried out.

(3) Demagnetization should be carried out first. Gradually increase the exciting voltage up to the value corresponding to 1.9 T or more of the magnetic flux density and then reduce it to zero. Voltage should be smoothly and

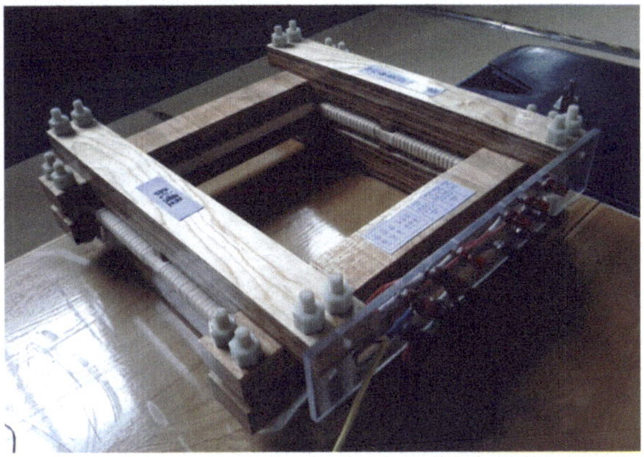

Fig. 9.7 Model C2

Table 9.2 Experimental system and related performances

Equipment	Specifications	Performance index
Contact voltage regulator (Shanghai)	TSGC2J (Regulated current $\leq$ 25 A)	Output voltage 0–430 V; output current: 0–27 A; rated capacity: 20 kVA; three phases; frequency 50–60 Hz
Harmonic power system (NF, Japan)	WF-1974 4520A	Output frequency: 0–20 kHz; output voltage: 0–240 V; rated capacity: 4 kVA; arbitrary waveform available
Precision power analyzer (YOKOGAWA)	WT3000	Maximum current measured: 30 A (rms); maximum voltage measured: 1000 V (rms); current and voltage reading accuracy: ±0.02%; range accuracy: ±0.04%; power accuracy: ±0.06%; frequency range: DC, 0.1 Hz–1 MHz

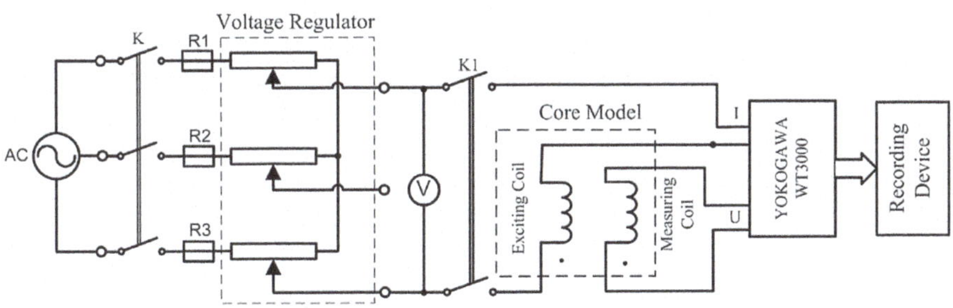

Fig. 9.8 Experimental circuit based on core models

continuously adjusted to avoid waveform jitter and recording erroneous transient losses caused by waveform jitter.

(4) Increase the exciting voltage from zero and regulated according to the estimated value in the log sheet, and record the parameters such as exciting voltage, current, exciting power, total loss and voltage waveform of search coil.
(5) Collate the experimental data, record in the corresponding tables, post-process, and plot.

9.3 Measurement Results of Magnetic Properties of Product-Level Core Model

9.3.1 Waveforms of Exciting Current and Voltage

The induced voltage of the search coil and the current waveform of the exciting coil are selected from the experimental data.

The induced voltage of the search coil and the current waveform of the exciting coil, when B_m of Model C1 is 1.0 and 1.7 T, are shown in Figs. 9.9 and 9.10.

It can be seen that as the magnetic flux density in the core model increases, the distortion degree of the exciting current waveform obviously increases too.

9.3.2 Experimental Data of Core Models

The experimental data of Model C1 are shown in Table 9.3. In the table, the voltage U and current I are both RMS; U_1 is the exciting voltage; I is the exciting current; S is the total exciting power of the core; P is the total iron core loss; U_2 is the induced voltage; u_1 is the measured voltage of the local search coil 1; and u_4 is the measured voltage of the local search coil 4, as shown in Fig. 9.4.

The experimental data of Model C2 are shown in Table 9.4. U_1 is the exciting voltage; I is the exciting current; S is the exciting power of the core; P is the total

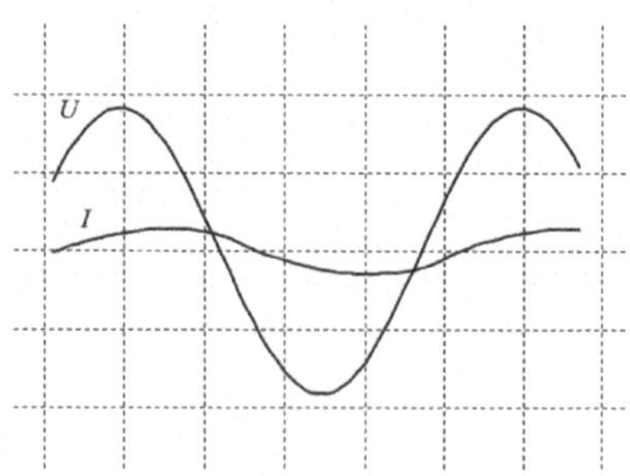

Fig. 9.9 Voltage and current waveforms (B_m: 1.0 T)

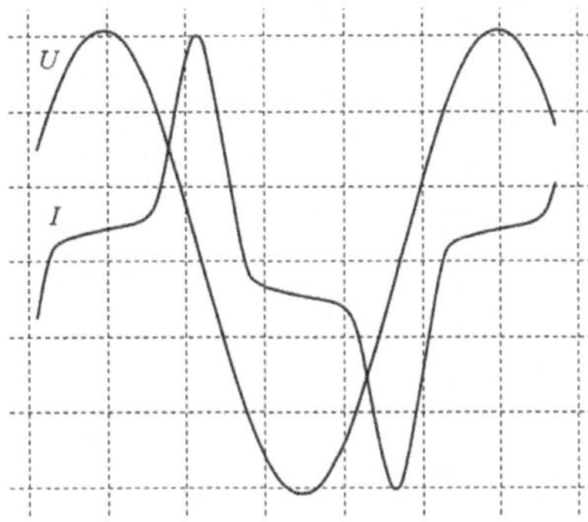

Fig. 9.10 Voltage and current waveforms (B_m: 1.7 T)

Table 9.3 Experimental data of Model C1

U_1 (V)	I (A)	S (VA)	P (W)	U_2 (V)	u_1 (V)	u_4 (V)
11.6630	0.0235	0.2741	0.1664	5.8269	0.4034	0.4046
13.0591	0.0258	0.3375	0.2054	6.5245	0.4518	0.4530
25.5822	0.0449	1.1486	0.7268	12.7815	0.8866	0.8877
38.6394	0.0620	2.3954	1.5890	19.3068	1.3394	1.3409
51.3662	0.0763	3.9174	2.7076	25.6655	1.7810	1.7828
64.1691	0.0889	5.7057	4.0674	32.0634	2.2253	2.2272
77.1740	0.1005	7.7553	5.6842	38.5645	2.6763	2.6786
89.9020	0.1108	9.9633	7.4938	44.9241	3.1178	3.1204
102.7290	0.1208	12.4113	9.5594	51.3345	3.5628	3.5657
109.3990	0.1260	13.7853	10.7244	54.6679	3.7943	3.7973
115.7410	0.1311	15.1690	11.8956	57.8378	4.0144	4.0175
122.5680	0.1367	16.7510	13.2188	61.2485	4.2512	4.2546
128.6850	0.1419	18.2600	14.4656	64.3058	4.4634	4.4670
134.9640	0.1476	19.9250	15.8322	67.4440	4.6812	4.6850
141.0020	0.1533	21.6200	17.1786	70.4610	4.8906	4.8948
147.7940	0.1604	23.7040	18.7846	73.8550	5.1263	5.1301
153.8690	0.1680	25.8460	20.3182	76.8910	5.3368	5.3411
160.5040	0.1779	28.5610	22.0646	80.2070	5.5668	5.5716
167.2160	0.1938	32.4130	23.9098	83.5610	5.7992	5.8049
173.4950	0.2182	37.8560	25.7070	86.6990	6.0159	6.0236
179.9090	0.2534	45.5940	27.5924	89.9040	6.2370	6.2471
186.8550	0.3046	56.9200	29.7262	93.3750	6.4758	6.4894
192.7140	0.3585	69.0970	31.6142	96.3040	6.6766	6.6940
199.0970	0.4279	85.1940	33.7706	99.4920	6.8949	6.9168
205.2080	0.5042	103.4590	35.9784	102.5450	7.1034	7.1304
211.8530	0.5998	127.0640	38.6640	105.8650	7.3297	7.3629
218.4010	0.7087	154.7870	41.7520	109.1360	7.5516	7.5919

(continued)

Table 9.3 (continued)

U_1 (V)	I (A)	S (VA)	P (W)	U_2 (V)	u_1 (V)	u_4 (V)
224.9600	0.8482	190.8200	45.9840	112.4120	7.7729	7.8215
231.3680	1.0607	245.4210	53.2200	115.6070	7.9881	8.0460
237.7030	1.4745	350.5010	64.2000	118.7610	8.2001	8.2672
244.9390	2.5487	624.2800	75.1800	122.3640	8.4413	8.5119
248.3550	3.4115	847.2600	80.0400	124.0630	8.5551	8.6388

Table 9.4 Experimental data of Model C2

U_1 (V)	I (A)	S (VA)	P (W)	U_2 (V)	u_1 (V)	u_3 (V)
11.5695	0.0166	0.1926	0.1170	5.7817	0.4011	0.4004
13.1315	0.0185	0.2430	0.1476	6.5623	0.4553	0.4547
25.6151	0.0321	0.8210	0.5212	12.8008	0.8882	0.8883
38.6901	0.0442	1.7084	1.1402	19.3363	1.3420	1.3420
51.6706	0.0547	2.8243	1.9564	25.8229	1.7927	1.7926
64.4281	0.0637	4.1041	2.9234	32.1991	2.2353	2.2354
76.9280	0.0718	5.5216	4.0358	38.4488	2.6691	2.6691
90.1970	0.0799	7.2096	5.3896	45.0801	3.1295	3.1295
102.5960	0.0872	8.9432	6.8260	51.2769	3.5598	3.5598
109.2000	0.0911	9.9487	7.6512	54.5779	3.7890	3.7892
115.8080	0.0952	11.0200	8.5384	57.8822	4.0184	4.0186
122.1010	0.0991	12.1050	9.4238	61.0269	4.2368	4.2371
128.7430	0.1036	13.3330	10.4052	64.3463	4.4673	4.4677
135.0030	0.1080	14.5740	11.3832	67.4760	4.6844	4.6851
141.4750	0.1128	15.9640	12.4572	70.7100	4.9090	4.9097
147.9380	0.1184	17.5120	13.5890	73.9410	5.1328	5.1343
153.9770	0.1245	19.1630	14.7214	76.9600	5.3423	5.3439
160.9500	0.1335	21.4930	16.1110	80.4450	5.5839	5.5681
167.3070	0.1453	24.3150	17.4478	83.6210	5.8041	5.8070
173.2960	0.1620	28.0660	18.7512	86.6150	6.0115	6.0148
180.5520	0.1911	34.4960	20.3626	90.2410	6.2623	6.2670
186.6670	0.2244	41.8940	21.7846	93.2970	6.4735	6.4795
192.6330	0.2646	50.9620	23.2360	96.2790	6.6793	6.6867
199.2500	0.3168	63.1280	24.9346	99.5860	6.9072	6.9168
205.3990	0.3729	76.6010	26.6166	102.6580	7.1189	7.1304
212.1020	0.4472	94.8600	28.7182	106.0070	7.3493	7.3637
217.9380	0.5219	113.7390	30.8626	108.9220	7.5496	7.5665
224.9130	0.6348	142.7750	34.3920	112.4070	7.7886	7.8089
231.0270	0.7889	182.2590	39.5640	115.4580	7.9973	8.0212
237.8390	1.1496	273.4200	48.0740	118.8520	8.2284	8.2578
244.6220	1.9381	474.1090	55.4560	122.2340	8.4572	8.4929
247.3800	2.4529	606.7900	58.2800	123.6100	8.5501	8.5883

iron core loss; U_2 is the induced voltage; u_1 is the measured voltage of the local search coil 1; and u_3 is the measured voltage of the local search coil 3, as shown in Fig. 9.5.

As can be seen from the measurement data of u_1, u_4 in Table 9.3 and u_1, u_3 in Table 9.4, except for the first few points with smaller values, the measured voltage of the local search coil in the geometric center of the middle uniform area is greater than that near the joint, which indicates that the average flux density of the middle uniform area in the core model is slightly greater than that near the joint.

9.3.3 Magnetic Properties of Core Models and Comparison with Material Properties

9.3.3.1 Magnetic Properties of Core Models

The results of exciting power and specific total loss of the two-core models are obtained by processing the measured experimental data, as shown in Table 9.5.

Based on the experimental data, B_m and H_m (the maximum magnetic field strength H on hysteresis loop corresponding to B_m) can be obtained by means of programming. The B_m–H_m data are listed in Table 9.6.

9.3.3.2 Magnetic Properties of Silicon Steel

The grain-oriented silicon steel sheet (27ZH100), which is the same batch as the core model, was cut into standard 25 cm Epstein specimens and then stress relief annealed. The results of B_m–H_m and B_m–P_{loss} curves measured at ambient temperature of 23 °C are listed in Table 9.7.

9.3.3.3 Comparison of Magnetic Properties

The comparison of magnetic properties of B_m–H_m and B_m–P_{loss} measured using core models and the material is shown in Figs. 9.11 and 9.12.

It can be seen from the comparison of the loss curves that the core model's loss curve is close to the core material's loss curve. At the same magnetic flux density B_m, the specific total loss of core material is the smallest, the specific total loss of Model C1 is slightly larger, and the specific total loss of Model C2 is the largest. This is reasonable because the assembly of cores will generate joints of higher loss density, and its influence on Model C1 is relatively small.

Table 9.5 Exciting power and specific total loss

B_m (T)	Exciting power S_e (VA/kg)		Specific loss P_{loss} (W/kg)	
	Model C1	Model C2	Model C1	Model C2
0.10	0.0075	0.0076	0.0046	0.0046
0.20	0.0268	0.0268	0.0170	0.0170
0.30	0.0552	0.0552	0.0366	0.0368
0.40	0.0908	0.0906	0.0627	0.0626
0.50	0.1326	0.1324	0.0945	0.0941
0.60	0.1792	0.1793	0.1312	0.1310
0.70	0.2309	0.2322	0.1736	0.1733
0.80	0.2872	0.2902	0.2211	0.2214
0.90	0.3508	0.3557	0.2750	0.2754
1.00	0.4216	0.4295	0.3338	0.3349
1.10	0.5047	0.5159	0.4011	0.4025
1.20	0.5993	0.6207	0.4712	0.4770
1.25	0.6606	0.6902	0.5107	0.5186
1.30	0.7476	0.7814	0.5531	0.5631
1.35	0.8701	0.9080	0.5943	0.6084
1.40	1.0434	1.0912	0.6373	0.6547
1.45	1.2830	1.3334	0.6833	0.7033
1.50	1.5867	1.6357	0.7319	0.7525
1.55	1.9595	2.0222	0.7822	0.8069
1.60	2.4079	2.4831	0.8368	0.8651
1.65	2.9403	3.0235	0.8972	0.9266
1.70	3.5936	3.7050	0.9701	1.0049
1.75	4.4217	4.5772	1.0670	1.1087
1.80	5.7395	5.9290	1.2404	1.2872
1.85	8.8559	9.1062	1.5300	1.5689
1.90	17.5288	17.8553	1.8215	1.8779

9.4 Separation of Exciting Power and Active Power Loss in the Joint Area and Middle Uniform Area of the Core

9.4.1 Separation of Exciting Power

The main flux of the core is generated by exciting current. When the core is saturated to a certain degree, leakage magnetic flux will be generated. As an approximate engineering approach, it can be considered that the total exciting power is consumed only in the core joint and in the limb and yoke outside the joint under no-load and within a certain magnetic flux density range [3, 4].

Table 9.6 B_m–H_m data of core models

Model C1		Model C2	
B_m (T)	H_m (A/m)	B_m (T)	H_m (A/m)
0.102	3.780	0.103	3.748
0.200	6.457	0.200	6.467
0.302	8.929	0.302	8.871
0.401	10.984	0.404	11.019
0.500	12.731	0.503	12.897
0.602	14.391	0.601	14.446
0.702	15.877	0.705	16.054
0.802	17.218	0.802	17.372
0.903	18.532	0.904	18.770
1.004	19.790	1.006	20.155
1.098	21.043	1.104	21.874
1.202	23.321	1.203	24.550
1.303	29.184	1.305	31.546
1.354	37.073	1.352	38.793
1.405	48.053	1.409	50.455
1.458	61.614	1.456	64.467
1.503	74.790	1.504	77.065
1.553	91.503	1.554	94.440
1.600	108.767	1.601	112.812
1.651	131.834	1.656	137.180
1.700	156.859	1.700	162.515
1.751	193.290	1.754	203.975
1.799	254.948	1.800	267.440
1.844	391.862	1.849	433.268
1.888	737.754	1.892	792.972
1.906	995.027	1.909	1021.436

The active and reactive components of core loss can be measured on a core model, i.e., the total exciting power and active core loss can be obtained. However, it is impossible to separate the total exciting power at the joint and at the limb and yoke on a single model by experimental approach. The separation of the total exciting power of the iron core is realized by means of the "two-core method" proposed by the authors, i.e., the total exciting power of the core is measured based on the two-core models with the same joint type and material but different structural dimensions, and the distribution of the total exciting power in the joint area and limb and yoke area of the core is determined according to the experimental results [3, 4].

In the separation of total exciting power, core joints can be treated by a different approach: (1) The core joint is treated as a 2D planar joint in a traditional way, i.e., characterized by the exciting power per unit area and (2) due to the adoption of the

Table 9.7 B_m–H_m and B_m–P_{loss} oriented silicon steel

B_m (T)	P_{loss} (W/kg)	H_m (A/m)
0.100	0.0044	4.40
0.200	0.0162	7.35
0.300	0.0347	9.88
0.400	0.0594	12.02
0.500	0.0895	13.85
0.600	0.1251	15.46
0.700	0.1657	16.96
0.800	0.2117	18.27
0.900	0.2631	19.50
1.000	0.3205	20.72
1.100	0.3832	21.86
1.200	0.4525	23.02
1.300	0.5274	24.09
1.400	0.6098	25.28
1.500	0.7032	28.14
1.600	0.8134	36.20
1.650	0.8801	44.66
1.675	0.9185	51.27
1.700	0.9632	60.44
1.725	1.0158	73.45
1.750	1.0797	92.81
1.775	1.1567	122.39
1.800	1.2510	170.40
1.825	1.3651	251.71
1.850	1.4944	394.16
1.875	1.6427	640.82
1.900	1.7935	1068.57
1.910	1.8315	1369.63
1.920	1.8968	1810.31
1.930	1.9725	2587.88

joint type of stepping lap, which is beyond the concept of "planar joint," and as the core saturation degree increases, the core joint can be expanded into a 3D area bounded by the envisaged "geometric joint," which can be called "joint influence domain (volume)," that is, expanded to both sides of the joint. In fact, this approach extends the scope of the traditional method and can bring about more reasonable results.

In this experiment, the second approach is adopted. Figure 9.13a, b shows that the joint influence domain of the core can be treated in two different ways, where x denotes the distance the joint influence domain extending outward from the

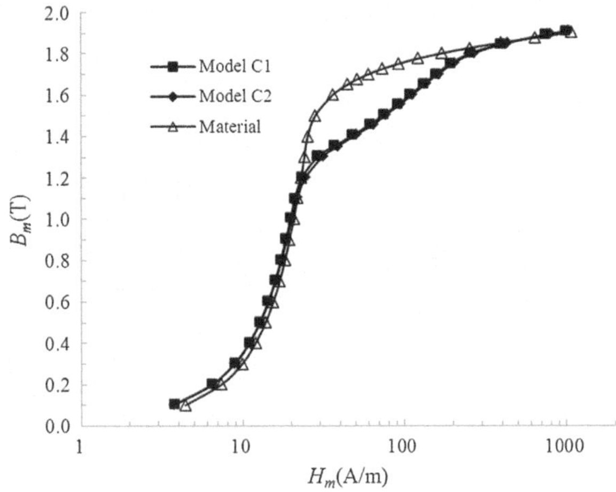

Fig. 9.11 Comparison of B_m–H_m curves

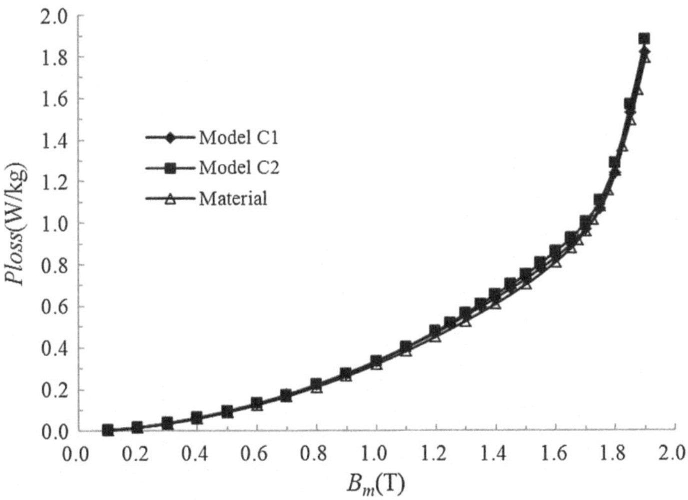

Fig. 9.12 Comparison of specific total loss curves

neighborhood of the core joint and indicates the volume change of the joint influence domain.

From the distribution characteristics of the flux in silicon steel materials, the processing method of Fig. 9.13a is relatively reasonable and is adopted in this experiment.

The distances corresponding to x in Fig. 9.11a are listed in Table 9.8.

The total exciting powers S of the two-core models (of the same joint and material but different structural dimensions) are measured respectively in a certain

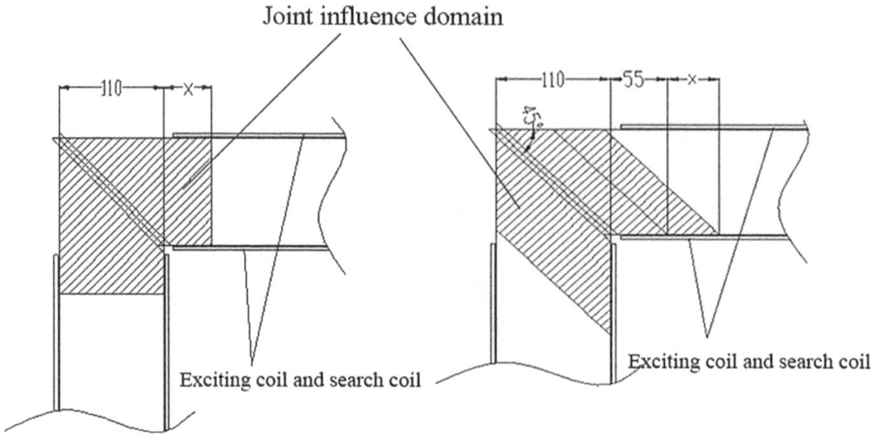

(a) Joint influence domain (I) (b) Joint influence domain (II)

Fig. 9.13 Schematic diagram of joint influence domain (hatched part)

Table 9.8 Distances corresponding to x

S/N	1	2	3	4	5	6	7	8
x (mm)	10	20	30	40	50	60	70	80

magnetic flux density range. Let the exciting power per unit weight in the joint influence domain and its outer middle uniform area be S_c and S_m, respectively, that can be expressed by the relations:

$$\rho V_{m1} S_m + \rho V_{c1} S_c = S_1 \tag{9.1}$$

$$\rho V_{m2} S_m + \rho V_{c2} S_c = S_2 \tag{9.2}$$

where $V_{m1} = (600 - 2x) \times 100 \times h \times 0.97 \times 4 \times 10^{-9}$ (m^3) denotes the core volume outside the joint influence domain of Model C1; $V_{m2} = (400 - 2x) \times 100 \times h \times 0.97 \times 4 \times 10^{-9}$ (m^3) denotes the core volume outside the joint influence domain of Model C2; $V_{c1} = (100 + 2x) \times 100 \times h \times 0.97 \times 4 \times 10^{-9}$ (m^3) denotes the volume of the joint influence domain of Model C1; $V_{c2} = (100 + 2x) \times 100 \times h \times 0.97 \times 4 \times 10^{-9}$ (m^3) denotes the volume of the joint influence domain of Model C2; h(20.7 mm) denotes the core thickness of Model C1 and Model C2, and the stacking factor is 0.97; S_1 denotes the total exciting power (VA) of Model C1 at a certain magnetic flux density measured by the power analyzer WT3000; and S_2 denotes the total exciting power (VA) of Model C2 at the same magnetic flux density measured by the power analyzer WT3000.

It should be noted that the joint influence domain will be expanded outward from both sides of the joint as the core saturation degree increases.

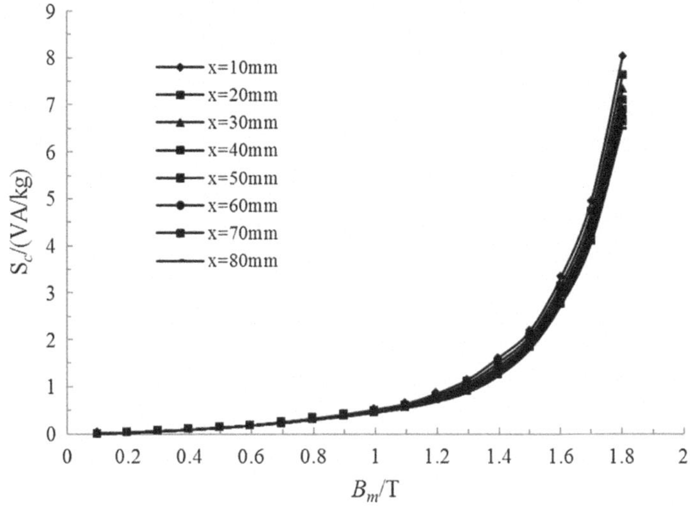

Fig. 9.14 B_m–S_c curve with the variation of extended distance from the joint to the middle uniform area of the core

The exciting power S_c per unit weight at the joint under different magnetic flux densities can be obtained by using experimental data and Eqs. (9.1) and (9.2), and the value of x has no influence on the exciting power S_m per unit weight in the middle uniform area of the core on the premise that the set joint influence domain of the two-core models is uniform.

The B_m–S_c curve showing the relationship between magnetic flux density and exciting power obtained in different extended distances from joint to the middle uniform area of the core is shown in Fig. 9.14. The separation results of exciting power with the variation of x are listed in Table 9.9.

It can be seen from the resulting separations of exciting power based on the joint influence domain that the unit weight exciting power in the joint area is larger than that in the middle uniform area of the core when $B_m > 0.5$ T, and the difference between the two would be more significant when the joint influence domain is smaller, e.g., where $B_m = 1.5$ T and $S_m = 1.46$ VA/kg, $S_c = 2.18$ VA/kg when x is 10 mm, $S_c = 2.00$ VA/kg when x is 30 mm, and $S_c = 1.79$ VA/kg when x is 80 mm.

9.4.2 Separation of the Active Power Loss

The active power loss can be separated by the same method as exciting power separation dividing the core into the joint influence domain and the middle uniform area. The B_m–P_c curves of joint areas with different joint influence domains under

Table 9.9 Separations results with the variation of extended distance from the joint to the middle uniform area of the core

B_m (T)	S_m (VA/kg)	S_c (VA/kg)							
		x = 10 mm	x = 20 mm	x = 30 mm	x = 40 mm	x = 50 mm	x = 60 mm	x = 70 mm	x = 80 mm
0.10	0.007	0.009	0.009	0.008	0.008	0.008	0.008	0.008	0.008
0.20	0.027	0.027	0.027	0.027	0.027	0.027	0.027	0.027	0.027
0.30	0.055	0.055	0.055	0.055	0.055	0.055	0.055	0.055	0.055
0.40	0.091	0.088	0.089	0.089	0.089	0.090	0.090	0.090	0.090
0.50	0.133	0.130	0.131	0.131	0.131	0.131	0.132	0.132	0.132
0.60	0.179	0.180	0.180	0.180	0.180	0.180	0.180	0.180	0.180
0.70	0.228	0.247	0.244	0.242	0.240	0.239	0.238	0.237	0.236
0.80	0.280	0.323	0.317	0.313	0.309	0.306	0.304	0.302	0.300
0.90	0.339	0.410	0.400	0.392	0.386	0.381	0.378	0.374	0.372
1.00	0.402	0.517	0.501	0.488	0.479	0.471	0.465	0.459	0.455
1.10	0.477	0.640	0.617	0.599	0.586	0.575	0.566	0.558	0.552
1.20	0.546	0.858	0.813	0.780	0.754	0.733	0.716	0.702	0.690
1.30	0.663	1.156	1.086	1.033	0.992	0.959	0.932	0.910	0.891
1.40	0.924	1.621	1.521	1.447	1.389	1.342	1.304	1.272	1.246
1.50	1.464	2.179	2.077	2.000	1.941	1.893	1.854	1.822	1.794
1.60	2.220	3.317	3.160	3.043	2.951	2.878	2.818	2.768	2.726
1.65	2.732	3.946	3.773	3.643	3.541	3.460	3.394	3.339	3.292
1.70	3.315	4.940	4.708	4.534	4.398	4.290	4.201	4.128	4.065
1.75	4.033	6.301	5.977	5.734	5.545	5.394	5.270	5.167	5.080
1.80	5.266	8.030	7.635	7.339	7.108	6.924	6.773	6.648	6.541
1.85	8.230	11.881	11.360	10.968	10.664	10.421	10.222	10.056	9.915
1.90	16.712	21.476	20.795	20.285	19.888	19.570	19.311	19.094	18.911

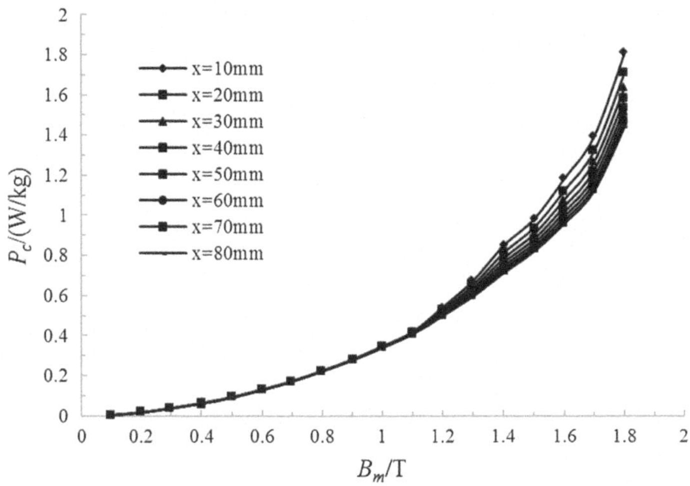

Fig. 9.15 B_m–P_c curve with the variation of extended distance from the joint to the middle uniform area of the core

various magnetic flux densities are shown in Fig. 9.15, and specific data of separations are shown in Table 9.10, where P_m is the specific total loss in the middle uniform area and P_c is the specific total loss in the corner joint area.

It can be seen from the loss separation results that the specific total loss in the joint area is larger than that in the middle uniform area, and the difference between them tends to decrease while the influence domain increases.

9.5 Specific Total Loss Calculation of the Middle Uniform Area with Two-Core Method

The measurement results above show that, based on the two-core models with the same joint type, magnetic material, and cross-section but different structural dimensions, the specific total loss of middle uniform area of the core can be obtained by means of the two-core method, and assuming that the joint influence domains of the two-core models are the same, the specific total loss of the middle uniform area is independent of the size of joint influence domain. The specific total loss P_e in the middle uniform area of the core can be calculated by the following equation:

$$P_e = \frac{P_{c1} - P_{c2}}{m_{c1} - m_{c2}} \tag{9.3}$$

Table 9.10 Separations results with the variation of extended distance from the joint to the middle uniform area of the core

B_m (T)	P_m (W/kg)	P_c (W/kg)							
		x = 10 mm	x = 20 mm	x = 30 mm	x = 40 mm	x = 50 mm	x = 60 mm	x = 70 mm	x = 80 mm
0.10	0.005	0.005	0.005	0.005	0.005	0.005	0.005	0.005	0.005
0.20	0.017	0.017	0.017	0.017	0.017	0.017	0.017	0.017	0.017
0.30	0.036	0.039	0.039	0.038	0.038	0.038	0.038	0.038	0.037
0.40	0.063	0.061	0.062	0.062	0.062	0.062	0.062	0.062	0.062
0.50	0.095	0.090	0.091	0.091	0.092	0.092	0.092	0.093	0.093
0.60	0.132	0.129	0.129	0.130	0.130	0.130	0.130	0.130	0.130
0.70	0.174	0.170	0.171	0.171	0.171	0.172	0.172	0.172	0.172
0.80	0.220	0.225	0.224	0.224	0.223	0.223	0.223	0.223	0.222
0.90	0.274	0.280	0.279	0.278	0.278	0.278	0.277	0.277	0.277
1.00	0.331	0.347	0.345	0.343	0.342	0.341	0.340	0.339	0.338
1.10	0.398	0.418	0.415	0.413	0.411	0.410	0.409	0.408	0.407
1.20	0.457	0.541	0.529	0.520	0.513	0.507	0.503	0.499	0.496
1.30	0.528	0.674	0.653	0.638	0.625	0.616	0.608	0.601	0.595
1.40	0.594	0.848	0.811	0.784	0.763	0.746	0.732	0.721	0.711
1.50	0.680	0.981	0.938	0.906	0.881	0.861	0.844	0.831	0.819
1.60	0.766	1.179	1.120	1.076	1.041	1.014	0.991	0.972	0.957
1.65	0.824	1.253	1.191	1.145	1.110	1.081	1.058	1.038	1.022
1.70	0.883	1.391	1.318	1.264	1.221	1.188	1.160	1.137	1.117
1.75	0.963	1.571	1.484	1.419	1.368	1.328	1.295	1.267	1.243
1.80	1.123	1.806	1.709	1.635	1.578	1.533	1.496	1.465	1.438
1.85	1.433	2.000	1.919	1.858	1.811	1.773	1.742	1.716	1.695
1.90	1.755	2.141	2.086	2.044	2.012	1.987	1.966	1.948	1.933

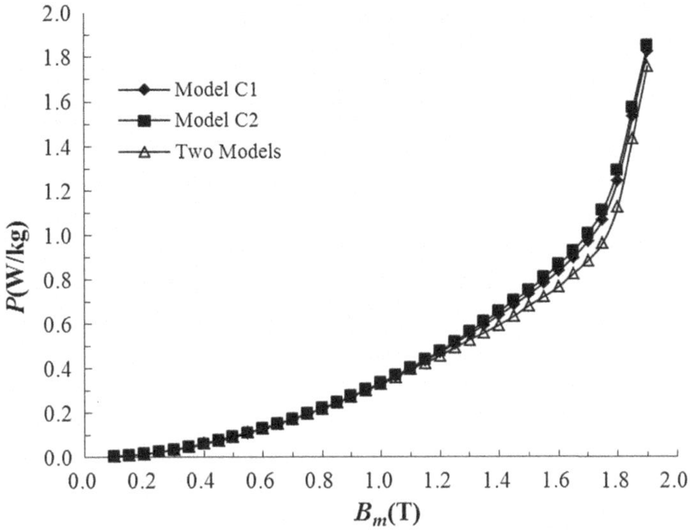

Fig. 9.16 Comparison of specific total loss curves

where P_{c1} is the measured total loss of Model C1, P_{c2} is the measured total loss of Model C2, m_{c1} is the total weight of Model C1, and m_{c2} is the total weight of Model C2.

The B_m–P_{loss} curves (P_{loss}: the specific total loss of core, including the loss in the joint area and middle uniform area of the core) obtained based on Model C1 and Model C2, respectively, and B_m–P_e curves (P_e: the specific total loss, excluding the loss in the joint influence domain) of middle uniform area obtained based on the two-core method are shown in Fig. 9.16 and Table 9.11.

It can be seen from the loss curves and data comparison that, at the same magnetic flux density B_m, P_e is smaller than the specific loss P_{c1} and P_{c2} of Model C1 and Model C2, which is reasonable. Moreover, although the joint types and materials of the two-core models are the same, the middle uniform area of Model C1 is larger than that of Model C2, and the joint area with higher loss density has relatively less influence on Model C1, and therefore, its specific total loss is relatively lower. It is also reasonable that the specific total loss of Model C1, P_{loss1}, is smaller than P_{loss2} of Model C2.

9.6 Determination of Building Factor of Core Model

Nowadays, the concept of building factor or loss factor [5] is used by transformer manufacturers when calculating the no-load loss of transformer core, to comprehensively consider the loss changes caused by the magnetic properties of the

Table 9.11 Comparison of specific total loss data

B_m (T)	Specific total loss		
	P_{loss1} (W/kg)	P_{loss2} (W/kg)	P_e (W/kg)
0.10	0.0046	0.0046	0.0045
0.15	0.0098	0.0098	0.0099
0.20	0.0170	0.0170	0.0169
0.25	0.0259	0.0260	0.0255
0.30	0.0366	0.0368	0.0360
0.35	0.0489	0.0490	0.0486
0.40	0.0627	0.0626	0.0629
0.45	0.0780	0.0777	0.0787
0.50	0.0945	0.0941	0.0955
0.55	0.1122	0.1119	0.1130
0.60	0.1312	0.1310	0.1317
0.65	0.1517	0.1514	0.1524
0.70	0.1736	0.1733	0.1743
0.75	0.1967	0.1967	0.1967
0.80	0.2211	0.2214	0.2204
0.85	0.2470	0.2473	0.2463
0.90	0.2750	0.2754	0.2738
0.95	0.3037	0.3043	0.3022
1.00	0.3338	0.3349	0.3311
1.05	0.3655	0.3679	0.3597
1.10	0.4011	0.4025	0.3974
1.15	0.4345	0.4390	0.4233
1.20	0.4712	0.4770	0.4567
1.25	0.5107	0.5186	0.4909
1.30	0.5531	0.5631	0.5279
1.35	0.5943	0.6084	0.5592
1.40	0.6373	0.6547	0.5940
1.45	0.6833	0.7033	0.6335
1.50	0.7319	0.7525	0.6805
1.55	0.7822	0.8069	0.7207
1.60	0.8368	0.8651	0.7660
1.65	0.8972	0.9266	0.8239
1.70	0.9701	1.0049	0.8833
1.75	1.0670	1.1087	0.9625
1.80	1.2404	1.2872	1.1233
1.85	1.5300	1.5689	1.4328
1.90	1.8215	1.8479	1.7555

Table 9.12 Building factor (BF)

B_m (T)	BF	
	Model C1	Model C2
0.10	1.032	1.035
0.20	1.045	1.047
0.30	1.052	1.058
0.40	1.056	1.054
0.50	1.056	1.051
0.60	1.049	1.048
0.70	1.047	1.045
0.80	1.044	1.045
0.90	1.045	1.047
1.00	1.042	1.045
1.10	1.047	1.050
1.20	1.041	1.054
1.30	1.049	1.068
1.40	1.045	1.074
1.50	1.041	1.070
1.60	1.029	1.064
1.65	1.019	1.053
1.70	1.007	1.043
1.75	0.988	1.027
1.80	0.991	1.029
1.85	1.024	1.050
1.90	1.016	1.030

grain-oriented silicon steels and core manufacturing processes; the building factor is used to correct the calculated no-load loss of transformer core. The purpose of this experiment is to study the building factor of the product-level core model based on the test results of the product-level transformer core model and the standard Epstein frame test system.

The building factor *BF* is the ratio of the measured total core loss (P_{total}) to the product of the specific total loss of core material ($P_{material}$) and total core weight (W_{core}), as shown in (9.4).

$$\mathrm{BF} = \frac{P_{\mathrm{total}}}{P_{\mathrm{material}} - W_{\mathrm{core}}} \tag{9.4}$$

The building factor BF is varying with the flux density of the core, which can be treated as a function of the peak value of flux density inside core (B_m). The B_m–BF curves for the two models (Model C1 and Model C2) with lamination factor of 0.97 are shown in Table 9.12.

The B_m–BF curves of two-core models are shown in Fig. 9.17.

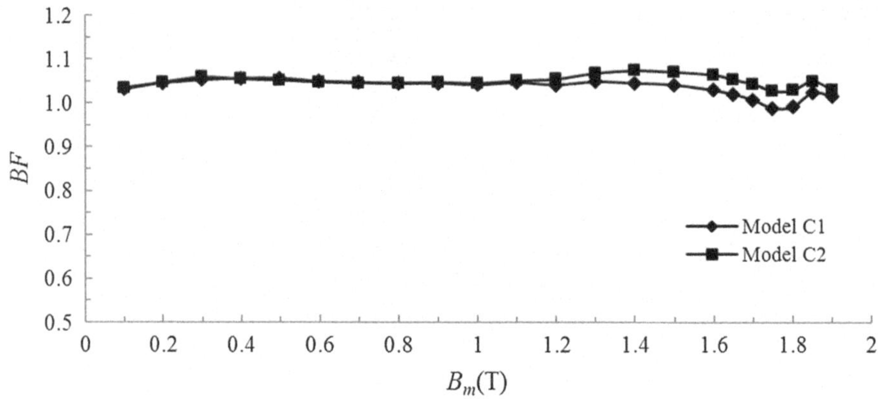

Fig. 9.17 B_m–BF curves

From Table 9.12 and Fig. 9.17, it can be seen that the variation trend of BF curves of the two-core models are basically consistent, and the BF value of Model C2 is slightly higher than that of Model C1.

Further investigation has shown that the BF of the laminated core is not only dependent on the manufacturing factors (e.g., core assembly, sheet-cutting), but also dependent on the actual operating conditions, involving the effect of non-uniform temperature, the additional core loss caused by rotating flux, inter-laminar flux, and extreme excitation condition, such as over-excitation, multi-harmonic, and/or DC bias [6].

9.7 Research on Magnetic Measurement of Transformer Core at Different Ambient Temperatures

When the power transformer is in operation, the temperature rises and the core will continuously operate under high-temperature conditions. According to China national standard, the maximum ambient temperature specified in the normal operating conditions of the power transformer is +40 °C, and the maximum temperature rise of the oil-immersed transformer core is 80 K [7], which indicate that the core may continuously operate at an ambient temperature of up to 120 °C. The typical data of magnetic properties of oriented silicon steel materials provided by silicon steel sheet manufacturers are generally obtained at the temperature specified by the measurement standard. The magnetic properties of oriented silicon steel vary at different ambient temperatures, indicating that there may be some errors in the design of transformer and electromagnetic simulation by using the material magnetic property data obtained at the standard temperature.

The magnetization curve and loss curve used in the simulation and modeling can be closer to the actual operation condition of transformer core by studying the variation of magnetic properties of oriented silicon steel under different ambient temperatures, which can provide reference for improving the simulation and design optimization of power transformers.

9.7.1 Experimental Setup and Process

The influence of temperature on the magnetic properties of oriented silicon steel material B27R095 (Baosteel, China) and core model is studied. The Epstein frame loaded with test specimen and the product-level core model made of the same material (design parameters are the same as Model C1) are respectively placed in a temperature-controlled chamber, and the magnetic properties of the oriented silicon steel material and the core model are measured under the ambient temperatures of 25 °C, 50 °C, 75 °C, and 105 °C, respectively.

First, the magnetic properties of materials at different ambient temperatures are tested by Epstein Frame Measuring System (Manufacturer: Hunan Tongdian, China) [8, 9]. The standard test specimens (48 pieces in total) are placed in the Epstein frame test system and fixed, then the Epstein frame is put into a temperature-controlled chamber, and the ambient temperature is regulated to 25 °C, 50 °C, 75 °C, and 105 °C, respectively. After reaching each set ambient temperature, the temperature must be maintained for at least 2 h before the test, in order to ensure that the material temperature is the same as the set ambient temperature. In addition, in order to ensure the validity of measurement data, it is necessary to test three times continuously and ensure that the measurement error between three tests is less than 1%. The Epstein Frame Measurement System and the temperature-controlled chamber are shown in Fig. 9.18.

Next, the core model is placed in the temperature-controlled chamber to be tested at 25 °C, 50 °C, 75 °C, and 105 °C, respectively. Similarly, after reaching each set ambient temperature, the temperature must be maintained for at least 2 h before the test in order to ensure that the material temperature is the same as the set ambient temperature. Figure 9.19 shows the core model in the temperature-controlled chamber.

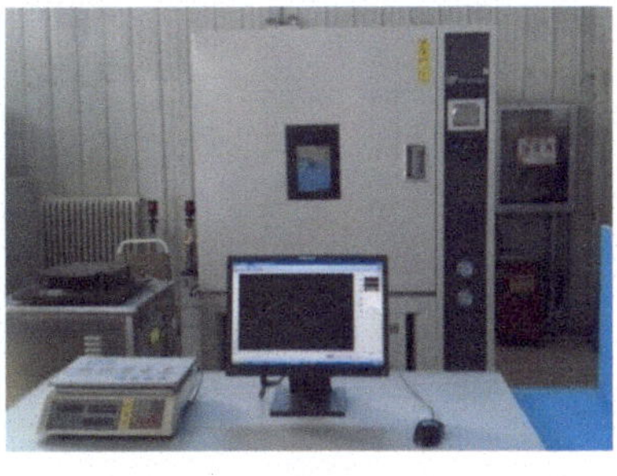

Fig. 9.18 Epstein frame and temperature-controlled chamber

Fig. 9.19 Core model in temperature-controlled chamber

9.7.2 *Measurement Results and Analysis*

9.7.2.1 Material Properties Under Different Temperatures

The specific total loss data of oriented silicon steel material measured under ambient temperature of 25, 50, 75, and 105 °C are listed in Table 9.13.

The specific total loss curves of B27R095 at four grades of temperature are shown in Fig. 9.20.

Table 9.13 Specific total loss data

B_m (T)	P_{loss} (W/kg)			
	25 °C	50 °C	75 °C	105 °C
0.10	0.0037	0.0036	0.0035	0.0034
0.20	0.0139	0.0135	0.0131	0.0127
0.30	0.0304	0.0295	0.0286	0.0277
0.40	0.0530	0.0516	0.0501	0.0485
0.50	0.0812	0.0790	0.0768	0.0744
0.60	0.1149	0.1120	0.1090	0.1057
0.70	0.1545	0.1507	0.1468	0.1424
0.80	0.2003	0.1955	0.1906	0.1849
0.90	0.2521	0.2463	0.2404	0.2334
1.00	0.3103	0.3030	0.2957	0.2871
1.10	0.3740	0.3652	0.3564	0.3461
1.20	0.4430	0.4327	0.4225	0.4102
1.30	0.5178	0.5059	0.4938	0.4798
1.40	0.5977	0.5843	0.5707	0.5546
1.50	0.6867	0.6720	0.6570	0.6393
1.60	0.7896	0.7737	0.7582	0.7392
1.65	0.8506	0.8352	0.8191	0.8010
1.70	0.9261	0.9112	0.8969	0.8794
1.75	1.0250	1.0120	0.9989	0.9841
1.80	1.1718	1.1630	1.1524	1.1393
1.90	1.6546	1.6423	1.6194	1.5895

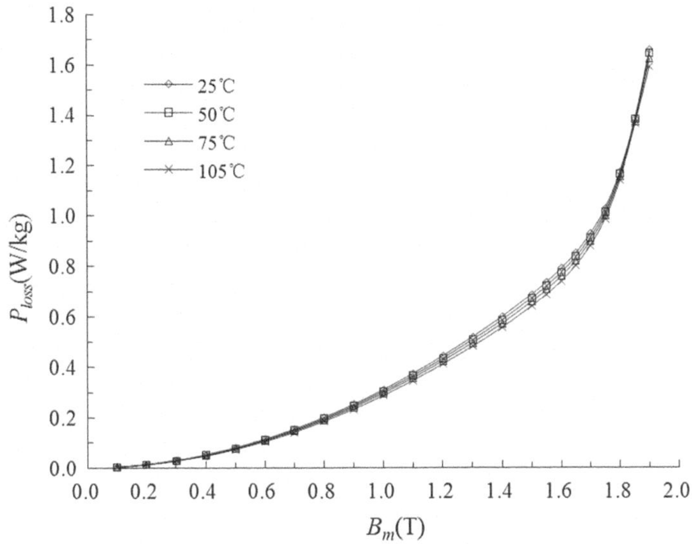

Fig. 9.20 Specific total loss curves at different ambient temperature

At each flux density points, the specific total losses have been measured at different ambient temperatures, such as 25, 50, 75, and 105 °C, referred to as P_{25}, P_{50}, P_{75}, and P_{105}, respectively. The increasements of the specific total loss $P_x - P_{25}$ (x = 50, 70 and 105) are shown in Table 9.14.

The percentages PP_x of the increasement of the specific total loss under different ambient temperatures ($P_x - P_{25}$) (x = 50, 75, and 105) to P_{25} can be expressed by (9.5), as shown in Table 9.15.

$$PP_x = \frac{P_x - P_{25}}{P_{25}} \times 100\% \tag{9.5}$$

It can be seen from the loss data and curves of B27R095 measured at different ambient temperatures that the specific total loss varies with the temperature at any fixed operating point B_{m} and decreases gradually with the increase of temperature, resulting in better loss properties.

The magnetization property data of B27R095 under ambient temperatures of 25, 50, 75, and 105 °C are shown in Table 9.16.

The magnetization curves of B27R095 measured at different ambient temperatures are shown in Fig. 9.21.

From the magnetization property data and curves of B27R095 at different ambient temperatures, it can be seen that when $B_{\mathrm{m}} \leq 1.2$ T, the H_{m} required to reach any flux density decreases to a certain extent with the increase of temperature, but the decrease is not significant; when $B_{\mathrm{m}} \geq 1.3$ T, the H_{m} required to reach any flux density increases with the increase of temperature, and the greater the flux

Table 9.14 Increasements of specific total losses

B_m (T)	$P_x - P_{25}$ (W/kg)		
	$P_{50} - P_{25}$	$P_{75} - P_{25}$	$P_{105} - P_{25}$
0.10	−0.0001	−0.0002	−0.0003
0.20	−0.0004	−0.0008	−0.0012
0.30	−0.0009	−0.0018	−0.0027
0.40	−0.0014	−0.0029	−0.0045
0.50	−0.0022	−0.0044	−0.0068
0.60	−0.0029	−0.0059	−0.0092
0.70	−0.0038	−0.0077	−0.0121
0.80	−0.0048	−0.0097	−0.0154
0.90	−0.0058	−0.0117	−0.0187
1.00	−0.0073	−0.0146	−0.0232
1.10	−0.0088	−0.0176	−0.0279
1.20	−0.0103	−0.0205	−0.0328
1.30	−0.0119	−0.0240	−0.0380
1.40	−0.0134	−0.0270	−0.0431
1.50	−0.0147	−0.0297	−0.0474
1.55	−0.0158	−0.0310	−0.0494
1.60	−0.0159	−0.0314	−0.0504
1.65	−0.0154	−0.0315	−0.0496
1.70	−0.0149	−0.0292	−0.0467
1.75	−0.0130	−0.0261	−0.0409
1.80	−0.0088	−0.0194	−0.0325
1.85	−0.0027	−0.0117	−0.0185
1.90	−0.0123	−0.0352	−0.0651

density, the greater the increase of H_m required. Considering the working point of 1.7 T, the magnetization property gradually deteriorates with the increase of temperature.

9.7.2.2 Magnetic Properties of Core Model Measured Under Different Temperatures

The specific total losses, P_{loss}, of the core model measured at the different ambient temperatures, such as 25, 50, 75, and 105 °C, are listed in Table 9.17.

The specific total loss curves of core model at different temperatures are shown in Fig. 9.22.

At each flux density points, the specific total losses have been measured at different ambient temperatures, such as 25, 50, 75, and 105 °C, referred to as P_{25}, P_{50}, P_{75}, and P_{105}, respectively. The increasements of the specific total loss $P_x - P_{25}$ (x = 50, 70, and 105) are shown in Table 9.18.

Table 9.15 Percentages of increasements of specific total losses

B_m (T)	PP_{50} (%)	PP_{75} (%)	PP_{105} (%)
0.10	−2.70	−5.41	−8.11
0.20	−2.88	−5.76	−8.63
0.30	−2.96	−5.92	−8.88
0.40	−2.64	−5.47	−8.49
0.50	−2.71	−5.42	−8.37
0.60	−2.52	−5.13	−8.01
0.70	−2.46	−4.98	−7.83
0.80	−2.40	−4.84	−7.69
0.90	−2.30	−4.64	−7.42
1.00	−2.35	−4.71	−7.48
1.10	−2.35	−4.71	−7.46
1.20	−2.33	−4.63	−7.40
1.30	−2.30	−4.63	−7.34
1.40	−2.24	−4.52	−7.21
1.50	−2.14	−4.33	−6.90
1.60	−2.01	−3.98	−6.38
1.65	−1.81	−3.70	−5.83
1.70	−1.61	−3.15	−5.04
1.75	−1.27	−2.55	−3.99
1.80	−0.75	−1.66	−2.77
1.90	−0.74	−2.13	−3.93

The percentages PP_x of the increasement of the specific total losses under different ambient temperatures ($P_x - P_{25}$) ($x = 50$, 75, and 105) to P_{25} can also be calculated by (9.5), as shown in Table 9.19.

It can be seen from the loss data and curves of core model at different ambient temperatures that the specific total loss varies with the temperature at any fixed working point B_m and decreases gradually with the increase of temperature, resulting in better loss properties.

The magnetization property data of the core model measured at ambient temperatures of 25, 50, 75, and 105 °C are shown in Table 9.20.

The magnetization curves of the core model measured at different ambient temperatures are shown in Fig. 9.23.

From the magnetization property data and curves of core model at different ambient temperatures, it can be seen that when $B_m \leq 1.2$ T, the H_m required to reach any flux density decreases to a certain extent with the increase of temperature, but the decrease is not significant; when $B_m \geq 1.3$ T, the H_m required to reach any flux density increases with the increase of temperature, and the greater the flux density, the greater the increase of H_m required. Considering the working point of 1.7 T, the magnetization property gradually deteriorates with the increase of temperature.

Table 9.16 Magnetization property data

B_m (T)	H_m (A/m)			
	25 °C	50 °C	75 °C	105 °C
0.10	3.13	3.07	3.01	2.96
0.20	5.43	5.32	5.22	5.12
0.30	7.48	7.35	7.20	7.06
0.40	9.37	9.20	9.03	8.84
0.50	11.00	10.82	10.63	10.42
0.60	12.52	12.31	12.11	11.88
0.70	13.96	13.74	13.51	13.26
0.80	15.34	15.10	14.86	14.60
0.90	16.68	16.42	16.19	15.92
1.00	18.00	17.73	17.50	17.24
1.10	19.32	19.06	18.86	18.66
1.20	20.78	20.61	20.53	20.52
1.30	22.86	22.95	23.14	23.41
1.40	26.29	26.74	27.24	27.87
1.50	32.14	33.05	33.96	35.10
1.55	36.68	37.89	39.16	40.70
1.60	43.37	45.16	47.00	49.24
1.65	54.30	57.33	60.18	63.79
1.70	74.83	79.85	85.37	92.43
1.75	117.61	128.08	138.92	154.90
1.80	209.34	235.37	263.21	302.74
1.85	427.98	490.09	552.92	648.98
1.90	894.83	1031.53	1181.97	1431.39

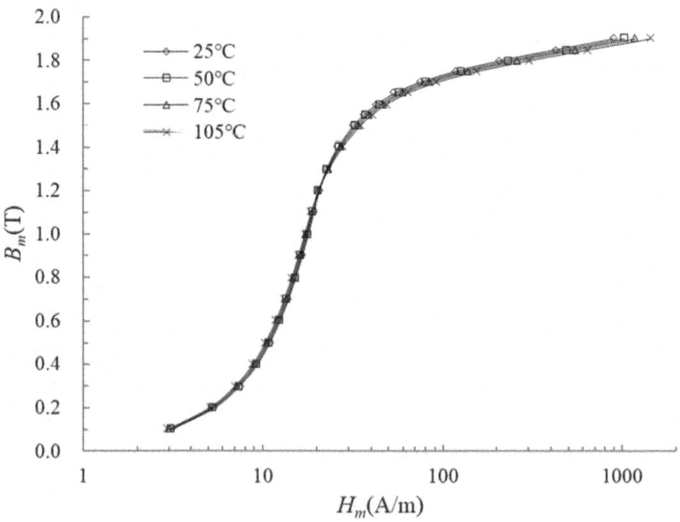

Fig. 9.21 Magnetization curves at different ambient temperature

Table 9.17 Specific total loss data of core model

B_m (T)	P_{loss} (W/kg)			
	25 °C	50 °C	75 °C	105 °C
0.1	0.0030	0.0029	0.0029	0.0030
0.2	0.0118	0.0115	0.0112	0.0110
0.3	0.0259	0.0255	0.0247	0.0240
0.4	0.0458	0.0442	0.0433	0.0420
0.5	0.0705	0.0683	0.0666	0.0646
0.6	0.0996	0.0972	0.0948	0.0923
0.7	0.1342	0.1315	0.1287	0.1248
0.8	0.1753	0.1717	0.1669	0.1626
0.9	0.2217	0.2169	0.2114	0.2066
1.0	0.2749	0.2692	0.2631	0.2572
1.1	0.3342	0.3261	0.3212	0.3126
1.2	0.4006	0.3916	0.3850	0.3748
1.3	0.4723	0.4625	0.4534	0.4404
1.4	0.5490	0.5379	0.5291	0.5127
1.5	0.6322	0.6208	0.6064	0.5929
1.6	0.7250	0.7123	0.6972	0.6816
1.7	0.8411	0.8216	0.8075	0.7962
1.8	1.0107	0.9972	0.9944	0.9881
1.9	1.3995	1.4012	1.3865	1.3794

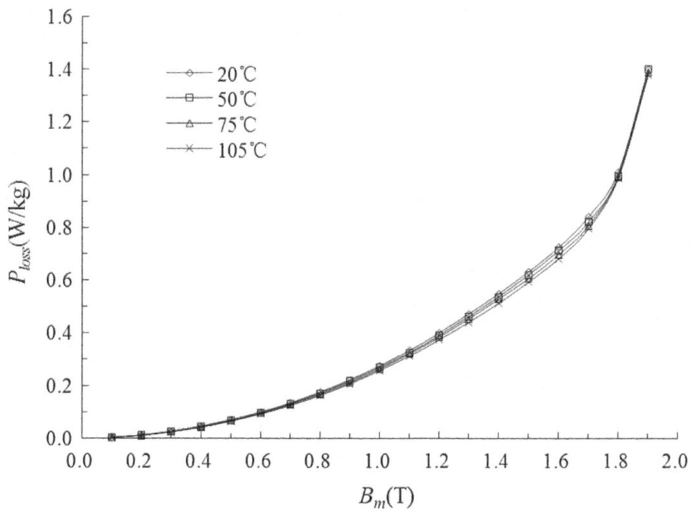

Fig. 9.22 Specific total loss curves of core model at different temperatures

Table 9.18 Increasements of specific total losses

B_m (T)	$P_x - P_{25}$ (W/kg)		
	$P_{50} - P_{25}$	$P_{75} - P_{25}$	$P_{105} - P_{25}$
0.1	−0.0001	−0.0001	−0.0001
0.2	−0.0003	−0.0006	−0.0007
0.3	−0.0005	−0.0013	−0.0019
0.4	−0.0016	−0.0025	−0.0038
0.5	−0.0022	−0.0039	−0.0059
0.6	−0.0024	−0.0048	−0.0073
0.7	−0.0026	−0.0054	−0.0094
0.8	−0.0036	−0.0084	−0.0127
0.9	−0.0048	−0.0103	−0.0151
1.0	−0.0056	−0.0118	−0.0177
1.1	−0.0081	−0.0130	−0.0216
1.2	−0.0090	−0.0156	−0.0257
1.3	−0.0098	−0.0188	−0.0319
1.4	−0.0111	−0.0199	−0.0363
1.5	−0.0114	−0.0258	−0.0393
1.6	−0.0126	−0.0278	−0.0434
1.7	−0.0195	−0.0336	−0.0449
1.8	−0.0134	−0.0163	−0.0225
1.9	−0.0039	−0.0131	−0.0202

Table 9.19 Percentage of increasements of specific total losses

B_m (T)	PP_{50} (%)	PP_{75} (%)	PP_{105} (%)
0.1	−2.85	−3.75	−1.98
0.2	−2.22	−4.94	−6.30
0.3	−1.79	−4.90	−7.48
0.4	−3.43	−5.46	−8.20
0.5	−3.09	−5.50	−8.34
0.6	−2.39	−4.79	−7.33
0.7	−1.97	−4.03	−7.00
0.8	−2.08	−4.81	−7.26
0.9	−2.16	−4.64	−6.82
1.0	−2.04	−4.29	−6.42
1.1	−2.42	−3.88	−6.45
1.2	−2.24	−3.88	−6.43
1.3	−2.07	−3.99	−6.76
1.4	−2.02	−3.62	−6.61
1.5	−1.80	−4.08	−6.21
1.6	−1.74	−3.83	−5.99
1.7	−2.32	−3.99	−5.33
1.8	−1.33	−1.61	−2.23
1.9	−0.28	−0.93	−1.44

Table 9.20 Magnetization property data

B_m (T)	H_m (A/m)			
	25 °C	50 °C	75 °C	105 °C
0.10	2.43	2.41	2.35	2.32
0.20	4.41	4.30	4.20	4.11
0.30	6.23	6.01	5.90	5.71
0.40	7.85	7.66	7.39	7.20
0.50	9.36	9.09	8.84	8.56
0.60	10.68	10.48	10.19	9.88
0.70	12.02	11.67	11.43	11.10
0.80	13.32	13.08	12.68	12.31
0.90	14.71	14.33	13.98	13.65
1.00	15.97	15.63	15.56	15.06
1.10	17.87	17.72	17.63	17.00
1.20	20.50	20.44	20.31	20.70
1.30	27.97	28.56	28.45	30.18
1.40	45.59	47.31	45.98	49.71
1.50	73.32	75.27	76.87	78.58
1.55	90.34	93.06	93.66	96.86
1.60	110.14	112.63	114.64	117.83
1.65	133.70	136.67	139.84	143.31
1.70	163.80	168.46	173.11	178.02
1.75	204.28	213.66	221.09	235.17
1.80	268.55	288.99	308.83	327.47
1.85	440.54	493.70	538.56	622.70
1.90	860.07	1043.50	1217.47	1593.44

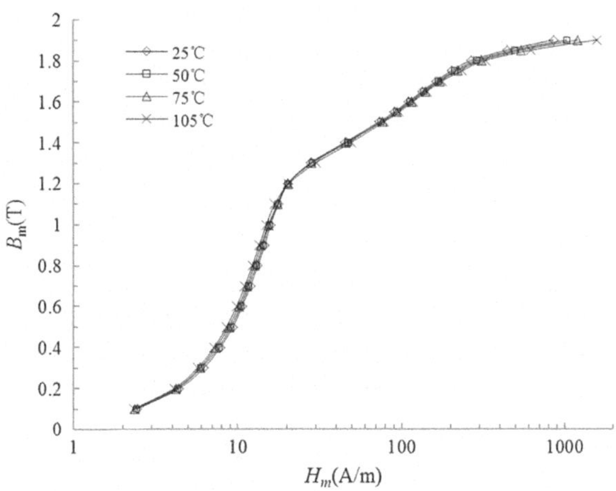

Fig. 9.23 Magnetization curves of the core model measured at four temperature levels

9.8 Magnetic Properties Modeling Based on Ring Cores Before and After Annealing

Unlike the laminated core of large power transformer, the ring core of current transformer must be annealed to eliminate internal stress after it is made and improve the average magnetic properties of the core [10]. Generally, stress-relief annealing of large transformer core is not carried out.

The difference of magnetic properties before and after annealing, including B_m–H_b curves (relation curve between peak flux density B_m and corresponding magnetic field strength H_b), B_m–P_{loss} curves (relation curve between peak flux density B_m and specific total loss P_{loss}), and B_m–S_e curves (relation curve between peak flux density B_m and specific exciting power S_e) of the ring core and the comparison with the magnetic properties of silicon steel materials are of concern to product developers and designers [11]. The selected sample is the core of the current transformer product, and the measured data are close to the properties of the core in real engineering environment.

The equipment, experimental circuit, and method used for measuring the magnetic properties of the ring core are the same as mentioned in Sect. 9.3 of this chapter.

The material of the ring core in magnetic properties experiment is oriented silicon steel 30P120 (POSCO, Korea) [12]. The properties data of silicon steel sheet are provided by the manufacturer, and the measurement refers to the relevant test methods of electrical steel specified in Japanese enterprise standard (JISC2553) or ASTM A343 [13, 14]. According to the standard, the Epstein frame specimens should be subject to stress-relief annealing at 840° ± 10° before measurement, which lasts for one hour under the 100% nitrogen gas protection.

It should be noted that the supplier of 30P120 silicon steel sheet used in the core model did not provide detailed magnetic property data (curves), but only the guaranteed properties value range of specimens.

Iron loss: 50 Hz/1.7 T: 0.97 ~ 1.02 (W/kg); 60 Hz/1.7 T: 1.30–1.35 (W/kg);

Flux density at the 800 A/m of magnetic field strength: 1.90–1.93 (T).

9.8.1 *Ring Core*

Two ring cores of a current transformer with the same design specification are randomly selected as test models and called models 1# and 2#, as shown in Fig. 9.24. The parameters of the two ring cores are listed in Table 9.21.

Fig. 9.24 Ring core model 1# (upper) and 2# (lower) of current transformer

Table 9.21 Ring core parameters

Ring core	Actual core size (mm)	Winding layers around core	Actual weight (kg)	Average magnetic path length (m)
1#	Ø 450.0/Ø 551.6 × 100.0	175	59.49	1.5733
2#	Ø 450.0/Ø 550.0 × 100.0	171	58.46	1.5708

Note The design weight of a ring core is 57.68 kg

9.8.2 Annealing Conditions

The ring core should be annealed according to the annealing conditions of current transformer core:

① Nitrogen protection is required during annealing.
② The heating rate is 60 °C/h from ambient temperature to 820 °C.
③ Keep the temperature at 820 °C for 2 h.
④ Reduce the temperature to about 600 °C with a cooling rate of 20–30 °C/h, then let it naturally reduce to room temperature.

9.8.3 Experimental Result

The magnetic properties of ring core before and after annealing, such as B_m–H_b curves, B_m–P_{loss} curves, and B_m–S_e curves, were obtained by processing the experimental data.

Based on the measurement data and the calculation results, the B_m–H_b, B_m–P_{loss}, and B_m–S_e curves before and after annealing of the ring core are shown in Figs. 9.25, 9.26, 9.27, 9.28, 9.29. It should be noted that the data used in each figure are valid with voltage distortion rate of less than 3%. The magnetic property curves of the ring core before and after annealing are shown in Appendix 9.1.

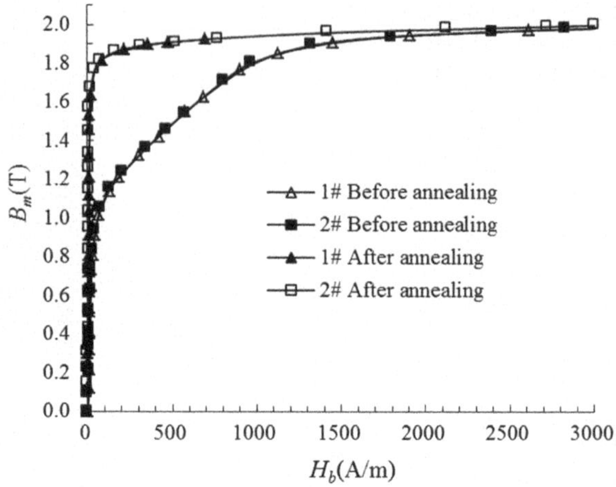

Fig. 9.25 B_m–H_b curves of ring core models 1# and 2# before and after annealing

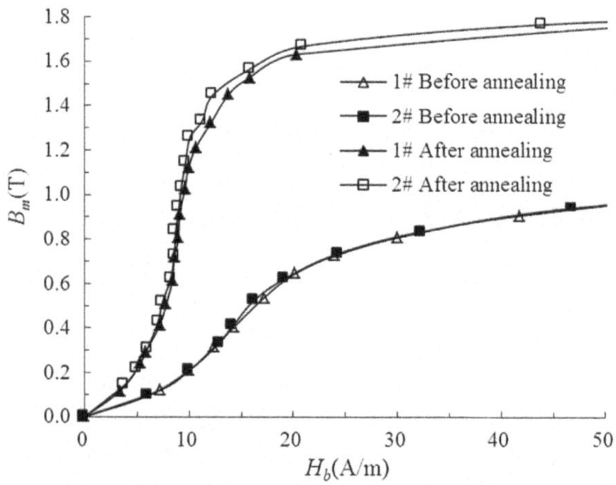

Fig. 9.26 B_m–H_b curves of ring core models 1# and 2# before and after annealing (partially enlarged)

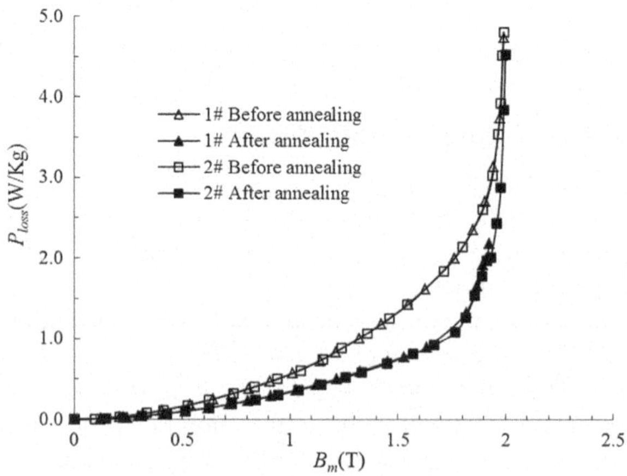

Fig. 9.27 B_m–P_{loss} curves of ring core models 1# and 2# before and after annealing

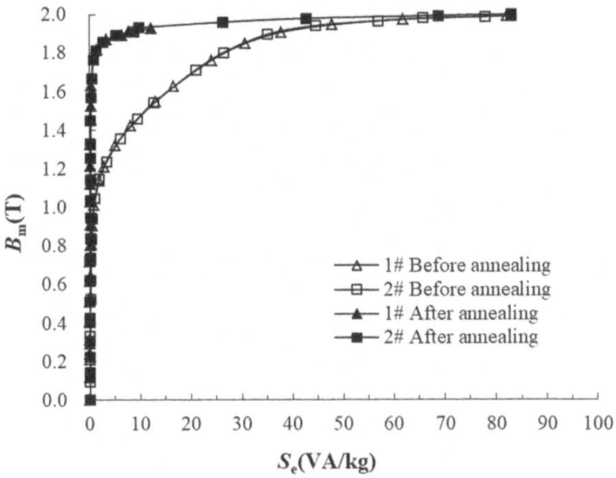

Fig. 9.28 B_m–S_e curves of ring core models 1# and 2# before and after annealing

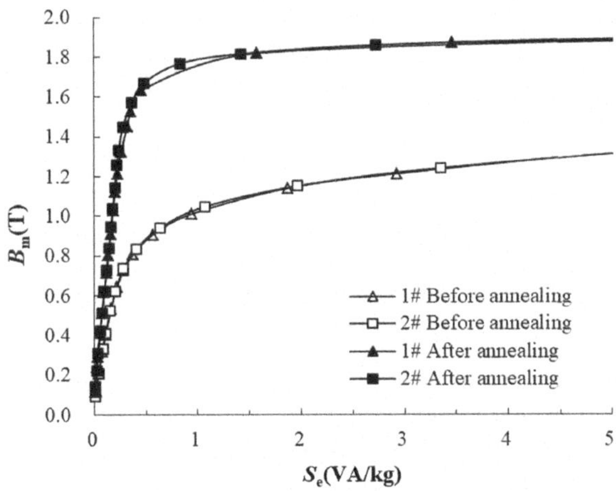

Fig. 9.29 B_m–S_e curves of ring core models 1# and 2# before and after annealing (partially enlarged)

9.8.3.1 B_m–H_b Curves of Ring Core Before and After Annealing

The B_m–H_b curves of ring core models 1# and 2# before and after annealing are shown in Fig. 9.25.

In order to observe the difference in magnetic permeability before and after annealing, the B_m–H_b curve in Fig. 9.25 is enlarged locally, as shown in Fig. 9.26.

Figures 9.25 and 9.26 show that the magnetic permeability properties of the annealed ring core are improved significantly. For example, when the H_b of the annealed ring core is 100 A/m, B_m is greater than 1.8 T; when the H_b of the ring core before annealing is 100 A/m, B_m is less than 1.0 T. The B_m–H_b curves of ring core model 1# and model 2# before and after annealing are almost the same.

9.8.3.2 B_m–P_{loss} Curves of Ring Core Before and After Annealing

The B_m–P_{loss} curves of ring core models 1# and 2# before and after annealing are shown in Fig. 9.27.

The results in Fig. 9.27 show that the loss of the ring core after annealing is significantly lower than that before annealing in the range of 0.5–1.9 T.

9.8.3.3 Exciting Power (B_m–S_e) Curves of Ring Core Before and After Annealing

The exciting power (B_m–S_e) curves of ring core models 1# and 2# before and after annealing are shown in Fig. 9.28. In order to observe the difference in magnetic permeability before and after annealing, the B_m–H_b curve in Fig. 9.28 is enlarged locally, as shown in Fig. 9.29.

It can be seen from the B_m–S_e curves of ring core models 1# and 2# before and after annealing that the exciting power of ring cores after annealing is significantly reduced in a relatively large magnetic flux density range, e.g., 1.1–1.9 T, than that before annealing.

The experimental results of magnetic properties of product-level ring core models before and after annealing show that:

① The magnetic properties of the ring core after annealing are obviously better than that before annealing.

② The magnetic properties of ring core models 1# and 2# measured under the same condition are almost the same.

③ After annealing, the specific total loss measured at 1.7 T (50 Hz) is in the range of 1.02–1.07 w/kg. See Appendix 9.1, Tables 9.24 and 9.25, the measured value is close to the typical value of 0.97–1.02 W/kg provided by the silicon steel sheet supplier.

9.9 Concluding Remarks

In this chapter, several product-level core models are designed and manufactured according to the actual structure and manufacturing process of transformer core, and "no-load" magnetic properties are measured under standard sinusoidal excitation based on the core models.

9.9.1 Separation of Exciting Power and Magnetic Loss Based on Laminated Core Models

The "two-core method" is proposed, and the separation of exciting power and magnetic loss between the joint area and middle uniform area of the core model is realized. The separation results of exciting power and loss show that:

① The exciting power (or loss) in the joint area is obviously higher than that in the middle uniform area, especially when the magnetic flux density is greater than 0.5 T.
② As the joint influence domain expands, the difference in exciting power (or loss) between joint area and middle uniform area decreases.
③ The separation of exciting power and power loss by "two-core method" can be used as a feasible engineering management method.

The specific total loss of the middle uniform area is obtained based on the "two-core method" in this chapter. On the premise that the joint influence domains of two-core models are identical, the specific total loss of the middle uniform area is independent of the size of the joint influence domain.

In addition, based on the core models, the building factor of transformer manufacture is studied, and the study results show that the building factor is a variable value that varies with the flux density.

9.9.2 Effect of Temperature on the Magnetic Properties

The influence of different ambient temperatures on the magnetic properties of oriented silicon steel materials and core models is also studied. The experimental results show that:

① The loss properties of oriented silicon steel material and core model have the same trend with the change of temperature. Under the same working flux density, the loss decreases with the increase of ambient temperature. Therefore, the loss properties of core model become better as the temperature increases;

② The magnetization properties of oriented silicon steel material and core model also have the same trend with the change of temperature. When the working flux density is greater than 1.2 T, the magnetization properties of the core model gradually become worse with the increase of the ambient temperature.

The study of magnetic properties of oriented silicon steel at different ambient temperatures is helpful to further understand the properties of materials and transformer products under working conditions and provide more data for the magnetic properties of materials which is close to the actual working conditions of products for electromagnetic simulation calculation; thus, it is helpful to the design and optimization of transformer products.

In addition, two ring cores of the current transformer of the same design specifications are selected as test models to study the changing trend of magnetic properties of the ring cores before and after annealing. The experimental results show that the magnetic properties after annealing become better than that before annealing.

Acknowledgements The experimental research was completed by Lanrong LIU, Yana FAN, Yang LIU, Zhigang ZHAO, Junjie ZHANG, Fulai CHE, Chen CHANG, Meilin LU, Guisheng HAN, Yong DU, and Xian ZHANG under the guidance of Zhiguang CHENG and Qifan HU and with the support of the leaders and colleagues of the relevant departments of Baobian Electric and Tebian Electric. The research project was supported in part by State Grid Corporation of China (no. sgri-wd-71-13-002).

Appendix 9.1: Magnetic Property Curves of the Ring Core Before and After Annealing

The magnetic properties of ring core models 1# and 2# measured before annealing are listed in Tables 9.22 and 9.23, respectively.

The magnetic properties of ring core models 1# and 2# after annealing are listed in Tables 9.24 and 9.25, respectively.

Table 9.22 Measurement results of the ring core model 1# obtained before annealing

B_m (T)	H_b (A/m)	P (W/kg)	S (VA/kg)	THD (%)
0.1219	7.2	0.01	0.02	
0.2097	10.0	0.03	0.04	
0.3170	12.4	0.07	0.08	
0.4029	14.2	0.11	0.11	
0.5362	17.2	0.18	0.16	
0.6431	20.1	0.26	0.22	
0.7269	24.0	0.32	0.28	
0.8085	30.1	0.38	0.38	
0.9067	41.8	0.47	0.57	
1.0096	65.5	0.58	0.94	
1.1395	126.0	0.73	1.87	
1.2106	184.9	0.84	2.93	
1.3202	300.4	1.01	5.16	
1.4224	420.9	1.18	8.20	
1.5455	575.9	1.43	13.01	
1.6246	683.9	1.62	16.67	
1.7615	905.1	2.00	24.08	
1.8474	1128.0	2.35	30.66	
1.9069	1450.4	2.71	37.82	
1.9448	1908.1	3.14	47.86	
1.9721	2616.0	3.74	61.95	3.4
1.9907	3291.7	4.74	82.25	5.4

Notes

1. The data with THD > 3% in the table are for reference only
2. THD = Total Harmonic Distortion, i.e., the ratio of total harmonic RMS to fundamental harmonic RMS, the same below
3. H_b is the magnetic field strength corresponding to the maximum flux density B_m, the same below
4. P specific total loss, S specific exciting power, the same below
5. Ambient temperature: 22 °C

Table 9.23 Measurement results of the ring core model 2# obtained before annealing

B_m (T)	H_b (A/m)	P (W/kg)	S (VA/kg)	THD (%)
0.0943	5.9	0.01	0.01	
0.2092	9.9	0.03	0.04	
0.3340	12.9	0.08	0.08	
0.4106	14.2	0.12	0.11	
0.5251	16.2	0.18	0.16	
0.6231	19.2	0.24	0.20	
0.7359	24.3	0.32	0.28	
0.8339	32.2	0.41	0.41	
0.9408	46.9	0.50	0.64	

(continued)

Table 9.23 (continued)

B_m (T)	H_b (A/m)	P (W/kg)	S (VA/kg)	THD (%)
1.0472	76.2	0.61	1.07	
1.1498	131.6	0.75	1.97	
1.2393	211.0	0.88	3.35	
1.3580	342.8	1.07	6.16	
1.4599	465.4	1.26	9.60	
1.5438	571.7	1.43	12.82	
1.7115	804.8	1.85	21.05	
1.8010	970.5	2.14	26.48	
1.8958	1328.0	2.60	35.02	
1.9403	1797.8	3.03	44.65	
1.9650	2398.8	3.54	56.94	3.3
1.9788	2824.0	3.92	65.76	4.3
1.9855	3169.1	4.52	77.93	5.0
1.9915	3465.8	4.81	83.00	5.5

Notes
1. The data with THD > 3% in the table are for reference only
2. Ambient temperature: 22 °C

Table 9.24 Measurement results of the ring core model 1# obtained after annealing

B_m (T)	H_b (A/m)	P (W/kg)	S (VA/kg)	THD (%)
0.1170	3.4	0.01	0.01	
0.2408	5.3	0.03	0.02	
0.2938	5.8	0.04	0.03	
0.4123	7.1	0.07	0.06	
0.5119	7.8	0.10	0.08	
0.6177	8.4	0.14	0.10	
0.7171	8.5	0.19	0.12	
0.8042	8.9	0.23	0.14	
0.9094	9.1	0.29	0.16	
1.0272	9.6	0.36	0.18	
1.1200	9.9	0.43	0.20	
1.2141	10.6	0.50	0.23	
1.3240	12.1	0.60	0.26	
1.4538	13.8	0.71	0.32	
1.5264	15.8	0.78	0.36	
1.6327	20.4	0.90	0.46	
1.8176	33.8	1.02	0.69	
1.8946	105.4	1.41	1.86	
1.8695	220.3	1.66	3.46	3.2
1.9093	481.4	1.99	7.85	7.3
1.9256	699.8	2.18	12.18	12.7

Notes
1. The data with THD > 3% in the table are for reference only
2. Ambient temperature: 23 °C

Table 9.25 Measurement results of the ring core model 2# obtained after annealing

B_m (T)	H_b (A/m)	P (W/kg)	S (VA/kg)	THD (%)
0.143	3.7	0.01	0.01	
0.222	5.0	0.02	0.02	
0.308	5.9	0.04	0.04	
0.426	7.0	0.07	0.06	
0.513	7.4	0.10	0.07	
0.622	8.2	0.14	0.10	
0.729	8.5	0.19	0.12	
0.838	8.7	0.25	0.14	
0.944	9.0	0.31	0.16	
1.036	9.2	0.36	0.18	
1.144	9.6	0.44	0.20	
1.257	10.0	0.52	0.22	
1.332	11.2	0.59	0.24	
1.449	12.2	0.69	0.28	
1.569	15.9	0.81	0.37	
1.668	20.8	0.92	0.48	
1.766	43.8	1.08	0.83	
1.814	79.6	1.27	1.42	
1.857	162.5	1.54	2.72	
1.891	319.1	1.78	5.23	
1.911	520.8	1.97	8.81	
1.930	772.9	2.01	9.95	
1.960	1422.7	2.44	26.27	3.3
1.978	2128.9	2.88	42.70	4.7
1.992	2717.1	3.84	68.86	6.7
1.999	2999.8	4.53	83.01	7.8

Notes

1. The data with THD > 3% in the table are for reference only
2. Ambient temperature: 23 °C

References

1. Z. Cheng, Q. Hu, C. Jiao, J. Zhang, Y. Du, X. Zhang, Y. Fan, L. Liu, and J. Wang, "Laminated core models for determining exciting power and saturation characteristics," Electromagnetic field problems and applications-I, vol.21, TSI press series, USA, 2008, pp. 415–418.
2. Y. Du, Z. Cheng, et al, "Magnetic Flux and Iron Loss Modeling at Laminated Core Joints in Power Transformers," *IEEE Trans. on Applied Superconductivity*, vol.20, no.3, pp. 1878–1882, 2010.
3. M. Zhang, "Electric Machinery (Volume 1)," Science Press, 1966:48.
4. Q. Hu, "No-load current and magnetic field," BTW research report, Mar. 26, 2007.
5. R. Girgis, "Core performance (ABB)," presented at IEEE/PES Transformers Committee, 24 Oct. 2009.

6. A. J. Moses, "Characterisation and performance of electrical steels for power transformers operating under extremes of magnetisation conditions," Presented at the International Colloquium Transformer Research and Asset Management, Cavtat, Croatia, Nov. 12–14, 2009.
7. Y. Xie, "Power Transformer Manual," China Machine Press, Beijing, 2003.
8. IEC 60404-2 Ed.3.1(2008): Magnetic Materials - Part 2: Methods of measurement of the magnetic properties of electrical steel sheet and strip by means of an Epstein frame; Amendment 1.
9. Z. Cheng, N. Takahashi, B. Forghani, A. Moses, P. Anderson, Y. Fan, T. Liu, X. Wang, Z. Zhao, and L. Liu, "Modeling of magnetic properties of GO electrical steel based on Epstein combination and loss data weighted processing," IEEE Trans. Magnetics, vol.50, no.1, 6300209, 2014.
10. Q. Qing, "Metallography and heat treatment," Metallurgical Industry Press, Beijing, 1996.
11. G. Zhao, "Discussion on Annealing to Eliminate Internal Stress in Wound Cores," Transformer, Volume 41, 2nd Issue, 2004: 22–24.
12. POSCO Pohang Works, Korea.
13. JIS C 2550-2000: Test methods for magnetic steel sheet and strip.
14. ASTM A343/A343 M-03(2008): Standard test method for alternating-current magnetic properties of materials at power frequencies using Wattmeter-Ammeter-Voltmeter method and 25-cm Epstein test frame.

Chapter 10
Rotational Magnetic Properties Measurement and Modeling

Yongjian Li

Abstract In the T-joints of power transformer core and behind the teeth of AC rotating machines, rotational magnetic field exists and is testified, especially in three-dimensional (3-D) distribution, which results in rotational core loss. A 3-D magnetic properties' tester with three orthogonal and symmetrical magnetization structure and combined *B–H* sensing structure is proposed and constructed. Based on that, 3-D rotational magnetic properties' measurements of the bulk SMC material and laminated silicon steels are performed under various excitation conditions and different frequencies. Magnetic hysteresis model is a key factor in electromagnetic design and strongly depends on magnetic properties' measurement. A hybrid vector hysteresis model associated with the classical Preisach and Stoner–Wohlfarth (S–W) model is presented in terms of magnetic energy equilibrium. Finally, the availability and practicality of the hysteresis model are verified by analyzing the magnetization process of SMC material under different excitations.

Keywords Rotational magnetic property · Magnetic measurement · 3-D tester · SMC material · Core loss · Hybrid vector hysteresis model

10.1 Development of Rotational Magnetic Properties Measurement

In 1896, Baily published the first paper about the quantitative investigation on rotational hysteresis loss of hard steel and soft iron, where he pointed out that under rotational field, the hysteresis loss shows a downward trend at relatively high magnetic flux density level [1]. Since then, rotational core loss of the soft magnetic materials has been extensively studied. In 1938, a qualitative explanation of rotational core loss phenomenon was proposed based on the domain theory of ferro-

Y. Li (✉)
State Key Laboratory of Reliability and Intelligence of Electrical Equipment,
Hebei University of Technology, Tianjin, China
e-mail: liyongjian@hebut.edu.cn

Z. Cheng et al. (eds.), *Modeling and Application of Electromagnetic and Thermal Field in Electrical Engineering*, https://doi.org/10.1007/978-981-15-0173-9_10

magnetism [2]. In 1961, Kaplan systematically measured the core loss of non-grain-oriented and grain-oriented silicon steel under various magnetic flux density ***B*** changing from pure alternating flux (only the magnitude of ***B*** varies with time) to a pure circular rotational flux (only the direction of ***B*** varies with time) [3]. In 1973, Moses studied the localized rotational magnetization and core loss in the T-joints of three-phase transformer [4]. In 1990, Fiorillo and Rietto proposed that the rotational core loss could also be divided into three parts, similar to the case of alternating one [5]. Nowadays, major research teams such as Sievert (Braunschweig), Moses (Cardiff), Fiorillo (Torino), Hempel (Aachen), Enokizono (Oita), Pillay (Concordia), and Y. J. Li (HEBUT) still continue the rotational magnetic properties' measurements and make great progress in this field.

10.1.1 Measurement Methods

In the early stage of rotational magnetic core loss measurement, torque-metric method with disk specimen was developed by some researchers [1, 2, 6, 7], such as Brailsford. In this method, the torque exerted on the fixed disk specimen in a rotational magnetic field (alternatively, the field is fixed while the specimen rotates) is measured over a full period for both clockwise and anticlockwise directions. Then, the difference between the two rotational directions represents the double-rotational core loss value. The advantages of this method are the direct reading of the measured torque, which corresponds to rotational core loss, and the ability to measure rotational core loss at high flux density. The pure torque-metric method is now almost out of date due to its tediousness and the limited information it can provide.

Thermometric method was proposed and developed during the 1960–1970s [8–10]. In this method, temperature of the specimen can be obtained by thermosensors (like thermocouples and thermos-viewers). The rotational core loss is proportional to the initial rate of the specimen temperature rise, if no cooling process is involved. This very general method applied to a variety of testing apparatus using different types of specimen, such as square, cross and round. The main drawbacks of this method are the difficulties of installation and calibration of the thermosensors. In addition, this method only provides the core loss data as the torque-metric method does.

From the 1970s, the field-metric method started becoming popular for rotational magnetic properties' measurement [11–15]. In this method, the rotational core loss is calculated from the measured B and H signals by using B and H sensing coils, as shown in Eq. (10.1). This method is characterized by high accuracy and generality. In addition, the measured instantaneous H and B values can yield more detailed information about the magnetic properties of the specimen, such as the loci of ***B*** and ***H*** vectors, and harmonics, permeability, and core loss [8]. The disadvantage of this method is that the calibration and installation of B and H sensors are complicated.

$$P_{\mathrm{t}} = \frac{f}{\rho}\int_{0}^{T}\left(\boldsymbol{H}\cdot\frac{\mathrm{d}\boldsymbol{B}}{\mathrm{d}t}\right)\mathrm{d}t = \frac{f}{\rho}\int_{0}^{T}\left(H_x\frac{\mathrm{d}B_x}{\mathrm{d}t}+H_y\frac{\mathrm{d}B_y}{\mathrm{d}t}\right)\mathrm{d}t \quad (\mathrm{W/kg}) \tag{10.1}$$

where f is the magnetization frequency in Hz, ρ is the mass density of the tested specimen in unit. H_x, H_y, B_x, B_y are the x and y components of $\boldsymbol{H}$ and $\boldsymbol{B}$ vectors in unit, respectively.

10.1.2 Measurement Apparatus in Field-Metric Method

The rotational magnetization apparatus, according to the magnetic circuit structure, can be mainly classified into horizontal structure, vertical structure and round-stack structure (including Halbach structure and traditional round structure) depending on specimen shapes, as shown in Fig. 10.1.

For the horizontal structure [12, 14, 16, 17], the main advantage is that it effectively saves the amount of magnetic circuit material and the supporting structure is easy to construct. The main problem is that the main flux may stray or get distorted due to the adjacent magnetic poles. In addition, this structure has air gap inevitably at the joint of the yokes and poles. This may cause the main flux pass through the non-rolling direction of the silicon steel sheet (the main material of the magnetic core and yoke). It increases magnetic circuit reluctance and results in asymmetrical flux distribution and increasing of the core loss of the magnetizer.

The advantage of vertical structure is that the magnetic circuit is independent and easy to control compared with the horizontal one [18, 19]. The major problem is that the asymmetrical single-yoke structure may also cause the asymmetrical field distribution in the magnetic circuit and measurement region, which may affect the measurement accuracy.

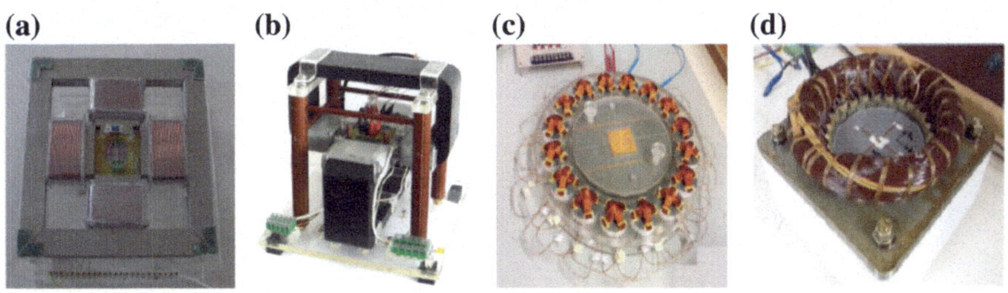

Fig. 10.1 Measurement apparatus **a** Horizontal structure. **b** Vertical structure. **c** Halbach structure. **d** Round structure

For Halbach and round structure magnetizers [20–22], which can be seen as a special electrical machine stator, the specimen is usually made round shape and surrounded by relatively deep yokes. Uneven magnetic flux exists around the specimen due to the yokes' cogging effect. Therefore, the specimen of this structure usually has larger size, compared with the horizontal and vertical structures, which increases the difficulty of excitation and the demand for power supply. Moreover, the magnetizers with this structure may have higher z-component of $\boldsymbol{H}$ (i.e., H_Z), which may interfere with the measurement because of the surrounding deep yokes.

10.1.3 Techniques for Measuring B *and* H *in Field-Metric Method*

In general, the magnetic flux density in the specimen is usually measured by sensing coil method or the needle method [3, 12–16]. The B sensing coil is generally wrapped around the outside of the specimen or two holes through the specimen, as shown in Fig. 10.2a and b. However, both ways would involve the inhomogeneous region of magnetic field caused by either the edge of specimen, where is usually considered non-uniform no matter what the shape of the specimen is, or the holes. Consequently, this will lead to the inaccurate measurement. In the needle method, as shown in Fig. 10.2c, each $\boldsymbol{B}$ component can be calculated by

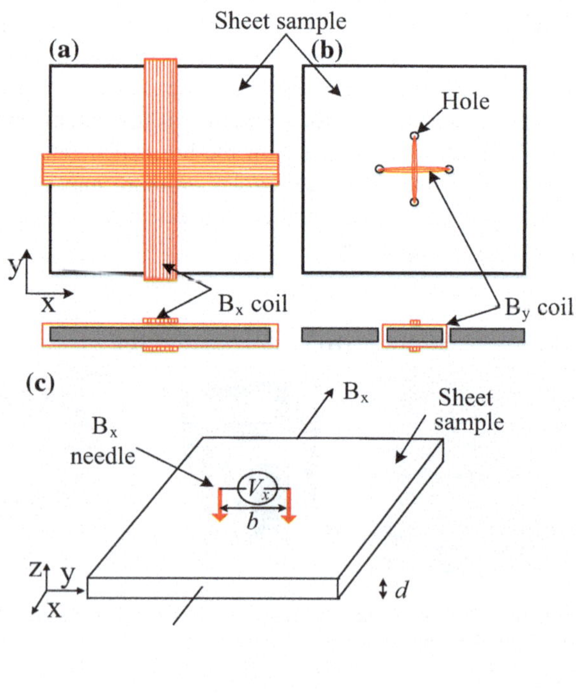

Fig. 10.2 *B* coils **a** Sensing coil around the specimen. **b** Sensing coil through the specimen by drilling. **c** *B* needle method

acquiring potential difference between the two points on the surface of a conducting sheet specimen without destroying it. The $\boldsymbol{B}$ component (such as B_x) can be determined using Eqs. (10.2)–(10.3). The B needle method is more suitable for batch measurements but only limited to conducting material.

$$V_x = \frac{bd}{2}\frac{\mathrm{d}B_x}{\mathrm{d}t} \tag{10.2}$$

$$B_x = \frac{2}{bd}\int V_x \mathrm{d}t \tag{10.3}$$

Generally, the magnetic field strength can be detected by, Hall sensor and sensing coil [18–23]. In terms of the Ampere's law, H is calculated from the magnetization current. This method is only suitable for alternating magnetization like toroid and Epstein frame where the effective magnetic circuit length is relatively easy to determine while it is not applicable to rotational magnetic properties measurement. By means of Hall Effect, Hall sensor detection method belongs to the so-called Spot-detection and the precision is low if the number of the sensors is not enough. Multi-sensors may cause the difficulty of installation and affect the testing field because they are active. The sensing coil method can be divided into three types, namely conventional H coil, double H coil, and Rogowski-Chattock potentiometer coil (RCP), as shown in Fig. 10.3. The first one can calculate the tangential component of $\boldsymbol{H}$ at the specimen surface by measuring the induced voltage across the H coil according to the magnetic field boundary condition. The tangential component of $\boldsymbol{H}$ can be calculated from (10.4). This method is commonly used in both alternating and rotational magnetization and is proven to have high accuracy as long as H coil is thin enough and closely attached to the specimen surface. The double H coil method is based on the hypothesis that $\boldsymbol{H}$ varies linearly

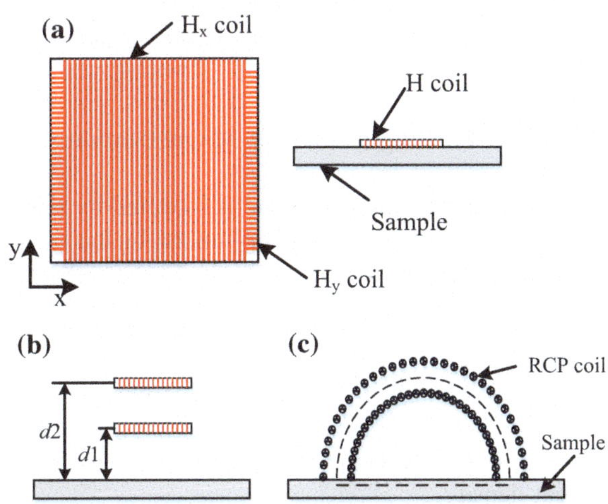

Fig. 10.3 *H* coils **a** Conventional *H* coil. **b** Double *H* coil. **c** RCP coil

with respect to the distance to the specimen surface. This method is excluded for rotational measurement because the distances of two H coils to the surface are hard to accurately determine, which may in return introduce excess error. In addition, the magnetic field intensity on the specimen surface is easily influenced by the surrounding electromagnetic environment, especially over the high-frequency range, which makes it uncertain whether $\boldsymbol{H}$ is linear with distance or not. Although the RCP method has high sensitivity, the complicated fabrication process and high symmetry requirement on both sides of the coil limit the expansion of its applications.

$$H = \frac{1}{\mu_0 K_{\mathrm{H}}} \int V_{\mathrm{H}} \mathrm{d}t \tag{10.4}$$

10.1.4 3-D Magnetic Testing System

10.1.4.1 3-D Excitation Structure

3-D Magnetization Structure

In order to smooth and balance the flux path in the three-dimensional (3-D) tester, and avoid structural anisotropy, a novel 3-D magnetization structure is designed. In Fig. 10.4, the new structure consists of three orthogonal "C-type" cores, and six multilayer excitation windings which are wound around the three pairs of orthogonal core poles. The "C-type" core is a laminated structure using HiB grain-oriented silicon steel, which can generate higher magnetic flux density and decrease core loss caused by leakage flux, compared with the core-yoke magnetization structure. To concentrate magnetic flux density and enhance the excitation field, terminals of the core are shaped in frustum of a square pyramid. A cubic laminated silicon steel specimen with B–H sensing coils is placed in the center of the tester [24].

10.1.4.2 Modeling of the 3-D Magnetization Circuit

The 3-D magnetization structure is modeled by means of finite element method to evaluate the magnetic flux distributions in the core. Magnetic vector potential is adopted to simulate the magnetic flux density distribution of the "C-type" core.

$$\nabla \times \boldsymbol{v} \nabla \times \boldsymbol{A} - \nabla \boldsymbol{v} \nabla \cdot \boldsymbol{A} = \boldsymbol{J}_S \tag{10.5}$$

$$\boldsymbol{B} = \nabla \times \boldsymbol{A} \tag{10.6}$$

where $\boldsymbol{A}$ is the magnetic vector potential, v is the magnetic reluctivity, $\boldsymbol{J}_S$ is the current density.

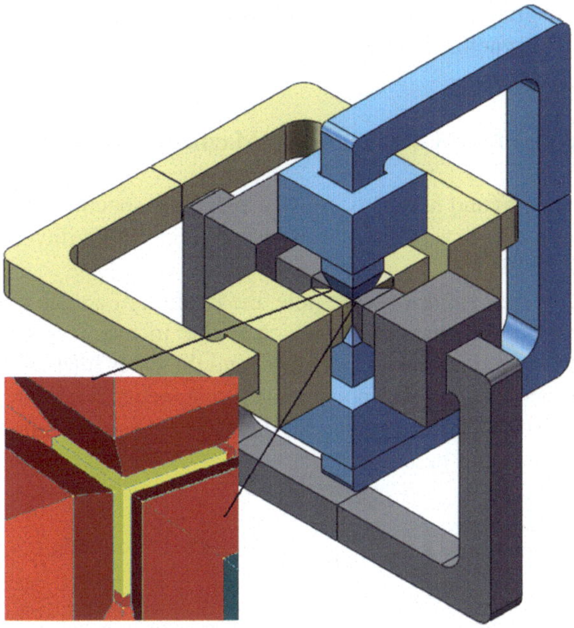

Fig. 10.4 Novel 3-D magnetization model

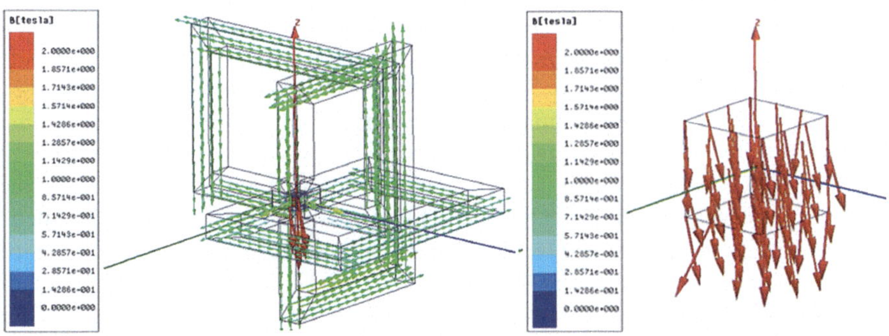

Fig. 10.5 Magnetic flux density distribution in the 3-D magnetization model

Grain-oriented direction is set along with the bulk model of the core, which can simulate the grain-oriented silicon steel lamination. The dimension of each core is about 1490 mm × 50 mm × 50 mm, and the air gap between the core poles, which can hold the cubic specimen, is 34 mm. Figure 10.5 shows the magnetic flux density distributions in the three cores at the same excitation field strength, and the magnetic flux density distributions in the cubic specimen. The magnetic flux distributions in the three "C-type" cores are nearly the same and are relatively uniform in each core of the magnetic circuit. It implies that 3-D excitation can be fulfilled in this symmetrical magnetization structure.

10.1.4.3 Shaping of the Core Poles and Magnetic Concentration

To generate a uniform and high flux density inside the specimen, the shape of core poles must be optimized; so that, it can also provide flexibility in design of the B sensing coils.

Compared with the cylindrical core pole, the frustum of a cone pole can generate a relatively large gap magnetic field, which is due to most lines of the magnetic force being concentrated across the pole face, and the other fraction passing through the conical surface. Therefore, concentrated magnetic flux density is formed through the frustum of a cone pole. In Fig. 10.6, the maximum magnetic flux density in the air gap can be expressed:

$$B_0 = \mu_0 M \left(1 - \frac{l_0}{\sqrt{l_0^2 + b^2}}\right) + M \sin^2\theta \cos\theta \ln\frac{a}{b} \tag{10.7}$$

where a is the radius of the bottom circle, b is the radius of the upper circle, l_0 is half length of the air gap, θ is the cone angle with the axis, and M is the magnetization intensity. If the pole has been uniformly magnetized to saturation, maximum magnetic field can be produced at $\theta = 54.5°$, yet the gap center of the magnetic field is not uniform. Engineering experience calculation shows that relatively high and uniform magnetic field between the frustum of the cone pole should satisfy $\theta = 39.5°$. In this 3-D tester, the laminated core poles are processed into frustum of a square pyramid shape by means of a wire-electrode, cutting though the cone shape, making it difficult to be processed like the laminated structure. The cone angle with the axis is about 40°. In addition, according to the design experience of single sheet tester (SST), the optimal performance in magnetic field strength measurement can be obtained when the core laminations are arranged vertically and the core poles are shaped in wedges. Therefore, concentrated magnetic flux density and uniform magnetic field can be obtained in the center of the magnetization structure.

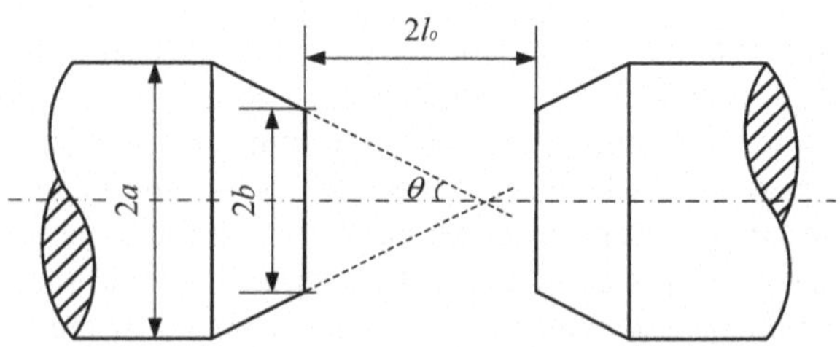

Fig. 10.6 Schematic structure of the frustum of a cone pole

10.1.4.4 Parameters of the Excitation Circuit

Parameters of the excitation circuit are important to the process of magnetization and should be designed in terms of the measurement frequency. The impedance of the coil will increase with increasing frequency because of the inductive reactance of the coil will increase. A series RLC resonance circuit is adopted in order to compensate the phase angle between AC voltage and current, and to obtain maximum excitation current. In Fig. 10.7, L is the inductance of the selected excitation coil, R_L is the DC resistance of the excitation coil, C is the AC capacitor, R is the compensation resistance for impedance of magnetization circuit matching for the power amplifier (PA) outputs. The wide-range-frequency experiment parameters are shown in Table 10.1 [25].

10.1.4.5 Excitation Model of the 3-D Magnetic Testing

Analysis of the 3-D Excitation Model

The 3-D magnetic testing system consists of three power amplifiers for x-, y-, z-directions, respectively, matching impedance, compensation capacitor, and excitation windings. The equivalent circuit of the 3-D tester is illustrated in Fig. 10.8. In order to analyze the 3-D rotational excitation feature, the transient voltage equations of the equivalent circuit can be written as

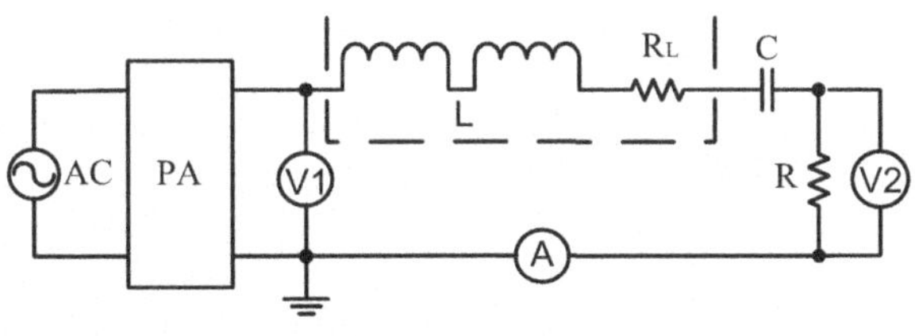

Fig. 10.7 Excitation circuit with RLC resonance

Table 10.1 Parameters of the wide-range-frequency experiments

f (Hz)	L (H)	N (Turns)	C (μF)	R_L (Ω)	R (Ω)
2	0.7742	1120	8180	3.6403	4
5	0.6178	980	1640	3.0537	4
20	0.2169	560	292	1.9384	4
50	0.1267	420	80	1.1927	2
100	0.1206	420	21	1.1927	2
200	0.0302	105	21	0.298	1
500	0.0291	105	3.48	0.2919	1
1000	0.0034	70	7.5	0.276	1

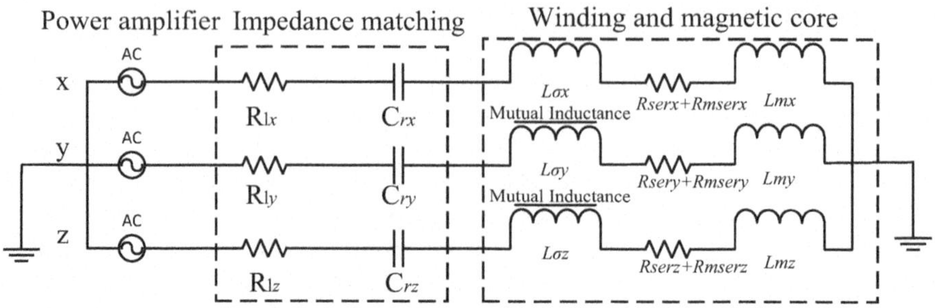

Fig. 10.8 Equivalent circuit of the 3-D magnetic tester

$$
\begin{aligned}
u &= R_1 i + C_r^{-1} \int i\mathrm{d}t + R_\omega i + \frac{\mathrm{d}\lambda}{\mathrm{d}t} \\
u_m &= R_{xyz} \frac{\mathrm{d}\lambda_m}{\mathrm{d}t}
\end{aligned}
\tag{10.8}
$$

where u and i denote the output voltage and current of the power amplifier, R_l is the compensation resistance of the series resonance, C_r is the resonant capacitance, R_w the diagonal element of the equivalent excitation winding resistance tensor, λ and λ_m the flux linkage through the main magnetic circuit and the specimen, respectively.

Since the cubic specimen cannot be perfectly located in the center of the three exciting magnetic poles, a resistance tensor in the coordinate system of the 3-D magnetic poles, R_{xyz}, is determined

$$
R_{xyz} = R_{xy} R_{xz} R_{yz} \tag{10.9}
$$

where R_{xy}, R_{xz}, and R_{yz} are the rotating transformation matrices in the xoy- , xoz- , and yoz-planes, respectively. They can be expressed as

$$
\begin{aligned}
R_{xy} &= \begin{pmatrix} \cos(\theta_{xy}) & -\sin(\theta_{xy}) & 0 \\ \sin(\theta_{xy}) & \cos(\theta_{xy}) & 0 \\ 0 & 0 & 0 \end{pmatrix} \\
R_{xz} &= \begin{pmatrix} \cos(\theta_{xz}) & 0 & -\sin(\theta_{xz}) \\ 0 & 0 & 0 \\ \sin(\theta_{xz}) & 0 & \cos(\theta_{xz}) \end{pmatrix} \\
R_{yz} &= \begin{pmatrix} 0 & 0 & 0 \\ 0 & \cos(\theta_{yz}) & -\sin(\theta_{yz}) \\ 0 & \sin(\theta_{yz}) & \cos(\theta_{yz}) \end{pmatrix}
\end{aligned}
\tag{10.10}
$$

where θ_{xy}, θ_{xz}, and θ_{yz} are the rotational angles of the specimen in the *xoy*- , *xoz*- , and *yoz*-planes, respectively. The main flux is defined as the magnetic flux crossing the specimen. The flux linkage equations can be expressed as

$$\begin{aligned} \lambda &= L_\sigma i + L_{\mathrm{m}} i + (R_{\mathrm{ser}} + R_{\mathrm{mser}}) \int_0^t i\mathrm{d}t \\ \lambda_m &= \left(L_{\mathrm{m}} i + R_{\mathrm{mser}} \int_0^t i\mathrm{d}t \right) \end{aligned} \tag{10.11}$$

where L_{m} and L_σ are the excitation and leakage inductances of the excitation windings, which strongly depend on the magnetic properties of the specimen. For a specimen of magnetically isotropic material, e.g., soft magnetic composite (SMC) material, L_{m} is a diagonal matrix. While for specimens of grain-oriented and non-oriented silicon steels and other magnetically anisotropic materials, L_{m} has nonzero off-diagonal elements. R_{ser} and R_{mser} are the equivalent resistors accounting for hysteresis losses in the main magnetic circuit and the specimen.

2-D Rotating Excitation Model

So far, there is no industrial standard for 2-D magnetic property testing. The most widely used 2-D rotating excitation model is circular or elliptic magnetization in a plane, which can be expressed as

$$\begin{aligned} B_x(t) &= B_{\mathrm{m}} \cos(2\pi ft) \\ B_y(t) &= \varepsilon_y B_{\mathrm{m}} \cos(2\pi ft + \pi/2) \end{aligned} \tag{10.12}$$

where f is the excitation frequency, B_{m} the magnitude of the rotating flux density vector $\boldsymbol{B}$, and ε_y the ratio of the major and minor axes. According to Faraday's law of electromagnetic induction, the time derivative of (10.12) is proportional to the back electromotive force (EMF) of the sensing coil and is used as the feedback signal in the feedback system.

However, the mutual inductance between the *x*- and *y*-axes exists due to the misalignment of the two opposite magnetic poles. In addition, the magnetic circuit parameters along the *x*- and *y*-axes are not identical to each other. The open-loop control method fails to satisfy the phases and magnitudes of the two directional waveforms and the resultant $\boldsymbol{B}$ locus is a distorted circular or elliptic one.

Spherical Excitation Model

The conventional magnetization techniques examine the overall hysteresis under 1-D and 2-D excitations, which cannot obtain experimental data in the direction perpendicular to the lamination plane. In fact, the nature of the magnetization process suggests that the magnetic properties should be studied under 3-D excitations. By setting the associated trigonometric equations in three directions, a ***B*** vector spiral spherical model is established as the following:

$$
\begin{aligned}
B_x(t) &= B_{\mathrm{m}} \cos(2\pi f_1 t) \cos(2\pi f_2 t) \\
B_y(t) &= \varepsilon_y B_{\mathrm{m}} \sin(2\pi f_1 t) \cos(2\pi f_2 t) \\
B_z(t) &= \varepsilon_z B_{\mathrm{m}} \sin(2\pi f_2 t)
\end{aligned} \tag{10.13}
$$

where ε_z is the ratio of the ***B*** magnitudes in the z- and x-axes, and ε_y the ratio of the ***B*** magnitudes in the y- and x-axes. The frequency f_2 is set very low to suppress the eddy current parallel to the plane of silicon steel laminations. If the eddy current in the plane of silicon steel sheets is significant, the magnetic field is extremely inhomogeneous and the measurement results will be unreliable. The envelope of the B_x and B_y waveforms are a sinusoidal waveform of frequency f_2. Unlike the 1-D and 2-D excitation model, the 3-D spiral spherical excitation model is involved at two different frequencies in one excitation axis [26].

10.1.5 B–H *Combined Sensing Structure*

In 3-D magnetic properties testing system, a novel-sensing structure with combined *B* and *H* sensing coils is designed and fabricated to guarantee the 3-D magnetic signals detection and improve the experimental precision. Figure 10.9 shows the schematic structure of the combined *B–H* sensing coils and the improved cubic sensing structure.

Each *H* sensing coil is wound around a 0.5-mm-thickness epoxy resin board with 200 turns. A thin (0.4 mm thickness) and round (5 mm average diameter) 60-turn *B* coil is embedded in the central hole of the epoxy resin board. To decrease measurement error and improve the accuracy, 0.05-mm enameled copper wire is used for the sensing coils. Two terminals of each coil are twisted to eliminate EMF induced by unwanted stray field. Each pair of *H* coils or *B* coils on the opposite

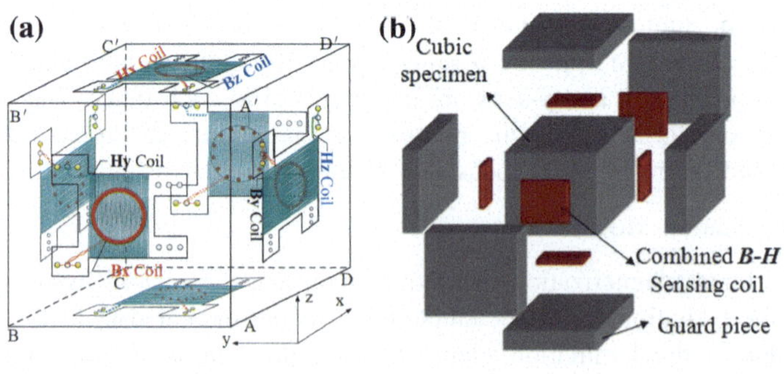

Fig. 10.9 3-D sensing structure **a** Schematic structure of the combined *B–H* sensing coils. **b** Cubic sensing box with specimen, sensing coils, and guard pieces

sides are connected in series to reduce the interference caused by the change of magnetic field or the slight movement of coil, which can balance the signal at the same time.

Six *B–H* combined coils are consolidated to form a cubic sensing box with internal space of $22 \times 22 \times 22\ \text{mm}^3$ so that a cubic specimen can be easily mounted into the sensing box and the surface of the specimen is as close as possible to the combined *B–H* sensing coils. In addition, a guard piece is placed between the *B–H* sensing coil and the pole to increase the field homogeneity within the specimen. The guard piece is $22 \times 22 \times 5\ \text{mm}^3$ and six guard pieces enclose the sensing box. The guard pieces are cut from the material around the specimen. Meanwhile, the guard pieces are arranged in such way that magnetic properties' direction of the guard piece is aligned with the direction of the specimen, such as rolling direction [27]. Therefore, the guard pieces are also functionally named homogeneous field core shoes.

10.1.6 Calibration and Compensation of the 3-D Tester

In the 3-D magnetic properties measurement, coefficients of the sensing coils must be calibrated through precise experiments to ensure measurement accuracy. Due to irregularities and flaws existing in the sensing coil, the measured voltage across the sensing coil is the overall voltage which is induced by the magnetic field from the three directions even though it is designed to measure the field perpendicular passing through the coil. In addition, there is an inevitable gap existing between adjacent windings, which is also an area responsible for induced voltage. Based on the sensing box structure, the coefficients of the 3-D sensing box are written in a 3×3 matrix. Each element can be determined by the calibration process.

$$K_{\mathrm{H}} = \begin{bmatrix} K_{\mathrm{H}xx} & K_{\mathrm{H}xy} & K_{\mathrm{H}xz} \\ K_{\mathrm{H}yx} & K_{\mathrm{H}yy} & K_{\mathrm{H}yz} \\ K_{\mathrm{H}zx} & K_{\mathrm{H}zy} & K_{\mathrm{H}zz} \end{bmatrix} \tag{10.14}$$

$$K_{\mathrm{B}} = \begin{bmatrix} K_{\mathrm{B}xx} & K_{\mathrm{B}xy} & K_{\mathrm{B}xz} \\ K_{\mathrm{B}yx} & K_{\mathrm{B}yy} & K_{\mathrm{B}yz} \\ K_{\mathrm{B}zx} & K_{\mathrm{B}zy} & K_{\mathrm{B}zz} \end{bmatrix} \tag{10.15}$$

The components of ***H*** and ***B*** vectors along each coordinate axis can be calculated from the induced EMF of the sensing coils by

$$H_i = \frac{1}{\mu_0 K_{\mathrm{H}_i}} \int U_{\mathrm{H}_i} \mathrm{d}t \quad (i = x,\ y,\ z) \tag{10.16}$$

$$B_i = \frac{1}{K_{\mathrm{B}_i}} \int U_{\mathrm{B}_i} \mathrm{d}t \quad (i = x,\ y,\ z) \tag{10.17}$$

where U_{Hi} and U_{Bi} are the induced voltages of the $\boldsymbol{H}_i$ and $\boldsymbol{B}_i$ sensing coils, K_{Hi} and K_{Bi} are the coefficients of the $\boldsymbol{H}_i$ and $\boldsymbol{B}_i$ coils obtained by calibration in a long solenoid which can generate linear and uniform magnetic field. The induced EMF of the sensing coils by alternating field is governed by Faraday's law; therefore, the coil coefficient K can be obtained:

$$K = \frac{E}{\sqrt{2}\pi f \mu_0 H_m} \tag{10.18}$$

where E is the RMS value of the induced EMF, $\mu_0 H_m$ is the peak value of the flux density in the center of the long solenoid, f is the excitation frequency. While calibrating the sensing box, the box should be aligned with the axis of the solenoid. In addition, the solenoid ought to be placed in an environment without external magnetic field interference in order to reduce system errors and ensure the calibration accuracy.

The off-diagonal element of the sensing box coefficient matrix is the key parameter for the accurate measurement of the 3-D rotational magnetic properties. It quantitatively describes the voltages induced by off-axis fields and effectively eliminates the errors caused by the coupling between magnetic poles in the measurement. The off-diagonal coefficients for B coils are ignored since they are very small compared to the diagonal values. Hence, the $\boldsymbol{B}$ vector can be simply calculated from (10.19). However, for H coils, the off-diagonal elements cannot be ignored because the field perpendicular to the laminar H coil also induces large EMF due to irregularity of the H coil. According to Faraday's law, the induced voltages crossing H and B coils can be expressed as

$$\begin{cases} U_{Bx} = K_{Bz} \dfrac{dB_z}{dt} \\ U_{By} = K_{Bx} \dfrac{dB_x}{dt} \\ U_{Bz} = K_{By} \dfrac{dB_y}{dt} \end{cases} \tag{10.19}$$

$$\begin{cases} U_{Hx} = \mu_0 \left(K_{Hxx} \dfrac{dH_x}{dt} + K_{Hxy} \dfrac{dH_y}{dt} + \dfrac{K_{Hxz}}{\mu_0} \dfrac{dB_z}{dt} \right) \\ U_{Hy} = \mu_0 \left(\dfrac{K_{Hyx}}{\mu_0} \dfrac{dB_x}{dt} + K_{Hyy} \dfrac{dH_y}{dt} + K_{Hyz} \dfrac{dH_z}{dt} \right) \\ U_{Hz} = \mu_0 \left(K_{Hzx} \dfrac{dH_x}{dt} + \dfrac{K_{Hzy}}{\mu_0} \dfrac{dB_y}{dt} + K_{Hzz} \dfrac{dH_z}{dt} \right) \end{cases} \tag{10.20}$$

The magnetic field strength and magnetic flux density components can then be worked out.

In order to further increase the accuracy of the experiment, the compensation and optimization of magnetic measurement should be considered as follows.

Unreasonable winding should be avoided in the design and fabrication for each sensing coil, and the winding should be as uniform and orderly as possible.

A thin-enameled wire should be used to reduce the gap between the sensing coil and the specimen. The leading wire should also be very thin and twisted tightly to eliminate the interference. The outer part of the leading wire should have a well-grounded shield layer.

The size of the combined *B–H* sensing coil should be minimized and be placed at the center of the specimen surface where the magnetic field is relatively uniform.

Installing homogeneous field core shoes with the same material enclosing the sensing box is equivalent to placing each sensing coil "inside" the specimen. It can significantly improve uniformity of the magnetic field at the specimen surface, hence improve the measurement accuracy. It can also effectively reduce the magnetic reluctance of the whole equivalent magnetic circuit and make the exciting current more efficient for magnetizing the specimen.

The comparison of the magnetic flux density distributions with and without homogeneous field core shoes is shown in Fig. 10.10. It can be seen that the magnetic field of the whole specimen surface is uniform after adding the shoes, which are very beneficial to the measurement of magnetic properties [28].

The 3-D magnetic circuit and sensing box are designed to keep the excitation and sensing box axes in the same direction. However, systematic deviations will occur in the experiment that can be compensated indirectly by means of rotating coordinate transformation. In the 3-D rotating coordinate system shown in Fig. 10.11, x, y, and z are orthogonal coordinate axes, and x', y', and z' are the deviation coordinates of the cubic sensor box. *B* signal is controlled in this system, in order to make compensation for the error in the *H* signal.

The transformation formula of magnetic field intensity is written as follow:

$$\begin{bmatrix} H_x' \\ H_y' \\ H_z' \end{bmatrix} = \begin{bmatrix} \cos\alpha_1 & \cos\beta_1 & \cos\gamma_1 \\ \cos\alpha_2 & \cos\beta_2 & \cos\gamma_2 \\ \cos\alpha_3 & \cos\beta_3 & \cos\gamma_3 \end{bmatrix} \begin{bmatrix} H_x \\ H_y \\ H_z \end{bmatrix} \tag{10.21}$$

where H_x', H_y', H_z' are measurement values, H_x, H_y, H_z are actual values, α_i, β_i, γ_i $(i = 1, 2, 3)$ are deviation angles.

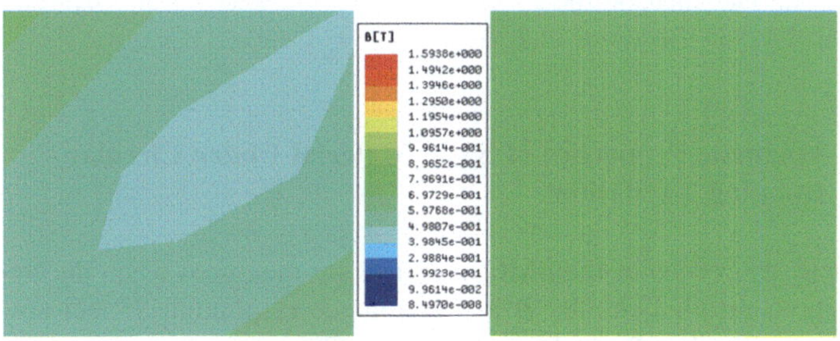

Fig. 10.10 Magnetic field distributions of the specimen without homogeneous field core shoes (left) and with homogeneous field core shoes (right)

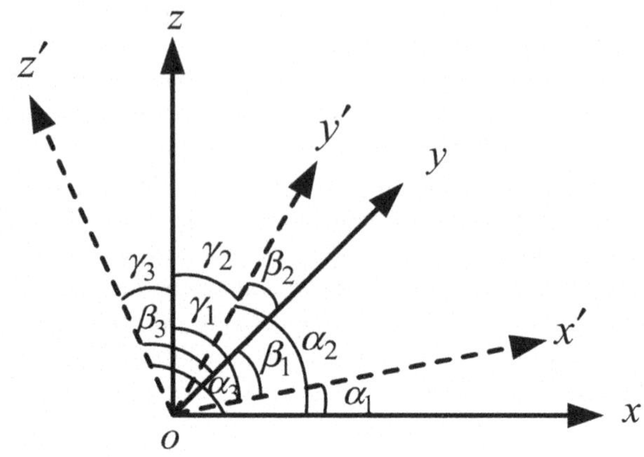

Fig. 10.11 Schematic rotation of 3-D coordinates

10.2 Measurement and Analysis of the Rotational Magnetic Properties

10.2.1 Magnetic Properties of the Soft Magnetic Composite Materials

Due to the intrinsic magnetic isotropy, SMC materials are applied in electrical machine with special structure, in which the magnetic flux distribution is 3-D. Conventional methods for analyzing the magnetic properties of SMC materials based on alternating excitation are not comprehensive due to the rotational magnetic flux in the rotating electrical machines. In this section, 3-D magnetic properties of the SMC material are measured using a 3-D testing system under complicated magnetizations, such as circular, elliptical, and spherical excitations. The magnetic flux density vector ***B*** loci, magnetic field strength vector ***H*** loci, and core losses over a wide excitation frequency range from 2 to 1000 Hz are measured and discussed. Experimental results of rotational core losses and alternating losses are compared. The measurements provide valuable data for the design and optimization of the 3-D electrical machines.

10.2.1.1 Magnetic Properties of SMC Material Under Circular Rotating Excitations

3-D magnetic properties of the SMC materials are measured when the frequency is at 50 Hz, the controlled circular ***B*** loci and the corresponding ***H*** loci are shown in Fig. 10.12 at increasing ***B*** amplitude (up to 1.6 T) in the *xoy*- , *yoz*- and *zox*-planes. It can be seen that the ***B*** and ***H*** loci lie in the same magnetization planes. The ***H*** loci in the *yoz*- and *zox*-planes evolve from ellipses into rectangular loops while the changing in the *xoy*-plane from circles to squares. It may imply that the particles are

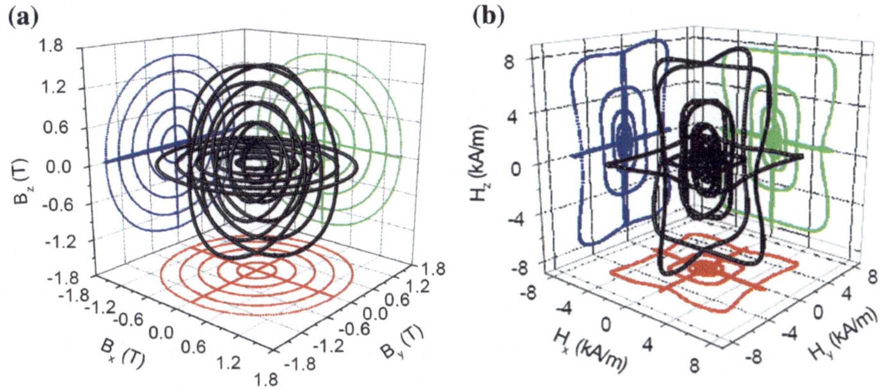

Fig. 10.12 Circular ***B*** loci and corresponding ***H*** loci in *xoy*- , *yoz*- , *zox*-planes and projections in three planes **a** Round ***B*** loci. **b** Corresponding ***H*** loci

much closer along *z*-axis (the compaction direction) than along *x*- and *y*-axes, i.e., high mass density and strong demagnetization field along the *z*-axis [29]. In other words, *z*-axis of this cubic SMC specimen is relatively harder to be magnetized than *x*- and *y*-axes, which is consistent with the alternating outputs. Also, when increasing the ***B*** amplitude in any plane, magnetic anisotropy becomes obvious due to movement of rotating domain [30].

A group of well-controlled circular ***B*** loci and corresponding experimental ***H*** loci in *xoy*-plane at 5, 20, and 500 Hz under rotational magnetizations are demonstrated and compared as shown in Fig. 10.13. At low frequency, such as 5 Hz, the outer ***H*** loci change from rectangular shape to saddle-like shape, which is close to the deep saturation state. If the frequency is higher, the ***B*** and ***H*** loci will be slightly deviated from the same magnetization plane due to the enhanced magnetic coupling among the 3-D cores. But the projection components do not contribute to the core losses due to ***B*** loci in *yoz*- and *zox*-planes being straight lines.

10.2.1.2 Magnetic Properties of SMC Material Under Elliptical Rotating Excitations

Figure 10.14 shows a series of ***B***, ***H*** loci and corresponding projections when ***B*** loci are controlled to be ellipses with increasing magnitudes up to 1.6 T at 50 Hz in the three orthogonal planes, respectively, where ε is the axis ratio of the minor axis to the major axis. When ε is 0, it is equal to the alternating property along the major axis. Similarly, it is equal to the circularly rotational property when ε is 1. Therefore, the outermost ***B*** and ***H*** loci are similar with that shown in Fig. 10.12. It can be seen that ***B*** and ***H*** loci lie in the same magnetization planes when the ***B*** loci are well controlled. Also, a slight anisotropy is found in this magnetization process, and *z*-axis is the hard direction to be magnetized.

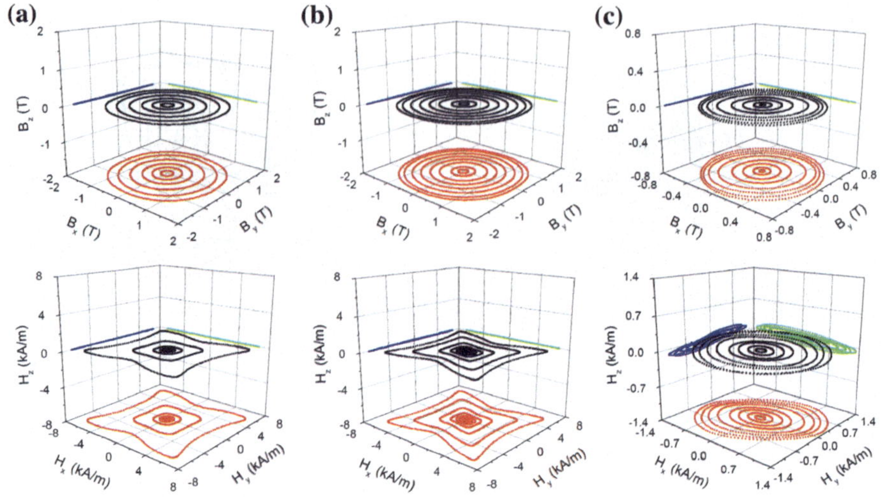

Fig. 10.13 Round $\boldsymbol{B}$ loci in *xoy*-plane (up), and corresponding $\boldsymbol{H}$ loci (down): **a** 5 Hz; **b** 20 Hz; **c** 500 Hz

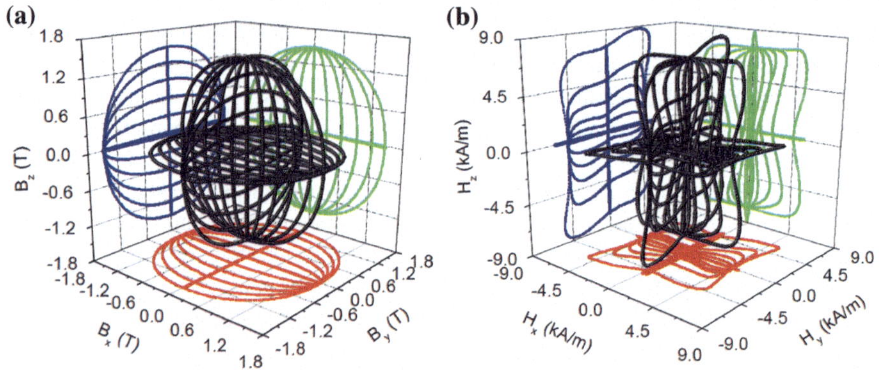

Fig. 10.14 Elliptical $\boldsymbol{B}$ loci and corresponding $\boldsymbol{H}$ loci in *xoy*- , *yoz*- , *zox*-plane and projections at 50 Hz **a** Elliptical $\boldsymbol{B}$ loci (the axis ratio $\varepsilon = 0, 0.2, 0.4, 0.6, 0.8, 1$), the major axis is x, y, and z, respectively. **b** Corresponding $\boldsymbol{H}$ loci

10.2.1.3 Magnetic Properties of SMC Material Under Spherical Rotating Excitations

In engineering applications, such as rotating machines, the magnetic material may be excited in any direction, i.e. 3-D excitation. In order to simulate the 3-D magnetic properties of the SMC materials under multi-directional excitation in space, it is necessary to construct 3-D excitation models, such as helical ellipsoid and helical spherical excitation models.

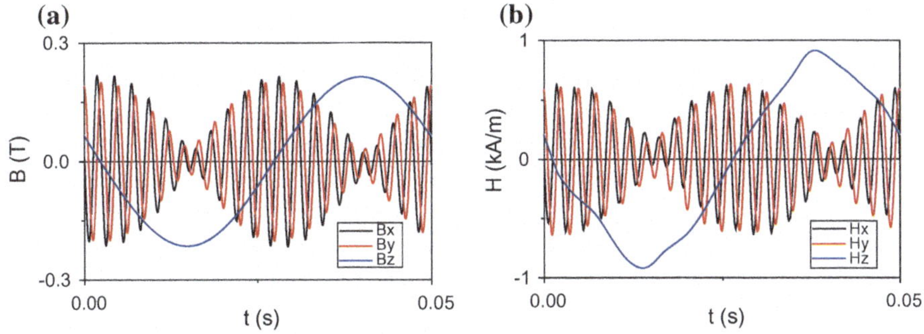

Fig. 10.15 Experimental waveforms along *x*- , *y*- and *z*-axes of ***B*** and ***H*** vectors under spherical model excitations: **a** Experimental waveforms of B_x, B_y and B_z, **b** The corresponding waveforms of H_x, H_y and H_z

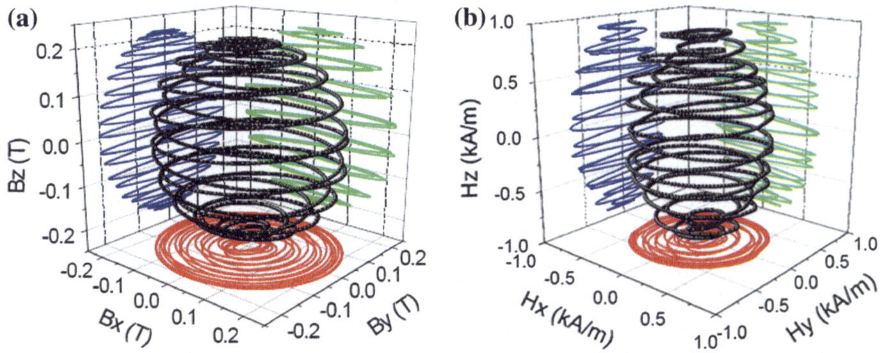

Fig. 10.16 Spherical ***B*** loci and corresponding ***H*** loci and projections in *xoy*- , *yoz*- , *zox*-planes **a** Spherical ***B*** loci. **b** Corresponding ***H*** loci

Figure 10.15 illustrates the experimental waveforms along *x*- , *y*- and *z*-axes of ***B*** and ***H*** vectors under spherical model excitations. It shows that the ***B*** component waveforms in three directions are well controlled to be sinusoidal and the corresponding ***H*** waveforms have some distortion, mainly reflected in *z*-axis along which the magnetization is relatively hard. The amplitude of H_z is obviously larger than that of H_x and H_y, and the waveform distortion is significant with the influence of third and fifth harmonics.

Figure 10.16 shows the experimental spherical ***B*** loci and corresponding ***H*** loci in 3-D space. It can be seen that the ***B*** loci are well controlled and the measured ***H*** loci change to be ellipsoidal contours. The projections in *xoy*-plane are approximate compared to ***B*** loci while the projections in *yoz*- and *zox*-planes look like stretched compared to ***B*** loci. It is evident that *z*-axis is the relatively hard magnetization direction and the slightly anisotropic phenomenon is consistent with that of alternating properties.

10.2.1.4 Harmonic Analysis Under Rotational Excitation

Figure 10.17 shows the H_z waveforms for SMC material under rotational excitation at 50 Hz (Circular $\boldsymbol{B}$ loci in *zox*-plane) with a different current excitation. It can be seen that the H_z waveform is sinusoidal at low current excitation, but it is distorted at high current excitation especially near the saturation as shown in the curve of H_{z5}. In order to analyze the influence of harmonics on the magnetic property curve, the H_z waveforms are expanded to Fourier series according to Formula (10.22).

$$H(t) = H_0 + \sum_{n=1}^{\infty} H_n \sin(n\omega t + \phi_n) \tag{10.22}$$

Since the waveform of the upper and lower half of the $\boldsymbol{H}$ loci is substantially symmetrical and no DC component is added, there are basically no even and zero-order harmonic components. In addition, harmonic analysis of the experimental H curves shows that the fundamental components are the dominant part while the odd harmonics constitutes the minor part. When the excitation current increases, the odd harmonics fraction increases rapidly.

Figure 10.18 shows the harmonic distributions of H_x, H_y, and H_z under 50 Hz rotational excitation. Amplitudes of the fundamental and higher harmonics along *z*-axis are greater than that of *x*-, *y*-axes. Moreover, this difference will be more significant as the excitation frequency increases. Therefore, the anisotropic characteristics can be intensified by increasing frequency. In addition, harmonics distributions of clockwise and anticlockwise in the same exciting conditions almost have no difference. In other words, the rotating direction of the magnetic field does not cause harmonic difference.

Figure 10.19 shows the relationship between the maximum flux densities, B_m, and corresponding maximum magnetic field strength values and fundamental components of vector $\boldsymbol{H}$ under rotational excitation at 50 Hz. The amplitude of fundamental component varies nonlinearly with the increase of B_m (radius of

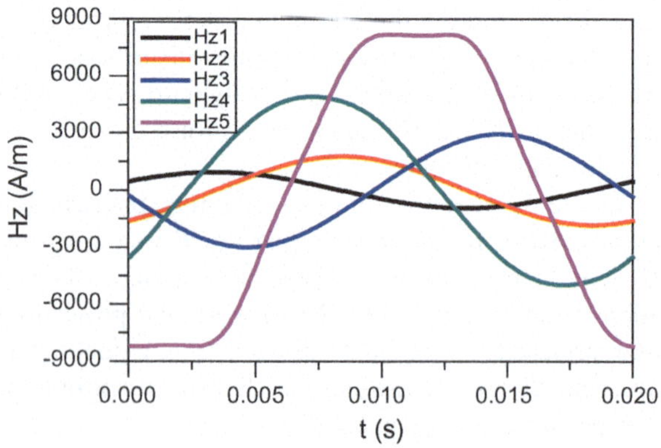

Fig. 10.17 H_z waveforms under rotational excitation at 50 Hz (Circular $\boldsymbol{B}$ loci in *zox*-plane)

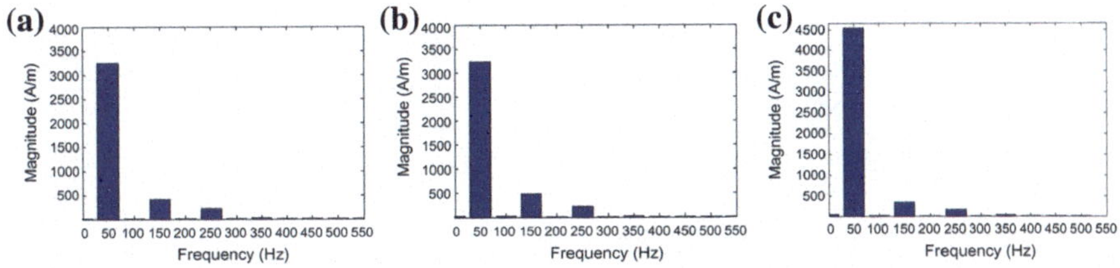

Fig. 10.18 Harmonic distributions of H_x, H_y, and H_z under 50 Hz rotational excitation: **a** H_x, **b** H_y, **c** H_z

Fig. 10.19 Relationship between the maximum flux densities and corresponding maximum values and fundamental components of H_x, H_y, and H_z under rotational excitation at 50 Hz

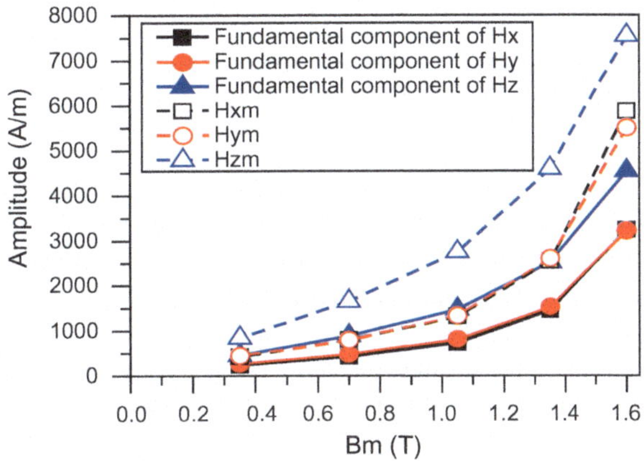

concentric circular ***B*** locus), and this difference will be more significant with the increase of magnetic flux density, as shown by solid line in Fig. 10.19. The same is true for experimental variations, as shown by dotted lines in Fig. 10.19. The variation trend of the characteristic curve of other frequency bands is similar to that of 50 Hz, but the nonlinear growth range is slightly different.

10.2.1.5 Measurement and Calculation of 3-D Rotational Core Loss

Rotational core losses are calculated by means of the Poynting's theorem from the measured ***B*** and ***H*** vectors. The Poynting vector integrated over the closed surface of the specimen represents the flow of energy into the specimen, and, averaged over one period.

Figure 10.20a shows the experimental rotational core losses when the round ***B*** loci are well controlled from 5 to 1000 Hz. For better comparison, the corresponding alternating (sinusoidal excitation along one axis) core loss features in the same frequency range are also presented. Obviously, the rotational core losses are greater than the corresponding alternating losses, in particular, at stronger magnetization and higher frequency. The comparison result is illustrated in Fig. 10.20b.

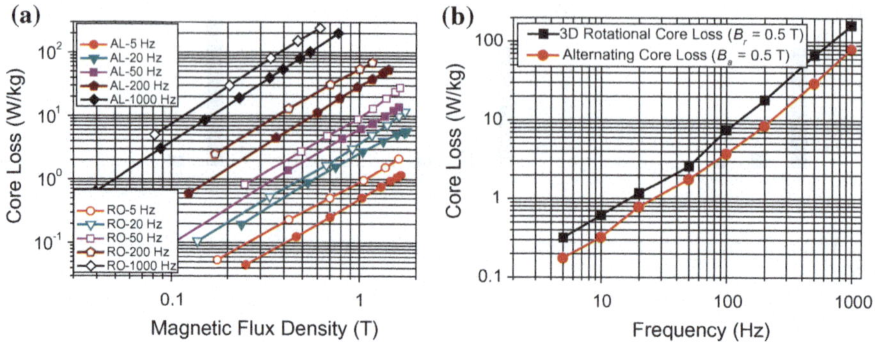

Fig. 10.20 Calculation of core loss for SMC material. **a** Comparison of alternating core losses (AL) and rotational core losses (RO) from 5 to 1000 Hz. **b** Relationship between core loss and frequency in AL and RO excitation

For example, the rotational core losses at 5, 10, 20, 50, 100, 200, 500, and 1000 Hz are 0.32, 0.62, 1.18, 2.6, 7.5, 18, 67, and 162 W/kg, respectively, when the magnitude of round ***B*** is 0.5 T, and the corresponding alternating core losses are 0.18, 0.33, 0.8, 1.8, 3.8, 8.6, 30, and 82 W/kg, respectively. The rotational loss is greater (about twice) than that of alternating core loss.

It is believed that the increasing loss is attributed to the domain rotating and domain wall motion, which cause the rotational hysteresis loss and anomalous loss steeply increasing. Due to the remarkable characteristics of low eddy current loss, the SMC material has better rotating core loss than the silicon steel in the middle and high-frequency bands, but no advantage for loss comparison in the low-frequency band.

10.2.2 Magnetic Properties of the Silicon Steels

Silicon steel sheets are widely used in the electrical equipment such as rotating machines (non-oriented steel, NO steel) and power transformers (grain-oriented steel, GO steel). Besides the common alternating magnetization, there are many situations, where the magnetic flux varies with time in terms of both magnitude and direction. For example, in the T-joints of multiphase transformer and behind the teeth of AC rotating machine, there exists a local 2-D circular or elliptical rotational magnetic flux density vector ***B***, which contributes to the rotational core loss. Extensive reports have shown that the rotational core loss can amount to more than 50% of the total core loss [24–26]. Therefore, it is of great interest to measure the magnetic properties of silicon steel under both alternating and rotational magnetization conditions.

10.2.2.1 Alternating Magnetic Properties Along Different Directions

There is a great deal of attention paid to the measurement of alternating magnetic properties along the rolling and transverse directions when the *B* signal is controlled to be sinusoidal alternating waveform by digital feedback. The magnetic properties of silicon sheets in the laminated direction can be obtained by 3-D magnetic properties tester, instead of 1-D and 2-D tester. There are also some differences in measurement between laminated direction and other two directions. Due to the larger eddy current and skin effect, the experiments should be completed in a short time to avoid excessive heat.

For the NO silicon steel, the hysteresis loops along the rolling, transverse, and laminated directions are measured and compared by improved 3-D tester with flexible multilayer windings at 50 Hz, as shown in Fig. 10.21. It shows that the hysteresis loops of NO silicon steel represent favorable coherence along the rolling and transverse directions. Due to the leading role of air permeability, the laminated direction is the most difficult magnetization direction and the permeability is approximately linear, which causes the hysteresis loops in this direction to be similar to ellipse.

Figure 10.22 shows the measured hysteresis loops along the rolling and transverse directions of GO silicon steel of GO 27ZH95 specimen. The rolling direction is the easiest magnetization direction, whose saturation magnetic flux density is approximately up to 1.8 T. The permeability of the specimen along the rolling direction is far greater than that along the transverse direction, which represents strong magnetic anisotropy [27]. In Fig. 10.23, the measured and calculated alternating core losses are obtained, respectively, where Rd-m-loss means the measured loss along the rolling direction and Td-p-loss means the predicted loss along the transverse direction. It is obvious that the alternating core loss of rolling direction is less than that of the transverse direction.

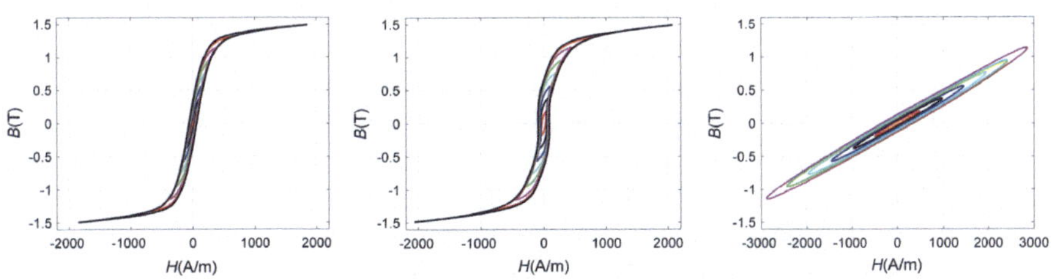

(a) rolling direction. (b) transverse direction. (c) laminated direction

Fig. 10.21 Hysteresis loops of the NO 35WW270 in three directions at 50 Hz

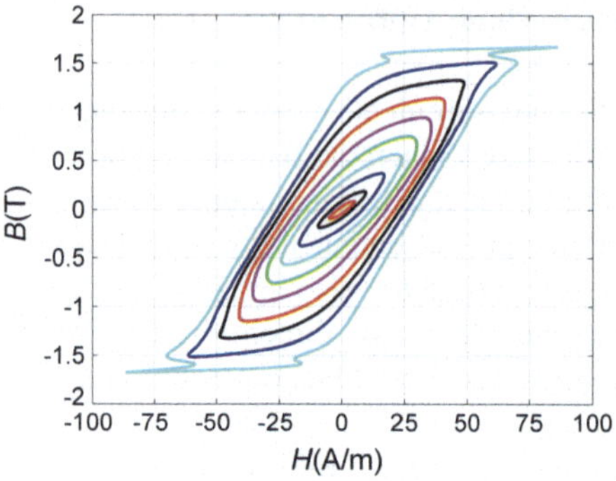

(a) Rolling direction.

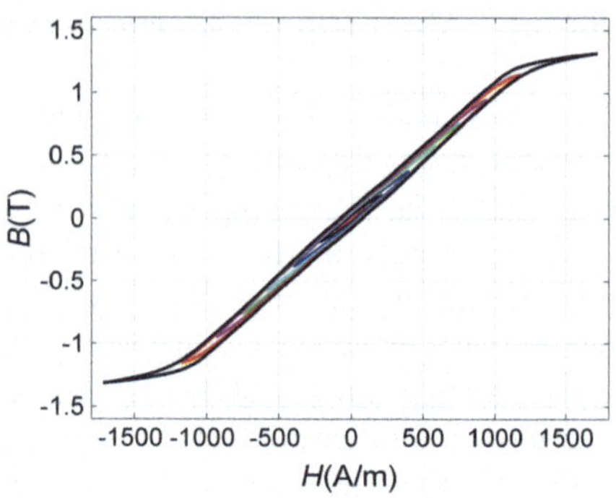

(b) Transverse direction.

Fig. 10.22 Hysteresis loops of the GO 27ZH95 specimen in rolling and transverse directions at 50 Hz

Fig. 10.23 Alternating core loss in the rolling and transverse directions at 50 Hz

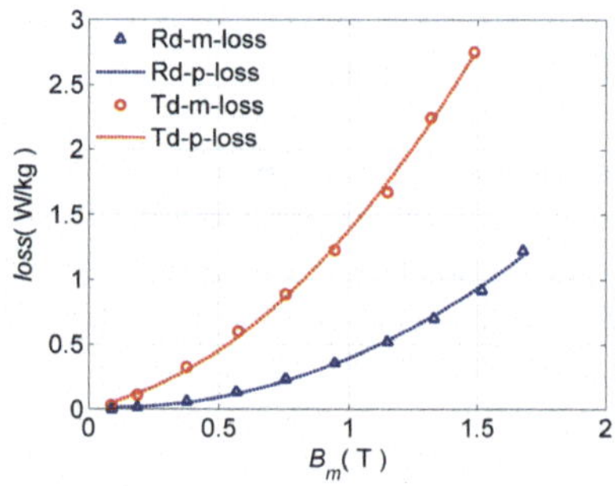

10.2.2.2 Rotational Magnetic Properties of the Silicon Steels

To analyze magnetic anisotropy comprehensively and provide reliable data to calculate core loss accurately, the rotating magnetic properties of laminated silicon steels are measured. When the magnetization loci are controlled to be circle, the ***B*** loci and the corresponding ***H*** loci are measured and compared in the rolling plane of the GO and NO silicon steels at 50 Hz as shown in Figs. 10.24 and 10.25.

For the GO 27ZH95 specimen, the x-direction is the rolling direction. When the magnetic flux density is less than 0.9 T, the ***H*** loci change to be a saddle-like shape with increasing excitation current. It is shown that the rolling direction of the GO silicon steel specimen is easier to be magnetized. When the magnetic flux density is larger than 0.9 T, the loci of ***H*** appear a sharp angle at the direction about 55° with respect to the x-axis. It means that the hard magnetization direction of Goss structure silicon steel is 55° with respect to x-axis. For the NO 35WW270 specimen, the y-axis is the rolling direction. It can be found that the components H_x and H_y of ***H*** have bigger difference at the lower excitation current while the difference is reduced progressively and tend to become stable with increasing excitation current.

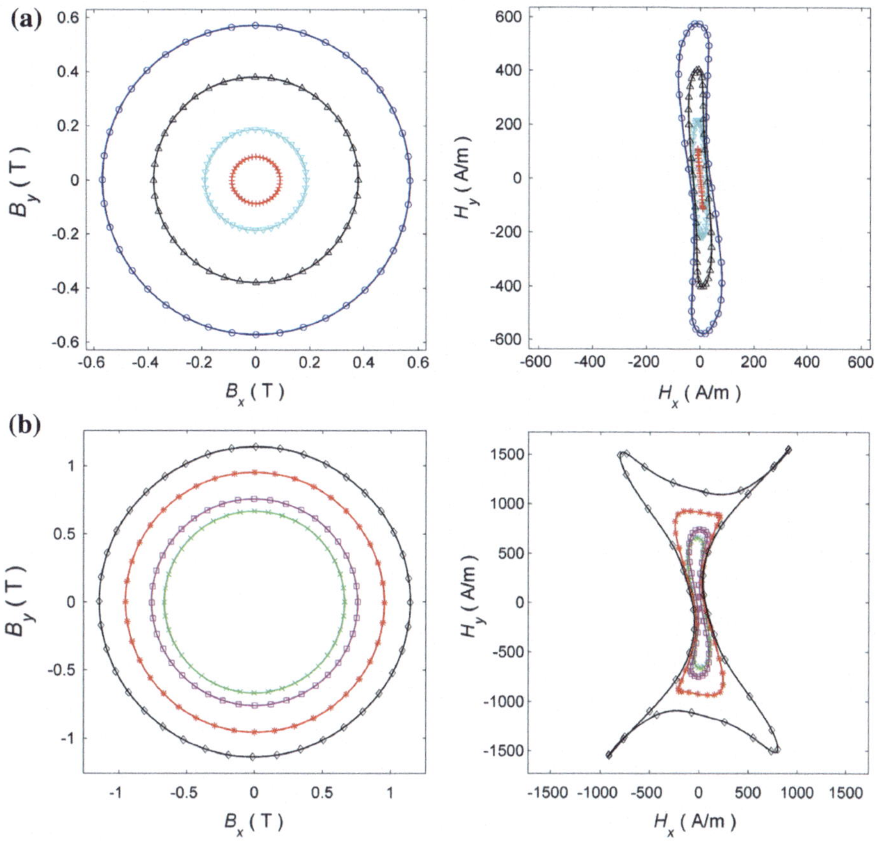

Fig. 10.24 ***B*** loci and corresponding ***H*** loci of the GO 27ZH95 in rolling plane. **a** Lower exciting current. **b** Higher exciting current

The above-observed macroscopic magnetic anisotropy is a joint result of crystal anisotropy and stress anisotropy, etc.

10.2.2.3 Core Loss Analysis

Alternating core losses along three directions and rotational core losses in the rolling plane for the NO 35WW270 are calculated and compared at 50 Hz, as shown in Fig. 10.26. It can be found that for the same ***B*** amplitude and frequency, the core loss measured for the laminated direction is far greater than rolling and transverse directions. The rotating core losses increase to peak value when the magnetic flux density is approximately 0.8 T and then gradually decrease. As the flux density increases, the magnetic domains expand and merge together. As a result, the total amount of the domain wall decreases, and hysteresis loss decreases.

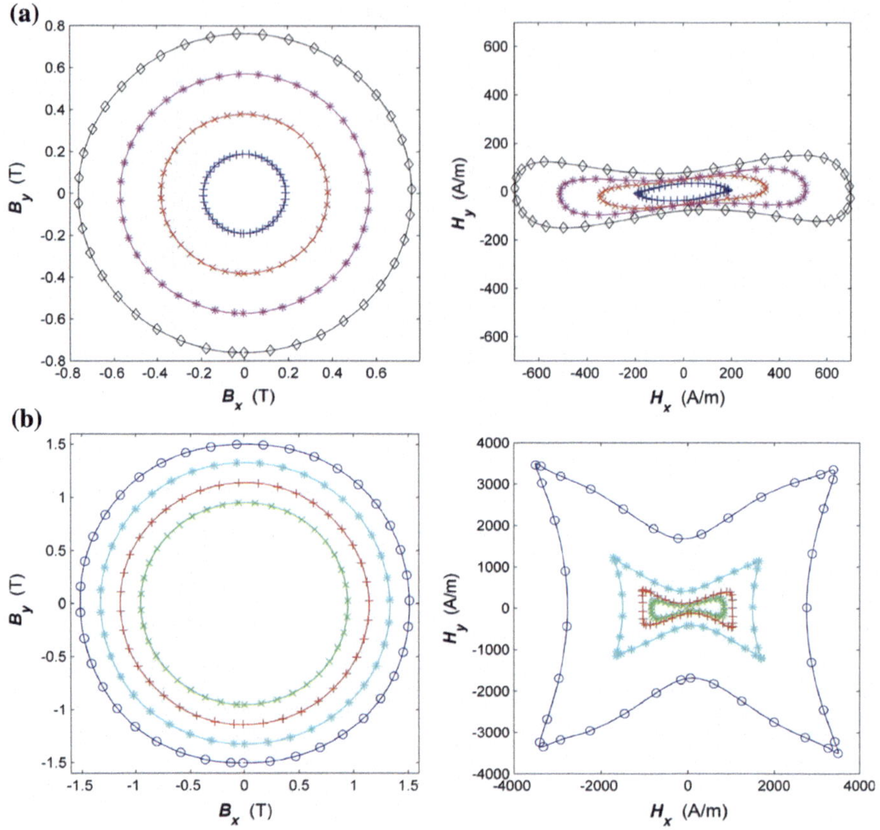

Fig. 10.25 ***B*** loci and corresponding ***H*** loci of the NO 35WW270 in rolling plane. **a** Lower exciting current. **b** Higher exciting current

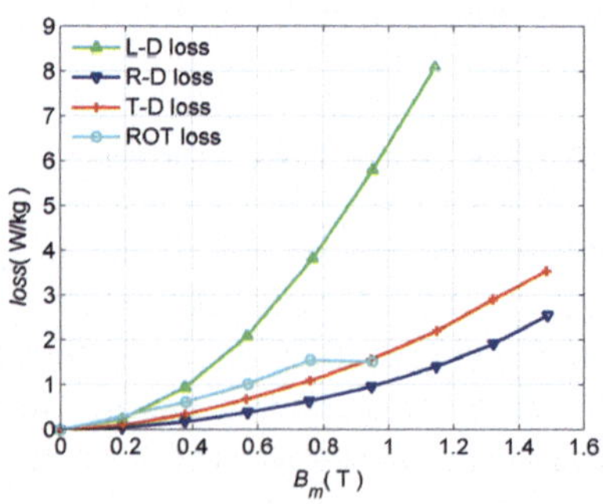

Fig. 10.26 Comparison of alternating and rotating core loss for the NO 35WW270 at 50 Hz

Fig. 10.27 Ratio of total loss to frequency for the NO 35WW270 at different frequencies

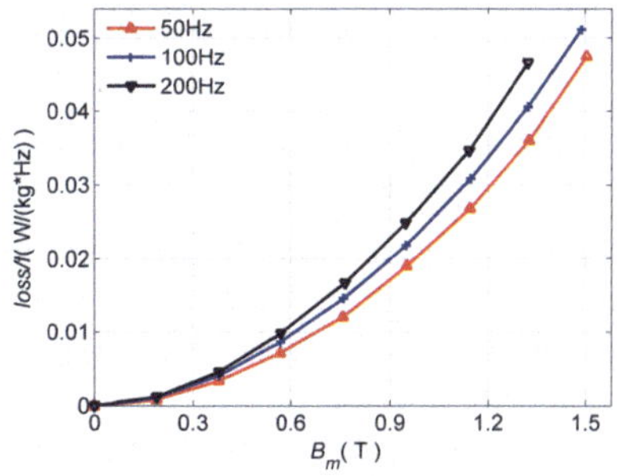

If the increase in the amount of the classical eddy current loss with increasing flux density is less than the decrease in the amount of hysteresis loss, the total rotating core loss decreases [30–32]. In Fig. 10.27, the ratio of total loss to frequency along the rolling direction for NO 35WW270 at 50, 100, and 200 Hz is plotted, which indicates that the total loss is nonlinear with frequency, and it increases faster than the increase in frequency.

10.3 Vector Hysteresis Model

Hysteresis modeling of the magnetic material is the major issue in accurate simulation of the magnetic characteristics. Since the scalar Preisach model presented by Preisach in 1935, various classical hysteresis models have been proposed successively [33]. In the last few decades, to overcome the drawbacks of a single model, hybrid models combining two or more classical models have been developed [34]. The classical Preisach model and the Stoner–Wohlfarth (S–W) model are combined into the Preisach–Stoner–Wohlfarth (PSW) model and the Della Torre Pinzaglia–Cardelli (DPC) model [35]. In the PSW model, the angle of the magnetization is defined by the net field, while in the DPC model, the magnetization direction is determined by the S–W model, and its magnitude can be calculated by the Preisach model.

In order to analyze the complex rotational magnetic properties of magnetic composite materials, a 2-D vector hybrid hysteresis model considering the advantages of the classical models is established. The hysteresis operator (hysteron), which is the significant modeling basis, is defined by the critical surface in the magnetic field. The critical surface function of hysteron is derived based on the S–W model.

10.3.1 Definition of the Vector Hysteron

10.3.1.1 Magnetic Particles and Equipotential Lines

Considering a small ellipsoidal magnetic particle with single-domain uniaxial anisotropy, which is saturated in a certain direction, we can describe its state by giving the orientation of its magnetization vector $\boldsymbol{M}$, or, equivalently, of the unit vector $\boldsymbol{m} = \boldsymbol{M}/\boldsymbol{M}_s$.

The behavior of the particle is governed by two energy terms, the uniaxial anisotropy energy and the interaction energy with the external field, $-\mu_0 \boldsymbol{M} \cdot \boldsymbol{H}_a$. We assume that the anisotropy axis is an easy magnetization direction. Under zero field, m is aligned to the easy axis. When we apply the external field $\boldsymbol{H}_a$, $\boldsymbol{m}$ rotates away from the easy axis, toward the field, by an angle depending on the relative strength of anisotropy and the field. Because of symmetry reasons, $\boldsymbol{m}$ will certainly lie in the plane containing the anisotropy axis and the external field. The configuration between uniaxial anisotropy axis, $\boldsymbol{M}$ and $\boldsymbol{H}_a$ is shown in Fig. 10.28. Where $\boldsymbol{\theta}$ is the angle between $\boldsymbol{M}$ and the easy axis, and $\boldsymbol{\theta}_H$ is the angle between $\boldsymbol{H}_a$ and the easy axis, of which one orientation is conventionally chosen as the positive one. $\boldsymbol{\theta}$ and $\boldsymbol{\theta}_H$ vary in the interval $(-\pi, \pi)$, with the convention that a positive (negative) angle means that the component of $\boldsymbol{M}$ or $\boldsymbol{H}_a$ perpendicular to the easy axis is positive (negative).

The system is described by a state variable θ, and two control variables, H_a and θ_H. The magnetization process is governed by the stability of the system under varying external field when the energy is minimized. The equipotential curves of constant energy are curves along which the differential $de_0 = 0$, where e_0 is the dimensionless quantity of free energy. The energy change along the equipotential

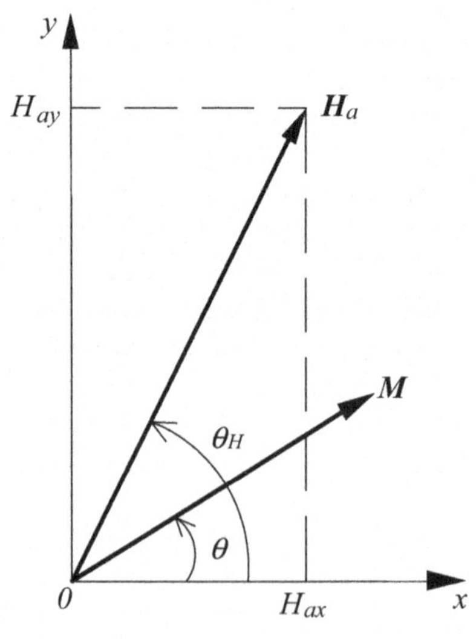

Fig. 10.28 Relations between applied field H_a, magnetization vector $\boldsymbol{M}$ and easy axis

curve is identical to zero. The energy e_{eq} of the equilibrium states is e_0 minima or maxima. When the applied field changes along the equipotential curve, the energy will change $de_{eq} = 0$. Therefore, the direction of magnetization is perpendicular to the equipotential curve.

10.3.1.2 Definition of the Vector Hysteron

The typical Preisach theory describes the hysteresis of a magnetic material with the help of an infinite set of magnetic dipoles. Each pair of dipoles is designated by a unique point in Preisach plane and has rectangular elementary hysteresis loops with different switching values of magnetic field strength. For a vector model, the changes in the magnitude and direction of the applied field should be considered. As in the scalar Preisach model, each hysteron is designated by a unique critical surface. Here, we limit the discussion to the case of an ellipsoidal critical surface. In this case, the easy axis of the hysteron is the minor axis and the hard axis is the major axis. Then, the hysteron is defined as an enclosed region surrounded by a specific equipotential line in the applied field plane.

A vector hysteron in 2-D plane with ellipsoidal critical surface is illustrated in Fig. 10.29. $\boldsymbol{\theta_H}$ is the angle between applied field and easy axis; H_α is the field at which an increasing applied field exits the surface, and H_β is the field at which a decreasing applied field exits the surface. $H_{\alpha x}$, $H_{\alpha y}$ and $H_{\beta x}$, $H_{\beta y}$ are the components of H_α and H_β along x- and y-axis, respectively. The magnetization of the vector hysteron is described by the normalized vector $\boldsymbol{m}$ (magnetization to the saturation value), which represents unit magnetization magnitude.

In order to obtain the magnetization curves with different θ_H, a hysteron is added in a negative saturation applied field, then the applied field varies along A–B–A in Fig. 10.29 with different θ_H. The hysteresis curves for alternating magnetizations with different θ_H are shown in Fig. 10.30. It is noted that the hysteresis loss of magnetization is proportional to θ_H when θ_H is increasing.

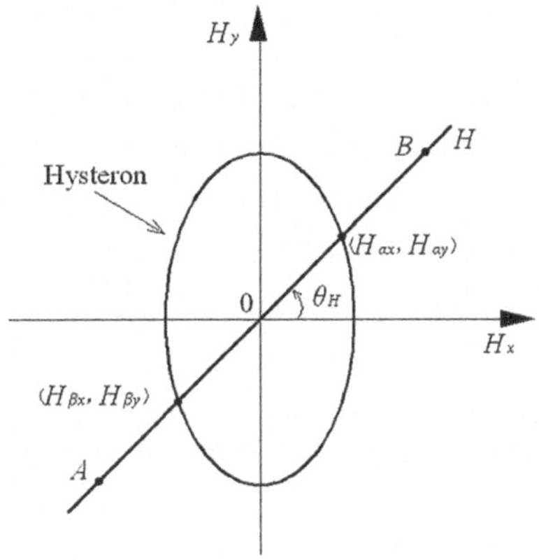

Fig. 10.29 A 2-D hysteron without interaction field

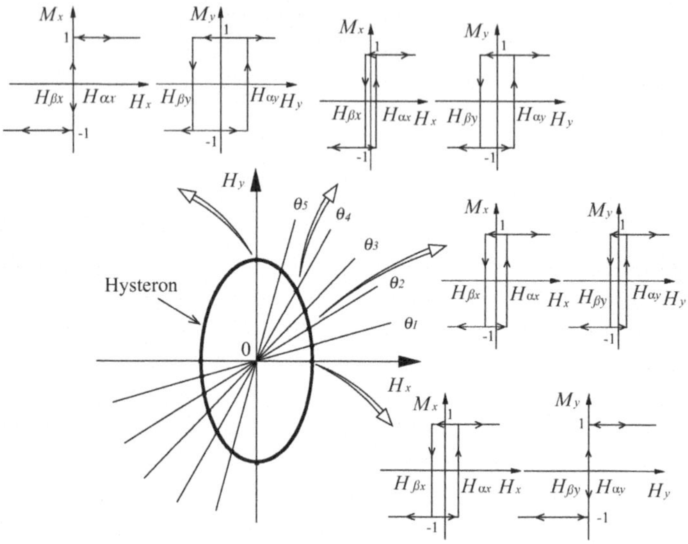

Fig. 10.30 Hysteresis loops of *x*- and *y*-component of magnetization for alternate magnetizations with different θ_H

10.3.1.3 Magnetization Direction of 2-D Vector Hysteron

The necessary and sufficient condition for the magnetized equilibrium state of points in the ***H***-plane is that H_x and H_y satisfy the first derivative of the energy equation to be equal to zero. All the external magnetic fields with angle of θ_0 between the magnetization vector ***m*** and the easy magnetization axis at the equilibrium points form a straight line in the H_a control plane, by writing Formula (10.23).

$$h_y = h_x \tan \theta_0 + \sin \theta_0 \tag{10.23}$$

The slope of the line is $\tan \theta_0$, and the intercept is $\sin \theta_0$. When θ_0 in the interval (0, π/2) that has $\tan \theta_0 > 0$ and $\sin \theta_0 > 0$. Figure 10.31a shows the contour plot of θ_0 in one quadrant of the applied field varying from 0 to π/2.

According to the definition of magnetic field line, the tangent direction of each point on the magnetic field line is consistent with the direction of magnetization at this point. All the points on the contour plot of θ_0 are magnetized along the line. Therefore, each contour plot of θ_0 is a magnetic field line.

Figure 10.31b shows the relationships between the contour plot of θ_0 and the critical surface of hysterons (θ_0 varies from 0 to π/2). The solid lines represent the critical surface of hysterons. The dotted lines refer to the contour plot. It can be concluded from Fig. 10.31b that all magnetic field lines are perpendicular to the critical curve of hysterons, and the magnetization direction of all points on the critical surface is outward along the magnetic field line at that point.

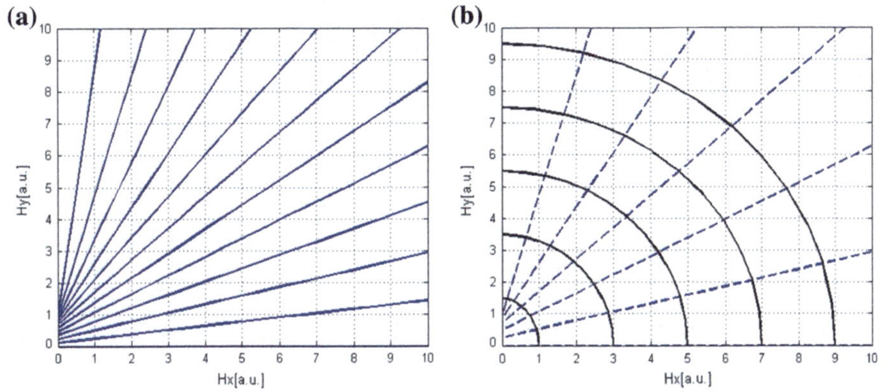

Fig. 10.31 **a** Contour plot of θ_0 in one quadrant of the applied field varies from 0 to $\pi/2$. **b** The relationship between contour plot of θ_0 and the critical surface of hysterons (θ_0 varies from 0 to $\pi/2$), solid lines represent the critical surface of hysterons, dotted lines represent the contour plot

10.3.1.4 Influence of Interaction Field Between Hysterons

Interaction field refers to an equivalent magnetic field formed by the interaction between adjacent magnetic particles due to their respective magnetization intensity. If the interaction field is taken into account, the effective field equals to the vector sum of the applied field and the interaction field. Figure 10.32 shows the effect of interacting field on the applied magnetic field of hysteron.

This method is used in both Jiles–Atherton (J–A) model and PSW model. But when using this method to calculate the magnetization process, it is necessary to calculate the value of $\boldsymbol{H}_e$ at each step before calculating the magnetization, which increases calculation time and complexity.

A simple method is used in the general vector hysteresis model (DPC model) and radial vector model (RVM model). The advantage of this method is that for one hysteron, it only needs to calculate the influence of the interaction field once.

Fig. 10.32 Effect of interacting field on the applied magnetic field of hysteron

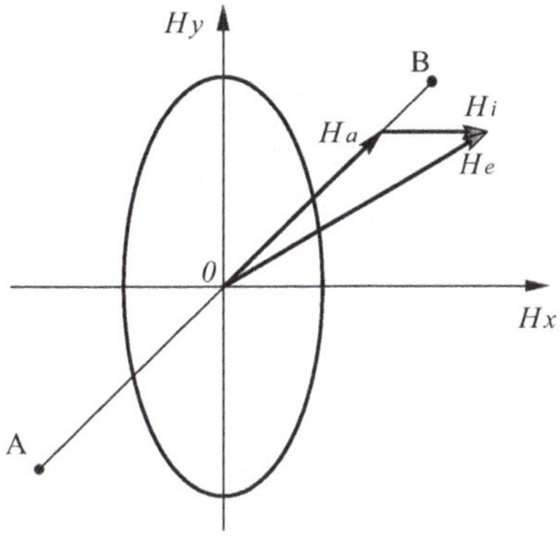

The hysteron is defined considering the effect of interaction field. The effect is to displace the critical curve of the hysteron in the vector direction of the interaction field. The critical surface equation of hysteron considering the influence of interacting fields can be described as formula (10.24).

$$\begin{cases} h_x = H_{ix} + \frac{1}{2}\cos^3\theta - \left(\frac{1}{2} + e\right)\cos\theta \\ h_y = H_{iy} - \frac{1}{2}\sin^3\theta + (1 - e)\sin\theta \end{cases} \tag{10.24}$$

10.3.2 *Modeling of the Vector Hysteresis Characteristics*

10.3.2.1 Magnetization Process of 2-D Hysterons

The magnetization direction of each point on the equipotential curve is along the normal direction outside the point. The magnetization is described by the unit magnetization vector ***m***, whose amplitude is equal to 1. Each hysteron is corresponding to a unique critical curve, which determines the magnetization characteristic. When the applied magnetic field changes along a trajectory, the variations of magnetization with the applied magnetic field are shown in Fig. 10.33.

It is assumed that, for the values of applied field outside the critical curve, the direction of magnetization is along the magnetic field lines. For fields inside the critical curve, the direction of magnetization is frozen in the magnetic field lines of the entrance point. When fields exit the critical curve, a Barkhausen jump would happen: the direction of magnetization is suddenly reoriented to the magnetic field line of the exit point. These rules allow this model to satisfy the saturation and the static loss property.

(1) The magnetization under alternating applied magnetic fields

In order to obtain the magnetization curves with different alternating applied magnetic fields, the hysteron is initially put in a negative saturation applied field, and the applied field is then varied. The switch field of the hysteron is described by

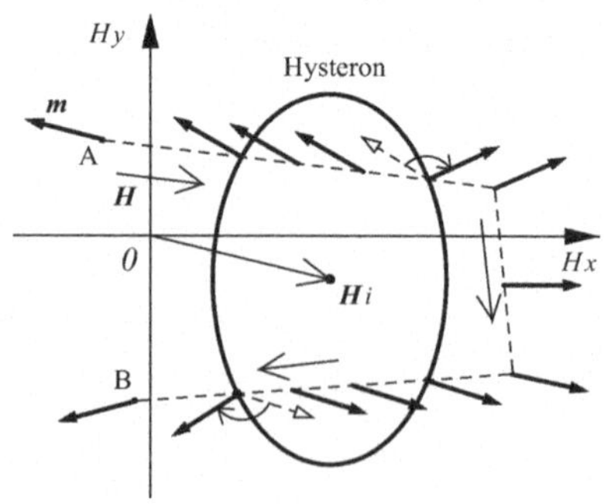

Fig. 10.33 Variations of magnetization with the applied magnetic field

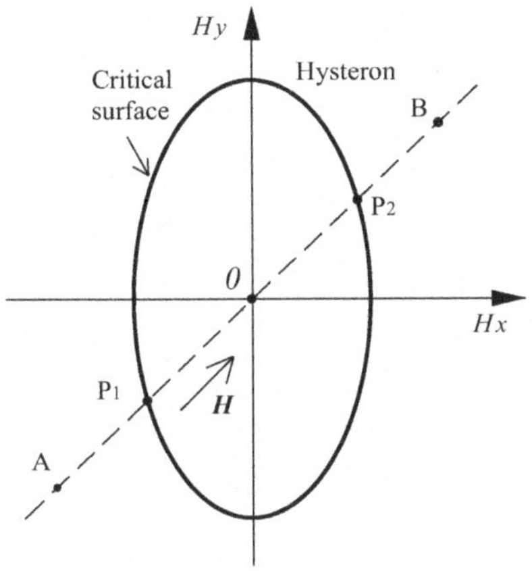

(a1) a hysteron without interaction field.

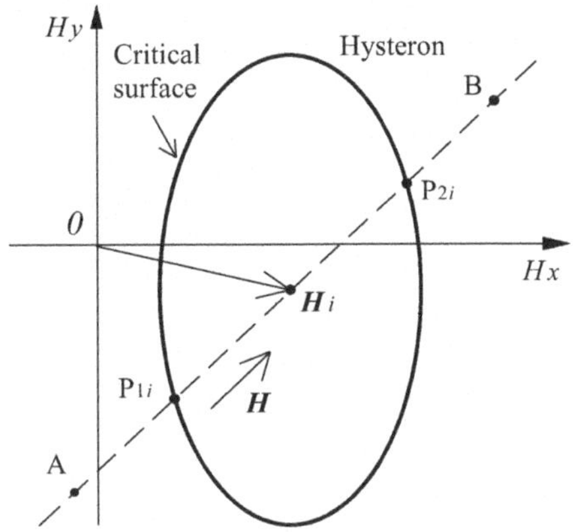

(a2) a hysteron with interaction field.

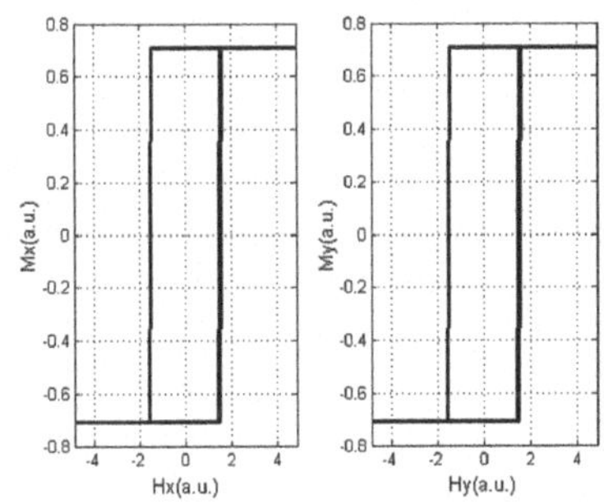

(b1) magnetization corresponding to (a1).

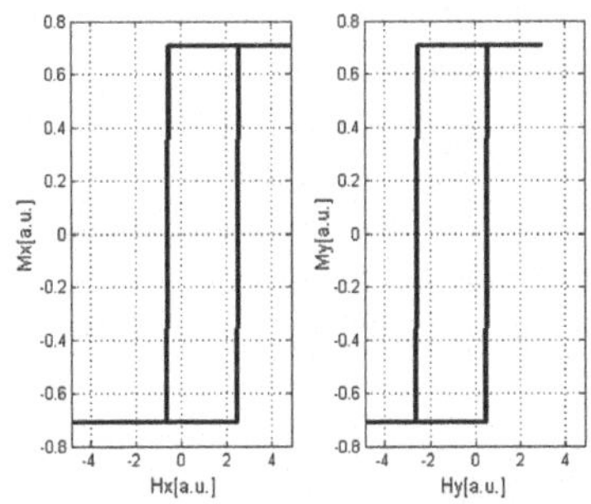

(b2) magnetization corresponding to (a2).

Fig. 10.34 M_x (H_x) and M_y (H_y) curves excited by alternating magnetization field

the intersection of the critical curve and the applied field. Figure 10.34b1, b2 shows the M_x (H_x) and M_y (H_y) curves of the hysteron excited by an alternate magnetization field with certain θ_H, such as the path A–B–A shown in Fig. 10.34a1, a2. Figure 10.34a1 is a hysteron without interaction field and (a2) is a hysteron with interaction field.

(2) The magnetization under rotating applied fields

Rotational applied field refers to an applied magnetic field $\boldsymbol{H}$ whose amplitude and phase angle change simultaneously. Consider a circular rotating magnetic field, where the amplitude of the $\boldsymbol{H}$ remains constant and the phase angle varies from 0 to 2π. According to the relationship between the magnitude of the external magnetic field and the critical surface of hysteron, the $\boldsymbol{H}$-plane can be divided into three regions, as shown in Fig. 10.35. Region 1 represents the amplitude of $\boldsymbol{H}$ less than the short axis of the critical surface. For Region 2, the amplitude of $\boldsymbol{H}$ is greater than

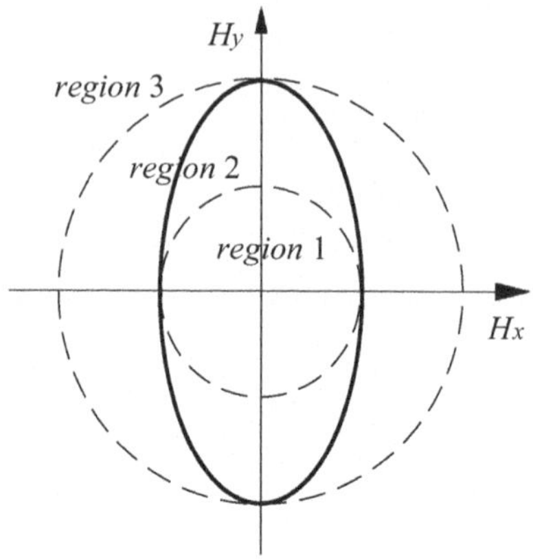

Fig. 10.35 Region diagram of hysteron under circular rotation magnetic fields

the short axis of critical surface but less than the long axis. The region where the amplitude of $\boldsymbol{H}$ is greater than the long axis of the critical surface is Region 3.

A hysteron without interaction field under circular rotational magnetic fields in Region 2 is shown in Fig. 10.36a. And, the magnetization curves where $\boldsymbol{H}$ rotating clockwise and anticlockwise are shown in Fig. 10.36b, c, respectively. The clockwise circular rotational magnetic field starts at $\theta_H = 0$ along the locus C in Fig. 10.36a. The anticlockwise circular rotational magnetic field starts at $\theta_H = 2\pi$ along the locus D in Fig. 10.36a.

10.3.2.2 Modeling of the Vector Hysteresis Characteristics

The vector hysteresis model is established based on the definition of 2-D vector hysteron. Based on that, each 2-D vector hysteron is described by three parameters: H_{ix}, H_{iy} and $\boldsymbol{e}$.

(1) The magnetization of hysteresis model under alternating magnetic field

As an example, a simple hysteresis model consists of three hysterons, as shown in Fig. 10.37a1. Firstly, the hysterons are put in the negative saturation $\boldsymbol{H}$. Then, $\boldsymbol{H}$ varies from negative to the positive saturation slowly. The switching field of the hysterons is described by the intersection of the critical curve and $\boldsymbol{H}$. Figure 10.37b1, b2 shows the M_x (H_x) and M_y (H_y) curves of the hysteron excited by an alternating magnetization field with certain θ_H, such as the path A–B–A shown in Fig. 10.37a1, a2. Figure 10.37a1 is a hysteron without interaction field and (a2) is a hysteron with interaction field.

(2) The magnetization of hysteresis model under rotational applied fields

The mathematical expressions in the $\boldsymbol{H}$-plane are curves composed of a group of continuous points, and the magnetization intensity of each point on the magnetization path can be calculated by using the vector hysteresis model. The total

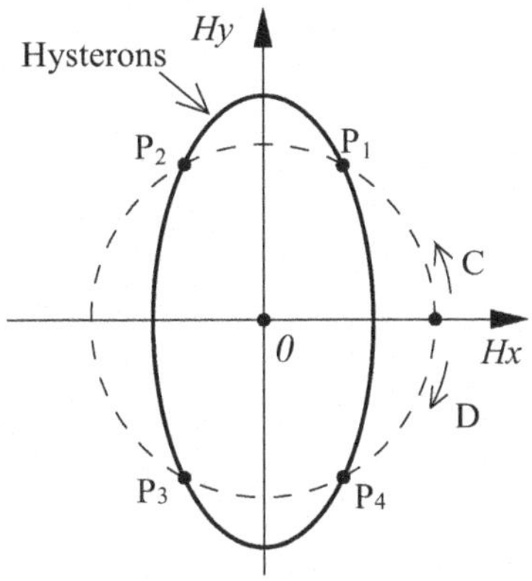

(a) A hysteron without interaction field under circular rotational magnetic fields.

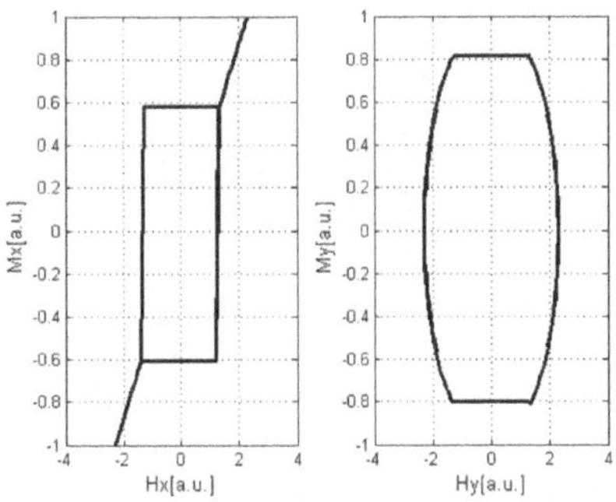

(b) magnetization corresponding to C

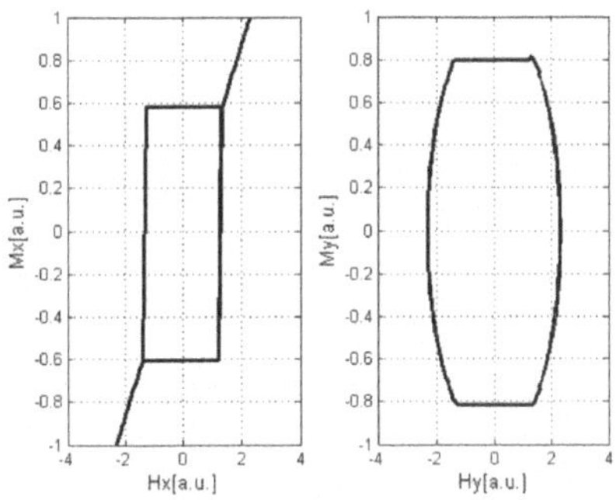

(c) magnetization corresponding to D

Fig. 10.36 $M_x(H_x)$ and $M_y(H_y)$ curves excited by rotation magnetization field

magnetization intensity of the vector hysteresis model is calculated as the vector addition of magnetization contributed by each hysteron. It is convenient to describe the parameters used in the numerical simulation in dimensionless form. Therefore, Formula (10.25) can be used to calculate the inverse normalized magnetization.

$$\boldsymbol{M}_{\text{sum}} = \boldsymbol{m}_{\text{sum}} \cdot M_{\text{s}} \tag{10.25}$$

where $\boldsymbol{M}_{\text{sum}}$ is the total magnetization intensity of the vector hysteresis model, $\boldsymbol{m}_{\text{sum}}$ is the total magnetization intensity before inverse normalization.

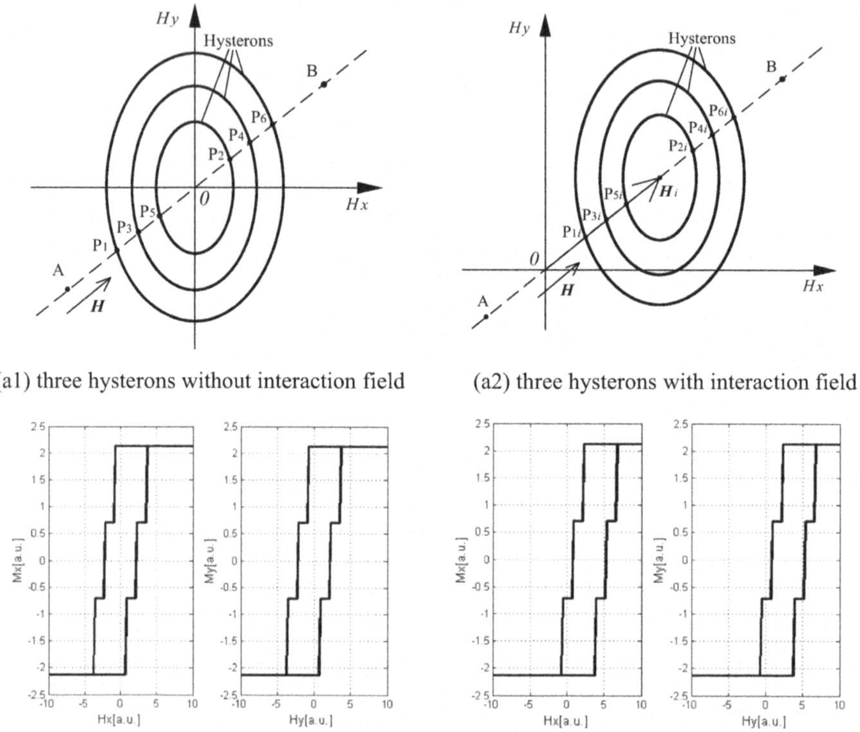

(a1) three hysterons without interaction field

(a2) three hysterons with interaction field

(b1) magnetization corresponding to(a1)

(b2) magnetization corresponding to (a2)

Fig. 10.37 $M_x(H_x)$ and $M_y(H_y)$ curves excited by rotation magnetization field

10.3.3 Magnetic Properties Prediction and Validation

The magnetization characteristics of SMC materials under alternating and rotational magnetic excitation field with different frequencies and amplitudes are simulated, and the simulation results are compared with the measured results.

In order to verify the practicability and validity of the vector model, the magnetic properties of SMC material are calculated under the alternating excitation field using the vector hysteresis model. The calculated results are compared with the measured results of a cubic SMC specimen (22 mm^3) using the improved 3-D tester with flexible excitation coils and novel-sensing coils. The specimen is the SMC-SOMALOY^{TM} 500.

10.3.3.1 Alternating Magnetic Properties Prediction and Validation

The magnetic properties of SMC-SOMALOY^{TM} 500 are measured under the given alternating experimental condition. And, the alternating magnetic properties of the

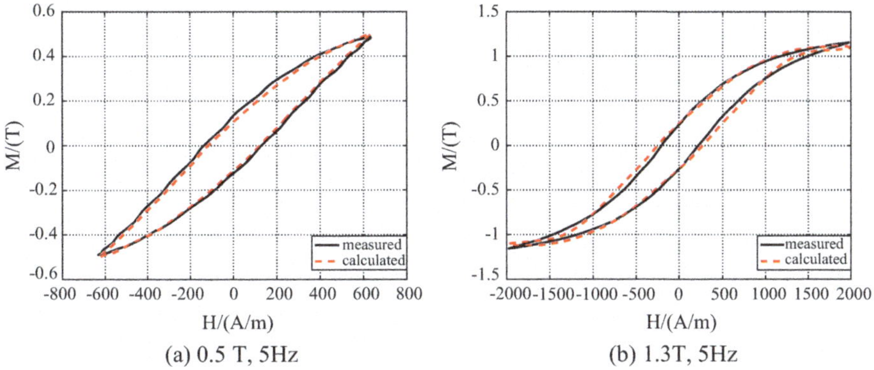

Fig. 10.38 Magnetic loops under the 5 Hz alternating applied field

SMC material are calculated by using the vector hysteresis model under the applied flux density of about 0.5 and 1.3 T when the frequency is 5 Hz; 0.5 and 1.5 T when the frequency is 20 Hz; 0.5 and 1.3 T when the frequency is 50 Hz, respectively.

The comparison between the calculated magnetization loci and measured ones is shown in Figs. 10.38a, b, 10.39a, b and 10.40a, b, respectively.

The solid lines represent the measured the magnetic loops, and the dotted lines refer to the computed ones. The data used in the calculations are based on the experimental measurement. Because of the measurement error, the densities of alternating applied field at different frequencies are slightly different. The magnetic field intensity measured near saturation at different frequencies was 1.3 T, 1.5 T, and 1.6 T, respectively. And, there have small flaws in the calculation curve due to the presence of measurement error.

From the comparison loci shown in Figs. 10.38a, 10.39a, and 10.40a, it can be concluded that the magnetic loops are matched well with the measured loops when the magnetization intensity of alternating applied field is 0.5 T at different frequencies. The results show that the influence of frequency is not significant when the magnetization intensity is 0.5 T. From the comparison loci shown in Figs. 10.38b, 10.39b, and 10.40b, it can be seen that the magnetic loops are matched well with the measured loops when the frequency is 5 Hz. But the magnetic loops are not matched well with the measured ones when the frequencies are 20 and 50 Hz. The results show that the frequency influence is obvious when the magnetization intensity is near to the saturation.

Comparing Figs. 10.38, 10.39, and 10.40, the calculated magnetic loops show good agreement with the measured ones for low magnetic field intensity and low excitation frequency. With the increase of external magnetic field, especially close to the saturation field, the consistency between the calculated results and the measurement results decreases. It is because the magnetic properties become nonlinear and anisotropic when approaching the saturation magnetization field.

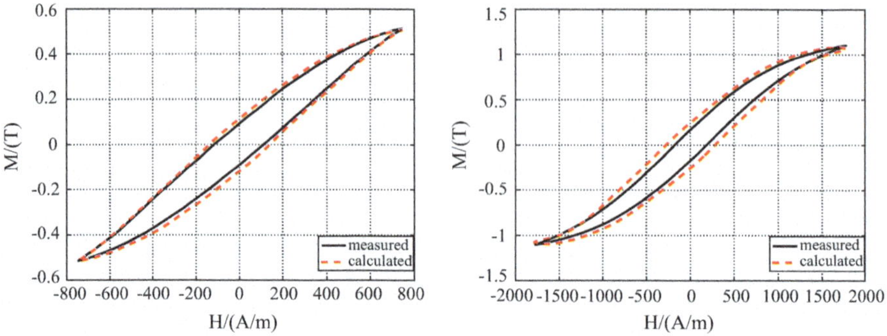

(a) 0.5 T, 20Hz (b) 1.3 T, 20Hz

Fig. 10.39 Magnetic hysteresis loops under the 20 Hz alternating applied field

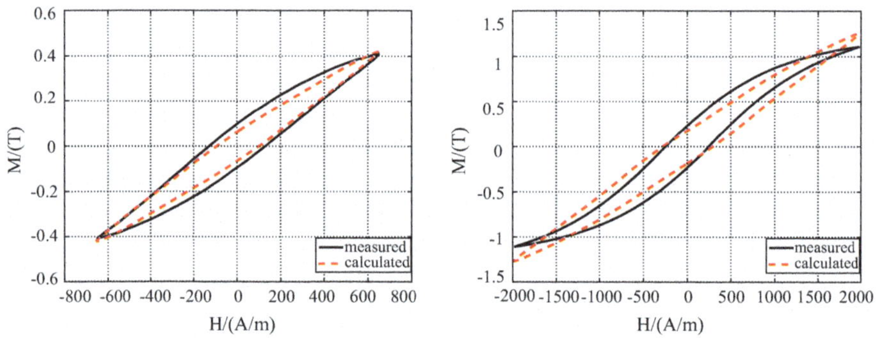

(a) 0.5 T, 50Hz (b) 1.3 T, 50Hz

Fig. 10.40 Magnetic hysteresis loops under the 50 Hz alternating applied field

Those influence factors in measurement are not considered in the numerical simulation, so the simulation precision is decreased when the frequency increases.

10.3.3.2 Rotational Magnetic Properties' Prediction and Validation

In order to describe the rotational properties of the SMC materials, more hysterons are taken into account. Take the specimen SMC-SOMALOY^{TM} 500 as example to verify the ability of the vector model to simulate the rotational properties. The data used in the calculation are the rotational properties, measured by the 3-D tester with flexible excitation coils and novel-sensing coils. According to the experimental data, five groups of data were selected at 5 and 10 Hz for the simulation. The comparison between the calculated magnetization loci and measured ones is shown in Figs. 10.41 and 10.42, respectively. The solid lines represent the measured

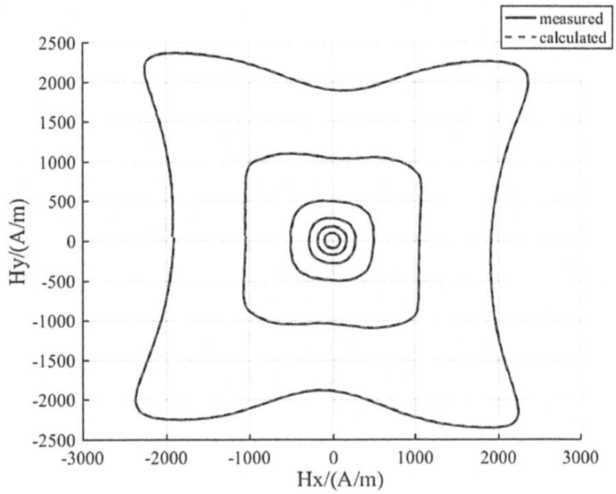

Fig. 10.41 Magnetic loci of SMC-SOMALOYTM 500 under 5 Hz rotating applied fields

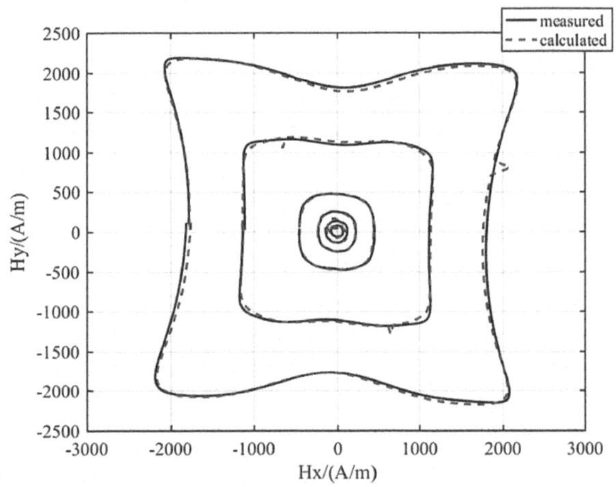

Fig. 10.42 Magnetic loci of SMC-SOMALOYTM 500 under 50 Hz rotating applied fields

magnetic field vector loci, and the dotted lines refer to the computed ones. From the comparison of the loci, it can be concluded that the rotational magnetic loci under low magnetic field strength are in better agreement as compared to the loops at high magnetic field strength. By controlling the current and the phase angle, the excitation signal is controlled as a circular rotating magnetic flux density with an amplitude of about 1.5–1.7 T in *xoy*-plane, which is slightly lower than the saturation magnetic flux density of the SMC material.

It is noted that the calculated results agree with the measured results when the external magnetic field is low. This is because of the increase of the excitation intensity, especially near the saturation state, the rotation and the movement of the magnetic domain wall is intensified, and the nonlinearity and anisotropy in the material characteristics gradually become obvious. In addition, the degree of anisotropy will increase with the increase of the excitation frequency, so that the difference between the simulation results and the experimental results will gradually increase, resulting in a reduced simulation accuracy.

It can be seen from the simulation results of the magnetization process of SMC materials under sinusoidal alternating and circular rotating excitation conditions that the calculated results are basically consistent with the measured results, but there are still differences. The possible reasons for such differences are as follows:

1. The hysteresis model is established by equivalent SMC material to an ideal isotropic material, but in fact, there are certain defects in SMC material, and certain inhomogeneity is caused in the process of material preparation, which will affect the magnetic properties of the material.
2. There is a certain difference between the calculation condition and the experimental measurement condition for the numerical simulation of hysteresis using the model. In the simulation process, the influence of temperature, stress, eddy current, and other factors were neglected, and the simulation was under the ideal conditions. However, in the actual measurement experiment, the influence of these factors existed, resulting in the difference between the simulation results and the measurement results.
3. When the intensity of excitation increases, especially close to the saturation state, the material will show nonlinearity and anisotropy. Influenced by these factors, the gap between simulation results and measurement results under high magnetic field increases.
4. The input signal of the model in the simulation process is the excitation signal measured in the experiment. The small error caused by various reasons in the measurement data will form a cumulative error in the iterative calculation process, resulting in the simulation deviation. Another possible reason for this phenomenon is that the simulation is carried out according to the magnetization law under certain conditions while the experimental data are obtained according to the average measurement data of clockwise and counterclockwise rotation.

In view of these problems, the model needs to be further modified. There is no simple linear relationship between material magnetization process and excitation frequency and field intensity. In order to establish a hysteresis model applicable to frequency and excitation intensity changes, it is necessary to have a deep understanding of the material's magnetization mechanism and measurement data, which is also the further work we will gradually carry out in the future.

In fact, the measured alternating hysteresis loops are matched better with the predicted loops than the rotational ones. The frequencies of the applied field have influenced on the simulation result, especially for rotating magnetic field. It can be described that the vector model can simulate the alternating and rotational magnetic properties under the low frequency applied field.

10.4 Summary

Rotational magnetic properties are tested in electrical equipment, such as electrical machines and power transformers, even in three dimensions. To investigate the rotational magnetic properties for soft magnetic materials, this chapter introduced a novel 3-D magnetic testing system and proposed a vector hysteresis modeling method. In the 3-D tester, the concentrated poles can generate relatively uniform and high magnetic field. The enclosed sensing structure with combined *B–H* coils and homogeneous field core shoes can ensure uniformity of the sensing field. Meanwhile, measurement precision is guaranteed by combining tensor coefficient and the positive compensation. By using the novel 3-D magnetic properties tester, comprehensive magnetic properties of the SMC material and laminated silicon steels are measured and analyzed. These data are very different from the basic manufacture data and important to the optimized design of the magnetic core, core loss prediction, and electromagnetic equipment operation with high reliability and low dissipation.

A vector magnetic hysteresis model is initially built considering anisotropic and isotropic characteristics in the magnetization process. The magnetic hysteresis properties of a SMC material under sinusoidal alternating and circular rotating excitations are simulated by using the hybrid vector hysteresis model. Comparing the simulation results with the experimental results, this model can efficiently simulate the hysteresis properties of the soft magnetic materials under relatively lower excitation magnetic field and lower frequency.

Acknowledgements This work was mainly supported by the Key Project of National Natural Science Foundation of China (No. 51237005), in part by the National Natural Science Foundation of China, (No. 51777055), and the National Key R&D Program of China (2017YFB0903904). Especially thanks to the Electromagentics team in State Key Laboratory of Reliability and Intelligence of Electrical Equipment, Hebei University of Technology.

References

1. F. G. Baily, "The hysteresis of iron and steel in a rotating magnetic field," *Philos. Trans. R. Soc.*, vol. 187, pp. 715–746, 1896.
2. F. Brailsford, "Rotational hysteresis loss in electrical sheet steels," *J. Inst. Elect. Eng.*, vol. 83, pp. 566–575, 1938.
3. A. Kaplan, "Magnetic core losses resulting from a rotating flux," *J. Appl. Phys.*, vol. 32, no. 3, pp. 370–371, Mar. 1961.
4. A. J. Moses and B. Thomas, "The spatial variation of localized power loss in two practical transformer t-joints," *IEEE Trans. Magn.*, vol. 9, no. 4, pp. 655–659, Dec. 1973.
5. F. Fiorillo and A. M. Rietto, "Rotational and alternating energy loss vs. magnetizing frequency in SiFe laminations," *J. Magn. Magn. Mater.*, vol. 83, pp. 402–404, 1990.
6. W. F. Archenhold, H. F. Sandham, and J. E. Thompson, "Rotational hysteresis loss in grain-oriented silicon-iron," Br. *J. Appl. Phys.*, vol. 11, no. 1, pp. 46–49, Jan. 1960.

7. P. J. Flanders, "The rotating-sample magnetometer," *J. Appl. Phys.*, vol. 38, no. 3, pp. 1293–1294, Mar. 1967.
8. C. R. Boon and J. E. Thompson, "Alternating and rotational power loss at 50c/s in 3% silicon-iron sheets," *Proc. Inst. Elect. Eng.*, vol. 112, no. 11, pp. 2147–2151, Nov. 1965.
9. F. Fiorillo and A. M. Rietto, "The measurement of rotational losses at I.E.N.: Use of the thermometric method," in Proc. *1st Int. Workshop Magnetic Properties Electr. Sheet Steel Under 2-D Excitation*, Physikalisch-Technische Bundesanstalt (PTB), Braunschweig, Germany, Sep. 1991, pp. 162–172.
10. A. J. Moses and B. Thomas, "Measurement of rotating flux in silicon iron laminations," *IEEE Trans. Magn.*, vol. 9, no. 4, pp. 651–654, Dec. 1973.
11. K. S. Tan, A. Datta, P. J. Flanders, and C. D. Graham, Jr., "Rotational loss in thin gage soft magnetic materials," *IEEE Trans. Magn.*, vol. 21, no. 5, pp. 1921–1923, Sep. 1985.
12. M. Enokizono, G. Shirakawa, T. Suzuki, and J. Sievert, "Two-dimensional magnetic properties of silicon steel sheet," (in Japanese) *J. Appl. Magn. Jpn.*, vol. 15, no. 2, pp. 265–270, 1991.
13. J. Sievert, "On measuring the magnetic properties of electrical sheet steel under rotational magnetization," *J. Magn. Magn. Mater.*, vol. 112, pp. 50–57, 1992.
14. J. G. Zhu, "Numerical modeling of magnetic materials for computer aided design of electromagnetic devices," Ph.D. thesis, Univ. Technology, Sydney, Australia, Jul. 1994.
15. Y. Alinejad-Beromi, A. J. Moses, and T. Meydan, "New aspects of rotational field and flux measurement in electrical steel," *J. Magn. Magn. Mater.*, vol. 112, no. 1–3, pp. 135–138, Jul. 1, 1992.
16. J. Sievert, "Recent advances in the one- and two-dimensional magnetic measurement technique for electrical sheet steel," *IEEE Trans. Magn.*, vol. 26, no. 5, pp. 2553–2558, Sep. 1990.
17. Y. Guo, J. G. Zhu, J. Zhong, H. Lu, and J. X. Jin, "Measurement and modeling of rotational core losses of soft magnetic materials used in electrical machines: A review," *IEEE Trans. Magn.*, vol. 44, no. 2, pp. 279–291, Feb. 2008.
18. J. Sievert, H. Ahlers, M. Enokizono, S. Kauke, L. Rahf, and J. Xu, "The measurement of rotational power loss in electrical sheet steel using a vertical yoke system," *J. Magnetism. Magn.. Mater.*, vol. 112, pp. 91–94, 1992.
19. L. Chen, Y. Wang, Z. Zhao, H. Zhao, C. Liu, and X. Yang, "A new magnetizer for measuring the two-dimensional magnetic properties of nanocrystalline alloys at high frequencies," *IEEE Magn. Lett*, vol. 8, 2017, Art. no. 6501705.
20. O. de la Barrière, C. Appino, F. Fiorillo, M. Lécrivain, C. Ragusa, and P. Vallade, "A novel magnetizer for 2D broadband characterization of steel sheets and soft magnetic composites," *Int. J. Appl. Electromagn. Mech.*, vol. 48, nos. 2–3, pp. 239–245, 2015.
21. J. Wanjiku and P. Pillay, "Design of a Sinusoidally Wound 2-D Rotational Core Loss Setup With the Consideration of Sensor Sizing," *IEEE Trans. Ind. Appl.*, vol. 52, no. 4, pp. 3022–3032, 2016.
22. J. Wanjiku and P. Pillay, "Shielding of the Z-Component of the Magnetic Field in a 2-D Magnetizer With a Deep Yoke," *IEEE Trans. Ind. Appl.*, vol. 52, no. 3, pp. 2289–3296, 2016.
23. International Electrotechnical Commission, "Magnetic materials, part3: Methods of measurements of the magnetic properties of magnetic sheet and strip by means of a single sheet tester," 404-3 IEC.
24. Y. Li, Q. Yang, J. Zhu, Z. Zhao, X. Liu, C. Zhang, "Design and analysis of a novel 3-d magnetization structure for laminated silicon steel." *IEEE Trans. Magn.*, vol. 50, no. 2, pp. 389–392, 2014.
25. Y. Li, Q. Yang, J. Zhu, Y. Guo, "Magnetic Properties Measurement of Soft Magnetic Composite Materials Over Wide Range of Excitation Frequency." *IEEE Trans. Ind. Appl.*, vol. 48, no. 1, pp. 88–97, 2012.
26. C. Zhang, Y. Li, J. Li, Q. Yang, J. Zhu, C, "Measurement of Three-Dimensional Magnetic Properties With Feedback Control and Harmonic Compensation." *IEEE Trans. Ind. Electron.*, vol. 64, no. 3, pp. 2476–2485, 2017.

27. Z. Lin, H. Lu, J. Zhu, et al. Vector characterization of soft magnetic materials. J. Appl. Phys., 2005, 97(10): 10R306-1-3.
28. Z. Lin, J. Zhu, Y. Guo, et al. Three-dimensional hysteresis of soft magnetic composite. J. Appl. Phys., 2006, 99(8): 08D909-1-3.
29. Z. Lin and J. Zhu, "Three-dimensional magnetic properties of soft magnetic composite materials," J. Magn. Magn. Mater., vol. 312, pp. 158–163, 2007.
30. V. Basso and G. Bertotti, "Hysteresis in soft magnetic materials," J. Magn. Magn. Mat., vol. 215, pp. 1–5, 2000.
31. C. Appino, F. Fiorillo, and C. Ragusa, "One-dimensional/two-dimensional loss measurements up to high inductions," *J. Appl. Phys.*, vol. 105, no. 7, pp. 07E718-1-07E718-3, 2009.
32. C. Appino, F. Fiorillo, C. Ragusa and B. Xie, "Magnetic losses at high flux densities in nonoriented Fe-Si alloys," *J. Magnetism. Magn. Mater*, vol. 320, pp. 2526–2529, 2008.
33. I. D. Mayergoyz. Mathematical Models of Hysteresis. New York: Springer Verlag, 1991.
34. E. Cardelli, E. Pinzaglia, E. D. Torre, Magnetic energy and radial vector model of hysteresis, J. Appl. Phys., 2006. 99: 08D703.
35. E. A Cardelli general hysteresis operator for the modeling of vector fields. IEEE Trans. Magn., 2011, 47(8): 2056–2067.

Chapter 11
Measurement and Prediction of Magnetic Property of GO Silicon Steel Under Non-standard Excitation Conditions

Zhenbin Du, Meilin Lu and Fulai Che

Abstract The working properties of both magnetic materials and components, instead of those measured under standard sinusoidal condition, are certainly required in electromagnetic analysis and design of large electrical equipment, operating under possible complex even extreme conditions. In this chapter, the magnetic properties of GO silicon steel are measured under non-standard conditions (e.g., harmonic and DC-bias), using an enhanced 1-D magnetic measurement system (Epstein frame and SST, Brockhaus, Germany). The integrated magnetic measurement system is used for measuring and predicting the working magnetic properties of the magnetic components (e.g., laminated core models) under harmonic and/or DC-bias conditions. Also, a new magnetic measurement method under AC–DC hybrid excitation is proposed, based on a specially designed test model, in which a DC bridge is installed in the center of the square core, with 5 mm gaps between the bridge ends and the yoke of the square core, to form a gapped-type three-limb core model, so that the multi-harmonic and DC excitations can be separately applied to the square core and DC bridge. Finally, the numerical analysis of magnetic loss under AC–DC hybrid excitation condition based on the proposed core models is validated.

Keywords GO silicon steel · Magnetic components · Working magnetic property · Measurement and prediction · Harmonic power supply · AC–DC hybrid excitation · Modeling and simulation

Z. Du (✉) · M. Lu · F. Che
Institute of Power Transmission and Transformation Technology,
Baobian Electric, Baoding, China
e-mail: sy16@btw.cn

Z. Cheng et al. (eds.), *Modeling and Application of Electromagnetic and Thermal Field in Electrical Engineering*, https://doi.org/10.1007/978-981-15-0173-9_11

11.1 Introduction

The magnetic property data measured by the standard means [1, 2] under ideal conditions cannot meet the strict requirements of analysis and design of very large electrical equipments, which are working usually under non-sinusoidal and extreme conditions [3–7]. In addition, there has been a consensus among the industrial and scientific communities that the magnetic material properties are always different from those of assembled magnetic components due to both the assembly and operating conditions of the devices [8]. Accordingly, the working properties obtained under the actual operating conditions of magnetic materials and components, and then the standardized measurements or predictions under such non-sinusoidal conditions, taking both the manufacturing and working factors into account, are in great demand [6].

This chapter includes the following three aspects: (1) the working properties of GO silicon steel are measured using the enhanced 1-D magnetic measurement system (Brockhaus, Germany), which are used in the numerical modeling and computation; (2) the electromagnetic behavior of the component-level laminated core under harmonic or DC-bias condition are modeled using a well-established integrated magnetic measure-bench, and the effects of every excitation parameter on the magnetic loss is examined; (3) the magnetic property of laminated core under AC–DC hybrid excitation conditions are measured based on a gapped-type three-limb core model proposed by the authors, in which harmonic and DC-bias excitation can be separately applied to the square frame and the individual limb ($L\Theta$, can be referred to as a DC bridge) placed in the frame's center, and there are narrow gaps between the square frame's yoke and $L\Theta$. Finally, the numerical analysis of magnetic loss based on the same core model is validated.

Note that a number of working property data of GO silicon steel and model-based magnetic measurements under non-standard conditions are also provided in Tables of the Appendix of this chapter to be used in further research and in-depth discussions on effects of various excitation conditions on both material and components properties.

11.2 1-D Magnetic Measurement Under Non-standard Conditions

In order to model the electromagnetic behaviors of electrical equipment under the non-sinusoidal excitations, the material property data used in numerical modeling and simulation should be measured under the identical non-sinusoidal conditions.

The 1-D magnetic measurement system (Brockhaus, Germany), including Epstein frame (standard 25 cm, and a scaled-down 20 cm Epsten customized according to a special requirement) and SST(specimen 500 × 500 mm), can be used under different frequency bands, such as high, medium, and low frequency

Fig. 11.1 1-D magnetic measurement system

(e.g., low to 3 Hz and up to 20 kHz), or AC–DC hybrid excitation, referred to as enhanced 1-D magnetic measurement system. See Fig. 11.1.

The magnetic properties of GO silicon steel sheets, such as 27ZH100 and B23R075, have been measured under different excitations, using this enhanced 1-D magnetic measurement system which will be used in the following modeling and simulation under AC–DC hybrid excitation conditions (see Sect. 11.4 of this chapter).

11.2.1 Measurement of Magnetic Loss Under Harmonic or DC-Bias Condition

The specific total losses (P_{loss}) of GO silicon steel sheet, 27ZH100 are measured under different harmonic conditions, i.e., different harmonic order and content, as shown in Figs. 11.2 and 11.3, which can also be referred to Table 11.6 in the Appendix of this chapter.

Note that all the phase angle of fundamental wave ϕ_1 in this chapter are 0°, and the phase angle of *n*th harmonic is expressed as ϕ_n.

In addition, it should be noted that in all tables and figures of this chapter, the phase angle difference, $\phi_n - \phi_1$, will be equal to ϕ_n (because ϕ_1 is zero in all measuring cases in this chapter), which is expressed in terms of angle difference (e.g., 0°, …, 180°) but not radian difference, for convenience only.

Figure 11.2 shows the loss results measured under fundamental wave (50 Hz) only or with 3rd harmonic of different content (0° phase angle difference).

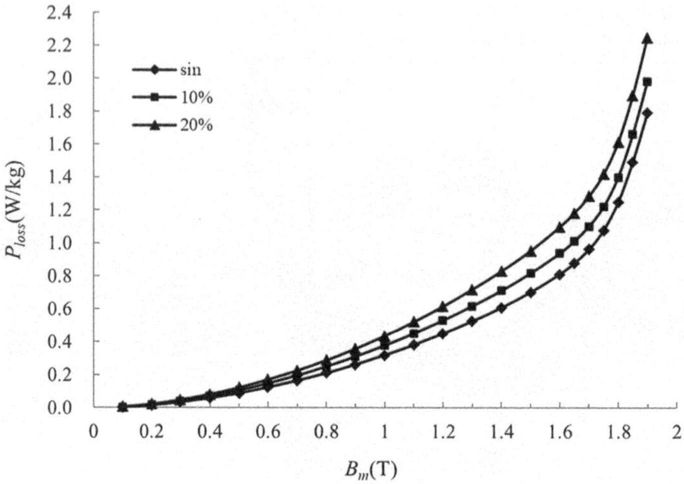

Fig. 11.2 Loss results measured under fundamental wave only or with 3rd harmonic of different content (0° phase angle difference)

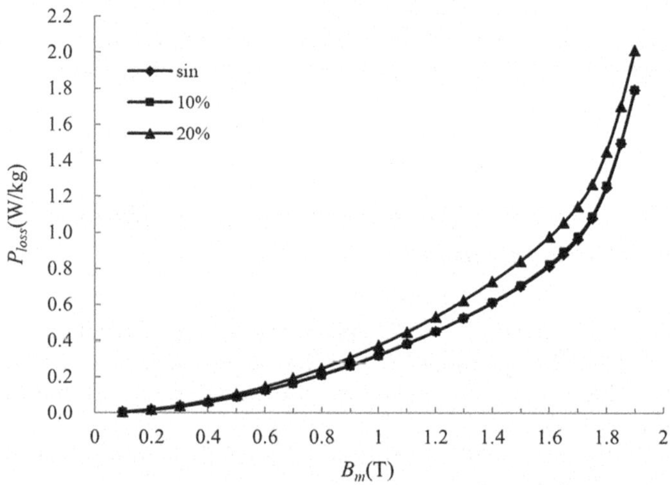

Fig. 11.3 Loss results measured under fundamental wave only or with 5th harmonic of different content (0° phase angle difference)

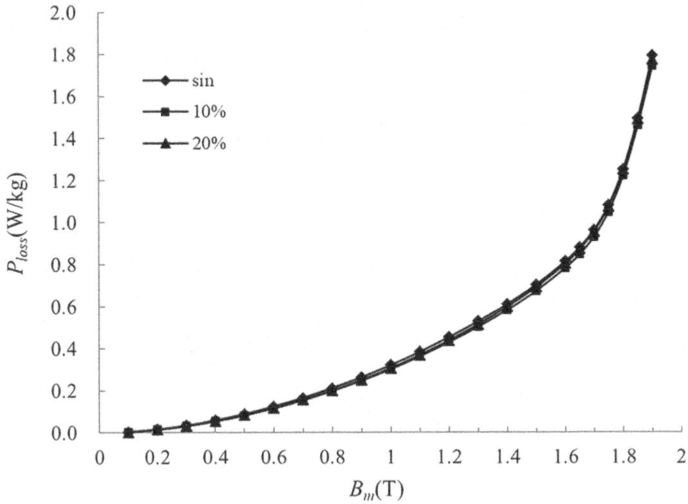

Fig. 11.4 Loss results measured under fundamental wave only or with 3rd harmonic of different content (180° phase angle difference)

Figure 11.3 shows the loss results measured under fundamental wave 50 Hz condition only or with 5th harmonic of different content (0° phase angle difference).

The loss results (P_{loss}) measured under different harmonic order and content conditions are shown in Figs. 11.4 and 11.5, which can also be referred to Table 11.7 in the Appendix of this chapter, and the phase angle difference here is 180° for the following two measurement cases.

Figure 11.4 shows the loss results measured under fundamental wave only or with 3rd harmonic of different content.

Figure 11.5 shows the loss results measured under fundamental wave only or with 5th harmonic of different content.

The loss results (P_{loss}) measured under fundamental wave only or with different DC-bias conditions (H_{dc}) are shown in Fig. 11.6, which can also be referred to Table 11.8 in the Appendix of this chapter.

11.2.2 Measurement of Magnetic Loss Under Harmonic and DC-Bias Condition

The loss results (P_{loss}) of GO silicon steel sheet, 27ZH100, measured under fundamental wave only or with 10% 3rd harmonic and different DC-bias (H_{dc}) conditions are shown in Fig. 11.7, which can also be referred to Table 11.9 in the Appendix of this chapter, and the phase angle difference is 0° for 3rd harmonic.

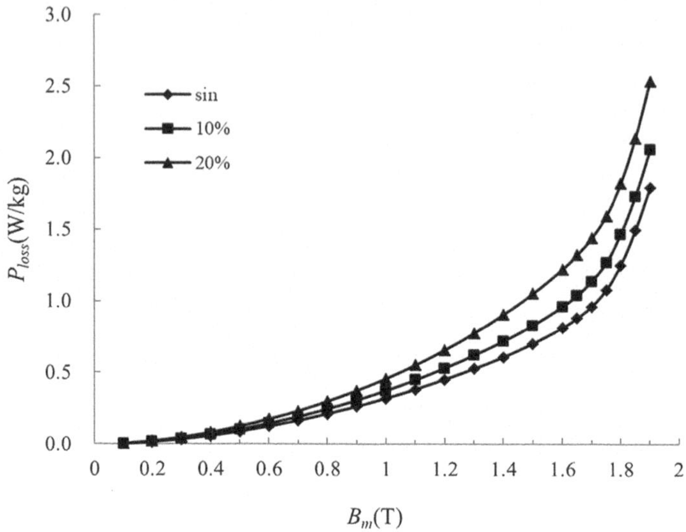

Fig. 11.5 Loss results measured under fundamental wave only or with 5th harmonic of different content (180° phase angle difference)

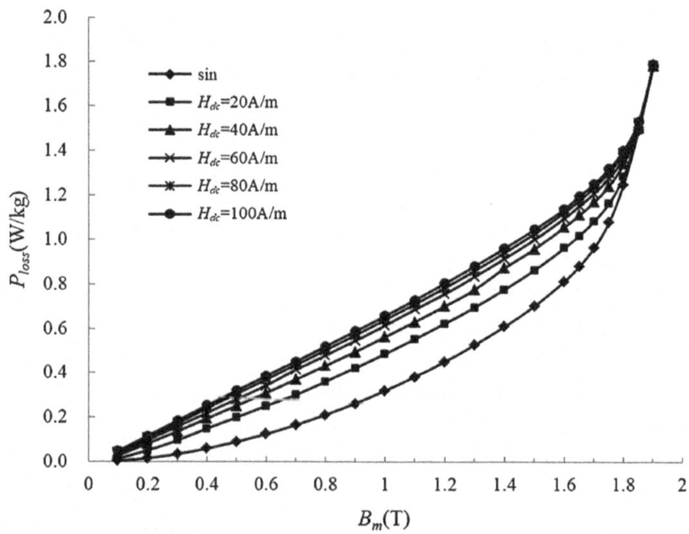

Fig. 11.6 Loss results measured under fundamental wave only or with different DC-bias (H_{dc}) conditions

The loss results (P_{loss}) measured under fundamental wave only or with 10% 3rd harmonic and different DC-bias (H_{dc}) conditions are shown in Fig. 11.8, which can also be referred to Table 11.10 in the Appendix of this chapter, and the phase angle difference is 180°(for 3rd harmonic).

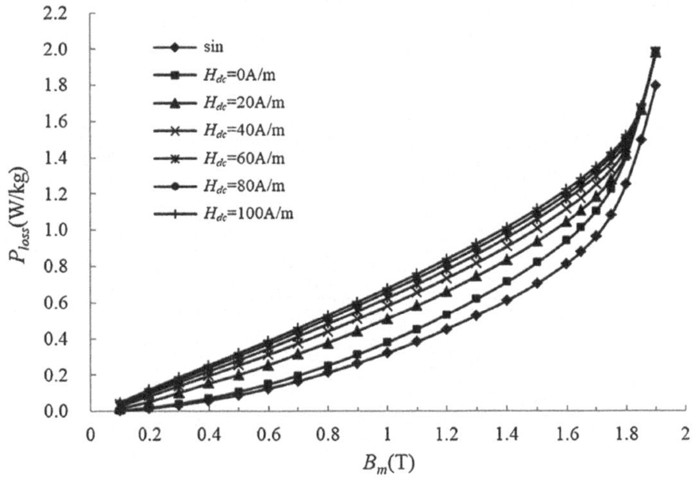

Fig. 11.7 Loss measurement results (P_{loss}) under fundamental wave only or with 10% 3rd harmonic and different DC-bias (H_{dc}) conditions, (0° phase angle difference)

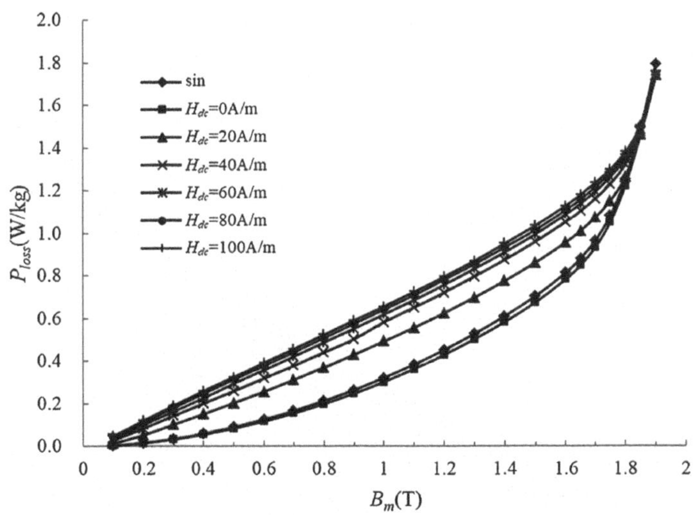

Fig. 11.8 Loss measurement results (P_{loss}) under fundamental wave only or with 10% 3rd harmonic and different DC-bias (H_{dc}) conditions, (180° phase angle difference)

The loss results (P_{loss}) measured under fundamental wave only or with 10% 5th harmonic and different DC-bias (H_{dc}) conditions are shown in Fig. 11.9, which can also be referred to Table 11.11 in the Appendix of this chapter, and the phase angle difference is 0° for 5th harmonic.

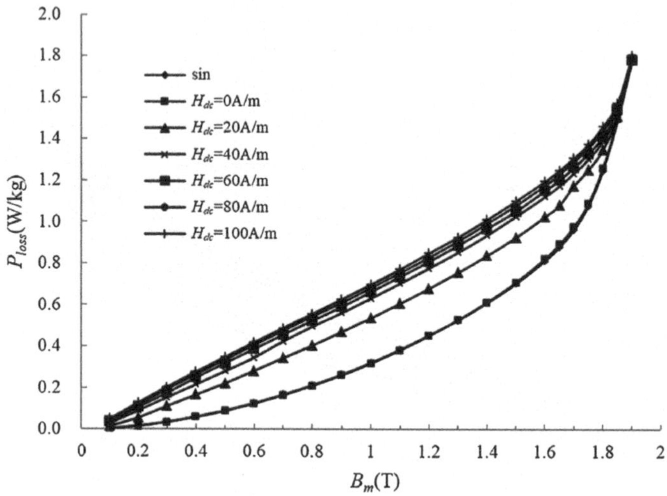

Fig. 11.9 Loss measurement results (P_{loss}) under fundamental wave only or with 10% 5th harmonic and different DC-bias (H_{dc}) conditions, (0° phase angle difference)

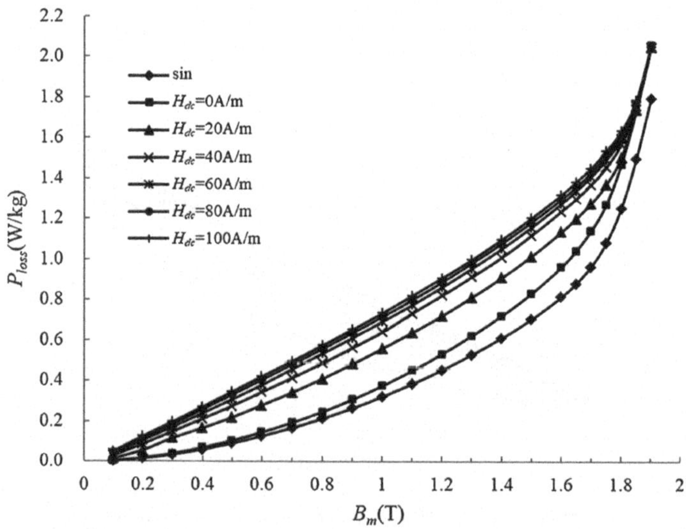

Fig. 11.10 Loss measurement results (P_{loss}) under fundamental wave only or with 10% 5th harmonic and different DC-bias (H_{dc}) conditions, (180° phase angle difference)

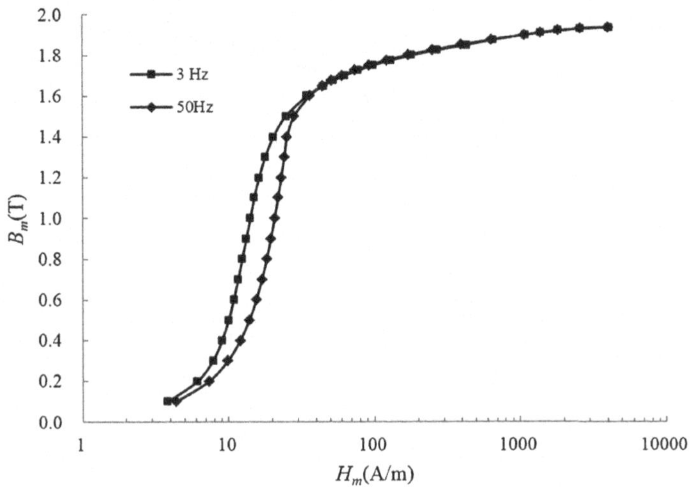

Fig. 11.11 *B–H* curve (27ZH100)

The loss results (P_{loss}) measured under fundamental wave only or with 10% 5th harmonic and different DC-bias (H_{dc}) conditions are shown in Fig. 11.10, which can also be referred to Table 11.12 in the Appendix of this chapter, and the phase angle difference is 180° for 5th harmonic.

Note that all the measured results of magnetic losses, as shown in Figs. 11.2, 11.3, 11.4, 11.5, 11.6, 11.7, 11.8, 11.9, and 11.10, under different excitation conditions, demonstrate the effects of the excitation parameters (e.g., the harmonic order, content, phase angle, the DC-bias content, and the different level of AC–DC hybrid excitation) on magnetic losses.

The measured results, as shown above and in the Tables of the Appendix of this chapter, enable us to further summarize the variation trends of the magnetic properties with different excitation parameters, from different perspectives, in the further research projects.

11.2.3 Measurement of Magnetization Property

In the numerical modeling and simulation of electromagnetic fields, under very complicated excitation conditions, the DC *B–H* curve is suggested to be used.

In this chapter, the *B–H* curves of GO silicon steel have been measured under very low frequency condition (3 Hz), which is used to approach its DC *B–H* curve.

The magnetization data of GO silicon steel 27ZH100 measured at 3 and 50 Hz are shown in Fig. 11.11, which can also be referred to Table 11.13 in the Appendix of this chapter.

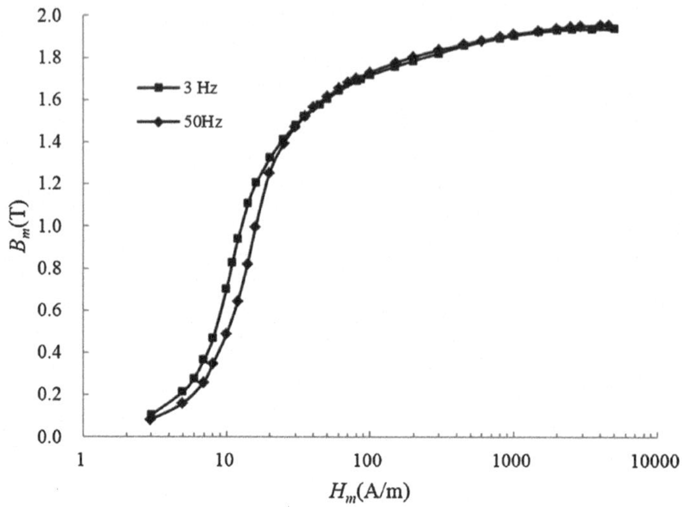

Fig. 11.12 *B–H* curve (B23R075)

The magnetization data of GO silicon steel, B23R075, measured at 3 and 50 Hz are shown in Fig. 11.12, which can also be referred to Table 11.14 in the Appendix of this chapter.

11.2.4 Measurement of B–H *Loop Under Harmonic and DC-Bias*

The *B–H* loops (27ZH100) have been measured under 3rd harmonic, different DC magnetic fields (H_{dc} = 0, 20 A/m), and different magnetic flux density levels (from 0.3 to 1.5 T) inside the laminated frame, as shown in Fig. 11.13.

11.3 Magnetic Measurement Under Non-standard Conditions Based on an Integrated Magnetic Measure-Bench

11.3.1 Magnetic Property Measure-Bench and Two Core Model Schemes

The harmonic power supply of the integrated magnetic measure-bench includes multi-functional generator of arbitrary waveforms (35 MHz, WF1974), three sets of

Fig. 11.13 *B–H* loop measured under harmonic and DC-bias (27ZH100)

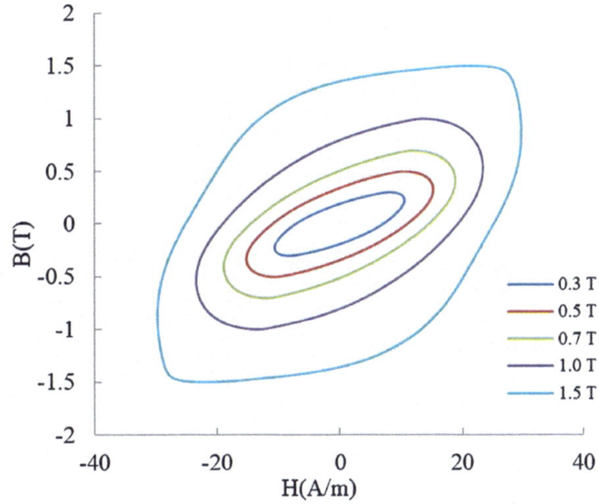

(a) H_{dc} = 0A /m

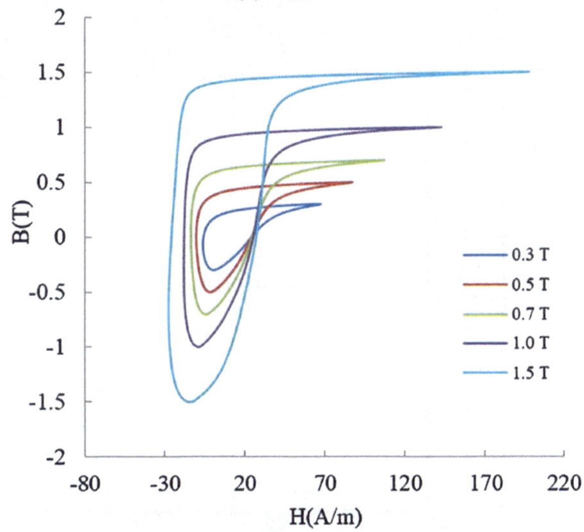

(b) H_{dc} = 20A /m

power amplifiers (4520 A) and boosters (4521 A), and power analyzer (WT3000E), as shown in Fig. 11.14a.

In the integrated magnetic measurement system, two core models are configured, which have the same material (27ZH100) and the same cross-sectional areas of the laminations and the same step-lap joints, and are fastened by wooden frame to keep fastening force as actual transformer core. The only difference between the two core models is in lengths of the laminated limbs, i.e., 70 cm for Model C70 and 50 cm for Model C50, as shown in Fig. 11.14a. See Sect. 9.2 of Chap. 9 for the detailed description of the laminated core models.

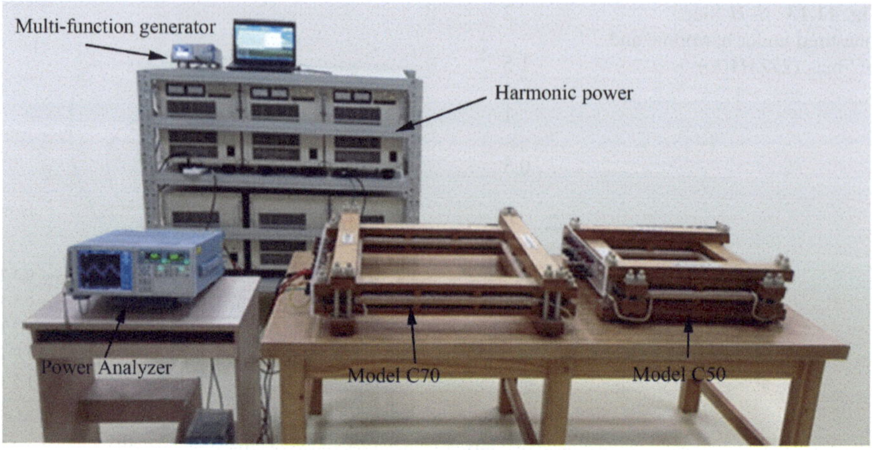

(a) Integrated measuring system

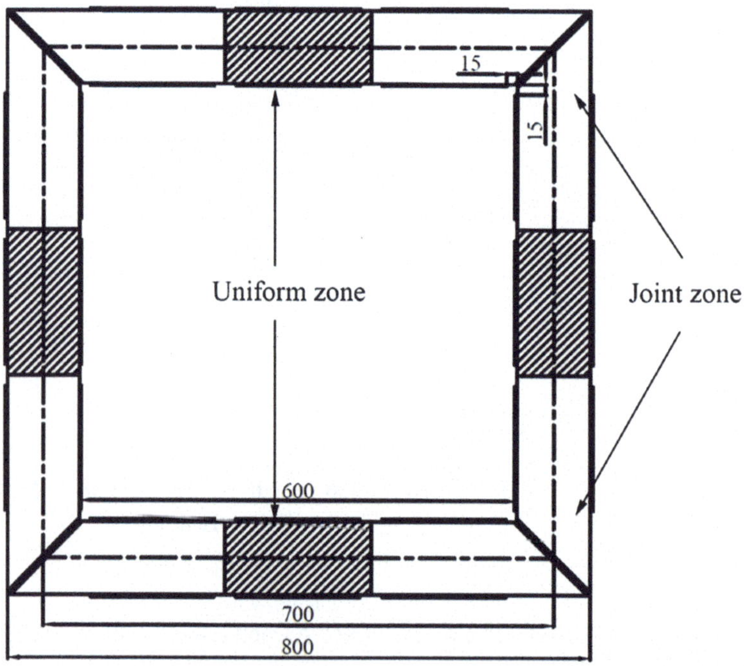

(b) Model (C70-C50)

Fig. 11.14 Integrated magnetic measure-bench

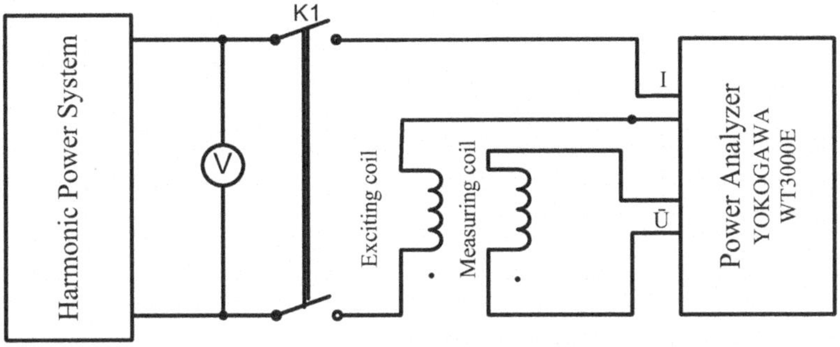

Fig. 11.15 Block diagram of measuring circuit

Obviously, the distribution of magnetic loss is not uniform inside the laminated core, like the non-uniform loss distribution in the Epstein frame. In this chapter, the two core model (2CM) schemes are used to reasonably determine and predict the magnetic properties in the laminated core. To achieve this purpose, the two following assumptions are made, as stated for the Epstein combination method in reference [9]: (I) the non-uniform magnetic field distribution and magnetic properties (such as specific total loss and exciting power) over the joint regions of both Model C70 and Model C50, are identical, despite the difference in the length of the limbs in the two core models; (II) the magnetic field and magnetic properties are uniform over the middle section of each limb. This can be imagined as a subtraction of two core models, i.e., Model (C70–C50) represents the uniform middle section (total volume of the Model (C70–C50): (200 × 20 × 100) × 4 mm^3) of the limb of Model C70, as shown in Fig. 11.14b [5, 6].

The block diagram of the measuring circuit is shown in Fig. 11.15.

11.3.2 Harmonic Excitation

The harmonic flux density B inside the core model can be expressed by (11.1),

$$B = \sum_{n=1}^{J} B_n \sin(n\omega t + \phi_n) \tag{11.1}$$

where ω is fundamental angle frequency, n—harmonic order, B_n and ϕ_n—amplitude and phase angle of nth harmonic component, respectively, j—maximum harmonic order.

In the measurement under harmonic excitations, the resulting magnetic flux density inside laminated core can be controlled. Based on the Faraday's law, after simple deduction, the harmonic flux density and the induced voltage $E(t)$ at the measurement coil have the following relationship, as (11.2)

$$E(t) = -NS\omega B_1 \sum_{n=1}^{J} nK_n \cos(n\omega t + \phi_n) \tag{11.2}$$

where N is number of turns of measurement coil, B_1 is fundamental amplitude, S is cross-sectional area of laminated core, and $K_n = B_n/B_1$ corresponds to the nth harmonic content (%).

11.3.3 Results and Discussions

The specific total losses under harmonic excitations (3rd and 5th harmonic applied individually or both) have been measured based on the laminated core models, Model C70 and Model C50, by using the magnetic measurement bench. The core models are made of GO silicon steel sheets (27ZH100), i.e., 0.27 mm thick and 1.00 W/kg (1.7 T/50 Hz).

11.3.3.1 Measurement of Specific Total Losses (Model C70)

The specific total losses are measured under harmonic excitations based on Model C70, all the phase angles are equal to 0°, which can be referred in Table 11.15 in Appendix of this chapter. The specific total losses are also measured under harmonic excitations based on Model C70, for the phase angles of 180°, which can be referred in Table 11.16 in Appendix of this chapter.

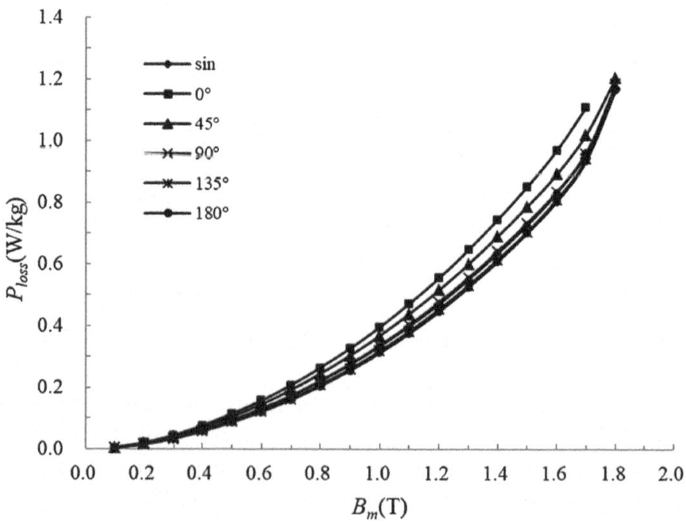

Fig. 11.16 Specific total losses (K_3 = 10%, different phase angle)

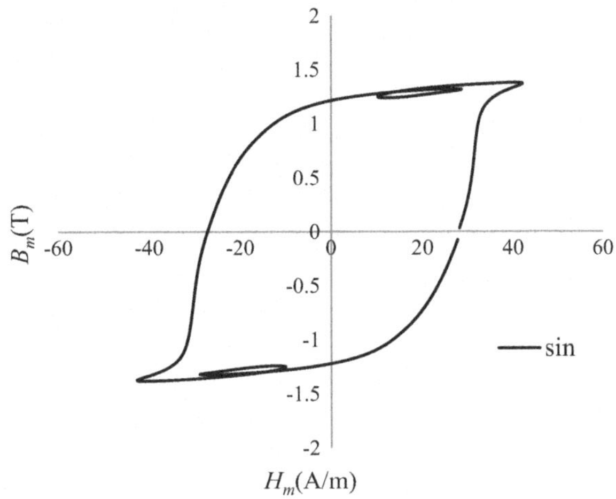

Fig. 11.17 *B–H* loop (K_3 = 20%, 0° phase angle difference, B_m = 1.38 T)

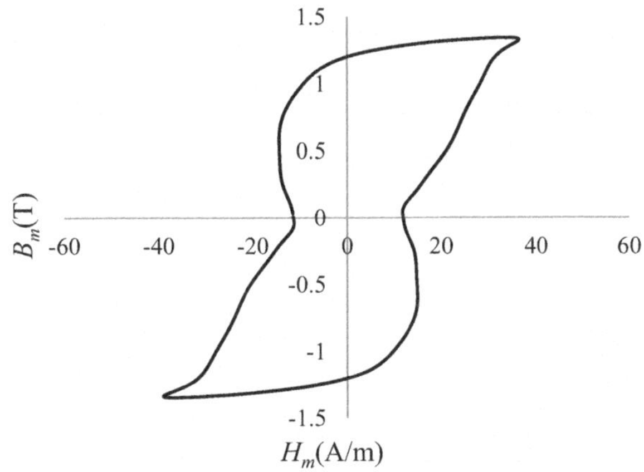

Fig. 11.18 *B–H* loop (K_3 = 20%, 180° phase angle difference, B_m = 1.34 T)

The specific total losses are measured under harmonic excitation (10% 3rd harmonic and different phase angles) based on Model C70, as shown Fig. 11.16, which can also be referred to Table 11.17 in the Appendix of this chapter.

It can be seen that specific loss reaches maximum at the 0° phase angle, which will be reduced with the phase angle increased.

Figures 11.17 and 11.18 show the *B–H* loops, including 5th harmonic, at the phase angle 0° and 180°, respectively.

It should be noted that the minor *B–H* loops appear at 0° phase angle, and the area of the *B–H* loop at 0° phase angle is greater than that at 180° phase angle. It results in the loss increase.

The specific losses are measured under the following harmonic parameters, i.e., 5th harmonic, 10% content, and different phase angles, as shown in Fig. 11.19, which can also be referred to Table 11.18 in the Appendix of this chapter.

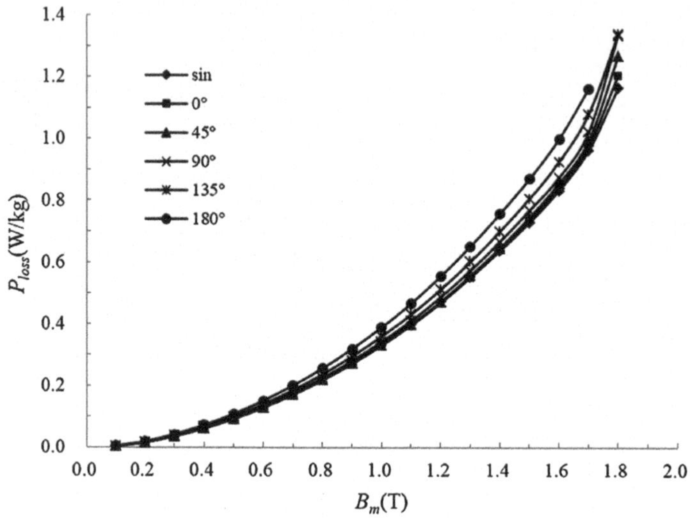

Fig. 11.19 Specific total losses (K_5 = 10%, different phase angle)

Fig. 11.20 *B–H* loop (K_5 = 20%, 0° phase angle, B_m = 1.36 T)

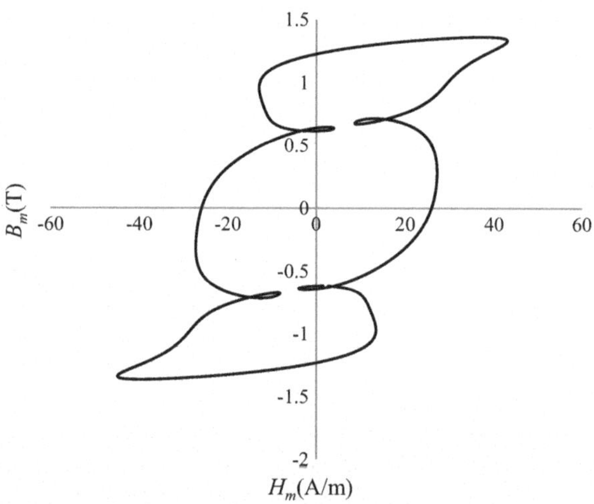

Note that the largest specific total loss is achieved at 180° phase, and the specific loss is reduced when the phase angle is reduced. This variation trend of the specific losses of 5th harmonic is different from that of 3rd harmonic.

Figures 11.20 and 11.21 show the *B–H* loops, including 5th harmonic, at the phase angles 0° and 180°, respectively.

Both the areas of the minor and major *B–H* loops are larger at 180° phase angle compared to that at 0° phase angle. This suggests an increase in the specific total loss.

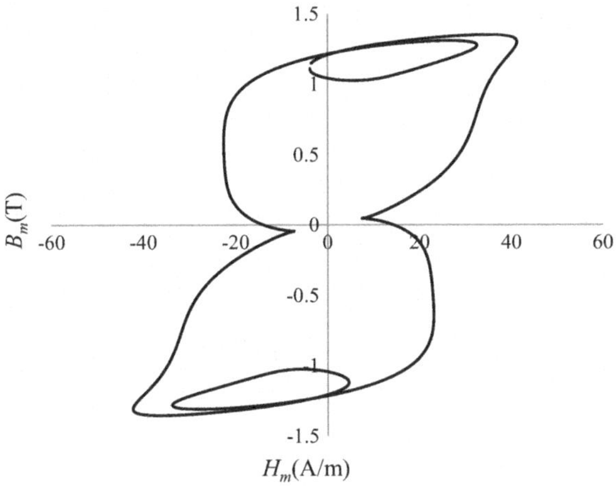

Fig. 11.21 *B–H* loop (K_5 = 20%, 180° phase angle, B_m = 1.31 T)

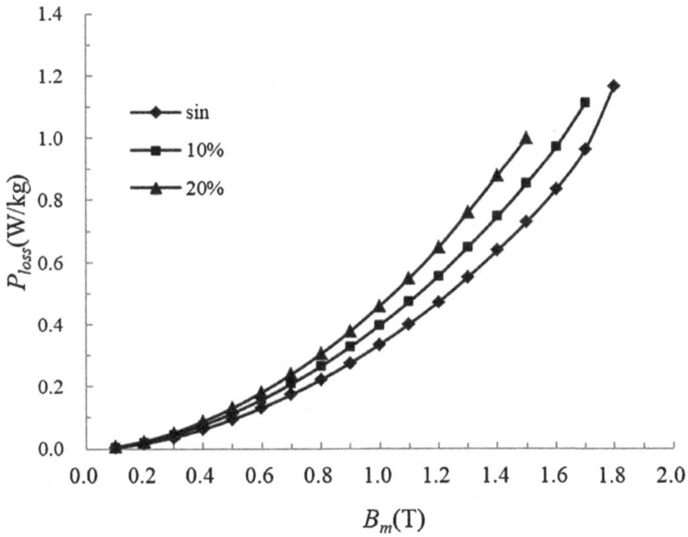

Fig. 11.22 Specific total loss (3rd, ϕ_3 = 0°, different content)

The specific total losses, with 3rd harmonic, 0° phase angle but different contents, are shown in Fig. 11.22.

The specific total losses, with 5th harmonic, 0° phase angle but different contents, are shown in Fig. 11.23.

It can be seen that the harmonic order affects the specific total loss, even when having the same phase angles and the same harmonic contents.

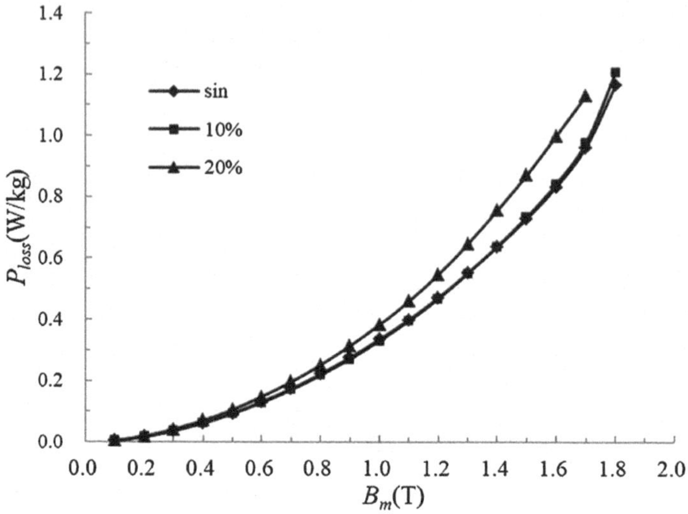

Fig. 11.23 Specific total loss (5th, $\phi_5 = 0°$, different content)

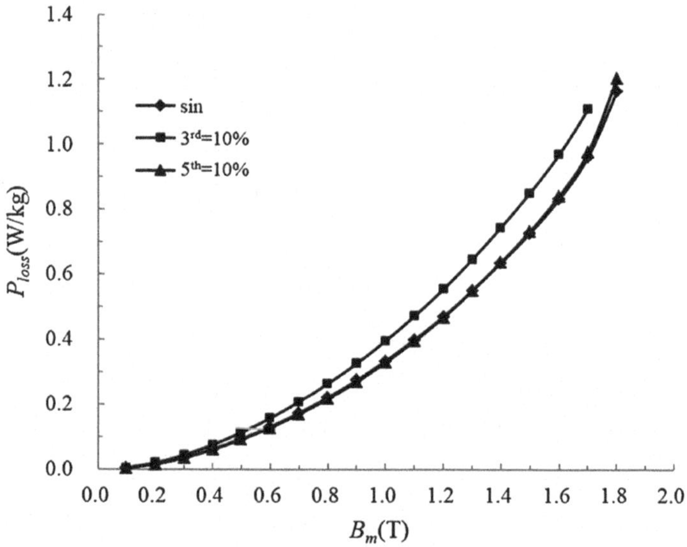

Fig. 11.24 Specific total loss ($K_3 = K_5 = 10\%$, $\phi_3 = \phi_5 = 0°$)

Figure 11.24 shows the effects of harmonic components (without harmonic, with 3rd and 5th harmonic of same contents, 0° phase angle difference) on specific total loss.

Figure 11.25 shows the effects of harmonic components (without harmonic, with 3rd and 5th harmonic of same contents, 0° phase angle difference) on specific total loss.

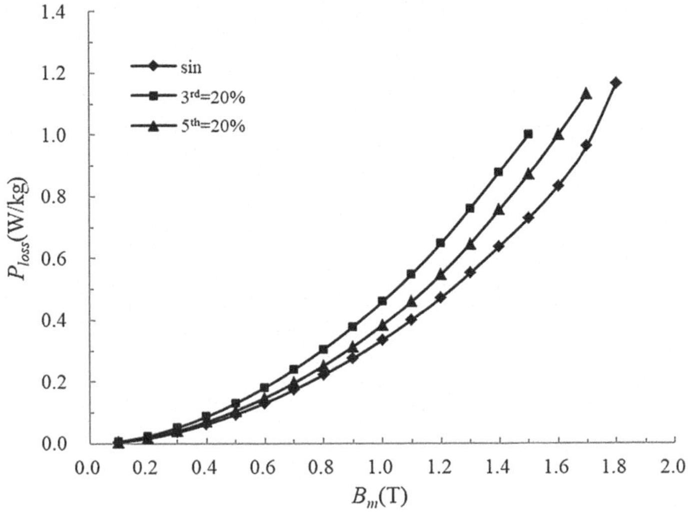

Fig. 11.25 Specific total loss ($K_3 = K_5 = 20\%$, $\phi_3 = \phi_5 = 0°$)

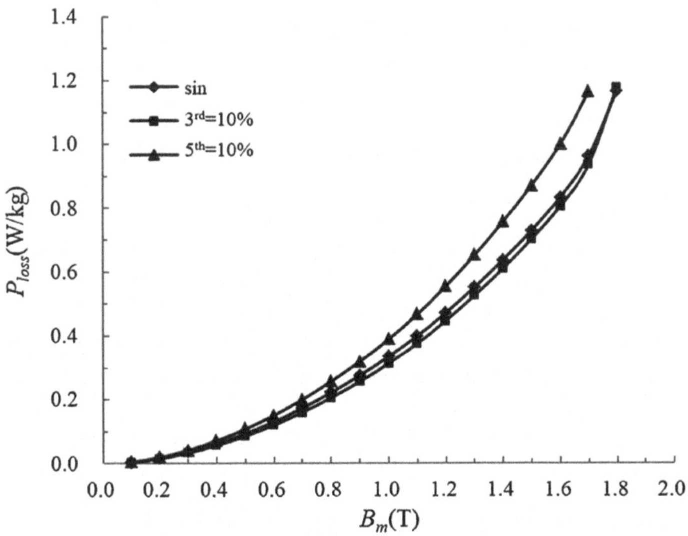

Fig. 11.26 Specific total loss ($K_3 = K_5 = 10\%$, $\phi_3 = \phi_5 = 180°$)

Figure 11.26 shows the effects of harmonic components (without harmonic, with 3rd and 5th harmonic with 10% contents, 180° phase angle difference) on specific total loss.

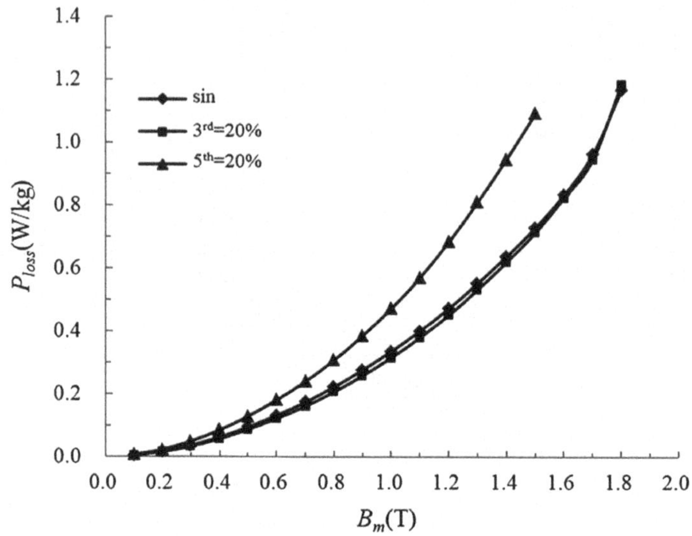

Fig. 11.27 Specific total loss ($K_3 = K_5 = 20\%$, $\phi_3 = \phi_5 = 180°$)

Figure 11.27 shows the effects of harmonic components (without harmonic, with 3rd and 5th harmonic with 20% contents, 180° phase angle difference) on specific total loss.

11.3.3.2 Measurement of Specific Total Loss (Model C50)

The specific total losses under different harmonic parameters (without harmonic, with different harmonic order, contents, and phase angle) are measured based on Model C50, which can be referred in Tables 11.19 and 11.20 in Appendix of this chapter.

It is shown that the variations of the specific total loss with the same harmonic parameters, measured based on either Model C70 or Model C50, indicate the same trend.

11.3.4 Specific Total Loss (Model (C70–C50))

The specific total loss at the middle section (200 mm in length) of Model C70 has been determined by using the two core model schemes, which can be referred to as the specific total loss of Model (C70–C50), as shown in Fig. 11.28, which can also be referred to Table 11.21 in the Appendix of this chapter.

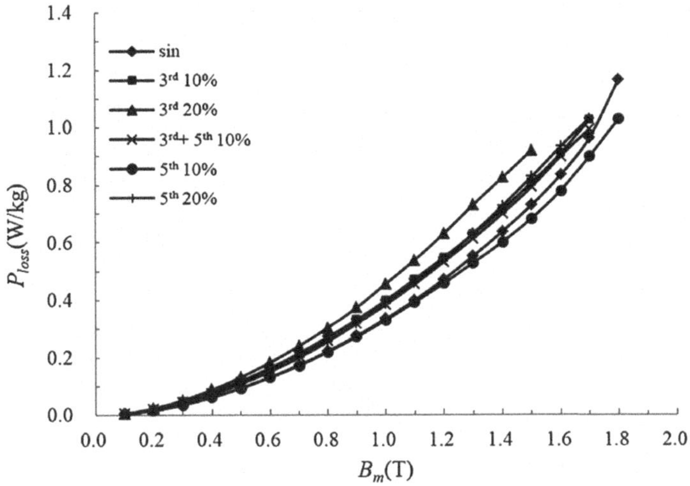

Fig. 11.28 Specific total loss at the middle section of Model C70

11.3.5 Comparisons Among Specific Total Losses Measured by Two Core Models

The specific total loss has been measured based on the two laminated core models, Model C70 and Model C50, under different harmonic parameters.

The following Figs. 11.29, 11.30, 11.31, 11.32, 11.33, and 11.34 and Tables 11.22, 11.23, 11.24, 11.25, 11.26, and 11.27 in the Appendix of this chapter

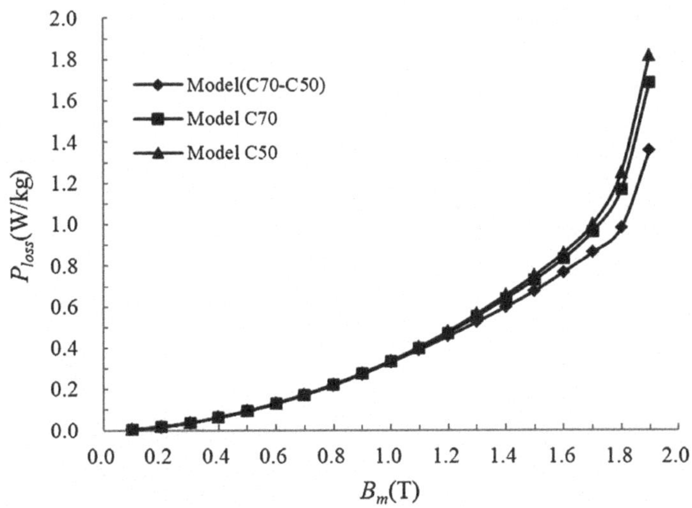

Fig. 11.29 Specific total loss (sin, 50 Hz)

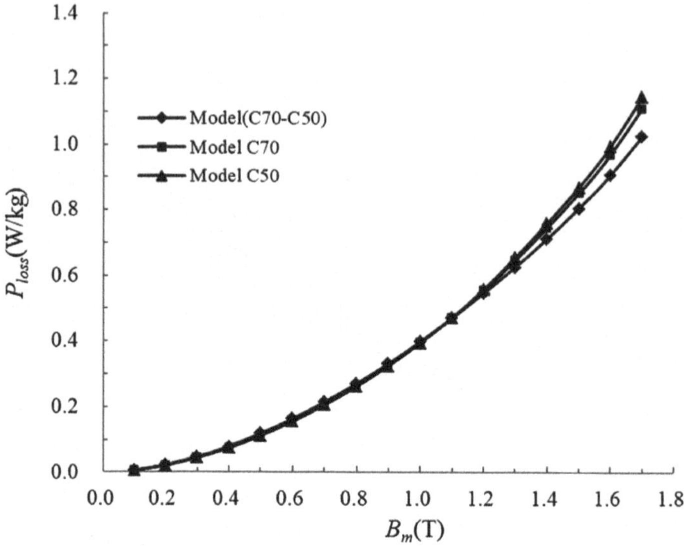

Fig. 11.30 Specific total loss (K_3 = 10%, ϕ_3 = 0°)

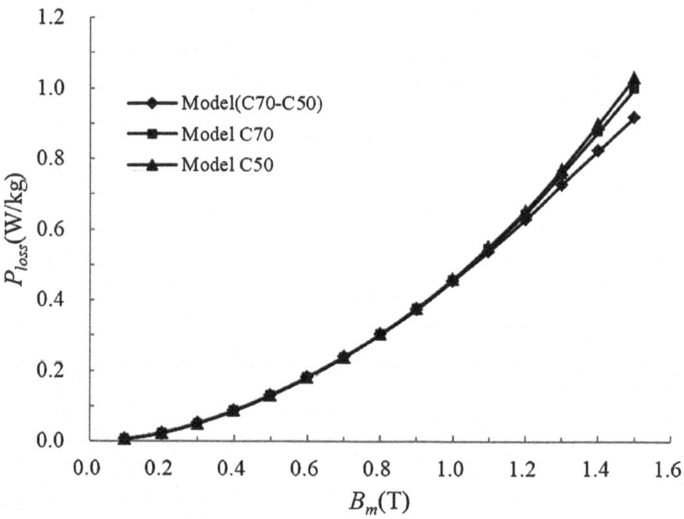

Fig. 11.31 Specific total loss (K_3 = 20%, ϕ_3 = 0°)

show the comparisons among specific total loss values determined by means Model C70, Model C50, and Model (C70–C50) at the same exciting condition.

As mentioned above, the specific total loss of Model (C70–C50) is easily determined based on the results of Model C70 and Model C50.

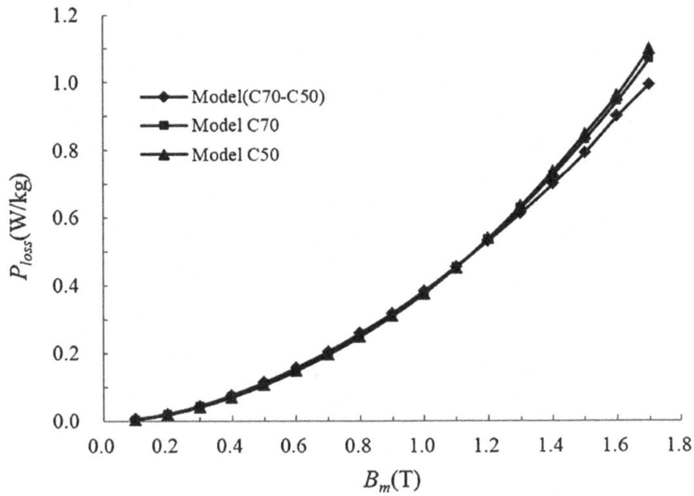

Fig. 11.32 Specific total loss ($K_3 = K_5 = 10\%$, $\phi_3 = \phi_5 = 0°$)

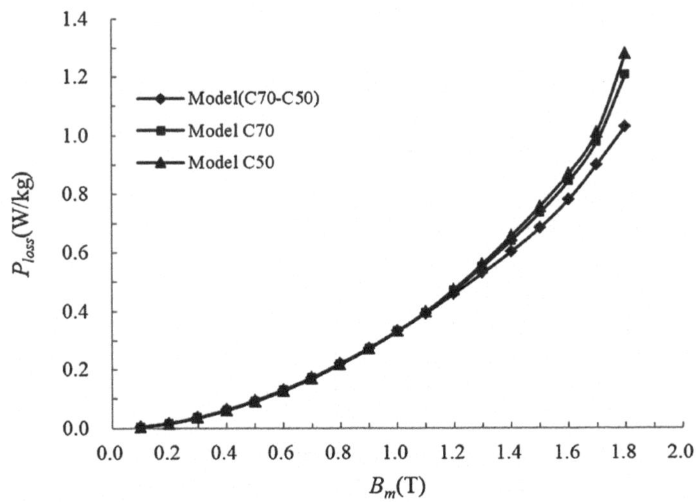

Fig. 11.33 Specific total loss ($K_5 = 10\%$, $\phi_5 = 0°$)

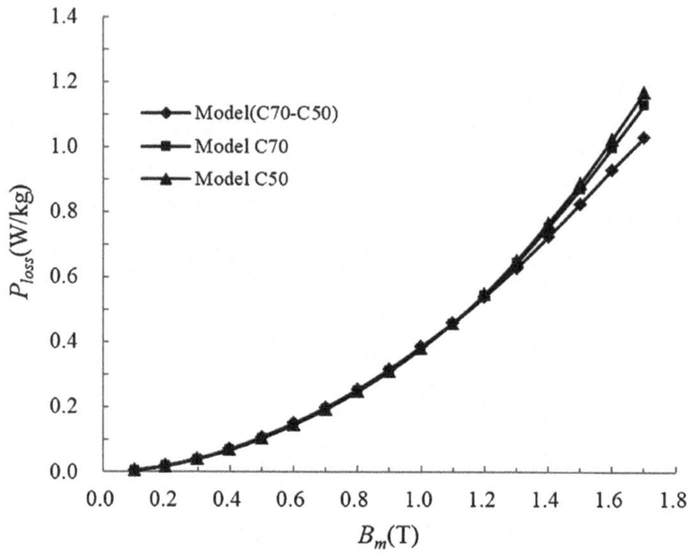

Fig. 11.34 Specific total loss ($K_5 = 20\%$, $\phi_5 = 0°$)

11.3.6 Comparison of Specific Total Loss Results (Using Core Models and Epstein Frame)

The specific total loss measured by different core models and by Epstein frame under sinusoidal excitation is shown in Fig. 11.35 (also can be referred in Table 11.28 of the Appendix of this chapter). It is noticed that:

(a) The Epstein loss is slightly lower than that of both Model C70 and Model C50, when the flux density is not higher than 1.6 T;
(b) The Epstein loss is in between those of Model C70 and Model C50, when flux density is equal or higher than 1.7 T, but higher than that of Model (C70–C50), i.e., the specific loss at the middle section of Model C70.

11.3.7 Exciting Power Inside Laminated Core

The distribution of the exciting power inside the laminated core (e.g., Model C70) is not uniform, i.e., the exciting power in the joint zone of the core is different from that of the middle section of each limb (Model (C70–C50)), in which the exciting power is uniform, as shown in Fig. 11.14b.

The two core model schemes, as stated already in Sect. 11.3.1, also enable us to separate the exciting power at different subregions. This is based on a basic fact, i.e., the Model C70 and Model C50 are made of the same material, have the same

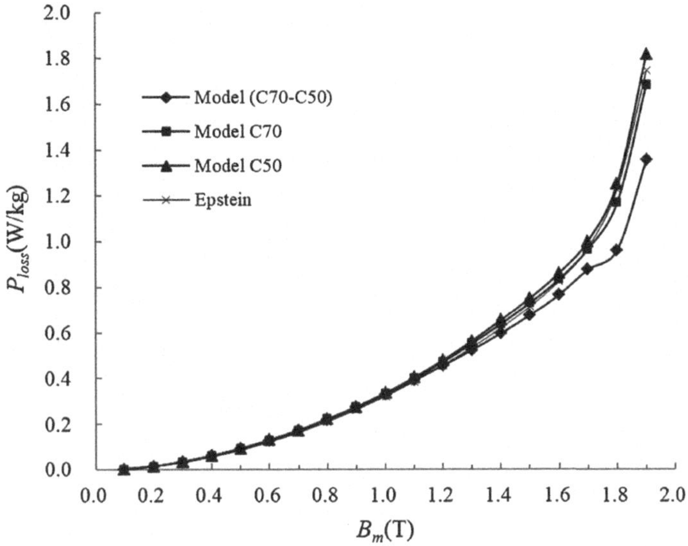

Fig. 11.35 Specific total loss (measured by core models and Epstein frame)

joint structure, and then have the same exciting power distribution under the same exciting level, at the joint zones.

In this way, the total exciting powers of Model C70 and Model C50, referred to as EP_{C70} and EP_{C50}, respectively, and the exciting power of each uniform zone of Model C70, denoted as $EP_{(C70-C50)}$, have the following relationship,

$$EP_{C70} - EP_{C50} = 4 \times EP_{(C70-C50)} \tag{11.3}$$

The comparisons among the measured results of the exciting power under different excitations (e.g., fundamental wave only or with harmonic) based on the core models (Model C70, Model C50, and Model (C70–C50)) are shown in Figs. 11.36, 11.37, 11.38, 11.39, 11.40, and 11.41, which can also be referred to Tables 11.29, 11.30, 11.31, 11.32, 11.33, and 11.34 in the Appendix of this chapter.

From Figs. 11.37, 11.38, 11.39, and 11.40, it can be found that the exciting power is dependent on the harmonic's phase angle even though the harmonic's order content is the same.

From Fig. 11.41, it can be seen how the exciting power is dependent on the harmonic combination (e.g., the fundamental excitation contains 3rd and 5th harmonics).

Figure 11.42 shows the effect of different harmonic excitations on the exciting power of Model C70, e.g., fundamental only, or fundamental with 10% of 3rd harmonic, or with 10% of (3rd + 5th) harmonic. It can also be referred in Table 11.35 in Appendix of this chapter.

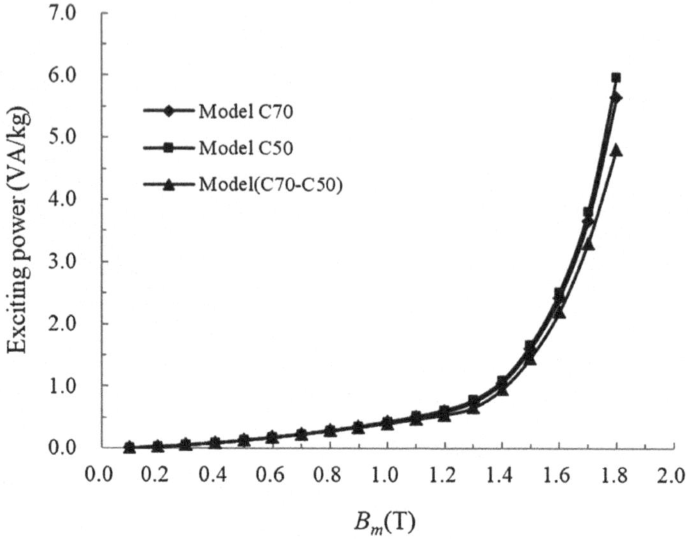

Fig. 11.36 Exciting power (sin, 50 Hz)

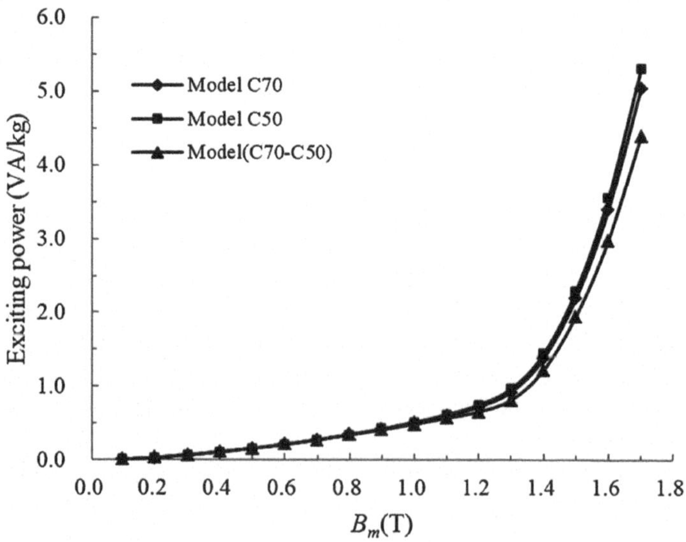

Fig. 11.37 Exciting power (K_3 = 10%, ϕ_3 = 0°)

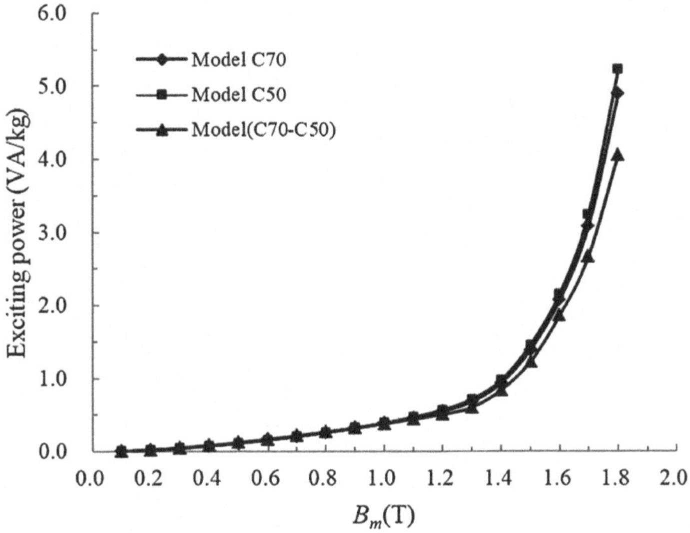

Fig. 11.38 Exciting power ($K_3 = 10\%$, $\phi_3 = 180°$)

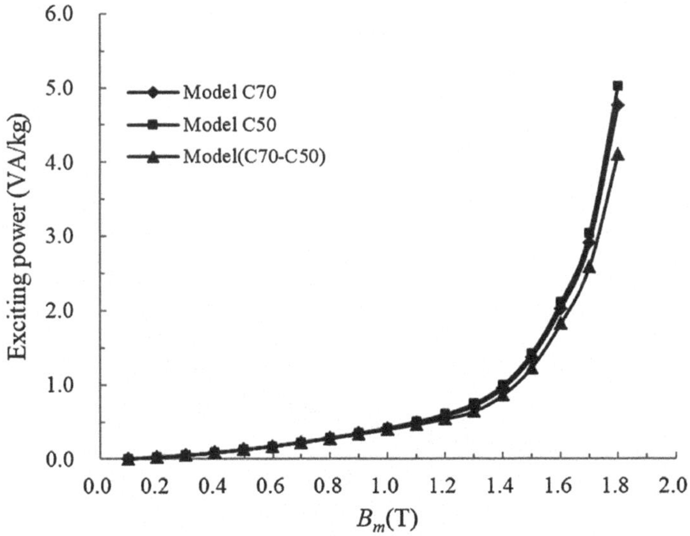

Fig. 11.39 Exciting power ($K_5 = 10\%$, $\phi_5 = 0°$)

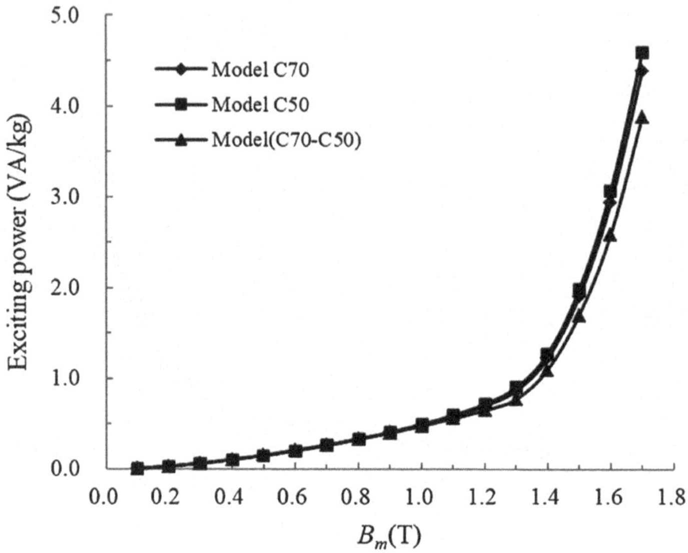

Fig. 11.40 Exciting power ($K_5 = 10\%$, $\phi_5 = 180°$)

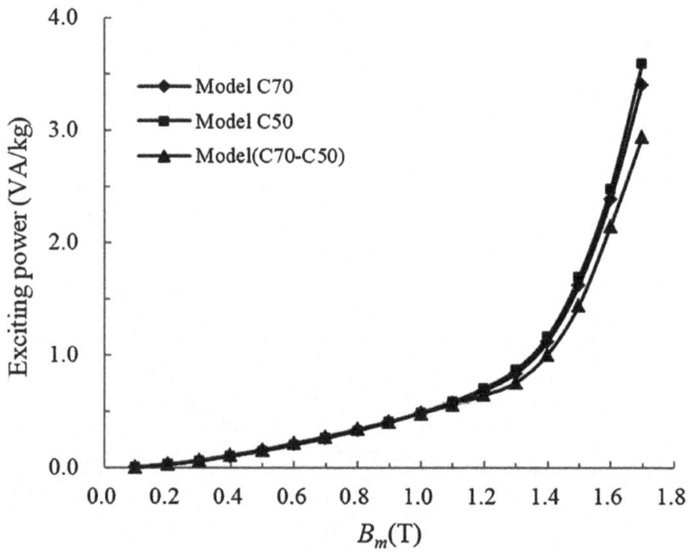

Fig. 11.41 Exciting power ($K_3 = K_5 = 10\%$, $\phi_3 = \phi_5 = 0°$)

From Fig. 11.42, it can be seen that 3rd harmonic results in the increase of exciting power; however, the combination of the 3rd and 5th harmonic does not increase the exciting power, compared to that of fundamental, when $\phi_3 = \phi_5 = 0°$.

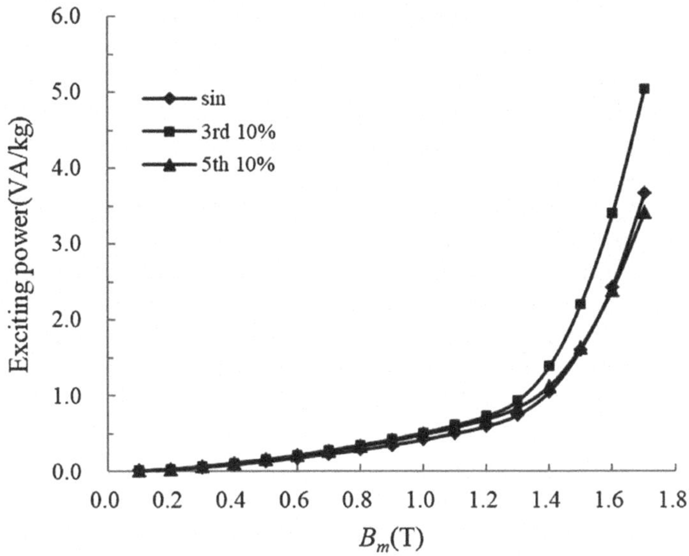

Fig. 11.42 Exciting power (Model C70, $\phi_3 = \phi_5 = 0°$)

11.3.8 Remarks 1

The specific total loss under different harmonic parameters (including the harmonic order, content, and phase angle) has been measured for two laminated core models, Model C70 and Model C50, using the magnetic measure-bench established in the laboratory of the Institute of Power Transmission and Transformation Technology, Baobian Electric.

(1) The magnetic measure-bench with an enhanced harmonic supply has been successfully used to model the magnetic component behavior under distorted excitations of non-standard waveforms, as required from the industry.
(2) The specific total loss determined based on the laminated core models is closely dependent on the harmonic parameters, i.e., the order, the content, and the phase angle of each harmonic; and the effect of the parameters of one harmonic on the specific total loss is probably different from that of another harmonic.
(3) The specific total loss of Model C70 is slightly greater than that of Model C50, and therefore, both results of specific total loss are practically in good agreement. However, the specific total loss of Model (C70–C50) is lower than that of both Model C70 and Model C50, especially, in the case of high flux densities. This is expected and reasonable, considering the impact of the non-uniform corner regions in the laminated core model.

(4) The specific total loss values measured by different core models and Epstein frame are compared against each other. A different loss behavior can be seen at different flux density levels.
(5) The effect of harmonic components (including the order, content, phase angle, and multi-harmonic combination) on exciting power has been examined, and as a result, 3rd harmonic results in the increase of exciting power; however, the combination of the 3rd and 5th harmonic does not increase the exciting power, compared to that of the fundamental.
(6) Further measurements of more magnetic properties, including the non-uniform exciting power, *B–H* loop, and the determination of building & working factor under multi-harmonic and/or DC-bias excitations, will be a future research project, due to the need for standardized measurements or predictions under non-sinusoidal conditions [5–8].

11.4 Measurement and Numerical Analysis of Magnetic Loss Under AC–DC Hybrid Excitation

11.4.1 Core Model Used for Magnetic Measurement Under Harmonic and DC-Bias Excitations

The magnetic measurement under multi-harmonic or DC-bias condition can be independently implemented based on the harmonic power and DC power supply, as shown in Fig. 11.43.

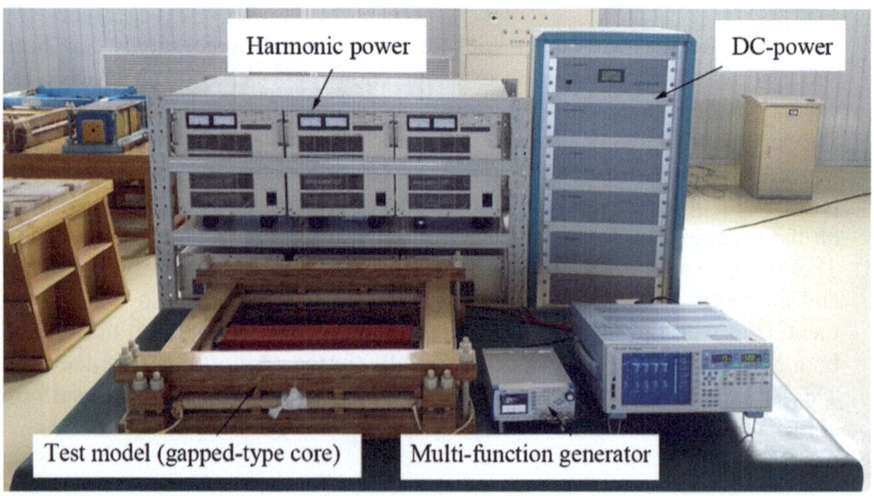

Fig. 11.43 Magnetic measurement system under complex excitations

However, in the case of special working condition of the converter transformers in HVDC transmission system, which involves both, multi-harmonics and DC excitation conditions, the magnetic measurement must be done under AC–DC hybrid excitation conditions. Certainly, the power supply's withstanding ability of unexpected DC and/or harmonic current components must be considered, because the harmonic power source may not withstand some DC components, or the DC power source may not withstand some AC components.

Accordingly, we proposed a new laminated core model, intending to implemented AC–DC hybrid excitation, in which a set of laminations *L*Θ is placed at the center of the square laminated frame (which has been used in other magnetic measurement, as a transformer-based square core model) to form a three-limb core. Note that there are two 5 mm gaps between the lamination *L*Θ and the core limbs, as shown in Fig. 11.44.

In this model, the lamination *L*Θ is excited by a DC exciting coil, and the square frame is excited by an AC exciting coil, including harmonic components, so the AC with multi-harmonics and DC excitations can be separately applied, and AC and DC exciting coils are not magnetically coupled. Therefore, in this test model, the laminations *L*Θ can be considered as a "DC bridge."

Note that the other experiment setup, in which the open-type SST and the Helmholtz coil are excited separately by AC and DC magnetic fields, respectively, has been proposed and used for magnetic loss measurement under DC-bias excitation [10]. See Chap. 7 (7.7.5 iron loss under DC-bias excitation).

The thickness of both lamination *L*Θ (B23R075) and laminated frame (27ZH100) is 20.7 mm. The number of the turns of the exciting coil wound on the

Fig. 11.44 Core model used for magnetic measurement under AC–DC hybrid excitations

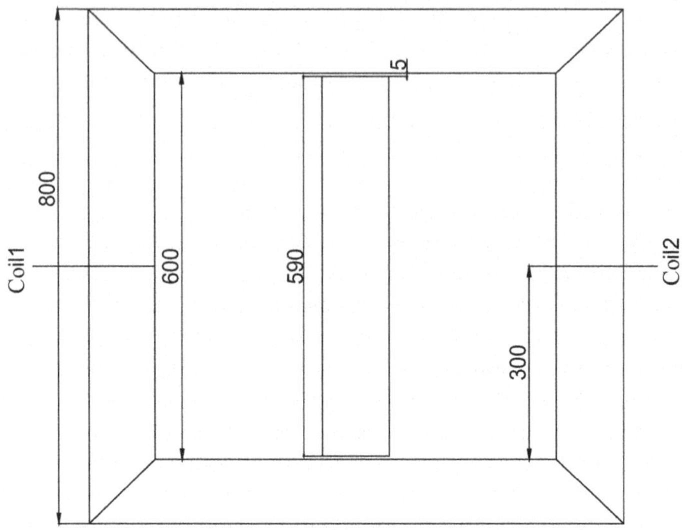

Fig. 11.45 Dimensions of the setup and search coils arrangement (unit: mm)

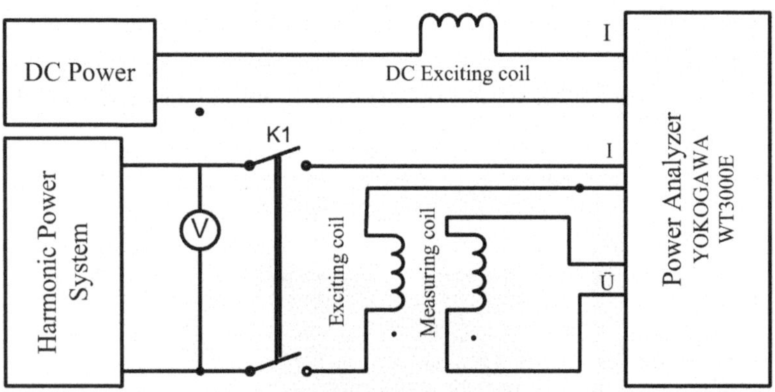

Fig. 11.46 Block diagram of measuring circuit

lamination $L\Theta$ is 327. The search coils (10 turns for each) are used for measuring the average magnetic flux densities inside the laminated frame. See Fig. 11.45.

Figure 11.46 shows the block diagram of measuring circuit under AC–DC hybrid excitation conditions.

11.4.2 *Feasibility of Magnetic Measurement Based on the New Core Model*

11.4.2.1 Effect of Lamination *L*Θ on Magnetic Loss in Square Laminated Frame

In order to examine the effect of the lamination *L*Θ on the magnetic loss inside the square laminated frame, the measurements with and without lamination *L*Θ have been done, and the results are shown in Table 11.1.

The loss results in Table 11.1 show that the effect of the no-load lamination *L*Θ (i.e., no DC applied to *L*Θ) on the magnetic loss inside the square laminated frame can be neglected, and relative errors are less than 1%. This means that almost no leakage flux passes through *L*Θ when the laminated frame is excited, even though the magnetic flux density is as high as 1.9 T, in this testing example.

Table 11.1 Measured loss results with and without lamination *L*Θ

B_m	P_{loss} without *L*Θ (W/kg)	P_{loss} with *L*Θ (W/kg)	Loss difference (W/kg)	Relative errors (%)
0.30	0.0366	0.0365	−0.0001	−0.30
0.40	0.0623	0.0624	0.0001	0.18
0.50	0.0938	0.0934	−0.0005	−0.51
0.60	0.1302	0.1292	−0.0010	−0.80
0.70	0.1718	0.1717	−0.0001	−0.05
0.80	0.2197	0.2200	0.0002	0.11
0.90	0.2734	0.2727	−0.0006	−0.24
1.00	0.3326	0.3333	0.0008	0.23
1.10	0.3981	0.3987	0.0006	0.16
1.20	0.4715	0.4734	0.0018	0.39
1.30	0.5523	0.5556	0.0034	0.61
1.40	0.6378	0.6421	0.0043	0.68
1.50	0.7315	0.7356	0.0041	0.56
1.55	0.7805	0.7865	0.0059	0.76
1.60	0.8346	0.8406	0.0060	0.72
1.65	0.8934	0.8985	0.0051	0.58
1.70	0.9633	0.9678	0.0045	0.46
1.75	1.0543	1.0579	0.0036	0.34
1.80	1.2245	1.2252	0.0007	0.06
1.85	1.5020	1.4966	−0.0054	−0.36
1.90	1.7862	1.7868	0.0006	0.04

11.4.2.2 Typical Measurement Results Under AC–DC Hybrid Excitation

As a measurement example, the DC (15.6 A) is applied in the exciting coil of the laminations *L*Θ, and the fundamental wave (50 Hz) and 3rd harmonic (10% content and 0° phase angle difference) are applied in the exciting coil of the square laminated frame.

The measurement results of total magnetic loss under the prescribed AC–DC hybrid exciting conditions are shown in Table 11.2.

The waveform of the exciting currents (fundamental with 3rd harmonic) of the square laminated frame is shown in Fig. 11.47, which can be used for computing magnetic field and loss.

The waveforms of the measured magnetic flux densities, using search coils 1 and 2, are shown in Fig. 11.48. The locations of the search coils can be found in Fig. 11.45.

Table 11.2 Measured magnetic loss under AC–DC hybrid excitation

Exciting conditions (A)		Measured magnetic loss (W)
Fundamental with 3rd harmonic (10%, 0 phase angle difference)	DC	
1.03	15.60	23.62

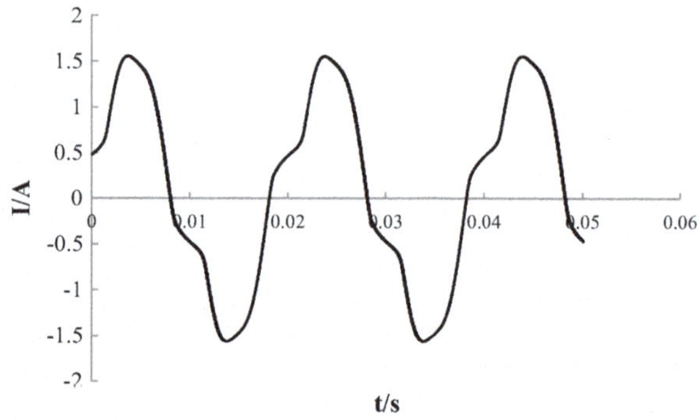

Fig. 11.47 Waveform of exciting currents of square laminated frame

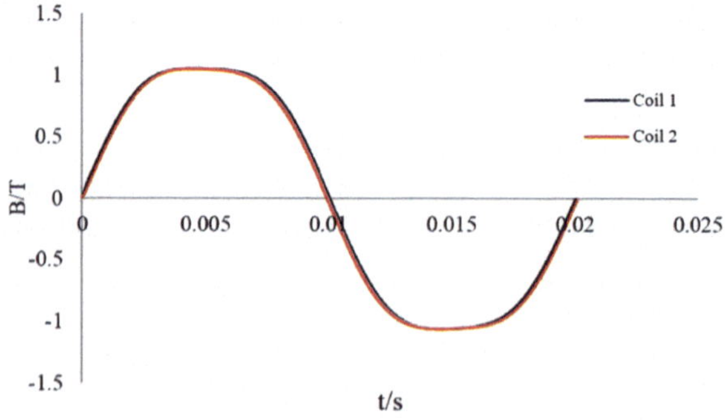

Fig. 11.48 Waveform of magnetic flux densities by using search coils

11.4.3 Numerical Calculation and Validation of Magnetic Loss Inside Square Laminated Frame Under AC–DC Hybrid Excitation Conditions

11.4.3.1 DC Magnetic Field

The DC magnetic field under the condition that only laminations *LΘ* are excited by DC (15.6 A) has been calculated, by Simcenter MAGNET, as shown in Fig. 11.49.

Figure 11.49 shows that the distribution of DC magnetic field inside the laminated square frame is not uniform. The specific total losses of the laminated frame are also different under different DC-bias levels. Thus, the multi-loss curves ($B_m - P$) are needed for the loss analysis under complex harmonic and DC-bias excitations, regarding the different DC-bias level and different AC working points.

The loss calculation, as a post-processing step, is performed by using the developed script. First, the determination of AC working point B_m and DC-bias field H_{dc} is based on the waveforms of magnetic flux density and magnetic field strength in each element, and then the specific total loss is calculated according to the determined B_m and H_{dc} and then summed up, using (11.4),

$$W = \rho \sum_{i=1}^{N_e} P^e\left(B_m^e,\ H_{dc}\right) V^e \tag{11.4}$$

where ρ is the material density; V^e is the element's volume.

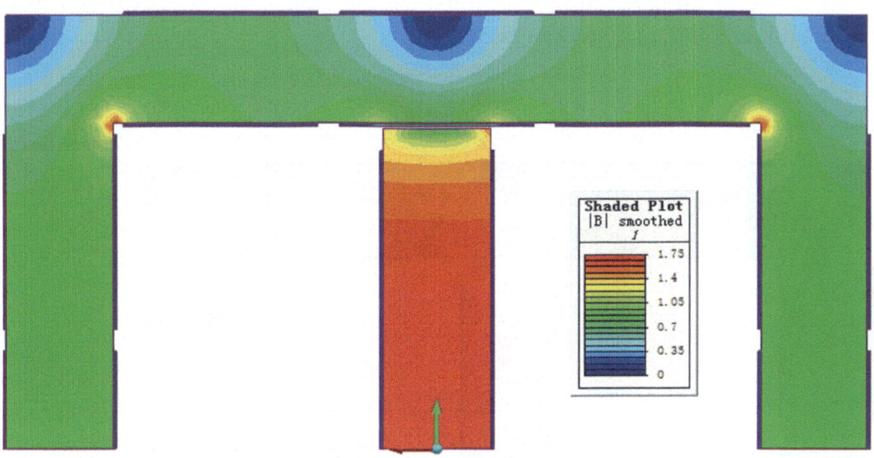

Fig. 11.49 DC magnetic field distribution

The AC working point B_m can be calculated using (11.5),

$$B_m = \frac{B_{max} - B_{min}}{2} \tag{11.5}$$

11.4.3.2 Zoning Scheme

It is shown that this loss calculation method is not applicable in engineering applications when the number of elements is very large and the time step is relatively small, due to the huge processing of B_m and H_{dc} within each transient period.

For this reason, a simplified zoning method, i.e., given the loss curves, $B_m - P$, for each prescribed sub-zone, regarding DC-biasing level, is proposed to calculate the magnetic losses inside each zone of the laminated frame.

The principle of sub-zone division is based on the local uniformity of DC-bias field distribution. The loss curve is determined according to the average value of H_{dc} of each sub-zone. In this way, the computational cost can be effectively reduced. Also, the magnetic loss can be directly computed using commercial software, e.g., Simcenter MAGNET.

Based on the zoning method, it is needed to rebuild the finite element model according to the zoned results, and assigning the property data to each sub-zone, and then solving the transient field.

In this chapter, the laminated square frame is divided into four sub-zones, as shown in Fig. 11.50. The H_{dc} value for each sub-zone is shown in Table 11.3.

The loss curves ($B_m - P$) for each sub-zone under harmonic (3rd, 10%, 0 phase angle difference) and different H_{dc} condition are shown in Table 11.4.

The comparison between the measured and calculated loss results is shown in Table 11.5, and the relative error is 3.51%.

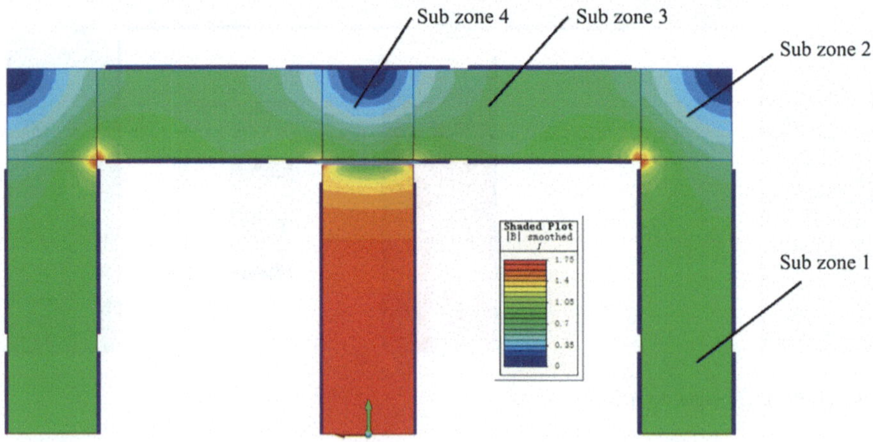

Fig. 11.50 Sub-zones in laminated square frame (DC: 15.6 A)

Table 11.3 H_{dc} of each sub-zone

Sub-zone	No. 1	No. 2	No. 3	No. 4
Average H_{dc} (A/m)	11.49	7.03	10.30	7.03

Table 11.4 Loss curves ($B_m - P$) for sub-zone

H_{dc} = 7.03 (A/m)		H_{dc} = 10.30 (A/m)		H_{dc} = 11.19 (A/m)	
B_m/T	P/(W/kg)	B_m/T	P/(W/kg)	B_m/T	P/(W/kg)
0.10	0.005	0.10	0.006	0.10	0.007
0.20	0.021	0.20	0.028	0.20	0.032
0.30	0.048	0.30	0.066	0.30	0.072
0.40	0.086	0.40	0.109	0.40	0.120
0.50	0.134	0.50	0.167	0.50	0.167
0.60	0.188	0.60	0.219	0.60	0.224
0.70	0.245	0.70	0.275	0.70	0.281
0.80	0.307	0.80	0.331	0.80	0.340
0.90	0.371	0.90	0.394	0.90	0.403
1.00	0.438	1.00	0.460	1.00	0.468
1.10	0.510	1.10	0.531	1.10	0.540
1.20	0.588	1.20	0.608	1.20	0.615
1.30	0.671	1.30	0.691	1.30	0.699
1.40	0.763	1.40	0.785	1.40	0.792
1.50	0.862	1.50	0.884	1.50	0.892
1.60	0.975	1.60	0.995	1.60	1.003
1.65	1.041	1.65	1.060	1.65	1.067
1.70	1.122	1.70	1.139	1.70	1.145
1.75	1.234	1.75	1.243	1.75	1.249
1.80	1.406	1.80	1.407	1.80	1.410
1.85	1.662	1.85	1.660	1.85	1.658
1.90	1.979	1.90	1.981	1.90	1.979
1.91	2.047	1.91	2.048	1.91	2.046

Table 11.5 Measured and calculated loss results

Excitations (A)		Measured magnetic loss (W)	Calculated magnetic loss (W)	Relative error (%)
Fundamental and 3rd harmonic (10%, 0 phase angle difference)	DC			
1.03	15.60	23.62	24.45	3.51

11.4.4 Remarks 2

The experiment setup, including harmonic and DC power supply and three-limb gapped-type core model, used for the magnetic measurement under harmonic and DC-bias hybrid excitation, is established, and some measured results are presented.

The magnetic losses inside the square laminated frame have been calculated based on the measured magnetic properties (by 1-D magnetic measurement system, Brockhaus, Germany), using the zoning scheme.

The calculated and measured results of magnetic losses are in good agreement. This shows the effectiveness of all the magnetic measurements and numerical computation.

11.5 Concluding Remarks

The enhanced 1-D magnetic measurement system (Brockhaus) has been used under both the standard and non-standard excitation conditions, including harmonic or/and DC-bias component. It is beneficial for industrial application.

In order to measure and predict the component-level magnetic properties under complex excitations, an integrated magnetic measure-bench has been established, including laminated core models. The effect of various harmonic and DC-bias parameters on the magnetic properties has been investigated in detail.

Especially, the magnetic measurement under the AC–DC hybrid excitations has been implemented based on a three-limb gapped-type core model proposed by the authors, in which the multi-harmonic components are included in the exciting coil of the square frame but the DC component is included in the exciting coils of the DC bridge.

The numerical computation of the magnetic loss inside the square frame under AC–DC hybrid excitation, using the zoning scheme, has been validated based on the test model.

Finally, it is emphasized that the importance of the fundamentals of magnetics and magnetic measurements [3, 11–13] should be fully understood, that the effects of non-standard or extreme excitation conditions on material microstructure of electrical steel laminations [3] and the standardized measurements or predictions under non-sinusoidal conditions [6] should be investigated, and that some limitations and shortcomings of the existing measurement methods, even though they have been used for many years [1, 2], should be correctly recognized, for possible improvements.

Acknowledgements The authors thank Prof. J. Sievert, Prof. A. J. Moses, and Dr. P. Anderson for helpful discussion and communications, and for their interesting lectures delivered at the joint seminar, Baobian Electric., September 2016, and thank all colleagues who contributed to the magnetic measurements and numerical simulation under non-standard conditions. This project was founded in part by the National Key R&D Program of China (no. 2016YFB0300300).

Appendix 1 Non-standard Magnetic Measurement Results

In this chapter, a number of non-standard magnetic properties have been measured under different excitation conditions, using the enhanced 1-D measuring system and the well-established measure-bench with AC–DC hybrid supply.

In order to quantifiably investigate the effects of all the excitation conditions, including multi-harmonic and/or DC-bias components, on the non-standard magnetic properties, and the effectiveness of numerical modeling and simulation based on the non-standard property data, the measured results are collected in the following tables in detail, which will be helpful for further research projects.

1.1 Magnetic Property of GO Silicon Steel Measured Under Non-standard Conditions

The specific total losses (P_{loss}) of GO silicon steel sheet, 27ZH100, measured under different harmonic conditions, i.e., different harmonic order and content, are shown in Table 11.6, in which the phase angle difference is zero for each measurement case.

Table 11.6 Measured loss results (0° phase angle difference)

B_m (T)	P_{loss} (W/kg)				
	Sin (50 Hz)	3rd 10%	3rd 20%	5th 10%	5th 20%
0.10	0.004	0.005	0.006	0.004	0.005
0.20	0.016	0.019	0.023	0.016	0.019
0.30	0.035	0.042	0.048	0.034	0.040
0.40	0.059	0.071	0.082	0.058	0.068
0.50	0.090	0.107	0.123	0.088	0.103
0.60	0.125	0.149	0.171	0.123	0.144
0.70	0.166	0.197	0.227	0.164	0.191
0.80	0.212	0.251	0.289	0.210	0.245
0.90	0.263	0.312	0.358	0.261	0.306
1.00	0.321	0.379	0.436	0.318	0.374
1.10	0.383	0.452	0.523	0.381	0.449
1.20	0.452	0.532	0.617	0.451	0.533
1.30	0.527	0.619	0.719	0.528	0.626
1.40	0.610	0.713	0.831	0.613	0.729
1.50	0.703	0.819	0.954	0.709	0.844
1.60	0.813	0.940	1.099	0.823	0.977
1.65	0.880	1.013	1.183	0.894	1.056
1.70	0.963	1.104	1.287	0.977	1.149
1.75	1.080	1.223	1.422	1.091	1.269
1.80	1.251	1.403	1.614	1.262	1.451
1.85	1.494	1.663	1.897	1.500	1.704
1.90	1.794	1.983	2.249	1.794	2.015

Table 11.7 Measured loss results (180° phase angle difference)

B_m (T)	P_{loss} (W/kg)				
	Sin (50 Hz)	3rd 10%	3rd 20%	5th 10%	5th 20%
0.10	0.004	0.004	0.004	0.005	0.006
0.20	0.016	0.015	0.015	0.019	0.022
0.30	0.035	0.032	0.033	0.040	0.048
0.40	0.059	0.055	0.056	0.069	0.082
0.50	0.090	0.083	0.084	0.104	0.125
0.60	0.125	0.117	0.118	0.145	0.175
0.70	0.166	0.154	0.156	0.192	0.232
0.80	0.212	0.198	0.200	0.246	0.299
0.90	0.263	0.247	0.250	0.306	0.374
1.00	0.321	0.302	0.306	0.374	0.458
1.10	0.383	0.362	0.368	0.449	0.552
1.20	0.452	0.429	0.437	0.531	0.657
1.30	0.527	0.502	0.513	0.621	0.775
1.40	0.610	0.582	0.598	0.720	0.907
1.50	0.703	0.674	0.693	0.831	1.052
1.60	0.813	0.782	0.804	0.961	1.222
1.65	0.880	0.848	0.873	1.040	1.323
1.70	0.963	0.931	0.956	1.139	1.442
1.75	1.080	1.049	1.074	1.270	1.596
1.80	1.251	1.223	1.252	1.468	1.823
1.85	1.494	1.463	1.492	1.732	2.140
1.90	1.794	1.745	1.776	2.058	2.536

The loss results (P_{loss}) measured under different harmonic orders and content conditions are shown in Table 11.7. Note that the phase angle difference here is 180° for each measurement case.

The loss results (P_{loss}) measured under fundamental wave and different DC-bias condition (H_{dc}) are shown in Table 11.8.

The loss results (P_{loss}) of GO silicon steel sheet, 27ZH100, measured under fundamental wave or with 10% 3rd harmonic and different DC-bias (H_{dc}) conditions are shown in Table 11.9, and the phase angle difference is 0° for 3rd harmonic.

The loss results (P_{loss}) measured under fundamental wave only or with 10% 3rd harmonic and different DC-bias (H_{dc}) conditions are shown in Table 11.10. Note that the phase angle difference here is 180° (for 3rd harmonic).

The loss results (P_{loss}) measured under fundamental wave only or with 10% 5th harmonic and different DC-bias (H_{dc}) conditions are shown in Table 11.11, and the phase angle difference is 0° for 5th harmonic.

Table 11.8 Loss results measured under different DC-bias conditions

B_m (T)	P_{loss} (W/kg)					
	Sin (50 Hz)	H_{dc} = 20 A/m	H_{dc} = 40 A/m	H_{dc} = 60 A/m	H_{dc} = 80 A/m	H_{dc} = 100 A/m
0.10	0.004	0.014	0.028	0.036	0.043	0.048
0.20	0.016	0.051	0.082	0.098	0.109	0.117
0.30	0.035	0.097	0.137	0.159	0.173	0.185
0.40	0.059	0.149	0.194	0.219	0.240	0.252
0.50	0.090	0.199	0.251	0.281	0.305	0.319
0.60	0.125	0.252	0.309	0.343	0.370	0.386
0.70	0.166	0.302	0.370	0.414	0.436	0.452
0.80	0.212	0.361	0.433	0.479	0.503	0.520
0.90	0.263	0.423	0.496	0.546	0.570	0.588
1.00	0.321	0.487	0.563	0.614	0.640	0.657
1.10	0.383	0.554	0.631	0.687	0.711	0.729
1.20	0.452	0.624	0.702	0.757	0.783	0.804
1.30	0.527	0.697	0.776	0.834	0.860	0.881
1.40	0.610	0.777	0.872	0.915	0.941	0.961
1.50	0.703	0.864	0.959	1.002	1.029	1.048
1.60	0.813	0.964	1.059	1.099	1.125	1.140
1.65	0.880	1.020	1.112	1.153	1.178	1.198
1.70	0.963	1.085	1.171	1.210	1.237	1.252
1.75	1.080	1.166	1.243	1.279	1.304	1.320
1.80	1.251	1.287	1.343	1.369	1.390	1.403
1.85	1.494	1.495	1.504	1.516	1.524	1.535
1.90	1.794	1.793	1.785	1.785	1.780	1.784

Table 11.9 Loss results under 3rd harmonic and DC-bias (0° phase angle difference)

B_m (T)	P_{loss} (W/kg)						
	Sin (50 Hz)	H_{dc} = 0 A/m	H_{dc} = 20 A/m	H_{dc} = 40 A/m	H_{dc} = 60 A/m	H_{dc} = 80 A/m	H_{dc} = 100 A/m
0.10	0.004	0.005	0.016	0.029	0.037	0.044	0.049
0.20	0.016	0.019	0.053	0.083	0.099	0.110	0.119
0.30	0.035	0.042	0.101	0.139	0.159	0.174	0.186
0.40	0.059	0.071	0.154	0.196	0.221	0.240	0.254
0.50	0.090	0.107	0.201	0.252	0.281	0.304	0.320
0.60	0.125	0.149	0.255	0.312	0.346	0.371	0.388
0.70	0.166	0.197	0.316	0.374	0.412	0.438	0.458
0.80	0.212	0.251	0.376	0.439	0.481	0.511	0.530
0.90	0.263	0.312	0.441	0.508	0.552	0.583	0.604
1.00	0.321	0.379	0.510	0.580	0.625	0.658	0.678
1.10	0.383	0.452	0.583	0.655	0.701	0.734	0.757

(continued)

Table 11.9 (continued)

B_m (T)	P_{loss} (W/kg)						
	Sin (50 Hz)	$H_{dc} = 0$ A/m	$H_{dc} = 20$ A/m	$H_{dc} = 40$ A/m	$H_{dc} = 60$ A/m	$H_{dc} = 80$ A/m	$H_{dc} =$ 100 A/m
1.20	0.452	0.532	0.660	0.733	0.781	0.815	0.839
1.30	0.527	0.619	0.745	0.817	0.865	0.899	0.924
1.40	0.610	0.713	0.835	0.908	0.958	0.991	1.015
1.50	0.703	0.819	0.935	1.007	1.058	1.089	1.114
1.60	0.813	0.940	1.045	1.117	1.169	1.197	1.222
1.65	0.880	1.013	1.109	1.176	1.229	1.256	1.282
1.70	0.963	1.104	1.181	1.248	1.296	1.326	1.347
1.75	1.080	1.223	1.277	1.330	1.375	1.404	1.423
1.80	1.251	1.403	1.421	1.452	1.482	1.504	1.520
1.85	1.494	1.663	1.661	1.662	1.670	1.672	1.682
1.90	1.794	1.983	1.979	1.976	1.974	1.972	1.971

Table 11.10 Loss results measured under 3rd harmonic and different DC-bias (180° phase angle difference)

B_m (T)	P_{loss} (W/kg)						
	Sin (50 Hz)	$H_{dc} = 0$ A/m	$H_{dc} = 20$ A/m	$H_{dc} = 40$ A/m	$H_{dc} = 60$ A/m	$H_{dc} = 80$ A/m	$H_{dc} =$ 100 A/m
0.10	0.004	0.004	0.014	0.028	0.037	0.044	0.049
0.20	0.016	0.015	0.052	0.084	0.100	0.111	0.120
0.30	0.035	0.032	0.102	0.142	0.163	0.180	0.190
0.40	0.059	0.055	0.151	0.200	0.225	0.247	0.259
0.50	0.090	0.083	0.203	0.258	0.293	0.313	0.325
0.60	0.125	0.117	0.255	0.318	0.358	0.377	0.392
0.70	0.166	0.154	0.312	0.378	0.422	0.442	0.457
0.80	0.212	0.198	0.369	0.440	0.487	0.507	0.523
0.90	0.263	0.247	0.430	0.503	0.551	0.574	0.590
1.00	0.321	0.302	0.492	0.582	0.618	0.641	0.656
1.10	0.383	0.362	0.557	0.651	0.686	0.709	0.725
1.20	0.452	0.429	0.624	0.721	0.756	0.781	0.794
1.30	0.527	0.502	0.698	0.796	0.832	0.853	0.868
1.40	0.610	0.582	0.775	0.874	0.908	0.932	0.950
1.50	0.703	0.674	0.861	0.959	0.993	1.013	1.034
1.60	0.813	0.782	0.955	1.052	1.087	1.108	1.128
1.65	0.880	0.848	1.010	1.103	1.138	1.158	1.179
1.70	0.963	0.931	1.073	1.161	1.194	1.216	1.236
1.75	1.080	1.049	1.150	1.230	1.263	1.282	1.298
1.80	1.251	1.223	1.263	1.322	1.348	1.364	1.381
1.85	1.494	1.463	1.462	1.474	1.483	1.496	1.505
1.90	1.794	1.745	1.740	1.738	1.740	1.736	1.741

Table 11.11 Loss results measured under 5th harmonic and different DC-bias (0° phase angle difference)

B_m (T)	P_{loss} (W/kg)						
	Sin (50 Hz)	H_{dc} = 0 A/m	H_{dc} = 20 A/m	H_{dc} = 40 A/m	H_{dc} = 60 A/m	H_{dc} = 80 A/m	H_{dc} = 100 A/m
0.10	0.004	0.004	0.015	0.029	0.038	0.045	0.050
0.20	0.016	0.016	0.055	0.089	0.106	0.117	0.127
0.30	0.035	0.034	0.111	0.152	0.174	0.192	0.202
0.40	0.059	0.058	0.166	0.215	0.248	0.265	0.276
0.50	0.090	0.088	0.221	0.279	0.318	0.335	0.348
0.60	0.125	0.123	0.281	0.344	0.386	0.405	0.418
0.70	0.166	0.164	0.342	0.423	0.454	0.475	0.486
0.80	0.212	0.210	0.403	0.494	0.523	0.544	0.557
0.90	0.263	0.261	0.469	0.562	0.591	0.612	0.627
1.00	0.321	0.318	0.537	0.632	0.664	0.683	0.699
1.10	0.383	0.381	0.607	0.703	0.736	0.755	0.771
1.20	0.452	0.451	0.679	0.778	0.809	0.829	0.850
1.30	0.527	0.528	0.756	0.853	0.888	0.908	0.927
1.40	0.610	0.613	0.838	0.936	0.968	0.992	1.011
1.50	0.703	0.709	0.925	1.027	1.052	1.080	1.101
1.60	0.813	0.823	1.026	1.122	1.152	1.177	1.197
1.65	0.880	0.894	1.082	1.175	1.204	1.232	1.250
1.70	0.963	0.977	1.177	1.237	1.265	1.293	1.310
1.75	1.080	1.091	1.254	1.304	1.332	1.360	1.376
1.80	1.251	1.262	1.350	1.393	1.413	1.442	1.461
1.85	1.494	1.500	1.512	1.529	1.544	1.560	1.576
1.90	1.794	1.794	1.784	1.786	1.786	1.796	1.805

The loss results (P_{loss}) measured under fundamental wave only or with 10% 5th harmonic and different DC-bias (H_{dc}) conditions are shown in Table 11.12, and the phase angle difference is 180° for the 5th harmonic.

The magnetization data of GO silicon steel 27ZH100 measured at 3 and 50 Hz are shown in Table 11.13.

The magnetization data of GO silicon steel, B23R075, measured at 3 and 50 Hz are shown in Table 11.14.

Table 11.12 Loss results measured under 5th harmonic and different DC-bias (180° phase angle difference)

B_m (T)	P_{loss} (W/kg)						
	Sin (50 Hz)	H_{dc} = 0 A/m	H_{dc} = 20 A/m	H_{dc} = 40 A/m	H_{dc} = 60 A/m	H_{dc} = 80 A/m	H_{dc} = 100A/m
0.10	0.004	0.005	0.016	0.030	0.039	0.046	0.051
0.20	0.016	0.019	0.056	0.087	0.105	0.116	0.125
0.30	0.035	0.040	0.116	0.149	0.171	0.185	0.196
0.40	0.059	0.069	0.166	0.211	0.238	0.257	0.270
0.50	0.090	0.104	0.218	0.274	0.307	0.330	0.344
0.60	0.125	0.145	0.278	0.342	0.378	0.403	0.419
0.70	0.166	0.192	0.341	0.412	0.455	0.479	0.495
0.80	0.212	0.246	0.409	0.485	0.533	0.556	0.573
0.90	0.263	0.306	0.481	0.561	0.610	0.635	0.652
1.00	0.321	0.374	0.557	0.639	0.692	0.716	0.734
1.10	0.383	0.449	0.637	0.733	0.774	0.799	0.820
1.20	0.452	0.531	0.721	0.822	0.862	0.888	0.907
1.30	0.527	0.621	0.812	0.912	0.954	0.980	0.997
1.40	0.610	0.720	0.908	1.008	1.051	1.077	1.096
1.50	0.703	0.831	1.014	1.117	1.157	1.182	1.202
1.60	0.813	0.961	1.134	1.236	1.274	1.296	1.318
1.65	0.880	1.040	1.201	1.298	1.339	1.358	1.383
1.70	0.963	1.139	1.275	1.369	1.408	1.428	1.450
1.75	1.080	1.270	1.369	1.455	1.492	1.514	1.535
1.80	1.251	1.468	1.504	1.569	1.595	1.614	1.633
1.85	1.494	1.732	1.737	1.742	1.756	1.767	1.783
1.90	1.794	2.058	2.050	2.048	2.045	2.046	2.052

Table 11.13 Magnetization data (27ZH100)

3 Hz		50 Hz	
H_m (A/m)	B_m (T)	H_m (A/m)	B_m (T)
3.85	0.100	4.40	0.100
6.12	0.200	7.35	0.200
7.82	0.300	9.88	0.300
9.03	0.400	12.02	0.400
10.00	0.500	13.85	0.500
10.86	0.600	15.46	0.600
11.67	0.700	16.96	0.700
12.42	0.800	18.27	0.800
13.24	0.900	19.50	0.900
14.12	1.000	20.72	1.000

(continued)

Table 11.13 (continued)

3 Hz		50 Hz	
H_m (A/m)	B_m (T)	H_m (A/m)	B_m (T)
15.03	1.100	21.86	1.100
16.20	1.200	23.02	1.200
17.86	1.300	24.09	1.300
20.36	1.400	25.28	1.400
24.69	1.500	28.14	1.500
34.69	1.600	36.20	1.600
44.82	1.650	44.66	1.650
52.37	1.675	51.27	1.675
62.57	1.700	60.44	1.700
76.89	1.725	73.45	1.725
97.93	1.750	92.81	1.750
129.90	1.775	122.39	1.775
181.19	1.800	170.40	1.800
270.78	1.825	251.71	1.825
432.38	1.850	394.16	1.850
650.00	1.875	640.82	1.875
1086.52	1.899	1068.57	1.900
1373.55	1.910	1369.63	1.910
1817.39	1.920	1810.31	1.920
2591.88	1.930	2587.88	1.930
4008.57	1.935	4003.57	1.935

Table 11.14 Magnetization data (B23R075)

3 Hz		50 Hz	
H_m (A/m)	B_m (T)	H_m (A/m)	B_m (T)
3.00	0.104	2.99	0.082
5.01	0.214	4.99	0.160
5.99	0.279	6.94	0.260
6.99	0.366	8.06	0.349
8.08	0.469	9.98	0.489
9.98	0.704	11.96	0.646
10.96	0.830	13.97	0.822
12.03	0.944	15.90	0.995
14.13	1.110	19.94	1.255
16.08	1.207	25.09	1.396
20.19	1.327	30.02	1.472
24.83	1.415	35.04	1.525
30.30	1.483	40.03	1.565
34.65	1.523	50.12	1.618

(continued)

Table 11.14 (continued)

3 Hz		50 Hz	
H_m (A/m)	B_m (T)	H_m (A/m)	B_m (T)
44.59	1.578	60.17	1.656
49.94	1.605	70.36	1.683
60.29	1.644	80.13	1.703
80.58	1.691	99.78	1.732
86.44	1.699	151.25	1.779
100.50	1.720	199.21	1.805
149.18	1.759	301.05	1.839
200.54	1.787	450.05	1.866
301.05	1.822	601.90	1.885
453.70	1.858	800.89	1.901
802.62	1.895	996.04	1.913
1002.32	1.908	1476.82	1.931
1505.66	1.925	1971.31	1.942
2008.48	1.932	2476.06	1.949
2533.65	1.938	2893.91	1.952
3517.41	1.938	4008.00	1.957
4988.40	1.940	4593.43	1.958

1.2 Magnetic Properties Measured Under Non-standard Conditions Based on Laminated Core Models

The specific total losses are measured under harmonic excitations ($\phi_3 = \phi_5 = 0°$, different content) based on Model C70, as shown in Table 11.15.

Table 11.16 shows the specific total losses under harmonic excitations ($\phi_3 = \phi_5 = 180°$, different content) based on Model C70.

The specific total losses measured ($K_3 = 10\%$, different phase angles) based on Model C70, are shown in Table 11.17.

The specific losses measured ($K_5 = 10\%$, different phase angle) based on Model C70, are shown in Table 11.18.

The specific total loss under different harmonic parameters (without harmonic, with different harmonic order, contents, and phase angle) measured based on Model C50, are shown in Tables 11.19 and 11.20.

The specific total loss at the middle section (200 mm in length) of Model C70 has been determined by using the two core model schemes, which can be referred to as the specific total loss of Model (C70–C50), as shown in Table 11.21.

Table 11.15 Specific total losses (Model C70, $\phi_3 = \phi_5 = 0°$)

B_m (T)	P_{loss} (W/kg)					
	Sin (50 Hz)	3rd 10%	3rd 20%	(3rd + 5th) 10%	5th 10%	5th 20%
0.1	0.005	0.006	0.006	0.005	0.005	0.005
0.2	0.017	0.021	0.024	0.020	0.017	0.019
0.3	0.037	0.045	0.051	0.042	0.036	0.041
0.4	0.063	0.076	0.087	0.072	0.061	0.070
0.5	0.094	0.114	0.131	0.108	0.092	0.106
0.6	0.131	0.158	0.182	0.150	0.129	0.148
0.7	0.174	0.209	0.239	0.198	0.171	0.196
0.8	0.222	0.265	0.304	0.252	0.218	0.251
0.9	0.275	0.327	0.377	0.312	0.271	0.313
1.0	0.335	0.396	0.459	0.379	0.330	0.383
1.1	0.400	0.472	0.548	0.454	0.396	0.460
1.2	0.471	0.555	0.647	0.536	0.468	0.547
1.3	0.552	0.648	0.760	0.628	0.551	0.646
1.4	0.637	0.746	0.878	0.726	0.639	0.756
1.5	0.729	0.852	1.000	0.831	0.735	0.872
1.6	0.833	0.970		0.944	0.843	0.999
1.7	0.963	1.111		1.067	0.978	1.131
1.8	1.165					
1.9	1.682					

Table 11.16 Specific total losses (Model C70, $\phi_3 = \phi_5 = 180°$)

B_m (T)	P_{loss} (W/kg)					
	Sin (50 Hz)	3rd 10%	3rd 20%	(3rd + 5th) 10%	5th 10%	5th 20%
0.1	0.005	0.004	0.004	0.005	0.005	0.006
0.2	0.017	0.016	0.015	0.019	0.020	0.023
0.3	0.037	0.034	0.034	0.042	0.042	0.050
0.4	0.063	0.058	0.058	0.071	0.072	0.085
0.5	0.094	0.087	0.087	0.107	0.109	0.129
0.6	0.131	0.121	0.121	0.149	0.151	0.180
0.7	0.174	0.161	0.161	0.198	0.201	0.239
0.8	0.222	0.205	0.206	0.253	0.256	0.307
0.9	0.275	0.256	0.257	0.314	0.319	0.384
1.0	0.335	0.313	0.315	0.383	0.389	0.471
1.1	0.400	0.376	0.378	0.460	0.467	0.570
1.2	0.471	0.446	0.450	0.546	0.554	0.683
1.3	0.552	0.526	0.531	0.640	0.652	0.809

(continued)

Table 11.16 (continued)

B_m (T)	P_{loss} (W/kg)					
	Sin (50 Hz)	3rd 10%	3rd 20%	(3rd + 5th) 10%	5th 10%	5th 20%
1.4	0.637	0.610	0.619	0.742	0.756	0.947
1.5	0.729	0.702	0.712	0.851	0.871	1.092
1.6	0.833	0.805	0.823	0.973	1.000	
1.7	0.963	0.935	0.945	1.123	1.163	
1.8	1.165	1.176	1.185			
1.9	1.682	1.768	1.847			

Table 11.17 Specific total losses (Model C70, $K_3 = 10\%$, different phase angle)

B_m (T)	P_{loss} (W/kg)					
	Sin (50 Hz)	0°	45°	90°	135°	180°
0.1	0.005	0.006	0.005	0.005	0.004	0.004
0.2	0.017	0.021	0.019	0.017	0.016	0.016
0.3	0.037	0.045	0.041	0.036	0.034	0.034
0.4	0.063	0.076	0.070	0.062	0.058	0.058
0.5	0.094	0.114	0.105	0.094	0.088	0.087
0.6	0.131	0.158	0.146	0.131	0.123	0.121
0.7	0.174	0.209	0.193	0.174	0.163	0.161
0.8	0.222	0.265	0.245	0.223	0.210	0.205
0.9	0.275	0.327	0.304	0.277	0.261	0.256
1.0	0.335	0.396	0.369	0.337	0.319	0.313
1.1	0.400	0.472	0.439	0.403	0.383	0.376
1.2	0.471	0.555	0.516	0.477	0.454	0.446
1.3	0.552	0.648	0.602	0.557	0.532	0.526
1.4	0.637	0.746	0.692	0.643	0.617	0.610
1.5	0.729	0.852	0.788	0.735	0.706	0.702
1.6	0.833	0.970	0.893	0.836	0.811	0.805
1.7	0.963	1.111	1.021	0.960	0.942	0.935
1.8	1.165		1.206	1.186	1.180	1.167
1.9	1.682					1.768

The specific total loss has been measured based on the two laminated core models, Model C70 and Model C50, under different harmonic parameters. The following Tables 11.22, 11.23, 11.24, 11.25, 11.26, and 11.27 show the comparisons among specific total loss values determined by means of Model C70, Model C50, and Model (C70–C50) under the same exciting condition. As mentioned above, the specific total loss of Model (C70–C50) is easily determined based on the results of Model C70 and Model C50. The specific total loss measured by different core models and by Epstein frame under sinusoidal excitation are shown in Table 11.28.

Table 11.18 Specific total losses (Model C70, K_5 = 10%, different phase angle)

B_m (T)	P_{loss} (W/kg)					
	Sin (50 Hz)	0°	45°	90°	135°	180°
0.1	0.005	0.005	0.005	0.005	0.005	0.005
0.2	0.017	0.017	0.017	0.017	0.018	0.020
0.3	0.037	0.036	0.036	0.037	0.039	0.042
0.4	0.063	0.061	0.061	0.064	0.067	0.072
0.5	0.094	0.092	0.093	0.096	0.101	0.109
0.6	0.131	0.129	0.129	0.134	0.140	0.151
0.7	0.174	0.171	0.171	0.177	0.186	0.201
0.8	0.222	0.218	0.219	0.226	0.237	0.256
0.9	0.275	0.271	0.272	0.281	0.295	0.319
1.0	0.335	0.330	0.332	0.342	0.360	0.389
1.1	0.400	0.396	0.398	0.411	0.431	0.467
1.2	0.471	0.468	0.472	0.487	0.511	0.554
1.3	0.552	0.551	0.557	0.573	0.602	0.652
1.4	0.637	0.639	0.646	0.666	0.700	0.756
1.5	0.729	0.735	0.745	0.767	0.806	0.871
1.6	0.833	0.843	0.857	0.875	0.926	1.000
1.7	0.963	0.978	0.994	1.022	1.081	1.163
1.8	1.165	1.206	1.268	1.336	1.341	
1.9	1.682					

Table 11.19 Specific total loss with different harmonic parameters (Model C50, $\phi_3 = \phi_5 = 0°$)

B_m (T)	P_{loss} (W/kg)					
	Sin (50 Hz)	3rd 10%	3rd 20%	(3rd + 5th) 10%	5th 10%	5th 20%
0.1	0.005	0.005	0.006	0.005	0.004	0.005
0.2	0.017	0.020	0.024	0.019	0.016	0.019
0.3	0.037	0.044	0.051	0.041	0.036	0.041
0.4	0.063	0.075	0.087	0.071	0.061	0.069
0.5	0.094	0.112	0.130	0.106	0.092	0.105
0.6	0.132	0.156	0.181	0.148	0.128	0.147
0.7	0.174	0.206	0.239	0.196	0.170	0.195
0.8	0.223	0.262	0.304	0.249	0.217	0.249
0.9	0.276	0.325	0.378	0.310	0.271	0.311
1.0	0.336	0.395	0.461	0.377	0.330	0.381
1.1	0.403	0.472	0.552	0.453	0.397	0.460
1.2	0.478	0.559	0.654	0.538	0.473	0.549
1.3	0.563	0.656	0.772	0.634	0.560	0.652
1.4	0.654	0.760	0.900	0.737	0.654	0.768
1.5	0.750	0.871	1.032	0.847	0.755	0.889
1.6	0.860	0.995		0.961	0.868	1.026
1.7	0.998	1.145		1.097	1.009	1.171
1.8	1.248				1.276	
1.9	1.814					

Table 11.20 Specific total loss with different harmonic parameters (Model C50, $\phi_3 = \phi_5 = 180°$)

B_m (T)	P_{loss} (W/kg)					
	Sin (50 Hz)	3rd 10%	3rd 20%	(3rd + 5th) 10%	5th 10%	5th 20%
0.1	0.005	0.004	0.004	0.005	0.005	0.006
0.2	0.017	0.015	0.016	0.019	0.019	0.023
0.3	0.037	0.033	0.034	0.041	0.042	0.049
0.4	0.063	0.057	0.057	0.070	0.071	0.085
0.5	0.094	0.085	0.087	0.105	0.107	0.128
0.6	0.132	0.119	0.121	0.147	0.150	0.179
0.7	0.174	0.158	0.161	0.195	0.199	0.238
0.8	0.223	0.202	0.207	0.250	0.255	0.306
0.9	0.276	0.253	0.258	0.311	0.318	0.383
1.0	0.336	0.309	0.316	0.380	0.388	0.471
1.1	0.403	0.373	0.381	0.458	0.468	0.570
1.2	0.478	0.447	0.455	0.546	0.557	0.688
1.3	0.563	0.530	0.542	0.645	0.660	0.819
1.4	0.654	0.621	0.635	0.752	0.771	0.968
1.5	0.750	0.719	0.733	0.869	0.892	1.116
1.6	0.860	0.829	0.849	0.998	1.030	
1.7	0.998	0.969	0.981	1.160	1.205	
1.8	1.248	1.257	1.297			
1.9	1.814	1.778	1.883			

Table 11.21 Specific total loss at the middle section of Model C70

B_m (T)	P_{loss} (W/kg)					
	Sin (50 Hz)	3rd 10%	3rd 20%	(3rd + 5th) 10%	5th 10%	5th 20%
0.1	0.005	0.006	0.006	0.006	0.005	0.005
0.2	0.017	0.022	0.024	0.021	0.017	0.019
0.3	0.037	0.047	0.052	0.045	0.036	0.042
0.4	0.063	0.079	0.088	0.076	0.063	0.072
0.5	0.095	0.119	0.132	0.114	0.094	0.109
0.6	0.131	0.164	0.183	0.157	0.131	0.151
0.7	0.173	0.215	0.240	0.205	0.172	0.200
0.8	0.220	0.271	0.304	0.259	0.220	0.256
0.9	0.273	0.332	0.375	0.319	0.272	0.318
1.0	0.330	0.399	0.457	0.384	0.330	0.387
1.1	0.391	0.472	0.539	0.455	0.393	0.460
1.2	0.455	0.547	0.630	0.532	0.458	0.541

(continued)

Table 11.21 (continued)

B_m (T)	P_{loss} (W/kg)					
	Sin (50 Hz)	3rd 10%	3rd 20%	(3rd + 5th) 10%	5th 10%	5th 20%
1.3	0.523	0.628	0.730	0.613	0.528	0.630
1.4	0.596	0.713	0.825	0.698	0.601	0.726
1.5	0.675	0.806	0.920	0.791	0.682	0.827
1.6	0.765	0.908		0.899	0.779	0.933
1.7	0.874	1.026		0.993	0.899	1.006
1.8	0.958				1.029	1.031
1.9	1.353					

Table 11.22 Specific total loss (sin, 50 Hz)

B_m (T)	P_{loss} (W/kg)		
	Model (C70–C50)	Model C70	Model C50
0.1	0.005	0.005	0.005
0.2	0.017	0.017	0.017
0.3	0.037	0.037	0.037
0.4	0.063	0.063	0.063
0.5	0.095	0.094	0.094
0.6	0.131	0.131	0.132
0.7	0.173	0.174	0.174
0.8	0.220	0.222	0.223
0.9	0.273	0.275	0.276
1.0	0.330	0.335	0.336
1.1	0.391	0.400	0.403
1.2	0.455	0.471	0.478
1.3	0.523	0.552	0.563
1.4	0.596	0.637	0.654
1.5	0.675	0.729	0.750
1.6	0.765	0.833	0.860
1.7	0.874	0.963	0.998
1.8	0.958	1.165	1.248
1.9	1.353	1.682	1.814

The two core model schemes enable us to separate the exciting power over the different subregions, i.e., the exciting power under different excitations over the core joint subregions of Model C70, and the rest can be represented by that of Model (C70–C50), i.e., the middle section of each limb, as shown in Tables 11.29, 11.30, 11.31, 11.32, 11.33, and 11.34.

Table 11.35 shows the effect of different harmonic excitations on the exciting power of Model C70, e.g., fundamental only, or fundamental with 10% of 3rd harmonic, or with 10% of (3rd + 5th) harmonic.

Table 11.23 Specific total loss (K_3 = 10%, ϕ_3 = 0°)

B_m (T)	P_{loss} (W/kg)		
	Model (C70–C50)	Model C70	Model C50
0.1	0.006	0.006	0.005
0.2	0.022	0.021	0.020
0.3	0.047	0.045	0.044
0.4	0.079	0.076	0.075
0.5	0.119	0.114	0.112
0.6	0.164	0.158	0.156
0.7	0.215	0.209	0.206
0.8	0.271	0.265	0.262
0.9	0.332	0.327	0.325
1.0	0.399	0.396	0.395
1.1	0.472	0.472	0.472
1.2	0.547	0.555	0.559
1.3	0.628	0.648	0.656
1.4	0.713	0.746	0.760
1.5	0.806	0.852	0.871
1.6	0.908	0.970	0.995
1.7	1.026	1.111	1.145

Table 11.24 Specific total loss (K_3 = 20%, ϕ_3 = 0°)

B_m (T)	P_{loss} (W/kg)		
	Model (C70–C50)	Model C70	Model C50
0.1	0.006	0.006	0.006
0.2	0.024	0.024	0.024
0.3	0.052	0.051	0.051
0.4	0.088	0.087	0.087
0.5	0.132	0.131	0.130
0.6	0.183	0.182	0.181
0.7	0.240	0.239	0.239
0.8	0.304	0.304	0.304
0.9	0.375	0.377	0.378
1.0	0.457	0.459	0.461
1.1	0.539	0.548	0.552
1.2	0.630	0.647	0.654
1.3	0.730	0.760	0.772
1.4	0.825	0.878	0.900
1.5	0.920	1.000	1.032

Table 11.25 Specific total loss ($K_3 = K_5 = 10\%$, $\phi_3 = \phi_5 = 0°$)

B_m (T)	P_{loss} (W/kg)		
	Model (C70–C50)	Model C70	Model C50
0.1	0.006	0.005	0.005
0.2	0.021	0.020	0.019
0.3	0.045	0.042	0.041
0.4	0.076	0.072	0.071
0.5	0.114	0.108	0.106
0.6	0.157	0.150	0.148
0.7	0.205	0.198	0.196
0.8	0.259	0.252	0.249
0.9	0.319	0.312	0.310
1.0	0.384	0.379	0.377
1.1	0.455	0.454	0.453
1.2	0.532	0.536	0.538
1.3	0.613	0.628	0.634
1.4	0.698	0.726	0.737
1.5	0.791	0.831	0.847
1.6	0.899	0.944	0.961
1.7	0.993	1.067	1.097

Table 11.26 Specific total loss ($K_5 = 10\%$, $\phi_5 = 0°$)

B_m (T)	P_{loss} (W/kg)		
	Model (C70–C50)	Model C70	Model C50
0.1	0.005	0.005	0.004
0.2	0.017	0.017	0.016
0.3	0.036	0.036	0.036
0.4	0.063	0.061	0.061
0.5	0.094	0.092	0.092
0.6	0.131	0.129	0.128
0.7	0.172	0.171	0.170
0.8	0.220	0.218	0.217
0.9	0.272	0.271	0.271
1.0	0.330	0.330	0.330
1.1	0.393	0.396	0.397
1.2	0.458	0.468	0.473
1.3	0.528	0.551	0.560
1.4	0.601	0.639	0.654
1.5	0.682	0.735	0.755
1.6	0.779	0.843	0.868
1.7	0.899	0.978	1.009
1.8	1.029		1.276

Table 11.27 Specific total loss (K_5 = 20%, ϕ_5 = 0°)

B_m (T)	P_{loss} (W/kg)		
	Model (C70–C50)	Model C70	Model C50
0.1	0.005	0.005	0.005
0.2	0.019	0.019	0.019
0.3	0.042	0.041	0.041
0.4	0.072	0.070	0.069
0.5	0.109	0.106	0.105
0.6	0.151	0.148	0.147
0.7	0.200	0.196	0.195
0.8	0.256	0.251	0.249
0.9	0.318	0.313	0.311
1.0	0.387	0.383	0.381
1.1	0.460	0.460	0.460
1.2	0.541	0.547	0.549
1.3	0.630	0.646	0.652
1.4	0.726	0.756	0.768
1.5	0.827	0.872	0.889
1.6	0.933	0.999	1.026
1.7	1.006	1.131	1.171
1.8	1.031		

Table 11.28 Specific total loss by core models and Epstein frame

B_m (T)	P_{loss} (Laminated core models) (W/kg)			P_{loss} (Epstein frame) (W/kg)
	Model (C70–C50)	Model C70	Model C50	
0.1	0.005	0.005	0.005	0.0043
0.2	0.017	0.017	0.017	0.0156
0.3	0.037	0.037	0.037	0.0338
0.4	0.063	0.063	0.063	0.0580
0.5	0.095	0.094	0.094	0.0878
0.6	0.131	0.131	0.132	0.1233
0.7	0.173	0.174	0.174	0.1643
0.8	0.220	0.222	0.223	0.2111
0.9	0.273	0.275	0.276	0.2635
1.0	0.330	0.335	0.336	0.3219
1.1	0.391	0.400	0.403	0.3862
1.2	0.455	0.471	0.478	0.4571
1.3	0.523	0.552	0.563	0.5341
1.4	0.596	0.637	0.654	0.6184
1.5	0.675	0.729	0.750	0.7126
1.6	0.765	0.833	0.860	0.8242
1.7	0.874	0.963	0.998	0.9690
1.8	0.958	1.165	1.248	1.2249
1.9	1.353	1.682	1.814	1.7422

Table 11.29 Exciting power (sin, 50 Hz)

B_m (T)	Exciting power (VA/kg)		
	Model C70	Model C50	Model (C70–C50)
0.1	0.008	0.008	0.008
0.2	0.027	0.027	0.027
0.3	0.055	0.055	0.055
0.4	0.091	0.091	0.091
0.5	0.132	0.132	0.133
0.6	0.179	0.179	0.179
0.7	0.231	0.231	0.230
0.8	0.288	0.290	0.283
0.9	0.351	0.355	0.341
1.0	0.421	0.428	0.403
1.1	0.502	0.514	0.471
1.2	0.597	0.619	0.541
1.3	0.743	0.777	0.656
1.4	1.048	1.088	0.948
1.5	1.598	1.657	1.452
1.6	2.426	2.515	2.203
1.7	3.662	3.805	3.302
1.8	5.638	5.967	4.813

Table 11.30 Exciting power (K_3 = 10%, ϕ_3 = 0°)

B_m (T)	Exciting power (VA/kg)		
	Model C70	Model C50	Model (C70–C50)
0.1	0.009	0.009	0.010
0.2	0.033	0.033	0.034
0.3	0.069	0.068	0.071
0.4	0.113	0.112	0.114
0.5	0.163	0.162	0.165
0.6	0.220	0.218	0.224
0.7	0.282	0.281	0.284
0.8	0.350	0.350	0.351
0.9	0.426	0.429	0.417
1.0	0.512	0.519	0.493
1.1	0.611	0.624	0.579
1.2	0.730	0.757	0.662
1.3	0.928	0.971	0.822
1.4	1.381	1.444	1.224
1.5	2.194	2.293	1.948
1.6	3.392	3.556	2.981
1.7	5.043	5.303	4.391

Table 11.31 Exciting power (K_3 = 10%, ϕ_3 = 180°)

B_m (T)	Exciting power (VA/kg)		
	Model C70	Model C50	Model (C70–C50)
0.1	0.007	0.007	0.007
0.2	0.024	0.024	0.025
0.3	0.050	0.049	0.052
0.4	0.082	0.080	0.086
0.5	0.120	0.117	0.126
0.6	0.163	0.160	0.170
0.7	0.211	0.208	0.218
0.8	0.264	0.262	0.269
0.9	0.323	0.323	0.323
1.0	0.389	0.392	0.383
1.1	0.465	0.472	0.446
1.2	0.552	0.570	0.509
1.3	0.684	0.715	0.605
1.4	0.937	0.974	0.842
1.5	1.393	1.459	1.228
1.6	2.075	2.160	1.862
1.7	3.079	3.244	2.665
1.8	4.892	5.224	4.062

Table 11.32 Exciting power (K_5 = 10%, ϕ_5 = 0°)

B_m (T)	Exciting power (VA/kg)		
	Model C70	Model C50	Model (C70–C50)
0.1	0.007	0.007	0.007
0.2	0.025	0.026	0.025
0.3	0.053	0.053	0.052
0.4	0.087	0.087	0.086
0.5	0.128	0.128	0.127
0.6	0.174	0.174	0.173
0.7	0.225	0.226	0.223
0.8	0.281	0.283	0.277
0.9	0.343	0.347	0.335
1.0	0.413	0.419	0.399
1.1	0.491	0.501	0.467
1.2	0.584	0.602	0.539
1.3	0.713	0.743	0.638
1.4	0.951	0.988	0.858
1.5	1.366	1.421	1.229
1.6	2.024	2.104	1.823
1.7	2.903	3.030	2.586
1.8	4.758	5.022	4.098

Table 11.33 Exciting power ($K_5 = 10\%$, $\phi_5 = 180°$)

B_m (T)	Exciting power (VA/kg)		
	Model C70	Model C50	Model (C70–C50)
0.1	0.009	0.009	0.009
0.2	0.031	0.031	0.032
0.3	0.064	0.063	0.066
0.4	0.106	0.104	0.109
0.5	0.154	0.152	0.158
0.6	0.208	0.206	0.214
0.7	0.269	0.267	0.273
0.8	0.335	0.335	0.337
0.9	0.409	0.411	0.403
1.0	0.492	0.498	0.478
1.1	0.588	0.598	0.562
1.2	0.702	0.722	0.651
1.3	0.865	0.901	0.773
1.4	1.222	1.272	1.097
1.5	1.903	1.983	1.701
1.6	2.931	3.071	2.582
1.7	4.385	4.588	3.879

Table 11.34 Exciting power ($K_3 = K_5 = 10\%$, $\phi_3 = \phi_5 = 0°$)

B_m (T)	Exciting power (VA/kg)		
	Model C70	Model C50	Model (C70–C50)
0.1	0.009	0.009	0.009
0.2	0.032	0.031	0.034
0.3	0.065	0.064	0.069
0.4	0.107	0.105	0.112
0.5	0.156	0.153	0.163
0.6	0.210	0.207	0.219
0.7	0.271	0.267	0.280
0.8	0.336	0.333	0.344
0.9	0.409	0.408	0.412
1.0	0.490	0.491	0.485
1.1	0.581	0.588	0.564
1.2	0.690	0.707	0.646
1.3	0.839	0.872	0.758
1.4	1.123	1.168	1.011
1.5	1.626	1.697	1.448
1.6	2.389	2.484	2.151
1.7	3.411	3.599	2.942

Table 11.35 Exciting power (Model C70, fundamental or with harmonic)

B_m (T)	Exciting power (Model C70) (VA/kg)		
	sin (50 Hz)	K_3 = 10%	$K_3 = K_5$ = 10%
0.1	0.008	0.009	0.009
0.2	0.027	0.033	0.032
0.3	0.055	0.069	0.065
0.4	0.091	0.113	0.107
0.5	0.132	0.163	0.156
0.6	0.179	0.220	0.210
0.7	0.231	0.282	0.271
0.8	0.288	0.350	0.336
0.9	0.351	0.426	0.409
1.0	0.421	0.512	0.490
1.1	0.502	0.611	0.581
1.2	0.597	0.730	0.690
1.3	0.743	0.928	0.839
1.4	1.048	1.381	1.123
1.5	1.598	2.194	1.626
1.6	2.426	3.392	2.389
1.7	3.662	5.043	3.411
1.8	5.638	/	/

References

1. IEC 60404–2 Ed.3.1(2008): Magnetic Materials - Part 2: Methods of measurement of the magnetic properties of electrical steel sheet and strip by means of an Epstein frame; Amendment 1.
2. IEC 60404-3Ed.2.2 (2010): Magnetic Materials–Part 3: Methods of measurement of the magnetic properties of electrical steel strip and sheet by means of a single sheet tester-Edition2.2; Consolidated Reprint.
3. A. J. Moses,"Relevance of microstructure and texture to the accuracy and interpretation of 1 and 2 directional characterisation and testing of grain-oriented electrical steels," *International Journal of Applied Electromagnetics and Mechanics,* **55**(2017), 3–13.
4. E. Barbisio, F. Fiorillo, and C. Ragusa, "Predicting Loss in Magnetic Steels Under Arbitrary Induction Waveform and With Minor Hysteresis Loops," *IEEE Trans Magn.*, **40**(2004), 1810–1819.
5. Z. Cheng, B. Forghani, X. Wang, L. Liu, T. Liu, Y. Fan, J. Zhang, X. Zhao, and Y. Liu, "Engineering-oriented investigation of magnetic property modeling and application," *International Journal of Applied Electromagnetics and Mechanics,* **55**(2017), 147–158.
6. A. J. Moses, "Characterisation and performance of electrical steels for power transformers operating under extremes of magnetisation conditions," Presented at the International Colloquium Transformer Research and Asset Management, Cavtat, Croatia, Nov. 12–14, 2009.
7. IEC TR 62383-2006 Determination of magnetic loss under magnetic polarization waveforms including higher harmonic components. Measurement, modeling and calculation methods.

8. T. Liu, L. Liu, F. Che, Z. Du, B. Forghani, Y. Li, Y. Liu, X. Zhao, and Z. Cheng, "Modeling of magnetic component behaviors under distorted excitations based on an integrated magnetic measure-bench," Presented at 1&2DM 2018, France.
9. Z. Cheng, N. Takahashi, B. Forghani, A. J. Moses, P. Anderson, Y. Fan, T. Liu, X. Wang, Z. Zhao, and L. Liu, "Modeling of magnetic properties of GO electrical steel based on Epstein combination and loss data weighted processing," *IEEE Trans Magn.,* **50** (2014), 6300209.
10. D. Miyagi, T. Yoshida, M. Nakano, and N. Takahashi, "Development of Measuring Equipment of DC-Biased Magnetic Properties Using Open-Type Single-Sheet Tester", *IEEE Trans. Magn.*, Vol. 42, No.10, pp. 2846–2848, 2006.
11. J. Sievert, "New perceptions of standardized measurement methods for high permeability electrical steel", Invited lecture at the Joint Seminar on Magnetics and Magnetic Measurement, 2016.09.22, Baobian Electric.
12. A. J. Moses, "Grain-oriented electrical steels domain structures, losses, magnetostriction and applications in transformer cores", invited lecture at the Joint Seminar on Magnetics and Magnetic Measurement, 2016.09.22, Baobian Electric.
13. P. Anderson, "Electrical steels research at Cardiff University", invited lecture at the Joint Seminar on Magnetics and Magnetic Measurement, 2016.09.22, Baobian Electric.

Part IV
Validation Based on a Well-Established Benchmarking System

Chapter 12
Establishment and Development of Benchmark Family (P21)

Zhiguang Cheng, Norio Takahashi, Behzad Forghani and Lanrong Liu

Abstract Testing Electromagnetic Analysis Methods (TEAM) Problem 21 (P21) was proposed by the author in November 1993 based on the industrial background of modeling stray-field loss in electrical equipment. Now, it has been extended to a TEAM Benchmark Family, including 5 sets of 16 member models, approved by the International Compumag Society (ICS). In the modeling and simulation of engineering electromagnetic field problems, it is necessary to properly deal with the nonlinear, anisotropic, and hysteretic behaviors of materials and components, multiply connected regions, multi-scale configurations, skin effect, and deep magnetic saturation behavior, etc., under standard (e.g., the excitation with sinusoidal waveform) and non-standard excitation conditions (e.g., when the excitation waveform contains higher harmonics and/or DC components). The confident validation of the modeling and simulation based on well-established benchmark models is certainly essential and challenging. This chapter briefly outlines the Problem 21-related benchmarking activities, the basic characteristics and the upgrade of the Problem 21 Family, the research on measurement and analysis method, and the model-based results. The significance of extending the scope of TEAM, establishing more realistic and challenging benchmark models, in combination with advanced material modeling, is emphasized. Further co-research projects related to Problem 21 Family are also proposed.

Keywords Benchmarking · Testing Electromagnetic Analysis Methods (TEAM) · Problem 21 Family · Leakage magnetic flux · Stray-field loss · Finite element (FE) · Grain-oriented (GO) silicon steel · Lamination · Magnetic steel plate · Magnetic property modeling

---

Z. Cheng (✉) · L. Liu
Institute of Power Transmission and Transformation Technology, Baobian Electric, Baoding, China
e-mail: emlabzcheng@yahoo.com

N. Takahashi (deceased)
Okayama University, Okayama, Japan

B. Forghani
Mentor Infolytica, a Siemens Business, Montreal, QC, Canada

Z. Cheng et al. (eds.), *Modeling and Application of Electromagnetic and Thermal Field in Electrical Engineering*, https://doi.org/10.1007/978-981-15-0173-9_12

12.1 Introduction

The original idea of the Testing Electromagnetic Analysis Methods (TEAM) came into being in March 1985 and was proposed by Sam Berk (Office of Fusion Energy, USA). During the Compumag Colorado conference in June 1985, consensus was reached on organizing a workshop aimed at testing electromagnetic analysis methods and software, and then, a planning meeting was held in November 1985 (ANL, USA). After a year of preparation, the first workshop was hosted by C.R.I. Emson at the Rutherford Appleton Laboratory, UK, on March 27, 1986. The early workshop was called "Eddy Current Workshop" to meet the need to validate eddy current analysis software, while the name "TEAM" was officially adopted in April 1987 [1–4].

Activities related to TEAM are conducted under the organization and leadership of the TEAM Board, which has approved a series of benchmark problems as models for testing numerical methods and software. The benchmark problem should have clear electromagnetic characteristics and typical scientific and industrial background. Although there is a certain degree of difficulty, the unknowns specified in the definition can be calculated and measured. The proponent must give a strict definition of the benchmark problem and publish the analysis and measured results of the problem. Researchers from all over the world may publish their results in the TEAM Workshop scheduled by the TEAM Board.

Different research groups have solved the same benchmark problem using different numerical methods and element types (e.g., node-based elements, edge-based elements, facet elements, volume elements, element-free) and have tested and compared the effectiveness and usefulness of their method, making continuous improvement; this activity has played an important role in promoting the development of computational electromagnetic (CEM) and industrial application technology. Practice indicates that testing analysis method on the benchmark model is the right way; otherwise, the purpose might not be reached if testing analysis methods and software are based on large and complex device, because it may not be clear whether the error is caused by unreasonably simplified calculation model or by the method itself. However, the usefulness and validity of software must be validated through physical models. For this purpose, until now, a series of benchmark problems (models) have been approved by the International Compumag Society (ICS) Board.

From 1986 to 2002, in addition to the TEAM Workshop following the Compumag conference, a series of regional TEAM Workshops have been held around the world with the approval of the TEAM Board. As an innovation in the field of international computational electromagnetics, these workshops have given a powerful impetus to the development and progress of international computational electromagnetic and industrial applications.

TEAM Workshops have been held several times in China. The first TEAM Workshop in China was held in Qiandao Lake, Zhejiang, in October 1992, sponsored by Zhejiang University [5]. The second TEAM Workshop was hosted by

Huazhong University of Technology in Yichang, Hubei, in 1996 [6]. The third TEAM Workshop was hosted by Hebei University of Technology in Chengde, Hebei, in 2000 [7]. These TEAM Workshops were all held in other places, immediately following the International Conference of Electromagnetic Field Problems and Applications (ICEF) organized in China. The first three TEAM Workshops held in China were chaired by Zhiguang Cheng, one of the authors of this chapter. At the TEAM Workshop held in Chengde, participants worldwide discussed the now and future of the TEAM with the topic "TEAM, the way ahead!" and put forward important suggestions [8]. TEAM activities in China have been supported by the founders, advocates, and active participants of the international TEAM, including overseas scholars C. W. Trowbridge, A. Kost, L. R. Turner, D. Lowther, T. Nakata, O. Biro, and domestic scholars Zhou Keding, Sheng Jianni, Tang Renyuan, Yan Weili, Ni Guangzheng, Fan Mingwu, and other predecessors in computational electromagnetics.

The International Compumag Society decided, starting in 2003, to incorporate the separate TEAM Workshop into the Compumag Conference program, as a topic of benchmarking (TEAM), based on the TEAM Workshop activities. Shenyang University of Technology held the Compumag Conference in Shenyang in June 2005. It was the first time that Compumag was held in China. An introduction to the early work of computational electromagnetics in China, its current direction and future, was presented by Xie Dexin and Tang Renyuan at the Compumag Shenyang in 2005 [9]. Compumag Shengyang was highly praised by colleagues, as an event attracting worldwide attention in the history of the CEM. During the conference, as a special arrangement, the TEAM Workshop was held again and co-chaired by Oszkar Biro of Graz University of Technology and Zhiguang Cheng of Baobian Electric. Since then, several benchmarking (TEAM) sessions were also set up in China, including the session held in the ICEF-2008 organized by Chongqing University [10].

The development and contribution of benchmarking (TEAM) have been widely recognized in the field of computational electromagnetics since the inception of TEAM. The authors firmly believe that any of the available electromagnetic field computation methods needs to be validated, and any worthwhile commercial software needs to pass repeated tests based on various typical models. It is inconceivable to carry out electromagnetic design without any proven analysis software. Nonetheless, the electromagnetic analysis software that cannot solve the benchmark problems approved by the ICS will not be trusted to solve complex engineering problems. A further problem is that even if the computation methods and software have been tested on benchmark model, it does not mean that they can be used blindly in product design. Therefore, the TEAM models in the future should be more engineering science-oriented, closer to the electrical engineering or great scientific engineering practices, more challenging, and focused more on testing the ability of modeling technology to solve complex problems. Of course, it is not easy to establish such benchmark problems.

Admittedly, it cannot be said that there is no product design without TEAM. Look at the power transformer engineering, e.g., it took more than 130 years from the world's first closed core type transformer (September 1884) to the first TEAM Workshop (March 1986), and the transformer's single-unit capacity, voltage level, variety of transformer products, total installed capacity, and so on have been greatly developed. Apparently, not all electromagnetic design calculation methods have been validated by the "TEAM" model defined many years ago to be applied in design and manufacture of products. The analysis and test of the 3D forced hot oil flow in the internal heating and cooling of transformers, the electromagnetic and thermal coupling in products, the vibration and noise in transformers, e.g., cannot be said to have been completely solved. In fact, the product design-oriented calculation method has been repeatedly tested and validated in the process of design, manufacture, test, operation, and maintenance for a long time in such a manner that it may be counted as a "TEAM" in the sense of practicing engineering, but the validation cost is really high and it is difficult to come to a strict and clear conclusion. It is obviously almost impossible to detect the effectiveness of electromagnetic analysis methods based on huge and complex electromagnetic devices.

It should be pointed out that solving a complex problem requires not only advanced and efficient analysis methods, but also the support of complete material property data. Therefore, the effectiveness of modeling and simulation depends on the analysis method and material property data used. This means that the international TEAM activities and the scope of TEAM need to be extended, continuously.

12.2 Development of TEAM Problem 21

12.2.1 Modeling and Prediction of Stray-Field Loss

Stray-field loss in electrical engineering has been a classic complex problem both for experimental research and numerical simulation. Taking the large power transformer as an example, stray-field loss is induced by the leakage electromagnetic field of the transformer in the conducting solid or laminated components. The modeling and prediction of the stray-field loss, hot-spot temperature rise, and shielding effect have become important topics in the R&D and design of power transformers. Although transformer theory, product design, and manufacture have a very long history, it is no exaggeration to say that the problem of 3D stray-field loss, under complex operating conditions, has not been completely solved, and it is still not easy to accurately solve the large-scale stray-field loss problems.

Stray-field loss is an important part of the total loss of the transformer, and various factors lead to excessive local density of stray-field loss, which may cause unallowable local overheating and endanger the safe operation of the transformers. For transformers with extra-large capacity, such as 1000 kV UHV AC transformers, the single-unit capacity is up to 1000 MVA, and also, the UHV DC transformer involves very complex operating conditions, which make the stray-field loss problem more and more complex and important, and any structural details that lead to increased stray-field loss cannot be ignored.

The total stray-field loss always includes various loss components, e.g., loss in exciting coils and loss in other conducting components, which cannot be measured directly, and is difficult to accurately separate from total stray-field losses.

On the other hand, the numerical modeling and prediction of local stray-field losses are actually inseparable from solving the whole 3D leakage electromagnetic field, which will inevitably lead to complications. Moreover, it is also a knotty problem to validate the effectiveness of the stray-field loss analysis under complex conditions.

12.2.2 Proposal and Updates to Problem 21

In 1993, the author proposed an engineering-oriented stray-field loss model, which was approved by the TEAM Board and ranked 21 among the previously accepted benchmark problems; hence, it is called Problem 21 [11].

Larry Turner, one of the international TEAM advocates, commented on Problem 21 as follows: "It has the properties desirable for a TEAM problem: computationally challenging but achievable, and also relevant to the design of transformers and other electrical machines." T. Nakata and his research team supported and collaborated in the establishment of Problem 21 and the long-term co-research that followed. At that time, Problem 21 was then known as the TEAM benchmark problem with high degree of difficulty [12].

For over more than 25 years of development, the definition of Problem 21 has been updated three times (1999, 2005, and 2009) since 1993; the scope has been expanded, and the number of benchmark models has increased from the original two models (Model A and Model B) to the Benchmark Family of 5 sets of 16 models: Problem 21^0/P21^a/ P21^b/P21^c/P21^d) [13–17]. The definition of the Problem 21 Benchmark Family provides the industrial background of the benchmark models, the description of the structure, material, and field quantities to be solved, as well as the partial measurement and calculation results of the magnetic field and stray-field loss of the benchmark model.

Scientists and engineers around the world have published numerous calculation results related to Problem 21 [18–48]. Even a "Problem 21 hot" emerged in the 1990s. Up to now, it is still of interest to the international computational electromagnetics and engineering circles and is continuously updated by the authors [49].

- In 1993, an engineering-oriented loss model proposed by the author was approved by the TEAM Board as TEAM Problem 21, at the International TEAM Workshop, Miami, USA [11].
- In 1996, a set of benchmark models with transformer core tie plate, as the direct engineering background, was proposed; the benchmark model and related results were published at TEAM Workshop—Yichang [13], China, 1996.

- In 1999, Problem 21 was revised for the first time and the first updated version of Problem 21 was released at International TEAM Workshop—Sapporo, Japan [14]. The *B–H* curve of A3 steel (used in Problem 21) was re-measured by the Electromagnetic Device Laboratory (Okayama University, Japan), and the relation between the specific hysteresis loss and magnetic flux density, i.e., $B_m - W_h$ curve, conductivity, and other material property data, was also accurately measured.
- In 2002, in order to investigate the different electromagnetic behaviors and loss distributions of different steel plate configurations, under the same excitation, three benchmark models were proposed, and the related results were presented at IEEE CEM 2002 [15], Bournemouth, UK.
- In 2005, four benchmark models for modeling electromagnetic and magnetic shielding were proposed and related results were published [16], presented at Benchmarking (TEAM) and International TEAM Workshop of Compumag 2005, Shenyang. The *B–H*, $B_m - W$, and $B_m - W_h$ curves, of the GO silicon steel sheet (30RGH120) used in Problem 21, were measured by the Electromagnetic Device Laboratory (Okayama University, Japan). The Problem 21 Benchmark Family composed of 13 models was proposed, as the second upgrade of Problem 21, which was approved by the ICS Board.
- In 2006, two models were proposed to model the loss and magnetic field inside the hybrid steel weldments (including ordinary A3 steel and non-magnetic steel 20Mn23Al); the models and related results were published at ACES—2006 [17], Miami, USA.
- In 2007, due to the fact that the ferromagnetic material in the Problem 21 model (exciting current 10 A, rms, 50 Hz) originally defined did not reach the desired magnetic saturation, the exciting currents were increased to model the loss and magnetic flux density in ordinary A3 steel under magnetic saturation or quasi-magnetic saturation conditions; The results were published at the Benchmarking(TEAM) of the Compumag 2007, held in Aachen, Germany.
- In 2008, based on the simplified magnetic shielding model in the Problem 21 Benchmark Family, i.e., the shielded ordinary steel plate is removed from the shielding model, the additional iron loss caused by leakage magnetic flux entering the silicon steel lamination vertically was investigated; simplified models and the related results were published at IEEE CEFC—2008, Athens, Greece. TEAM Problem 21 (V. 2009) was approved by the ICS Board.

The Problem 21 Benchmark Family (V.2009) is its latest version. Table 12.1 shows the characteristics and development of the Benchmark Family.

The Problem 21 Benchmark Family follows the core purpose of International TEAM and, like all TEAM Benchmark problems, can be used to test the effectiveness of electromagnetic analysis methods. In addition, Problem 21 has been extensively tested and continues to evolve in response to new engineering requirements. Benchmark models derived from different engineering backgrounds have been proposed one after another, and the excitation intensity has been increased so that ferromagnetic materials in the models can reach different

Table 12.1 Problem 21 Benchmark Family (V.2009)

Benchmark sets	Member models	Characteristics of field problems	Proposed at
$P21^0$	$P21^0$-A $P21^0$-B	3D nonlinear eddy current-hysteresis problems, including multiply connected domains, isotropic materials, and accounting for skin effect	TEAM, Miami, USA, **1993**
$P21^a$	$P21^a$-0 $P21^a$-1 $P21^a$-2 $P21^a$-3	3D linear eddy current problems, including multiply connected domains, isotropic materials	TEAM, Yichang, China, **1996**
$P21^b$	$P21^b$-MN $P21^b$-2 M $P21^b$-2 N	3D nonlinear eddy current–hysteresis problems, different (or same kind of) steel plates placed separately, isotropic materials, and accounting for skin effect	IEE CEM, Bournemouth, UK, **2002**
	$P21^b$-MNM $P21^b$-NMN	3D nonlinear eddy current-hysteresis problem, different steel plates welded as a whole, isotropic materials	ACES, Miami, USA, **2006**
$P21^c$	$P21^c$-M1 $P21^c$-M2 $P21^c$-EM1 $P21^c$-EM2	3D nonlinear eddy current–hysteresis problems, magnetic steel plates shielded by copperplates or by GO silicon steel lamination	Compumag Shenyang, China, **2005**
$P21^d$	$P21^d$-M	3D nonlinear eddy current–hysteresis problems, GO silicon steel lamination, and accounting for additional iron loss	IEEE CEFC-2008, Athens, Greece

saturation levels. Now, we can examine the deviation between the measured and calculated results of interlinkage flux inside the conductor, as well as, comparing the measured and calculated results of leakage magnetic flux at a specified position outside the conductor (in the air). The conducting (magnetic or non-magnetic) components used in benchmark models also changed from an isotropic "solid" plate structure to an anisotropic "laminated" structure. The Problem 21 Benchmark Family has the following basic characteristics and engineering science significance:

(1) Each set of member models in the Benchmark Family corresponds to a clear engineering background and is a "microcosm" of the typical power transformer product structure. It is mainly used to study the stray-field loss problems in electrical engineering in-depth and systematically and covers various control techniques over stray-field losses. Benchmarking based on Problem 21 has always been guided by the idea that models come from engineering and results serve aims of engineering.

(2) The possibility of comparative study between different member models within a set can be provided by each set (e.g., $P21^0$, $P21^a$, $P21^b$, and $P21^c$), which enables detailed investigation of the electromagnetic behavior of typical structures and material configurations. The regular results and conclusions obtained are valuable in engineering application.
(3) The numerical analysis and test of a plurality of benchmark models (which involves materials having different structures, electromagnetic properties, and different types of problems) are helpful to the R&D and test of efficient and feasible electromagnetic analysis technology and to the reasonable establishment of calculation and simulation models.
(4) The magnetic steel plate in $P21^c$-M1 is removed to become $P21^d$, which makes it possible to study, in detail, the magnetic field and loss distribution in the lamination and the additional iron loss caused by the leakage magnetic flux entering the lamination vertically.
(5) The excitations of all the current 16 benchmark models are sinusoidal current sources. It should be pointed out that because the magnetic characteristics under complex excitation conditions (e.g., including higher harmonic and/or DC components) in today's power system have created the need for modeling and simulation under corresponding conditions, the establishment of new benchmark models under non-sinusoidal excitation (including two types of typical magnetic components: solid magnetic steel plate and silicon steel lamination) has become a further project of the authors in TEAM research [50].

Note that all the development of Problem 21 Benchmark Family are certainly inseparable from the concern and support from international computational electromagnetics. In addition, the authors have carried out continuous and in-depth research [51–62] on Problem 21 Family for more than 25 years. Prof. J. Sykulski highly evaluates the contribution of Problem 21: "The TEAM Problem 21 is well-established and recognized worldwide and has been successfully used by the computational electromagnetic community both in academia and in industry" [63].

12.3 Definition of Problem 21 Benchmark Family

Scholars around the world have published the calculated results about the TEAM Problem 21 at TEAM Workshops and in publications after it was approved in 1993. It should be noted that, according to the preliminary calculated results, most of the calculated results of the magnetic flux density in the air at the specified position are consistent with each other, but the calculated losses inside the magnetic steel plate are different. Further loss analysis and experimental study show that the effect of hysteresis loss in magnetic steels must be adequately taken into account, and the definition of this problem was revised at the TEAM Workshop held in Sapporo, Japan, in 1999 [14]. Later, other TEAM benchmark models with more specific engineering background, and closer to typical engineering problems, have been put forward, including three sets of models: Problem 21^+ [13], modeling core tie plate of transformer; Problem 21^* [15, 17], modeling hybrid steel structure of transformer oil tank; and Problem $21'$ [16], modeling electromagnetic and magnetic shields of transformer. Problems 21^+, Problems 21^*, and Problems $21'$ were renamed as Problems 21^a, Problems 21^b, and Problem 21^c, respectively, in abbreviation referred to as $P21^a$, $P21^b$, and $P21^c$ in the updated definition, for convenience of expression, at the suggestion of O. Biro, as shown in Table 12.1. Problem 21 is named as a Benchmark Family because all "member models" share a common excitation source but with different electromagnetic and structural characteristics.

It should be pointed out that Problem 21 Benchmark Family is provided with a clear engineering background, and the 16 member models reflect the measures of reducing loss and avoiding hazardous local overheating in transformer engineering, such as slotting solid or laminated components, to reduce eddy currents ($P21^a$), reasonably designing hybrid steel structures ($P21^b$), adopting electromagnetic and magnetic shields ($P21^c$), which can effectively reduce power consumption and improve reliability. These are the crystallization of the wisdom of designers and manufacturers. However, to quantitatively analyze and predict the effectiveness of these measures, describe the electromagnetic behavior of typical structures, or provide further information needed for design, effective numerical modeling and simulation of electromagnetic fields are a requirement. As pointed out earlier, TEAM was given the purpose of examining the effectiveness of electromagnetic analysis methods. Meanwhile, the applicability of the analysis methods to be used for solving large-scale engineering electromagnetic filed problems is also emphasized below.

The TEAM benchmark problems need to be strictly defined and then approved by the ICS Board. Moveover, any changes to the definition of TEAM benchmark problems and any additions and updates to the published results related to the problems must be approved by the ICS Board.

Up to now, 34 benchmark problems have been approved by the ICS Board and posted on the ICS Web site, www.compumag.org/team.

12.3.1 Benchmark Models

1. **Problem 21^0(1993)**

TEAM Problem 21 Family is based on stray-field loss problems in power transformer engineering, and all member models take transformer components as industrial background, such as oil tank, core tie plate, shieldings. The V.1993 of the Problem 21 includes two member models: Model A and Model B, which are the two original models of the Problem 21 Benchmark Family, called $P21^0$-A and $P21^0$-B for short. $P21^0$-A consists of two exciting coils of the same specification and two magnetic steel plates with the same overall dimensions, however, one of which is provided with a rectangular hole in the center to form a multiply connected region. Current flows in opposite directions in the two exciting coils to strengthen the leakage magnetic flux into the model steel plate.

P21^0-A includes two steel plates so that the difference of eddy current and loss in the same material with and without holes and having the same external dimensions, under the same external field excitation condition, can be investigated. Note that, for numerical testing, the optimal algorithm can be implemented in different sub-regions, e.g., using $\boldsymbol{T} - \psi - \psi$ in the singly connected steel plate region of Model A and $\boldsymbol{A} - V - \boldsymbol{A}$ in the multiply connected steel plate region [23, 64, 65]. Thus, the advantages of different algorithms can be complemented. The author has completed the numerical implementation of the above combinatorial algorithm [66].

The air-core exciting coils are used in all the member models of Problem 21, which is to focus on solving the stray-field loss problems in the excited conductors without adding more variables, i.e., avoiding the 3D, grain-oriented, and nonlinear laminated core structure to simplify the concerned problems.

P21^0-B has the same exciting coils as model P21^0-A, but only a steel plate of the same material without hole. The structural design and photographs of the models are shown in Fig. 12.1.

2. **Problem 21^a(1996)**

The direct engineering background of Problem 21^a (hereinafter referred to as P21^a) is the structural design of transformer core tie plate. The typical laminated core and the core tie plate of large transformer are shown in Fig. 12.2. For the tie plate as a load-bearing component connected to the transformer core, quantitative analysis is required for the given transformer capacity, involving magnetic or non-magnetic materials to be selected, the structural size of the tie plate, the type and number of slots, or the means to separate narrow steel strips, as well as for eddy current distribution and loss, etc. It is obvious that the solution to local eddy current problems is related to the whole leakage electromagnetic field.

As a member of the Problem 21 Benchmark Family, P21^a has 4 member models, represented by P21^a-0, P21^a-1, P21^a-2, and P21^a-3, respectively. The numbers 0–3 in the defined symbols refer to the number of slots in non-magnetic steel plates. Each non-magnetic steel plate is excited by the same exciting coils, and the structural design and photographs of the models are shown in Fig. 12.3. The length, width, and height of the slot in the non-magnetic steel plate in the model are 660 mm, 10 mm, and 10 mm, respectively.

It should be noted that, for Model P21^a-3, there are three slots in the non-magnetic steel plate, and the distance between each two adjacent slots is equal in model design, as shown in Fig. 12.3(d-1); however, from the corresponding photograph, it can be seen that the distance between each two adjacent slots in the non-magnetic plate is not equal, as shown in Fig. 12.3(d-2), so the eddy current loss will be different.

Section 12.5.2 of this chapter shows the calculated loss results and eddy current distribution in the cases of equidistant and non-equidistant slots.

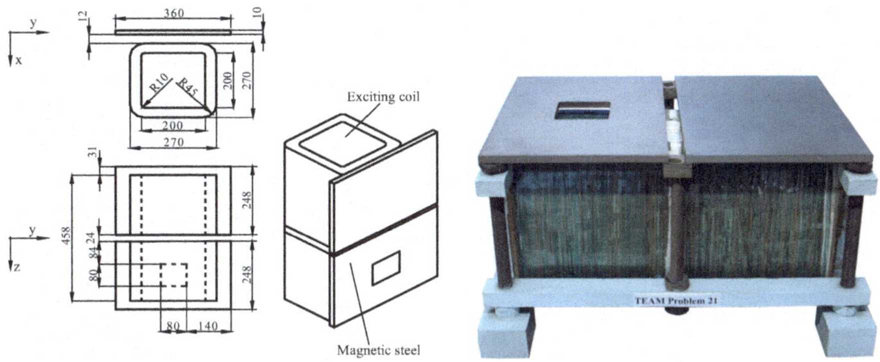

(a-1) (a-2)

(a) Model P21^0-A

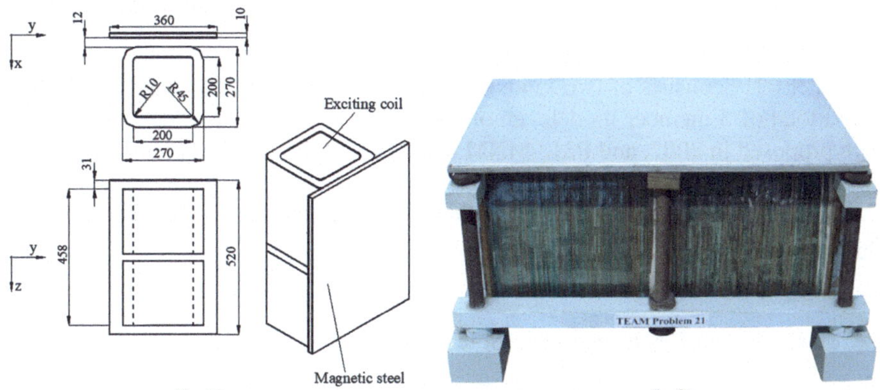

(b-1) (b-2)

(b) Model P21^0-B

Fig. 12.1 Problem P21^0

3. **Problem 21^b(2002, 2006)**

The direct engineering background of Problem 21^b (hereinafter referred to as P21^b) is modeling of the magnetic and non-magnetic hybrid steel structure in the transformer oil tank; that is, the transformer oil tank is basically made of ordinary A3 steel, except that non-magnetic steel is used in the vicinity where the heavy current leads pass to reduce eddy current losses and avoid possible concentration of loss density in the tank. Figure 12.4a shows the hybrid steel structure of the oil tank of the large transformer.

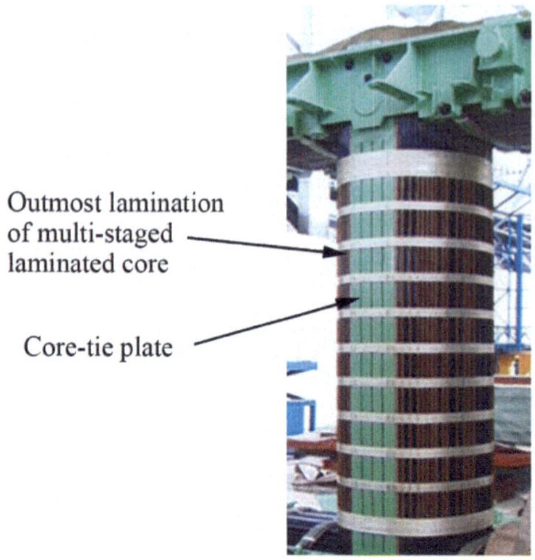

Fig. 12.2 Large transformer core. *Note* The laminated core includes multi-staged laminations and core tie plate

Model P21^{b} consists of two kinds of steel plates separated and welded together, with a total of 5 member models, among which P21^{b}-MN, P21^{b}-2M, and P21^{b}-2N were proposed in 2002, and P21^{b}-MNM and P21^{b}-NMN, 2006. "M" and "N" in the defined symbols represent magnetic and non-magnetic steel plates, respectively. The structural design and photograph of the 5 member models of P21^{b} are shown in Fig. 12.4b–e.

4. **Problem 21^{c}(2005)**

① Magnetic and Electromagnetic Shielding

In order to control stray-field loss, in addition to optimizing the structure of the winding, reasonably distributing the ampere-turns and selecting components' materials, one can also arrange electromagnetic shields (used to prevent magnetic flux from entering, called screening) and magnetic shields (used to add a magnetic shunt, called shunting) in the leakage electromagnetic field regions appropriately, so as to reduce stray-field losses in windings and other components, to avoid excessive local heating hazard, and to reduce short-circuit crushing forces in the windings. In this case, the path and distribution of the leakage magnetic flux are changed, so the stray-field loss becomes controllable.

The electromagnetic shielding of transformer has been used for many years. It is generally designed as a copperplate (e.g., with the thickness of 4–6 mm) or an aluminum plate (e.g., with the thickness of 7–8 mm). Magnetic shields, on the other hand, are safer from the point of view of preventing possible heating. However, the magnetic shield creates a magnetic shunt and is placed perpendicular or parallel to the tank wall. If it is not properly placed, or if the shielding structure is not properly sized, it may be counterproductive.

Nowadays, both the electromagnetic and magnetic shields are widely used in power transformers to optimize the distribution of leakage electromagnetic field and

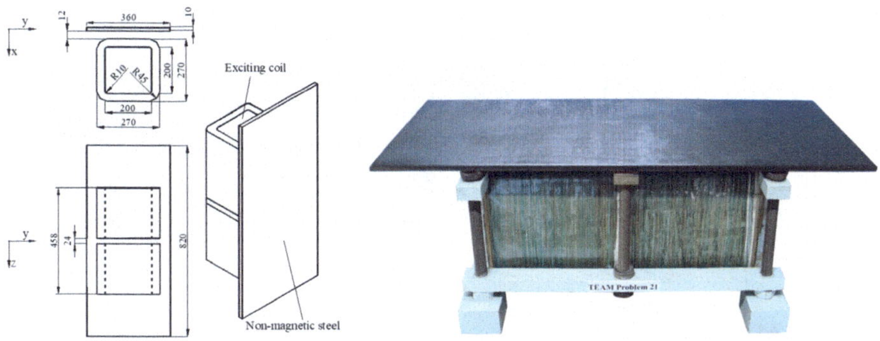

(a-1) (a-2)

(a) P21[a]-0

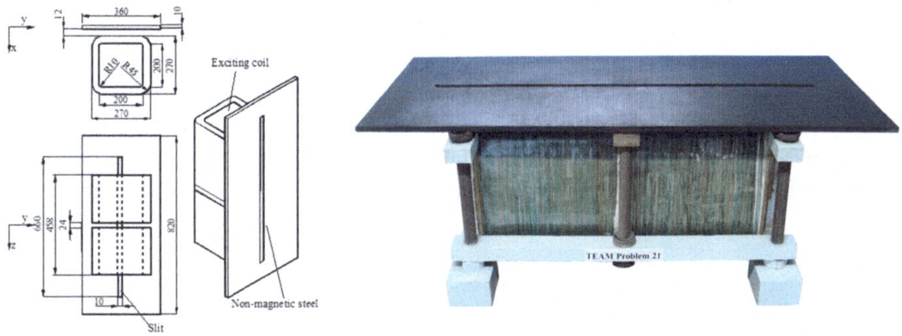

(b-1) (b-2)

(b) P21[a]-1

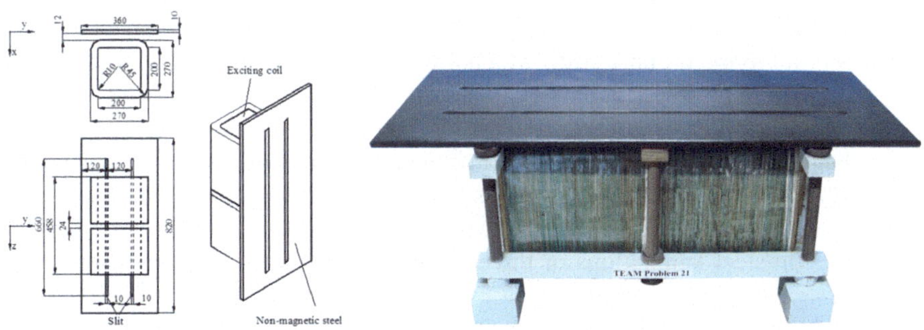

(c-1) (c-2)

(c) P21[a]-2

Fig. 12.3 Problem 21[a]

(d-1) (d-2)

(d) P21^a-3

Fig. 12.3 (continued)

to reduce losses. The design optimization of the shield structure, the correct selection of different kinds of shields, at different locations, or the different purpose for, and the accurate modeling and simulation of electromagnetic-thermal behavior of the shields are important research projects.

② Model P21^c

The direct engineering background of Problem 21^c (hereinafter referred to as P21^c) is the magnetic shields and the electromagnetic shields widely used in large transformers; the shields are made of anisotropic silicon steel sheets and copperplates, respectively. Figure 12.5a shows the magnetic shields installed on the sidewall of the oil tank of a large transformer. In fact, there are two ways to fix the magnetic shields on the oil tank: ① The wider silicon steel sheets are laminated and placed parallel to the oil tank wall, which are fixed at the designed positions of the tank wall; ② the narrow silicon steel sheets are laminated and vertically fixed on the tank wall in order to further reduce eddy current loss in shielding.

The benchmark shielding models of Problem 21^c can be divided into two types: Type 1 (P21^c-M1 or P21^c-EM1) includes only one set of silicon steel lamination or one whole copperplate; Type 2 (P21^c-M2 and P21^c-EM2) includes three sets of narrow silicon steel laminations or three narrow copperplates separated from each other.

Note that the laminated sheets in magnetic shields (P21^c-M1 and P21^c-M2) and the corresponding copperplates in the electromagnetic shields (P21^c-EM1 and P21^c-EM2) have the same external structural dimensions, respectively. See the structural design and photographs shown in Fig. 12.5b–d.

In addition, it should be noted that

(1) The silicon steel sheet material of the magnetic shielding model is anisotropic. However, the leakage electromagnetic field in the Model P21^c-M1/M2 is

(a) An example of hybrid steel structure in the oil tank of large transformer

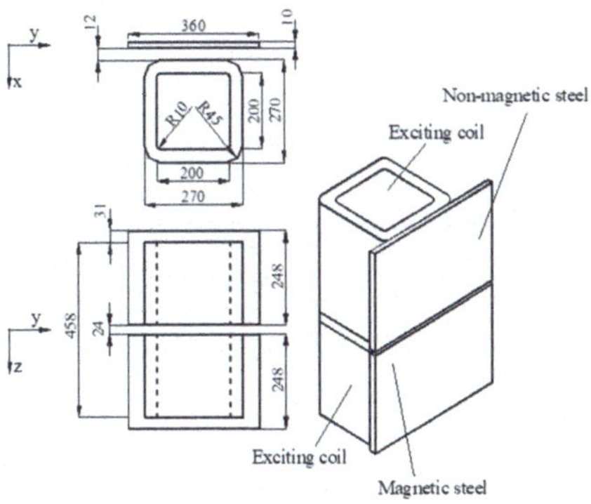

(b-1)

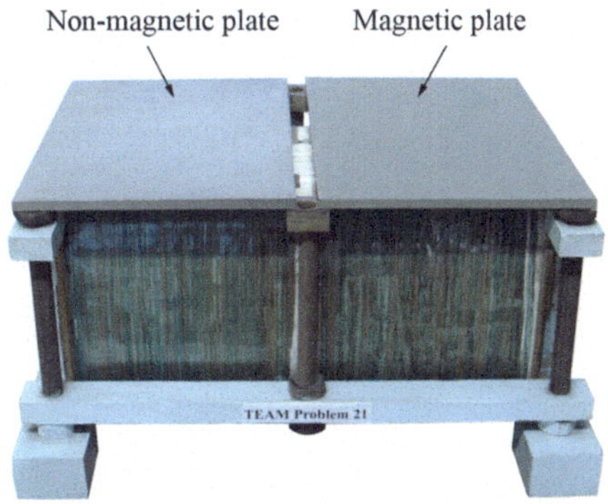

(b-2)

(b) P21^{b}-MN(2002)

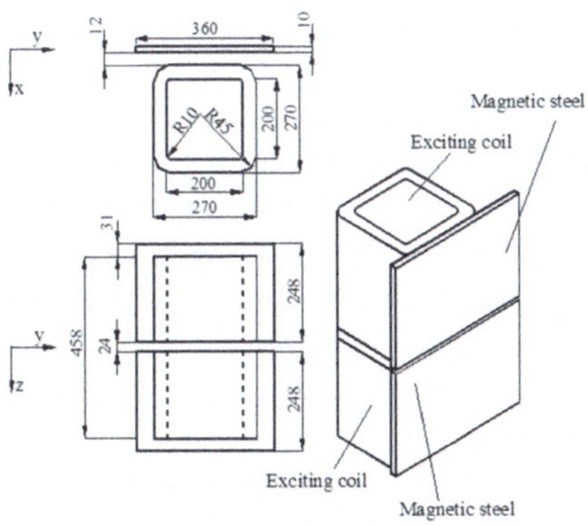

(c-1)

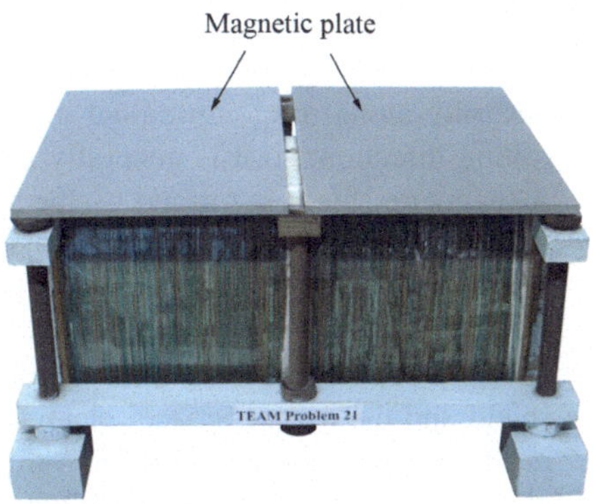

(c-2)

(c) P21^{b}-2M(2002)

Fig. 12.4 Problem 21^{b}

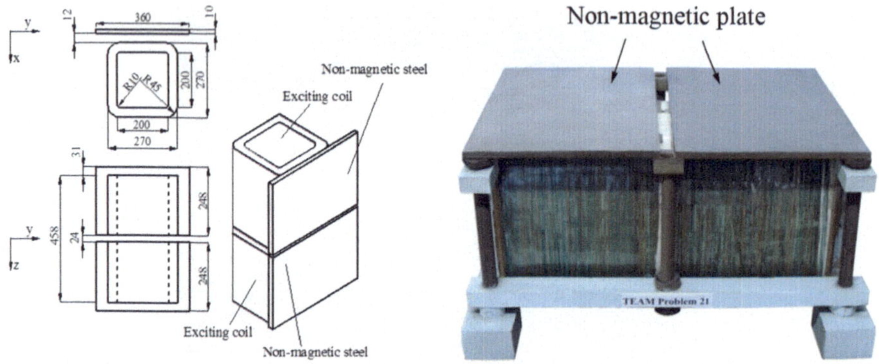

(d-1) (d-2)

(d) P21^b-2N(2002)

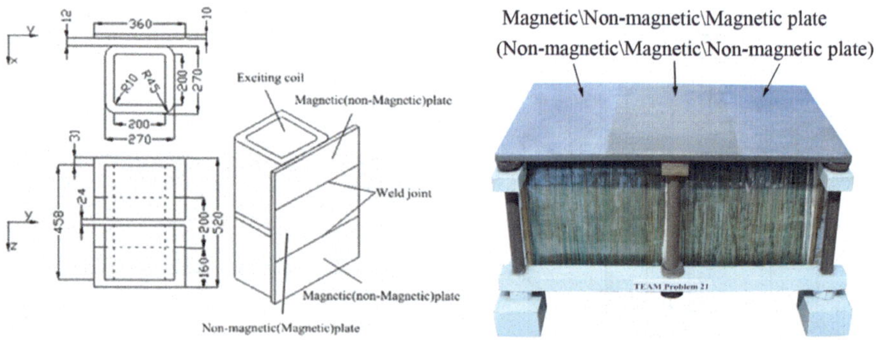

(e-1) (e-2)

(e) P21^b-MNM/NMN (2006)

Fig. 12.4 (continued)

basically along the longitudinal direction of the shielding lamination, i.e., the rolling direction, so it is generally a problem of weak anisotropy.

(2) A magnetic shield is formed by the silicon steel sheets stacked together, and a thin insulating film (on the order of microns) is attached to the surface of the silicon steel sheet to ensure good insulation between the magnetic shield and the shielded magnetic steel plate (A3). However, the electromagnetic shield is made of copperplate (6 mm thick) and there is no insulating film between the shield and the magnetic steel plate, so it is necessary to consider adding a thin insulating material (e.g., 0.05 mm thick) between the copperplate and the steel plate to separate them from each other during calculation and measurement. To keep the structural size consistent between the magnetic and electromagnetic shielding models, the same thickness of insulating material should be added between the silicon steel sheet and the steel plate in the magnetic shielding model.

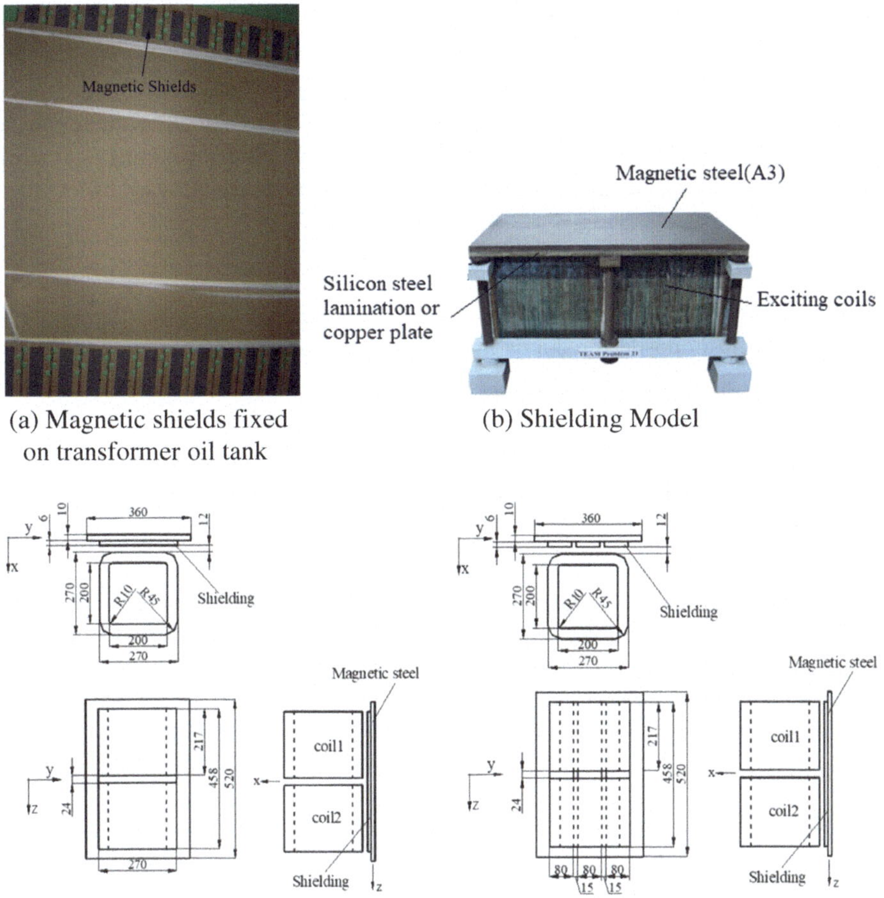

(a) Magnetic shields fixed on transformer oil tank

(b) Shielding Model

(c) P21^{c}-EM1 & M1

(d) P21^{c}-EM2 & M2

Fig. 12.5 Problem 21^{c}-EM and M

5. Problem 21^{d}(2008)

The leakage magnetic flux enters the GO silicon steel lamination vertically, which induces eddy currents in the lamination and produces eddy current losses. In order to study such additional iron loss, the simplified magnetic shielding model was proposed at IEEE CEFC 2008, Athens, Greece, i.e., the shielded magnetic steel plate was removed from P21^{c}-M1, referred to as P21^{d}-M, which was then approved by the ICS Board as a benchmark model of TEAM Problem 21 Family (V.2009).

The numerical analysis and experimental verification of additional iron loss based on simplified models will be detailed in Chap. 13.

12.3.2 Benchmark Family Data

The data for the Problem 21 Benchmark Family includes two parts: the common exciting coils for each model, and the different kinds of conducting components excited by the exciting coils, such as the magnetic steel plates, non-magnetic steel plates, copperplates, and silicon steel lamination, which are used in various member models.

1. **Excitation Source**

In order to enhance the leakage flux into the conductor component, the exciting currents flow in opposite directions in two exciting coils of the same size.

The number of turns in each coil: 300;
Exciting current: ±10 A (50 Hz, rms value);
Specification of bare copper wire for coil: 2.0 × 6.7 mm (net cross-sectional area of wire: 13.04 mm^2);
Density of copper wire: 8.9 × 10^3 kg/m^3;
Conductivity of copper wire: $\sigma = 5.7143 \times 10^7$ S/m.

It should be pointed out that the authors have developed an enhanced excitation device to further validate the effectiveness of modeling and simulation of stray-field losses in components, which are at different magnetic saturation level or under complex excitation conditions (e.g., the excitation source contains harmonics and/or DC components). See Sect. 12.8 of this chapter.

2. **Conducting parts**

① Magnetic steel plate (Q235A steel, or A3 steel, used in $P21^0$/$P21^b$/$P21^c$)

Thickness of steel plate: 10 mm;
Conductivity: $\sigma = 6.484 \times 10^6$ S/m;
Density: 7.8 × 10^3 kg/m^3.
The data for *B–H*, $B_m - W_h$ curves of isotropic magnetic steel plates are listed in Appendix 12.1.

② Non-magnetic steel plate (20Mn23Al used in $P21^a$/$P21^b$)

Thickness of steel plate: 10 mm;
Conductivity: $\sigma = 1.3889 \times 10^6$ S/m;
Relative permeability: $\mu_r = 1$.

③ Shielding material (for $P21^c$)

Problem 21^c includes two types of benchmark models, Problem 21^c-EM ($P21^c$-EM) and Problem 21^c-M ($P21^c$-M), as shown in Fig. 12.5.

As a set of electromagnetic shield models, $P21^c$-EM includes two member models, $P21^c$-EM1 and $P21^c$-EM2. In $P21^c$-EM1, a rectangular copperplate is fastened to the magnetic steel plate of $P21^0$-B (i.e., the magnetic steel plate is the same as the steel plate in Model B of Problem 21), and the copperplate is disposed

Table 12.2 Copperplates for electromagnetic shielding

Model	Size (mm)	Qty	Production requirements
P21^{c}-EM1	458 × 270 × 6	1	Plates must be cut by machines; size and flatness shall meet the requirements
P21^{c}-EM2	458 × 80 × 6	3	

between the exciting coil and the magnetic steel plate, whereas there are three narrow copperplates parallel to and separate from each other in P21^{c}-EM2.

As a set of magnetic shield (shunting) models, P21^{c}-M has 2 member-models, P21^{c}-M1 and P21^{c}-M2. In P21^{c}-M1, a set of silicon steel sheets is fixed on the magnetic steel plate (which is also the same as the magnetic steel plate in Model B of Problem 21), while P21^{c}-M2 includes three sets of narrow silicon steel sheets parallel to and separate from each other.

(1) *Problem 21^{c}-EM*

In P21^{c}-EM1, an entire 458 × 270 × 6 mm copperplate is installed between the magnetic steel plate and the exciting coil, but P21^{c}-EM2, as mentioned earlier, is composed of three copperplates of 458 × 80 × 6 mm. See Table 12.2 for the specifications and quantities. The physical properties of the copperplate are as follows:

Conductivity: $\sigma = 5.7143 \times 10^{7}$ S/m;
Relative permeability: $\mu_r = 1$;
Density: 8.9×10^{3} kg/m^{3};
Models P21^{c}-EM1 and P21^{c}-EM2 are shown in Fig. 12.5.

(2) *Problem 21^{c}-M*

Except for the difference in electromagnetic properties of the shielding materials (GO silicon steel sheet), the size and structure of the Problem 21^{c}-M are the same as those of Problem 21^{c}-EM, as shown in Fig. 12.5. The detailed dimensions and physical properties of the anisotropic silicon steel sheet used in Problem 21^{c}-M are as follows:

Silicon steel sheet: 30RGH120;
The laminated sheets of P21^{c}-M1 and P21^{c}-M2 have the same length and thickness, 458 mm and 0.3 mm, respectively, with different widths: 270 mm (1 laminated sheet) and 80 mm (3 laminated sheets) respectively;
Conductivity: $\sigma = 2.22 \times 10^{6}$ S/m;
Density: 7.65×103 kg/m^{3}.

For P21^{c}-M1 and P21^{c}-M2, the number of silicon steel sheets of each lamination is 20, as shown in Table 12.3.

Table 12.3 Silicon steel sheets used in magnetic shielding

Model	Size (mm)	Qty	Production requirements
P21^{c}-M1	458 × 27 × ×0.3	20	Cut with special machine; sampling direction and size must be strictly guaranteed
P21^{c}-M2	458 × 80 × 0.3	60	

The *B–H*, B_m-W_h, and B_m-W curves (along the rolling direction and the vertical rolling (or transverse) direction) are illustrated in Appendix 12.2. The B_m-W_h curve indicates that hysteresis loss is a function of the maximum magnetic flux density B_m, while B_m-W curve shows the relationship between total losses (including hysteresis loss, classical eddy current loss, and excess losses) and the maximum magnetic flux density B_m in silicon steel sheet.

The *B–H*, B_m-W_h, and B_m-W curves of 30RGH120 silicon steel sheets are measured and provided by the Electromagnetic Device Laboratory (Okayama University, Japan).

Note that Appendix 12.3 also provides a quick search of Problem 21 data.

12.3.3 Field Quantities to Be Calculated

In order to test the effectiveness of the electromagnetic analysis methods, it is required to compare the local field quantity (e.g., magnetic flux density or eddy current density at designated positions) and integral quantity (e.g., the total loss and interlinkage flux), which can be accurately calculated and measured based on the benchmark model. Therefore, some field quantities to be calculated are clearly specified for Problem 21.

① Magnetic Flux Density at Specified Position

It is necessary to calculate the magnetic flux densities at the specified positions, including those entering and exiting both sides of the conductor (magnetic flux enters from one side and exits from the other side of the conductor), as shown in Table 12.4.

② Losses in Conducting Materials

Losses in conducting materials, such as losses caused in magnetic steel plates, non-magnetic steel plates, and electromagnetic and magnetic shields (such as copperplates and silicon steel sheets), including hysteresis, eddy current, or total losses need to be calculated, as shown in Table 12.5.

Table 12.4 Calculated B_x

z (mm)	Flux entry side		Flux exit side	
	$x = 5 + \delta$ (or $11 + \delta$) (mm)		$x = -5 - \delta$ (mm)	
z1				
z2				
z3				
.				
.				

Notes
(1) δ (where $\delta = 0.76$ mm) is half of the thickness of the magnetic probe used
(2) The measuring position $x = 11 + \delta$ (mm) on the side where the flux enters the steel plate is only for the shielding model, while for other models, $x = 5 + \delta$ (mm)

Table 12.5 Calculated results of losses

Conductor parts	Loss (W)		
	Eddy current loss	Hysteresis loss	Total losses
Magnetic steel			
Non-magnetic steel		–	
Shield			–

Notes
(1) The 3D distribution of eddy currents in the model is shown to provide an insight into the electromagnetic behavior of typical structures
(2) The losses in the silicon steel sheets in the magnetic shielding model should be calculated based on the specific total loss curve (B_m–W)

12.4 Numerical Analysis and Measurement

Great progress has been made in electromagnetic design and analysis technology, advanced material modeling, nonlinear and anisotropic processing technologies, efficient solvers of large algebraic equations, and extensive industrial application. 3D eddy current analysis and model-based experimental research are inseparable and mutually reinforcing. The following is a brief summary of some of the work done by the authors, which certainly will be continuously improved by other researchers.

12.4.1 On Eddy Current Analysis Method

Various eddy current solvers have been developed by the authors, based on different potential sets, including $\boldsymbol{A} - V - \boldsymbol{A}$ ($\boldsymbol{A}$ is the magnetic vector potential for the entire region; V is the electric scalar potential for conducting regions only);

$\boldsymbol{A}_{\mathrm{r}} - V - \boldsymbol{A}_{\mathrm{r}}$ ($\boldsymbol{A}_r$ is the reduced magnetic vector potential for the entire region; total magnetic vector potential $\boldsymbol{A} = \boldsymbol{A}_r + \boldsymbol{A}_s$; $\boldsymbol{A}_s$ represents the contribution of the excitation source) and $\boldsymbol{T} - \varphi - \varphi$ ($\boldsymbol{T}$ is the current vector potential for the entire conducting region; φ is the reduced magnetic scalar potential for the entire region) [23]; using two types of finite elements, node-based element and edge-based element; time-stepping, time-periodic finite element, or time-harmonic methods to deal with the time differential terms; the nonlinear or quasi-nonlinear approach (QNA) for magnetic steel plates.

The formulation based on potential sets, $\boldsymbol{A} - V - \boldsymbol{A}$ and $\boldsymbol{T} - \varphi - \varphi$ [23, 64–66], and the numerical implementation are outlined in Chap. 2 of this book. The corresponding solvers developed by the authors have all been tested for a long time and in various ways.

The results show that hysteresis loss in magnetic steel plate must be considered in magnetic loss analysis. A practical way to solve the problem of hysteresis loss has been proposed by the authors; i.e., hysteresis loss W_h is regarded as a function of the peak value of magnetic flux density B_{m}, and $B_{\mathrm{m}}{-}W_{\mathrm{h}}$ curve of specified materials can be measured in advance, like measuring other magnetic property curves [51]. Based on the analysis results of the magnetic field, hysteresis loss W_{h} can be calculated by the following equation:

$$W_{\mathrm{h}} = \sum_{e=1}^{N_e} W_{\mathrm{h}}^{(e)}\left(B_{\mathrm{m}}^{(e)}\right)\rho V^{(e)} \tag{12.1}$$

where $W(e)\ h$ is the hysteresis loss of element (W/kg);
$B(e)\ m$ is the peak value of magnetic flux density;
ρ is the density of steel plate;
$V^{(e)}$ is element volume;
N_e is the total number of elements.

In P21^{c}-M, the magnetic shield is composed of GO silicon steel sheets (30RGH120). To solve the anisotropy problem of silicon steel lamination, the *B–H*, and $B_{\mathrm{m}}{-}W_{\mathrm{h}}$ curves have been measured in the rolling direction and the vertical rolling direction, respectively.

Coulomb gauging is imposed on the governing equation of eddy current based on the $\boldsymbol{A} - V - \boldsymbol{A}$ (or $\boldsymbol{A}_r - V - \boldsymbol{A}_r$) potential set [23]. In the $\boldsymbol{A} - V - \boldsymbol{A}$ algorithm based on node-based elements, the equations of the sub-regions of eddy current are as follows:

$$\nabla \times \frac{1}{\mu}\nabla \times \boldsymbol{A} - \left[\nabla\frac{1}{\mu_c}\nabla\cdot\boldsymbol{A}\right] + \sigma\left(\frac{\partial \boldsymbol{A}}{\partial t} + \nabla\frac{\partial V}{\partial t}\right) = 0 \tag{12.2}$$

The penalty function term within square brackets in Eq. (12.2) is used to force the zero divergence condition to be satisfied. In addition, the permeability μ_{c} in penalty function term is treated as isotropic to ensure the symmetry of finite element matrix system. A simple method is to average the local permeability (at inspected points) in different directions.

The thickness of silicon steel lamination in a magnetic shielding model is as thin as 0.3 mm. In the finite element analysis of large-scale 3D eddy current field, the multi-scale calculations become very difficult due to the thin finite element mesh layers. It is also important to note that the characteristics of a single silicon steel sheet are different from those of an entire silicon steel lamination shield. For the reasons above, effective permeability is introduced herein. According to the continuity condition of $\boldsymbol{B}$ and $\boldsymbol{H}$ between silicon steel sheet and the air, and considering that the permeability of silicon steel sheet is much higher than that of air, the relationship between the effective permeability $\underline{\mu}_x$, $\underline{\mu}_y$, $\underline{\mu}_z$ and the corresponding measured permeability can be expressed as follows:

$$\begin{aligned} \underline{\mu}_x &= \mu_0/(1 - C_{\mathrm{p}}) \\ \underline{\mu}_y &= C_{\mathrm{p}}\mu_y \\ \underline{\mu}_z &= C_{\mathrm{p}}\mu_z \end{aligned} \tag{12.3}$$

where C_{p} is the packing factor and μ_0 is the permeability of vacuum.

So far, the authors have published a number of measured and calculated results related to the magnetic flux density and loss based on TEAM Problem 21 benchmark models, which will be shown in Sect. 12.5 of this chapter.

Note that a number of P21-based numerical results have been obtained, using independently developed solvers by the authors, or using well-established commercial software based on international cooperative projects, such as indicated in the corresponding sections. The measured results concerning with the specified field quantities based on the benchmark models were completed at Baobian Electric, China, and the magnetic property data of magnetic materials were provided by the Electromagnetic Device Laboratory (Okayama University, Japan).

It is hoped that the benchmarking (TEAM) encourages the continuous adoption of new methods to obtain results for further improvement and to test the effectiveness and usefulness of more methods, as expressed in TEAM's purpose.

12.4.2 Measurement of Magnetic Flux Density and Interlinkage Flux at Specified Positions

The measurement and numerical analysis of the field quantity specified in the definition of benchmark problem are important aspects of the TEAM and are mutually validated and supported.

Clear requirements are included in the definition of the magnetic flux density and interlinkage flux measurement, at specified positions in the air. In a model of Problem 21, the component of the magnetic flux density in the direction perpendicular to the surface of the conductor, such as B_x, is measured point by point, using a Gaussmeter probe, at a designated position on the surface (the side facing the excitation source or the other side facing away from the excitation source) of the conductor (e.g., steel plates or oriented silicon steel lamination in the benchmark model) excited by the leakage electromagnetic field (in the definition, the x-axis is the normal direction of the conductor surface). In fact, the distance from the centroid of the sensor of the probe, that is close to the conductor surface, is 1/2 of the thickness of the probe.

In the specified positions of the conducting components in benchmark models, the search coil is wound with a thin wire against the surface of the conductor. A potential is induced on the conductor when the alternating flux is interlinked with the wound search coil. Based on the collected induced potential, the interlinkage flux at the measured position can be obtained through data processing.

Gaussmeter is very flexible and convenient for measuring the magnetic flux density at a designated position. The waveform of the magnetic flux density and the interlinkage flux at the specified cross section can be obtained by processing the data collected by Gaussmeter.

The Gaussmeter (F. W. Bell 9200) used by the authors in the magnetic field measurement, at the specified positions, is briefly introduced below, as shown in Fig. 12.6.

It is a wide-range portable Gaussmeter with a built-in rechargeable battery (lasts for 8 h under normal operating condition) and can be connected to AC power supply. Such Gaussmeter can be used with various measuring probes. Communication interface: RS—232.

Measurement range: 200 μT–20 T (Tesla)

Resolution is related to range, as shown in Table 12.6.

Accuracy:

(a) Accuracy under DC

The accuracy of the given output mode and range under DC is shown in Table 12.7.

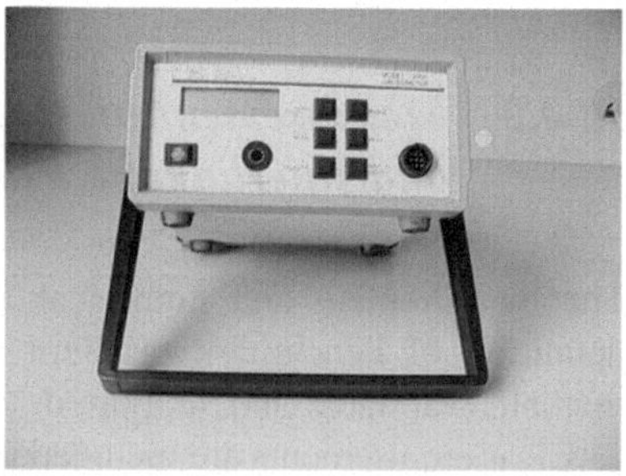

Fig. 12.6 Gaussmeter (F. W. Bell 9200)

Table 12.6 Range and resolution of Gaussmeter

Range		Resolution	
Tesla (T)	Gauss (G)	Tesla (T)	Gauss (G)
200×10^{-6}	2	0.1×10^{-6}	1×10^{-3}
2×10^{-3}	20	1×10^{-6}	10×10^{-3}
20×10^{-3}	200	10×10^{-6}	100×10^{-3}
200×10^{-3}	2000	100×10^{-6}	1
2	20,000	1×10^{-3}	10
20	200,000	10×10^{-3}	100

Table 12.7 Accuracy under DC

Output mode	Range	
	20 (Gauss)	200–200,000 Gauss
Screen display	±5% (reading)	5% (reading)
Analog output (%)	0.5	0.5

Table 12.8 Accuracy under AC

Frequency (Hz)	±% (full range)	Remarks
DC-9	1.5	Analog output
10–49	1.5	
50–99	2.0	
100–499	2.5	
500–999	3.0	
1000–10,000	4.0	

(b) Accuracy under AC

The accuracy of the given output mode and range under AC is shown in Table 12.8.

The transverse measuring probe used with the Gaussmeter is shown in Fig. 12.7.

Specifications
Model: HTB92-0618
Total probe length: 20″
Probe width: 0.180″

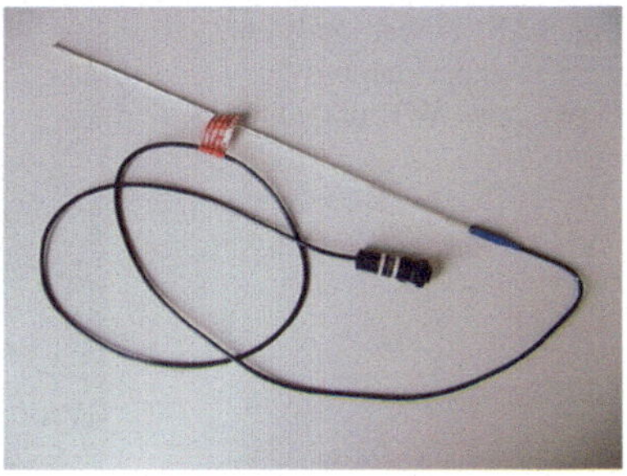

Fig. 12.7 Transverse measuring probe (F. W. Bell)

Probe thickness: 0.060″
The distance from the centroid of the sensing element to the edge of the probe: 0.145″
Operating conditions
Frequency: DC—2000 Hz
Temperature range: 0–75 °C.

12.4.3 Indirect Determination of Loss in Conducting Components

The structures of the benchmark models of Problem 21 Family are seemingly simple, only including a set of air-core exciting coils and the conducting components with simple shape, which are composed of solid conducting plates and/or laminated sheets. However, the determination of the loss inside conducting components is difficult, like solving all the stray-field loss problems.

Taking Model P21^0-A as an example, the measured total loss W_t includes the loss W_{steel} produced in the magnetic steel plate (including eddy current loss and hysteresis loss) and the loss W_{coil} produced in the exciting coils (including eddy current loss W_{coile} and resistance loss W_{coilr}). Thus, the total loss W_t measured in the P21^0-A is

$$W_t = W_{steel} + W_{coil} \quad (12.4)$$

Since it is not possible to directly measure the stray-field loss W_{steel} in the steel plate, the eddy current loss in the coil can only be obtained through calculation, so W_{steel} can only be determined indirectly, i.e.,

$$W_{steel} = W_t - (W_{coile} + W_{coilr}) \quad (12.5)$$

The total loss of all the member models of Problem 21 Family has been measured using the Precision Power Analyzer (WT3000/WT3000E, Yokogawa, Japan). Scc Fig. 12.8.

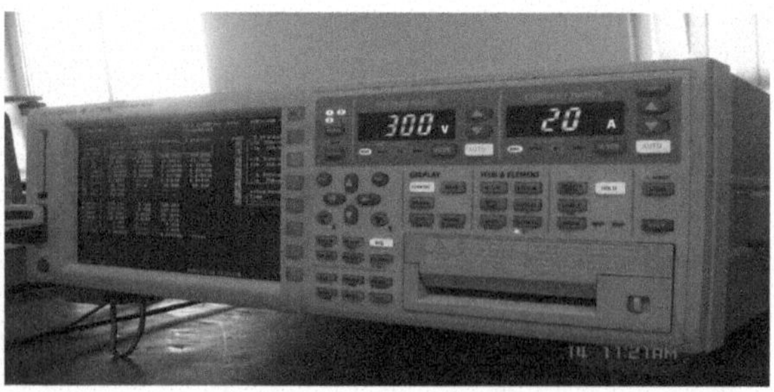

Fig. 12.8 Power analyzer used in experiments (Yokogawa WT3000, Japan)

12.4.4 Determination of Upper and Lower Bounds of Losses

① Upper and Lower Bounds of Eddy Current Losses in Exciting Coils

To calculate the eddy current loss in exciting coils, it is necessary to know exactly the distribution of the leakage electromagnetic field at the coil regions. Also, the effect of magnetic components on the leakage flux and then the eddy currents inside the exciting coils must be taken into account. It is difficult to determine the eddy current loss in the exciting coils accurately.

However, the upper and lower bounds of eddy current losses in the exciting coils can be obtained, which is helpful in evaluating the loss in the magnetic components. The two extremes that affect the transverse leakage flux occur when:

(1) The silicon steel lamination or the magnetic steel plate is assumed to be removed from the benchmark model, the effect of magnetic components on the leakage magnetic field is completely eliminated, and then, both the transverse leakage flux and $\mathrm{W_{coile}}$ reach the minimum $\mathrm{W_{coile}^{min}}$.
(2) The silicon steel lamination or the magnetic steel plate is assumed to be removed from the benchmark model; however, a set of mirror coils (with respect to the exciting coils) is placed on the surface of the original magnetic component (e.g., magnetic steel plate or lamination); this is equivalent to having a magnetic component with infinite permeability. In this case, the transverse leakage flux and W_{coile} reach the maximum $W_{\mathrm{coile}}^{\max}$.

As an example, the upper and lower bounds of total eddy current losses in the exciting coil of Problem 21 are as follows:

$$\begin{aligned} W_{\mathrm{coile}}^{\max} &= 3.333\ (\mathrm{W}) \\ W_{\mathrm{coile}}^{\min} &= 2.673\ (\mathrm{W}) \end{aligned} \quad (12.6)$$

② Upper and Lower Bounds of Magnetic Losses in Magnetic Components

The upper and lower bounds of eddy current losses in the exciting coils (the resistance loss of the exciting coils W_{coilr} is unchanged since the exciting currents remain the same) and the total loss of the magnetic shield model measured correctly are helpful in evaluating the magnetic loss in magnetic components:

(1) The lower bound of the magnetic loss in the magnetic steel, $W_{\mathrm{steel}}^{\min}$, corresponds to the upper bound of eddy current loss in the exciting coils, $W_{\mathrm{coile}}^{\max}$, i.e.,

$$W_{\mathrm{steel}}^{\min} = W_{\mathrm{t}} - \left(W_{\mathrm{coile}}^{\max} + W_{\mathrm{coilr}}\right) \quad (12.7)$$

(2) The upper bound of the magnetic loss in the magnetic steel W_{steel}^{max} corresponds to the lower bound of the eddy current loss in the exciting coils, W_{coile}^{min}, i.e.,

$$W_{steel}^{max} = W_t - \left(W_{coile}^{min} + W_{coilr}\right) \tag{12.8}$$

Clearly, the upper and lower bounds of the total loss in the magnetic steel help to determine the validity of the loss results.

Note that Sect. 12.8 of this chapter shows the upgraded P21 models, including a set of magnetic flux compensation coils (or the mirror coils mentioned above, see Fig. 12.42), which can provide the upper and lower bounds of total loss (i.e., eddy current losses and resistance loss) in exciting coils.

12.4.5 *Eddy Current Losses in Exciting Coils*

The following simplified assumptions are made for the calculation of eddy current losses in the exciting coil:

(1) The fillet of the exciting coil of the "racetrack-shaped" cross section of Problem 21 is not considered;
(2) The exciting current density is uniformly distributed in the whole space in which coils are located.

The eddy current loss in the exciting coil is induced by the leakage electromagnetic field in the copper wire. It can be found that the eddy current loss W_{coile} in the coil can be divided into two parts: the loss $W_{coile}^{(l)}$ caused by the transverse leakage flux perpendicular to the wide surface (e.g., 6.7 mm high in member models of P21) of the copper wire of exciting coil and the loss $W_{coile}^{(s)}$ caused by the axial leakage flux perpendicular to the narrow surface (e.g., 2 mm wide in member models of P21) of the copper wire, i.e.,

$$W_{coile} = W_{coile}^{(l)} + W_{coile}^{(s)} \tag{12.9}$$

Further analysis shows that $W_{coile}^{(l)}$ is the major component of total eddy current losses in the exciting coils caused by the transverse leakage flux.

The eddy current loss in the exciting coil is calculated by the following equation based on the above two cases:

$$W_{coil} = \frac{\sigma}{24}\left(\omega B_r h\right)^2 V \tag{12.10}$$

where

σ conductivity
ω angular frequency
B_r transverse or axial leakage flux
h the size of the conductor in the transverse or axial direction
V conductor volume (of exciting coil) at corresponding spot of B_r.

It should be noted that the accurate 3D modeling and simulation of eddy current loss in exciting coils of Problem 21 have been carried out by the authors, even under complex excitation condition. The stray-field loss and flux distribution inside the magnetic steel plate, under harmonic excitation, have been investigated in some detail [61].

12.5 Typical Calculated and Measured Results

The scholars from all over the world have published a number of results based on the Problem 21 Benchmark Family. In this section, some typical measured and calculated results obtained by the authors are presented below to detail the model-based electromagnetic behavior and the validity of modeling and simulation. The further results of Problem 21 Family and the related discussions can also be found in the updated definition of TEAM Problem 21 at www.compumag.org/team, or in the authors' research articles [51–61].

It is hoped that new and more effective computational methods will be developed to further improve the calculation results which will benefit industrial applications.

12.5.1 Problem 21^0 (P21^0-A and P21^0-B)

For TEAM Problem 21^0 (P21^0-A and P21^0-B), the measured B_x (normal component of magnetic flux density $\boldsymbol{B}$ on the steel plate surface of the model) at the designated position ($x = 5.76$ mm, $y = 0.0$ mm) is shown in Table 12.9, and Fig. 12.9 shows the comparison between the measured and calculated results (B_x).

The calculated loss results (including eddy current and hysteresis loss) inside the magnetic plate of Problem 21 Model A, according to different treatments of magnetic properties of material (A3 steel), are listed in Table 12.10.

Based on the calculated results shown in Table 12.10, it can be seen that the deviation between the calculated (by different processing of magnetic material property, e.g., using linear permeability, $\mu_r = 1000$ or 5000, or quasi-nonlinear approach) and the measured results is significant. Therefore, the nonlinearity of the magnetic material must be taken into account when calculating the magnetic loss,

Table 12.9 Measured B_x (P21^0-A and B, presented in 1993, at TEAM-Miami, USA)

z (mm)	P21^0-A (Model A)	z (mm)	P21^0-B (Model B)
	Measured ($\times 10^{-4}$ T)		Measured ($\times 10^{-4}$ T)
218.0	102.50	227.5	122.47
197.0	68.20	196.0	70.56
176.0	41.20	162.5	24.18
136.0	7.41	130.0	8.96
96.0	51.80	97.5	45.82
75.0	82.60	65.0	93.62
54.0	118.70	32.5	156.70
0.0	127.60	0.0	215.67
−54.0	109.90	−32.5	(1/4 symmetric)
−75.0	69.70	−65.0	–
−96.0	39.30	−97.5	–
−136.0	1.21	−130.0	–
−176.0	49.60	−162.5	–
−197.0	77.90	−196.0	–
−218.0	113.80	−227.5	–

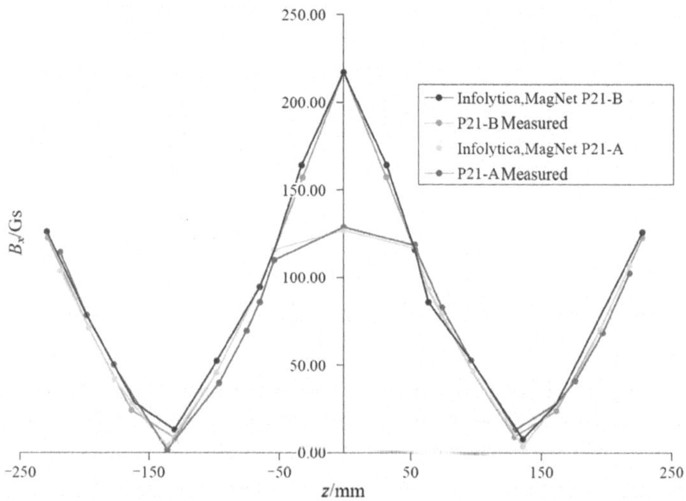

Fig. 12.9 Magnetic flux densities at designated position of P21^0-A/B (calculated results by Simcenter MAGNET)

even though the magnetic flux density in the air is smaller near the surface of the steel plate.

Additional calculation results of the losses of Model A and Model B are shown in Fig. 12.10, in which the solver Simcenter MAGNET (Mentor Infolytica, Canada)

Table 12.10 Eddy current and hysteresis loss of Problem 21 Model A

Computation method	Calculated results (W)				Measured results (W)
	P_e	P_h	P_t	P_h/P_t (%)	
Linear (jω) $\mu_r = 1000$	5.92	0.29	6.21	4.7	9.28
Linear (jω) $\mu_r = 5000$	3.26	4.16	7.42	56.1	
Quasi-nonlinear	6.87	2.24	9.11	24.6	

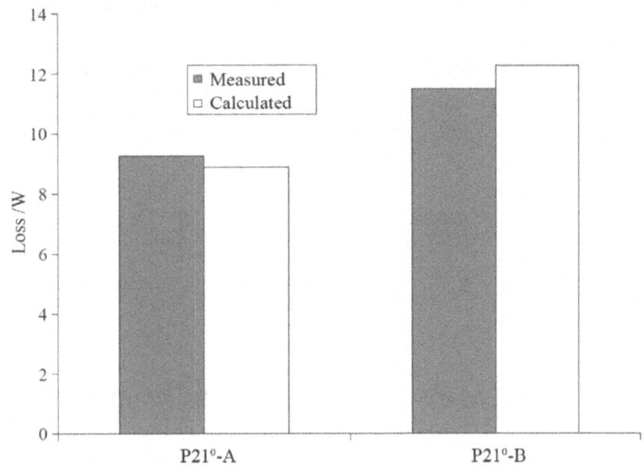

Fig. 12.10 Stray-field loss in magnetic steel plate ($P21^0$-A and B)

is used and the magnetic steel is treated as a nonlinear material, which achieves satisfactory accuracy compared with the measured results.

The magnetic flux density distribution on both sides (i.e., the side facing the exciting coil and the side facing away from the exciting coil) of steel plate in Model A and Model B is shown in Fig. 12.11. The shielding effect of magnetic steel plate on leakage magnetic flux is obvious, and the difference of magnetic flux distribution between $P21^0$-A and $P21^0$-B steel plates is also illustrated.

12.5.2 Problem 21^a

For TEAM Problem 21^a-2, the results of B_x measured and calculated (based on $\boldsymbol{A_r} - V - \boldsymbol{A_r}$) at the designated positions ($x = \pm 5.76$ mm, $y = 0.0$ mm) are shown in Table 12.11 and Fig. 12.12.

The measured and calculated loss results, using different 3D eddy current analysis methods, including the 2D calculated results, are shown in Table 12.12. The eddy current distributions of $P21^a$-3 model are shown in Fig. 12.13, in which the actual asymmetry of the slots is also taken into account.

A comparison of the eddy current distribution calculated according to the actual asymmetric slot ($P21^a$-3) is shown in Fig. 12.13. The difference between

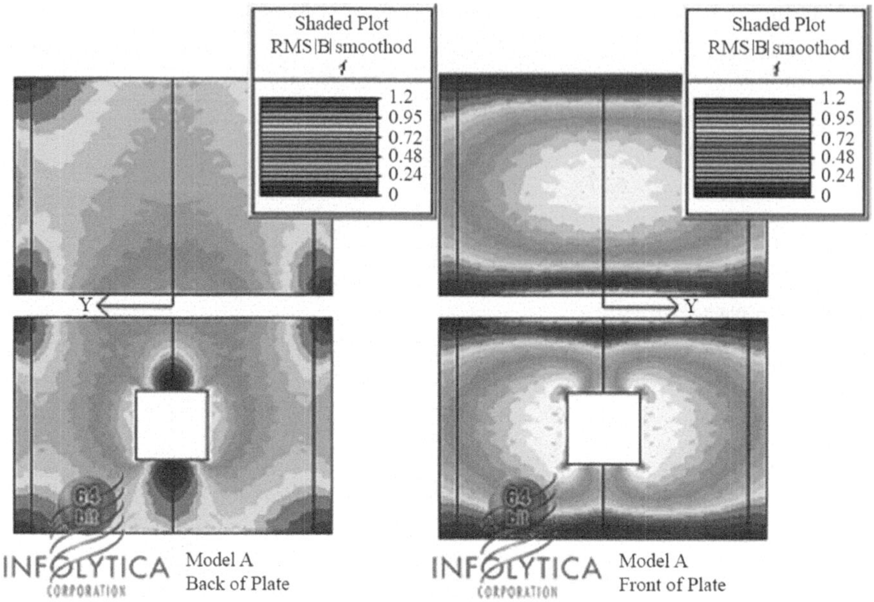

(a) Magnetic flux density distribution (Model A, or P21^{0}-A)

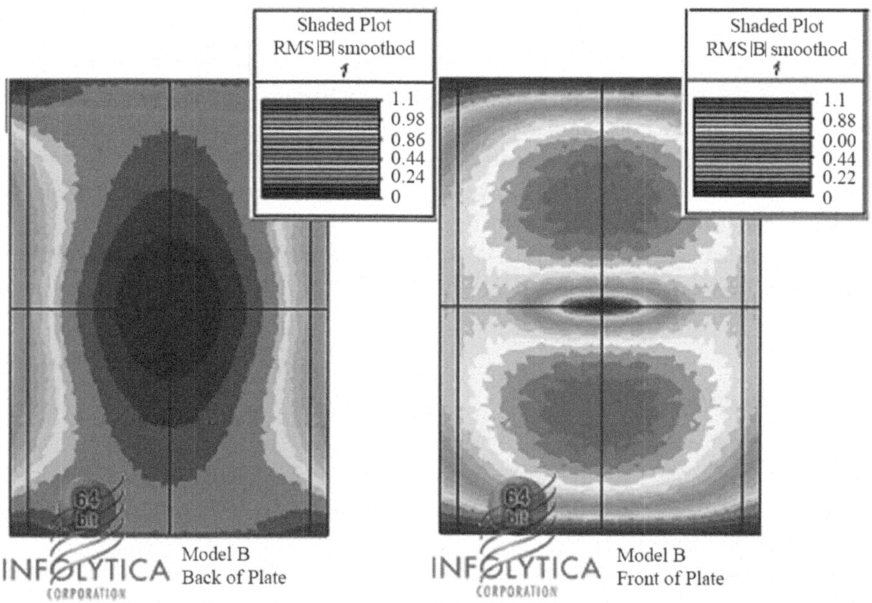

(b) Magnetic flux density distribution (Model B, or P21^{0}-B)

Fig. 12.11 Magnetic flux density distribution in magnetic steel (by Simcenter MAGNET, Mentor Infolytica)

Table 12.11 Results of magnetic flux density (P21[a]-2, presented at TEAM-Yichang, 1996, China)

z (mm)	x = 5.76 mm		x = −5.76 mm	
	Measured ($\times 10^{-4}$ T)	Calculated ($\times 10^{-4}$ T)	Measured ($\times 10^{-4}$ T)	Calculated ($\times 10^{-4}$ T)
6.0	81.25	81.70	64.60	65.60
30.0	63.80	63.10	53.40	54.28
66.0	34.70	37.50	31.60	34.63
102.0	16.20	20.20	14.90	17.79
139.0	1.00	1.96	1.30	0.43
175.0	−15.50	−15.04	−12.50	−13.02
211.0	−35.10	−30.73	−27.50	−25.25
230.0	−42.00	−41.72	−32.20	−31.66
246.0	−37.20	−36.87	−29.90	−29.53
280.0	−23.70	−24.26	−20.70	−21.90
313.0	−15.30	−15.10	−13.60	−14.36
344.0	−11.20	−10.18	−9.40	−9.80
372.0	−8.70	−7.17	−6.80	−7.07
398.0	−7.60	−5.27	−5.20	−5.25

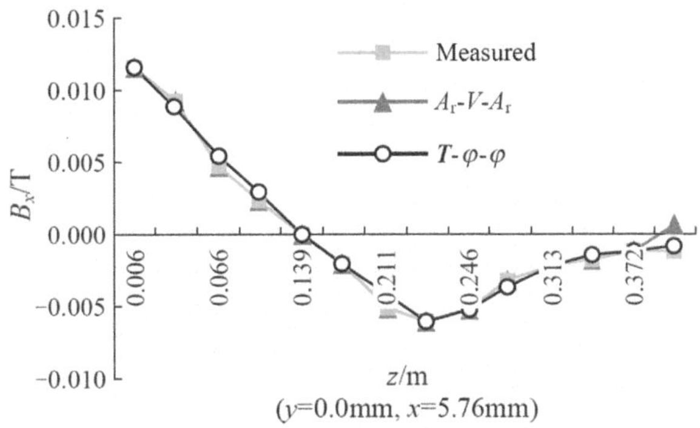

Fig. 12.12 Distribution of B_x (P21[a]-0, x = 5.76 mm, y = 0.0 mm)

symmetrical and asymmetrical slots can only be quantitatively explained in terms of eddy current distribution and loss by 3D eddy current analysis.

Figure 12.14 shows that the loss in non-magnetic steel (P21[a]) decreases rapidly with the increase of the number of slots, and the calculated and measured loss results are in good agreement. In addition, if the non-magnetic steel in Model P21[a] is replaced by the magnetic steel without changing the excitation conditions, the corresponding losses are increased greatly, as shown in Fig. 12.14.

The comparison between the calculated and measured eddy current losses of P21[a] can be summarized as follows:

Table 12.12 Calculated and measured eddy current loss (P21[a])

Model	Measured results (W)	Calculated results (W)			
		$T-\varphi-\varphi$	A_r-V-A_r	$A-V$*** (crack element)	2D calculation
P21[a]-0	9.17	9.50	9.31	9.22	14.75
P21[a]-1	3.40	3.37	3.34	3.35	6.23
P21[a]-2	1.68	1.67	1.66	1.68	3.07
P21[a]-3	1.25	0.95*(1.15**)	0.94*(1.14**)	1.00*	1.86

Notes

(1) *represents the calculated results in the case of the three slots being equidistant (i.e., three-slot symmetry)

(2) **represents the calculated results in the case of the three slots being non-equidistant (i.e., three-slot asymmetry)

(3) *** represents the calculated results using crack elements for narrow slots in large components. The potentials on both sides of the slot are set to 0 and unknown, respectively, and the condition of no current passing through the slot needs to be imposed [29]

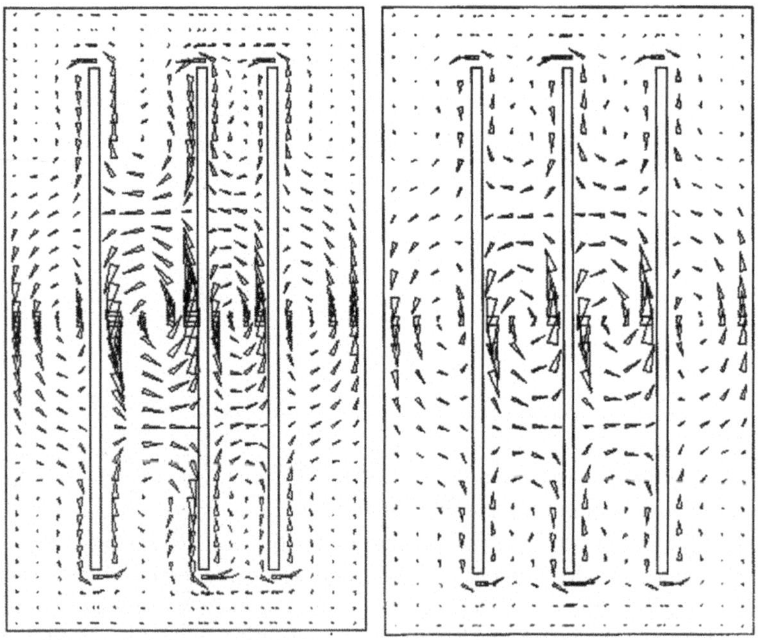

(a) 3-slot asymmetry (b) 3-slot symmetry

Fig. 12.13 Eddy current distribution in non-magnetic steel plate with symmetry or asymmetry three-slot structures

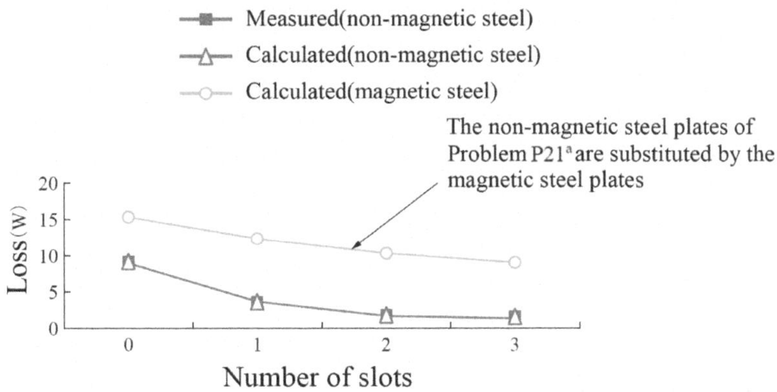

Fig. 12.14 Loss in steel plate varying with number of slots

(1) The results given by 3D eddy current analysis software formulated by different "potential sets," based on $\boldsymbol{A}_r - V - \boldsymbol{A}_r$ and $\boldsymbol{T} - \varphi - \varphi$ and $\boldsymbol{A} - V$ (using crack element), are satisfactory. Further numerical analysis shows that there is almost no difference in the calculated results of 2 and 5 layers of 10-mm-thick non-magnetic steel plate when "crack element" is used.
(2) The loss in non-magnetic steel ($P21^a$) decreases rapidly with the increase in the number of slots, and the calculated and measured results agree well with each other. In addition, without changing the excitation conditions and the structural design of the non-magnetic steel, the corresponding loss increases significantly if the non-magnetic steel in $P21^a$ is replaced by the magnetic steel. In such a case, however, the loss in the magnetic steel also decreases rapidly with the increase of the number of slots.
(3) It is a common understanding in industrial applications that 3D method is not recommended to solve the problems that can be solved by the 2D method, to meet the engineering needs. However, the analysis of Models $P21^a$ and the comparison from the experimental results show that the calculated loss based on the 2D eddy current field is not applicable even when calculating the eddy current loss in linear material such as non-magnetic steel.

12.5.3 Problem 21^b

For $P21^b$-MN, the results of magnetic flux density measured and calculated at the designated positions ($x = \pm 5.76$ mm, $y = 0.0$ mm) are shown in Table 12.13.

Figures 12.15, 12.16, 12.17, and 12.18 show that the results of the magnetic flux density calculated and measured at specified positions (including both sides of the steel plate) of each $P21^b$ model agree well with each other. Also, the difference in

Table 12.13 Results of magnetic flux densities (P21[b]-MN, presented at IEE CEM, 2002, UK)

z (mm)	x = 5.76 mm		x = −5.76 mm	
	Measured ($\times10^{-4}$ T)	Calculated ($\times10^{-4}$ T)	Measured ($\times10^{-4}$ T)	Calculated ($\times10^{-4}$ T)
−252	−48.3	−50.3	26.3	35.5
−218	−51.0	−56.3	9.1	8.3
−197	−35.4	−36.6	7.0	7.1
−186	−25.5	−27.6	6.4	6.9
−176	−20.6	−20.1	6.0	6.6
−162	−13.4	−10.9	5.6	6.3
−136	0.0	4.13	5.1	6.2
−109	20.7	19.7	5.1	6.1
−100	26.6	25.2	5.8	6.5
−89	34.9	32.7	6.6	6.8
−75	43.8	43.3	6.8	6.7
−47	74.1	72.0	5.3	4.4
−33	91.1	91.5	−9.0	−2.0
−18	107.0	114.0	−36.1	−25.8
−9	121.0	123.0	−13.2	−4.1
−3	128.0	129.0	46.2	47.2
0	133.0	140.0	59.8	47.7
3	134.0	150.0	63.2	68.8
9	204.0	223.0	73.0	77.3
18	283.0	271.0	76.8	78.7
33	181.0	198.0	70.7	72.2
47	119.0	123.0	60.4	61.1
75	95.9	98.1	39.3	39.3
89	76.3	71.4	30.2	31.0
100	61.6	58.4	23.9	24.0
109	44.6	42.1	18.8	18.2
136	−3.58	−3.86	4.8	4.7
162	−32.4	−26.0	−8.2	−11.1
176	−46.6	−43.0	−15.6	−16.5
186	−60.4	−57.0	21.0	−22.6
197	−75.8	−73.5	−38.4	−29.7
218	−111.0	−112.0	−41.4	−41.4
252	−146.0	−133.0	−22.8	−40.1

the magnetic flux density distribution and magnitude, between the steel plates facing and facing away from the exciting coil, is quantitatively demonstrated.

Figure 12.19 shows the loss density distribution at specified positions in the steel plate (for Models P21[b]-MNM and P21[b]-NMN, x = 4.6 mm, y = 0.0 mm).

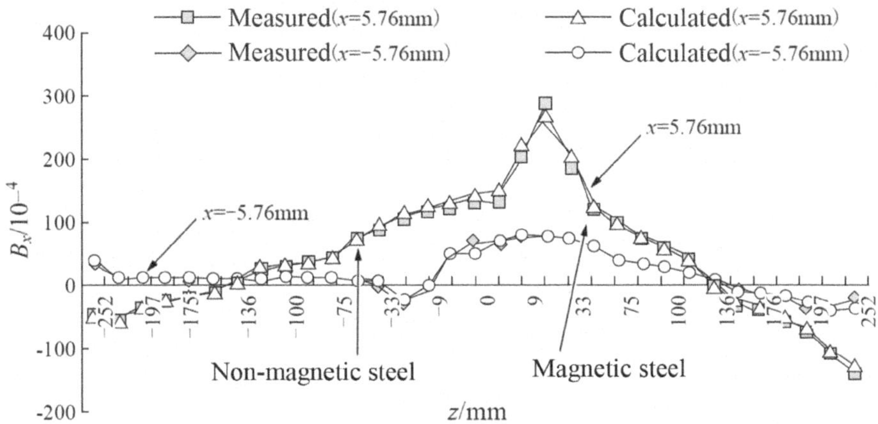

Fig. 12.15 Distribution of B_x (P21^b-MN, $x = \pm 5.76$ mm, $y = 0.0$ mm)

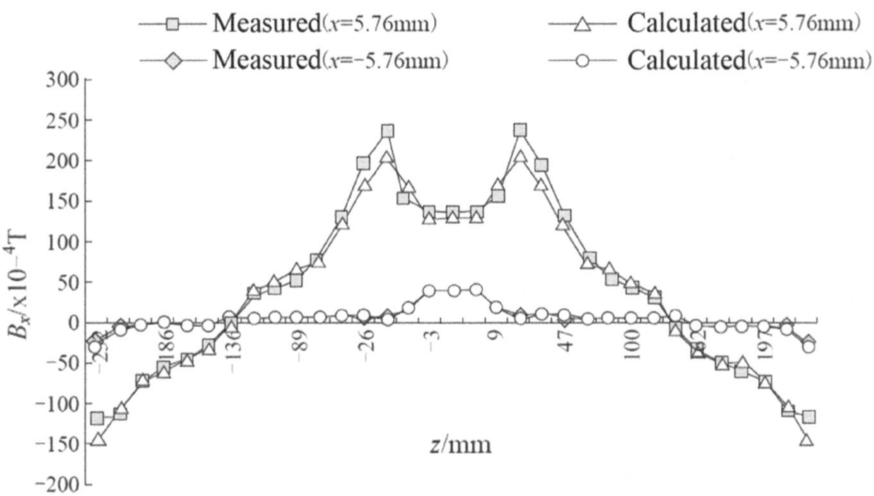

Fig. 12.16 Distribution of B_x (P21^b-2 M, $x = \pm 5.76$ mm, $y = 0.0$ mm)

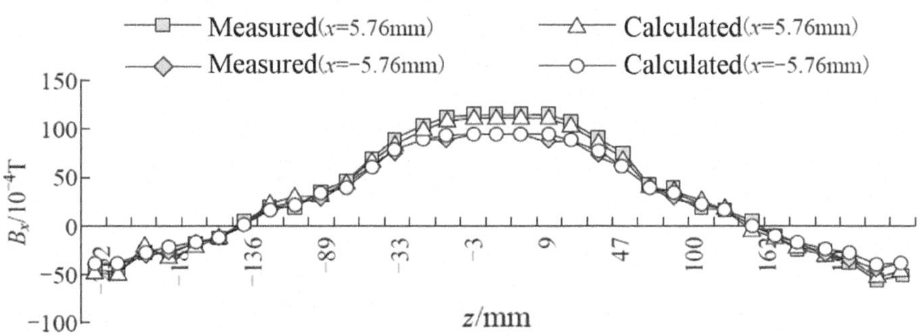

Fig. 12.17 Distribution of B_x (P21^b-2 N, $x = \pm 5.76$ mm, $y = 0.0$ mm)

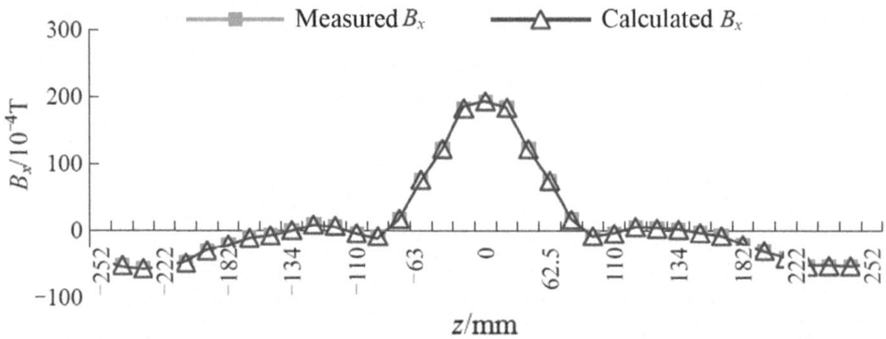

Fig. 12.18 Distribution of B_x (P21^b-NMN, x = 5.76 mm, y = 0.0 mm). *Note* This figure shows only 1/2 of Model P21^b-MNM/NMN in the length direction (i.e., Z-direction in the global coordinate)

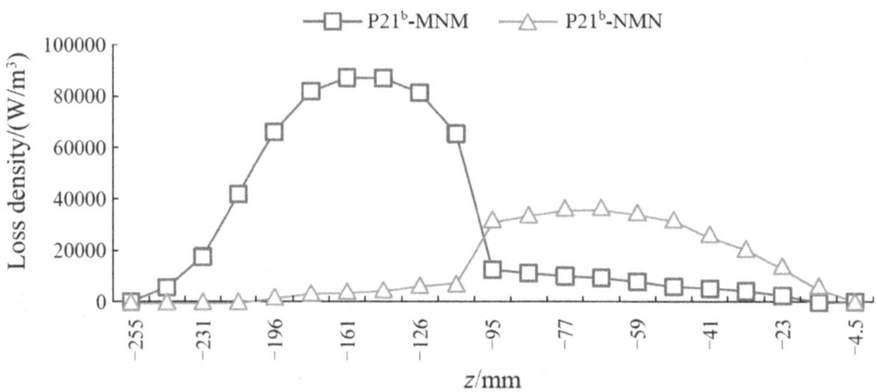

Fig. 12.19 Loss density at specified position (P21^b-MNM/NMN, x = 4.6 mm, y = 0.0 mm)

The above numerical simulation and experiment can be summarized as follows:

(1) From the experiment, simulation, and validation based on the benchmark models, it can be seen that the modeling of the magnetic flux density inside and outside the composite steel plate and the loss in the steel is not only feasible, but also can achieve satisfactory accuracy.
(2) The quantitative analysis of the loss density distribution and total loss of specific composite steel structure will provide evaluation basis for the structural designs.

In addition, the 2D magnetic field distributions on the symmetry planes of the models P21^b-MN, P21^b-MNM, and P21^b-NMN, and the eddy current distribution inside different welding components, in models P21^b-MNM and P21^b-NMN, have been obtained by the authors, to further investigate the effects of the structural configuration of magnetic and non-magnetic steel plates on their magnetic field, eddy currents, and stray-field loss.

12.5.4 Problem 21^c

For P21^c-M1, the results of magnetic flux density measured and calculated at the designated positions are shown in Table 12.14 and Figs. 12.20, 12.21.

Based on the results of the magnetic flux density, measured and calculated at the designated positions, shown in Figs. 12.20 and 12.21, it can be seen that the measured and calculated results of the magnetic flux density of each P21^c model

Table 12.14 Results of magnetic flux density (P21^c-M1, presented at Compumag 2005, Shenyang, China)

z (mm)	x = 11.76 mm, y = 0.0 mm		x = −5.76 mm, y = 0.0 mm	
	Measured ($\times 10^{-4}$ T)	Calculated ($\times 10^{-4}$ T)	Measured ($\times 10^{-4}$ T)	Calculated ($\times 10^{-4}$ T)
0.0	222.31	213.99	2.69	8.17
−6.0	220.31	214.00	2.68	8.15
−12.0	213.76	214.00	2.67	8.13
−18.2	201.81	187.65	2.66	6.88
−30.6	169.85	187.72	2.67	6.73
−43.0	138.95	137.19	2.69	5.62
−55.4	111.30	137.40	2.72	5.32
−67.8	89.10	93.35	2.74	4.44
−80.2	68.02	58.44	2.80	4.14
−92.6	51.05	58.90	2.88	3.64
−105.0	34.65	29.94	2.97	3.65
−117.4	20.19	30.43	3.09	3.13
−129.8	6.50	4.33	3.22	3.30
−142.2	9.20	4.72	3.42	2.84
−154.6	−22.17	−21.19	3.67	3.10
−167.0	−37.83	−50.04	4.02	3.13
−179.4	−55.86	−50.39	4.44	3.14
−191.8	−75.09	−82.263	5.07	3.27
−204.2	−98.64	−83.17	5.94	3.67
−216.6	−130.04	−143.74	7.27	4.83
−229.0	−178.97	−145.31	9.62	5.52
−241.4	−95.32	−86.86	14.12	11.10
−253.8	−100.20	−96.06	28.57	37.88
−258.0	−96.38	−95.96	45.61	38.14

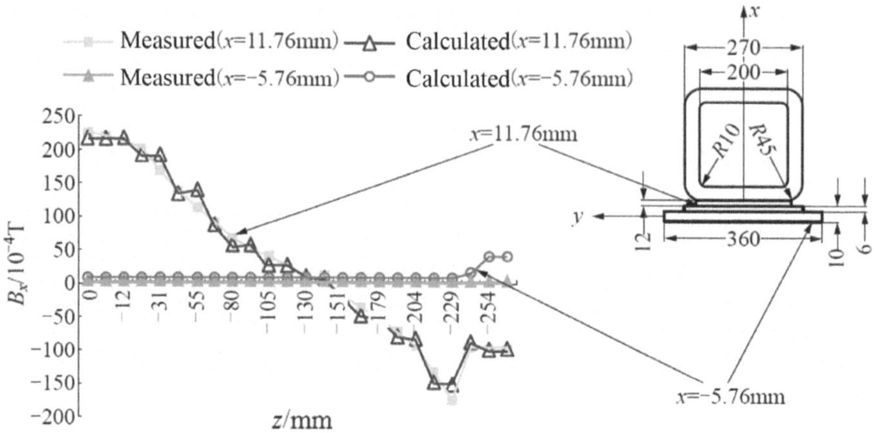

Fig. 12.20 Distribution of B_x (P21[c]-M1, $x = -5.76$ mm, $x = 11.76$ mm, $y = 0.0$ mm)

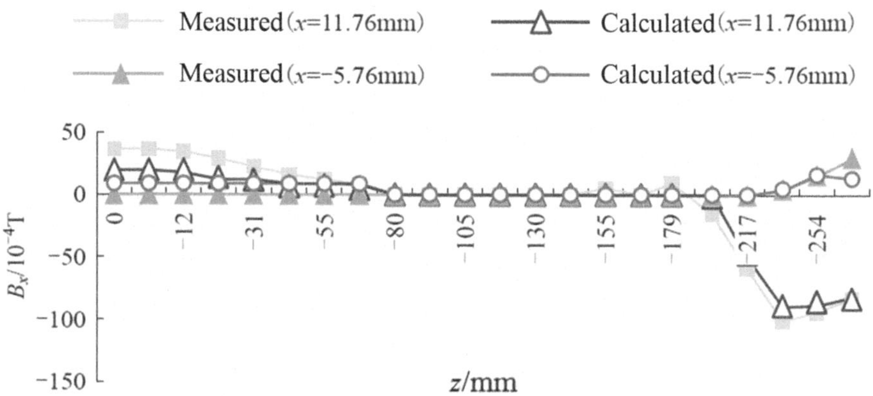

Fig. 12.21 Distribution of B_x (P21[c]-EM1, $x = -5.76$ mm, $x = 11.76$ mm, $y = 0.0$ mm)

agree well with each other. The effect of shielding on the leakage magnetic flux of the exciting coil can be quantitatively demonstrated.

It should be noted that the designated positions here are somewhere on the shielding surface S1 facing the exciting coil and on the surface S2 (i.e., the back of the magnetic steel plate) of shielded magnetic steel plate (A3) that faces away from the exciting coil. It goes like this: The leakage magnetic flux of the exciting coil first enters the magnetic or electromagnetic shield through S1, then enters the shielded magnetic steel plate, then passes through the magnetic steel plate, and finally enters the free space through S2. Therefore, it is not easy for the measured and calculated magnetic flux densities at the designated positions on both sides to be in good agreement. This is a good example for validating the numerical modeling and simulation.

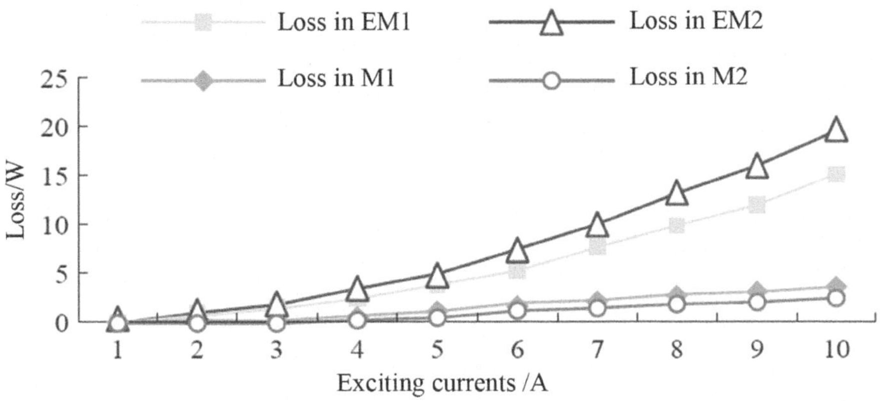

Fig. 12.22 Variation of losses with exciting current (P21^{c}-M1/M2/EM1/EM2)

The variation of total losses inside conducting components of both the magnetic shield (P21^{c}-M1 and M2) and the electromagnetic shield (P21^{c}-EM1 and EM2) with exciting currents is shown in Fig. 12.22, which also demonstrates the different shielding effects.

The related calculated and measured results, based on Model P21^{d}-M, a member model simplified from P21^{c}-M1, for calculating the iron loss in laminations in detail, can be seen in Sect. 12.5.5 of this chapter, as shown in Table 12.15 and Fig. 12.23.

12.5.5 Loss Spectrum of Problem 21 Benchmark Family

The measured and calculated losses of the Problem 21 Benchmark Family include 16 member models (exciting currents: 10 A, rms, 50 Hz), as shown in Table 12.15. The arrangement of the losses of each member model, in the proposed sequence in the Benchmark Family, forms a loss "spectrum" [53], as shown in Fig. 12.23.

All the measured and calculated losses of the Problem 21 Benchmark Family can be summarized as follows:

(1) Under the defined excitation condition (10 A, 50 Hz, rms), the calculated and measured losses of the 16 benchmark models are in good agreement with each other, indicating that the measurement and computation methods are effective.
(2) It can be seen from the comparison and analysis of the losses of each set of models (e.g., P21^{0}, P21^{a}, P21^{b}, P21^{c}, and P21^{d}) that:

    (a) P21^{0}: The steel plates in Model A and Model B have the same external dimensions, but the steel plate in Model A is split into two parts; i.e., a "through slot" divides the steel plate into two halves, and one of them has a hole, so the loss is reduced.

Table 12.15 Measured and calculated loss of P21 Models

Models	Measured loss (W)	Calculated loss (W)			
		Total losses	Losses in magnetic and/or non-magnetic steel plates		Losses in laminated sheets or Cu plate
			Eddy current	Hysteresis	
$P21^0$-A	9.28	9.11	6.87	2.24	
$P21^0$-B	11.97	12.04	8.10	3.94	
$P21^a$-0	9.17	9.31	9.31	–	
$P21^a$-1	3.40	3.34	3.34	–	
$P21^a$-2	1.68	1.66	1.66	–	
$P21^a$-3	1.25	1.14	1.14	–	
$P21^b$-MN	7.03	6.83	5.30	1.53	
$P21^b$-2M	9.34	9.88	7.44	2.44	
$P21^b$-2N	1.38	1.37	1.37	–	
$P21^b$-MNM	10.53	10.04	8.96	1.08	
$P21^b$-NMN	7.44	7.88	6.84	1.04	
$P21^c$-M1	3.72	3.79	0.90	0.30	2.59
$P21^c$-M2	2.64	3.16	1.65	0.68	0.83
$P21^c$-EM1	15.24	16.22	3.87	1.50	10.85
$P21^c$-EM2	20.07	20.11	8.45	2.35	9.31
$P21^d$-M	2.20	2.33	–	–	2.33

(b) $P21^a$: The loss in non-magnetic steel decreases rapidly with the increase in the number of slots.

(c) $P21^b$: When the magnetic and the non-magnetic steel plates are separated, conceivably, $P21^b$-2 N has the lowest loss, and $P21^b$-2 M has the highest, while the loss of $P21^b$-MN is average.

The loss of the hybrid steel part $P21^b$-MNM is greater than that of $P21^b$-NMN. Therefore, it can be concluded that the total loss in hybrid steel part depends on the proportion of the magnetic steel in its total volume.

(d) $P21^c$:

① **Magnetic shield**

The loss in the separate laminated magnetic shield ($P21^c$-M2) is lower than that in the integral laminated magnetic shield ($P21^c$-M1), which is mainly a result of the increase of eddy current loss in the Model $P21^c$-M1.

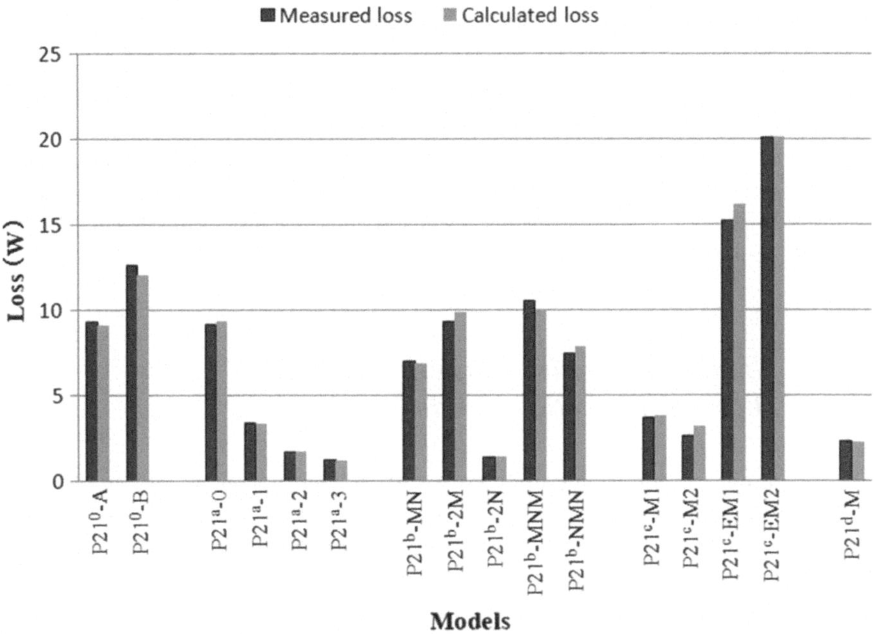

Fig. 12.23 Comparison between calculated and measured losses of Problem 21 Benchmark Family

② **Electromagnetic shield**

Unlike the case of the magnetic shield, the total loss of the electromagnetic shield, composed of separated copperplates ($P21^c$-EM2), is larger than that of the whole copperplate ($P21^c$-EM1). Further analysis shows that this is mainly a result of the poor shielding effect of the separate Cu screen (in $P21^c$-EM2) on leakage flux. Although the eddy current loss in the separated copperplate (in $P21^c$-EM2) is slightly smaller than that in the whole copperplate (in $P21^c$-EM1), the loss in the shielded steel plate in the separate Cu screen model ($P21^c$-EM2) increases significantly, resulting in an increase in the total loss. See Table 12.15.

(e) $P21^d$:

The calculated and measured loss results caused in the laminated sheets of Model $P21^d$ are also in good agreement. Moreover, it becomes possible to accurately study the electromagnetic behavior and the magnetic loss at the single sheet level. See Chap. 13 for additional results.

12.6 Problem 21 in Magnetic Saturation

In this section, Problem 21 is extended based on Model B(P21^0-B). The exciting current limit of 10 A, originally defined, is exceeded, and the exciting current is gradually increased to such an extent that the Problem 21 is at different degrees of magnetic saturation. Instead of calculating the magnetic flux density at designated positions in the air, the interlinkage flux and the magnetic flux density in the magnetic steel are calculated.

12.6.1 Nonlinear Iterative Convergence Process Under Different Excitation Conditions

If the exciting current is not higher than 10 A (rms, 50 Hz), the Problem 21 Benchmark Family, e.g., the magnetic steel plate in Model B (P21^0-B), is unsaturated; the nonlinearity of the magnetic steel must be considered in the loss analysis although the magnetic flux density in the air is relatively low. However, the nonlinear convergence can be achieved quickly under unsaturated conditions, without using accelerating convergence factors, as shown in Fig. 12.24.

Both Figs. 12.25 and 12.26 show that a convergent solution can be obtained, under weak excitation, in 10 nonlinear iterations by using $\boldsymbol{T} - \varphi - \varphi$ and $\boldsymbol{A}_r - V - \boldsymbol{A}_r$ solvers developed by the authors to calculate the loss and the interlinkage flux in the steel plate in the Problem 21 model, and the results of the two methods are consistent.

Figure 12.27 shows the calculated and measured loss in steel under weak excitation (10 A), which shows that the two are in good agreement.

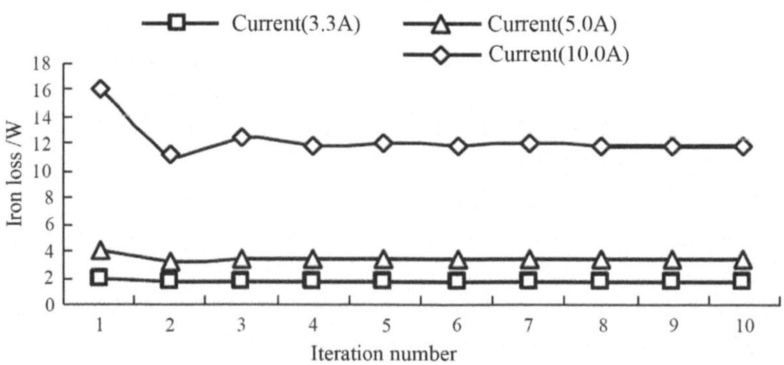

Fig. 12.24 Nonlinear convergence process (P21^0-B)

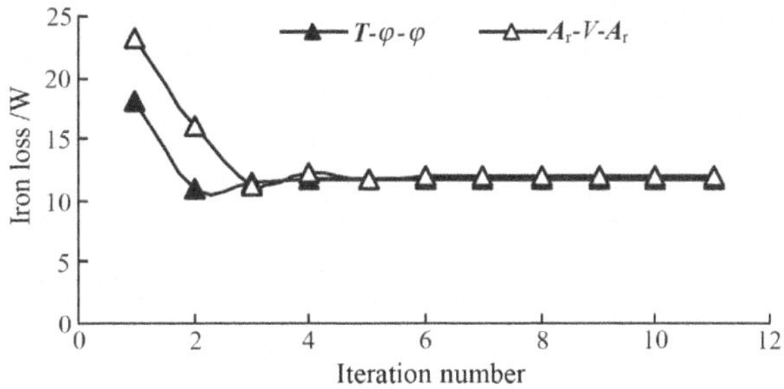

Fig. 12.25 Iron loss calculated using two methods (P21^0-B, 10 A)

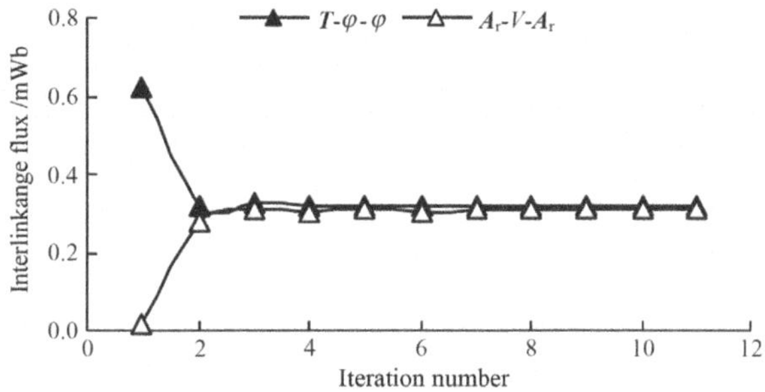

Fig. 12.26 Interlinkage magnetic flux density calculated using two methods (P21^0-B, 10 A)

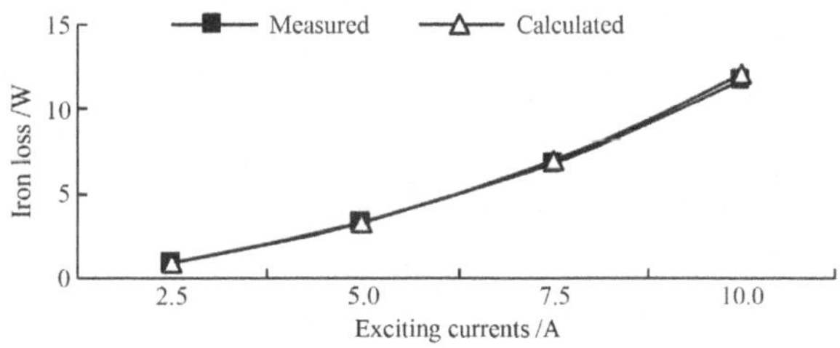

Fig. 12.27 Measured and calculated iron loss

However, the nonlinear convergence becomes more and more difficult with the increase of exciting current. For example, if the exciting current is 25 A (rms, 50 Hz), continuous oscillation occurs in the nonlinear iterative process of loss calculation. Notice that the main factor causing the oscillation is the eddy current loss, as shown in Fig. 12.28.

12.6.2 Introduction of Magnetic Saturation Factor

Magnetic components exhibit linearity under the influence of a small exterior excitation. The saturation degree of the magnetic components gradually increases with the increase of the exciting current therein. A saturation factor is introduced to

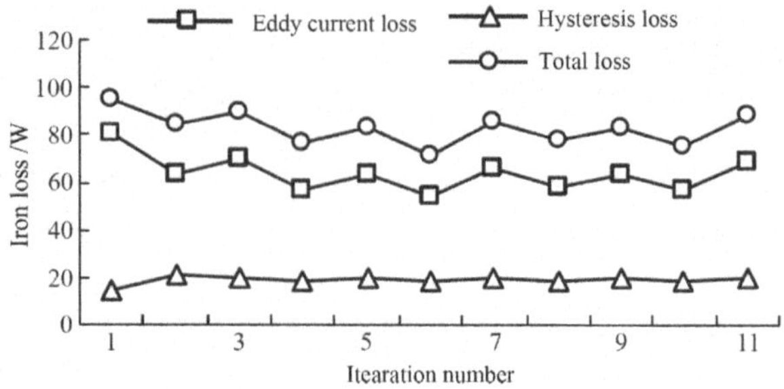

Fig. 12.28 Oscillation in nonlinear iteration (25 A)

characterize the saturation degree. Taking Model B (P21^0-B) of Problem 21 as an example, the saturation degree of magnetic plate is examined using the ratio C_s of the average magnetic flux density (B_{aver}) to the 0.5 of the maximum magnetic flux density (B_{m}) on the cross section at the designated position (z = 120.5 mm) in the model, i.e.,

$$C_{\text{s}} = \frac{B_{\text{aver}}}{0.5B_{\text{m}}} \tag{12.11}$$

It is quasi-saturated when $C_{\text{s}} \leq 0.5$. Of course, there can be different ways to define saturation degrees. The findings from P21^0-B are shown in Figs. 12.29 and 12.30.

It can be seen from the simulation results that P21^0-B can hardly be deeply saturated because the leakage magnetic flux does not form a closed circuit in the model's magnetic steel, and the exciting current cannot be increased indefinitely due to the limited current-carrying capacity of the exciting coil, so it might as well be called "quasi-saturation."

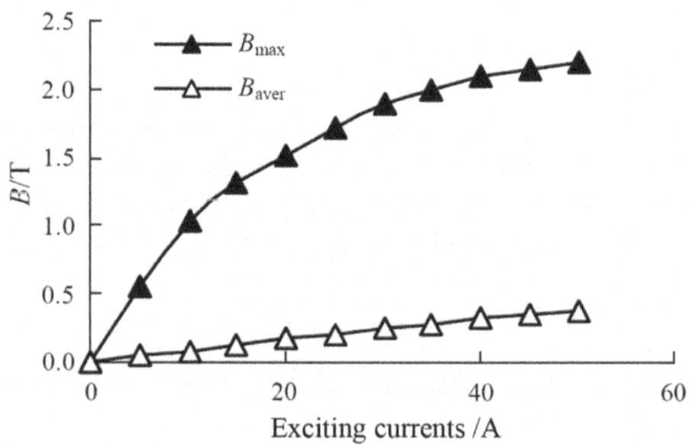

Fig. 12.29 Variation of average magnetic flux density and maximum magnetic flux density with exciting current at designated position (z = 120.5 mm)

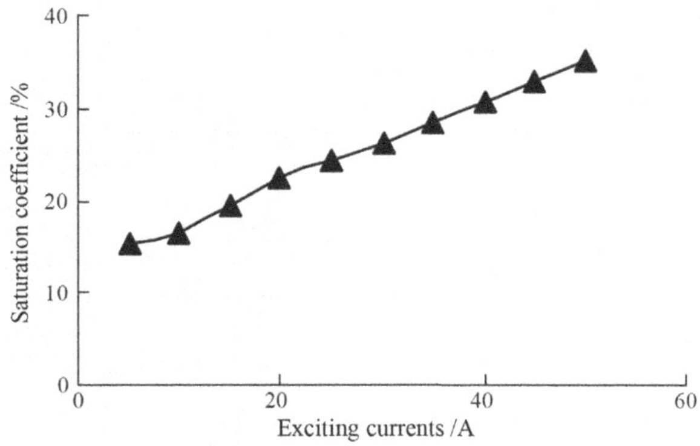

Fig. 12.30 Variation of saturation degree with exciting current

12.6.3 Analysis of Quasi-saturation

1. **Epitaxial *B–H* Curve**

When the magnetic material is saturated, the maximum magnetic flux density of the element may exceed the maximum magnetic flux density of the existing *B–H* curve in the nonlinear iterative process, so the epitaxial processing is required for the existing curve. In Model B of Problem 21, the epitaxy result of *B–H* of ordinary A3 steel can be expressed by Eq. (12.12).

$$\begin{aligned} & B = \mu_0 H - 1.9538 \times 10^{-10} H^2 + 1.9043 \times 10^{-5} H + 1.5729 \\ & 1.85\,\boldsymbol{T} < B \leq 2.1\,\boldsymbol{T} \\ & B = \mu_0 H + 2.0368 \\ & B > 2.1\,\boldsymbol{T} \end{aligned} \tag{12.12}$$

2. **Quasi-nonlinear Approach**

Quasi-nonlinear approach (QNA) is widely used in low saturation problems, including in commercial electromagnetic field software. As a matter of fact, it is a time-harmonic approach. Strictly speaking, such approach can only be used to solve linear problems, but the influence of nonlinearity can be considered in certain situations; i.e., the element's permeability is updated based on the traditional nonlinear *B–H* curve. However, more accurate and effective calculations must be studied in case of strong nonlinearity and deep saturation.

The waveform changes of the exciting current and the interlinkage flux in magnetic steel in the model with the increase of the exciting current are studied in detail by the authors, based on P21^0-B. For this purpose, a search coil is wound with thin copper wire (20 turns) at the designated position (z = 120.5 mm) of the steel plate of P21^0-B, as shown in Fig. 12.31. The induced potential on the search coil is measured under a certain exciting current, and then, time integration is

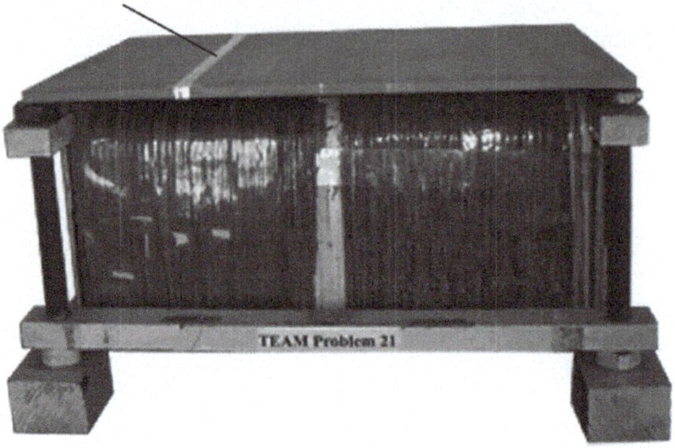

Fig. 12.31 Measurement of interlinkage flux based on P21^0-B

performed to obtain the interlinkage flux through the search coil according to Faraday's law.

The waveforms of the measured exciting current and induced potential when the exciting current rises to 50 A (rms, 50 Hz) are shown in Figs. 12.32 and 12.33. The waveform of the exciting current is sinusoidal. It can be seen that the waveform of the exciting current is still sinusoidal, but the induced potential waveform is already distorted to some extent and exhibits hysteresis when the exciting current increases to 50A, as shown in Fig. 12.34. However, the total harmonic distortion (THD) is less than 10%. It is considered that P21^0-B is still in quasi-saturated state and can be solved by quasi-nonlinear method.

As the exciting current increases, magnetic materials gradually become saturated, and nonlinear iteration has difficulty to converge. An under-relaxation factor α needs to be introduced.

$$x^{(\text{new})} = x^{(\text{old})} + \alpha\left(x^{(\text{now})} - x^{(\text{old})}\right) \tag{12.13}$$

where $x^{(\text{now})}$ is the current solution, $x^{(\text{old})}$ is the last solution, and $x^{(\text{new})}$ is the "new" solution after under-relaxation. The under-relaxation factor α is between 0 and 1, depending on the degree of saturation of the material. How to determine the

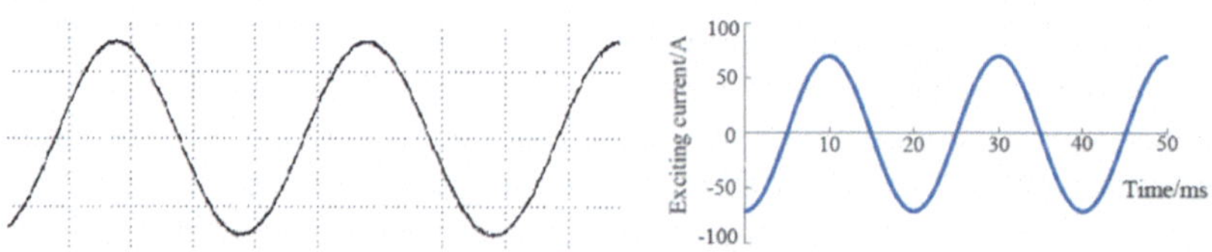

(a) Measured waveform (b) Sinusoidal waveform

Fig. 12.32 Waveform of exciting current (50 A, rms, 50 Hz)

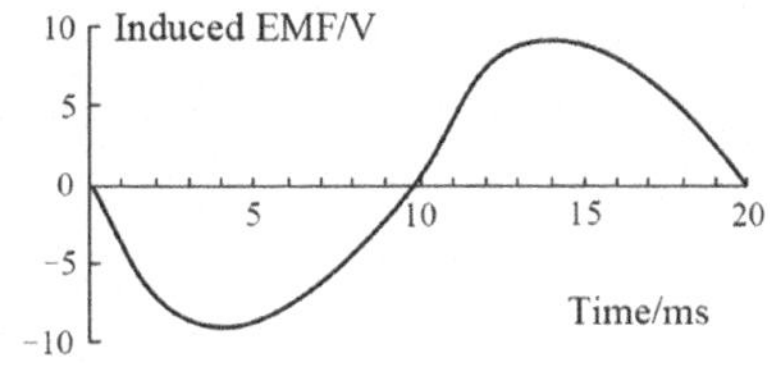

Fig. 12.33 Waveform of induced potential (50 A, rms, 50 Hz)

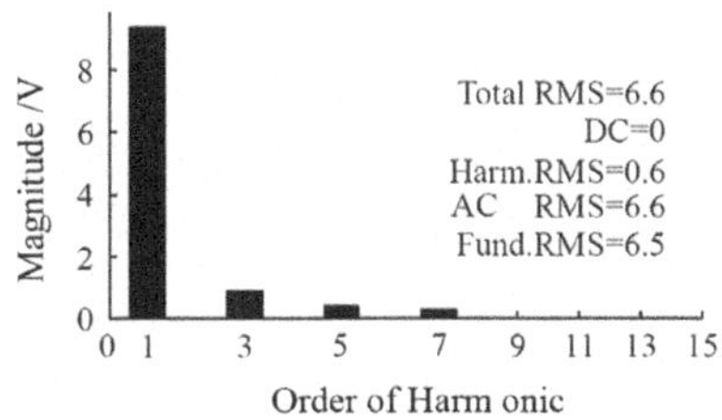

Fig. 12.34 Harmonic analysis of induced potential waveform (50 A, rms, 50 Hz)

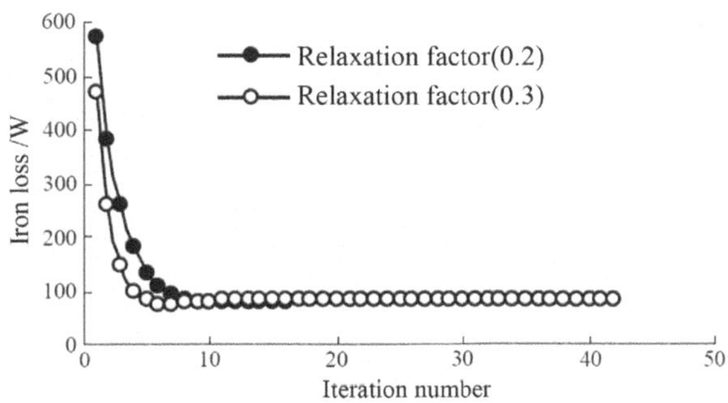

Fig. 12.35 Nonlinear convergence after introducing under-relaxation factor ($P21^0$-B, 25 A)

appropriate relaxation factor is a problem that needs further study. An example of calculating magnetic loss of P210-B is given, in which the exciting current is 25 A (rms, 50 Hz) and the relaxation factors are 0.2 and 0.3, respectively. The nonlinear convergence is shown in Fig. 12.35.

3. **Measured and Calculated Iron Loss and Interlinkage Flux under Quasi-saturation**

It is of great interest to computational electromagnetics to investigate the effectiveness of the method for calculating the interlinkage flux in magnetic steel and the iron losses under saturation conditions. What is discussed here is the situation where the exciting current gradually increases from 10 A, originally defined, to 50 A and the magnetic steel (A3 steel) is saturated to some extent, i.e., the state of quasi-saturation.

Table 12.16 Measured and calculated iron loss (P21^0-B)

Exciting current (A, rms, 50 Hz)	Measured $P_{meas.}$ (W)	Calculated $P_{calcu.}$ (W)	$(P_{calcu.} - P_{meas.})/P_{meas.}$ (%)
5	3.30	3.30	0.0
10	11.97	12.04	0.1
15	26.89	27.12	0.9
20	49.59	50.92	2.7
25	82.39	84.78	2.9
30	123.70	128.67	4.0
35	179.10	183.15	2.3
40	248.00	250.45	1.0
45	330.00	330.91	0.3
50	423.00	425.07	0.5

Table 12.17 Measured and calculated interlinkage flux (P21^0-B)

Exciting current (A, rms, 50 Hz)	Measured results $\Phi_{meas.}$ (Wb)	Calculated results $\Phi_{calc.}$ (Wb)	$(\Phi_{calc.} - \Phi_{meas.})/\Phi_{meas.}$ (%)
5	1.576E−4	1.511E−4	−4.1
10	3.183E−4	3.060E−4	−3.9
15	4.775E−4	4.578E−4	−4.1
20	6.175E−4	6.053E−4	−2.0
25	7.703E−4	7.498E−4	−2.7
30	9.360E−04	8.901E−4	−2.7
35	1.064E−03	1.024E−3	−4.1
40	1.206E−03	1.152E−3	−4.6
45	1.357E−03	1.276E−3	−6.0
50	1.486E−03	1.396E−3	−6.1

The measured and calculated total loss in steel and interlinkage flux at designated position (z = 120.5 mm) are given in Tables 12.16, 12.17. Power analyzer, Yokogawa WT 3000, Japan, and oscilloscope, Tektronix TPS 2024, are used in measurement.

The interlinkage flux ψ_{in} at the designated position of P21^0-B steel plate is calculated according to Eq. (12.14) on the basis of the magnetic field calculation result,

$$\psi_{in} = \sum_{i=1}^{N^e} B_{zi}^{(e)} \cdot S_i^{(e)} \tag{12.14}$$

where $B_{zi}^{(e)}$ and $S_i^{(e)}$ are the z-component of the magnetic flux density and the area of the element "e" in the z-direction, respectively, and N^e is the cross-sectional area of the designated position.

12.7 Further Co-research for Problem 21 Family

12.7.1 New Proposal of Problem 21 Family

As mentioned above, in the original planning meeting for the TEAM, the goal for the TEAM Workshops was stated as: "The ultimate goal is to show the effectiveness of numerical techniques and associated computer codes in solving electromagnetic field problems and to gain confidence in their predictions. The workshops should also provide cooperation between workers, leading to an interchange of ideas" [2].

It should be noted again that the effectiveness of the computational electromagnetic modeling and simulation is dependent on not only efficient electromagnetic analysis method and computational software, but also sufficient and applicable material property data. Consequently, the material property modeling, especially, under complex, even extreme working conditions, is also very important for effective modeling and simulation.

In the current version of Problem 21 Family, V.2009, posted on the ICS official Web site (www.compumag.org/team), all the exciting currents have a sinusoidal waveform, even though the magnetic saturation can be achieved.

Accordingly, further extension of the Problem 21 Family should handle the extreme excitation condition; i.e., the magnetic components of the member models of Problem 21 Family are excited by a non-sinusoidal supply which may contain a number of harmonics and DC components. Table 12.18 describes two newly proposed models, $P21^e$-B(NS) and $P21^e$-M(NS), based on member models of Problem 21 Family, $P21^0$-B and $P21^d$-M [50].

Table 12.18 New proposal of Problem 21 Family with extreme excitations

Proposed new models	Electromagnetic features	Industry background	Remarks
$P21^e$-B (NS)	3D nonlinear transient field in solid magnetic plate under multi-harmonic and/or DC bias excitations	Magnetic loss in solid magnetic components under extreme conditions	Upgraded $P21^0$-B with magnetic flux compensation
$P21^e$-M (NS)	3D nonlinear transient field in laminated sheets under multi-harmonic and/or DC bias excitations	Magnetic loss in laminated magnetic components under extreme conditions	Ungraded $P21^c$-M1 with different types of laminated sheets

Fig. 12.36 Upgraded TEAM P21's power supply

Meanwhile, an enhanced excitation supply with leakage flux complementary coils has been established by the authors [59], which can be used for magnetic measurement under non-sinusoidal conditions:

(1) The enhanced exciting coils are built up; i.e., the number of the turns of the exciting coil is increased from 300 to 400, and the dimension of the copper wire is also upgraded from 6.7 × 2 mm to 9 × 3 mm.
(2) In order to keep the leakage magnetic field of the exciting coils almost the same, when the magnetic components (such as the magnetic plate or GO silicon steel lamination) are removed (referred to as no-load case), i.e., keeping the loss generated in the exciting coils almost the same, when the magnetic components of high permeability are removed, two complementary coils (called the C-coil1 and C-coil2) are utilized, which have exactly the same specifications as excitation coils (E-Coils) and can move in parallel tracks. The enhanced measuring setup is helpful to evaluate the total loss (i.e., resistance and eddy current loss) caused in exciting coils, and then indirectly determine the magnetic loss inside the magnetic components (GO silicon steel lamination or steel plate). See Fig. 12.36.

Note that additional results concerning with the stray-field loss and flux distribution inside magnetic steel plate under multi-harmonic excitations have been presented by the authors [61], based on the proposed benchmark Model P21^{e}-B (NS). An improved method for accurately determining the stray-field loss in the magnetic steel components is also proposed. The effects of the harmonic order and phase angle on the stray-field loss in magnetic steel components are examined in detail.

12.7.2 Improved Method to Determine Stray-Field Loss

As mentioned above, there are some difficulties to determine the stray-field loss in conducting components of P21 models, mainly including:

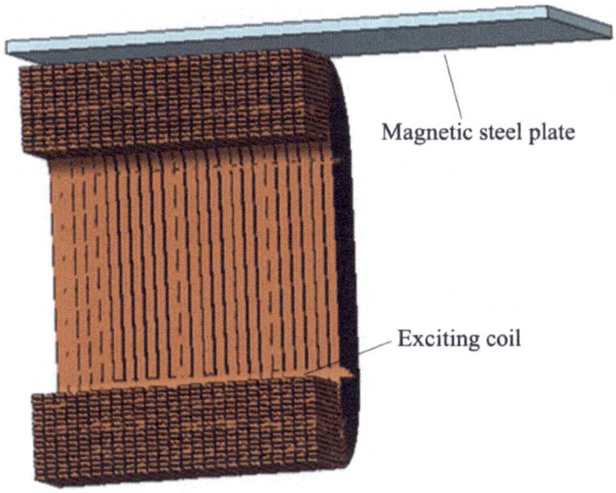

Fig. 12.37 FE model for calculating eddy current loss in exciting coil

(1) The measured total loss of P21 model includes the stray-field loss in the conducting component and the total loss caused in the exciting coils, which cannot be experimentally separated;
(2) The total loss in the exciting coils, especially for the eddy current loss therein, is closely dependent on the leakage flux distribution, which is different when the excitation is applied on the P21 model with or without the conducting components.

Fortunately, the measurement methodology of the total loss of P21 models, with or without conducting components, referred to as load and no-load cases, respectively, has been well-established. Also, the eddy current loss of the exciting coils under load or no-load conditions can be accurately calculated. Figure 12.37 shows an example of 3D FE model of eddy current loss calculation based on Model P21^{e}-B(NS).

Based on the combination of advanced magnetic measurement and numerical simulation, an improved method to indirectly determine the stray-field loss in conducting components P_x has been proposed and implemented by the authors [61], i.e.,

$$P_x = P_{\mathrm{load(m)}} - P_{n\mathrm{load(m)}} - \Delta \tag{12.15}$$

$$\Delta = P_{\mathrm{coil}} - \mathrm{load(c)} - P_{\mathrm{coil}-n\mathrm{load(c)}} \tag{12.16}$$

where $P_{\mathrm{load(m)}}$ and $P_{n\mathrm{load(m)}}$ denote the measured total loss of P21 model under the load and no-load conditions, respectively; Δ denotes the variation of eddy current loss of the exciting coils due to conducting component; $P_{\text{coil-load(c)}}$ and $P_{\text{coil-}n\text{load(c)}}$ denote the calculated eddy current loss of the exciting coil under the load and no-load conditions, respectively.

12.8 Summary and Outlook

12.8.1 Summary on Problem 21 Family

After more than 25 years of development, TEAM Problem 21 has been upgraded from only two benchmark models, at the very beginning, to a Benchmark Family, including 5 sets of 16 member models (V.2009), with typical and practical engineering background. In the P21-based benchmarking (TEAM), the stray-field losses and the 3D eddy current distributions in conducting components specified in all benchmark model definitions, as well as the magnetic fields on the conductor surface, in the air and at designated positions, inside the conducting parts, have been calculated and measured. A number of valuable results, with engineering scientific significance, have been obtained, which will be beneficial in helping to make a reasonable selection of materials and optimizing the structure configuration in electromagnetic design and deepen the research and engineering application of stray-field loss problems in electrical engineering. It is briefly summarized as follows:

(1) There may be great differences between the analysis results of 2D and 3D eddy current fields for a particular problem; e.g., the eddy current loss results in $P21^a$ models. It can be seen from these differences that the loss results calculated from 2D current field are not desirable even for linear problems and that 3D eddy current field analysis should be carried out; i.e., different analysis methods are needed for different problems.
(2) Based on the fact that linear and nonlinear processing techniques lead to great differences when being used to calculate the magnetic loss of benchmark models, the nonlinearity of material must be considered in the calculation of magnetic loss, even if the leakage electromagnetic field in the air nearby the components is relatively low.
(3) A clear engineering background is provided for each set of benchmark models, and comparability exists between member models. By modeling the electromagnetic behaviors of the benchmark models (eddy current, magnetic field, loss distribution, skin effect, etc.), it can be found that quantifying the regularity of those behaviors contained in typical electromagnetic structures is helpful to the optimal design of the product structure. Such optimization may involve the type and number of slots in the core tie plate of transformer, the basis for using hybrid steel structure in oil tank design and the structural design, and optimization of shields.
(4) The effectiveness of the benchmark models used to research, develop, and validate various engineering electromagnetic field computation methods and the corresponding software has been verified, while it is almost impossible to carry out such validation on large electromagnetic devices.
(5) All the benchmarking (TEAM) results of 16 member models of the Problem 21 Benchmark Family show that the measured and calculated magnetic flux densities and losses of each model agree well with each other within the

definition of benchmark problems, which verifies that the proposed and implemented analysis method for eddy current–hysteresis loss problems is practical and effective.

(6) The definition of the Problem 21 benchmark problem can be exceeded to increase the exciting current to gradually saturate the magnetic material. The comparison between experimental and calculated results shows that the improved quasi-nonlinear approach can be used to obtain satisfactory results when under quasi-saturation. However, for the problems of deep saturation and strong nonlinearity, accurate and effective approaches are needed, which requires further study.

(7) The Problem 21 Benchmark Family can be extended to change the direction and effective value of the exciting current in the two exciting coils, so as to study the influence of the change in excitation condition on the loss in the conducting components. For instance, when the current flows in one direction, the loss decreases due to the significant reduction of eddy current loss.

(8) A P21-based benchmarking platform is well-established, which can be used for further electromagnetic analysis and design. Simplifying the magnetic shielding model, e.g., can be used for in-depth study on the additional iron loss caused by alternative leakage magnetic flux entering silicon steel sheets vertically. Additional details of the models, experiments, and simulations can be found in Chap. 13.

12.8.2 Outlook on the Future Co-research

With the rapid development of power systems and the construction of UHV AC/DC transmission and transformation projects and the complex operating conditions, serious challenges have been posed to the R&D, analysis and design of very large UHV power equipment. The measurement of material properties, the multi-physics and multi-scale modeling and simulation, and the experimental validation, etc., under non-sinusoidal (including multiple harmonics and DC bias) excitation, have become a series of key topics.

In this chapter, the further co-research project related to Problem 21 Benchmark Family is proposed, considering the realistic complexity of engineering field problems, such as stray-field loss under non-sinusoidal excitation conditions, i.e., establishing new benchmark models involving complex excitation conditions. Meanwhile, the extended TEAM study and material modeling need to be closely combined. This is because the effectiveness of modeling and simulation involves not only electromagnetic analysis methods, but also related material properties. The validation of modeling and simulation under complex operating conditions is particularly important [50, 61].

Acknowledgements The authors would like to express their thanks to the ICS Board for approving the proposal (1993) and updates (1999–2009) of the Problem 21 Benchmark Family, to T. Nakata, A. Kost, L. R. Turner, J. Sykulski, and O. Biro, for their support and help, to all colleagues, including Sheng Gao, Mingxun Wu, Junjie Zhang, Yana Fan, Xiaoyan Wang, Zhigang Zhao, Yang Liu, Sumei Yang, Yong Du, Qingyi Kong, et al., for their long-term cooperation of this benchmarking research, and to many postgraduate students for their contributions to the relevant experimental studies and numerical calculations in this book over the years. This project was supported in part by the Natural Science Foundation of China under No. 59277296, No. 59924035, and No. 50777042, Hebei Natural Science Foundation under No. E2006001036, and Hebei Provincial Government's special fund for talent training. The authors thank all related leaders of Baobian Electric. and thank D. Lowther and his research team for great support in P21-based numerical analysis and for all comments and advices from computational electromagnetic colleagues worldwide.

Appendix 12.1: Characteristics of Magnetic Steel Plates Used in Problem 21 Family

The *B–H*, and $B_m - W_h$ curves of the magnetic steel plate (A3) used in Problem 21 Family have been measured by the Electromagnetic Device Laboratory (Okayama University, Japan). As shown in Table 12.19, the conductivity of the steel plate is $\sigma = 6.484 \times 10^6$ S/m.

Table 12.19 $B_m - W_h$ curve of magnetic steel plate

B (T)	*H* (A/m)	W_h (W/kg)	*B* (T)	*H* (A/m)	W_h (W/kg)
0.049	115	0.02	1.449	1965	11.86
0.101	171	0.11	1.500	2506	12.80
0.150	196	0.24	1.550	3291	13.77
0.200	214	0.41	1.600	4430	14.68
0.299	245	0.82	1.639	5599	15.28
0.399	279	1.32	1.670	6698	15.73
0.499	316	1.88	1.701	7926	16.22
0.601	359	2.51	1.729	9251	16.84
0.700	405	3.21	1.760	10,792	17.32
0.801	461	3.98	1.781	11,930	17.51
0.899	528	4.82	1.800	13,106	17.57
1.001	616	5.77	1.830	14,949	17.75
1.099	732	6.81	1.850	16,290	17.72
1.200	898	8.01	1.875	18,002	17.82
1.300	1154	9.39	1.900	19,942	17.85
1.401	1606	10.99			

Appendix 12.2: Characteristics of Silicon Steel Sheets Used in Problem 21 Family

The B_m–H_b, B_m–W_h, and B_m–W curves of anisotropic silicon steel sheets (30RGH120) used in Problem 21 Family have been measured by the Electromagnetic Device Laboratory (Okayama University, Japan), as shown in Tables 12.20, 12.21, 12.22, 12.23, 12.24 and Figs. 12.38, 12.39, 12.40, 12.41, 12.42, 12.43.

It should be noted that the magnetic field strength at the maximum magnetic flux density (B_m) is Hb. It is applied at B_m and H_b because they provide an instant when the eddy current is zero. This means that the measured B_m–H_b curve is very close to the DC B–H curve. W represents the total loss per kilogram of silicon steel sheet in Figs. 12.42, 12.43, and Table 12.24.

Table 12.20 B_m–H_b of silicon steel sheet in rolling direction

B_m (T)	H_b (A/m)	B_m (T)	H_b (A/m)
0.0500	1.5757	1.0496	9.6248
0.1000	2.5912	1.0998	9.8313
0.1501	3.4846	1.1497	10.2613
0.2001	4.2920	1.1998	10.6696
0.2501	4.9280	1.2497	11.2566
0.3001	5.5596	1.2997	12.0681
0.3501	6.0624	1.3496	0.1694
0.4002	6.3726	1.3997	15.0969
0.4502	6.8566	1.4496	17.6901
0.5001	7.0721	1.4997	21.2103
0.5503	7.4201	1.5494	26.2651
0.6000	7.6989	1.5994	33.4483
0.6503	7.8836	1.6495	46.9983
0.7004	8.1211	1.6995	69.1130
0.7504	8.3246	1.7494	105.8169
0.8004	8.6006	1.7994	182.9606
0.8504	8.7677	1.8491	357.9689
0.9004	8.9253	1.8991	742.4458
0.9497	9.1544	1.9490	1667.8985
1.0001	9.3758	1.9990	9081.6648

Table 12.21 B_m–H_b curve of silicon steel sheet in a direction perpendicular to the rolling direction

B_m (T)	H_b (A/m)	B_m (T)	H_b (A/m)
0.0501	85.3664	1.0489	189.4970
0.1000	120.8005	1.0999	196.7729
0.1493	135.6965	1.1490	208.2999
0.2006	144.5633	1.1998	232.3968
0.2489	150.0459	1.2492	289.9317
0.3002	154.0995	1.2997	433.3158
0.3494	157.5049	1.3491	736.1902
0.4004	159.7633	1.3997	1366.9074
0.4494	161.6266	1.4492	2361.8001
0.5001	163.9006	1.4996	3698.2184
0.5493	165.3869	1.5491	5293.3038
0.6004	166.8455	1.5993	7186.9947
0.6492	168.3099	1.6493	9316.4081
0.7000	169.6325	1.6990	11650.1128
0.7491	171.4374	1.7492	14193.6353
0.8002	173.1495	1.8001	16952.0366
0.8492	175.3864	1.8501	19834.8760
0.8999	178.1123	1.9001	22893.7254
0.9490	181.0530	1.9502	26145.8958
0.9998	184.8982	2.0002	29665.5458

Table 12.22 B_m–W_h curve of silicon steel sheet (in rolling direction)

B_m (T)	W_h (W/kg)	B_m (T)	W_h (W/kg)
0.0500	0.0007	1.0498	0.1693
0.1000	0.0022	1.0998	0.1839
0.1501	0.0046	1.1498	0.2028
0.2001	0.0077	1.1997	0.2224
0.25011	0.0122	1.2495	0.2431
0.3002	0.0166	1.2996	0.2588
0.3501	0.0222	1.3493	0.2793
0.4001	0.0284	1.3995	0.3035
0.4502	0.0370	1.4495	0.3313
0.5002	0.0435	1.4995	0.3536
0.5502	0.0521	1.5489	0.3885
0.6002	0.0610	1.5988	0.4280
0.6502	0.0726	1.6485	0.4811
0.7003	0.0800	1.6989	0.5439
0.7501	0.0918	1.7492	0.6348
0.8002	0.1022	1.7992	0.7610
0.8501	0.1165	1.8493	0.9071
0.9001	0.1279	1.8993	1.0725
0.9500	0.1434	1.9491	1.2449
1.0000	0.1561	1.9992	1.3170

Table 12.23 B_m–W_h curve of silicon steel sheet (in a direction perpendicular to the rolling direction, 50 Hz)

B_m (T)	W_h (W/kg)	B_m (T)	W_h (W/kg)
0.0501	0.0204	0.8498	0.9289
0.1003	0.0648	0.8999	0.9896
0.1500	0.1187	0.9496	1.0490
0.2001	0.1774	0.9997	1.1130
0.2499	0.2369	1.0495	1.1784
0.3001	0.2993	1.0995	1.2481
0.3499	0.3589	1.1494	1.3293
0.4002	0.4191	1.1994	1.4254
0.4500	0.4771	1.2494	1.5539
0.4999	0.5358	1.2996	1.7322
0.5499	0.5932	1.3493	1.9579
0.6000	0.6502	1.3994	2.2127
0.6499	0.7058	1.4494	2.4482
0.6999	0.7613	1.4993	2.6532
0.7499	0.8164	1.5492	2.8002
0.7999	0.8725	1.5993	2.9182

Table 12.24 B_m–W_h curve (50 Hz)

Rolling direction		Perpendicular to rolling direction	
B_m (T)	W (W/kg)	B_m (T)	W (W/kg)
0.0500	0.0011	0.0500	0.0290
0.1000	0.0040	0.1000	0.0922
0.1501	0.0088	0.1493	0.1685
0.2001	0.0151	0.2006	0.2526
0.2501	0.0226	0.2489	0.3344
0.3001	0.0324	0.3002	0.4224
0.3501	0.0436	0.3494	0.5048
0.4002	0.0549	0.4004	0.5910
0.4502	0.0704	0.4494	0.6719
0.5001	0.0850	0.5001	0.7566
0.5503	0.1012	0.5493	0.8374
0.6000	0.1202	0.6004	0.9218
0.6503	0.1426	0.6492	0.9994
0.7004	0.1626	0.7001	1.0838
0.7504	0.1886	0.7491	1.1651
0.8004	0.2090	0.8002	1.2501
0.8504	0.2352	0.8492	1.3342
0.9004	0.2580	0.8999	1.4246
0.9497	0.2946	0.9490	1.5123
1.0001	0.3213	0.9998	1.6041

(continued)

Table 12.24 (continued)

Rolling direction		Perpendicular to rolling direction	
B_m (T)	W (W/kg)	B_m (T)	W (W/kg)
1.0496	0.3557	1.0489	1.6978
1.0998	0.3861	1.0999	1.8012
1.1497	0.4234	1.1490	1.9122
1.1998	0.4503	1.1998	2.0386
1.2497	0.4977	1.2492	2.1965
1.2997	0.5355	1.2997	2.4087
1.3496	0.5806	1.3491	2.6844
1.3997	0.6227	1.3996	2.9984
1.4496	0.6770	1.4492	32.752
1.4997	0.7234	1.4996	3.5174
1.5494	0.7834	1.5491	3.6992
1.5994	0.8487	1.5993	3.8528
1.6495	0.9241	1.6493	3.9627
1.6995	1.0100	1.6990	4.0424
1.7494	1.1262	1.7492	4.0971
1.7994	1.2803	1.8001	4.1714
1.8491	1.4580	1.8501	4.2433
1.8991	1.6727	1.9001	4.3202
1.9490	1.9030	1.9502	4.4122
1.9990	2.1173	2.0002	4.5229

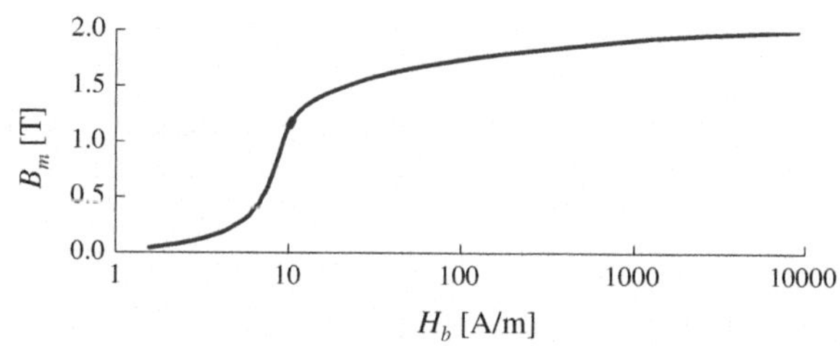

Fig. 12.38 B_m–H_b curve of silicon steel sheet (in rolling direction)

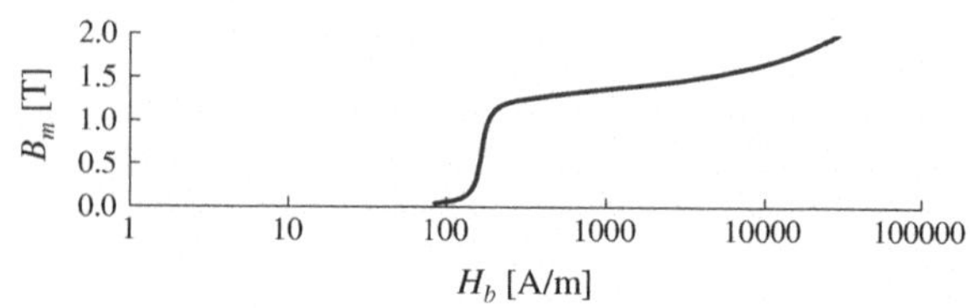

Fig. 12.39 B_m–H_b curve of silicon steel sheet (in a direction perpendicular to the rolling direction)

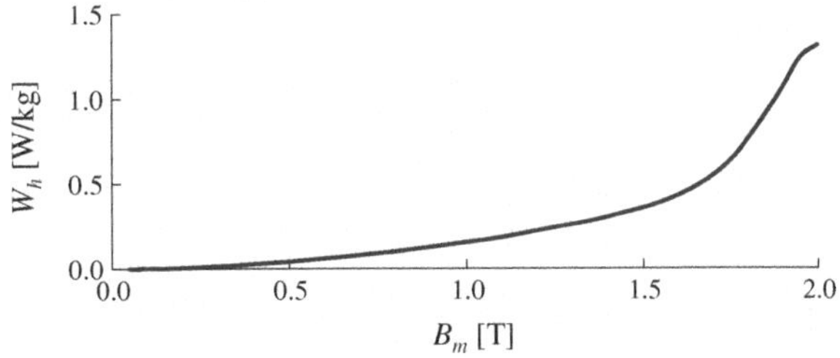

Fig. 12.40 B_m–H_b curve of silicon steel sheet (in rolling direction)

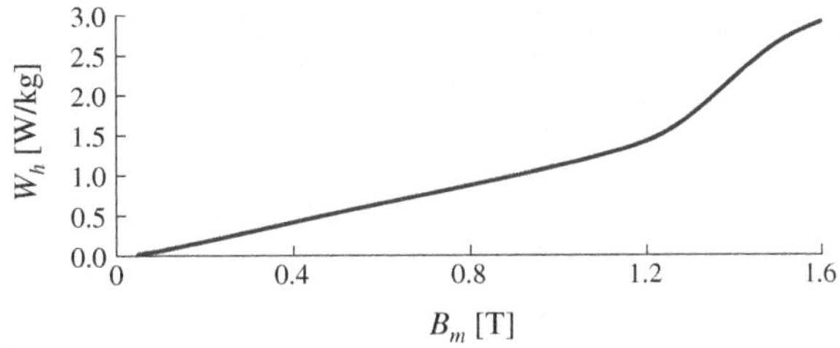

Fig. 12.41 B_m–H_b curve of silicon steel sheet (in a direction perpendicular to the rolling direction, 50 Hz)

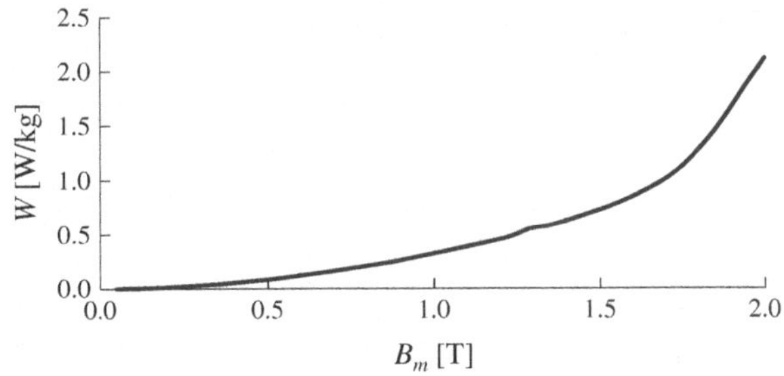

Fig. 12.42 B_m–W curve of silicon steel sheet (in rolling direction, 50 Hz)

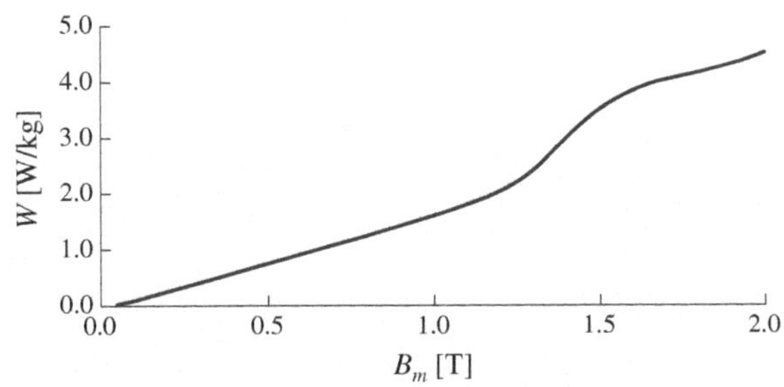

Fig. 12.43 B_m–W curve of silicon steel sheet (in a direction perpendicular to the rolling direction, 50 Hz)

Appendix 12.3: Reference Data of Problem 21 Family

Some input about Problem 21 Family has been collected into Tables 12.25 and 12.26 for quick reference.

Table 12.25 Conductor specifications used in Problem 21

Models	Magnetic steel plate (A3)		Non-magnetic steel (20Mn23Al)		Silicon steel sheet (30RGH120)		Copperplate	
	Size (mm)	Qty	Size (mm)	Qty	Size (mm)	Qty	Size (mm)	Qty
$P21^0$-A	360 × 248 × 10	2						
$P21^0$-B	360 × 520 × 10	1						
$P21^a$-0			360 × 820 × 10	1				
$P21^a$-1			360 × 820 × 10	1				
$P21^a$-2			360 × 820 × 10	1				
$P21^a$-3			360 × 820 × 10	1				
$P21^b$-MN	360 × 248 × 10	1	360 × 248 × 10	1				
$P21^b$-2M	360 × 248 × 10	2						
$P21^b$-2N			360 × 248 × 10	2				
$P21^c$-M1	360 × 520 × 10	1			270 × 458 × 0.3	20		
$P21^c$-M2	360 × 520 × 10	1			80 × 458 × 0.3	20 × 3		
$P21^c$-EM1	360 × 520 × 10	1					270 × 458 × 6	1
$P21^c$-EM2	360 × 520 × 10	1					80 × 458 × 6	3

Table 12.26 Conductor physical parameters used in Problem 21

Performance parameters	Magnetic steel plate (A3)	Non-magnetic steel plate (20Mn23Al)	Silicon steel sheet (30RGH120)	Copperplate
Conductivity (S/m)	×	×	×	×
Density (kg/m^3)			×	×
B–H curve (isotropic)	Appendix 12.1 (Table 12.19)			
B_m-W_h curve (isotropic)	Appendix 12.1 (Table 12.19)			
B–H curve (in rolling direction)			Appendix 12.2 (Table 12.20)	
B–H curve (in a direction perpendicular to the rolling direction)			Appendix 12.2 (Table 12.21)	
B_m-W_h curve (in rolling direction)			Appendix 12.2 (Table 12.22)	
B_m-W_h curve (in a direction perpendicular to the rolling direction)			Appendix 12.2 (Table 12.23)	
B_m-W curve (in rolling direction)			Appendix 12.2 (Table 12.24)	
B_m-W curve (in a direction perpendicular to the rolling direction)			Appendix 12.2 (Table 12.24)	

References

1. TEAM Workshops: Test problems(ed. by L. R. Turner), April 1988.
2. L. R. Turner, "New directions for the TEAM Workshops," Verification of Software for 3-D Electromagnetic Field Analysis," International Academic Publishers, Proc. of Asian TEAM Workshop (ed by Z. Cheng, K. Jiang, and N. Takahashi), Qiandaohu, China, 1992, pp. 6–15.
3. N. Ida, "The TEAM Workshops: fourth round," Proc. of TEAM Workshop on computation of applied electromagnetics in materials, 30 Jan., 1993, pp. 9–12.
4. Proc. of European TEAM Workshop and international seminar on electromagnetic field analysis, (ed. by Y. Crutzen, N. J. Diserens, C. R. I. Emson, and D. Rodger), Oxford, England, pp. 23–25, April 1990.
5. Proc. of TEAM Workshop-Qiandaohu, China(ed. by Z. Cheng, K. Jiang and N. Takahashi), 1992.
6. Proc. of ICEF-1996, Electromagnetic Field Problems and Application (ed. by K. Zhou), International Academic Publishers, 1997.
7. Proc. of ICEF-2000, Electromagnetic Field Problems and Application (ed. by W. Yan), 2001.
8. C. W. Trowbridge, "The future of TEAM: some personal thoughts(invited talk)," TEAM Workshop, Chengde, September 2000.
9. D. Xie and R. Tang, "Development of Computational Electromagnetics in China," *IEEE Trans. on Magn.*, vo.42, no.4, 2005, pp. 509–514.
10. Proc. of ICEF-2008, Electromagnetic Field Problems and Application-II, vol.22(ed. by M. Chen, S. Yang, D. Xiao and X. Zhang),TSI Press Series, USA, 2008, pp. 357–366.

11. Z. Cheng, Q. Hu, S. Gao, Z. Liu, C. Ye, M. Wu, J. Wang and H. Zhu, "An engineering-oriented loss model (Problem 21)," Proc. of the International TEAM Workshop, Miami, 1993, pp. 137–143.
12. T. Nakata and K. Muramutsu, "Has the 3D magnetic field analysis come into practical application?" (invited), CBMAG, 1995, pp. 7–12.
13. S. Gao, M. Wu, H. He, J. Wang, Z. Liu, Q. Hu and Z. Cheng, "Problem 21^+: slotted non-magnetic steel plate driven by Problem 21's source," Proc. of the ICEF & TEAM Workshop, Yichang, 1996,pp. 366–370.
14. Z. Cheng, N. Takahashi, S. Gao and T. Sakura, "Loss analysis based on revised version of TEAM Problem 21," Presented by N. Takahashi at TEAM-Sapporo, 1999.
15. Z. Cheng, N. Takahashi, Q. Hu and C. Fan, "TEAM-based benchmark family: Problem 21/ $21^+/21^*$," Presented by Z. Cheng at 4th IEE CEM, UK, 2002.
16. Z. Cheng, N. Takahashi, S. Yan, T. Asano, Q. Hu and X. Ren, "Proposal of Problem 21-based shielding model (Problem 21^*)," Proc. of TEAM Workshop, Shenyang, 2005, pp. 15–20.
17. Z. Cheng, N. Takahashi, S. Liu, S. Yang, C. Fan, Qi. Hu, L. Liu, M. Guo, and J. Zhang, "Benchmarking-based approach to engineering stray-field loss problems," presented by S. Liu at ACES-2006, Miami, USA.
18. O. Biro and K. Preis, "Solution of TEAM benchmark problem 21 (an engineering-oriented loss model)," Proceeding of the International TEAM Workshop, Aix-Les-Bains, 1994, pp. 25–27.
19. Z. Cheng, Q. Hu, S. Gao, Z. Liu, C. Ye and M. Wu, "On the benchmark solution of a typical engineering loss problem," 10th Annual Review of Progress in Applied Computational Electromagnetics, Monterey, 1994, pp. 335–342.
20. Z. Cheng, Q. Hu, S. Gao, Z. Liu, C. Ye and M. Wu, "Supplementary report on problem 21," Proc. of the International TEAM Workshop, Aix-Les-Bains, France, 1994, pp. 37–47.
21. T. Nakata, N. Takahashi, K. Fujiwara, H. Ohashi and H. Zhu, "Analysis of an engineering-oriented loss model(Problem 21)," Proc. of the TEAM Workshop, Seoul, 1994, pp. 51–55.
22. T. Nakata, N. Takahashi, K. Fujiwara, H. Ohashi and H. Zhu, "Analysis of flux and eddy current distribution of 3-D stray-field lossmodel(Problem 21)," Proc. of the International TEAM Workshop, Aix-Les-Bains, 1994, pp. 21–24.
23. O. Biro, K. Pries, and K. R. Richter, "Various FEM formulation for the calculation of the calculation of transient 3D eddy currents in nonlinear media," *IEEE Trans. on Magn.*, vol.31, no.3, 1995, pp. 1307–1312.
24. J.P. Bastos, N. Ida, R.C. Mesquita and J. Hector, "Problem 21: solution using edge elements and a vector potential **A**," Proc. of the International TEAM Workshop, Berlin,1995, pp. 71–75.
25. Z. Cheng, Q. Hu, S. Gao, Z. Liu and M. Wu, "Stray-field loss analysis and measurement (Problem 21)," Proc. of the International TEAM Workshop, Berlin, 1995, pp. 62–66.
26. N. Takahashi, T. Nakata, K. Fujiwara, K. Muramatsu and T. Torii, "Time periodic finite element analysis of Problem 21," Proc. of the International TEAM Workshop, Berlin, 1995, pp. 56–61.
27. J. Gu, W. Yan and Q. Yang, "Computation of flux and eddy current distributions of an engineering-oriented loss model (Problem 21)," Proc. of the International TEAM Workshop, Berlin, 1995, pp. 52–55.
28. L. Li, X. Cui and J. Yuan, "Numerical calculation of 3D eddy current loss and magnetic field (Problem 21)," Proc. of the International TEAM Workshop, Berlin, 1995, 67–70.
29. N. Takhashi, K. Fujiwara, and T. Torill, "Analysis of eddy current in slotted plate utilizing crack element," Electromagnetic field problems and applications (ed. by Zhou Keding), ICEF, 1996, pp. 371–375.
30. Z. Cheng, S. Gao, Q. Hu, Z. Liu, and M. Wu, "Summary of results for Problem 21(3-D stray-field Loss model)," Proceedings of the international TEAM Workshop, Berlin, July 14,1995, pp. 76–83.

31. J. Fetzer, S. Kurz, and G. Lehner, "The solution of TEAM Workshop Problem 21 using BEM-FEM coupling," Proc. of the International TEAM Workshop, Okayama, 1996, pp. 19–22.
32. C. Golavanov, Y. Marechal and G. Meunir, "Solution of TEAM Workshop problem 21," Proc. of the International TEAM Workshop, Okayama, 1996, pp. 26–30.
33. N. Takahashi, K. Fujiwara, and T. Torii, "Analysis of eddy current in slotted plate utilizing crack element," Proc. of TEAM Workshop, Yichang, Electromagnetic Field Problems and Application (ed. by Zhou Keding)pp. 371–375, 1996.
34. P. J. Leonard, D. Rodger, G. Ciuprina and D. Loan, "Solution of TEAM Problem No. 21," Proc. of TEAM Workshop, Graz, 1996, pp. 17–19.
35. Y. Yao, D. Xie and B. Bai, "The application of the modified impedance boundary condition to the Problem 21," Electromagnetic Field Problems and Applications (ed. by Zhou Keding), ICEF-1996, pp. 376–379.
36. Y. Yao, D. Xie, B. Bai, and L. Zeng,"Nonlinear transient analysis and nonlinear surface impedance method of TEAM Problem 21," Proceedings of ICEF-Tianjin and TEAM-Chengde, 2000.
37. Y. Hu, and R. Tang, "T-φ-φ_{m} + GTM method for 3-D unbounded eddy current analysis (Problem 21)," Proceedings of ICEF-Tianjin and TEAM-Chengde, 2000.
38. H. Lin, "Solution of TEAM Workshop Problem 21 with improved sub-structure technique," Electromagnetic Field Problems and Applications(ed. by Zhou Keding), ICEF, 1996, pp. 380–382.
39. N. Ida and J. P. A. Bastos, *Electromagnetics and calculation of fields*, Springer(second edition), 1997, pp. 477–480.
40. Z. Cheng, N. Takahashi, S. Gao, and Z. Liu, "Some new developments of Problem 21," Proc. of the international TEAM Workshop, Rio de Janeiro, Brazil, November 1997, pp. 21–24.
41. N. Takahashi, "Calculation of iron loss of engineering oriented loss model (Problem 21)," Proc. of the international TEAM Workshop, Rio de Janeiro, Brazil, November 1997, pp. 25–26.
42. N. Takahashi, K. Fujiwara, K. Sugiyama, and J. Takehara, "Verification of eddy current analysis of engineering oriented loss model(Problem 21)," Proc. of the ACES, USA, 1998, pp. 779–786.
43. Z. Cheng, N. Takahashi, S. Gao, and Q. Hu, "Hysteresis loss analysis in steel based on W_h-$B_m$curve," Proc. of CICEM, Xian, China,1999, pp. 454–457.
44. I. Sebestyen and M. Kuczmann, "Investigation of different approaches for iron loss prediction methods using TEAM Problem 21," Record of the 15th Compumag Conference on the Computation of Electromagnetic Fields, Shenyang, 2005, vol.2,pp. 268–269 (OD-3); Proc. of TEAM Workshop, Shenyang, 2005, pp. 5–6.
45. K. Ishibashi, Z. Andjelic, and D. Pusch, "Nonlinear eddy current analysis by BEM equation utilizing adaptive technique," presented at IEEE CEFC-2008, Athens.
46. S. Gao, Z. Liu, M. Wu, and Z. Cheng, "Problem 21: loss modelling and validation," Proc. of TEAM Workshop, Okayama, 1996. pp. 23–25.
47. Z. Cheng, N. Takahashi, S. Yang, Y. Du, J. Zhang, L. Liu, Y. Fan, J. Wang, and Q. Hu," Analysis and measurement of iron loss and linkage flux inside nonlinear magnetic steel," Electromagnetic field problems and applications-I, vol.22, TSI press series, USA, pp. 363–366, 2008.
48. Z. Cheng, S. Gao, J. Wang, H. He, Z. Liu, M. Wu, H. Li and Q. Hu, "Loss evaluation of non-magnetic tie-plates in transformers," COMPEL, vol.17, no.1/2/3, 1998, pp. 347–351.
49. Z. Cheng, N. Takahashi, B. Forghani, X. Wang, et al, "Extended progress in TEAM Problem 21 family," COMPEL, 33, 1/2, pp. 234–244, 2014.
50. Z. Cheng, B. Forghani, X. Wang, L. Liu, T. Liu, Y. Fan, J. Zhang, X. Zhao, and Y. Liu, "Engineering-oriented investigation of magnetic property modeling and application," invited speech at the 1&2DM2016, *International Journal of Applied Electromagnetics and Mechanics,* **55**(2017), 147–158.

51. N. Takahashi, T. Sakura, and Z. Cheng, “Nonlinear analysis of eddy current and hysteresis losses of 3-D stray field loss model (Problem 21),” *IEEE Trans. on Magn.*, vol.37, no.5, 2001, pp. 3672–3675.
52. Z. Cheng, R. Hao, N. Takahashi, Q. Hu and C. Fan, “Engineering-oriented benchmarking of Problem 21 family and experimental verification,” *IEEE Trans. on Magn.,* vol. 40, no.2, 2004, pp. 1394–1397.
53. Z. Cheng, N. Takahashi, S. Yang, T. Asano, Q. Hu, S. Gao, X. Ren, H. Yang, L. Liu, L. Gou, “Loss spectrum and electromagnetic behavior of problem 21 family”, *IEEE Trans. Magn.*, vol.42, no.4, pp. 1467–1470, 2006.
54. Z. Cheng, N. Takahashi, S. Yang, C. Fan, M. Guo, L. Liu, J. Zhang and S. Gao, “Eddy current and loss analysis of multi-steel configuration and validation,” *IEEE Trans. On Magn.,* vol.43, no.4, 2007, pp. 1737–1740.
55. Z. Cheng, N. Takahashi, B. Forghani, G. Gilbert, J. Zhang, L. Liu, Y. Fan, X. Zhang, Y. Du, J. Wang, and C. Jiao, “Analysis and measurements of iron loss and flux inside silicon steel laminations,” *IEEE Trans. on Magn*etics, vo.45, no.3, pp. 1222–1225, 2009.
56. Z. Cheng, N. Takahashi, B. Forghani, et al, “Effect of excitation patterns on both iron loss and flux in solid and laminated steel configurations,” *IEEE Trans. on Magnetics,* vol.46, no.8, pp. 3185–3188, 2010.
57. Z. Cheng, N. Takahashi, B. Forghani, Y. Du, Y. Fan, L. Liu, and H. Wang, “Effect of variation of B-H properties on both iron loss and flux in silicon steel lamination,” *IEEE Trans. on Magnetics,* vol.47, no.5, pp. 1346–1349, 2011.
58. W. Zheng, and Z. Cheng, “Efficient finite element simulation for GO silicon steel laminations using inner-constrained laminar separation,” *IEEE Trans. on Magn.*, vol.48, no.8, pp. 2277–2283, 2012 (IEEE regular paper).
59. Z. Cheng, N. Takahashi, B. Forghani, L. Liu, Y. Fan, T. Liu, J. Zhang, and X. Wang, “3-D finite element modeling and validation of power frequency multi-shielding effect,” *IEEE Trans. on Magnetics*, vol.48, no.2, pp. 243–246, 2012.
60. Z. Zhao, Z. Cheng, B. Forghani, F. Liu, Y. Li, and L. Liu, “Analytical study and corresponding experiments for ironloss inside laminated core under ac-dc hybrid excitation,” *International Journal of Applied Electromagnetics and Mechanics,* **55**(2017), 159–167.
61. X. Zhao, F. Meng, Z. Cheng, L. Liu, J. Zhang, and C. Fan, “Stray-field loss and flux distribution inside magnetic steel plate under harmonic excitation, COMPEL, 36, 6, pp. 1715–1728, 2017.
62. Z. Cheng, N. Takahashi, and B. Forghani, TEAM Problem 21 Family(V.2009), Approved by the International Compumag Society Board at Compumag-2009, Florianópolis, Brazil.
63. T. Nakata, A. Kost, J. Sykulski, L. Turmer, O. Biro, Private communications on TEAM Problem 21, 1992–2012.
64. O. Biro, K. Preis, W. Renhart, K. R. Richter, and G. Vrisk, “Performance of different vector potential formulations in solving multiply connected 3-D eddy current problems,” *IEEE Trans. on Magn.*, vol.26, no.2, 1990, pp. 438–441.
65. J. P. Webb and B. Forghani, “T-Omega method using hierarchal edge elements,” *IEE Proc.-Sci. Meas. Technol.*, Vol. 142, No. 2, March 1995.
66. Z. Cheng, S. Gao, and L. Li. *Eddy Current Analysis and Validation in Electrical Engineering* (Supported by National Natural Science Foundation). Higher Education Press, China, ISBN 7-04-009888-1, 2001 (in Chinese).

Chapter 13
Analysis and Validation of Additional Iron Loss Based on Benchmark Models

Zhiguang Cheng, Chen Chang and Dongjie Wang

Abstract The eddy current loss caused by the leakage magnetic flux entering the laminated sheets perpendicularly becomes, in some special cases, a significant portion of the total iron loss, which is usually referred to as additional iron loss. It is very challenging to compute the additional iron loss accurately due to the non-uniform distribution of loss and field in laminated sheets, and the multi-scale problems, involving very thin silicon steel sheets, the very small depth of penetration, and the insulation film insulating the sheets from each other. Some practical measures, in order to reduce additional iron loss and then avoid the unacceptable local overheating, are considered in the power transformer design. In this chapter, a detailed numerical analysis of the iron loss and the flux in GO silicon steel laminations is presented. The corresponding verification experiments are carried out based on the simplified magnetic shield models of TEAM Problem 21 Family (www.compumag.org/team), $P21^d$-M, and newly proposed P21-based models $P21^d$-M2, and $P21^d$-M3. This makes it possible to accurately analyze the loss and flux at the single-sheet level.

Keywords Additional iron loss · GO silicon steel lamination · 3-D leakage magnetic flux · Induced eddy currents · Benchmark model

13.1 Introduction

The magnetic flux in the transformer core includes the alternating flux in the limb and yoke, and the rotating flux in the T-joint area of the iron core together they cause the total core loss. Some effective modeling methods, such as homogenization, have been proposed for the calculation of core loss [1, 2]. However, the loss inside the iron core, caused by the leakage magnetic flux entering perpendicularly,

Z. Cheng (✉) · C. Chang · D. Wang
Institute of Power Transmission and Transformation Technology,
Baobian Electric, Baoding, China
e-mail: emlabzcheng@yahoo.com

Z. Cheng et al. (eds.), *Modeling and Application of Electromagnetic and Thermal Field in Electrical Engineering*, https://doi.org/10.1007/978-981-15-0173-9_13

is different from the "standard" core loss caused by the main magnetic flux, and therefore can be called additional core loss or surprising loss [3]. In particular, the considerable eddy current loss due to the leakage magnetic flux of transformer coil is produced inside the outermost layer of multi-stage laminated core of large transformers.

Two questions, therefore, can be raised: (1) How to effectively reduce eddy current loss inside the outermost layer of the laminated core and the adjacent tie plate, caused by leakage magnetic flux? (2) How to accurately calculate the additional core loss?

Problem 1, mentioned above, encountered in engineering design, has been solved by experienced designers; the outmost layer of core lamination (refer to Fig. 12.2 of Chap. 12 of this book) that is directly affected by leakage flux can be "slotted" or cut into "narrow strips" to reduce eddy current losses there. However, Problem 2 for calculating the loss in such lamination layer, penetrated by the leakage electromagnetic field, is not as simple. The loss in the lamination layer can no longer be simply calculated using the existing loss curves provided by the silicon steel supplier, because they do not include the eddy current loss induced by the normal leakage flux entering the lamination perpendicularly [4, 5].

In order to study quantitatively the above problems and establish simplified models, the magnetic shielding models P21^{c}-M1 and P21^{c}-M2 of the international TEAM Problem 21 Benchmark Family [6] are simplified, i.e., the shielded magnetic steel plate (10 mm thick, ordinary A3 steel plate) is removed from the shielding models. It is used to accurately investigate the variation of both the iron loss and the magnetic flux inside silicon steel sheets (30RGH120) with the increase of the number of sheets and exciting currents. The magnetic properties (loss and flux) of two types of simplified magnetic shields are also compared.

In this chapter, the eddy current losses caused by alternating flux entering the laminated silicon steel sheets perpendicularly, and its portion in the total core loss, are quantitatively investigated through the accurate magnetic measurement and numerical analysis based on the simplified benchmark models.

13.2 Model Structure and Design Data

13.2.1 Structure and Dimension of the Models

In the version 2009 of TEAM Problem 21 Family, P21^{d}-M is simplified from P21^{c}-M1, that is, the magnetic steel in P21^{c}-M1 is removed, leaving only the exciting coil and 20 silicon steel sheets (30RGH120) [6]. For the same purpose, the magnetic plate in the member-model P21^{c}-M2 of Problem 21 Family is also removed, the simplified model is suggested to be referred to as P21^{d}-M2. See Figs. 13.1 and 13.2. However, P21^{d}-M2 is not a member-model of Problem 21 Family until now.

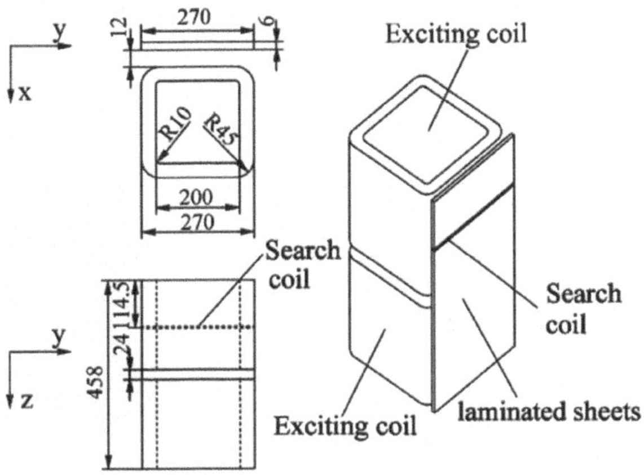

Fig. 13.1 P21^d-M model

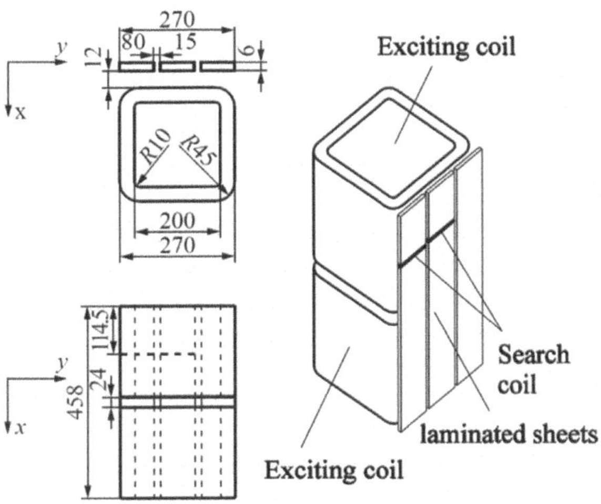

Fig. 13.2 P21^d-M2 model

Note that P21^d-M2 comprises three narrow laminated sheets (30RGH120), separated from each other, instead of one wide laminated sheet. The magnetic property data of the silicon steel sheet (30RGH120) can be found in the definition of Problem 21 benchmark family [6].

13.2.2 Locations of the Search Coil

Figures 13.4 and 13.5 schematically show the specific locations and numbering of the search coils. There are 20 turns in each search coil, and the diameter of the wire used in the search coils is 0.056 mm. The search coil is wound closely to the silicon steel sheet, and it is assumed that there is no air gap between the search coil and the wound silicon steel sheets.

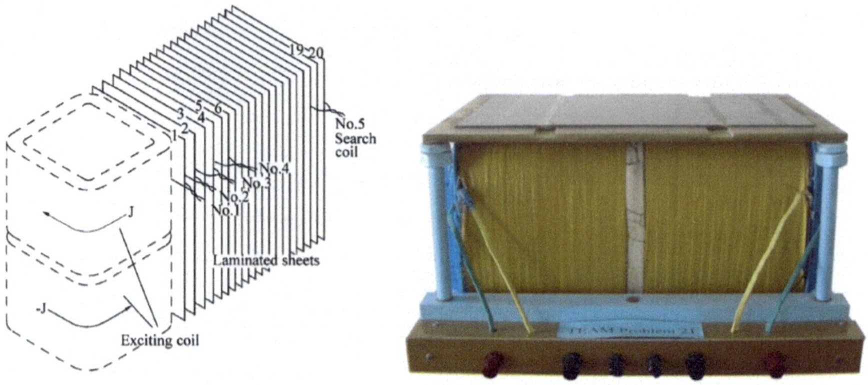

(a) Locations of search coils(P21^{d}-M) (b) Model P21^{d}-M

Fig. 13.3 P21^{d}-M and the locations of search coils

The P21^{d}-M model is used as an example to illustrate the locations of the search coils:

(1) Search coil No. 1 is wound around the designated location of the first silicon steel sheet facing the exciting coil, z = 114.5 mm, x–y–z coordinates, refer to Fig. 13.1, and the locations of the search coil, refer to Fig. 13.3.
(2) Search coil No. 2 is wound at the corresponding locations of silicon steel sheet No. 2 arranged in sequence.
(3) The silicon steel sheets wound by search coil No. 3 are silicon steel sheets No. 3 and No. 4 (2 sheets).
(4) The silicon steel sheets wound by search coil No. 4 are silicon steel sheets No. 5 and No. 6 (2 sheets).
(5) The silicon steel sheets wound by search coil No. 5 are silicon steel sheets No. 19 and No. 20 (2 sheets).

The locations of search coils No. 2–5 can be seen in Figs. 13.1 and 13.3.

Similarly, search coils No. 1–8 are arranged at predetermined positions in P21^{d}-M2 (see Figs. 13.2 and 13.4). As mentioned above, the difference between P21^{d}-M and P21^{d}-M2 is that the laminated shielding of P21^{d}-M2 is composed of three separate stacks of "laminated sheets" (20 sheets per stack) placed in parallel, while P21^{d}-M is an integral (non-separated) stack of laminated sheets and also contains 20 silicon steel sheets of the same material.

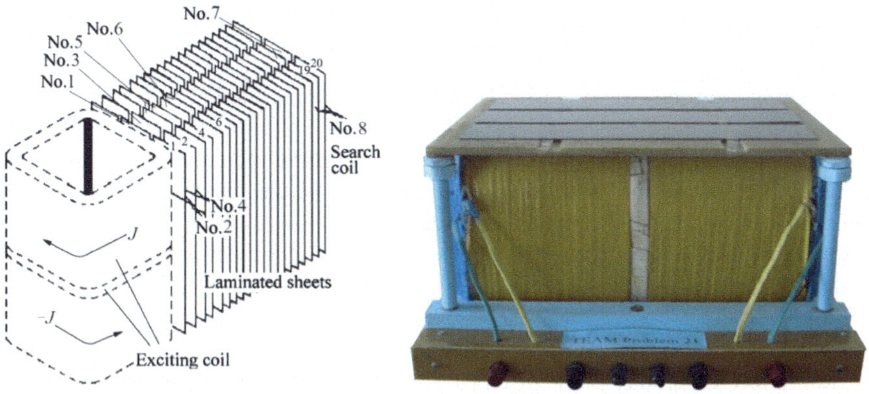

(a) Locations of search coils (P21^d-M2)

(b) Model P21^d-M2

Fig. 13.4 P21^d-M2 and the locations of search coils

13.3 Experimental Method and Targets

The magnetic loss and interlinkage flux in the laminated sheets under different lamination layers and excitation conditions are measured based on P21^d-M and P21^d-M2 models.

13.3.1 Experimental Circuit

The experimental circuit based on the test models P21^d-M and P21^d-M2 is shown in Fig. 13.5.

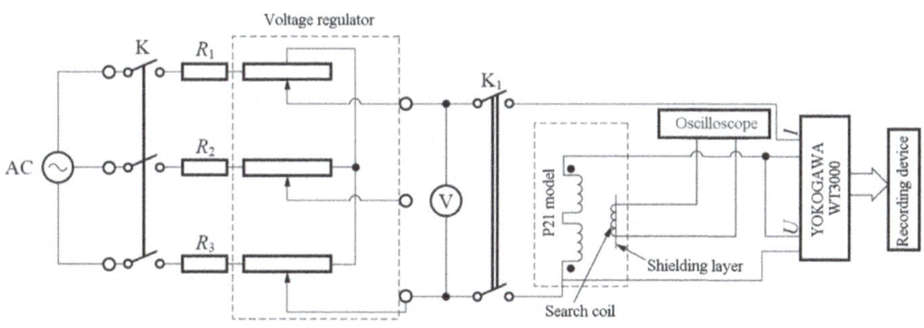

Fig. 13.5 Experimental circuit (schematic)

13.3.2 Measurement Procedure

The exciting currents of 10, 15, 20, and 25 A are applied, respectively (rms, 50 Hz).

As the number of layers of the shielding lamination increases under different excitation conditions, the induced voltage waveform of the search coil and the loss under corresponding conditions are recorded.

The laboratory instruments used, as well as the specifications, are shown in Table 13.1.

1. **Determining the Interlinkage Flux**

According to the Faraday's law of electromagnetic induction, the transient voltage in search coil (voltage waveform may be distorted) measured in the experiment is integrated over sampling time to determine the maximum value of the interlinkage flux in the silicon steel sheet wound by the search coil, which is then divided by the effective area of the silicon steel sheet contained in the coil. Thus, the maximum average magnetic flux density (B_{m}) at the corresponding location is obtained.

It should be noted that flux cannot be measured directly because it varies over time. As a matter of fact, a rate of variation of flux over time is measured as induced e.m.f., according to the Faraday's law of electromagnetic induction, i.e.,

$$e(t) = -N\frac{\mathrm{d}\psi}{\mathrm{d}t} \tag{13.1}$$

Each cycle (20 ms for 50 Hz) in the measurement is divided into 200 time steps. Transient induced voltage and exciting current are collected at the same time, and the acquired voltage is integrated over sampling time. For example, the time t_i corresponding to the ith time step:

Table 13.1 Instruments used in experiments and the related performance

Equipment	Specifications	Performance index
Power analyzer (Japan)	WT3000, YOKOGAWA	Maximum current measured: 30 A (rms); current and voltage reading accuracy: ±0.02%; range accuracy: ±0.04%; power accuracy: ±0.06%; frequency range: DC, 0.1 Hz–1 MHz
Infrared thermometer (Jiangsu)	68 IR THERMOMETER	Temperature range: −32 ~ + 760 °C; wavelength: 630–670 nm; emissivity: digital and adjustable; step size: 0.01
Contact voltage regulator (Shanghai)	TSGC2J (Regulated current ≤ 25 A)	Rated input voltage: 380 V; rated output voltage: 0–430 V; rated capacity: 20 kVA; output current: 27 A; number of phases: 3; frequency: 50–60 Hz; insulation class: A

$$\psi(t_i) = \psi(0) + \frac{1}{N}\int_{0}^{t_i} e(t)\mathrm{d}t \text{ (Let the initial integration timet} = 0) \tag{13.2}$$

The maximum value Ψ_{m} is obtained after 200 time steps $\Psi_{i,}$, i.e., the maximum value of the induced flux is obtained after the integration of 200 time steps is completed. Further, the average magnetic flux density $B_{\mathrm{m}}^{(\mathrm{average})}$ of the cross section S passing through the location of the search coil can be obtained.

$$B_{\mathrm{m}}^{(\mathrm{average})} = \frac{\psi_{\mathrm{m}}}{\mathrm{SN}} \tag{13.3}$$

2. **Experimental Procedure**

The experimental procedure is illustrated with the test of P21^{d}-M as an example:

(1) The search coil is wound;
(2) wiring is completed according to Fig. 13.5, and the first layer of silicon steel sheet is placed on the model according to Figs. 13.3 and 13.4; the test shall not be started until the wiring is checked;
(3) a current of 10A is applied, so that the voltage signal waveform of the search coil can be recorded; WT3000 is used to record the loss generated (load condition);
(4) the voltage regulator is set to zero and the magnetic shield (laminated sheets) is removed; then the voltage regulator is quickly adjusted to a specified exciting current; WT3000 is used to record the loss thus generated as no-load data (no-load condition); due to the short time interval between steps (3) and (4), the actual magnetic loss in silicon steel sheets can be considered as the difference between "load" and "no-load" losses without considering the influence of temperature difference.
It should be noted that this is an approximate method for determining the magnetic loss. The authors have studied the method for further precise determination of magnetic loss in magnetic components, regarding the variations of both leakage flux and temperature, under complex excitation conditions [7];
(5) the current is increased to 15 A, 20 A, and 25 A, respectively, and the steps (3)–(4) are repeated;
(6) the second layer of silicon steel sheet is added to the magnetic shield according to Fig. 13.3, and the steps (3)–(5) are repeated;
(7) the third and fourth layers of silicon steel sheet continue to be added to the original magnetic shield, as shown in Fig. 13.3, and the steps (3)–(5) are repeated;
(8) the fifth and sixth layers of silicon steel sheet continue to be added to the original magnetic shield, as shown in Fig. 13.3, and the steps (3)–(5) are repeated;

(9) then four, five, and five layers of silicon steel sheets continue to be added to the original magnetic shield, respectively, according to Fig. 13.3, and the steps (3)–(5) are repeated.

As for P21^{d}-M2, the experimental procedure is almost the same except that the locations of the search coils may be different, and that each layer of the lamination is no longer an integral (each layer comprises three narrow strips of equal width that are separated from each other). Also note that when silicon steel sheets are added, the corresponding number of sheets is added on each stack of laminated sheets as required.

13.4 Measurement Results

13.4.1 Measured Loss Results of P21^{d}-M

The measured losses of P21^{d}-M under different numbers of silicon steel sheets and different exciting currents (A, rms, 50 Hz) are shown in Table 13.2 and Fig. 13.6.

The analysis of measured loss shows that when a current is applied and the laminated sheets reach a certain number, the loss of laminated sheets hardly further increases with the increase of the number of sheets. This indicates the shielding effect of the laminated silicon steel sheets on the impressed field. It can be seen from further analysis that the eddy current loss caused by the leakage flux, entering the laminated sheets perpendicularly, possibly becomes a "significant" portion of the total iron loss.

13.4.2 Measured Loss Results of P21^{d}-M2

The measured losses of P21^{d}-M2 under different numbers of silicon steel sheets and different exciting currents (rms, 50 Hz) are shown in Table 13.3 and Fig. 13.7.

Table 13.2 Variation of losses in laminated sheets with numbers of sheets and exciting currents (P21^{d}-M, Unit: W)

Number of laminated sheets	Exciting currents (A, rms, 50 Hz)			
	10	15	20	25
1	1.3	2.2	2.3	3.4
2	1.9	4.0	5.5	6.7
4	1.9	4.8	9.2	14.3
6	2.0	4.9	9.7	16.5
10	2.1	5.0	10.2	16.9
15	2.1	5.3	10.3	17.2
20	2.1	5.3	10.4	17.2

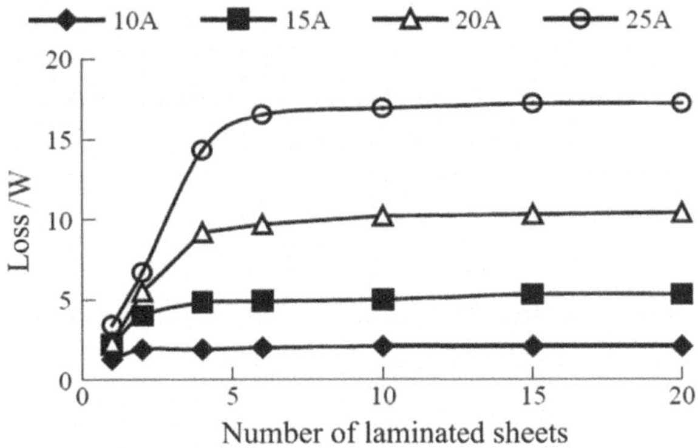

Fig. 13.6 Losses in laminated sheets under different numbers of sheets and different exciting currents (P21[d]-M)

Table 13.3 Variation of losses in laminated sheets with numbers of sheets and exciting currents (P21[d]-M2, Unit: W)

Number of laminated sheets	Exciting currents (A, rms, 50 Hz)			
	10	15	20	25
1	1.0	1.2	1.3	1.5
2	1.2	2.0	2.8	3.6
4	1.1	2.3	4.1	5.1
6	1.2	2.4	4.4	6.2
10	1.2	2.5	4.3	6.4
15	1.2	2.4	4.4	6.3
20	1.2	2.6	4.5	6.4

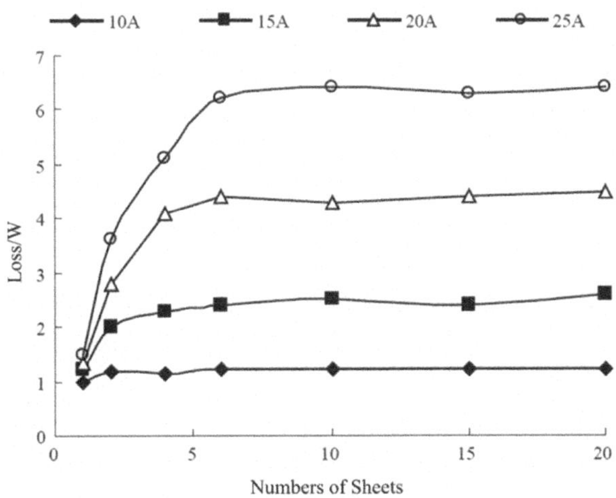

Fig. 13.7 Losses in laminated sheets under different numbers of sheets and different exciting currents (P21[d]-M2)

It is indicated from the measured losses of the two models (P21^d-M and P21^d-M2) that when an exciting current, e.g., 25 A (rms, 50 Hz), is applied and the silicon steel sheets is increased to a certain number, for example, when the number of sheets increases to six or more under the current specific test conditions, the loss hardly increases with the increase of the number of silicon steel sheets.

13.4.3 Flux Waveforms Obtained by Search Coils in Models P21^d-M and P21^d-M2

In the Model P21^d-M, for example, when exciting current of 25A (rms, 50 Hz) is applied, the waveforms of the interlinkage flux of the search coils No. 1–4 can be seen in Fig. 13.8a.

Furthermore, the waveform of the sum of flux interlinking with the first 6 layers of silicon steel sheets is also obtained. It is shown that the sum of the flux interlinking with the first 6 layers accounts for the vast majority of, and even close to, the total interlinkage flux of the 20 layers, and is basically sinusoidal (THD = 4.7%) [8]. All the waveforms of flux under different conditions are shown in Fig. 13.8a.

Similarly, in P21^d-M2, for example, when exciting current of 25 A is applied, the waveforms of the interlinkage flux of the search coils No. 1–4 are given in Fig. 13.8b, and the waveform of the sum of flux interacting with the first 6 layers of silicon steel sheets is also obtained. It is also shown that the sum of the flux interlinking with the first 6 layers accounts for the vast majority of, and even close to, the total interlinkage flux of the 20 layers, and is basically a sine wave (THD = 4.9%), as shown in Fig. 13.8b.

Moreover, it can be found that in P21^d-M2, the flux of two stacks (of total 3 stacks) was measured under the condition of a total of 20 layers, where one stack in the middle and another stack on the side are counted as Φ_1 and Φ_2, respectively, resulting in the total interlinkage flux of 20 layers in P21^d-M2 of $\Phi_{M2} = \Phi_1 + 2 \times \Phi_2$, which is known to be similar to the total interlinkage flux of the 20 layers in P21^d-M.

13.4.4 Average Flux Density Waveforms in the Laminated Sheets of Models P21^d-M and P21^d-M2

Based on the resulting flux linkage, the average magnetic flux density B in the lamination of P21^d-M and P21^d-M2 can be calculated according to Eq. (13.3).

The packing factor is taken into account, and the film thickness of the silicon steel sheet surface is removed in the calculation of net cross-sectional area *S*. The waveform of the average magnetic flux density is shown in Fig. 13.9.

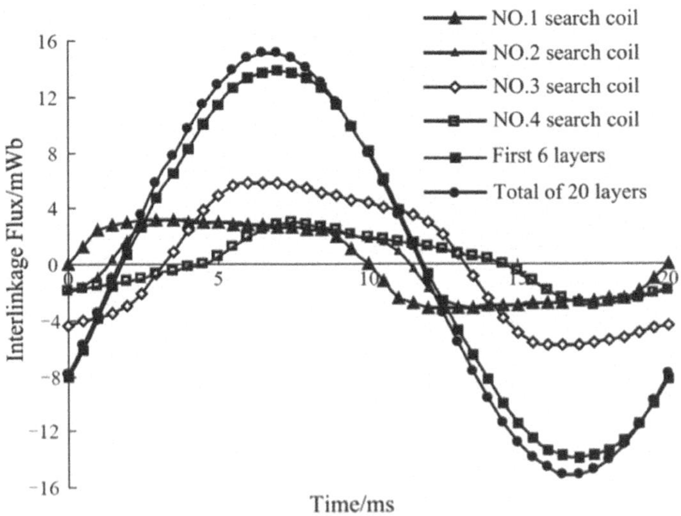

(a) Interlinkage flux of each search coil (25A, P21^{d}-M)

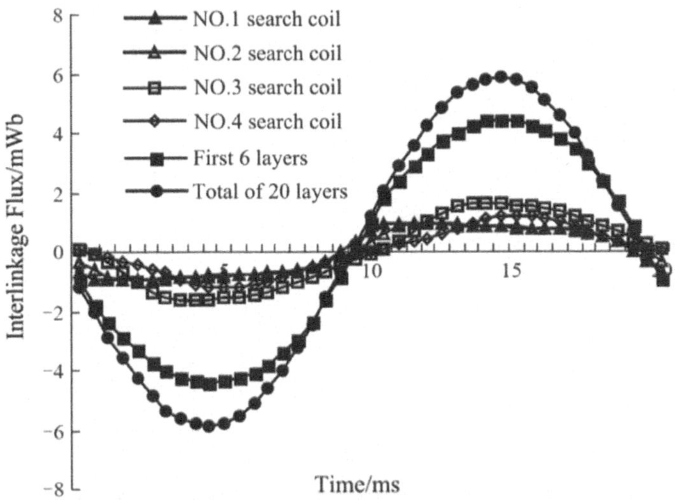

(b) Interlinkage flux of each search coil (25A, P21^{d}-M2)

Fig. 13.8 Interlinkage flux interlinked with search coils under various conditions

It should be noted that when the silicon steel sheets are strongly saturated, the flux interlinking with the search coil is no longer strictly equal to the flux in the silicon steel sheet, due to leakage flux and/or interlaminar flux, etc.

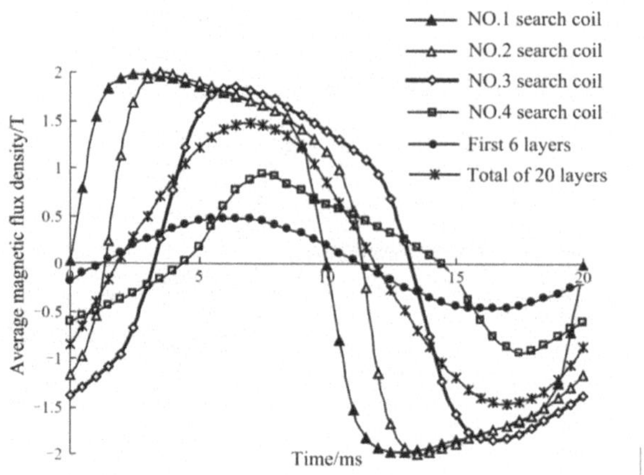

(a) Average magnetic flux density interlinking with each search coil in P21^d-M

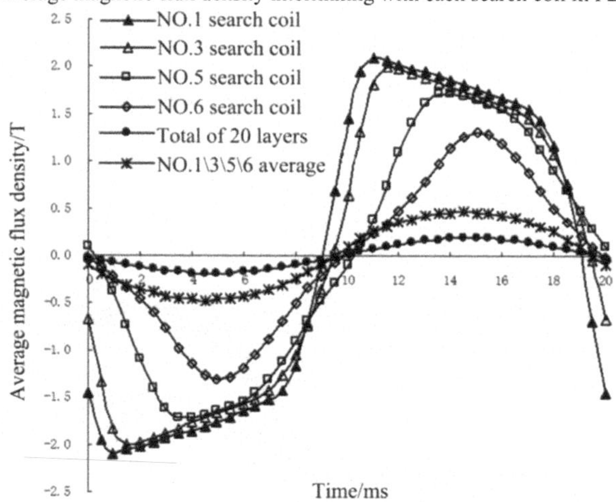

(b) Average magnetic flux density interlinking with each search coil in P21^d-M2

Fig. 13.9 Waveform of magnetic flux density interlinking with search coils under various conditions

13.4.5 Determination of Maximum Values of Flux Density Based on P21^d-M and P21^d-M2

The maximum average magnetic flux density in the silicon steel sheet is further calculated based on the interlinkage flux with the search coil. The calculated results of the maximum average magnetic flux density as a function of the exciting current corresponding to each search coil in P21^d-M are listed in Table 13.4 and illustrated in Figs. 13.10, 13.11, 13.12, 13.13, and 13.14. The locations of search coils in P21^d-M can be found in Fig. 13.4.

Table 13.4 Maximum average magnetic flux density (B_m) in the laminated sheets (P21[d]-M, Unit: T)

Number of laminated sheets	Search coil No.	Exciting currents(A, rms, 50 Hz)			
		10	15	20	25
1	No. 1	2.0946	2.1730	2.1880	2.2122
2	No. 1	1.8549	2.0082	2.0847	2.1054
	No. 2	1.8668	2.1115	2.1657	2.2187
4	No. 1	1.7981	1.9036	1.9772	2.0211
	No. 2	1.4022	1.8481	1.9798	2.0307
	No. 3	0.3263	1.0634	1.9065	2.0524
6	No. 1	1.7688	1.8846	1.9590	1.9971
	No. 2	1.3870	1.8287	1.9467	1.9954
	No. 3	0.2944	0.9559	1.6601	1.9241
	No. 4	0.0677	0.1502	0.4357	1.1846
10 (6 + 4)	No. 1	1.7364	1.8625	1.9408	1.9802
	No. 2	1.3968	1.8114	1.9550	2.0181
	No. 3	0.3028	0.9622	1.6194	1.8522
	No. 4	0.0504	0.1170	0.3558	0.9807
15 (6 + 9)	No. 1	1.7364	1.8596	1.9391	1.9818
	No. 2	1.2622	1.8083	1.9443	2.0087
	No. 3	0.3046	0.9402	1.6071	1.8510
	No. 4	0.0696	0.1187	0.3582	0.9469
20 (8 + 12)	No. 1	1.6301	1.8508	1.9387	1.9800
	No. 2	0.8408	1.7966	1.9480	2.0201
	No. 3	0.1836	0.9468	1.5939	1.8449
	No. 4	0.0382	0.1157	0.3728	0.9614
	No. 5	0.0342	0.0495	0.0618	0.0691

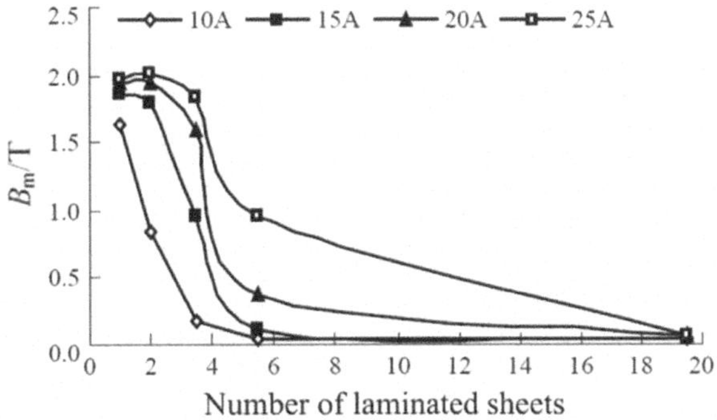

Fig. 13.10 Maximum average magnetic flux density in each test sheet as a function of current when 20 silicon steel sheets are used (P21[d]-M)

Fig. 13.11 Variation of the maximum average magnetic flux density in the silicon steel sheet corresponding to search coil No. 1 with the increase of the current and number of silicon steel sheets in the shielding layer (P21[d]-M)

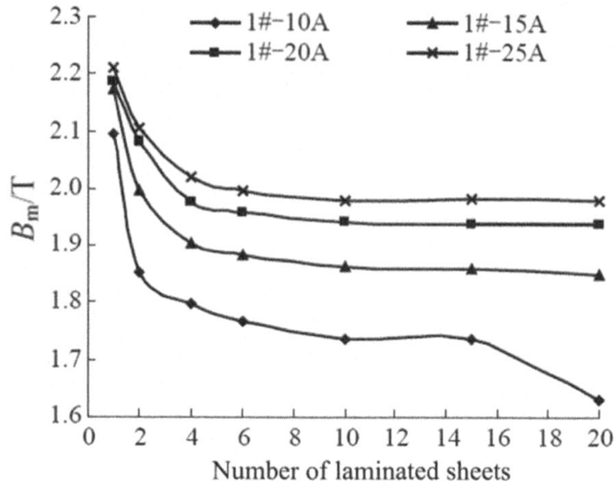

Fig. 13.12 Variation of the maximum average magnetic flux density in the silicon steel sheet corresponding to search coil No. 2 with the increase of the current and number of silicon steel sheets in the shielding layer (P21[d]-M)

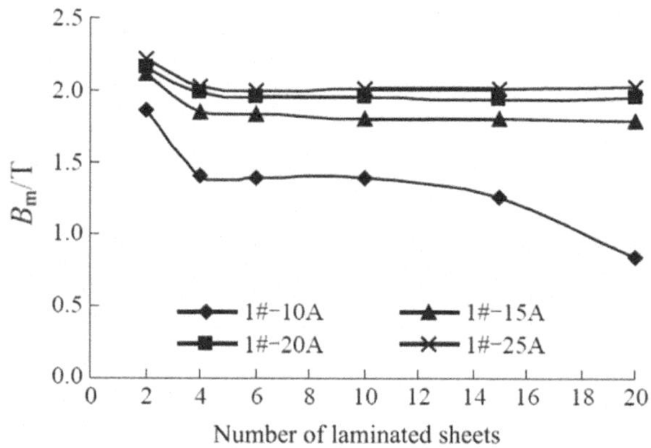

Fig. 13.13 Variation of the maximum average magnetic flux density in the silicon steel sheets corresponding to search coil No. 3 with the increase of the current and number of silicon steel sheets in the shielding layer (P21[d]-M)

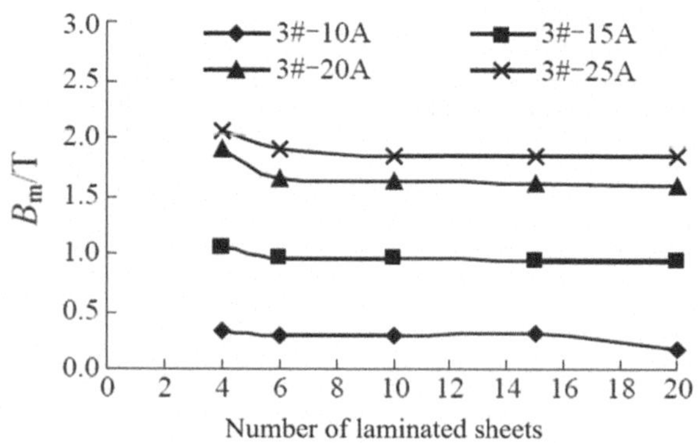

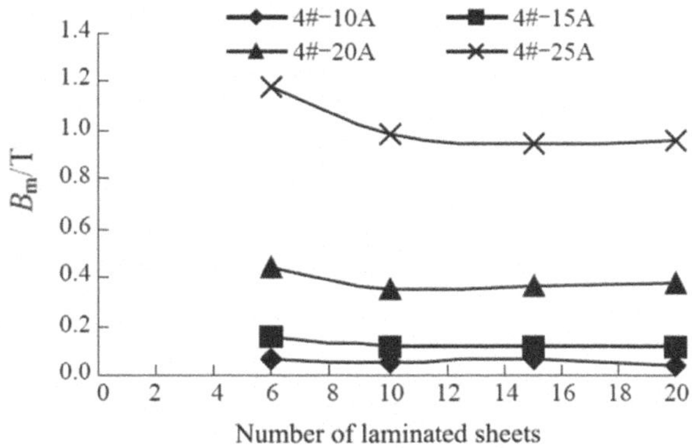

Fig. 13.14 Variation of the maximum average magnetic flux density in the silicon steel sheets corresponding to search coil No. 4 with the increase of the current and number of silicon steel sheets in the shielding layer ($P21^d$-M)

The calculated results of the maximum average magnetic flux density (peak value) of $P21^d$-M2 are listed in Table 13.5 and illustrated in Figs. 13.15, 13.16, 13.17, 13.18, 13.19, and 13.20. The locations of search coils in $P21^d$-M2 are shown in Fig. 13.5. The leakage flux outside the silicon steel sheet shall be taken into account in the calculated magnetic flux density in Table 13.5 when the magnetic flux density is highly saturated (those marked with ** in the Table should be corrected).

13.4.6 Remarks on the Measured Results

All the measured results based on the simplified models presented in this section (Sect. 13.4) can be conclusively remarked as follows:

(1) With a given total number of silicon steel sheets (e.g., less than 4 sheets), the average magnetic flux density B_z (in the rolling direction) in each laminated sheet increases with the increasing exciting current; under the same current, the interlinkage flux in the laminated sheet close to the exciting coil will decrease as the sheets are gradually added, as shown in Tables 13.4 and 13.5.

(2) When the laminated sheets are added to a certain number, the iron loss will no longer increase with the increasing number of sheets, for example, the exciting current is 10 A, the number of silicon steel sheets is more than just 2 or 3, the iron loss will no longer increase; where the exciting current is 25 A, the number of silicon steel sheets is more than just 6 or 7, the iron loss will no longer increase.

(3) The sum of the flux interlinking with the first 6 sheets (i.e., 1st–6th sheets) accounts for the vast majority of the total interlinkage flux of the 20 sheets The waveform thereof is close to a sinusoid such as the one for the total interlinkage flux of 20 sheets, even though sine wave is not produced with the flux interlinking with each layer of silicon steel sheets. However, sine wave is not

Table 13.5 Maximum average magnetic flux density (B_m) of the test sheets in P21[d]-M2 (Unit: T)

Number of laminated sheets	Search coil No.	Exciting currents (A, rms, 50 Hz)			
		10	15	20	25
1	No. 1	2.2247	2.2970**	2.3400**	2.4130**
	No. 2	2.0493	2.1105	2.1487	2.2307
2	No. 1	2.0405	2.1742	2.2419	2.2950
	No. 2	1.8917	2.0725	2.1515	2.1945
	No. 3	2.1991	2.3336	2.4045	2.4432
	No. 4	1.8422	1.9917	2.0893	2.1143
4	No. 1	1.8849	2.0145	2.0644	2.1298
	No. 2	1.4467	1.8234	1.9440	2.0557
	No. 3	1.5814	2.0526	2.2020	2.2971
	No. 4	0.9022	1.7867	1.9752	2.0392
6	No. 1	1.8601	2.0049	2.0700	2.1085
	No. 3	1.3067	1.9072	2.0158	2.0811
	No. 5	0.5769	1.1697	1.7116	1.8548
	No. 6	0.4374	0.8033	1.5973	2.0246
10 (6 + 4)	No. 1	1.8273	1.9860	2.0557	2.1051
	No. 3	1.2197	1.8440	1.9663	2.0262
	No. 5	0.4736	0.9919	1.5720	1.7634
	No. 6	0.2299	0.3910	0.7590	1.4041
15 (6 + 9)	No. 1	1.8118	1.9711	2.0465	2.0904
	No. 3	1.2014	1.8535	1.9719	2.0339
	No. 5	0.4390	0.9417	1.5330	1.7491
	No. 6	0.1891	0.3358	0.6786	1.3363
20 (8 + 12)	No. 1	1.7963	1.9743	2.0395	2.0837
	No. 3	1.1891	1.8370	1.9535	2.0167
	No. 5	0.4389	0.9587	1.5205	1.7361
	No. 6	0.1776	0.3082	0.6683	1.3165
	No. 7	0.0950	0.1453	0.2070	0.2627
	No. 8	0.1162	0.1755	0.2449	0.3153

produced with the total flux interlinking with the 20 silicon steel sheets in the strict sense because leakage flux may be generated.

(4) The variation of iron loss in the laminations and interlinkage flux at the designated position with the increasing exciting current are investigated in detail, and the corresponding results from measurement and calculation are in good agreement.

(5) The measured and calculated magnetic flux densities in the air and at the prescribed positions on the lamination surface are also in good agreement.

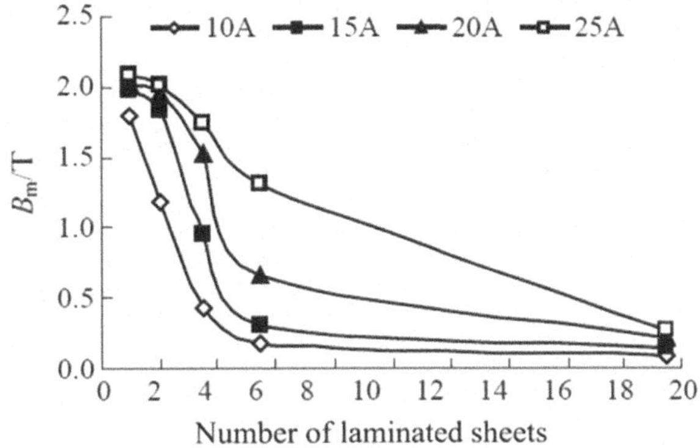

Fig. 13.15 Maximum average magnetic flux density in each test sheet as a function of current when 20 layers of silicon steel sheets are used (P21[d]-M2)

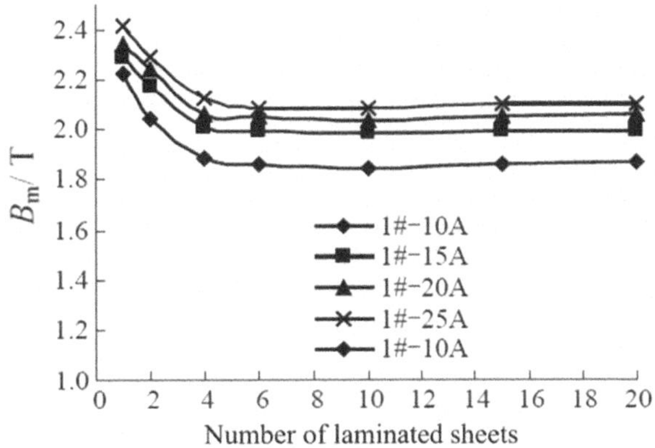

Fig. 13.16 Variation of the maximum average magnetic flux density in the silicon steel sheet corresponding to search coil No. 1 with the increase of the current and number of silicon steel sheets in the shielding layer (P21[d]-M2)

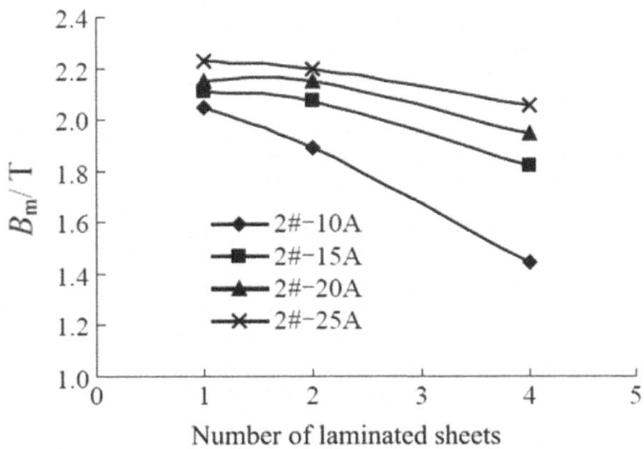

Fig. 13.17 Variation of the maximum average magnetic flux density in the silicon steel sheet corresponding to search coil No. 2 with the increase of the current and number of silicon steel sheets in the shielding layer (P21[d]-M2)

Fig. 13.18 Variation of the maximum average magnetic flux density in the silicon steel sheet corresponding to search coil No. 3 with the increase of the current and number of silicon steel sheets in the shielding layer (P21d-M2)

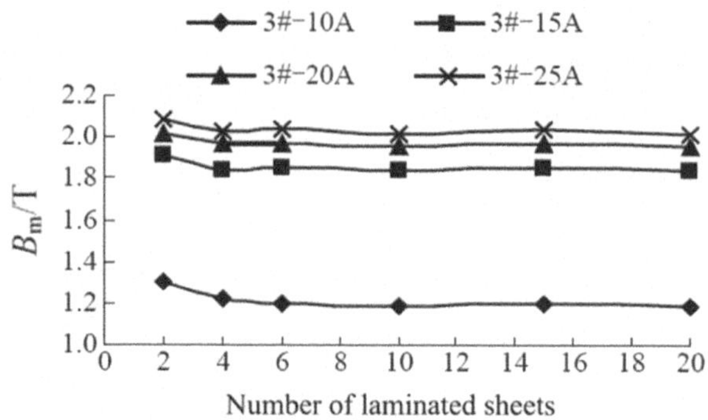

Fig. 13.19 Variation of the maximum average magnetic flux density in the silicon steel sheet corresponding to search coil No. 5 with the increase of the current and number of silicon steel sheets in the shielding layer (P21d-M2)

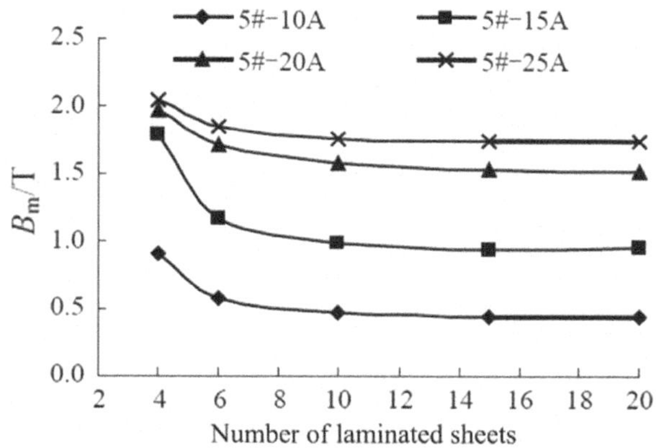

Fig. 13.20 Variation of the maximum average magnetic flux density in the silicon steel sheet corresponding to search coil No. 6 with the increase of the current and number of silicon steel sheets in the shielding layer (P21d-M2)

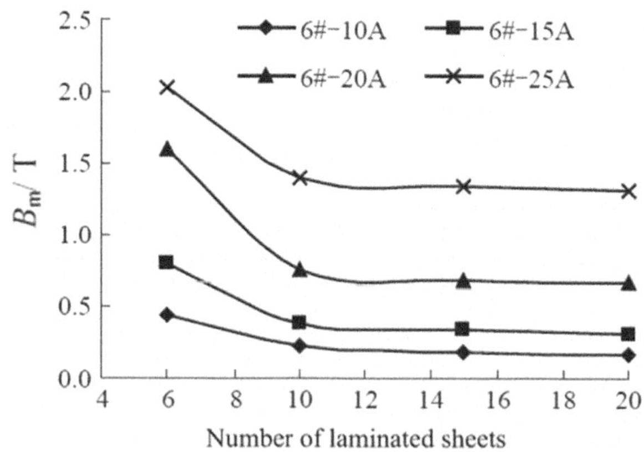

13.4.7 3-D Finite Element Computation of Additional Iron Loss

13.4.7.1 3-D Finite Element Model

The different zoning schemes for computing the total iron loss inside laminated sheets have been investigated by the authors [9, 10], including the additional iron loss caused by the normal leakage flux, and the standard iron loss based on the already measured loss curve under standard condition without a normal-direction excitation. The following magnetic loss analysis based on the benchmark model P21^d-M has been carried out, which has 20 GO silicon steel sheets (30RGH120) with the dimensions of 270 × 458 mm for each sheet.

The GO silicon laminations of 20 sheets are portioned to two parts in its thickness direction: the surface layer facing to the normal-direction flux is subdivided as several single sheets, and the rest is treated as bulk, as shown in Fig. 13.21. In order to solve the eddy currents induced by the normal-direction leakage flux, 2 practical simplifications are taken as follows:

(1) In the entire surface layer, the anisotropic conductivity is treated as that, the $\sigma_y = \sigma_z$, and $\sigma_x = 0$. This means that induced eddy currents by the normal-direction flux are limited in single sheet in *Y*–*Z* plane(2-D), resulting in additional iron loss.
(2) Note that the resulting magnetic field inside the laminated sheets of Model P21^d-M is almost in one direction (along *z*-axis), making it a weak magnetic anisotropy problem, and the orthogonal anisotropic permeability is assigned to all the laminations.

The iron loss and magnetic field generated inside the GO laminations have been computed based on the results of the eddy current field as part of the finite element post processing. It is shown that the additional iron loss P_a caused by the flux perpendicularly entering the laminated sheets must be fully considered. The total iron loss P_t is divided into 2 parts [9, 10], i.e.,

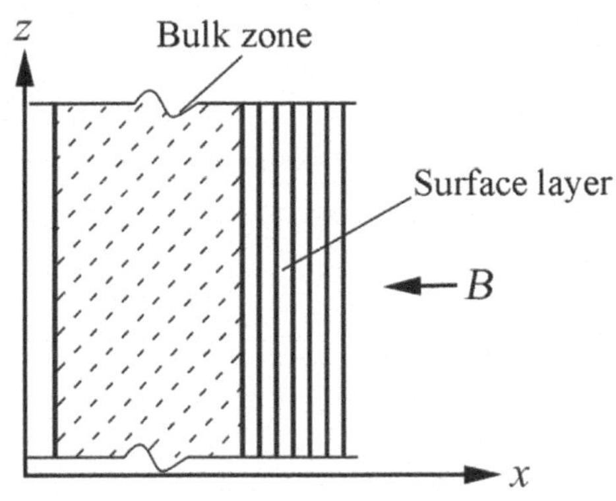

Fig. 13.21 Simplification of laminated sheets (P21^d-M)

$$P_t = P_a + P_s \tag{13.4}$$

where P_s, as mentioned above, can be determined based on the measured loss curve $B_m - W_t$, and P_a can be calculated based on the field results of 2-D eddy current $\boldsymbol{J}_{yz}$ flowing in the plane of the lamination by (13.5),

$$P_a = \int \frac{\overline{|J_{yz}|_{rms}}^2}{\sigma_{yz}} dv \tag{13.5}$$

The standard specific iron loss P_s includes all the loss components, such as the classical eddy current loss $P_e^{classic}$, the anomalous eddy current loss P_e^{anomal}, and the hysteresis loss P_h,

$$P_s = P_e^{classic} + P_e^{anomal} + P_h \tag{13.6}$$

However, P_s does not include the additional iron loss P_a.

The magnetic flux inside GO silicon sheets can be obtained by integrating the calculated flux densities over the specified cross-sectional areas or based on the measurements using the search coils.

13.4.8 Measured and Calculated Results of Iron Loss and Magnetic Flux

Two test models (P21^d-M and P21^d-M2) are used for measurement and numerical analysis of magnetic loss inside laminated sheets.

1. **Model P21^d-M**

The measured and calculated (using Simcenter MAGNET) results of both total iron losses and magnetic flux inside the laminations of 20 laminated sheets, under different exciting currents, are shown in Tables 13.6 and 13.7 [9].

Tables 13.6 and 13.7 show that the measured and calculated results are practically in good agreement. Figure 13.22 shows the further comparison among the

Table 13.6 Total iron loss inside laminations under different exciting currents (P21^d-M)

Exciting currents (A, rms, 50 Hz)	Calculated iron losses (W)	Measured iron losses (W)
10	2.33	2.20
15	5.04	5.30
20	10.32	10.20
25	16.33	16.80

Table 13.7 Flux inside 20 sheets under different exciting currents (P21^{d}-M)

Exciting currents (A, rms, 50 Hz)	Calculated flux (mWb)	Measured flux (mWb)
10	0.311	0.297
15	0.447	0.444
20	0.594	0.589
25	0.702	0.738

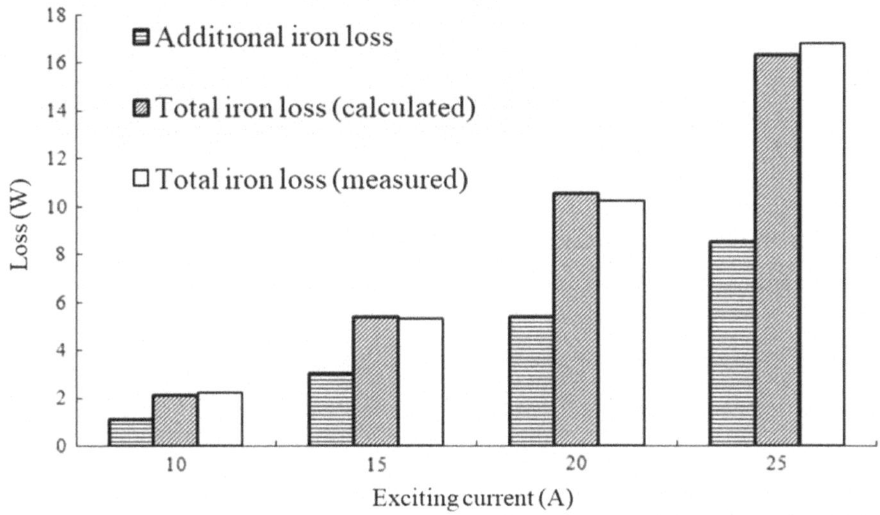

Fig. 13.22 Measured and calculated total iron losses in laminations (P21^{d}-M)

measured and calculated loss results. It can be found that the additional iron loss becomes the considerable part of the calculated total iron loss in this case study.

2. **Model P21^{d}-M3**

In order to further demonstrate the noticeable contributions of additional iron loss caused by the normal flux to the total iron loss, a further simplified test model is designed by the authors based on the original power supply of P21^{d}-M, in which the dimensions of the laminations (30P120) are enlarged from 270 × 458 mm to 500 × 500 mm, but the number of the laminated sheets is reduced from 20 to 6 [10], referred to as Model P21^{d}-M3. See Fig. 13.23.

This is a more prominent example regarding additional iron loss. The measured and calculated iron loss inside the laminations of P21^{d}-M3, and the comparison between the additional and total iron losses are shown in Tables 13.8 and 13.9, respectively [10].

Note that in this special thin lamination model P21^{d}-M3 the additional iron loss caused by normal flux becomes dramatically the major part of the total iron loss inside the 6-sheet magnetic component.

Fig. 13.23 Model P21^{d}-M3

Table 13.8 Total iron loss inside laminations (P21^{d}-M3)

Exciting currents (A, rms, 50 Hz)	Calculated iron losses (W)	Measured iron losses (W)
10	2.54	2.52
15	6.66	7.12
20	13.37	13.7
25	23.68	23.8

Table 13.9 Components of total iron loss (P21^{d}-M3)

Exciting currents (A, rms, 50 Hz)	Additional iron loss P_a (W)	Standard iron loss P_s (W)	$P_a/(P_s + P_a)$ (%)
10	1.73	0.81	68.1
15	5.20	1.46	78.1
20	11.15	2.22	83.4
25	21.22	2.46	89.6

In addition, the authors have obtained a number of measured and calculated results of both iron losses and magnetic fluxes inside solid steel configurations based on other benchmark models, which have been presented in [9].

All the measured and calculated results presented in this section can be summarized as follows:

(1) The additional eddy current loss P_a induced by the normal component of the 3-D magnetic flux inside the laminated sheets can be a considerable portion of the total iron loss and is not included in the standard total iron loss P_s, and P_a should be separately calculated based on the field results.

(2) All the calculated and measured results of both iron loss and flux for different test cases are practically in good agreement. This proves that the proposed computation method of the additional iron loss is effective.

13.4.9 Comparison Between Waveforms of Measured and Calculated Flux

When an exciting current, up to 25 A (50 Hz, rms), is applied, the laminated sheets corresponding to search coils No. 1–4 (shown in Fig. 13.4), as the surface layer of laminated sheets, penetrated directly by the leakage magnetic flux in P21^{d}-M, are saturated, and the waveforms of flux and induced voltage are distorted. See Figs. 13.24, 13.25, 13.26, 13.27, 13.28, 13.29, 13.30, and 13.31, respectively [9].

Note that "Layer 1" in Fig. 13.24 refers to the first silicon steel sheet that faces the exciting coil and is surrounded by the first search coil (No. 1), as shown in Fig. 13.4. In a similar fashion, "Layer *x*" in the following refers to the silicon steel sheet surrounded by the *x*th search coil. Only Layer 3 and 4 contain two silicon steel sheets each, while Layer 1 and 2 contain one.

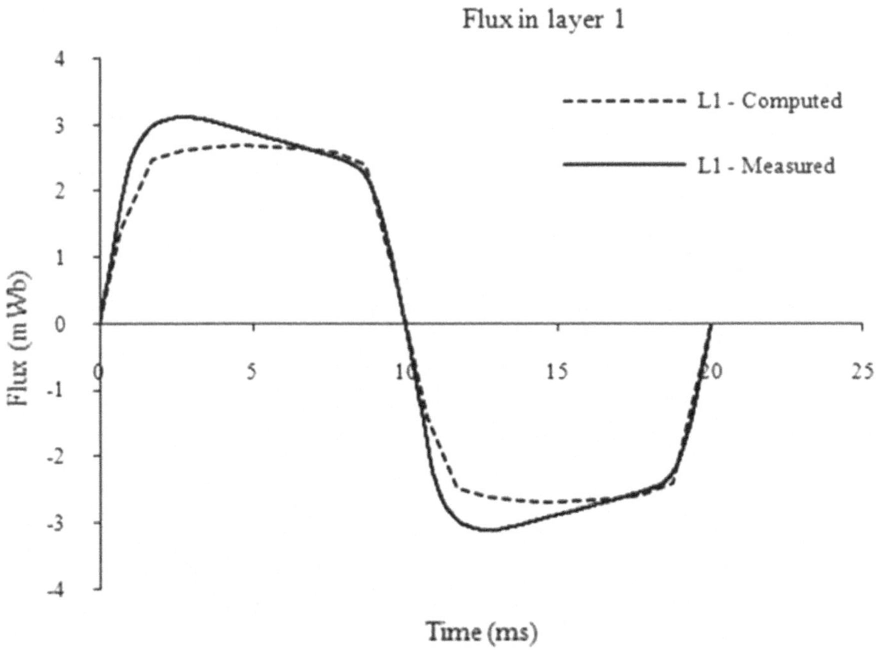

Fig. 13.24 Waveform of flux in layer 1

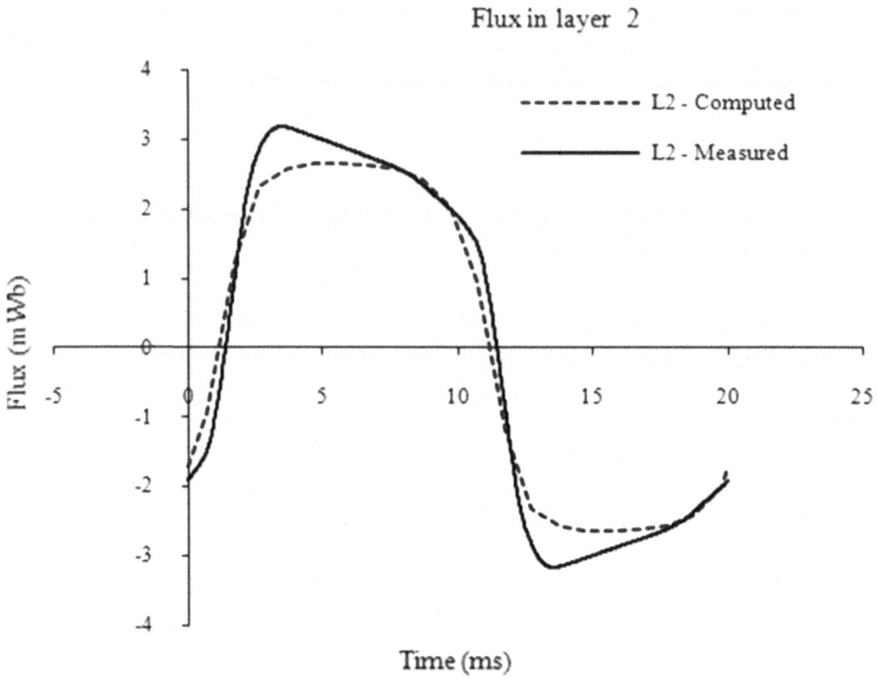

Fig. 13.25 Waveform of flux in layer 2

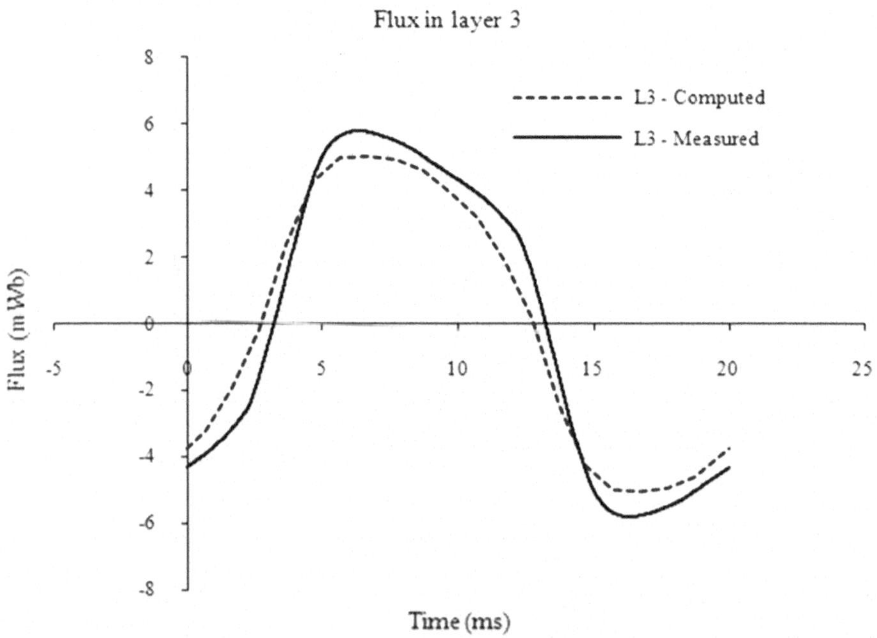

Fig. 13.26 Waveform of flux in layer 3

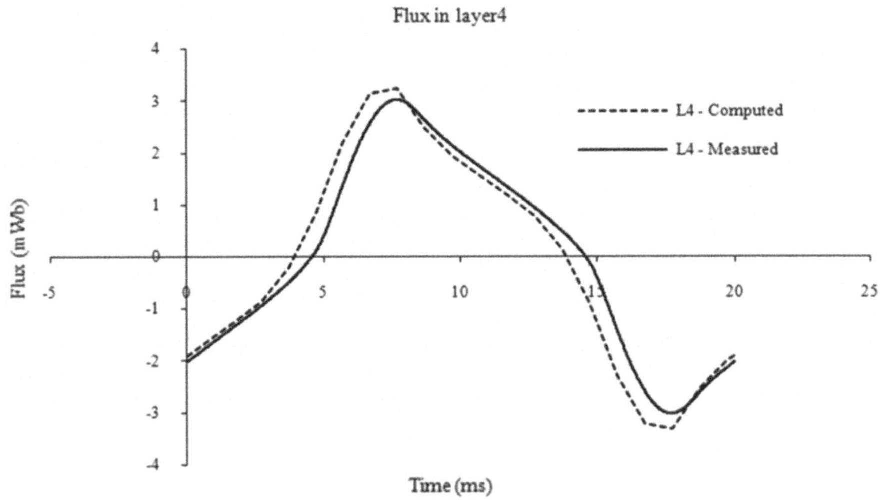

Fig. 13.27 Waveform of flux in layer 4

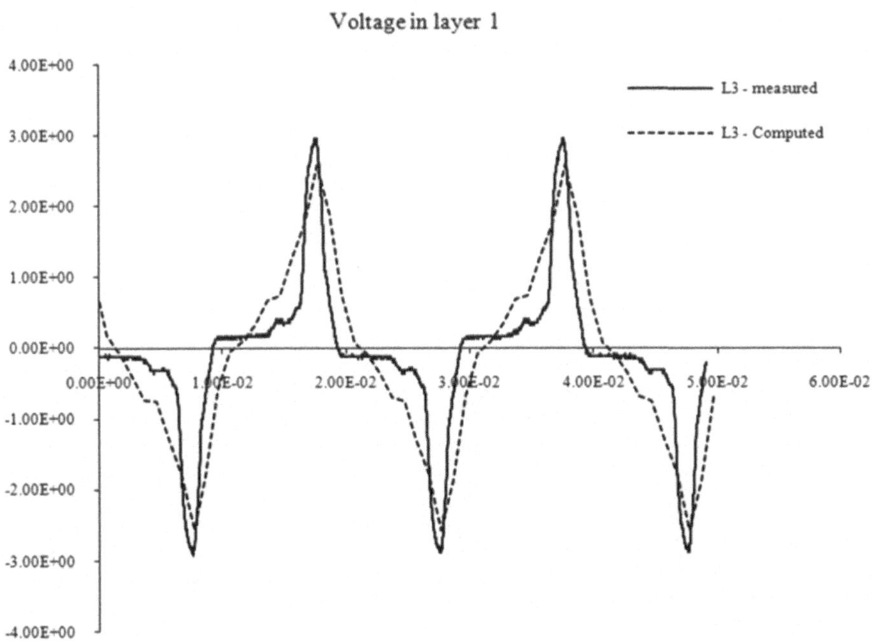

Fig. 13.28 Waveform of voltage in layer 1

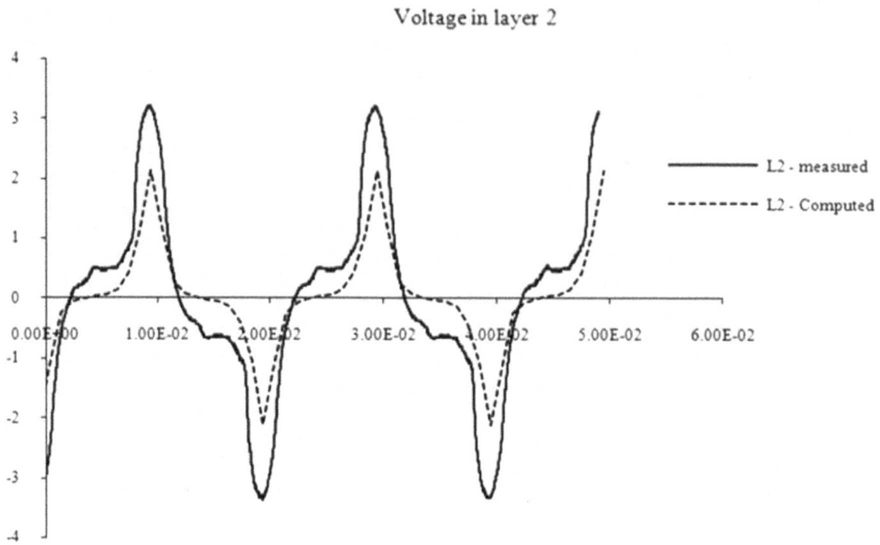

Fig. 13.29 Waveform of voltage in layer 2

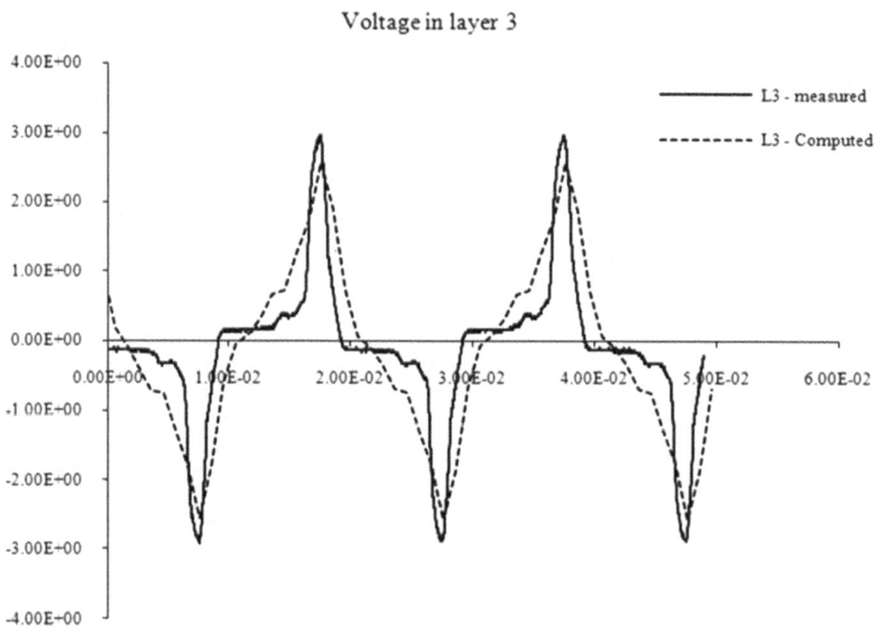

Fig. 13.30 Waveform of voltage in layer 3

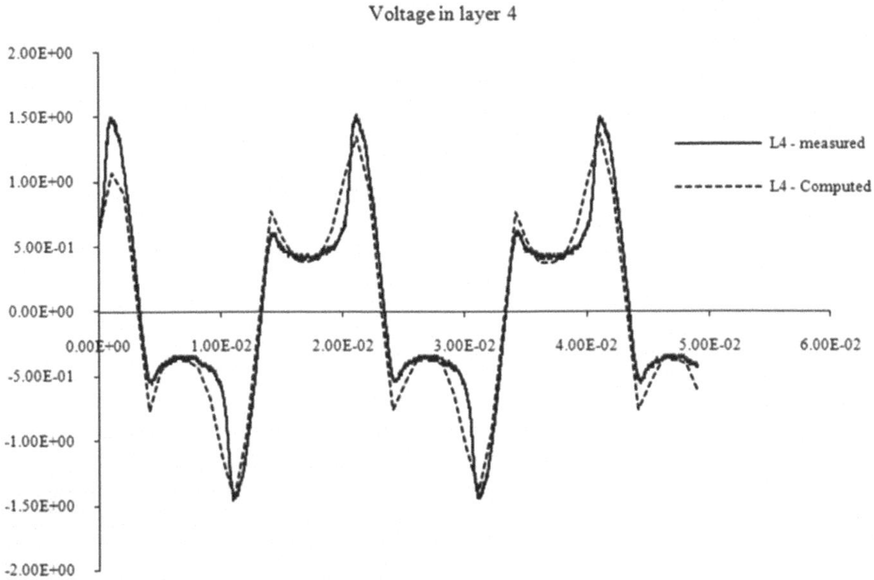

Fig. 13.31 Waveform of voltage in layer 4

The interlinkage flux of the search coils No.1–4 in P21^{d}-M has been obtained using Simcenter MAGNET transient solver for longer than 3 cycles of excitation source. As indicated, the calculated and measured waveforms of flux and induced voltage show satisfactory correlation.

13.4.10 Measured and Calculated Flux Densities at Specified Positions

The normal magnetic flux density (B_x) at the specified positions on both sides of the lamination of P21^{d}-M) is measured using a Gaussmeter (F. W. Bell, In.), under the applied exciting current of 10 A (50 Hz, rms), and compared with the corresponding calculation values, to further confirm the validity of the results, as shown in Fig. 13.32. See Fig. 13.1 for the specified measurement positions.

It is observed that the two are in good agreement.

Note: the half-thickness of the Gaussmeter's sensor: 0.76 mm.

Based on the simplified models of laminated sheets, the comparison between the calculated and measured results of several field quantities, including the magnetic flux densities at specified positions in the air, the flux in individual laminated sheets, and the waveforms of the induced voltage on the search coils, show a satisfactory agreement.

The measuring system is shown in Fig. 13.33.

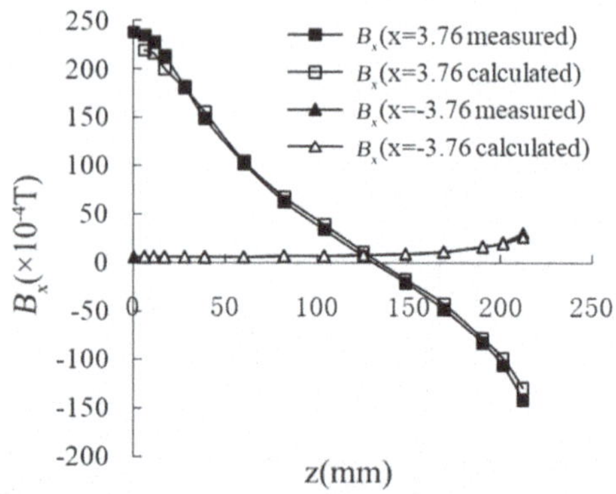

Fig. 13.32 B_x distribution on both sides of lamination (P21^d-M)

Fig. 13.33 Measurement and prediction of magnetic behaviors in laminated sheets

13.4.11 Question and Discussion

In this chapter, the additional iron loss induced in the laminated sheets, as an increment of the standard specific total loss measured under standard condition, is investigated based on simplified models.

One question still worth considering is whether the leakage flux entering the laminations vertically only increases some eddy current losses, or may affect the magnetic microstructure of GO silicon steel [11], thereby changing its specific total loss. If so, the specific total loss after the change will no longer be equal to the standard specific total loss obtained by standard 1-D or 2-D magnetic measurement method. Is the stronger the leakage flux, the more prominent the problem?

The further research is needed to solve this problem.

13.5 Concluding Remarks

The detailed numerical simulation and measurement of the additional iron loss based on laminated models can be briefly summarized as follows:

(1) The measured and calculated results based on simplified models $P21^d$-M, $P21^d$-M2, and $P21^d$-M3, including the total and additional iron loss, the interlinkage flux at designated positions of the silicon steel laminations, and magnetic flux densities at the specified positions in the air under various conditions (e.g., different numbers of silicon steel sheets and/or different exciting currents), are practically in good agreement, which demonstrates the effectiveness of modeling and computation.

(2) From the benchmarking results, it can be seen that the additional iron loss probably becomes a considerable portion of total iron loss in the laminated sheets, depending on the normal-direction leakage flux condition.

(3) Based on the simplified models ($P21^d$-M, $P21^d$-M2, and $P21^d$-M3), the electromagnetic behaviors, including the variations of iron loss and magnetic flux with exciting currents and laminated configurations, are studied in detail. This is helpful for optimum structural design and for preventing and solving local overheating, which may be caused by additional iron loss.

Moreover, Appendix 13.1 shows some calculated and measured results, which demonstrate how the magnetic loss and flux inside the laminated sheets, also in magnetic steel plate, are closely dependent on excitation patterns [9].

Acknowledgements The authors thank the colleagues of the Electromagnetic Laboratory of Institute of Power Transmission and Transformation Technology, Baobian Electric, for their cooperation in the model-based experiment and numerical analysis. Some of the calculated results shown in this chapter are provided by using Simcenter MAGNET. This project was supported in part by the Natural Science Foundation of China (NSFC) (no. 50677016 and no. 50777042) and a key project of the Hebei Natural Science Foundation (no. E2006001036).

Appendix 13.1 Magnetic Loss and Flux Under Different Excitation Patterns

In Problem 21 Family, the exciting source used in all the member-models is composed of two exciting coils with the same structural specifications, and the exciting currents of the two coils flowing in opposite directions.

To further examine the effect of different excitation patterns on magnetic loss and flux, based on P21^d-M, three test cases depending on the different exciting currents **J** in coils 1 and 2 are proposed, as shown in Table 13.10. Figure 13.34 shows the 2-D magnetic field distributions for three cases, i.e., Cases I–III.

13.1.1 Benchmarking Results Based on P21^d-M

The results of magnetic loss and linked flux measured and calculated under different excitation pattern and different exciting currents, based on P21^d-M, are shown in Tables 13.11 and 13.12.

The results shown in Tables 13.11 and 13.12 can be summarized as follows:

(1) The calculated results of both magnetic loss and linked flux agree well with the measured ones under each excitation pattern and different exciting currents;
(2) The effect of the excitation pattern on magnetic loss in laminated sheets is remarkable. The magnetic loss of the Case I is greater than those of both Case II and Case III under different exciting currents. Noticed that the magnetic loss of Case II becomes the lowest, with the increase of excitation currents.
(3) However, the effect of the excitation pattern on linked magnetic flux is not like the situation with the magnetic loss: The results of three cases are not so much different, and the values of magnetic flux of the Case II become higher compared to the other two cases.

Table 13.10 Different excitation patterns

Cases	Exciting currents (50 Hz)		Main property of magnetic flux
	In Coil 1	In Coil 2	
I	**J**	−**J**	Mostly perpendicular to steel plate (or lamination)
II	**J**	**J**	Mostly parallel to steel plate (or lamination)
III	**J**	**0**	Partly perpendicular, partly parallel

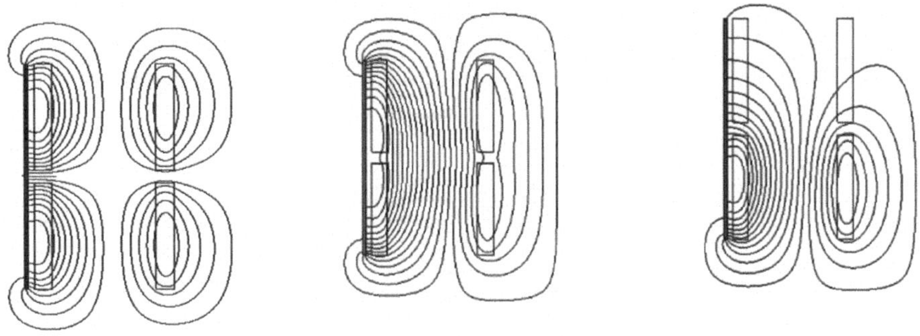

(a) Case I (b) Case II (c) Case III

Fig. 13.34 2-D magnetic flux distributions under different excitation patterns

Table 13.11 Magnetic loss in laminated sheets (P21^{d}-M)

Exciting currents (A, rms, 50 Hz)	Case I (W) (by $T-\Omega$)		Case II (W) (by $T-\Omega$)		Case III (W) (by $T-\Omega$)	
	Meas.	Calc.	Meas.	Calc.	Meas.	Calc.
10	2.20	2.33	0.66	0.62	0.59	0.57
15	5.30	5.04	1.43	1.33	1.39	1.26
20	10.20	10.32	2.71	2.45	2.99	2.76
25	16.80	16.33	4.72	4.49	5.19	4.91

Table 13.12 Linked magnetic flux in laminated sheets (P21^{d}-M)

Exciting currents (A, rms, 50 Hz)	Case I (mWb) (by $T-\Omega$)		Case II (mWb) (by $T-\Omega$)		Case III (mWb) (by $T-\Omega$)	
	Meas.	Calc.	Meas.	Calc.	Meas.	Calc.
10	0.297	0.311	0.357	0.381	0.329	0.323
15	0.444	0.447	0.532	0.569	0.490	0.501
20	0.589	0.594	0.708	0.707	0.652	0.672
25	0.738	0.702	0.886	0.893	0.817	0.832

Note The position of search coil can be seen in Fig. 13.3

13.1.2 Benchmarking Results Based on P21^{0}-B

In order to compare the effect of the excitation pattern on magnetic properties inside solid magnetic steel plate and laminated sheets, the measured and calculated results of magnetic loss and flux under different excitation pattern and different exciting currents, based on P21^{0}-B with solid magnetic steel plate, have been obtained by the authors, as shown in Tables 13.13 and 13.14.

Table 13.13 Magnetic loss in magnetic steel plate (P21^0-B)

Exciting currents (A, rms, 50 Hz	Case I (W) (calc. by $\boldsymbol{A}_r - \boldsymbol{V} - \boldsymbol{A}_r$)		Case II (W) (calc. by $\boldsymbol{T} - \Omega$)		Case III (W) (calc. by $\boldsymbol{T} - \Omega$)	
	Meas.	Calc.	Meas.	Calc.	Meas.	Calc.
10	11.97	12.04	11.61	10.92	6.04	6.57
15	26.89	27.12	26.52	25.80	13.48	14.43
20	49.59	50.92	47.16	48.32	24.35	25.49
25	82.39	84.78	74.40	72.36	39.44	40.12
30	123.70	128.67	107.60	101.12	58.90	58.64
35	179.10	183.15	155.00	150.76	83.27	81.32
40	248.00	250.45	205.00	197.32	114.10	108.72
45	330.00	330.91	258.00	249.92	146.37	152.92
50	423.00	425.07	335.50	316.82	189.00	190.42

Table 13.14 Linked magnetic flux in laminated sheets (P21^0-B)

Exciting currents (A, rms, 50 Hz	Case I (mWb) (calc. by $\boldsymbol{A}_r - \boldsymbol{V} - \boldsymbol{A}_r$)		Case II (mWb) (calc. by $\boldsymbol{T} - \Omega$)		Case III (mWb) (calc. by $\boldsymbol{T} - \Omega$)	
	Meas.	Calc.	Meas.	Calc.	Meas.	Calc.
10	0.318	0.306	0.341	0.356	0.326	0.318
15	0.478	0.458	0.513	0.538	0.492	0.481
20	0.618	0.605	0.679	0.687	0.652	0.637
25	0.770	0.750	0.834	0.841	0.808	0.780
30	0.936	0.890	0.983	0.970	0.967	1.003
35	1.064	1.024	1.129	1.141	1.113	1.108
40	1.206	1.152	1.270	1.300	1.256	1.260
45	1.357	1.276	1.404	1.410	1.398	1.413
50	1.486	1.396	1.540	1.580	1.537	1.500

Note The position of search coil can be seen in Fig. 12.35 of Chap. 12 of this book

The results shown in Tables 13.13 and 13.14 can be summarized as follows:

(1) The calculated results of both magnetic loss and linked flux agree well with the measured ones under each excitation pattern and different exciting currents;
(2) The effect of the excitation pattern on magnetic loss in solid magnetic plate is much different from that in the laminated sheets. The magnetic losses in both Case I and Case II are much higher than that in Case III.
(3) Also the effect of excitation pattern on linked magnetic flux is different from that of magnetic loss. The magnetic fluxes of both Case II and Case III are slightly higher than that of Case I.

13.1.3 Remarks

The effect of excitation pattern on both magnetic loss and linked flux has been investigated based on $P21^d$-M and $P21^0$-B, as two member-models of Problem 21 Family.

All results shown in Tables 13.11, 13.12, 13.13, and 13.14 indicate that the effects of excitation patterns on both magnetic loss and linked flux inside laminated sheets and magnetic steel plate are different, and the laminated sheets and solid magnetic plate have different performances under the same excitation patterns.

Accordingly, the effect of the excitation patterns on magnetic properties should be fully taken into account in magnetic measurements, and further in-depth research is needed, especially under complex (even extreme) excitation conditions.

References

1. I. Sebestyen, S. Gyimothy, J. Pavo, and O. Biro: "Calculation of losses in laminated ferromagnetic materials," *IEEE Trans. Magn.*, Vol.40, No.2, pp. 924–927, 2004.
2. H. Kaimori, A. Kameari, and K. Fujiwara, "FEM computation of magnetic field and iron loss in laminated iron core using homogenization method," *IEEE Trans. Magn.*, Vol. 43, No. 4, pp. 1405–1408, 2007.
3. J. Turowski, X. M. Lopez-Fernandez, A. Soto, and D. Souto, "Stray losses control in core- and shell type transformers," Presented at Advanced Research Workshop in Transformers (ARWtr-2007), Baiona, Spain.
4. K. Muramatsu, T. Okitsu, H. Fujitsu, and F. Shimanoe: "Method of nonlinear magnetic field analysis taking into account eddy current in laminated Core," *IEEE Trans. Magn.*, Vol.40, No.2, pp. 896–899, 2004.
5. Z. Cheng, N. Takahashi, B. Forghani, G. Gilbert, J. Zhang, L. Liu, Y. Fan, X. Zhang, Y. Du, J. Wang, and C. Jiao, "Analysis and measurements of iron loss and flux inside silicon steel laminations," *IEEE Trans. Magn.*, vo.45, no.3, pp. 1222–1225, 2009.
6. TEAM Problem 21:Benchmark Family, [Online]. Available: www.compumag.org/team.
7. X. Zhao, F. Meng, Z. Cheng, L. Liu, J. Zhang, and C. Fan, "Stray-field loss and flux distribution inside magnetic steel plate under harmonic excitation, COMPEL, 36, 6, pp. 1715–1728, 2017.
8. National Standard of China: Quality of Electric Energy Supply/Harmonics in Public Supply Network, GB/T 14549-93.
9. Z. Cheng, N. Takahashi, B. Forghani, G. Gilbert, Y. Du, Y. Fan, L. Liu, Z. Zhai, W. Wu, and J. Zhang, "Effect of excitation patterns on both iron loss and flux in solid and laminated steel configurations," *IEEE Trans. Magn.*, vol.46, no.8, pp. 3185–3188, 2010.
10. Z. Cheng, N. Takahashi, B. Forghani, Y. Du, Y. Fan, L. Liu, and H. Wang, "Effect of variation of B-H properties on both iron loss and flux in silicon steel lamination," *IEEE Trans. Magn.*, vol.47, no.5, pp. 1346–1349, 2011.
11. A. J. Moses, "Relevance of microstructure and texture to the accuracy and interpretation of 1 and 2 directional characterization and testing of grain-oriented electrical steels," IJAEM (International Journal of Applied Electromagnetics and Mechanics), vol.55, Supplement 1, pp. 3–13, 2017.

Part V
Transformer-Related Electromagnetic and Thermal Modeling and Application

Chapter 14
Electromagnetic and Thermal Modeling Based on Large Power Transformers

Lanrong Liu, Jie Li and Fulai Che

Abstract To study the variation of metal material conductivity and other related performance parameters under the real working conditions, a measuring system for the electromagnetic and thermal properties of metal materials commonly used in transformer engineering is well established. Meanwhile, the measurement methodology and devices used for determining the material surface coefficient of heat transfer are also developed, through which the corresponding working properties of some key materials are obtained. The effectiveness of electromagnetic and thermal modeling and simulation, using the measured material property data, are experimentally verified based on the proposed test models. Finally, the engineering usefulness of both the large-scale electromagnetic and thermal modeling, and the structural optimization to reduce stray-field loss, to prevent the local overheating, are validated based on larger power transformers.

Keywords Electromagnetic and thermal simulation · Finite element (FE) · Power transformer · Material property measurement · Conductivity · DC magnetization curve · Surface heat transfer coefficient

14.1 Introduction

With the development and construction of China's state grid, the voltage grades and capacities of transformers are increasing. By now, the capacity of a single-phase transformer has reached 1500 MVA; the AC and DC rated voltages have reached 1000 kV and ±1100 kV, respectively, thus the leakage magnetic field of large power transformers and stray-field losses of metal components have also increased. Under normal operation, the stray-field losses in the metal components of a large power transformer can reach 5–8%, even more, of load losses. Excessive stray-field

L. Liu (✉) · J. Li · F. Che
Institute of Power Transmission and Transformation Technology,
Baobian Electric, Baoding, China
e-mail: liulanrong@btw.cn

Z. Cheng et al. (eds.), *Modeling and Application of Electromagnetic and Thermal Field in Electrical Engineering*, https://doi.org/10.1007/978-981-15-0173-9_14

loss [1, 2] can easily cause local overheating of metal components, such as the tank, clamping, core tie-plate, tensile belt, hoist seat and other metal components, which is the result of the leakage magnetic field caused by the windings, the active parts and the heavy current leads [3, 4]. The stray-field loss in metal components must not be ignored, because local overheating of these components causes insulation aging, anomalous oil chromatogram and other hazards, affecting the normal operation of transformer, which could easily lead to transformer faults. Nowadays, the stray-field loss calculation and local overheating prevention have become key technologies for large power transformers.

Compared with the traditional analytical method and empirical formula, the electromagnetic and thermal simulation has the advantages of not only high accuracy of the calculated results, but also visual observation of the simulated field distributions, and therefore has become an indispensable scientific tool to study stray-field loss distribution characteristics and for controlling the local temperature. Electromagnetic and thermal simulation technology provides support for reducing stray-field loss, improving operating efficiency, saving materials and reducing consumption, optimizing structure of transformers and ensuring safe operation.

In this chapter, the measurement devices used for determining the material surface coefficient of heat transfer and obtaining the corresponding working properties of some key materials are developed. Meanwhile, the electromagnetic and thermal properties of metal materials are measured using the established experimental setup and regarding the actual operation condition. The engineering usefulness of large-scale electromagnetic and thermal modeling and simulation, and the effectiveness of the structural optimization to reduce loss and prevent local overheating are verified based on larger power transformers.

14.2 Measurement of Electromagnetic and Thermal Properties of Commonly Used Metal Materials

A transformer essentially comprises iron core, winding, active-part insulation, leads, transformer oil, tank, heat radiator and bushing. The metal materials used in these components are mainly magnetic steel plates, non-magnetic steel plates, copper and aluminum plates, which are either magnetic or non-magnetic conducting materials. It is found that the standard parameter values in the manual are usually directly used by many designers to solve the transformer's electromagnetic and thermal problems and may cause great errors. A platform is setup to study the variation of metal material conductivity and other parameters under the real service condition of the transformer and to determine the optimum parameters for simulation. The conductivity tests of common materials of large oil-immersed power transformers, such as magnetic steel plates Q235B and Q345B, non-magnetic steel plate 20Mn23Al, non-magnetic high strength round steel 50Mn, copper plate T2Y, aluminum plate 1200, and DC magnetization curve measurements of magnetic steel plates Q235B and Q345B, are given below.

14.2.1 Conductivity Measurement

The DC resistance loss of transformer and stray-field loss of components are all temperature dependent, the reference temperature of oil-immersed power transformer stipulated in IEC 60076-1 is 75 °C, so the load loss value needs to be corrected to the reference temperature [2, 5, 6]. The efficiency of magneto-thermal coupling simulation of stray-field losses can be greatly reduced by converting the conductivity of metal materials to the operating temperature [7]. A platform for measuring material parameter characteristics is built to study the variation of metal material conductivity and other parameters under the real service condition of transformers. Through repeated tests using the platform, the conductivity of the material at 75 °C is determined and applied to the simulation to improve accuracy.

14.2.1.1 Test Specimen

For the metal plate, the dimensions of the specimen are 300 × 30 × 10 mm. For the round steel, the length of the specimen is 300 mm. However, the actual dimensions of the specimens to be used must be accurately redetermined before property measurement.

14.2.1.2 Measurement Method

The resistance of the metal specimen is measured by Kelvin four-terminal detecting method, i.e., using a DC constant current power to generate a DC voltage drop U on the tested resistance R. The advantage of using the "four-terminal" (current terminals, potential terminals) method is that the errors caused by leads and contact resistances can be eliminated.

Equation (14.1) below is used to calculate conductivity after the specimen resistance value is obtained:

Equation for conductivity:

$$\sigma = \frac{1}{\rho} = \frac{1}{\frac{RS}{L_C}} = \frac{L_C}{RS} \tag{14.1}$$

where

σ is conductivity (S/m);
L_C is distance between two potential measuring points (m);
R is resistance (Ω);
S is cross-section area of test specimen (m^2).

14.2.1.3 Test Temperature Control

The specimens are placed in a high-low temperature test chamber, heated to the corresponding constant temperature, as shown in Fig. 14.1.

14.2.1.4 DC Resistance Tester

DC resistance tester (HIOKI 3541, Japan) is recommended (as shown in Fig. 14.2), the specifications can be found in Table 14.1.

14.2.1.5 Measured Conductivity

The conductivities measured at 20 °C and at 75 °C are shown in Table 14.2.

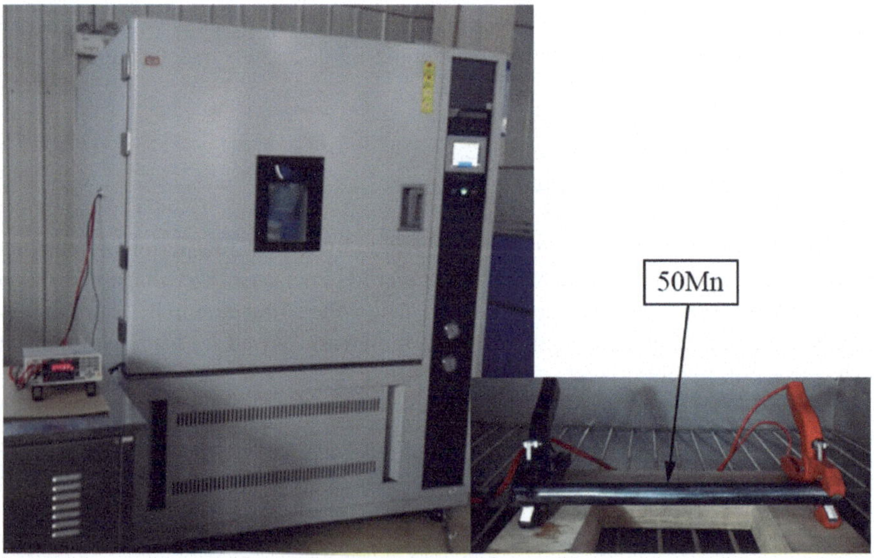

Fig. 14.1 Conductivity test platform for metal specimens

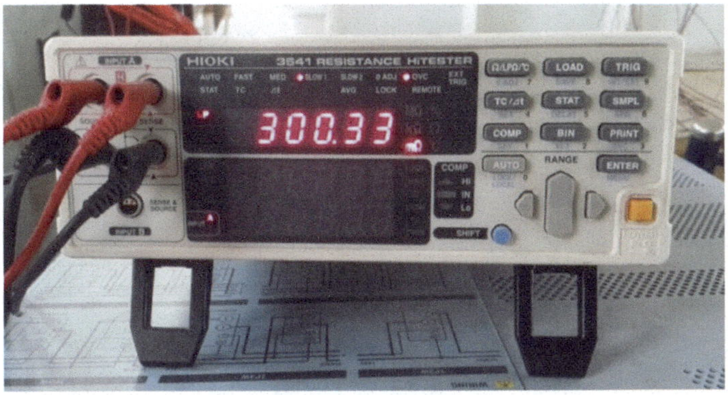

Fig. 14.2 DC resistance tester (HIOKI 3541, Japan)

Table 14.1 Specifications of DC resistance tester (HIOKI 3541, Japan)

Range	20 m/200 mΩ, 2/20/200/2 k/20 k/100 kΩ, 1 M/10 M/100 MΩ; minimum resolution: 0.1 μΩ
Display	Maximum display value: 200,000 (20 m–20 kΩ), LED
Accuracy	(SLOW2) ± 0.007% rdg. ± 15dgt. (2 k/20 kΩ)

Table 14.2 Conductivity of commonly used metal materials

S/N	Specimen material	Conductivity (S/m)		Remarks
		20 °C	75 °C	
1	Q235B	5.7898×10^6	4.9134×10^6	Magnetic steel plate
2	Q345B	6.6355×10^6	5.4683×10^6	Magnetic steel plate
3	20 Mn23Al	4.2693×10^6	3.7552×10^6	Non-magnetic steel plate
4	50Mn	1.4543×10^6	1.3985×10^6	Non-magnetic high strength round steel
5	T2Y	5.8101×10^7	4.8107×10^7	Copper plate
6	1200	3.4767×10^7	2.8548×10^7	Aluminum plate
7	14Mn	4.055×10^6	3.5622×10^6	Magnetic steel plate

14.2.2 Measurement of DC Magnetization Curve of Magnetic Steel Plate

14.2.2.1 Ring-Shaped Specimens

The DC magnetization curve (*B–H* curve) of magnetic steel plate is measured using ring-shaped specimens which are made according to IEC604-4 and GB/T 13012.

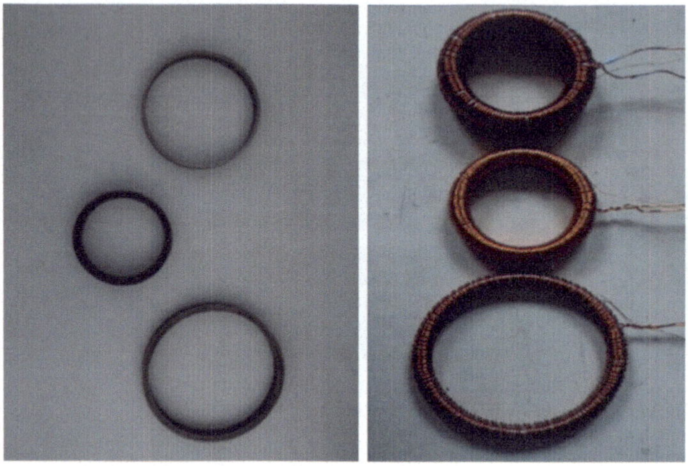

Fig. 14.3 Ring-shaped specimens for *B–H* curve of magnetic steel plate

The outer surface of the specimens is wound with the measuring windings and the exciting windings (as shown in Fig. 14.3).

14.2.2.2 Measured Results of DC Magnetization Curve of Magnetic Steel Plate

The DC magnetization curves are measured by the DC magnetic tester (Metron MTR2488, Japan). The measured results and the *B–H* curves are shown, respectively, in Fig. 14.4 and Table 14.3.

14.2.3 Surface Heat Transfer Coefficient of Steel Plate

Simcenter MAGNET (including a magnetic field simulation software) and Simcenter MAGNET Thermal (a thermal field simulation software) are widely used in electromagnetic and thermal simulation of power transformers. The surface heat transfer coefficient given in the process of the magnetic and thermal coupling simulation must be accurate, otherwise the results will have great errors, which even lead to erroneous conclusions. A series of models and devices for measuring the surface heat transfer coefficient of materials such as steel plates are put forward and made to obtain the parameters used for engineering application. These parameters are tested and validated to help and improve the accuracy of the electromagnetic and thermal simulation.

There are two kinds of media, transformer oil and air, used to dissipate heat from steel components of oil-immersed power transformers, in which the components

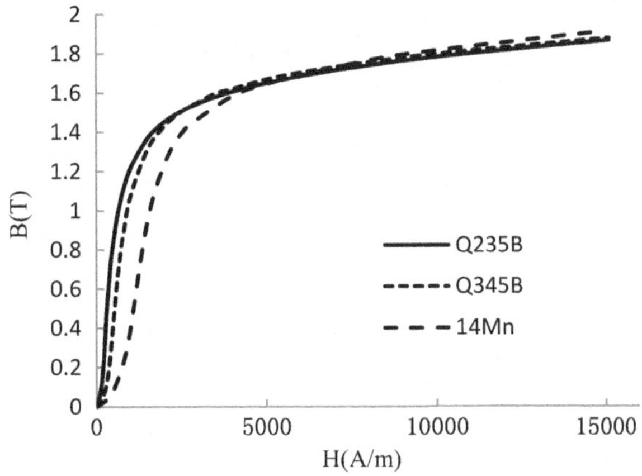

Fig. 14.4 *B–H* curve of magnetic steel plate

Table 14.3 Test data for *B–H* curve of magnetic steel plate

Measure point	Q235B		Q345B		14Mn	
	H (A/m)	*B* (T)	*H* (A/m)	*B* (T)	*H* (A/m)	*B* (T)
B1	100.46	0.058	100.45	0.021	99.61	0.011
B2	201.15	0.192	200.96	0.055	/	/
B3	301.63	0.485	301.67	0.119	298.5	0.038
B4	402.25	0.701	402.23	0.251	/	/
B5	502.8	0.848	502.81	0.442	497.57	0.085
B6	603.43	0.958	603.45	0.628	/	/
B7	703.06	1.044	704.02	0.778	/	/
B8	803.45	1.112	803.59	0.896	796.67	0.229
B10	1004.47	1.217	1004.37	1.074	995.9	0.395
B15	1507.04	1.370	1506.95	1.314	1494.08	0.945
B20	2011.44	1.454	2009.39	1.433	1992.38	1.234
B25	2514.44	1.510	2511.22	1.508	2486.36	1.392
B30	3017.43	1.549	3017.36	1.556	2983.25	1.468
B35	3520.33	1.582	3505.74	1.598	/	/
B40	4023.26	1.608	4018.97	1.622	3978.08	1.582
B50	5029.76	1.652	5025.06	1.670	4973.66	1.645
B60	6036.39	1.685	6043.42	1.705	/	/
B70	7038.33	1.716	7030.89	1.733	/	/
B80	8043.92	1.741	8028.2	1.757	7961.85	1.765
B100	10,054.57	1.785	10,052.87	1.800	9952.48	1.816
B150	15,083.86	1.868	15,083.8	1.880	14,930.13	1.914

inside the tank (e.g., core tie-plate, core clamping, inside wall of tank) are completely immersed in transformer oil and the components outside the tank (e.g., support for cooling equipment, outside wall of tank) are completely exposed to air. The surface heat transfer coefficient is a key parameter and must be accurate, otherwise the result will have great errors, which even lead to erroneous conclusions. The device made by the team can measure the surface heat transfer coefficients of steel plate either in the transformer oil or air.

14.2.3.1 Models for Measuring Surface Temperature Property of the Steel Plate

A square steel plate with electrothermal pipes embedded therein is used to achieve a test model with approximately uniform heat dissipation on the surface. The heating elements of the electrothermal pipes are completely and uniformly embedded in the steel plate, and the lead-out wire is sealed with sealant. The reliable insulation is provided by a ceramic tube configured between electrothermal pipes and the steel plate. Thermoelectric couples are buried in the steel plate surface according to certain rules for measuring the surface temperature of the steel plate. The steel plate model and test model are shown in Figs. 14.5 and 14.6.

14.2.3.2 Measurement Methods and Instruments

The terminals of the electrothermal pipes are connected in parallel, and the ends of the model are connected to the DC power supply (WT3000 power analyzer). After power on, the Joule heat is transmitted inside the steel plate and radiated to the cooling medium (air or transformer oil) via the surface of the steel plate. It is convenient to measure the power output from DC power supply to the electrothermal pipes, that is to say, it is easy to obtain the power of the heat source.

The TP700 data recorder is used to monitor the surface temperature of the steel plate by the thermoelectric couple, the surface temperature is recorded when it reaches a stable level. The surface temperature of the model is scanned by the

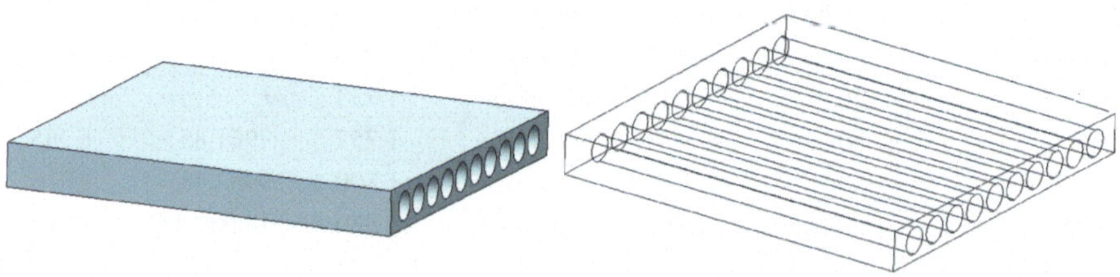

Fig. 14.5 Test model for steel plate

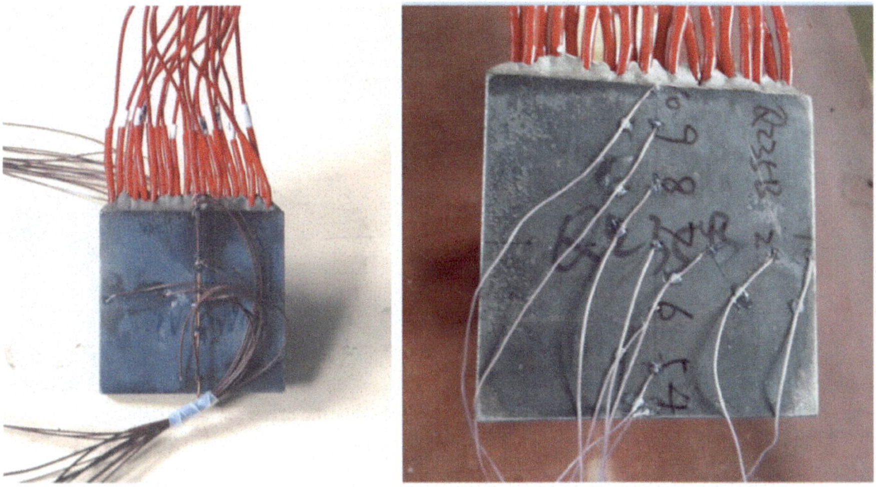

Fig. 14.6 Test model for steel surface coefficient of heat transfer

infrared thermal imager and recorded. The photographs, in which coefficients are measured in air and the thermal image, are shown in Figs. 14.7 and 14.8, respectively. The measurement model is placed on three-tapered iron frames to reduce the heat transfer from the support, as shown in Fig. 14.7a.

The internal average loss density of the steel plate, P_0, is calculated as follows:

$$P_0 = \frac{P_a}{V_s} \tag{14.2}$$

where P_a is the applied power and V_s is the total volume of the steel plate.

Thermal simulation is carried out on the test specimen with Simcenter MAGNET Thermal, and the loss density can be obtained by Eq. (14.2). The surface heat transfer coefficient of the steel measured under such condition is obtained by adjusting the surface heat transfer coefficient of the model to make the surface hot-spot temperature of the test model equal to the measured value.

14.2.3.3 Measurement Results

The surface heat transfer coefficients of the steel obtained by the method of combining test and simulation [7] are shown in Tables 14.4, 14.5 and 14.6.

(a) The upper and lower surfaces of the test

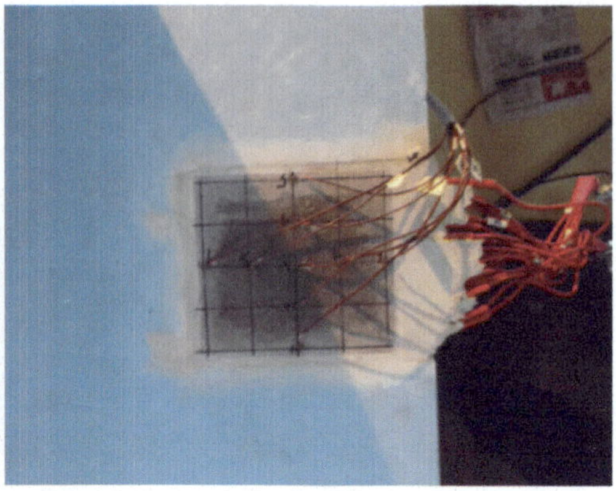

(b) Only the upper surface dissipates heat model dissipate heat at the same time

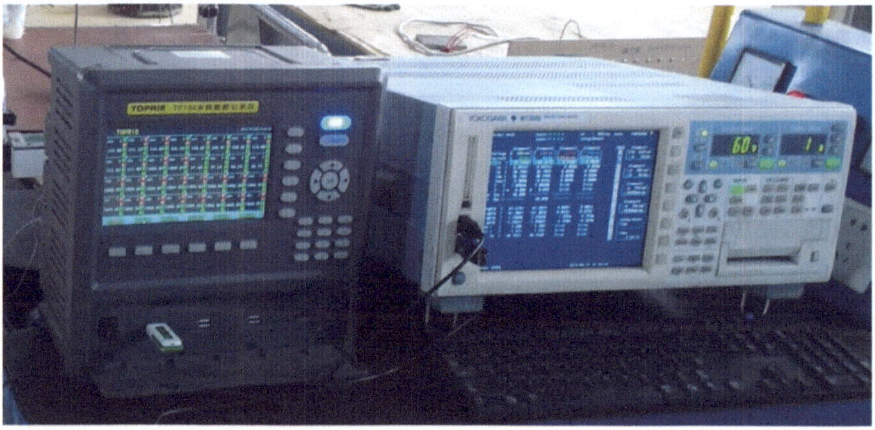

(c) Test Instrument

Fig. 14.7 Measuring the steel surface coefficient of heat transfer measured in air

14.3 Validation of Modeling and Simulation of Loss and Surface Hot-Spot Temperature of Steel Plate

Another test model and platform are built to further determine the correctness and credibility of the parameters, such as the conductivity and surface heat transfer coefficient. Such models can be placed in air or transformer oil.

The winding used here is the excitation source, the internal space of which generates a magnetic field when the model is powered on, and the built-in steel specimen may generate loss and heat under the action of the magnetic field. The test platform is an approach closer to the actual service conditions, and the parameters obtained are more accurate.

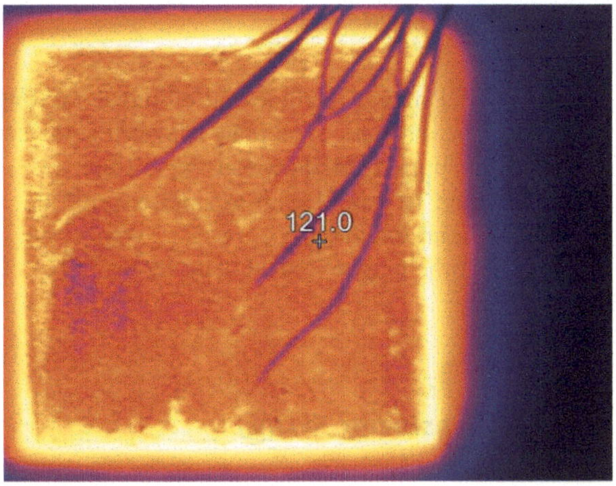

Fig. 14.8 Surface temperature measurement of magnetic steel plate Q235B (using thermal infrared imager, fluke TI32)

Table 14.4 Surface heat transfer coefficient of magnetic steel measured in air

Ambient temperature (°C)	Loss density P_0 (W/m^3)	Orientations	Surface hot-spot temperature (°C)	Surface heat transfer coefficient (W/m^2 °C)
20	16,000	Horizontal	69.8	11.9
20	26,500	Horizontal	121.0	16.3
20	16,000	Vertical	70.4	12.1
20	26,500	Vertical	112.4	17.2

Table 14.5 Surface heat transfer coefficient of magnetic steel measured in oil

Oil temperature (°C)	Loss density (W/m^3)	Orientations	Surface hot-spot temperature (°C)	Surface heat transfer coefficient (W/m^2 °C)
20	1,000,000	Horizontal	72.9	99.3
20	1,880,000	Horizontal	99.8	121.8
20	1,000,000	Vertical	73.1	98.9
20	1,880,000	Vertical	103.8	120.3

Table 14.6 Surface heat transfer coefficient of non-magnetic steel measured in oil

Oil temperature (°C)	Loss Density (W/m^3)	Orientations	Surface hot-spot temperature (°C)	Surface heat transfer coefficient (W/m^2 °C)
20	1,000,000	Horizontal	73.4	98.6
20	1,880,000	Horizontal	101.7	120.3
20	1,000,000	Vertical	73.8	99.7
20	1,880,000	Vertical	104.9	121.9

14.3.1 Test Model

The test model mainly includes an elliptical exciting coil and a square magnetic steel plate. The magnetic steel plate is placed at the center of the coil and is fixed by an insulating support, as shown in Fig. 14.9.

The winding is wound with a paper-covered flat copper conductor and has 24 discs in the axial direction.

The size of the magnetic steel plate shown in Fig. 14.9 is 500 mm × 300 mm × 10 mm.

14.3.2 Measuring System

Thermoelectric couples are spot welded on the surface of the steel plate, as shown in Fig. 14.10, to allow temperature measurement. The measuring system setup in the air is shown in Fig. 14.11. The laboratory instruments and equipment required as well as the specifications are listed in Table 14.7.

14.3.3 Results and Validation

When the transformer oil, for example, is used as the heat dissipation medium, the test model is simulated with Simcenter MAGNET, as shown in Fig. 14.12, and validated by the test.

Fig. 14.9 Test model

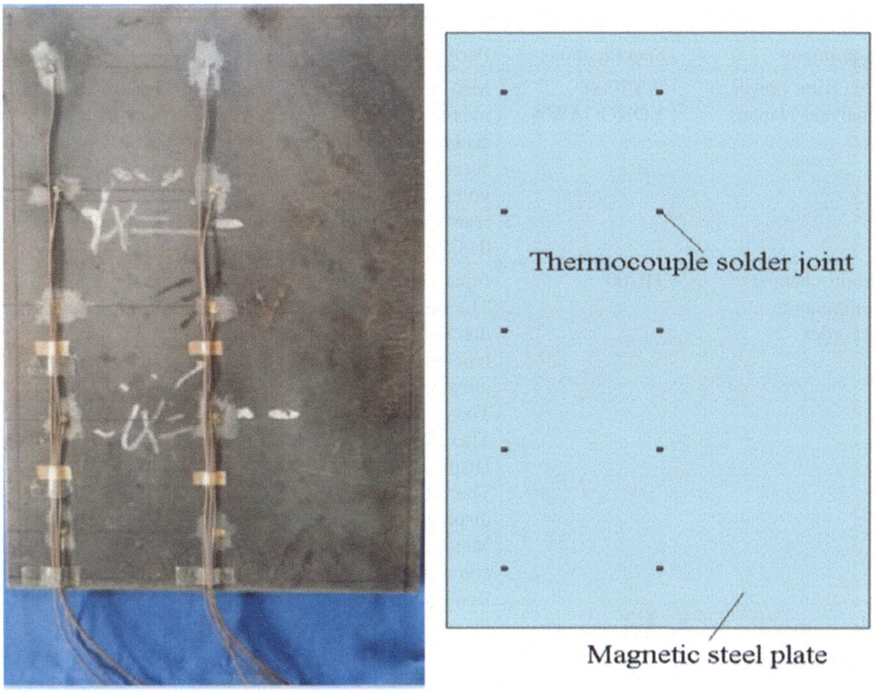

Fig. 14.10 Location of thermoelectric couples buried in steel plate surface

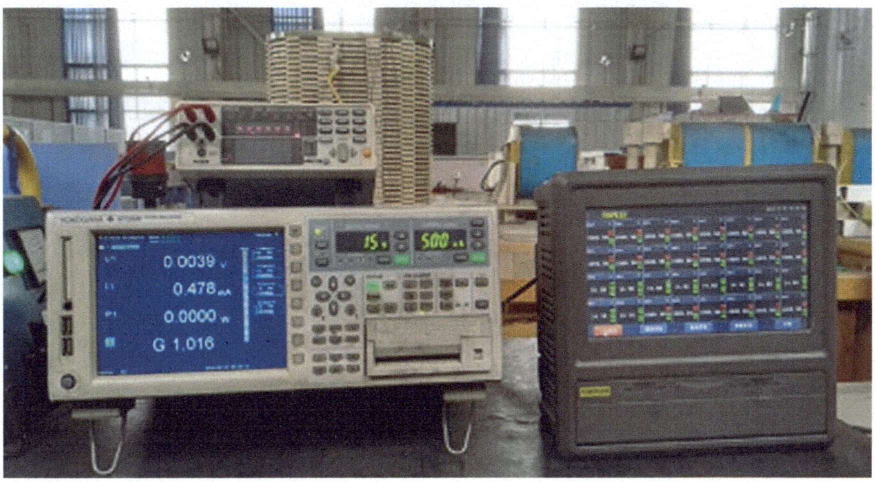

Fig. 14.11 Measuring system

Table 14.7 Performance of laboratory instruments and equipment

Equipment	Specifications	Performance Index
Precision power analyzer (Japan)	WT3000, YOKOGAWA	Measuring DC and AC signals; maximum current measured: 30 A (rms); current and voltage reading accuracy: ±0.02%; range accuracy: ±0.04%; power accuracy: ±0.06%; the internal resistance of the voltage measuring terminal: 10 MΩ; the internal resistance of the current measuring terminal: 0.0055 Ω; frequency range: DC, 0.1 Hz–1 MH
Multi-channel temperature recorder	TP700	Input signal: The input signal includes DC current, DC voltage, thermal resistance, thermoelectric couple and transmissible pressure gauge. Isolation input without jumper Thermal resistance: Pt100, Cu50 Thermoelectric couple: K, S, R, B, N, E, J, T Display: The intrinsic error is less than ±0.2% F S; digital display range is: −999.9 to ~1999.9 Measurement resolution: 1/60,000, 24-bit AD conversion Real-time curve inter-record gap: 1–9999 s
DC resistance tester (Japan)	3541, HIOKI	See Table 14.1

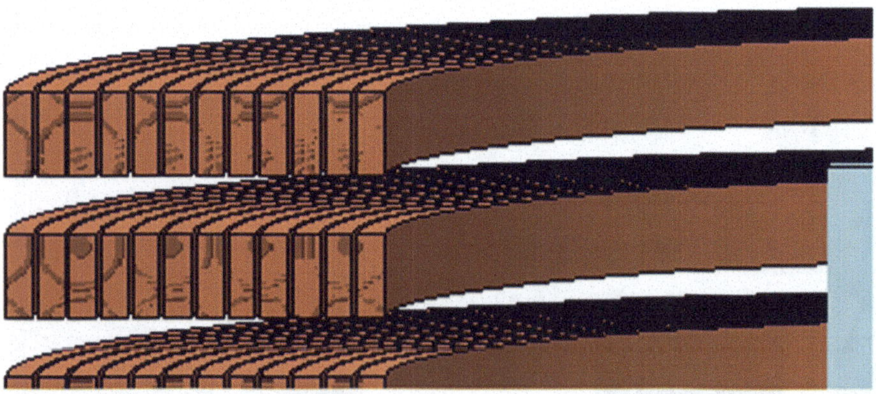

Fig. 14.12 3-D model of winding and steel plate

Parameter setting:

The winding current, conductor temperature and other parameters are the same as the test model; the electromagnetic and thermal property parameters of the steel plates are set in reference with Sects. 14.2.1, 14.2.2 and 14.2.3; the winding and steel plate losses are obtained by the magnetic field and loss simulation.

The winding loss P_{coil} includes DC resistance loss $P_{\mathrm{coil-r}}$ and eddy-current loss $P_{\mathrm{coil-e}}$. The real winding distribution and the cross section and length of conductor can be obtained, if the winding is modeled as a single turn, so that the accurate simulation and the real winding eddy-current loss can also be obtained.

The simulation results obtained from the Simcenter MAGNET are directly coupled to Simcenter MAGNET Thermal. The surface heat transfer coefficient is set according to the condition of air, and the temperature distribution and hot-spot temperature of the steel plate surface are obtained by simulation.

Based on the results obtained from the test model, the measurement value of steel plate loss $P_{\mathrm{steel}\text{-}m}$, is calculated based on the measured total loss of model P_{total} and the calculated total loss of windings P_{coil}, by Eqs. (14.3) and (14.4).

$$P_{\mathrm{coil}} = P_{\mathrm{coil-r}} + P_{\mathrm{coil-e}} \tag{14.3}$$

$$P_{\mathrm{steel-m}} = P_{\mathrm{total}} - P_{\mathrm{coil-m}} \tag{14.4}$$

where

P_{coil} is total loss of winding, W;
$P_{\mathrm{coil-r}}$ is DC resistance loss of winding, I^2R, W; I is exciting winding current;
R is winding thermal resistance at outage;
$P_{\mathrm{coil-e}}$ is eddy-current loss of windings obtained by simulation;
$P_{\mathrm{steel-m}}$ is loss of steel plate, W;
P_{total} is total loss of the model, as test results, in W.

The Maximum element size of the steel plate is 10 mm. Mesh layers assigned to the steel plate face are 0.5, 1 and 2 mm from outside to inside, respectively. Setting polynomial order is 2 for model calculation [3]. The results from measurement and simulation of the model are shown in Tables 14.8 and 14.9, while the temperature distribution on the steel surface is shown in Fig. 14.13.

The conductivity, *B–H* curve, surface heat transfer coefficient and other parameters of metal material obtained from the test are used to carry out the magnetic-thermal coupling simulation by Simcenter MAGNET and Simcenter MAGNET Thermal on the model in Fig. 14.9, and the errors between the results from simulation and measurement are within the engineering allowance.

Table 14.8 Comparison between the measured and simulated losses of Q235B in transformer oil

Current applied (A)	Test results				Simulation results		Error (%)
	P_{total} (W)	R (Ω)	$P_{\mathrm{coil-r}}$ (W)	$P_{\mathrm{steel-m}}$ (W)	$P_{\mathrm{coil-e}}$ (W)	$P_{\mathrm{steel-c}}$ (W)	
12	640.70	0.03029	43.98	583.74	12.98	610.76	4.63
16	1137.10	0.03041	78.14	1036.24	22.72	1040.68	0.43
20	1764.90	0.03074	123.00	1606.81	35.09	1561.26	−2.83

$P_{\mathrm{steel-c}}$ is calculated value of steel plate loss

Table 14.9 Comparison between the measured and simulated temperature of hot spot on the surface of Q235B

Current applied (A)	Measured temperature of hot spot on the steel plate surface (K)	Simulated temperature of hot spot on the steel plate surface (K)	Error (%)
12	23.3	25.3	8.6
16	39.1	41.9	7.2
20	55.5	59.2	6.7

Note Ambient temperature: 20 °C

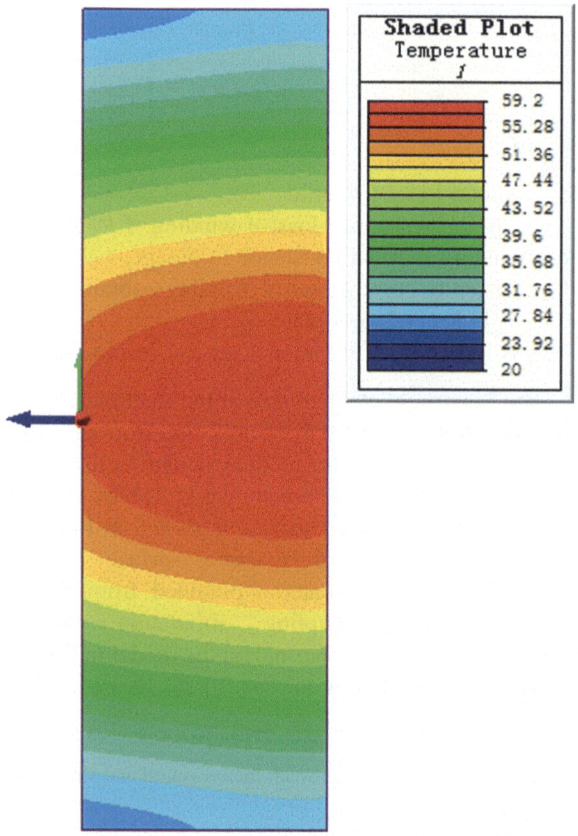

Fig. 14.13 Temperature distribution on the steel plate surface (current 20 A)

14.4 3-D FE Model for Simulating Transformer Component Loss

14.4.1 3-D Mesh for Magnetic Steel Plate

Compared to non-magnetic steel, the core stiffening plate, supporting plate, foot pad, tank and other components in large transformers are generally made of magnetic steel plates, whose *B–H* and *B–W* curves are nonlinear, which makes it much more difficult to accurately compute loss. The principle to select the mesh approach of magnetic steel plates for engineering simulation is introduced through a detailed research on meshing the magnetic steel plates, considering accuracy and efficiency.

14.4.1.1 Determining the Standard Loss for Simulation of Magnetic Steel Specimen

A magnetic steel (Q235B) specimen (size: 12 mm × 150 mm × 300 mm) is made and the loss in such magnetic plate is measured in the same way as in Sect. 14.2.3.

Simcenter MAGNET is used to calculate the loss inside the above specimen. Considering the symmetry of the solved field problem, its 1/4 FE model is built, as shown in Fig. 14.14, to reduce the amount of computation.

In order to ensure computation accuracy, the mesh layer near to the steel plate's surface should be fine (e.g., in this FE model, 0.2 mm thick for 15 layers + 0.4 mm thick for 5 layers), at the same time the maximum element size of the magnetic steel specimen is 10 mm and the second-order polynomial is used in the simulation process. The calculated loss results are shown in Table 14.10.

Table 14.10 shows that the error between the simulation results and the measured results is less than 5%. Therefore, the simulated losses in Table 14.10 are used as the standard values to compute the deviations of simulated losses obtained using different mesh approaches.

14.4.1.2 Analysis of Simulated Losses and Errors Obtained by Different Mesh Approaches

The magnetic flux density and loss density of the magnetic steel plate in the thickness direction (y-direction) are obtained inside the steel specimen (near the center of winding), respectively, as shown in Figs. 14.15 and 14.16. The penetration depths of the magnetic steel specimen [8–11] into the magnetic field are about 1.9, 2.8 and 3.8 mm corresponding to the exciting current of 10, 20 and 40 A. When such depths are exceeded, the magnetic flux density and loss density in the magnetic steel plate are close to zero.

Fig. 14.14 Simulation model

Table 14.10 Measured and simulated losses of magnetic steel specimen

Exciting current (A)	Measured loss (W)	Simulated loss (W)	Error (%)
10	32.53	33.89	4.2
20	143.2	149.6	4.5
40	590.8	619.7	4.9

The losses in the magnetic steel plate are mainly concentrated in the surface layer, as shown in Fig. 14.16. Therefore, in the loss simulation, the magnetic steel surface must be layered, usually the more the number of layers, the higher the accuracy. However, with the increase of the number of layers, the number of subdivision units also increases, and the simulation time will be significantly increased. As such, with the unchanged mesh size, the surfaces with one layer and with two layers are compared to achieve a cost-effective layering.

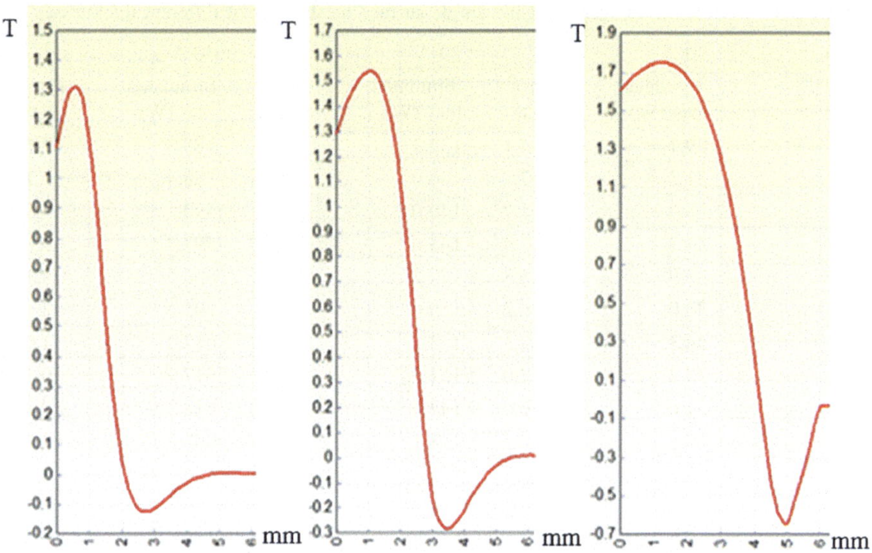

(a)10A (b)20A (c)40A

Fig. 14.15 By distribution of magnetic steel specimen in thickness direction (0–6 mm: center to surface of steel specimen)

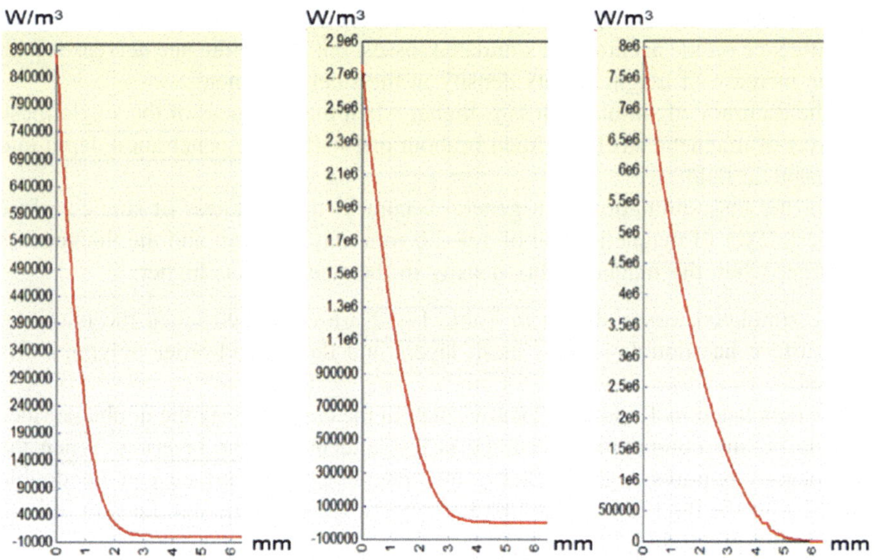

(a)10A (b)20A (c)40A

Fig. 14.16 Loss density distribution of magnetic steel specimen in thickness direction (0–6 mm: surface to center of steel specimen)

Table 14.11 Calculated loss results using single-surface mesh layer of different thicknesses

Layer thickness (mm)	10 A		20 A		40 A	
	Simulated loss (W)	Error (%)	Simulated loss (W)	Error (%)	Simulated loss (W)	Error (%)
0.4	29.1	−14.1	132.2	−11.6	585.3	−5.6
0.6	31.3	−7.7	136.0	−9.1	588.4	−5.1
0.8	33.3	−1.7	140.0	−6.5	592.1	−4.5
1.0	34.7	2.4	143.6	−4.0	597.2	−3.6
1.2	35.9	5.8	147.2	−1.6	600.5	−3.1
1.5	36.0	6.2	150.9	0.9	605.5	−2.3
2.0	34.7	2.5	153.1	2.3	612.3	−1.2
2.5	33.2	−2.0	151.3	1.1	617.9	−0.3
3.0	31.6	−6.6	147.0	−1.8	618.2	−0.2
4.0	28.4	−16.1	138.1	−7.7	608.3	−1.8

The magnetic losses inside magnetic steel plate under different thickness of the single-surface mesh layer and different exciting current are calculated, as shown in Table 14.11, which demonstrated the effect of the thickness of the surface mesh layer on total magnetic loss. In addition, the second-order polynomial [10] was adopted in this loss calculation.

Table 14.11 shows that:

(1) The thickness of single-surface mesh layer of magnetic steel has a great influence on the accuracy of simulated losses, but such influence decreases with the increase of magnetic flux density in the steel specimen;
(2) The accuracy of simulated loss is higher when the thickness of the single mesh layer of magnetic steel specimen is about one-half of the penetration depth into magnetic field;
(3) For the loss simulation with general accuracy, the thickness of single-surface mesh layer of magnetic steel of 1.0–2.0 mm may be used, and the thickness is larger when the magnetic flux density in the steel plate is higher.

The simulated losses shown in Table 14.12 are computed when the magnetic steel surface has double-surface mesh layers and the second-order polynomial is adopted.

The data listed in Table 14.12 shows that, in the case of using the double-surface mesh layer, the computation accuracy is less affected by the layering. When the computation requires high accuracy, the magnetic steel surface can be double layered with the thickness of the first layer being 0.8–1.0 mm and the sum of both layers being preferably 2.5–3.0 mm.

By comparing the values listed in Tables 14.11 and 14.12, the accuracy of calculated loss in the magnetic steel plate with double-surface mesh layer is higher than that with single-surface mesh layer.

Table 14.12 Calculated loss results using double-surface mesh layer of different thicknesses

Layer thickness (mm)	10 A		20 A		40 A	
	Simulated loss (W)	Error (%)	Simulated loss (W)	Error (%)	Simulated loss (W)	Error (%)
0.8–2.2	33.6	−1.0	148.6	−0.7	618.4	−0.2
0.8–2.0	33.7	−0.5	150.0	0.3	619.1	−0.1
0.8–1.6	34.0	−0.3	152.0	1.6	616.2	−0.6
0.8–1.0	35.1	3.5	152.6	2.0	609.8	−1.6
0.6–2.4	32.9	−3.1	147.7	−1.3	617.6	−0.3
0.6–2.0	33.5	−1.2	151.0	0.9	618.8	−0.2
0.6–1.5	34.2	0.9	152.8	2.1	613.5	−1.0
0.6–1.2	34.9	3.0	152.5	1.9	609.8	−1.6

14.4.2 3-D Mesh for Non-magnetic Steel Plate

The simulation accuracy is significantly affected by the mesh size in the simulation of large transformer losses. For the non-magnetic steel plate, the relative permeability μ_r generally equals to 1, and only the eddy-current loss caused by the magnetic field is considered. In simulation, the smaller the mesh, the higher is the accuracy of simulated loss. Nevertheless, it will take up a lot of computer memory and simulation time if the mesh is too small. Therefore, the choice of mesh is one of the key problems in large transformer loss simulation.

A single-leg simulation model is established and validated by meshing the non-magnetic steel plate, based on the key parameters of a large power transformer. The simulated losses of the core tie-plate and the web of core clamping for different meshes are shown in Table 14.13.

Table 14.13 Simulated losses of core tie-plate and web of core clamping for different meshes

Mesh (mm)	Tie plate loss		Web of core clamping loss		Memory occupied (G)	Time (min)
	P (W)	Error (%)	*P* (W)	Error (%)		
6	623.6	0.00	2498.0	0.00	37.7	305.0
8	616.8	−1.09	2498.0	−0.01	13.5	81.0
10	612.0	−1.86	2497.0	−0.08	7.3	42.0
15	601.6	−3.53	2478.0	−0.80	3.9	21.0
20	566.6	−9.14	2475.0	−0.92	2.9	12.0
30	509.4	−18.3	2403.0	−3.80	1.6	7.0
40	498.3	−20.09	2385.0	−4.52	1.5	6.0
50	73.8	−24.02	2327.0	−6.85	1.4	5.0
60	434.7	−30.29	2320.0	−7.13	1.3	5.0

The error in the simulated tie plate loss increases rapidly with the mesh size, for example, when the mesh size = 20 mm, the error is about 10%, as can be seen from Table 14.13. Therefore, the mesh size of non-magnetic steel tie plate is generally about 1/2 of the plate thickness in the simulation. The same happens to the error in the simulated web of core clamping loss which also increases with mesh size. Due to larger thickness and area, the base value of the web of core clamping loss is larger and its increase rate of error is smaller than that of the tie plate. Therefore, the mesh of the non-magnetic steel web of core clamping can be slightly larger, which is 1/2 of the web of core clamping thickness.

The loss density distributions of the tie plate and the web of core clamping are shown in Figs. 14.17 and 14.18, respectively. The nonuniform loss [10] distribution is found, in that, the tie plate loss is mainly concentrated at the position corresponding to the winding ends, and the web of core clamping loss, at the upper and

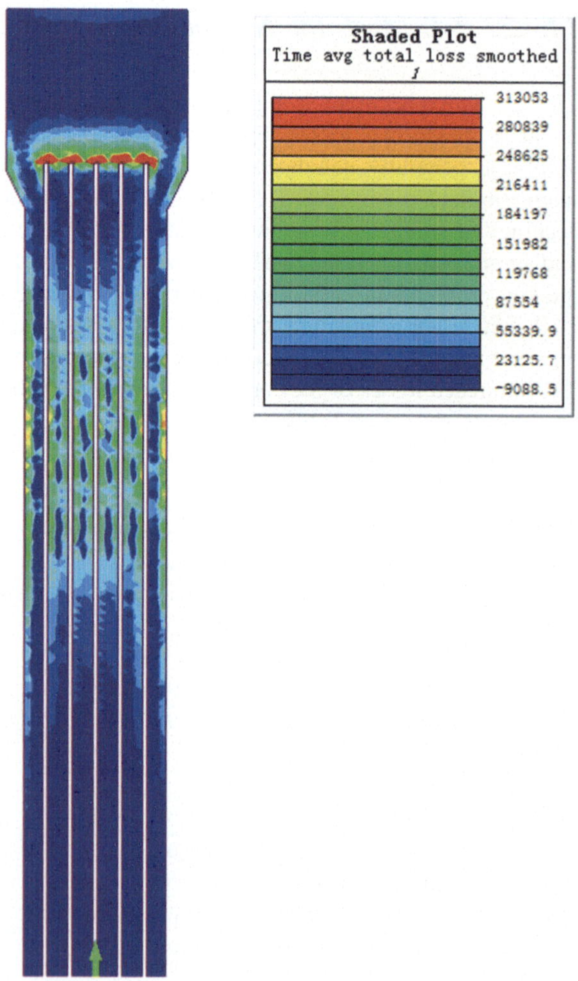

Fig. 14.17 Loss density distribution in tie plate

Fig. 14.18 Loss density distribution in web of upper clamping

Fig. 14.19 Local fine meshing of tie plate

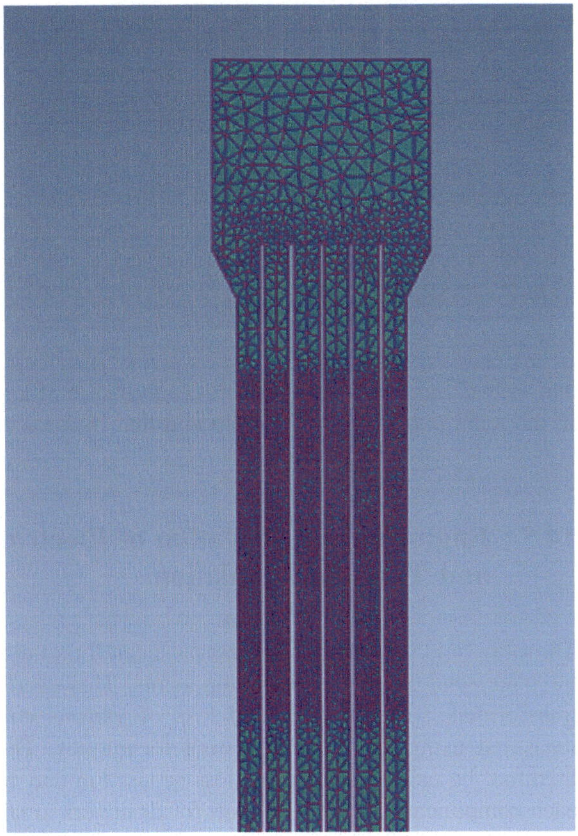

lower edges. Accordingly, accuracy can be further improved by local fine meshing, as shown in Figs. 14.18, 14.19, 14.20 and Table 14.14.

Table 14.14 shows that accurate simulation results can be obtained through a smaller computing scale by means of local mesh refinement. Therefore, local mesh refinement is recommended for components, such as non-magnetic steel tie plate

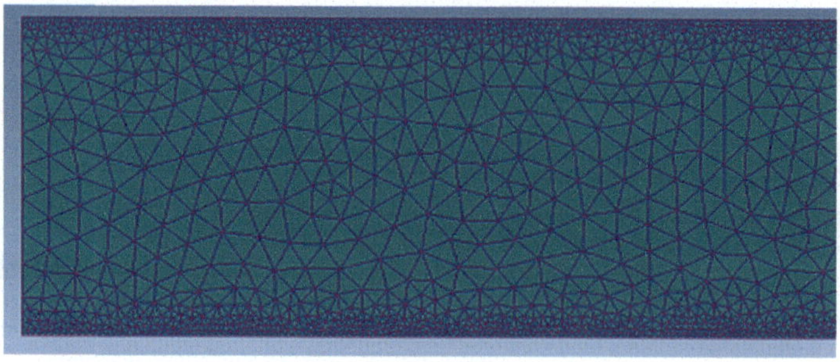

Fig. 14.20 Local fine meshing of web of upper clamping

Table 14.14 Calculated loss of tie plate and web of core clamping with local meshing

Component	Mesh (mm)	Simulated loss		Memory occupied (G)	Time (min)
		P (W)	Error (%)		
Tie plate	Main part 40—local part 10	591.7	−5.12	3.2	13.0
Web of core clamping	Main part 50—local part 15	2480	−0.72		
Tie plate	40	498.3	−20.09	1.5	6.0
Web of core clamping	50	2327	−6.85	1.4	5.0

and web of core clamping, in the loss simulation of large transformer, with the mesh in the main part being 40–50 mm and that in the local part being ¼–1/3.

14.5 Engineering Application of Electromagnetic and Thermal Simulation

The stray-field loss of transformers has been found in winding, tank, core clamping, tie plate, and among other components. However, the total stray-field loss of transformers and the stray-field loss of various components cannot be directly measured using the existing testing techniques. The simulation technology can therefore be used to realize the loss separation and compute the stray-field loss in each component providing support for designers to take corresponding measures to reduce the stray-field loss.

The model-based parameters facilitate the electromagnetic and thermal simulation accuracy, which has been applied in a 700 MVA/750 kV single-phase autotransformer. Note that in order to improve the transformer structure, the different design options were carefully evaluated based on the electromagnetic and thermal analysis results, by using Simcenter MAGNET and Simcenter MAGNET Thermal, at the design stage.

14.5.1 Large Single-Phase Autotransformer (700 MVA/750 KV)

14.5.1.1 The Specifications

The related technical parameters of the single-phase autotransformer 700 MVA/750 kV are as follows:

Rated capacity: 700/700/233 MVA;
Voltage combination: $\frac{765}{\sqrt{3}} / \frac{345}{\sqrt{3}} \pm 2 \times 2.5\% \ /63$ kV;
Rated phase current: 1584.9/3514.3/3698.4 A;
Short-circuit impedance: 18%/56%/36%;
Rated frequency: 50 Hz;
It is a single-phase four-leg structure. Figure 14.21 shows a structural diagram of the active part.

14.5.1.2 Magnetic Field Simulation of Winding

The magnetic field of winding is simulated by the Simcenter MAGNET and the magnetic field on the short-axis section of the tank is shown in Fig. 14.22.

The short-circuit impedance values obtained by the magnetic field simulation can be found in Table 14.15. The deviation between the values from simulation and measurement of short-circuit impedance is less than 0.7%.

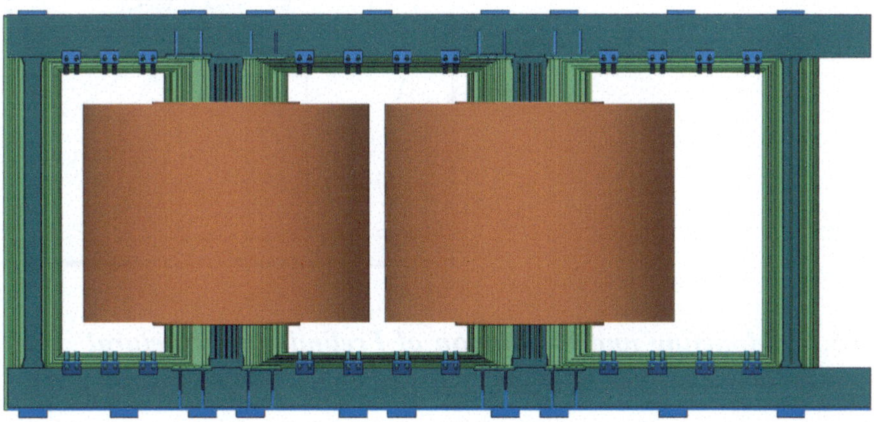

Fig. 14.21 Structural diagram of the active part of 700 MVA/750 kV

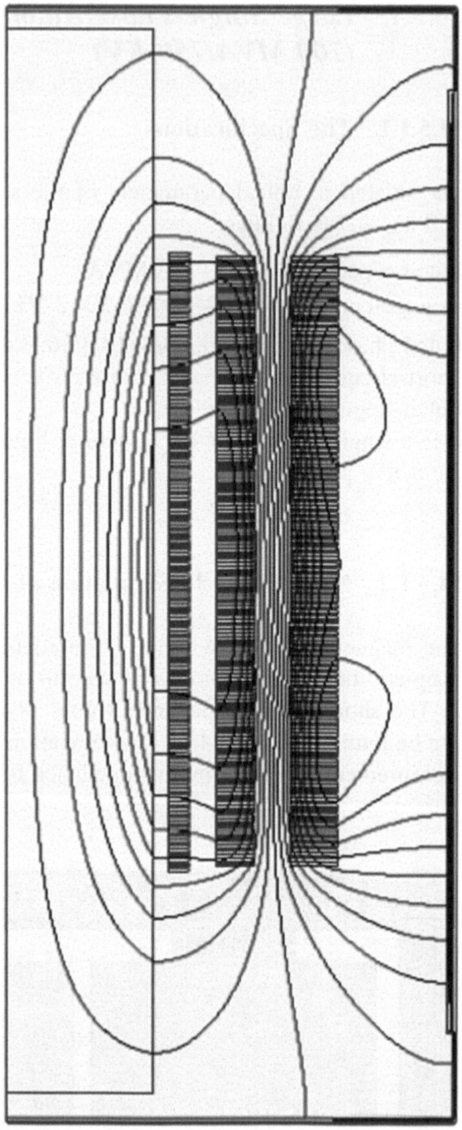

Fig. 14.22 Magnetic field map of HV-MV operating winding

14.5.2 *Modeling and Simulation of Preliminary Structural Design*

As one of the design options (D01), the stray-field losses in the key components of the autotransformer (700 MVA/750 kV) are calculated, and the losses in different components are given in Table 14.16.

Table 14.15 Values from simulation, measurement and design, as well as the deviations, of short-circuit impedance (700MVA/750 kV)

Operation mode	Measurement value (%)	Design value		Simulation value	
		Design value (%)	Deviation from measurement value (%)	Simulation value (%)	Deviation from measurement value (%)
HV-MV	17.96	18	0.22	18.02	0.36
HV-LV	57.89	56	−3.26	58.22	0.57
MV-LV	35.85	36	0.42	**36.09**	0.67

Table 14.16 Losses in key components in percentages of total calculated stray-field loss before structural optimization (700 MVA/750 kV, D01)

Components	Percentage (%)	Remarks
Windings	35.6	2-D computation; eddy-current loss
Oil tank	26.7	Taking account of difference between low- and high-voltage sides
Partial core laminations (affected by leakage flux)	1.0	Additional loss caused by the leakage flux inside partial core laminations
Core clampings	26.2	
Tie plate	3.3	
Tensile belt	0.2	
Pad footing	3.8	
Bracing plate	3.2	

Note In order to reduce the computation cost, the eddy-current losses of the winding are calculated by a 2-D solver

As shown in Table 14.16, the stray-field losses of the transformer components can be separately calculated. Except for the transformer windings, the stray-field losses of all the transformer components shown in Table 14.16 were calculated by using the 3-D solver. The stray-field loss distributions in the key components of the autotransformer (700 MVA/750 kV, D01) are shown in Figs. 14.23, 14.24 and 14.25.

14.5.3 Thermal Field Simulation of Components in the Active Part

The thermal field simulation of the key components of the autotransformer (700 MVA/750 kV, D01) is carried out with Simcenter MAGNET Thermal under the magnetic-thermal coupling mode [12–15]. The calculated temperature results of

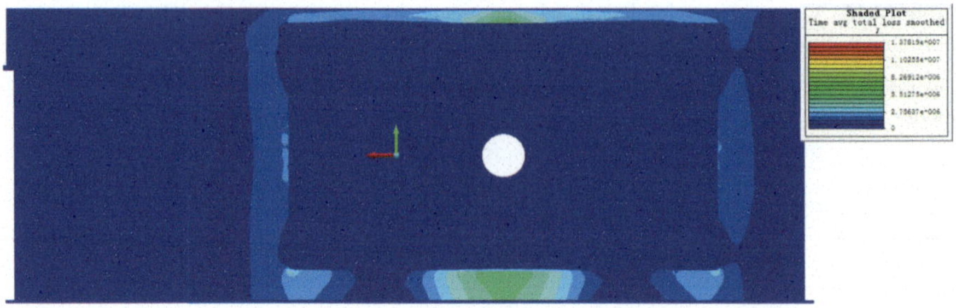

Fig. 14.23 Loss density distribution in the tank wall at high-voltage side (700 MVA/750 kV, D01)

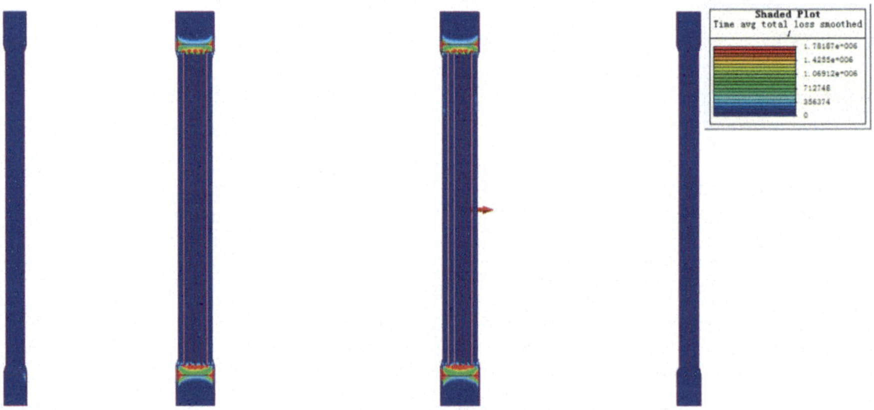

Fig. 14.24 Loss density distribution in the tie plate (700 MVA/750 kV, D01)

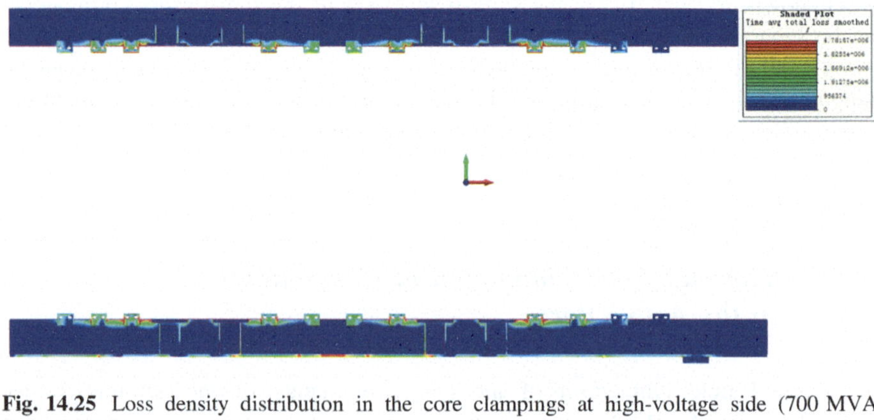

Fig. 14.25 Loss density distribution in the core clampings at high-voltage side (700 MVA/750 kV, D01)

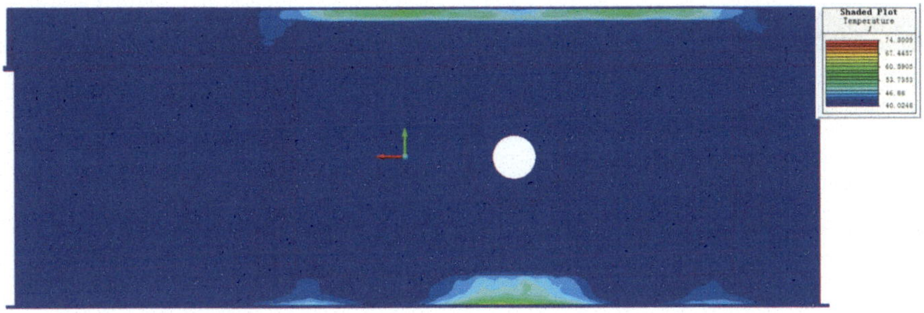

Fig. 14.26 Temperature distribution in the tank at high-voltage side (700 MVA/750 kV, D01)

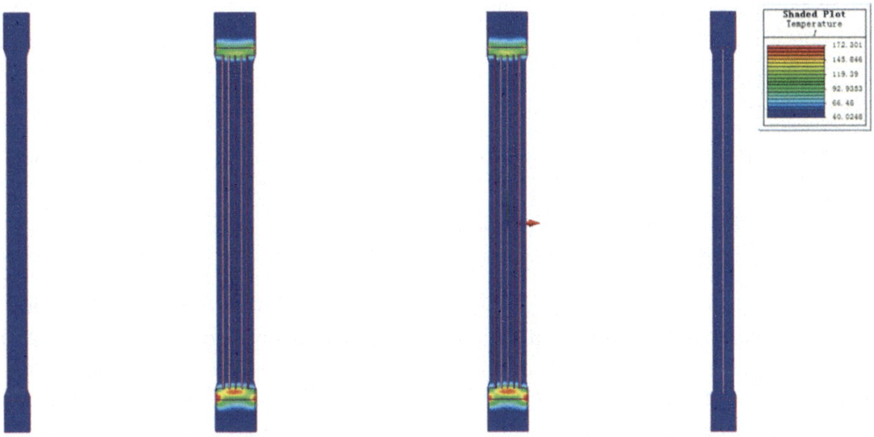

Fig. 14.27 Temperature distribution in the core tie-plate (700 MVA/750 kV, D01)

the key components (oil temperature: 40 °C) show that the hot-spot temperatures and temperature rise of some components are much higher than the warning values, including clamping at both HV and LV sides, and core tie-plate. The corresponding temperature distributions are shown in Figs. 14.26, 14.27 and 14.28.

The simulation results indicate an excessive local heating hazard in the preliminary transformer structure design (D01).

14.5.4 Modeling and Simulation of Optimized Structures

The structural improvements are made against the hot spots found by the preliminary simulation (D01), which is referred to as the optimized structure design, D01, mainly including:

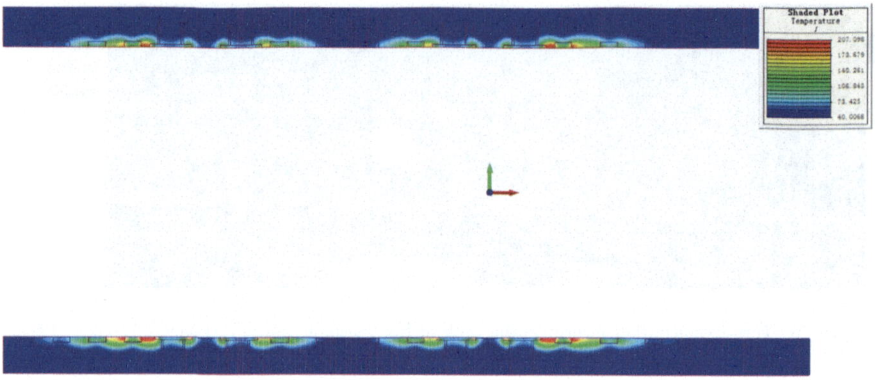

Fig. 14.28 Temperature distribution in the clamping at high-voltage side (700 MVA/750 kV, D01)

(1) The area covered by magnetic shields at the lower end of the active part is enlarged.
(2) The structure of the magnetic shields onto the tank is optimized.
(3) The structure of the core tie-plate is improved.

The calculated results after the structural optimization, referred to as D02, show the reductions of both stray-field loss and the hot-spot temperature. The loss and temperature of each component can be found in Tables 14.17 and 14.18, respectively. The distributions of both loss density and temperature of the components are emphatically improved, as shown in Figs. 14.29, 14.30, 14.31 and 14.32.

It is concluded that the stray-field loss, the temperature and temperature rise of all the components are considerably reduced to meet the engineering requirements. See Chap. 4 of this book for the coupled electromagnetic thermal field analysis in detail.

14.5.5 Discussion

The calculated total loss without structure optimization, P_c-D01, and the measured and calculated total loss based on the autotransformer (700 MVA/750 kV) with structure optimization, P_m-D02 and P_c-D02 are compared, as shown in Table 14.19.

From Table 14.19, it can be seen that both the calculated total losses (P_c-D01 and P_c-D02) are higher than the measured total loss (P_m-D02). The calculated total loss (P_c-D02) is in good agreement with the measured total loss (P_m-D02), and 4.4% lower than P_c-D01.

Table 14.17 Losses in key components in percentages of total measured loss after structural optimization (700 MVA/750 kV, D02)

Components	Percentages (%)	Remarks
Windings (DC)	76.9	DC resistance loss
Windings (Eddy Current)	9.3	eddy-current loss
Oil tank	7.1	Taking account of difference between low- and high-voltage sides, stray-field loss
Partial core laminations (affected by leakage flux)	0.32	Additional core loss caused by the leakage flux inside partial core laminations
Core clampings	4.85	
Tie plate	0.83	
Tensile belt	0.04	
Pad footing	0.89	
Bracing plate	0.83	

Table 14.18 Thermal simulation results of key components after structural optimization (700 MVA/750 kV, D02)

Component	Hot-spot temperature (°C)	Hot-spot temperature rise (K)	Warning value of temperature rise (TR_{limit}): 75 K
Clampings at high-voltage side	88.6	48.6	Low than TR_{limit}
Clampings at low-voltage side	84.7	44.7	Low than TR_{limit}
Core tie-plate	76.2	36.2	Low than TR_{limit}
Neutral-point lead	86.3	46.3	Low than TR_{limit}

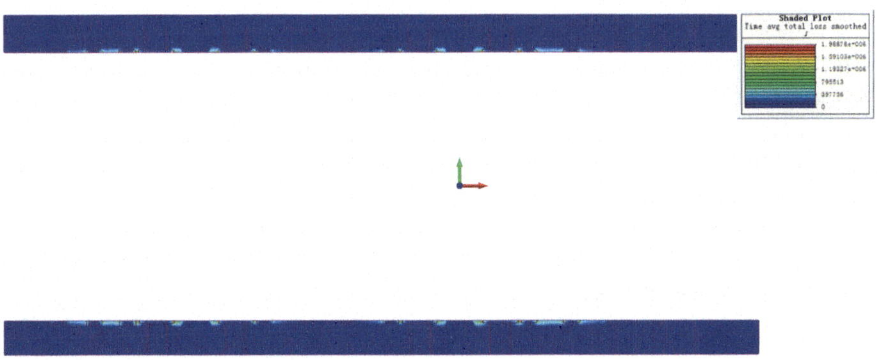

Fig. 14.29 Loss density distribution in the core clamping at high-voltage side (700MVA/750 kV, D02)

Fig. 14.30 Loss density distribution in the tie plate (700 MVA/750 kV, D02)

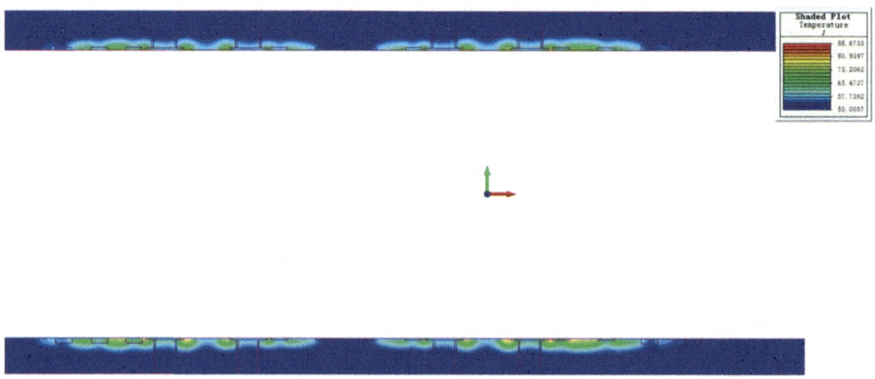

Fig. 14.31 Temperature distribution in the tank at high-voltage side (700 MVA/750 kV, D02)

It should be noted that the structure optimization of the autotransformer (700 MVA/750 kV) not only reduced the total loss, but also improved the loss distribution, and consequently avoided the unallowable local overheating. Therefore, the numerical modeling and simulation are really important to evaluate different design options.

Note that nine sets of single-phase autotransformer (700 MVA/750 kV) with the above-optimized structure were connected to the power grid in 2016 and have been in good condition ever since.

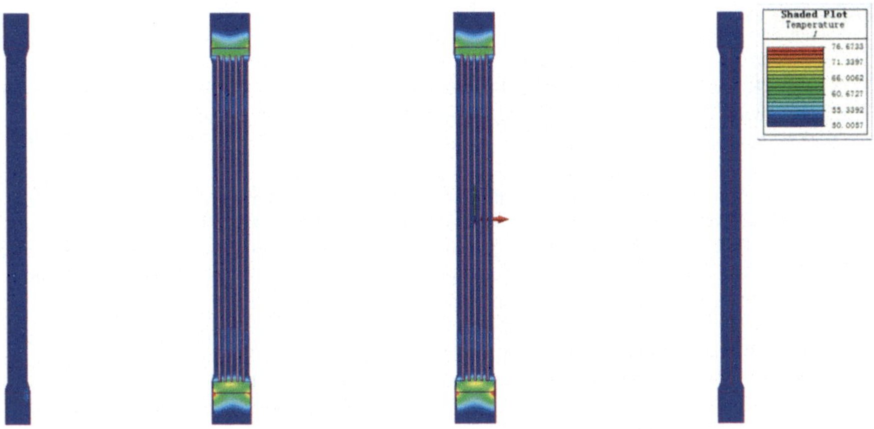

Fig. 14.32 Temperature distribution in the core tie-plate (700 MVA/750 kV, D02)

Table 14.19 Comparison of total loss with and without structural optimization

Components	Calculated total loss (P_c-D01) (%)	Calculated total loss (P_c-D02) (%)	Measured total loss (P_m-D02) (%)
Total	105.5	101.1	100.0

14.6 Summary

The key material properties, regarding the actual operating conditions, are of prime importance for accurately calculating the stray-field loss of large power transformers and preventing the hazardous local overheating. The experimental setup and test models are well established, including the practical and useful measurement methodology, which are used for measuring and predicting the electromagnetic and thermal properties of the key materials in power transformers. The valuable property data are obtained and used in the numerical modeling and simulation.

Using the electromagnetic and thermal properties obtained by the measuring system proposed by the authors, the engineering effectiveness and usefulness of both large-scale modeling and simulation have been verified by the structure optimization of a large autotransformer, which provides guidance for designers to carry out the simulation and the structural optimization in a more accurate and efficient way.

Acknowledgements This research project was supported by the National Natural Science Foundation of China (no. 11771057). The authors thank the colleagues of the National Engineering Research Center for Silicon Steel for their support in magnetic property measurement, thank Junjie Zhang for his guidance in modeling and simulation, and thank Aihua Guo and other colleagues for their supports in property measurements.

References

1. X. Yu, "Research on Eddy Current Loss Calculation and Temperature Rise Distribution of Large Power Transformer," Shenyang University of Technology, 2016.
2. Y. Xie, et al, "Power Transformer Manual," China Machine Press, ISBN 7–111-11082-X, 2014.
3. D. Liu, J. Zhang, et al. "3D Eddy Current Field Calculation and Partial Overheat Protection Technology Research of UHV Large Transformer," Transformer, 2009, 46(9):1–5.
4. J. Zhang, D. Liu, X. Zhang, D. Zhang, and F. Zhao. "The Simulation of Leakage Magnetic Field in the Oversize Generator Transformer Considering the Heavy Current Leads," Transaction of China Electrotechnical Society, pp. 157–161, 2013.
5. GB1094.1-2013. Power Transformers-Part 1: General.
6. JB/T501-2006. Test Guide for Power Transformers.
7. K. Preis, O. Bíró, G. Buchgraber, and I. Ticar, "Thermal-Electromagnetic coupling in the finite-element simulation of power transformers," *IEEE Trans. on Magn*, pp. 999–1002, 2006.
8. Z. Cheng, S. Gao, and L. Li. "Eddy Current Loss Analysis and Validation in Electrical Engineering," Higher Education Press, 2001.
9. L. Li, Y. Li, et al, "Research of Leakage Magnetic Field and Stray Loss Calculation in Power Transformer,"Transactions of China Electrotechnical Society, 2013, 28(s2):122–128.
10. J. Wang, C. Jing, et al, "Engineering Research on 3D Eddy current Field and Stray Loss of structural Parts in Large Transformer," *Transformer*, 2012,49(2):28–32.
11. Bogdan Cranganu-Cretu, and M. Schneider. "Coupled Electromagnetic-Thermal Analysis for ABB Power Transformers," Presented at International Colloquium Transformer Research and Asset Management, Cavtat, Croatia, November 12–14, 2009.
12. Z. Yao, Y. Qi, and J. Li. "Simulation and Research on Partial Overheat of Structural Parts for Large Transformer," *Transformer*, 2015,52(7):30–33.
13. Y. Zhang. "Research on the temperature field of single core ultra high voltage (UHV) shunt reactor," Shandong University, 2017.
14. Y. Kang. "Calculation and Analysis of the Temperature Rise and Eddy Current Loss in Power Transformer," Shenyang University of Technology, 2007.
15. E. Mechkov, R. Tzeneva, V. Mateev, and I. Yatchev. "Thermal analysis using 3D FEM model of oil-immersed distribution transformer," 2016 19th International Symposium on Electrical Apparatus and Technologies (SIELA). 29 May-1 June 2016. INSPEC Accession Number: 16229994. https://doi.org/10.1109/siela.2016.7543027.

Chapter 15
Engineering-Oriented Modeling and Experimental Research on DC-Biased Transformers

Mansheng Guo

Abstract Due to routine preventative maintenance of equipment and operation commissioning of converter station, the direct current (DC) transmission system may operate in a monopolar ground return mode or an unbalanced dipole mode, which may cause a large direct current to flow into the neutral grounded power transformers. The asymmetrical saturation of the core, the loud noise, the serious vibrations, the local loss density concentration, and the local overheating may occur in a DC-biased transformer when the DC is flowing through the transformer winding. In this chapter, the electromagnetic properties of transformer core, the exciting current under DC bias condition, and the harmonic distribution are experimentally investigated based on two sets of verification transformer models. Furthermore, a large-scale numerical analysis of the 3-D electromagnetic fields and the power loss are carried out. The different DC injection tests are also performed on a 500 kV autotransformer to further demonstrate the harmonic characteristic, no-load loss, and noise level under DC bias. Finally, an estimation method of the abilities to withstand the DC bias is proposed.

Keywords DC bias · Power transformer · Magnetic property · Experimental research · Modeling and simulation

15.1 Introduction

With the development of DC transmission technology, extra high voltage (EHV) and ultra-high voltage (UHV) DC transmission lines are rapidly increasing in China. China Southern Power Grid Co., Ltd. (CSG), for example, already has several EHV or UHV DC transmission lines that either have been put into operation or planned to be constructed soon. There are the Tianshengqiao—Guangzhou Beijiao, Anshun, Guizhou—Zhaoqing, Guangdong (Guizhou—Guangdong I),

M. Guo (✉)
Design Department, Baobian Electric, Baoding, China
e-mail: smggg@163.com

Z. Cheng et al. (eds.), *Modeling and Application of Electromagnetic and Thermal Field in Electrical Engineering*, https://doi.org/10.1007/978-981-15-0173-9_15

Xingren, Guizhou—Shenzhen, Guangdong (Guizhou—Guangdong II), Three Gorges—E'cheng, Guangdong, and other ±500 kV DC transmission lines as well as the Yunnan—Guangdong (Chuxiong, Yunnan—Suidong, Guangdong) ±800 kV DC transmission lines to be constructed soon. There are also several EHV DC transmission lines in operation in the Yangtze River Delta area, including Gezhouba—Nanqiao, Three Gorges—Changzhou, Three Gorges—Shanghai, and other ±500 kV DC transmission lines that have been put into operation, and Xiangjiaba—Nanhui, Shanghai, ±800 kV DC transmission lines that are about to be built soon.

With the increasing convenience of power transmission, the influence of DC transmission line on the operation of grounding transformer in AC system is becoming more and more serious. Due to the equipment failures in DC transmission system, routine preventive maintenance and monopole operation and commissioning at the initial stage of converter station construction, the probability of operating a DC transmission system in a monopolar ground return mode or an unbalanced dipole mode is greatly increased. The abnormal operations of EHV DC transmission lines may produce DC bias which will affect the safe operation of neutral grounded transformers in the AC system. The situation is getting worse with more EHV DC transmission lines put into operation in China. Therefore, problems arising from DC bias have been widely of concern to various grid companies, transformer manufacturers, scientific research institutes, and other relevant departments.

15.1.1 DC Bias Phenomenon on Power Transformers

Through research and practical measurements [1–4], it is found that the geomagnetic storm or DC transmission system is the main reason for the generation of DC flowing through the neutral point.

The interaction between solar flares and the geomagnetic field produces auroral electrojet, which causes transient fluctuations of the geomagnetic field. When this phenomenon is serious enough, it is called a geomagnetic storm. The earth is a conductive sphere. When a geomagnetic storm occurs, the transient fluctuation of the geomagnetic field puts a part of the earth in such a time-varying magnetic field and causes surface-induced potential, which can reach 1.2–6 V/km when the geomagnetic storm is strong in areas with high soil resistivity. Geomagnetic storms have been observed many times in the northern hemisphere, such as Canada, the USA, Japan, and Finland. From March 2001 to October 2002, the 500 kV main transformer of Shanghe Substation in Huai'an, Jiangsu Province, experienced anomalous noise increase five times consecutively due to geomagnetic storms (which were confirmed later). Geomagnetic storm is one of the reasons that cause DC bias phenomenon on grounding transformers in the AC system. The neutral direct current of the transformer due to the geomagnetic storm can reach up to 300 A.

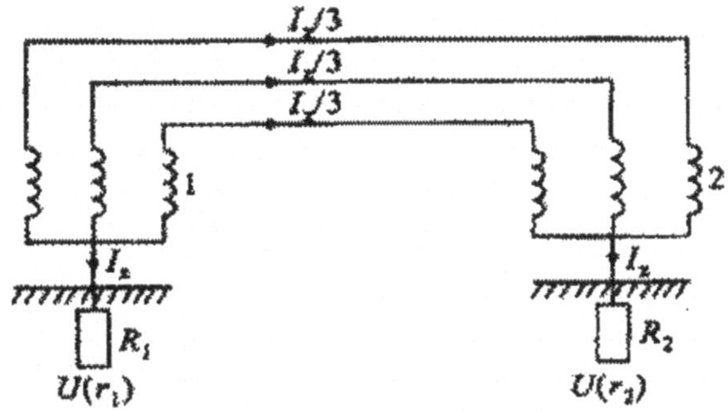

Fig. 15.1 Schematic diagram of DC generation mechanism

At present, the DC bias problem that we usually face is mainly caused by the DC transmission system. When DC transmission system operates in monopolar ground return mode or unbalanced dipole mode, DC will be injected into the ground through the grounding electrode and flows to the other end of the DC system. A large amount of DC flowing through the earth will generate potential difference on the path, especially in the vicinity of the grounding electrode, as shown in Fig. 15.1. The magnitude of such potential depends on the geological conditions of the earth. The higher the earth resistivity, the higher the potential and the broader the influenced range. The high potential of the DC grounding electrode acts on the grounding point of the AC substation and causes the DC to flow through the neutral grounded transformer, which shunts a part of the DC transmission current.

The results of the survey conducted in the Three Gorges—Changzhou DC transmission system in China show that the noise originated from two groups of 500 kV autotransformers in Wunan Substation (Changzhou) close to the Zhengping (Changzhou) DC converter station in Jiangsu Province increased, when the Three Gorges—Changzhou DC transmission system operated in monopole ground return mode. The measurement results show that the maximum direct current on the neutral point in Wunan Substation is 10.4 A when the DC transmission reaches 1540 MW (earth current 3320 A). The comprehensive measurement results from other substations in the grid show that about 2% of the direct current in Zhengping Converter Station flows back through the AC system. Similar problems have also been found in China's Tianshengqiao—Guangzhou, Three Gorges—Guangzhou, and Guizhou—Guangzhou HV DC transmission systems. The main transformers of Ling Ao Nuclear Power Station and Daya Bay Nuclear Power Station belonging to China General Nuclear Power Group were the most affected. It is estimated through preliminary analysis that about 5% of direct current flows into the AC system when the Tianshengqiao—Guangzhou or Three Gorges—Guangzhou DC system operates in ground return mode. The ratio of direct current flowing through the transformers of the above two power stations to the total direct current is less than 4%. During the monopole commissioning of the Guizhou—Guangdong DC transmission system, it is estimated that 2% of the direct current may flow into the AC system based on the measured results in the monopolar ground return mode.

If a direct current flows through the transformer winding, especially when the value of direct current is beyond the tolerance of the transformer, it will have a certain impact on the operation of the transformer in the following aspects:

- Increased transformer loss, aggravated temperature rise, and local overheating
  When there is a DC flowing through the transformer winding, the peak value of exciting current is greatly enlarged; the magnetic flux of the transformer is saturated, and the eddy current losses of the transformer winding, core, tank, clamping, and other structural parts are increased, resulting in increases of the temperature rise in the region of top oil and winding temperature rise in the transformer; when the direct current is sustained for a long time, local overheating will occur. The temperature difference measured between the upper end of the tie plate and the top oil is 52 K after a DC bias duration of 1 h [5] when a 370 MVA single-phase autotransformer with a voltage of 735 kV is injected with a DC of 75 A, which have been researched by Canadian scholars.
- Noise increase
  When there is a DC flowing through the transformer winding, the magnetic flux in the transformer core is saturated and the harmonic component is increased, resulting in significant magnetostriction and loud noise. Noise spectrum analysis shows that there are both even and odd harmonics. For example, the anomalous noise of the main transformer (500 kV, 750 MVA) under DC bias condition in Shanghe Substation (Jiangsu) appears at 50, 100, 150, and 200 Hz, while anomalous noise appears at 100 and 200 Hz when transformer operates under normal condition [6]. The transformer noise caused by DC bias during commissioning in Wunan Substation is up to 91.4 dB [7].
- Vibration intensified
  When there is a DC flowing through the transformer winding, the exciting current in the transformer is distorted, resulting in an increase of magnetostriction of the core, while the increase in leakage magnetic flux leads to an increase of the electrodynamic force of the winding, which to some extent aggravates the transformer vibration. For instance, when the neutral direct current in Wunan Substation in Jiangsu is 15.8 A, the highest vibration value of the oil tank wall is 194 μm [7].
  The influence of DC bias on transformer includes aging of the transformer insulation, decrease of operational reliability, and shorter service life.
  In addition, DC bias has a greater impact on the grid, as shown in the following aspects:
- Occurrence of harmonics
  Due to DC bias, there are not only odd harmonics but also even harmonics in the exciting current. The transformer becomes a harmonic source in the AC system, causing distortion of system voltage waveform, filter overload, protective relaying malfunction, continuous overvoltage generated when the no-load line is closed, increase of secondary arc current during single-phase reclosing, increase of circuit breaker recovery voltage, etc.

- Increase of reactive loss
  The transformer saturation caused by DC bias will lead to surges of the exciting current and the increase of reactive loss. Accordingly, the reactive power compensation device of the system is overloaded, or the system voltage drops, and even the entire grid may collapse in severe case.
- Failure of relay protection system
  The geomagnetic storm may cause relay protection failures, such as protection system malfunction, differential protection failure, and action delay.
- Corrosion of substation grounding grid caused by DC.

15.1.2 Brief Overview of Investigation on DC-Biased Problem

Up to now, there have been many references about the research on DC bias [8–19]. The research on the influence of geomagnetic induction on power system and electrical equipment has been going on since the 1980s. Researchers have studied the saturation process of transformer under DC bias and the consequently generated harmonics, which focused on the harm of DC bias to the system. There have also been studies on the local overheating of transformer due to saturation under DC bias and the determination of the bearing capacity of transformer, which focused on the damage of DC bias to the transformer. The relevant works published can be roughly divided into two categories: (1) The power system or research institution studies the influence of geomagnetic disturbance on AC transmission system and impact of DC grounding electrode on transmission system, from the perspective of safe operation of system, and (2) the simulation and calculation of the exciting current of the transformer under the condition of geomagnetic disturbance, and carrying out tests on transformers of different specifications through DC injection at neutral point. Analytical method and equivalent magnetic circuit approach are usually used in simulations and calculations. In 1992, Xiangheng Wang and Boxiong Xu published their results in *Transformer* Magazine and put forward a method to estimate the DC flux numerically [8]. In 1994, Toshiba, Hitachi, and Mitsubishi in conjunction with Tokyo Electric Power Company conducted tests on a transformer model to investigate the DC-biased transformer caused by geomagnetically induced current, and the relative results show that the structure of transformer has great influence on the magnitude and characteristics of exciting current under the effect of geomagnetically induced current (quasi-DC). The sensitivity of a single-phase three limbs, three-phase five limbs, and three-phase three limbs to DC bias current decreases successively. The experiment applying 200/3 A DC to a 30 MVA core-type transformer shows that the local temperature of the tie plate reaches 110 °C in a short time. The temperature of the tie plate made of non-magnetic steel of the same transformer is about 10 °C. In 1997, P. Picher and L. Bolduc et al. from Canada cooperated with ABB and conducted a two-dimensional simulation and experimental research on a small model and transformer products, for the ability of power transformers to withstand DC bias. Results show that the temperature rise of the tie plate to the top oil is 52 °C after applying DC 75 A to 370 MVA single-phase transformer for 1 h. In 2004, Zhiqiang Ma and Guihua Mei, et al. focused on the influence of DC transmission on AC system transformers by numerical simulation [9]. Chun Shang et al. proposed measures to suppress the influence of HV DC ground current on AC transformers [12]. Shuo Liu et al. have studied on DC biasing magnetic field and modeling of magnetic material property [13].

Despite many accomplished works on DC bias, there are still many unsolved problems, which are listed as follows:

- The influence of DC bias on the system and equipment is mostly studied from the view of the power system, without consideration of the internal detailed structure of the transformer, such as the number of coil turns, the length of magnetic circuit, and the rated working flux density of the core and other elements closely related to DC bias flux. Therefore, the conclusion is not definite. However, not only quantitative but also qualitative judgment should be made on the ability of a transformer to withstand DC bias.
  From the references, there are few reports on the relationship between the detailed structure inside the transformer and the ability of the transformer to withstand DC bias, such as the number of turns of the transformer coil, the size and material of the core structure, the structure of the pressbeam of clamping and tie plates, and the structure of the tank. The DC magnetic potential generated by the DC applied to the transformer varies with the layout design and materials of the structural components of the transformer. Therefore, the influence of the same DC on a different transformer may be very different, and the loss and heat distribution under the DC bias may also be very different; hence, it is important to understand the internal structural parameters of a transformer before carrying out the simulation analysis of the transformer under DC bias.
- There is a lack of practical and effective engineering-oriented methods to calculate exciting current of the DC-biased transformer.
- The calculation and analysis of the transformer's magnetic field and loss under DC bias are still at the primary stage, and the treatment of the transformer is relatively simplified. There is no in-depth and systematic analysis and research on the *B*–*H* curve, hysteresis loop, DC flux, $\Phi - I$ curve, and loss characteristics of the transformer's core under DC bias. Therefore, thc three dimensional nonlinear eddy current transient field has not been used to simulate and calculate the magnetic field loss and other characteristics of the transformer under DC bias condition. These efforts have to be made if we want to investigate DC-biased transformer intensively.
- There is a lack of experimental research on the DC-biased transformer, so there is a lack of sufficient basis for the evaluation of transformer's ability to withstand DC.
- It is required to study and draw up the standard for transformer's ability to withstand DC.

15.1.3 Key Research Projects

According to the typical requirements for the manufacturing and operating a power transformer, engineering-oriented research DC-biased problem is carried out in two aspects: simulation and model test. The effective methods to calculate the transformer leakage magnetic field under DC bias condition are studied, and the engineering methods to solve problems are explored. Also, the analysis and calculation of the internal magnetic field of the transformer are completed, the core performance and the loss of structural components are calculated, and analysis is carried out to evaluate the temperature rise and the ability to withstand DC bias based on the calculated results for the magnetic field. The transformer model to study the influence of DC bias on transformer characteristics is designed for manufacturing, and the test of transformer-related performance parameters under different DC bias conditions is completed. The simulation results are compared with the measured results to guide the test, and the validity of the simulation results is verified by the experimental results. Some achievements are as follows:

- The method of adopting different loss curves under different bias currents is proposed according to the basic characteristics of transformer under DC bias condition from an engineering point of view. Based on the transformer model of product level, the average *B–H* curve, *Φ–I* curve, loss curve, etc., of "laminated" ferromagnetic materials under DC bias condition are studied in depth, and their validity is verified by experimental study to solve the problems related to material properties that are relied on in simulation and calculation.
- Programming is conducted to obtain the exciting current of transformer and to calculate the DC flux under DC bias condition. Accordingly, the principle of determining limit of DC that the transformer can withstand is proposed.
- The parameters of the single-phase three-limb (and other types of) transformer model under DC bias condition have been determined, and the basic approach is developed to simulate and analyze the transformer under different DC bias conditions by utilizing Simcenter MAGNET for nonlinear three-dimensional eddy current field analysis (commercial electromagnetic analysis package).
- The test system for DC bias test on transformer is developed, and the technical requirements of DC bias test are determined to perform experimental study on relevant models.
- The experiment items including the no-load loss, harmonic characteristics, and sound level have been studied by applying a DC bias to a 500 kV autotransformer for the evaluation of transformer's ability to withstand DC bias.
- The curve representing relation between DC magnetic flux density and AC flux density under different DC bias conditions is established based on the DC flux density (ΔB), and then for transformer with a certain material, structure, and flux density, the ability to withstand DC bias can be presented as follows. That is, based on the $\Delta B - B_{\mathrm{m}} - H_{\mathrm{dc}}$ (I_{dc}) curve, the performance parameters of

transformer such as the temperature rise of the tie plate, sound level and loss of transformer body, and the allowable harmonic increment of the excitation current under DC bias, the allowable H_{dc} of the transformer can be obtained, and then the allowable DC of the transformer can be obtained by Eq. (15.1).

$$I_{dc} = H_{dc} \times \frac{L}{N} \tag{15.1}$$

15.2 Magnetic Properties of Product-Level Laminated Core Under DC Bias Condition

15.2.1 Φ–I *Curve and* B−H *Curve of Transformer Core*

To provide analytical assessment on the behaviors of transformer under DC bias, it is very important to know the magnetic characteristics of transformer core under DC bias, including the average *B–H* curve, *Φ–I* curve, and loss curve $B_m - W$ of the transformer core. The acquisition of the *Φ–I* curve is a prerequisite for the simulation and calculation of exciting current, while the *B–H* curve is a basic material parameter that is indispensable for the calculation of DC-biased magnetic field in transformer.

Generally, the parameters used in engineering are only those provided by various core material manufacturers, at power frequency and without DC bias, as shown in Figs. 15.2 and 15.3.

The research on DC-biased transformers from some universities is only limited to laboratory models (e.g., electronic transformer level), and the problem of magnetic field characteristics of cores when power transformers are subjected to DC bias is not explored extensively from an engineering point of view. Figures 15.4 and 15.5 show the *B–H* curve of ferromagnetic materials completed by Electromagnetic Device Laboratory (Okayama University, Japan), under DC-biased magnetization at a certain rated flux density [19].

The materials in Figs. 15.4 and 15.5 are very different from those used in the actual transformer, and their working flux density is too low to explain the actual situation of the power transformer under DC bias.

To study the calculation of DC-biased transformers, the *B–H* curve, *Φ–I* curve, and loss characteristics of the "laminated" ferromagnetic materials under DC bias must be obtained accurately.

In order to obtain the material properties of actual transformer products under DC bias condition and to further study the DC-biased transformer, a transformer model is designed and manufactured for the analysis of the influence of DC on different transformer core structures. The model is adopted with a single-phase three-limb transformer structure and a three-phase five-limb transformer structure

(a)

(b)

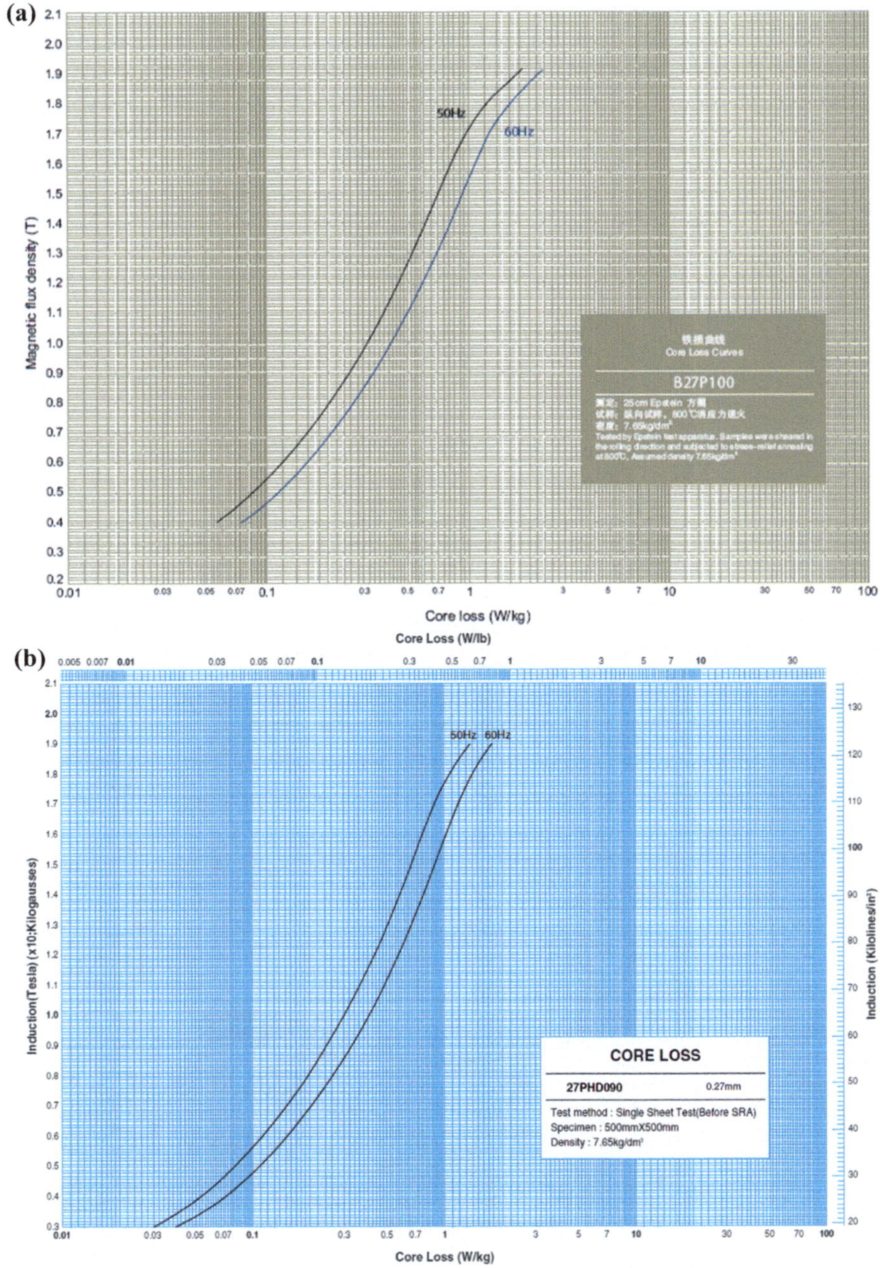

Fig. 15.2 **a** Loss curve of silicon steel sheet (B27P100). **b** Loss curve of silicon steel sheet (27PHD090)

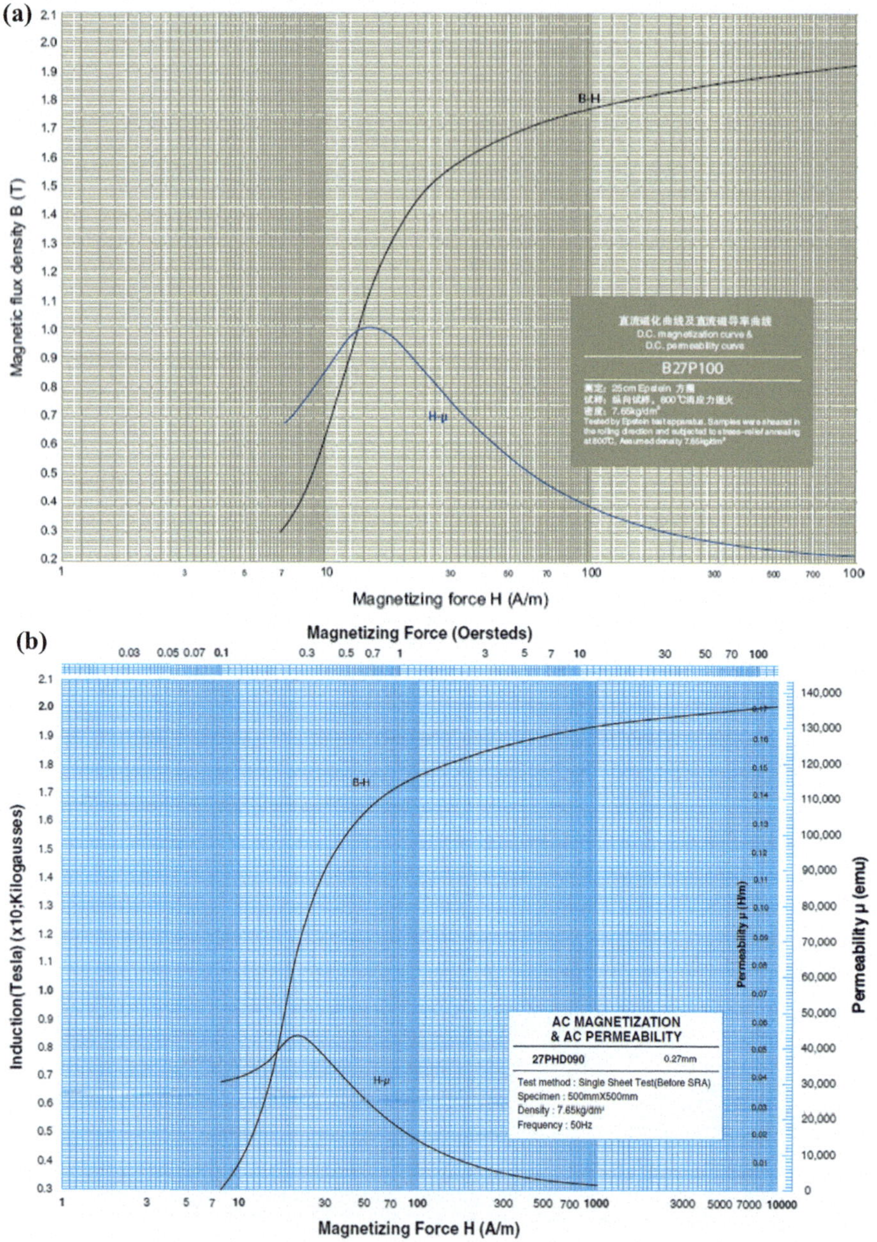

Fig. 15.3 **a** Magnetization curve of silicon steel sheet (B27P100). **b** Magnetization curve of silicon steel sheet (27PHD090)

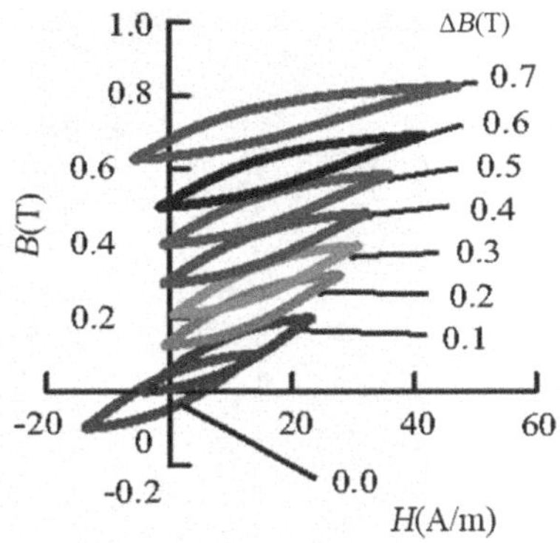

Fig. 15.4 *B–H* curve under DC-biased magnetization (6.5% Si–Fe, $B_m = 0.1$T)

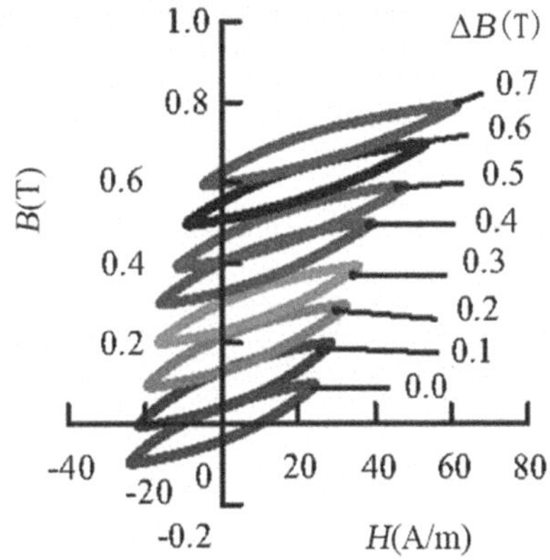

Fig. 15.5 *B–H* curve under DC-biased magnetization (50A290, $B_m = 0.1$T)

which are seriously affected by DC bias. To objectively reflect the situation of the large transformer working under direct current, the lamination, the core, and the clamping structure are all modeled with reduced scale on the structure of large transformer, and in order to obtain the *B–H* curve of the core material used in common transformers under DC bias, 30RGH120 silicon steel sheet is selected in manufacturing the transformer model. In this chapter, the research of single-phase three-limb transformer structure most affected by DC bias is mainly introduced.

The product-level three-phase and single-phase transformer model for DC bias test developed by the author is shown in Fig. 15.6, and the relevant technical parameters are shown in Table 15.1.

The following methods are adopted to obtain the *Φ–I* and *B–H* curves of the core [20].

- Different DCs and the corresponding DC biasing magnetic intensity are selected. According to the actual situation of the DC source, the DCs are selected as $I_{dc} = 0$ A, $I_{dc} = 1.26$ A, $I_{dc} = 2.53$ A, $I_{dc} = 3.2$ A, and the corresponding DC biasing magnetic intensity is $H_{dc1} = 0$ A/m; $H_{dc2} = 100$ A/m; $H_{dc3} = 150$ A/m; and $H_{dc4} = 200$ A/m.

Fig. 15.6 Transformer model for DC bias test

Table 15.1 Key parameters of single-phase transformer model

Voltage (V)	(231 ± 8%)/231
Rated current (A)	216.5/216.5
Connection symbol	I i0
No-load loss (kW)	0.2
No-load current (%)	0.85 (1.84 high-voltage side)
Load loss (kW)	1.5
Short-circuit impedance (%)	7.85
Turns (HV/LV)	64/64
Type of silicon steel sheet for core	30RGH120
Length of main magnetic circuit (m)	1.62
Sectional area of middle-column core (cm^2)	96.05
Sectional area of return yoke (cm^2)	50.35
Rated flux density of core (*T*)	1.691

- Under DC bias condition, 31 working conditions with the voltage from about 15 to 300 V are applied to the transformer while the DC I_{dc} is kept constant. The corresponding working flux densities are $B_{m1} = 0.5T$, $B_{m2} = 0.8T$, $B_{m3} = 1.0T$, $B_{m4} = 1.3T$, $B_{m5} = 1.5T$, $B_{m6} = 1.6T$, $B_{m7} = 1.62T$, $B_{m8} = 1.64T$, $B_{m9} = 1.66T$, $B_{m10} = 1.68T$, $B_{m11} = 1.7T$, $B_{m12} = 1.71T$, $B_{m13} = 1.72T$, $B_{m14} = 1.73T$, $B_{m15} = 1.74T$. For each working condition, the voltage applied to the winding, the induced voltage of the magnetic coil, the current flowing into the winding, and the power are measured. A total of 1002 sets of voltage and current data are collected by the power analyzer.

The law of electromagnetic induction is as follows:

$$e = -N\frac{\mathrm{d}\Phi}{\mathrm{d}t} \tag{15.2}$$

where N is the number of turns of the exciting coil. The flux Φ can be obtained by integrating the collected voltage over time.

$$\Phi = -\frac{1}{N}\int e\,\mathrm{d}t \tag{15.3}$$

The Φ–I data pair is obtained via the data processing and calculation based on the collected voltage e and current i.

- The data pair is converted into a B–H data pair. The B–H curve can be obtained from the Φ–I curve in accordance with the following formulations:

$$B = \frac{\Phi}{S} \tag{15.4}$$

$$\int H \cdot \mathrm{d}l = IN \tag{15.5}$$

- For a DC biasing magnetic intensity H_{dc}, two pairs of B_m–H_b data corresponding to the two vertices of the hysteresis loop can be obtained corresponding to a certain working flux density B_m and expressed as $B_{m+} - H_{b+}$ and $B_{m-} - H_{b-}$. The measured points $B_{m+} - H_{b+}$ at different working flux densities are connected together to form the B–H curve corresponding to the exciting current in the positive axis; the $B_{m-} - H_{b-}$ points measured at different working flux densities are connected together to form the B–H curve corresponding to the exciting current in the negative axis.

The above steps will be described with specific curves given as follows. First, as U = 15, 100, 231 V, the obtained Φ–I hysteresis loop is shown in Fig. 15.7.

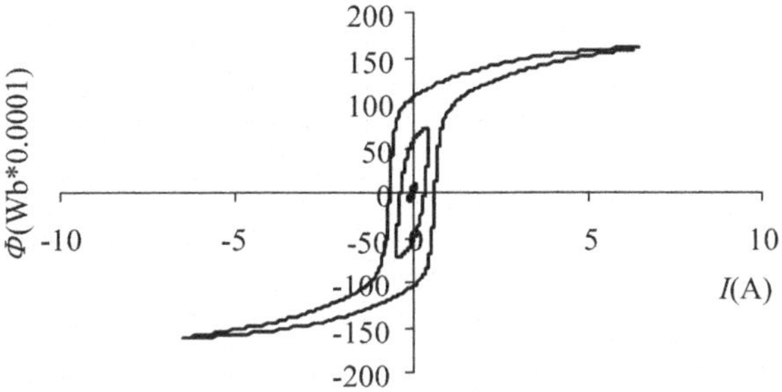

Fig. 15.7 Φ–I hysteresis loop of the model as I_{dc} = 0 A, U = 15, 100, 231 V

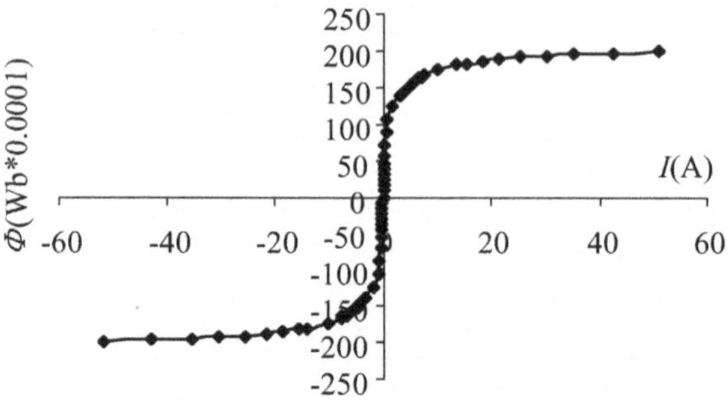

Fig. 15.8 *Φ–I* magnetization curve of the model as $I_{dc} = 0$ A

The vertices of the hysteresis loop are connected together to form the basic magnetization curve of the measured transformer under no-load operation, as shown in Fig. 15.8.

The *B–H* curve of the transformer core is obtained from the above *Φ–I* curve, as shown in Fig. 15.9.

The curve data, shown in Fig. 15.9, is given in Table 15.15, Appendix 15.1. To distinguish the difference of magnetic properties between a single piece of steel sheet and laminated core model, the *B–H* curve of the transformer model under DC bias condition is compared with the *B–H* curve of the same silicon steel sheet (30RGH120) used in international TEAM Problem 21, tested by Electromagnetic Device Laboratory, Okayama University, Japan, as shown in Fig. 15.10.

The *Φ–I* hysteresis loops of the transformer core are not symmetric any more when $I_{dc} = 1.26$ A, as shown in Fig. 15.11.

Similarly, the *Φ–I* hysteresis loops of the transformer core when $I_{dc} = 2.53$ A are shown in Fig. 15.12.

When $I_{dc} = 3.2$ A, the *Φ–I* hysteresis loops of transformer core are shown in Fig. 15.13.

For each DC bias current, the typical magnetization curve loops above are obtained only as $U = 15$ V, 100 V, 231 V, respectively. The *Φ–I* curves can be

Fig. 15.9 *B–H* magnetization curve of model as $I_{dc} = 0$ A

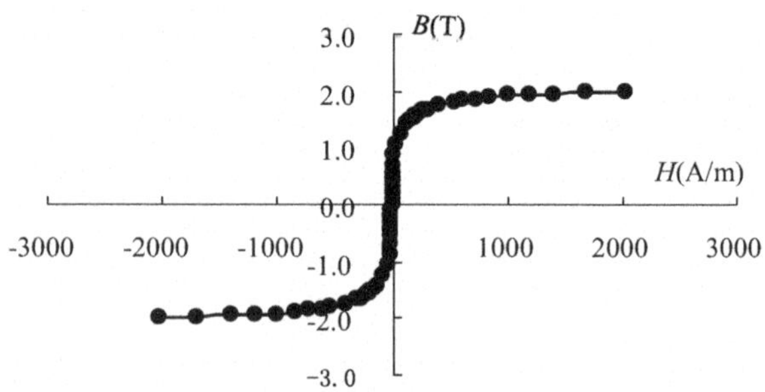

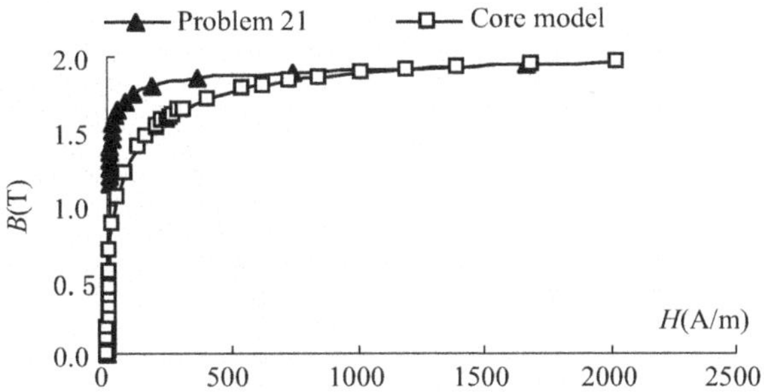

Fig. 15.10 Comparison of *B–H* curves between transformer core model and the silicon steel sheet (30RGH120) used in TEAM Problem 21 [21]

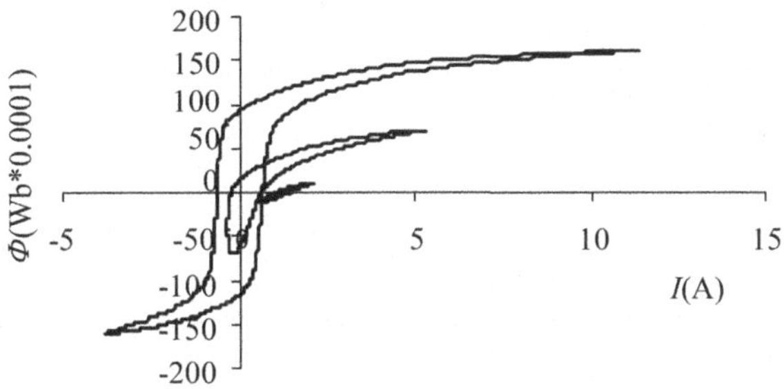

Fig. 15.11 Φ–I hysteresis loops of the model as I_{dc} = 1.26 A, U = 15, 100, 231 V

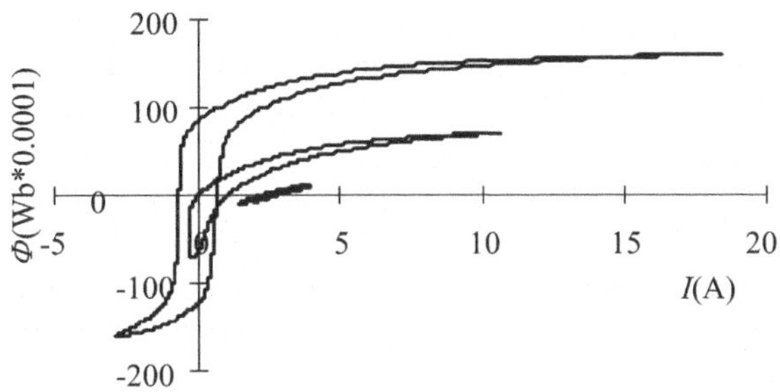

Fig. 15.12 Φ–I hysteresis loops of the model as I_{dc} = 2.53 A, U = 15, 100, 231 V

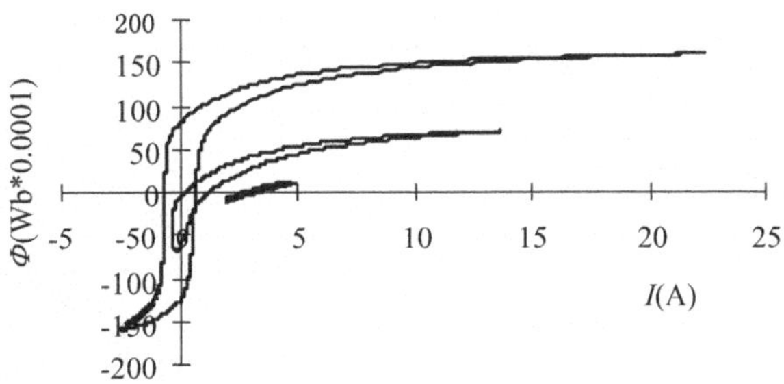

Fig. 15.13 Φ–I hysteresis loops of the model as I_{dc} = 3.2 A, U = 15, 100, 231 V

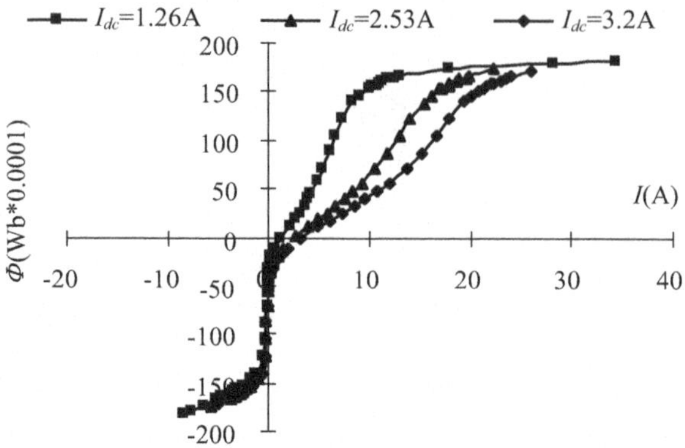

Fig. 15.14 Φ–I magnetization curve of the model as I_{dc} = 1.26 A, 2.53 A, 3.2 A (DC flux is not considered)

obtained by connecting the vertices of the hysteresis loop under different DC bias conditions (I_{dc} = 1.26, 2.53, and 3.2 A), as shown in Fig. 15.14.

Then, the *B*–*H* curve of the corresponding transformer core can be obtained, as shown in Fig. 15.15.

Since DC flux cannot be measured, the *B*–*H* curve in Fig. 15.15 is a magnetization curve without consideration of DC flux density. In practice, AC flux density must be added with a ΔB due to the hybrid field of AC and DC under DC bias condition. Furthermore, for the same DC, the DC flux ΔB in the transformer core is different from each other under different alternating voltage excitations. Therefore, in order to obtain the accurate flux density under practical DC bias condition, the following methods are adopted to obtain ΔB beforehand.

Firstly, with a given direct current I_{dc}, there are three parameters that can be obtained from the measured exciting current, including the positive peak value, the negative peak value, and the DC component. Secondly, based on the principle of iterative method, exciting current can be obtained when positive peak value and DC

Fig. 15.15 *B*–*H* curve as I_{dc} = 1.26 A, 2.53 A, 3.2 A (DC flux density is not considered)

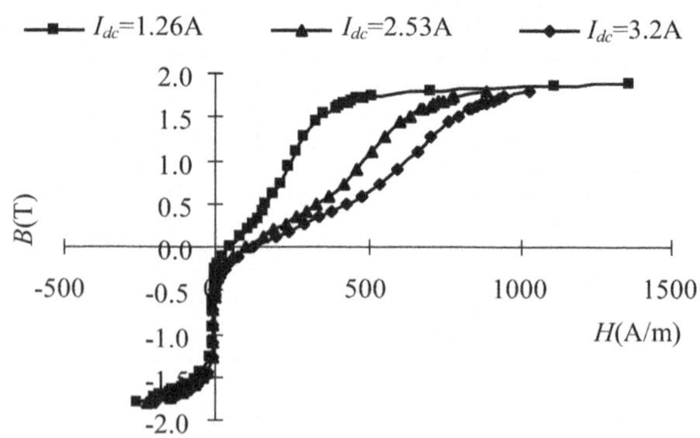

component of the exciting current obtained from simulation are consistent with the corresponding two measured parameters, respectively, by adjusting $\Delta\Phi$ interactively (notice that the negative peak value of the exciting current can hardly be consistent with the measured ones), the $\Delta\Phi$ is the DC flux corresponding to the working point, and ΔB can be obtained from $\Delta\Phi$. Finally, all other working conditions can be treated similarly to obtain a set of ΔB, a new B is obtained by adding each ΔB to the B corresponding to AC magnetization curve shown in Fig. 15.15, and the correlation curve between the new B and the H is the magnetization curve which AC and DC work together in the actual DC bias condition.

This iterative method is applied to working condition as $I_{dc} = 3.2$ A. Figure 15.16 shows the family of Φ–I curves based on a total of 22 voltage excitations from $U = 15$ V to $U = 250$ V at $I_{dc} = 3.2$ A. The Φ–I curve family, when DC bias flux is considered, is obtained by simulation and calculation corresponding to each working voltage by using the above iterative method, as shown in Fig. 15.17.

The average magnetization curve under DC bias condition is obtained to connect all the vertices of the hysteresis loops in Fig. 15.17, as shown in Fig. 15.18.

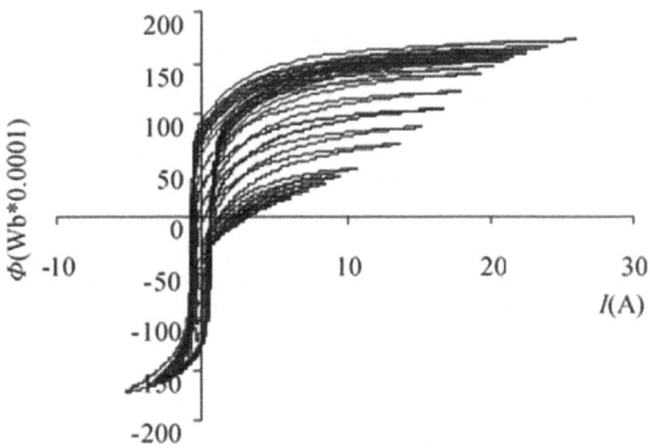

Fig. 15.16 Family of hysteresis loops as $I_{dc} = 3.2$ A ($\Delta\Phi$ is not considered)

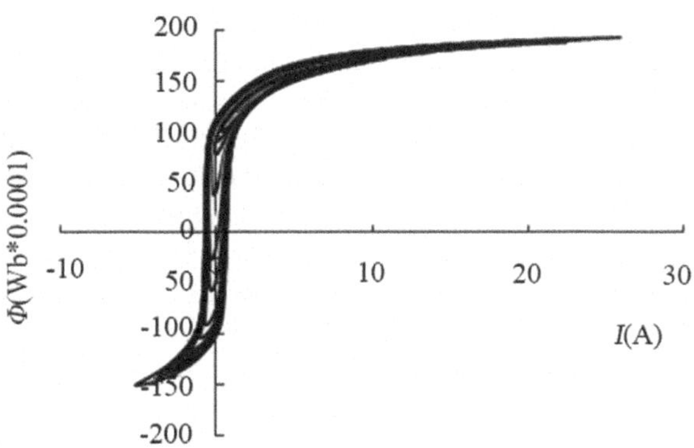

Fig. 15.17 Family of hysteresis loops as $I_{dc} = 3.2$ A ($\Delta\Phi$ is considered)

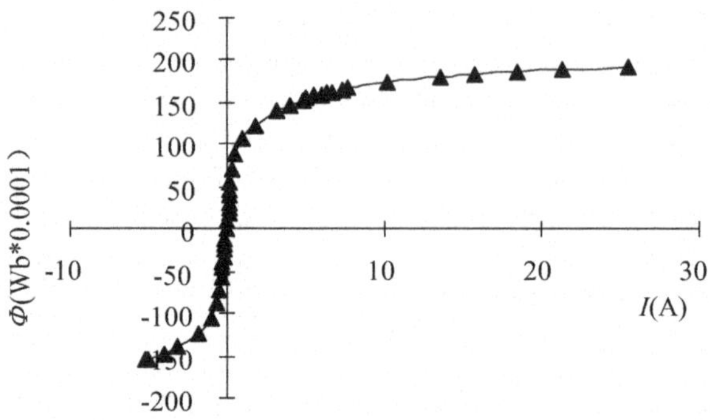

Fig. 15.18 Φ–I curve as $I_{dc} = 3.2$ A ($\Delta\Phi$ is considered)

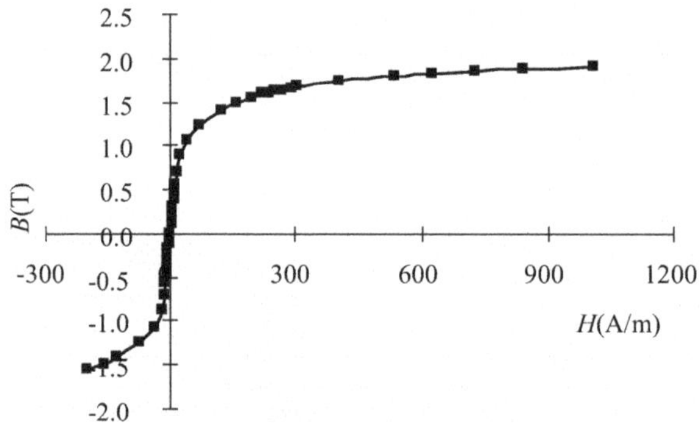

Fig. 15.19 Average B–H curve as $I_{dc} = 3.2$ A ($\Delta\Phi$ is considered)

The data sheet of the Φ–I curve for the single-phase transformer model under a specific DC bias condition ($I_{dc} = 3.2$ A) is shown in Table 15.16, Appendix 15.1.

Figure 15.19 shows the B–H curve obtained from the Φ–I curve in Fig. 15.18. Compared with the B–H curve measured in the absence of DC flux, one feature of such curve is that it is not symmetrical in the first and third quadrants, and another one is that the projection of the curve onto B-axis is asymmetrical about the H-axis. This curve is the basis for calculating exciting current.

The B–H curve above is the basic magnetization curve under DC bias condition; however, the hysteresis loop is not considered for the following reasons.

Note that, the shape and area (the area of the loop corresponds to the hysteresis loss) of hysteresis loop are different under different DC bias conditions (i.e., Φ_m or B_m is different). With the increase of flux density, the extreme value of magnetic field intensity tends to increase, leading to increase of the area of the hysteresis loop. These two factors directly affect the waveform of the exciting current. The factor that determines the magnitude of exciting current is the extreme value of magnetic field intensity. In practical transformer products, high-quality cold-rolled silicon steel sheets are usually used, hence the working flux density is generally about 1.7T, and the area of hysteresis loop is actually small and has little influence on analyzing the direct current under DC bias condition.

On the other hand, the calculation of magnetic field and loss based on the basic magnetization curve can meet mostly the demand on accuracy in practical engineering application. Therefore, in order to solve the DC biasing magnetic field, it is proposed to establish the "fundamental" *Φ–I* (or *B–H*) curve of the core which can reflect the basic characteristics of transformer core under DC bias condition, and compile program to correct the flux under different DC bias conditions, so as to obtain the relation curve between the flux and the current under AC/DC hybrid excitation (contribution of DC flux is considered), which is reasonable and necessary in engineering.

15.2.2 $B_m - W$ *Curve of Transformer Core*

According to the needs of engineering calculation, the curves of the total loss of transformer under no-load and non-bias and no-load bias are obtained through model experiments, as shown in Figs. 15.20 and 15.21, respectively. The area of hysteresis loop under each working condition is obtained through program processing, as shown in Figs. 15.22 and 15.23, respectively. As a comparison, Fig. 15.20 also lists the results tested by Electromagnetic Device Laboratory (Okayama University, Japan) [21]. It is worth noting that magnetic properties of the laminated core made of 30RGH120 are measured by Baobian Electric, while the material properties of silicon steel sheet on 30RGH120 are measured directly by Electromagnetic Device Laboratory (Okayama University, Japan).

The loss curve in this section is used to calculate the loss of transformer under DC bias condition and has important engineering value.

The data sheet of various losses of single-phase transformer model under DC bias condition is shown in Tables 15.17 and 15.18 of Appendix 15.1.

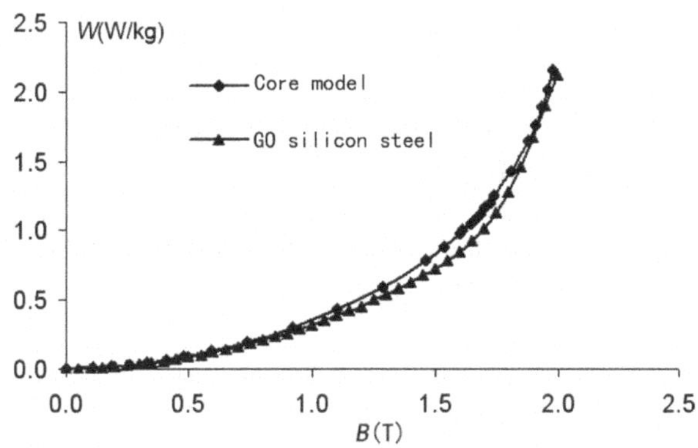

Fig. 15.20 No-load loss curve under DC bias condition

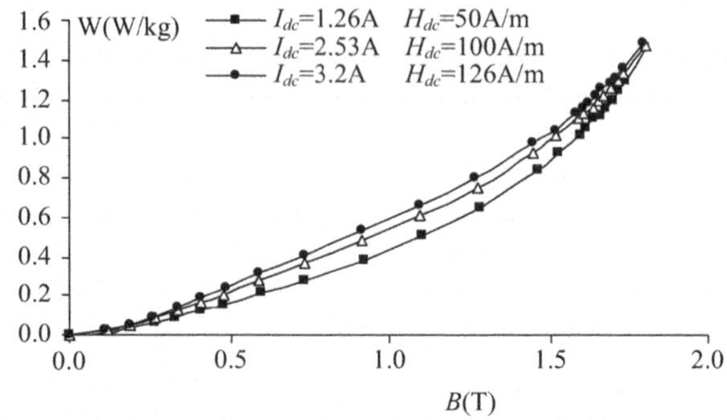

Fig. 15.21 Loss curves under different DC bias conditions

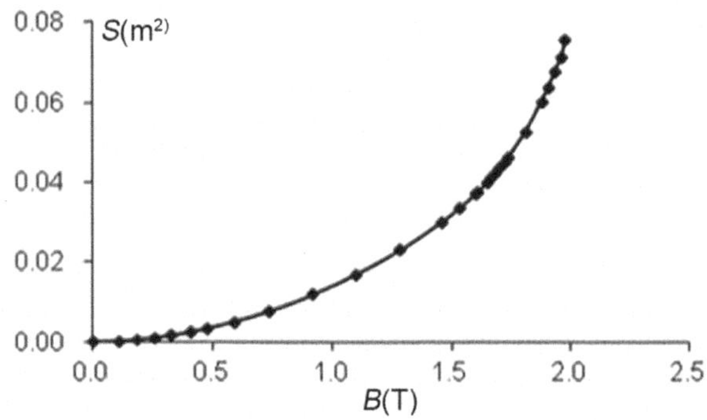

Fig. 15.22 Curve between hysteresis loop area and magnetic flux density under no-load condition

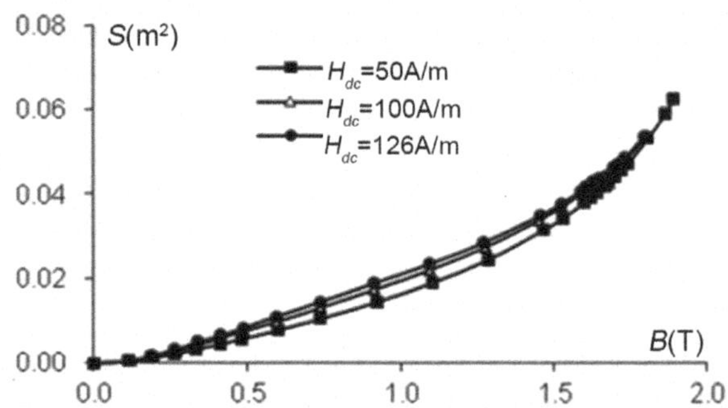

Fig. 15.23 Curve between hysteresis loop area and magnetic flux density under DC bias condition

15.3 Calculation of the Exciting Current Under DC Bias Condition

15.3.1 Principle of Simple Iteration to Determine DC Flux in Transformer Core

The exciting current in the winding and the waveform of the flux in the core change when the transformer is subjected to DC bias. The half-wave saturation

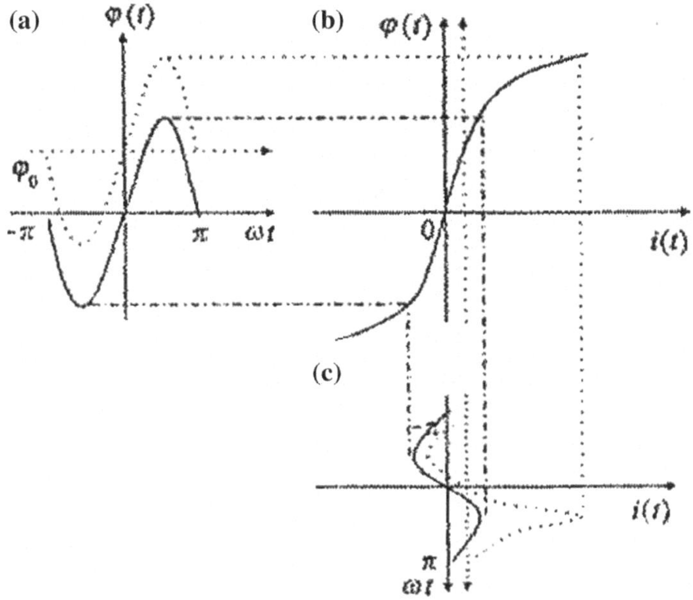

Fig. 15.24 Schematic diagram showing influence of DC bias on the exciting current of transformer

phenomenon causes the peak value of excitation current to increase rapidly in the positive half cycle, and the waveform of exciting current is distorted seriously.

In large power transformers, the core is made of high-quality silicon steel sheets, and hysteresis loops have little influence on exciting current. Therefore, it is feasible to calculate the positive and negative magnitude of the exciting current under DC bias condition using the normal magnetization curve (hysteresis characteristics are not considered).

It can be seen that the exciting current in the winding and the waveform of the flux in the core change as shown in Fig. 15.24 when the transformer is subjected to DC bias. The magnetization characteristics of the transformer, i.e., $\Phi = F\ (i)$, and the relationship between the DC bias and the DC component of exciting current are given.

Based on the above ideas, the iterative method is adopted to obtain exciting current in transformer winding from the engineering perspective.

The method is developed based on the following principle, an appropriate amount of bias ΔB is given, and the corresponding exciting current waveform $f(t)$ which is periodic but non-sinusoidal and asymmetrical is obtained according to the Φ–I curve of magnetization characteristics of the transformer core; the DC component I_{dc}^{x} can be determined according to Eq. (15.6),

$$I_{dc}^{x} = \frac{1}{T}\int_{0}^{T} f(t)\mathrm{d}t \tag{15.6}$$

where I_{dc}^{x} is compared with the known bias current I_{dc}, and the amount of bias ΔB is corrected according to the target that the two currents are equal; the bias ΔB and exciting current waveform $f(t)$ can be eventually determined upon repeated calculation and comparison.

Based on the principle of iteration, the corresponding program is compiled in matlib. The basic steps of the method are described in detail as follows:

- The Φ–I curve data is prepared:

The Φ–I curve data for a transformer is listed in Table 15.2.

The curve can be obtained by using static magnetic field analysis of the power transformer, and the curves are different for the transformer with different structures and materials.

- The rated flux data of the product, and the actual DC and the initial DC flux $\Delta\Phi$ are given.
- I_{dc} is obtained by calculation and should be examined whether it is consistent with the target I_{dc}, and if so, the calculation will be stopped; otherwise, the $\Delta\Phi$ will be adjusted and calculation is repeated until the results meet the requirements.
- Destination file is saved.

The results obtained in this example are as follows: Fig. 15.25 shows the waveform of exciting current, Fig. 15.26 shows the spectrum of exciting current, and Fig. 15.27 shows the relationship between the exciting current and the flux.

With the Φ–I curve of a transformer provided, the exciting current waveform and frequency spectrum can be obtained with a given DC using the principle of simple iteration.

15.3.2 Validation of Simple Iteration Method

To validate the effectiveness of this method, the exciting current of the transformer model is measured and compared with the simulated results.

15.3.2.1 No-Load Exciting Current Without DC Bias (I_{dc} = 0 A)

The waveform of the measured no-load current of the transformer model is shown in Fig. 15.28.

Table 15.2 *Φ–I* curve data for a transformer

Φ (Wb)	*I* (A)	*Φ* (Wb)	*I* (A)
−1.99E+02	−5.13E+01	0.00E+00	0.00E+00
−1.97E+02	−4.26E+01	1.07E+01	9.48E−02
−1.95E+02	−3.51E+01	1.79E+01	1.39E−01
−1.93E+02	−3.02E+01	2.50E+01	1.53E−01
−1.91E+02	−2.55E+01	3.17E+01	1.69E−01
−1.88E+02	−2.13E+01	3.92E+01	2.06E−01
−1.86E+02	−1.84E+01	4.60E+01	2.08E−01
−1.83E+02	−1.59E+01	5.68E+01	2.78E−01
−1.80E+02	−1.35E+01	7.08E+01	3.46E−01
−1.74E+02	−1.02E+01	8.83E+01	5.09E−01
−1.67E+02	−7.67E+00	1.06E+02	9.63E−01
−1.65E+02	−7.29E+00	1.23E+02	1.84E+00
−1.63E+02	−6.74E+00	1.40E+02	3.14E+00
−1.61E+02	−6.34E+00	1.47E+02	4.00E+00
−1.60E+02	−6.07E+00	1.54E+02	4.97E+00
−1.58E+02	−5.56E+00	1.55E+02	5.12E+00
−1.55E+02	−5.12E+00	1.58E+02	5.56E+00
−1.54E+02	−4.97E+00	1.60E+02	6.07E+00
−1.47E+02	−4.00E+00	1.61E+02	6.34E+00
−1.40E+02	−3.14E+00	1.63E+02	6.74E+00
−1.23E+02	−1.84E+00	1.65E+02	7.29E+00
−1.06E+02	−9.63E−01	1.67E+02	7.67E+00
−8.83E+01	−5.09E−01	1.74E+02	1.02E+01
−7.08E+01	−3.46E−01	1.80E+02	1.35E+01
−5.68E+01	−2.78E−01	1.83E+02	1.59E+01
−4.60E+01	−2.08E−01	1.86E+02	1.84E+01
−3.92E+01	−2.06E−01	1.88E+02	2.13E+01
−3.17E+01	−1.69E−01	1.91E+02	2.55E+01
−2.50E+01	−1.53E−01	1.93E+02	3.02E+01
−1.79E+01	−1.39E−01	1.95E+02	3.51E+01
−1.07E+01	−9.48E−02	1.97E+02	4.26E+01
		1.99E+02	5.13E+01

Based on the magnetization curve of the core measured experimentally, the waveform of calculated no-load current is shown in Fig. 15.29.

As shown in Fig. 15.30, the calculated and measured exciting currents agree well with each other.

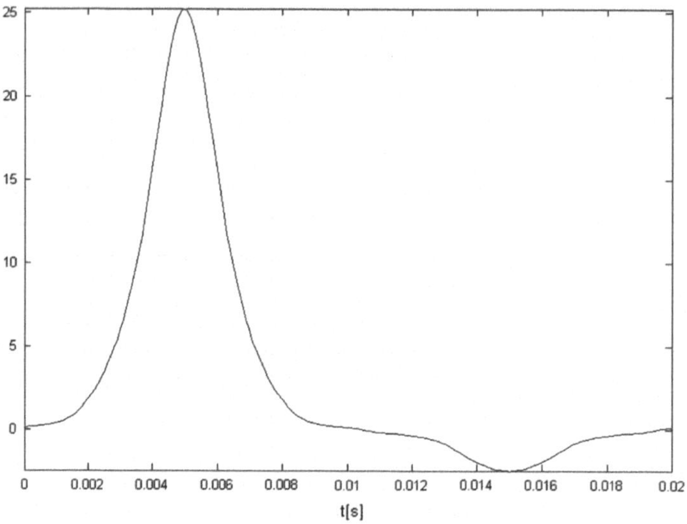

Fig. 15.25 Waveform of exciting current

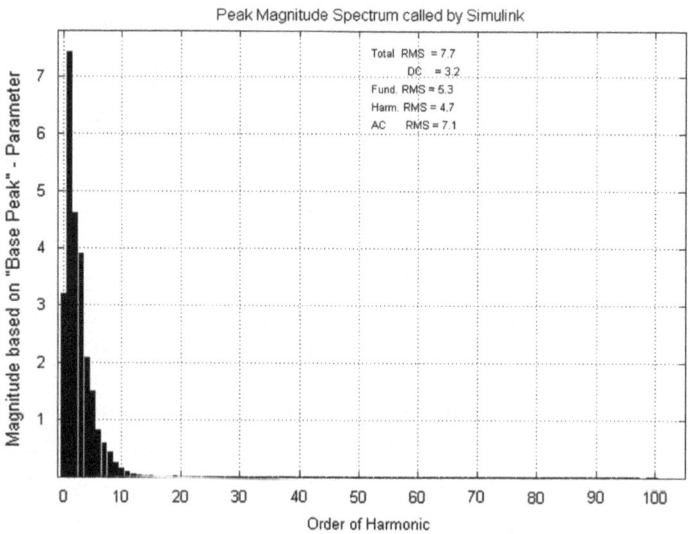

Fig. 15.26 Spectrum of exciting current

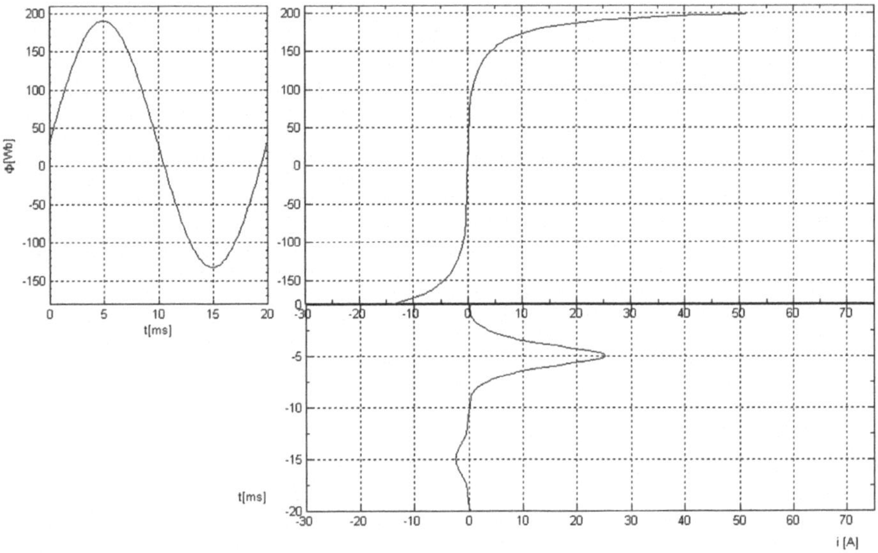

Fig. 15.27 Relationship between the exciting current and the flux under DC bias condition

Fig. 15.28 Measured no-load exciting current of single-phase transformer model under DC bias condition

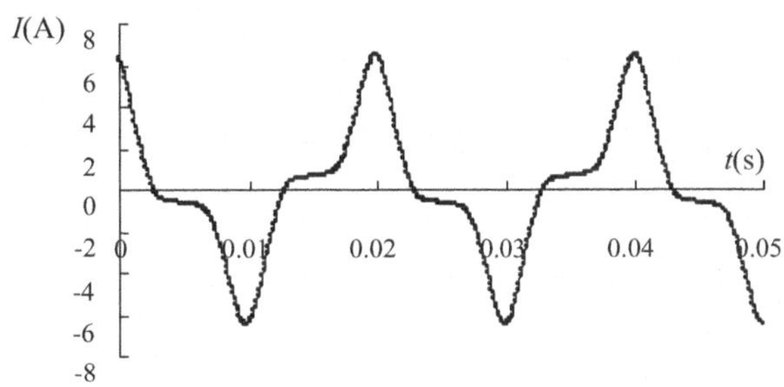

Fig. 15.29 Calculated no-load exciting current of single-phase transformer model under DC bias condition

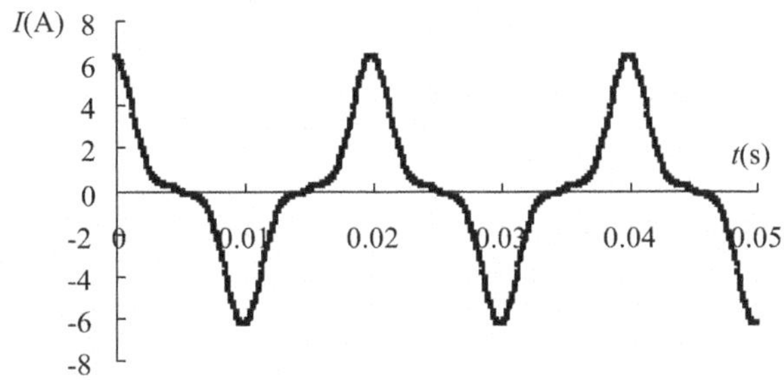

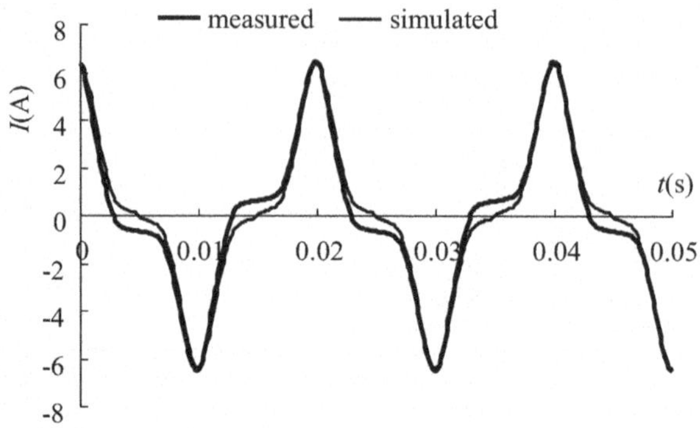

Fig. 15.30 Comparison of measured and simulated no-load exciting currents under DC bias

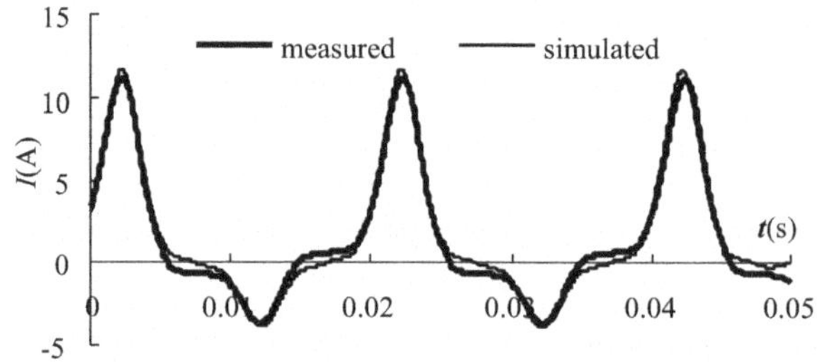

Fig. 15.31 Comparison of the measured and simulated exciting currents as $I_{dc} = 1.26$ A

15.3.2.2 No-Load Exciting Current Under DC Bias Condition ($I_{dc} = 1.26$ A)

The measured and calculated exciting currents with $I_{dc} = 1.26$ A are compared in Fig. 15.31.

15.3.2.3 No-Load Exciting Current Under DC Bias Condition ($I_{dc} = 2.53$ A)

The measured and calculated exciting currents with $I_{dc} = 2.53$ A are compared in Fig. 15.32.

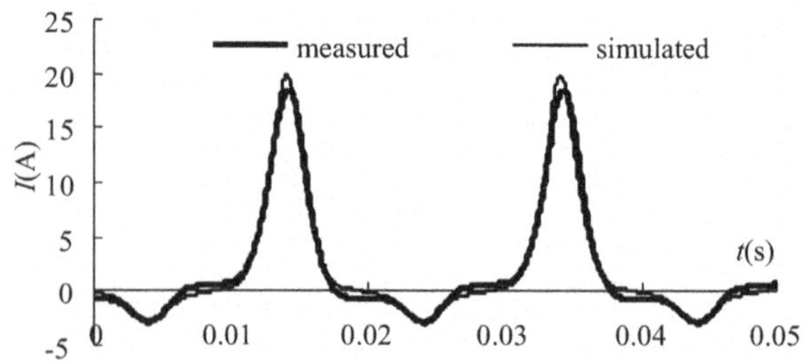

Fig. 15.32 Comparison of the measured and simulated exciting currents as $I_{dc} = 2.53$ A

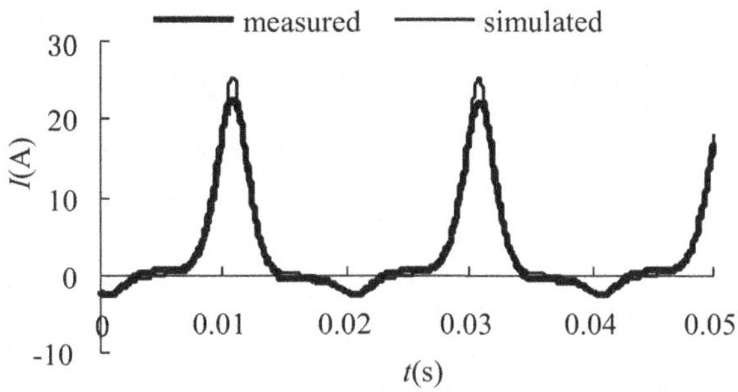

Fig. 15.33 Comparison of the measured and simulated exciting currents as I_{dc} = 3.2 A

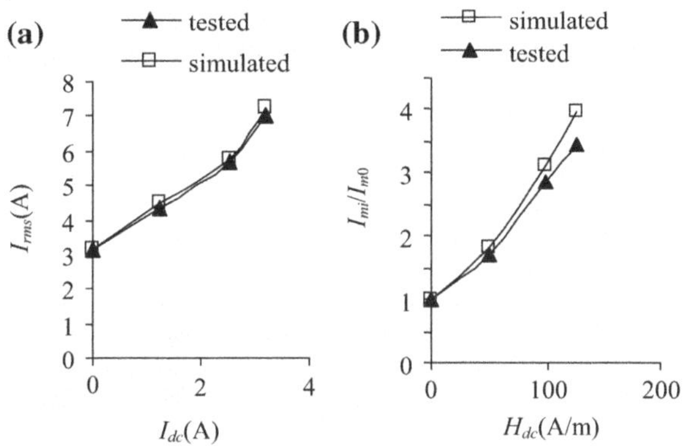

Fig. 15.34 Influence of DC bias on RMS value and magnitude of exciting current

15.3.2.4 No-Load Exciting Current Under DC Bias Condition (I_{dc} = 3.2 A)

The measured and calculated exciting currents with I_{dc} = 3.2 A are compared in Fig. 15.33.

The waveforms in Figs. 15.30, 15.31, 15.32, and 15.33 are analyzed to obtain the curves of RMS value and magnitude of exciting current, as shown in Fig. 15.34. The comparison between simulated and measured exciting currents in terms of RMS values and peak values under different DC bias conditions is shown in Tables 15.3 and 15.4.

From the curves given above, it can be seen that the computational error is within 5%, which meets the engineering needs.

15.3.3 Harmonic Analysis of Exciting Current [22]

The harmonic distribution of exciting current under DC bias condition shows obvious characteristics. The spectrum comparison between measured and simulated

Table 15.3 Comparison of simulated and measured RMS values of exciting current under different DC bias conditions

DC I_{dc} (A)	0.0	1.26	2.53	3.2
Simulated value I_{rms} (A) RMS	3.17	4.47	5.73	7.26
Tested value I_{rms} (A) RMS	3.18	4.35	5.69	7.03
Deviation (%)	−0.31	2.76	0.70	3.27

Table 15.4 Peak value of exciting currents under different DC bias conditions

DC I_{dc} (A)	0.0	1.26	2.53	3.2
DC magnetic potential H_{dc} (A/m)	0.0	49.78	99.95	126.42
Exciting current simulated I_{mi}/I_{m0}	1.0	1.7	2.8	3.5
Exciting current tested I_{mi}/I_{m0}	1.0	1.8	3.1	4.0

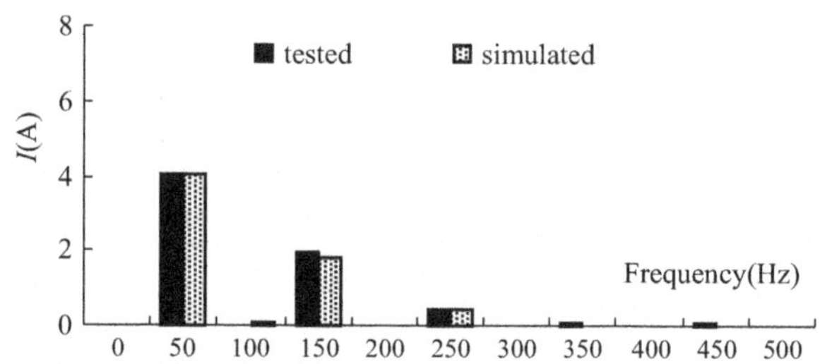

Fig. 15.35 Spectrum of no-load exciting current

exciting currents of the single-phase transformer model under different DC bias conditions is shown in Figs. 15.35, 15.36, 15.37, and 15.38.

The relationship between total harmonic distortion (THD) and DC is shown in Fig. 15.39.

With the increase of I_{dc}, THD increases slowly. When I_{dc} = 3.2 A, THD reaches 84.5% as the maximum value.

The harmonic components of exciting current under different DC bias conditions are compared as shown in Fig. 15.40. To further illustrate the problem, the relationship between the amplitude ratio of each higher harmonic to the fundamental

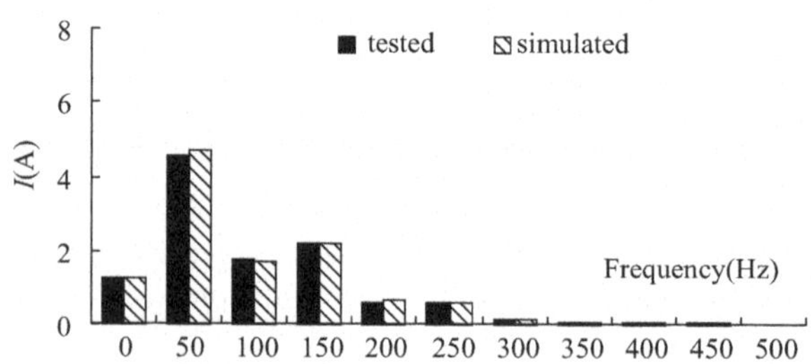

Fig. 15.36 Spectrum of exciting current as I_{dc} = 1.26 A

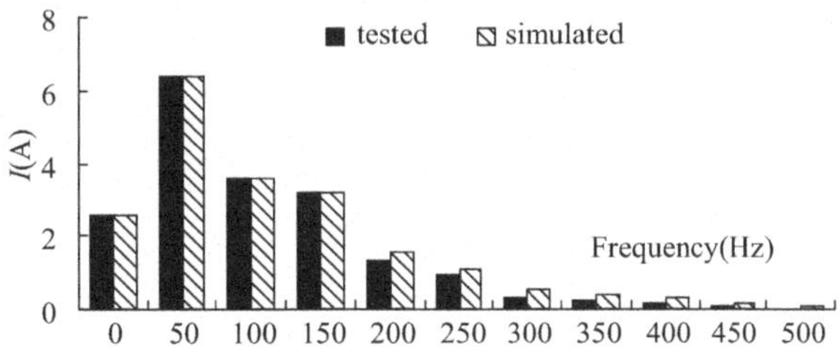

Fig. 15.37 Spectrum of exciting current as $I_{dc} = 2.53$ A

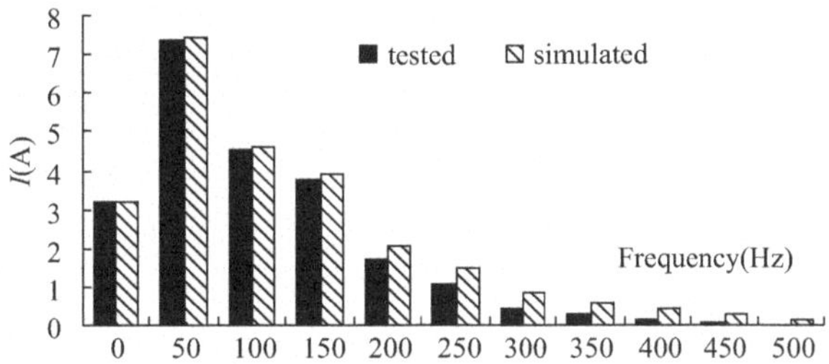

Fig. 15.38 Spectrum of exciting current as $I_{dc} = 3.2$ A

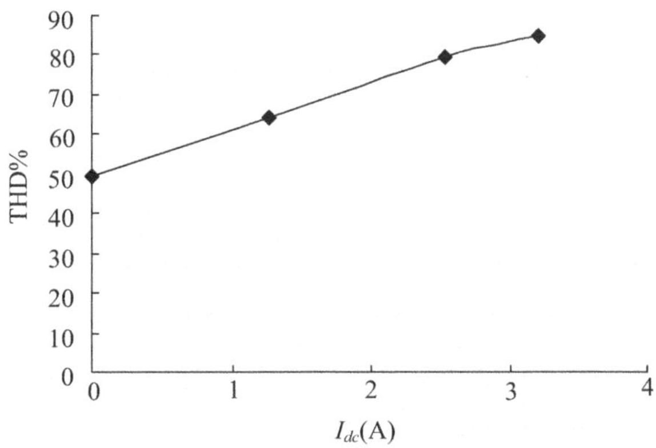

Fig. 15.39 Influence of DC bias on THD

component and the DC is drawn, as shown in Fig. 15.41, from which we can find that even harmonics increase significantly, especially the second harmonic increases fastest with the increase of bias current.

From the curves given above, it can be seen that with the increase of DC, all harmonic components increase to a certain extent. In addition, it can be seen that the amplitude ratios of the even harmonics are much larger than those of the odd harmonics.

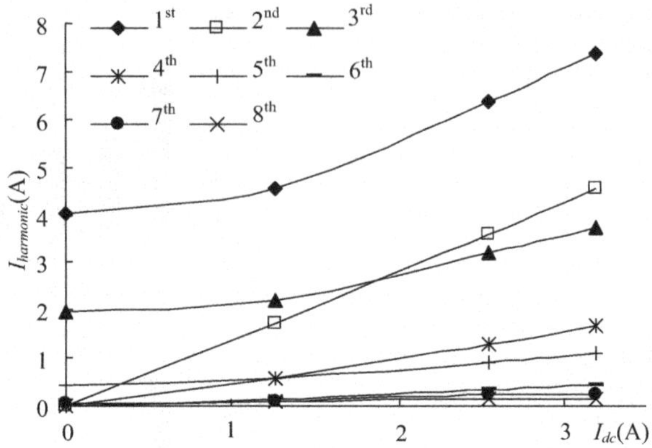

Fig. 15.40 Harmonic components of exciting current under different DC bias conditions

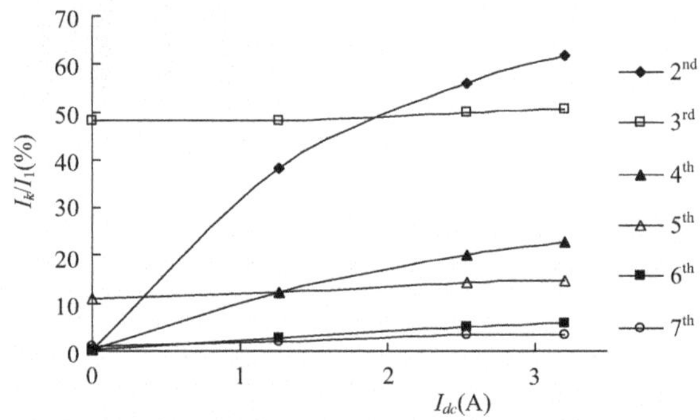

Fig. 15.41 Amplitude ratio of high-order harmonic to the fundamental harmonic under different DC bias conditions

15.4 Modeling and Computation of Magnetic Field and Loss Under DC Bias Condition

15.4.1 Some Key Factors in Modeling Under DC Bias Condition

The exciting current of transformer under DC bias is non-sinusoidal, and the positive and negative half cycles are asymmetric. With the increase of DC, the higher harmonic components of the excitation current increase significantly. Therefore, the problem related to DC bias is, as a matter of fact, a problem of nonlinear and asymmetric transient fields. To calculate the magnetic field and loss of transformer under DC bias, it is necessary to clearly understand the magnetic field characteristic parameters of transformer core under DC bias condition, including *B*–*H* curve and loss curve $B_{\mathrm{m}} - W$ of transformer core under DC bias condition, which have been described in Sect. 15.2.

Another key issue is the effectiveness and usefulness of simulation tools and computational method, because of the complexity of three-dimensional DC biasing magnetic field of transformer and large computational cost on computer memory and CPU time. In this chapter, Simcenter MAGNET was used to solve the three-dimensional DC biasing magnetic field of transformer based on its excellent computational efficiency. The efficient modeling and simulation of both the core loss and the stray-field loss in various components under DC bias conditions are difficult and investigated worldwide [23–30].

In order to calculate the total loss W in the component by finite element method, the element's loss can be calculated based on the measured loss curve $B_{\mathrm{m}} - W$ (W/kg) of the material and the calculated field result (i.e., B_{m} for each element) under DC bias condition, and then sum up all the element's loss, as shown in 15.7,

$$W = \rho \sum_{i=1}^{N_e} W^e(B^e_{\mathrm{m}})\, V^e W \tag{15.7}$$

where

ρ	is the material density;
N_e	is the total number of elements;
V^e	is the element volume;
*B*e m	is the magnitude of flux density in the finite element;
W^e	is the loss in the finite element.

In this section, single-phase transformer with the parameters listed in Table 15.1 is taken as an example to illustrate the calculation of the magnetic field and loss in the transformer under DC bias condition. First, the product or model needs to be modeled. In this model, 30RGH120 silicon steel sheet is used for the transformer core and magnetic shielding of tank, 20Mn23Al is used for the tie plate, A3 magnetic steel is used for the tank and clamping, etc. In the computational process, the material nonlinearity is considered, while the anisotropy of the oriented material is not. To save memory and computational time, 1/4 part of the model is modeled for computation considering the symmetry of the whole model.

Three-dimensional transient field (skin effect is considered) for single-phase transformer model is calculated by Simcenter MAGNET, the tie plate and clamping are layered at 0.5/1.5 mm in the thickness direction of the steel plate, the maximum grid is 15 mm during finite element mesh generation, and the tie plate, clamping, core, and magnetic shield are calculated by second-order elements The time steps are 1 and 0.5 ms, respectively. The model established is shown in Fig. 15.42.

Some mesh generation and computing information are as follows:

No. of tetrahedra = 732,742;	No. of nodes = 135,283;
No. of unary faces = 31,198;	No. of binary faces 0;
No. of edges = 883,623;	No. of unary edges = 0;
No. of binary edges = 0;	No. of scalar tets = 676,030;
No. of vector tets = 56,712;	No. of void tets = 0;

(continued)

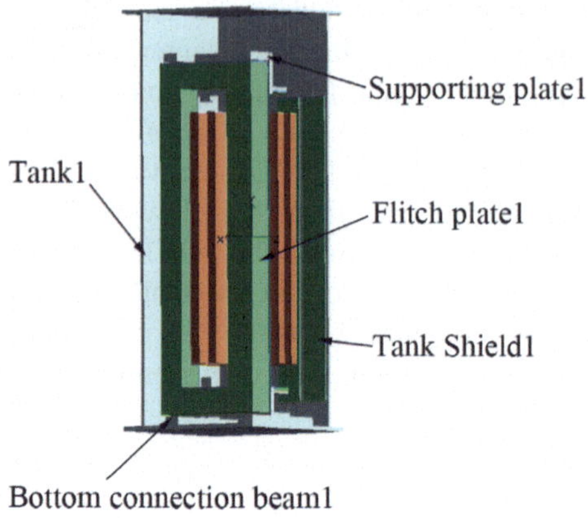

Fig. 15.42 Computational model of DC-biased transformer

(continued)

No. of scalar edges = 834,616;	No. of vector edges = 49,007;
No. of void edges = 0;	No. of field nodes = 419,508
No. of binary Fnodes(pairs) = 0;	CG tolerance = 0.0001%
Newton tolerance = 1%;	No. of nonzeros = 8,567,461。

15.4.2 Computation of Magnetic Field and Loss Under No-Load and DC Bias Condition

15.4.2.1 Results Under No-Load Condition with I_{dc} = 0 A

Figure 15.43 shows the waveform of exciting current used under no-load and non-bias condition. The flux linkage waveform is shown in Fig. 15.44.

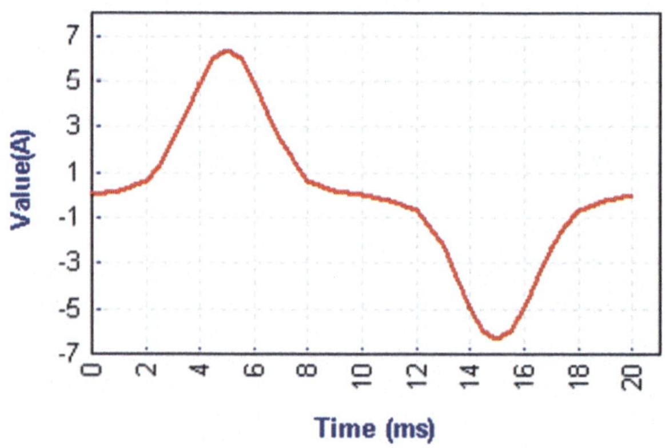

Fig. 15.43 Waveform of no-load exciting current

Fig. 15.44 Waveform of no-load exciting flux linkage

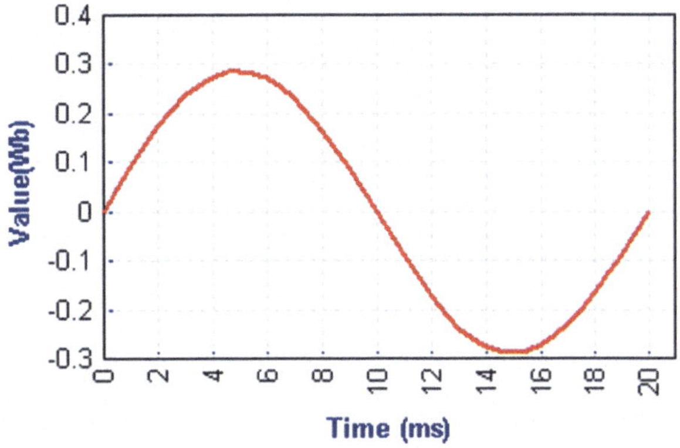

Three-dimensional and two-dimensional distributions of magnetic field provided by Simcenter MAGNET are shown in Figs. 15.45 and 15.46.

The maximum flux density in the limb is 1.7T, as shown in Fig. 15.46.

The magnetic field strength of a single-phase transformer model as I_{dc} = 0 A and t = 5 ms is shown in Fig. 15.47.

The losses in the components of a single-phase transformer model calculated under no-load and non-bias condition are listed in Table 15.5.

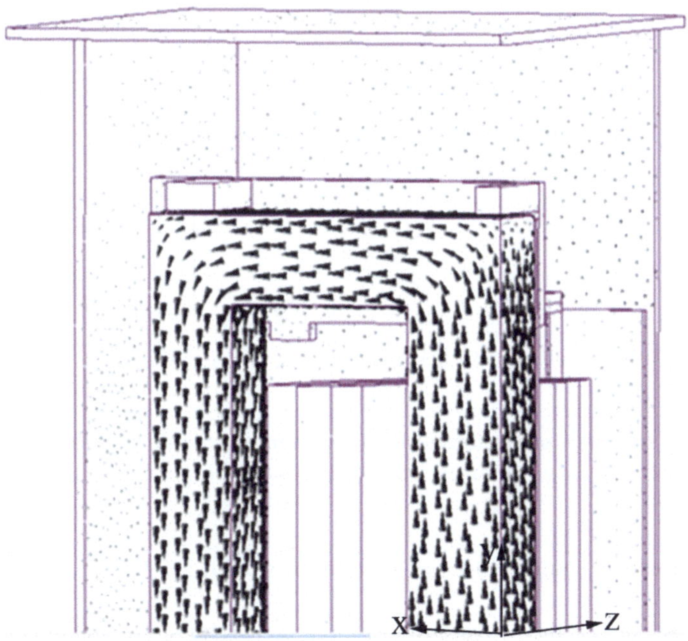

Fig. 15.45 Distribution of flux density as I_{dc} = 0 A and t = 5 ms

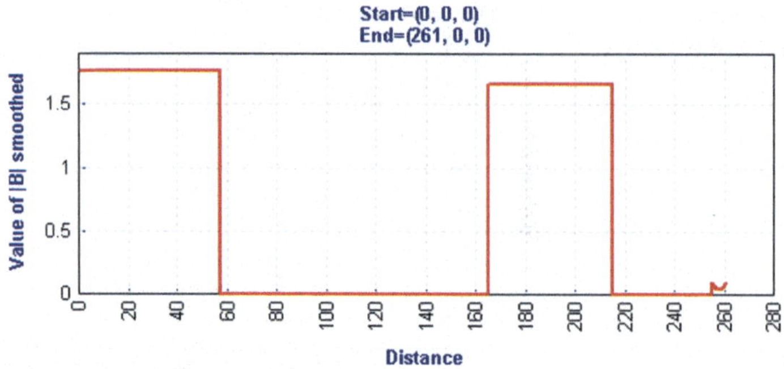

Fig. 15.46 Distribution of field quantity $|B|$ along centerline of x-axis as $I_{dc} = 0$ A and $t = 5$ ms (abscissa unit: mm; ordinate unit: tesla)

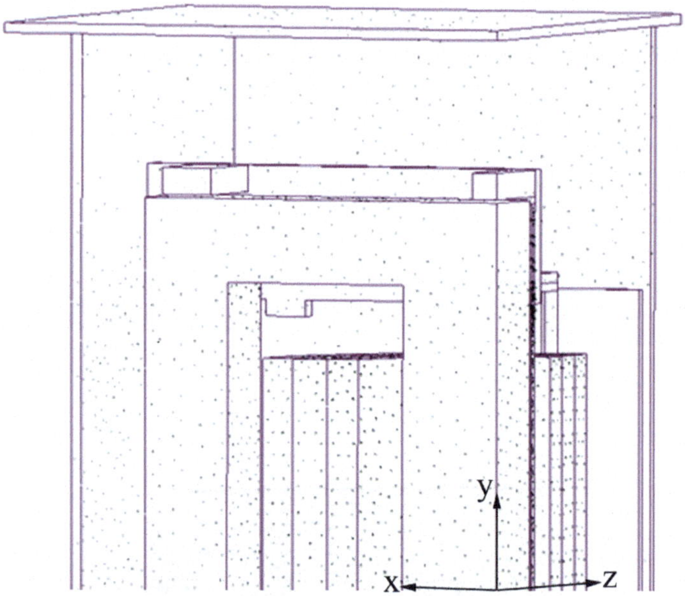

Fig. 15.47 Distribution of magnetic field strength as $I_{dc} = 0$ A and $t = 5$ ms

It can be seen that under the no-load and non-bias condition, the core loss is a larger part of total loss, while the stray-field loss in components is smaller.

15.4.2.2 Results Under No-Load Condition with $I_{dc} = 3.2$ A

Figure 15.48 shows the waveform of exciting current used under no-load and non-bias. The flux linkage waveform is shown in Fig. 15.49.

Table 15.5 Calculated losses of components under no-load condition with $I_{dc} = 0$

Component	Calculated losses (W)
Core	3.51E+01
Bottom connection beam 1	2.47E−02
Bottom connection beam 2	7.35E−03
Tie plate 1	4.86E−03
Tie plate 2	5.27E−03
Angle steel 1	2.53E−03
Angle steel 2	3.82E−03
Lower limb plate	2.15E−03
Shield of lower limb plate	2.68E−04
Tank shield 1	3.91E−03
Tank shield 2	8.45E−03
Upper tank	1.92E−02
Middle tank 1	1.22E−01
Middle tank 2	1.90E−01
Lower tank	5.09E−02
Pressbeam of upper clamping	4.95E−02
Upper limb plate	4.48E−04
Shield of upper limb plate	2.62E−04
Pressbeam of lower clamping	8.61E−02
Supporting plate 1	1.39E−03
Supporting plate 2	5.87E−04
Total	3.57E+01

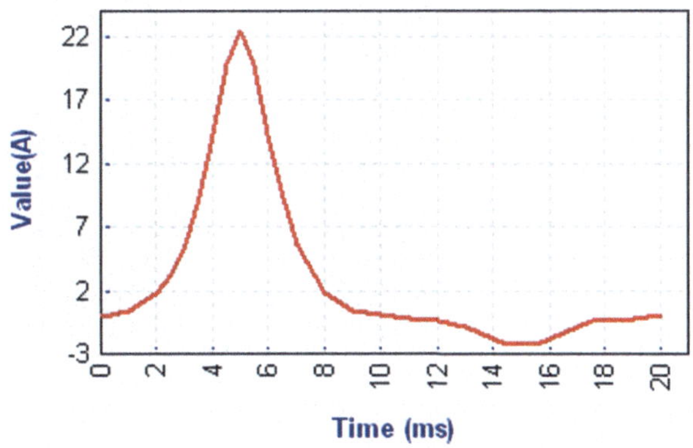

Fig. 15.48 Waveform of exciting current as $I_{dc} = 3.2$ A

Flux density distribution under no-load bias ($I_{dc} = 3.2$ A) is shown in Figs. 15.50 and 15.51.

The flux density distributions displayed in Fig. 15.50 when $t = 5$ ms and in Fig. 15.51 when $t = 15$ ms are different due to DC bias. The flux density when $t = 5$ ms is significantly higher than that when $t = 15$ ms. For clear illustration, a

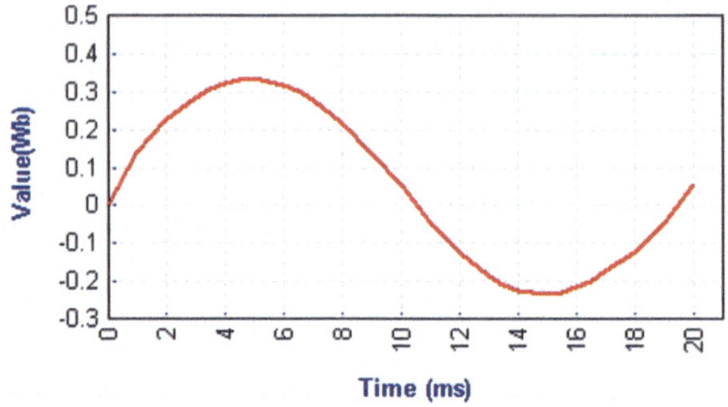

Fig. 15.49 Waveform of flux linkage as $I_{dc} = 3.2$ A

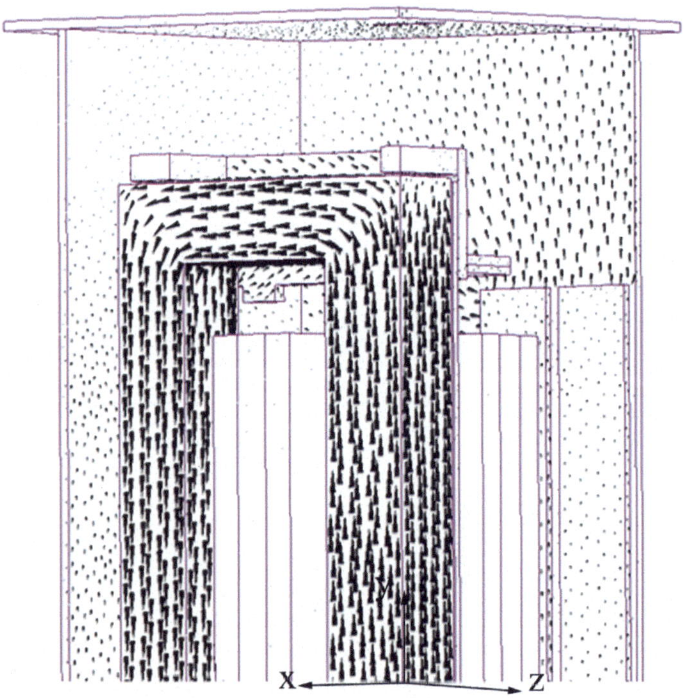

Fig. 15.50 Distribution of flux density as $I_{dc} = 3.2$ A and $t = 5$ ms

straight line is made along the centerline of the x-axis to obtain the distribution of flux density when $I_{dc} = 3.2$ A, $t = 5$ ms and $I_{dc} = 3.2$ A, $t = 15$ ms, as shown in Figs. 15.52 and 15.53, respectively.

In addition, the flux density distributions along the surface layer of the tie plate, the centerline of the magnetic shield, and the surface of the clamping under no-load

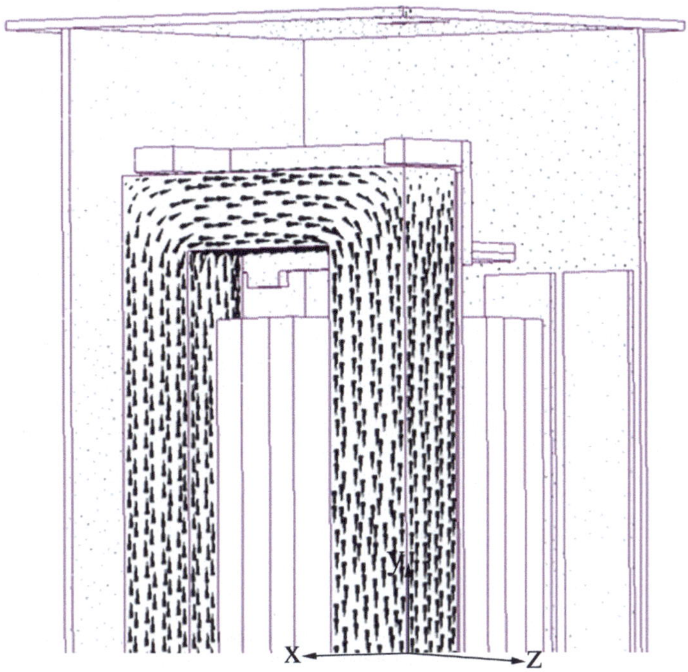

Fig. 15.51 Distribution of flux density as I_{dc} = 3.2 A and t = 15 ms

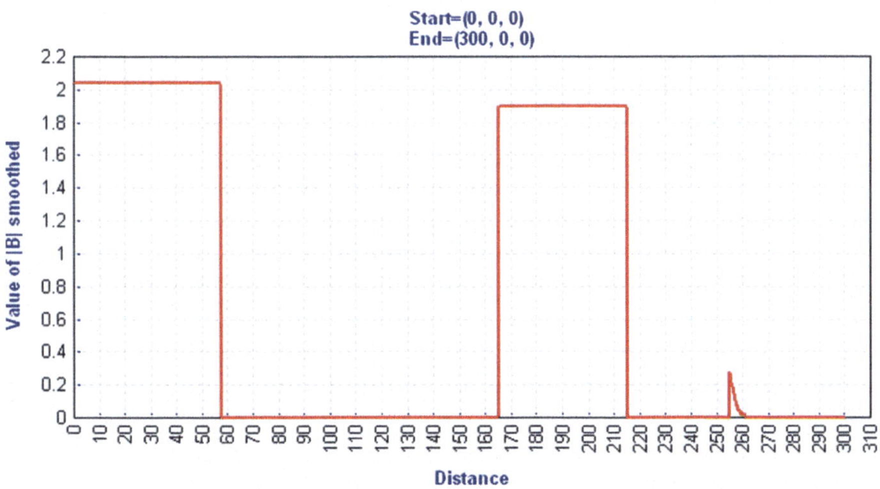

Fig. 15.52 Distribution of field quantity $|B|$ along centerline of x-axis as I_{dc} = 3.2 A and t = 5 ms (abscissa unit: mm; ordinate unit: tesla, the same below)

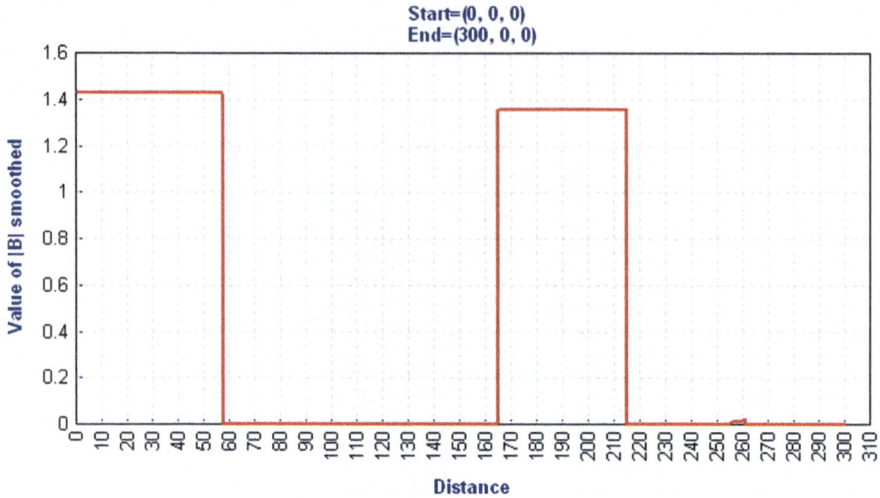

Fig. 15.53 Distribution of field quantity $|B|$ along centerline of x-axis as I_{dc} = 3.2 A and t = 15 ms

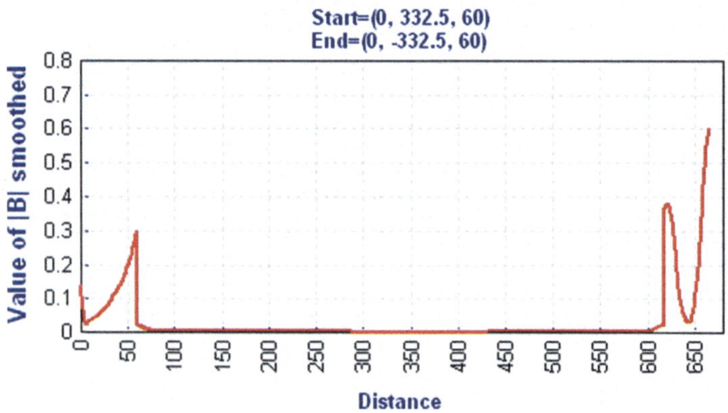

Fig. 15.54 Distribution of field quantity $|B|$ along z-axis of outer surface of tie plate as I_{dc} = 3.2 A and t = 5 ms

and DC bias condition are shown in Figs. 15.54, 15.55, 15.56, and 15.57, respectively.

The calculated losses inside the components of a single-phase model under no-load DC bias condition (I_{dc} = 3.2 A) are listed in Table 15.6.

Losses inside each component when I_{dc} = 0 A are compared with that when I_{dc} = 3.2 A under no-load condition, as shown in Fig. 15.58.

As shown in Fig. 15.58, compared with the no-load and non-bias condition, the loss increases in components including tank and pressbeam of clamping under DC bias condition. The hysteresis loss of each component has a large proportion in the

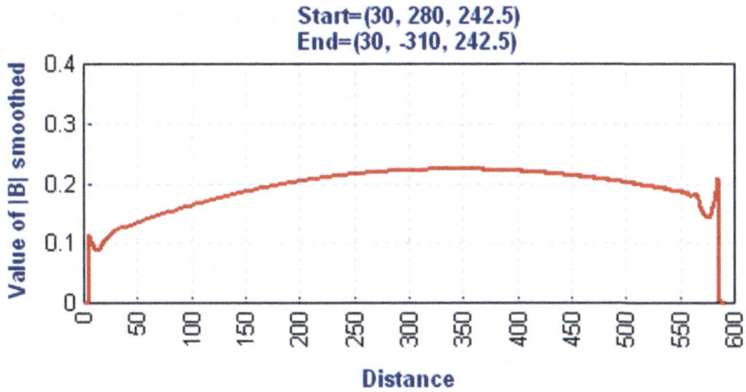

Fig. 15.55 Distribution of $|B|$ along y-direction of the center of tank shield 1 as I_{dc} = 3.2 A and t = 5 ms

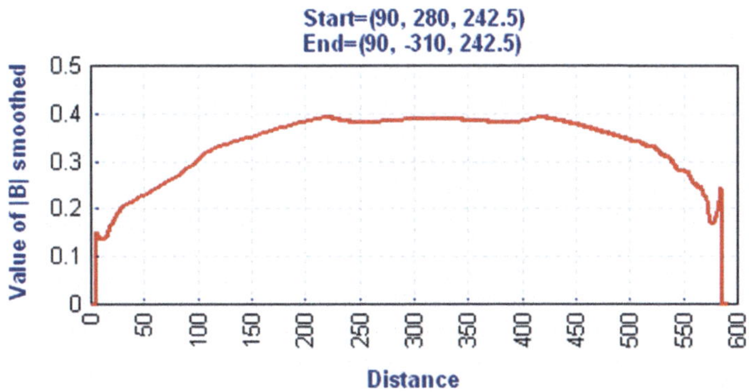

Fig. 15.56 Distribution of $|B|$ along y-direction of the center of tank shield 2 as I_{dc} = 3.2 A and t = 5 ms

total loss under no-load condition. The proportion of hysteresis loss when I_{dc} = 0 A and I_{dc} = 3.2 A is shown in Fig. 15.59.

The hysteresis loss accounts for 99.38 and 97.97% of the total loss when I_{dc} = 0 A and I_{dc} = 3.2 A, respectively.

The hysteresis loss and eddy current loss inside each component increase in varying degrees under no-load condition when I_{dc} = 3.2 A, compared with that under no-load and non-bias condition, as shown in Fig. 15.60.

Figure 15.60 shows that the hysteresis loss of the middle tank 1 and the eddy current loss of the middle tank 2 increase greatly. The eddy current loss distributions in the pressbeam of upper clamping, lower clamping, middle tank, and tie plate are shown in Figs. 15.61, 15.62, 15.63, and 15.64, respectively.

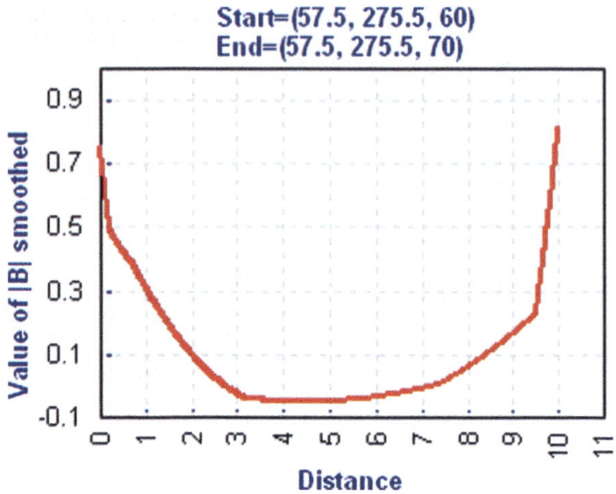

Fig. 15.57 Distribution of $|B|$ along clamping surface as $I_{dc} = 3.2$ A and $t = 5$ ms

Table 15.6 Calculated losses of components under no-load DC bias condition with $I_{dc} = 3.2$ A

Component	Calculated losses (W)
Core	3.68E+01
Bottom connection beam 1	8.74E−02
Bottom connection beam 2	2.53E−02
Tie plate 1	2.39E−02
Tie plate 2	1.96E−02
Angle steel 1	9.05E−03
Angle steel 2	1.17E−02
Lower limb plate	5.69E−03
Shield of lower limb plate	1.63E−03
Tank shield 1	2.40E−02
Tank shield 2	7.25E−02
Upper tank	7.45E−02
Middle tank 1	5.08E−01
Middle tank 2	6.58E−01
Lower tank	1.66E−01
Pressbeam of upper clamping	2.11E−01
Upper limb plate	1.02E−03
Shield of upper limb plate	1.67E−03
Pressbeam of lower clamping	2.99E−01
Supporting plate 1	5.03E−03
Supporting plate 2	1.03E−03
Total	3.90E+01

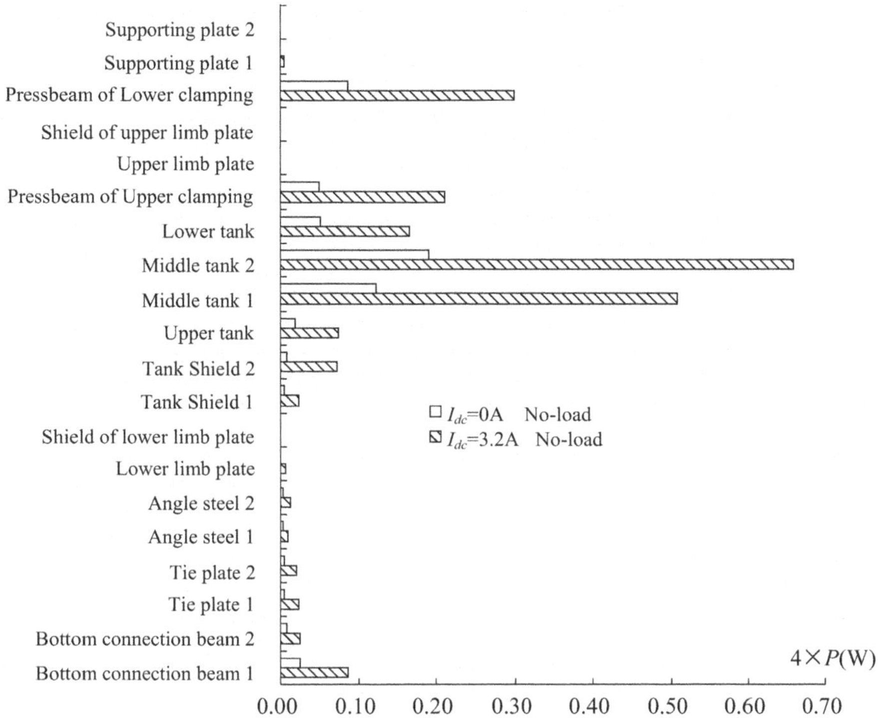

Fig. 15.58 Losses of components as $I_{dc} = 0$ A compared with that as $I_{dc} = 3.2$ A under no-load conditions

15.4.3 Computation of Magnetic Field and Loss Under Load and DC Bias Condition

15.4.3.1 Results Under Load Condition with $I_{dc} = 0$ A

Figures 15.65 and 15.66 show the exciting current applied on the low-voltage and high-voltage coils and the corresponding flux linkage under load condition with $I_{dc} = 0$ A.

The flux density (***B***) distribution in transformer under load (non-bias) condition obtained from the computational results by Simcenter MAGNET is shown in Fig. 15.67, and the distribution of magnetic field intensity (***H***) is shown in Fig. 15.68.

There is not only main flux but also leakage flux under load condition in a transformer, which is different from that under no-load condition. The comparison

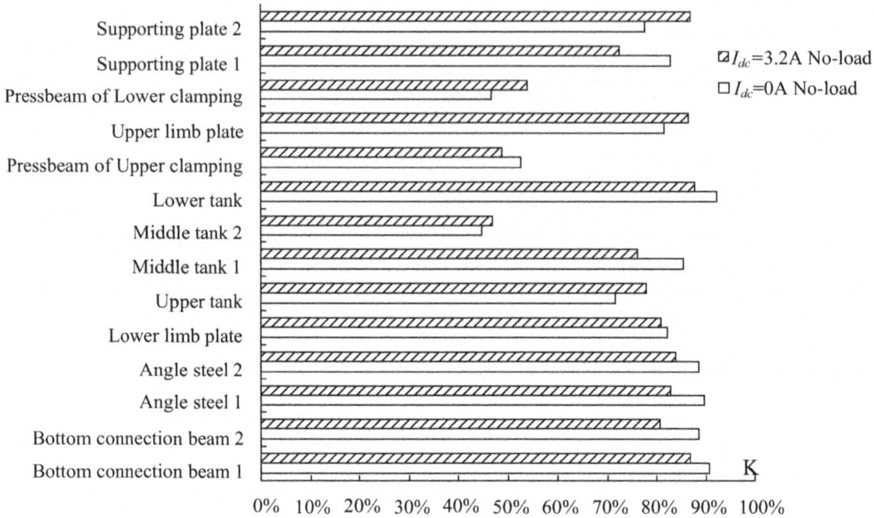

Fig. 15.59 Proportion of hysteresis loss of components accounting for total loss under no-load condition as I_{dc} = 0 A and 3.2 A, respectively

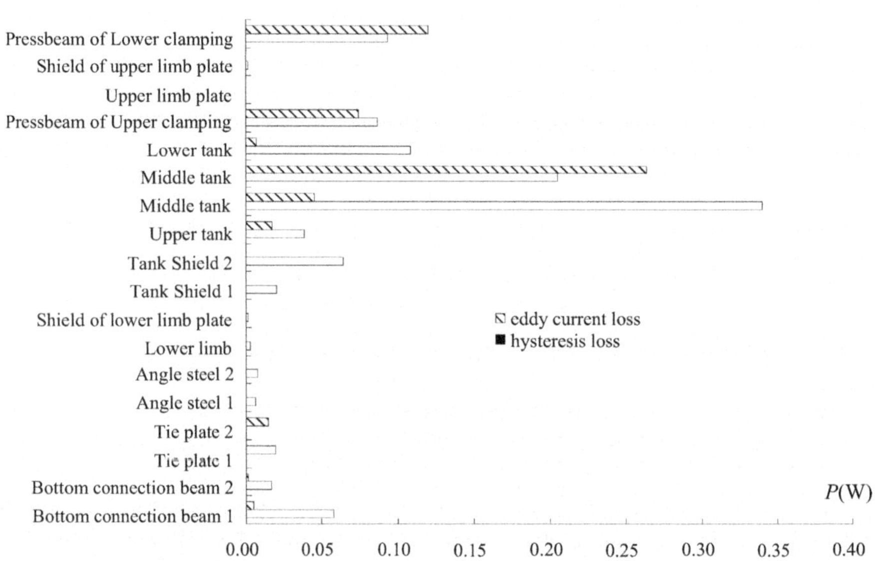

Fig. 15.60 Increment of hysteresis loss and eddy current loss under no-load condition when I_{dc} = 3.2 A

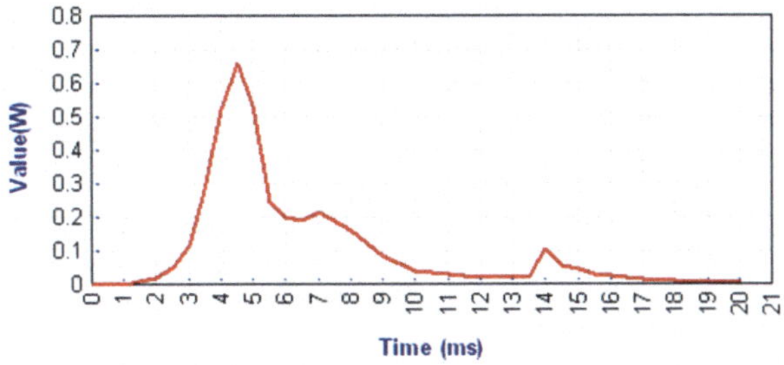

Fig. 15.61 Variation of loss inside pressbeam of upper clamping with time when I_{dc} = 3.2 A

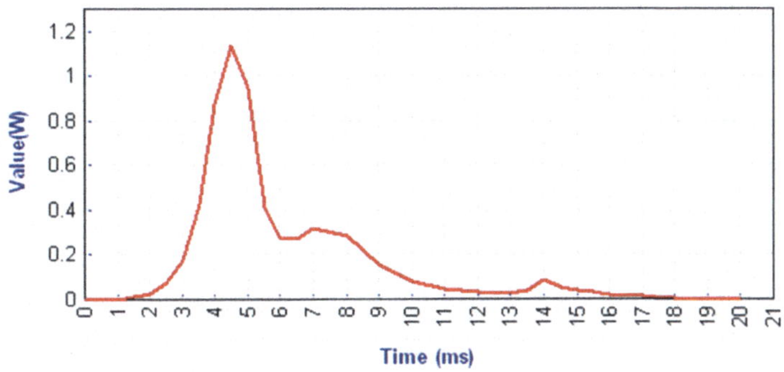

Fig. 15.62 Variation of loss inside pressbeam of lower clamping with time when I_{dc} = 3.2 A

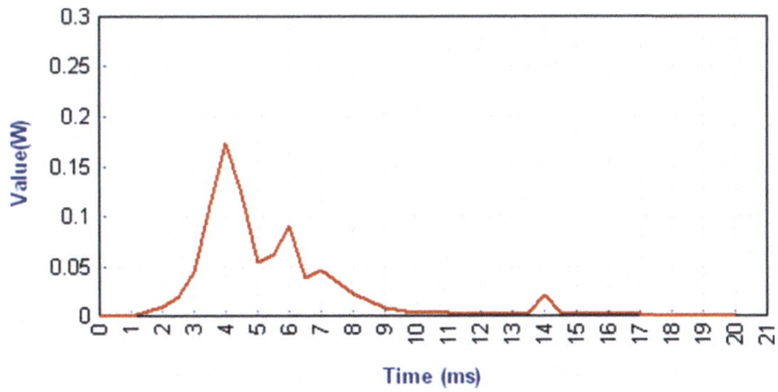

Fig. 15.63 Variation of the loss inside tie plate 1 with time when I_{dc} = 3.2 A

between Figs. 15.68 and 15.47 shows that the magnetic field intensity H reflects the distribution of transformer leakage electromagnetic field under load condition. To reflect the distribution of leakage flux and main flux, the $|B|$ distribution along the centerline of x-axis is shown in Fig. 15.69.

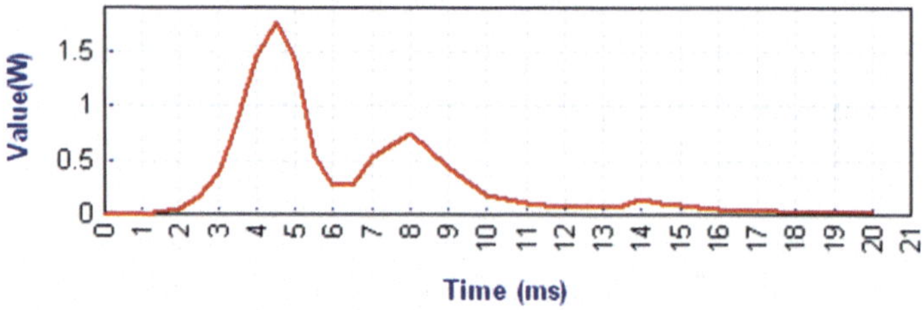

Fig. 15.64 Variation of the loss inside middle tank 2 with time when I_{dc} = 3.2 A

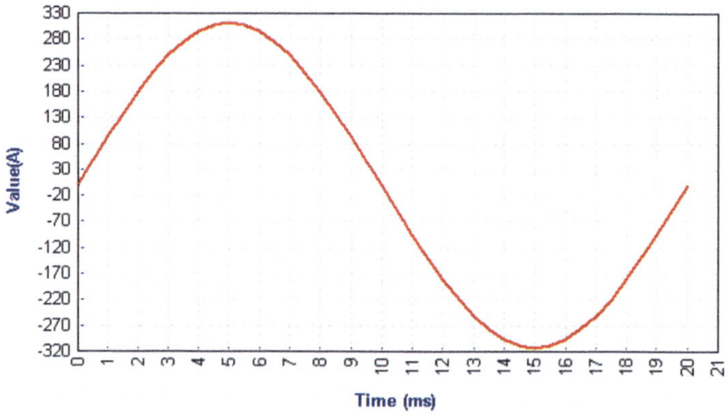

(a) Exciting current in low-voltage coil

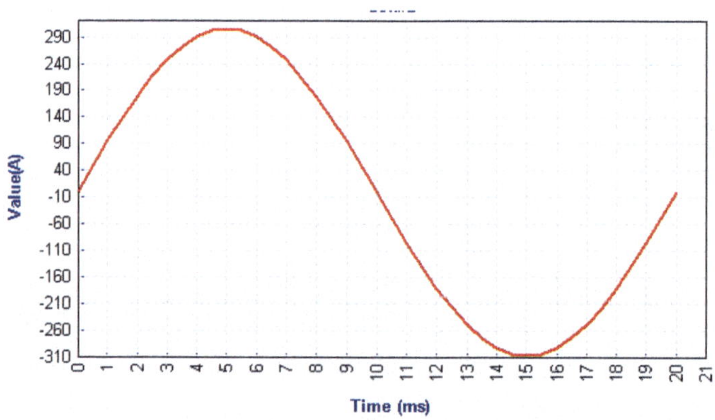

(b) Exciting current in high-voltage coil

Fig. 15.65 Exciting current waveform of single-phase transformer model under load condition

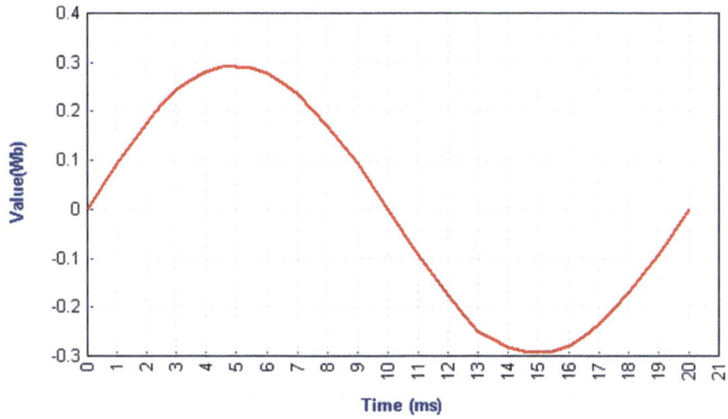

(a) Flux linkage of low-voltage coil

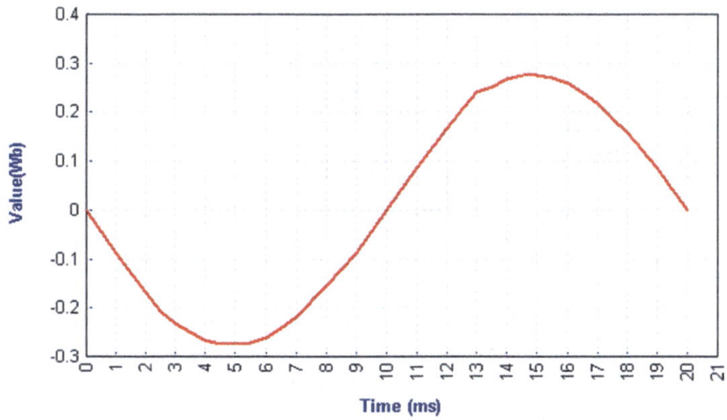

(b) Flux linkage of high-voltage coil

Fig. 15.66 Exciting flux linkage waveform of single-phase transformer model under load condition

The average flux density in the core is 1.703T, and the leakage flux density is 516 Gs. However, due to the deviation in modeling and simulation of the yoke, the calculated flux density in the core yoke may have some errors.

The calculated losses inside components of the single-phase model under rated load condition with I_{dc} = 0 A are listed in Table 15.7.

15.4.3.2 Results Under Load Condition with I_{dc} = 3.2 A

The applied exciting current under rated load condition when I_{dc} = 3.2 A is shown in Fig. 15.70. The corresponding flux linkage waveform is shown in Fig. 15.71.

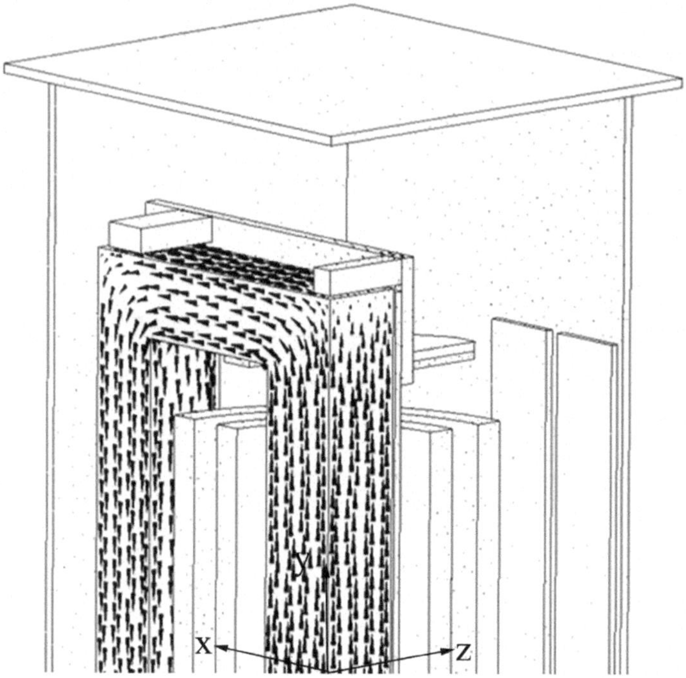

Fig. 15.67 Flux density distribution under load condition when $I_{dc} = 0$ A and $t = 5$ ms

Similarly, the flux density distribution under load condition with $I_{dc} = 3.2$ A is shown in Figs. 15.72 and 15.73 by using Simcenter MAGNET.

From the flux density distribution shown in Fig. 15.74, it can be seen that DC bias has a great influence on magnetic field in transformer core under load condition. The average flux density in the core is 1.9435T, and the leakage flux density is 0.0521T. Compared with the results when $I_{dc} = 0$ A, the flux density of the core increases by 14%, while the leakage flux density increases by only about 1%. The calculated losses of components in single-phase model under load condition with $I_{dc} = 3.2$ A are listed in Table 15.8.

For comparison, the loss inside each component when $I_{dc} = 0$ and $I_{dc} = 3.2$ A is shown in Fig. 15.75.

As shown in Fig. 15.75, the losses inside components excluding pressbeam of upper and lower clamping increase compared to that under load condition with $I_{dc} = 0$ A. From $I_{dc} = 0$ A to $I_{dc} = 3.2$ A, the proportion of the total hysteresis loss inside components is shown in Fig. 15.76. Regardless of the coil, the hysteresis loss accounts for 98.87 and 98.35% of total loss when $I_{dc} = 0$ A of and 3.2 A, respectively.

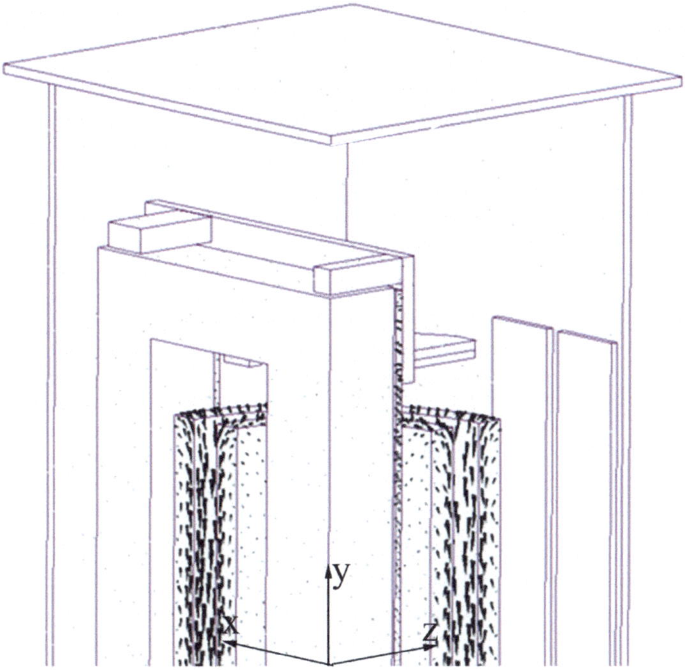

Fig. 15.68 Magnetic field intensity distribution under load condition when $I_{dc} = 0$ A and $t = 5$ ms

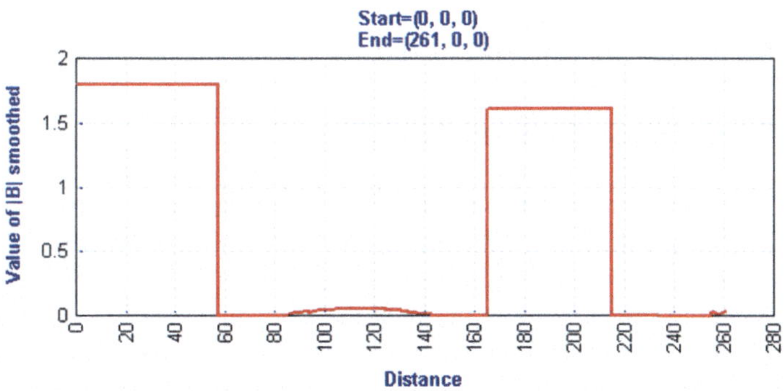

Fig. 15.69 Distribution of field quantity $|B|$ along the centerline of x-axis under rated load as $t = 5$ ms

The hysteresis loss and eddy current loss inside each component increase compared with that under load condition with $I_{dc} = 0$ A, as shown in Fig. 15.77. It can be seen that the increment of hysteresis loss and eddy current loss in the upper limb plate and supporting plate 1 increases significantly.

Table 15.7 Calculated losses of components under load DC bias condition with $I_{dc} = 0$ A

Component	Calculated losses (W)
Core	3.47E+01
Bottom connection beam 1	2.19E−02
Bottom connection beam 2	1.21E−03
Tie plate 1	5.31E−02
Tie plate 2	1.63E−02
Angle steel 1	2.17E−04
Angle steel 2	2.58E−03
Lower limb plate	4.85E−04
Shield of lower limb plate	4.96E−04
Tank shield 1	1.11E−03
Tank shield 2	1.35E−03
Upper tank	7.03E−03
Middle tank 1	2.84E−02
Middle tank 2	3.79E−02
Lower tank	1.23E−02
Pressbeam of upper clamping	3.00E−01
Upper limb plate	1.15E−03
Shield of upper limb plate	6.39E−04
Pressbeam of lower clamping	1.69E−01
Supporting plate 1	2.92E−03
Supporting plate 2	6.14E−04
Total	3.54E + 01

15.4.4 Influence of DC Bias on Loss

15.4.4.1 Results Under No-Load Condition

The calculated and measured losses of the single-phase transformer model under no-load conditions with $I_{dc} = 0$ A and $I_{dc} = 3.2$ A are shown and compared in Table 15.9. The core loss is analyzed separately, which is shown in Table 15.10.

Table 15.9 shows the deviation between the simulated results, measured results are within 5% and the simulation result is effective and accurate. However, with the increase of DC, the proportion of core loss in total loss becomes smaller, while loss proportion of components becomes larger, as shown in Table 15.10.

15.4.4.2 Results Under Load Condition

For the losses under load condition with $I_{dc} = 0$ A and load condition with $I_{dc} = 3.2$ A, the relative change with DC is listed in Table 15.11. The simulated core losses are shown in Table 15.12.

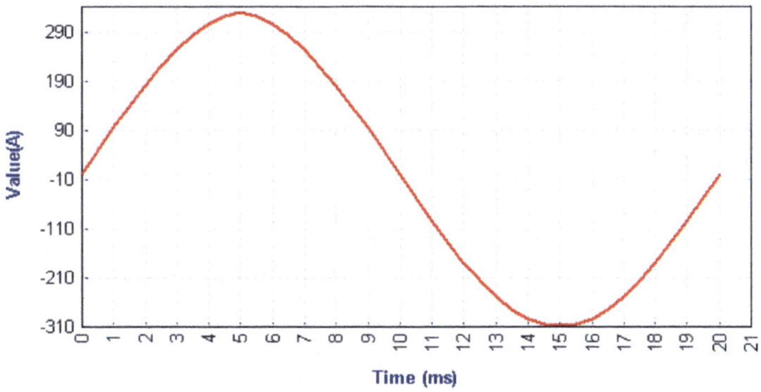

(a) Exciting current in low-voltage coil

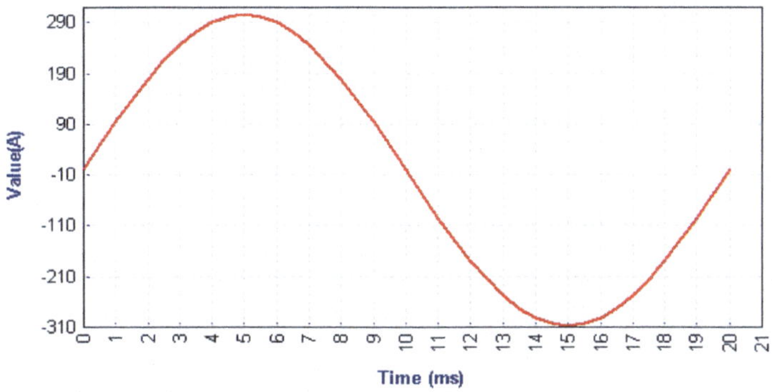

(b) Exciting current in high-voltage coil

Fig. 15.70 Load exciting current waveform of single-phase transformer model

It can be seen from Table 15.11 that the variation of calculated loss with DC is consistent with that of measured loss under load condition. Table 15.12 shows that the proportion of core loss in total loss decreases with the increase of DC.

15.4.4.3 Influence of Bias on No-Load and Load Loss

From the simulation, it can be shown that the total no-load loss as I_{dc} = 3.2 A is 9.15% higher than that as I_{dc} = 0 A, while the load loss as I_{dc} = 3.2 A is 5.72% higher than that as I_{dc} = 0 A. Measured results show that when the direct current up to 3.2 A is applied, the no-load loss increases by 11.6% and the load loss increases by 2.5%; thus, the influence of DC bias on loss under no-load condition is more significant than that under load condition.

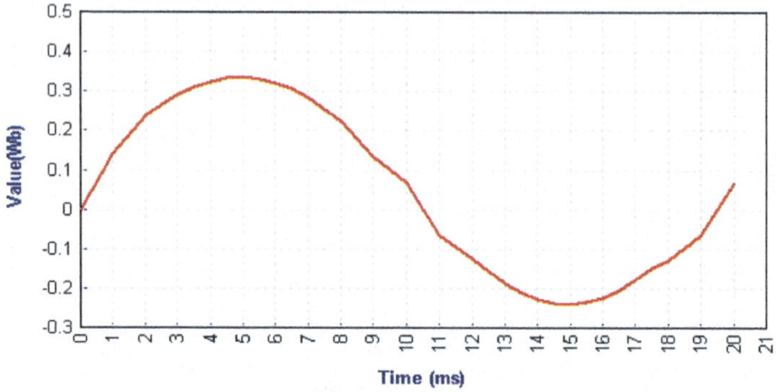

(a) Flux linkage of low-voltage coil

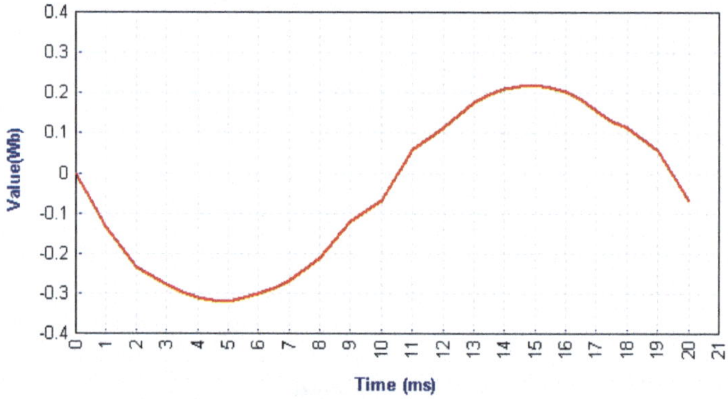

(b) Flux linkage of high-voltage coil

Fig. 15.71 Flux linkage waveform of single-phase transformer model under load condition

15.5 The Experimental Research on the DC-Biased 500 KV Autotransformer

DC bias tests with different direct current injections are carried out to further study the harmonics characteristic, no-load loss, and noise-level large power transformer under DC bias. The product for test is shown in Fig. 15.78. The electric circuit is shown in Fig. 15.79.

The relevant technical parameters for the transformer are as follows:

Capacity: 250/250/80 MVA
Voltage ratio: $(525/\sqrt{3})/(230/\sqrt{3} \pm 2 \times 2.5\%)/63$ kV
Connection symbol: YN a0 d11 (three phases)
Type of cooling: ONAF/ONAN (100%/70%)
Sheet type: 30ZH120 single-phase double frame; step-lap joint
Coefficient: lamination 0.97

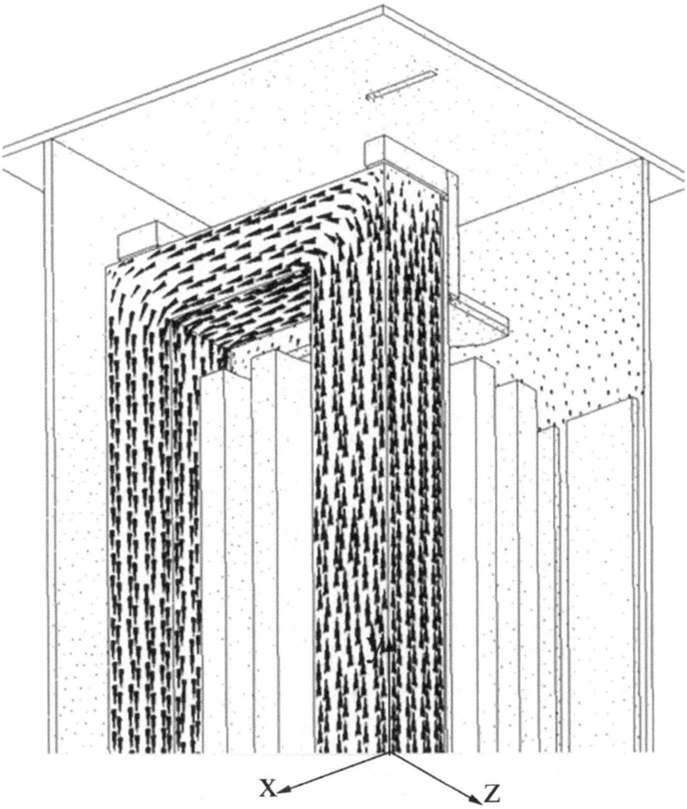

Fig. 15.72 Flux density distribution under rated load condition as I_{dc} = 3.2 A and t = 5 ms

15.5.1 No-Load Loss Measurement Under DC Bias Condition

When the DC I_{dc} equals to 0, 0.9, 1.9, and 3.4 A, respectively, the no-load loss of the transformer is measured. No-load loss of the transformer without DC applied is defined to 100%, the ratio of measured no-load loss under different DC bias conditions to that when I_{dc} = 0 is represented by the symbol K given in Fig. 15.80.

For the 500 kV autotransformer, the measured peak value of no-load exciting current is 0.83 A, and the measured no-load loss is 74.55 kW. When I_{dc} is larger than 4 amperes,the voltage is seriously distorted; hence, the measured maximum value of I_{dc} is 4 A, which is about 4.8 times of the RMS values of no-load exciting current. From above curve, it can be found that the no-load loss increases with the increase in I_{dc} until I_{dc} is equal to 4 A, and the no-load loss is 1.065 times of that without DC bias excitation.

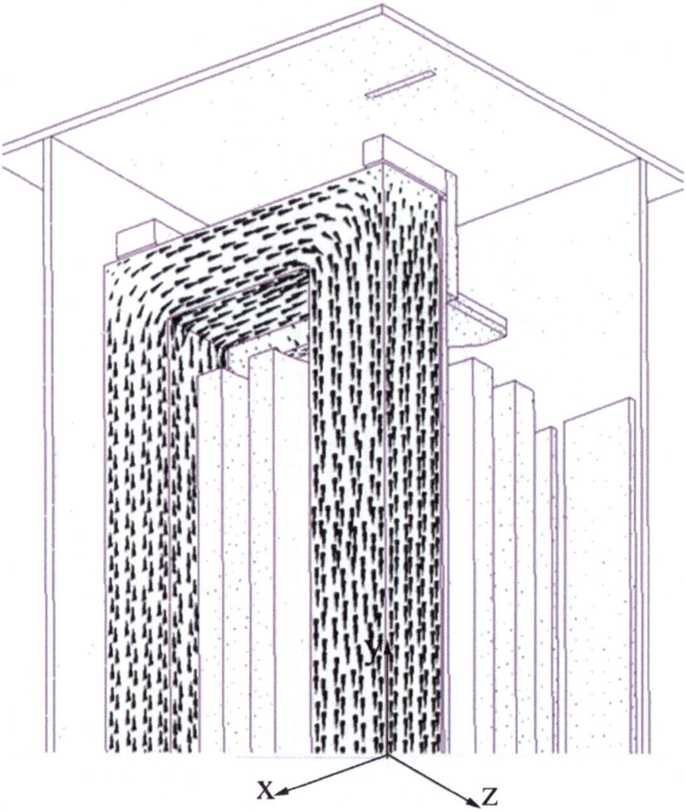

Fig. 15.73 Distribution of flux density under rated load as I_{dc} = 3.2 A and t = 15 ms

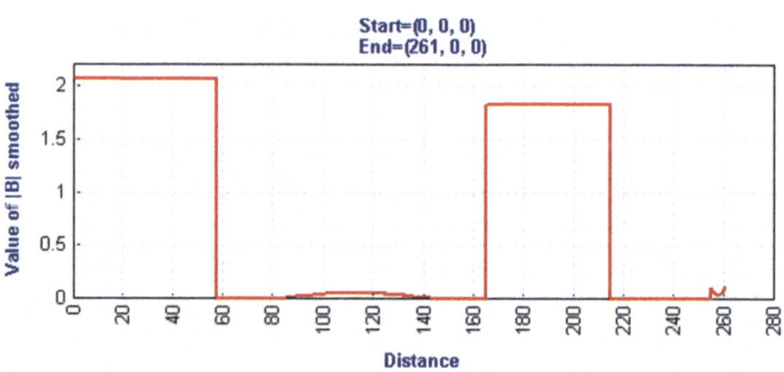

Fig. 15.74 Distribution of field quantity $|B|$ along the centerline of x-axis under rated load t = 5 ms (abscissa unit: mm; ordinate unit: tesla)

Table 15.8 Calculated losses of components under load DC bias condition with I_{dc} = 3.2 A

Component	Calculated losses (W)
Core	3.63E+01
Bottom connection beam 1	5.86E−02
Bottom connection beam 2	3.79E−03
Tie plate 1	6.84E−02
Tie plate 2	2.49E−02
Angle steel 1	5.98E−04
Angle steel 2	5.25E−03
Lower limb plate	6.20E−04
Shield of lower limb plate	1.95E−03
Tank shield 1	6.12E−03
Tank shield 2	1.18E−02
Upper tank	1.09E−02
Middle tank 1	5.80E−02
Middle tank 2	1.58E−01
Lower tank	2.98E−02
Pressbeam of upper clamping	5.15E−03
Upper limb plate	3.73E−01
Shield of upper limb plate	8.39E−04
Pressbeam of lower clamping	2.46E−03
Supporting plate 1	2.54E−01
Supporting plate 2	5.27E−03
Total	3.74E+01

15.5.2 Harmonics Analysis of Exciting Current

Analysis is performed based on the measured results. When I_{dc} equals to 0, 0.9, 1.9, 3, and 4 A (the corresponding ratio of direct current to RMS values of no-load current is about 0, 108.4, 228.9, 361.4, and 481.9%), respectively, and meanwhile 30, 100, and 113% of the rated voltage are applied on the transformer, harmonic spectrum variation of the exciting current is shown in Figs. 15.81, 15.82, and 15.83.

From harmonic spectrums corresponding to three typical alternating voltage excitations applied on a transformer, it can be seen that when no DC is injected, there are mainly odd harmonics which show an upward trend with the increase of applied voltage. When DC is applied, there are even harmonics in addition to odd harmonics, and all harmonic components increase with the increase of the applied voltage. However, when the applied voltage is not higher than the rated voltage, odd harmonics such as fundamental and third harmonics and even harmonics such as second and fourth harmonics increase significantly compared with those when no

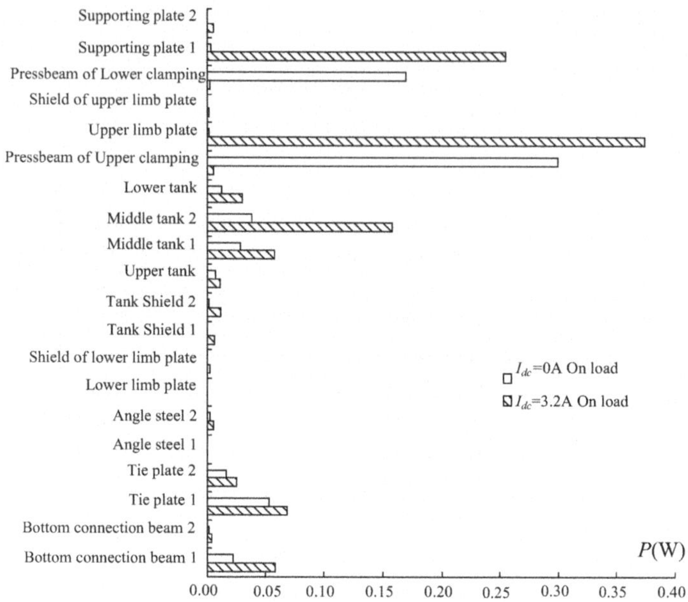

Fig. 15.75 Losses inside component when I_{dc} = 0 A compared with that when I_{dc} = 3.2 A, under load conditions

DC is applied. In the case of overvoltage excitation, the increment of odd harmonic component becomes smaller, while the even harmonic components only change a little. Figures 15.84, 15.85, and 15.86 show the variation of each harmonic component under different DC bias conditions.

It can be seen that with the increase of DC, the fundamental, second, third, and fourth harmonic components of the exciting current show an overall upward trend, while the rest of the harmonic components show little change and even downward trend. When the applied voltage is lower than the rated voltage, the order of growth speed is as follows: fundamental–second–third–fourth harmonic; when the applied voltage is equal to or smaller than the rated voltage, the second and fourth harmonic components increase significantly. Furthermore, harmonic spectrum with different applied voltages is shown in Figs. 15.87, 15.88, 15.89, 15.90, and 15.91.

The total harmonic distortion (THD) for exciting current of the transformer is shown in Table 15.13.

The THD and the ratio of high-order harmonic components to the fundamental component, such as I_2/I_1, I_3/I_1 under different DC bias conditions, are shown in Figs. 15.92, 15.93, and 15.94.

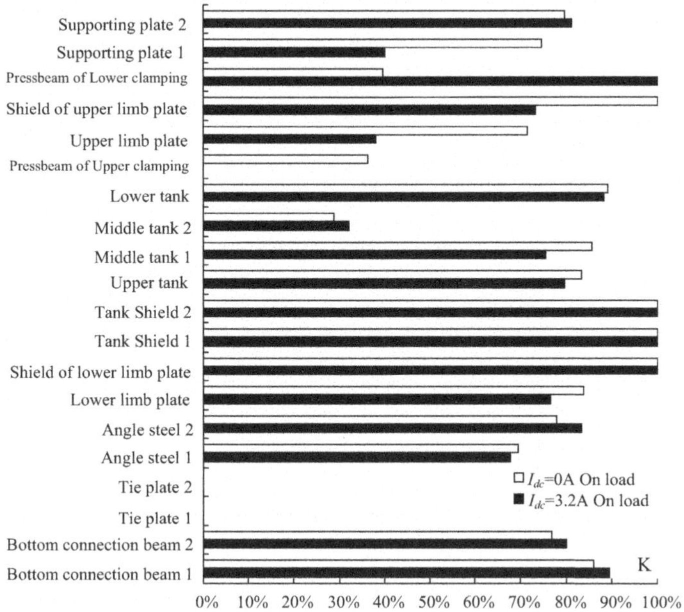

Fig. 15.76 Proportion of hysteresis loss inside components accounting for total loss when $I_{\mathrm{dc}} = 0$ A, 3.2 A under load condition

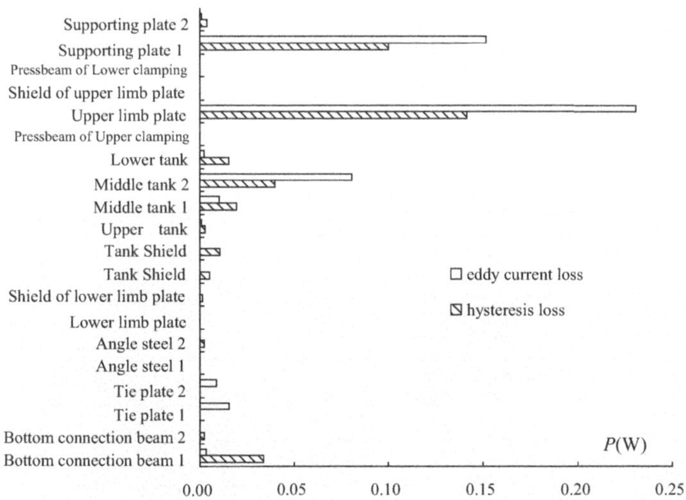

Fig. 15.77 Increment of hysteresis loss and eddy current loss under load condition when $I_{\mathrm{dc}} = 3.2$ A

Table 15.9 Comparison of the losses of single-phase transformer model under no-load and bias conditions 1

DC (A)	Measured results (W)	Calculated results (W)	Deviation (%)
0	143.47	142.89	−0.40
3.2	160.12	155.96	−2.60

Table 15.10 Comparison of the losses of single-phase transformer model under no-load and bias conditions 2

DC (A)	Core loss (W)	Percentage of core loss to total loss (%)	Loss of other components (W)	Percentage of loss of other components to total loss (%)
0	140.40	98.26	2.49	1.74
3.2	147.20	94.38	8.76	5.62

Table 15.11 Comparison of the losses of single-phase transformer model under load and DC bias condition (1)

DC (A)	Measured results (W) (%)	Calculated results (W) (%)
0	100	100
3.2	102.53	105.67

Table 15.12 Comparison of the losses of single-phase transformer model under load and DC bias condition (2)

DC (A)	Total losses (W)	Core loss (W)	Percentage of core loss to total loss (%)
0	141.6	138.8	98.02
3.2	149.7	145.2	97.0

Fig. 15.78 Product outline for a 500 kV autotransformer

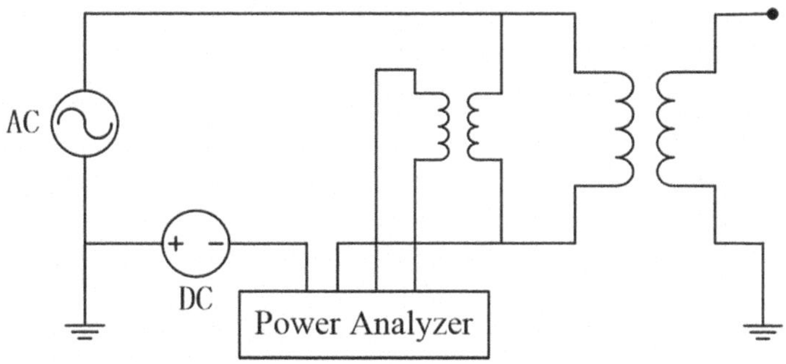

Fig. 15.79 Schematic diagram of circuit for test under DC bias

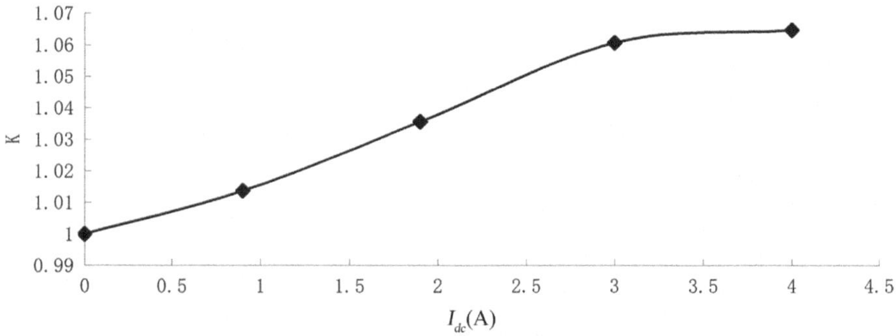

Fig. 15.80 Ratio of measured loss under DC-biased condition to the loss under no-load condition

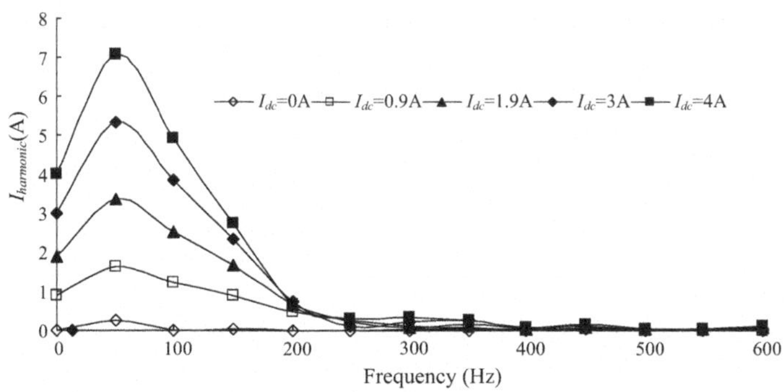

Fig. 15.81 Spectrum of exciting current under different DC bias conditions when 30% U_n of voltage is applied

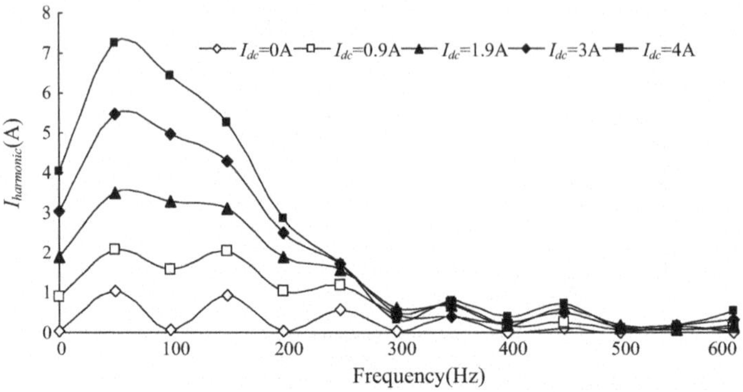

Fig. 15.82 Spectrum of exciting current under different DC bias conditions when 100% rated voltage is applied

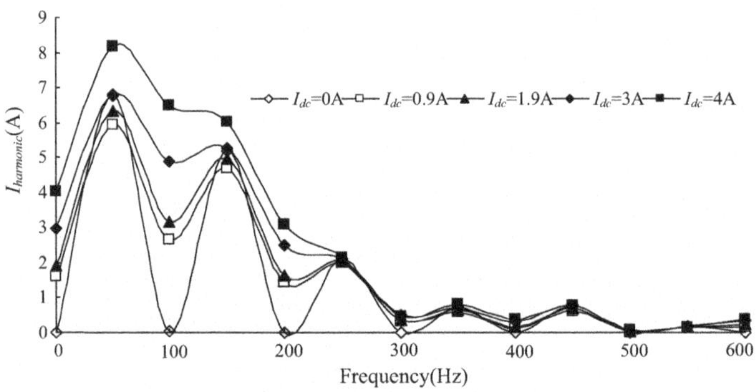

Fig. 15.83 Spectrum of exciting current under different DC bias conditions when 113% rated voltage is applied

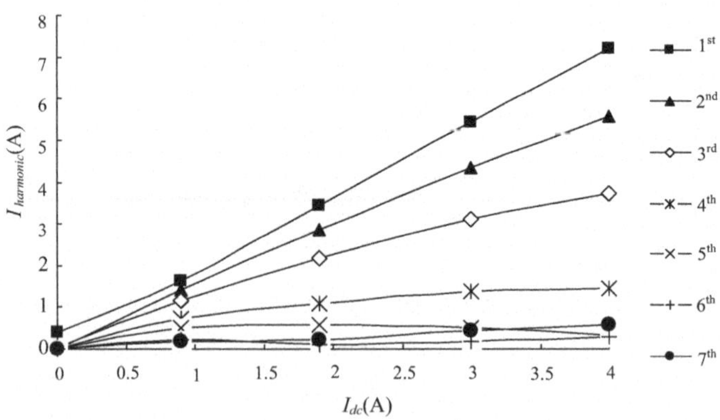

Fig. 15.84 Spectrum of exciting current DC bias condition when 30% rated voltage is applied

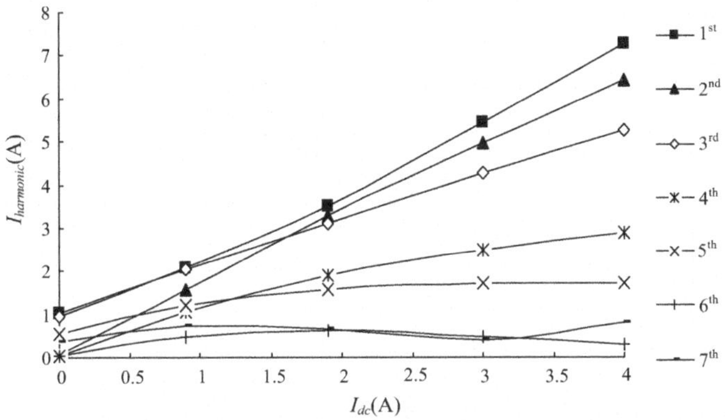

Fig. 15.85 Spectrum of exciting current DC bias condition when 100% rated voltage is applied

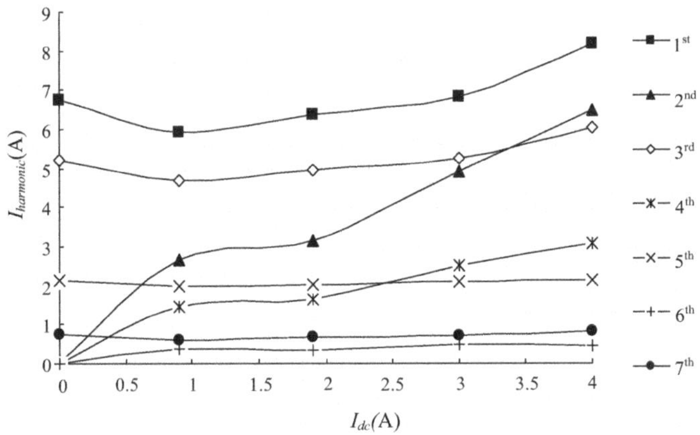

Fig. 15.86 Spectrum of exciting current DC bias condition when 113% rated voltage is applied

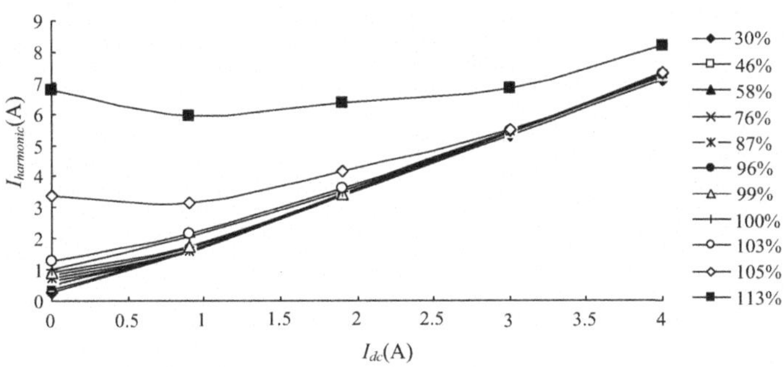

Fig. 15.87 Fundamental component under different DC bias conditions

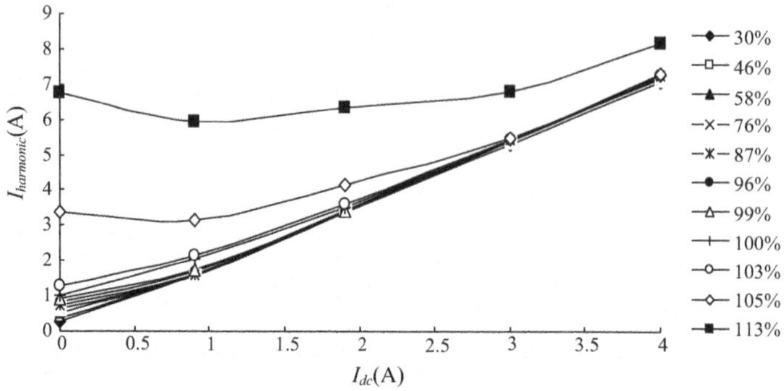

Fig. 15.88 Second harmonic component under different DC bias conditions

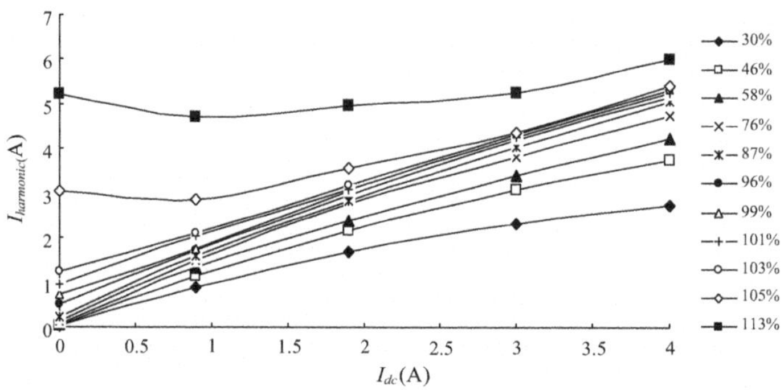

Fig. 15.89 Third harmonic component under different DC bias conditions

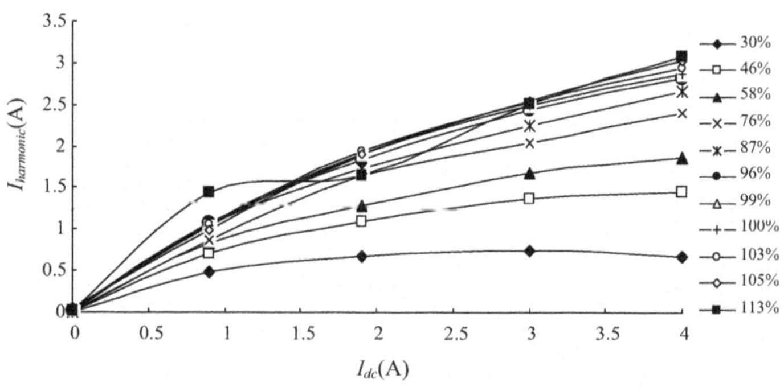

Fig. 15.90 Fourth harmonic component under different DC bias conditions

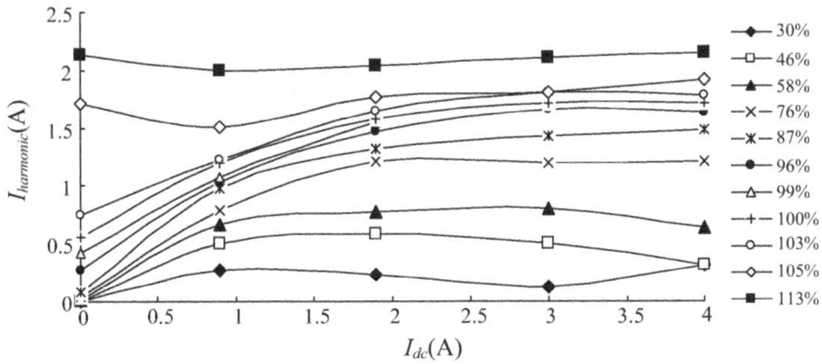

Fig. 15.91 Fifth harmonic component under different DC bias conditions

Table 15.13 THD for single-phase 500 kV autotransformer

U_n (%)	0 A (%)	0.9 A (%)	1.9 A (%)	3 A (%)	4 A (%)
30	20	101	93	86	81
46	22	126	112	103	96
58	23	142	121	112	104
76	29	159	141	124	118
87	39	175	147	131	124
96	78	172	153	137	128
99	102	168	155		129
100	115	153	153	138	131
103	122	149	152	140	133
105	107	121	138	139	
113	85		104	121	123

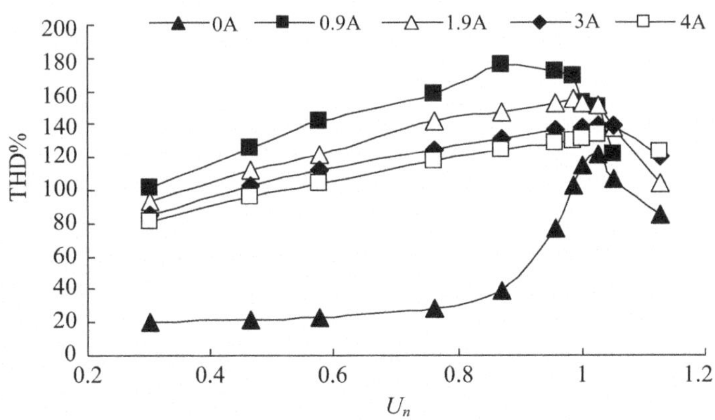

Fig. 15.92 Variation of THD with applied voltage under different DC bias conditions

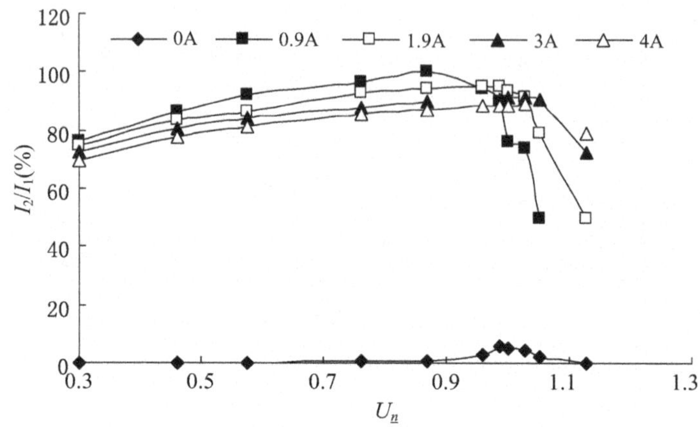

Fig. 15.93 Variation of I_2/I_1 with applied voltage under different DC bias conditions

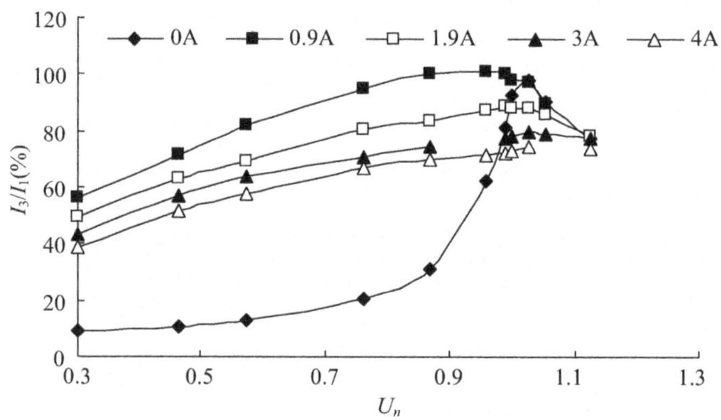

Fig. 15.94 Variation of I_3/I_1 with applied voltage under different DC bias conditions

It can be seen that THD, I_2/I_1, and I_3/I_1 are largest from $U = 80\%U_n$ to $U = 100\%U_n$. When $I_{dc} = 0.9$ A (the ratio of direct current to RMS values of no-load current is 108.4%), THD reaches up to 175%; in addition, I_2/I_1 and I_3/I_1 reach up to 100%.

Each harmonic component under DC bias condition is compared with that when $I_{dc} = 0$, as shown in Figs. 15.95, 15.96, 15.97, and 15.98 (k is the ratio of each harmonic component in RMS value in the figure). The results under DC bias condition with rated voltage applied are shown in Fig. 15.99.

It can be seen from the comparison that the increase in the second harmonic is much larger than that of the fundamental and third harmonics; the second harmonic increases by 131% when $I_{dc} = 4$ A at rated voltage.

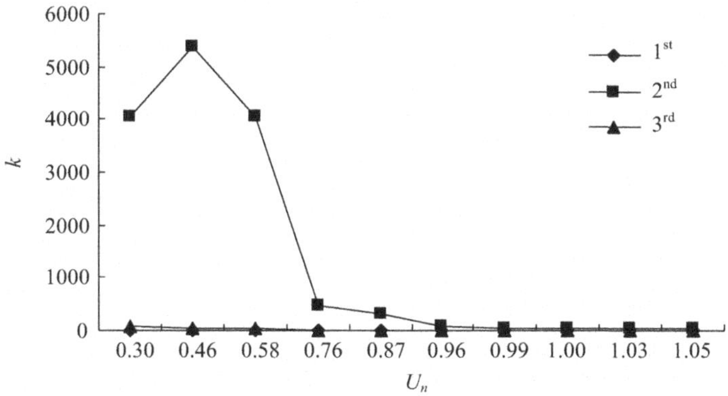

Fig. 15.95 Comparison of RMS values of fundamental, second and third harmonics at $I_{dc} = 0.9$ A

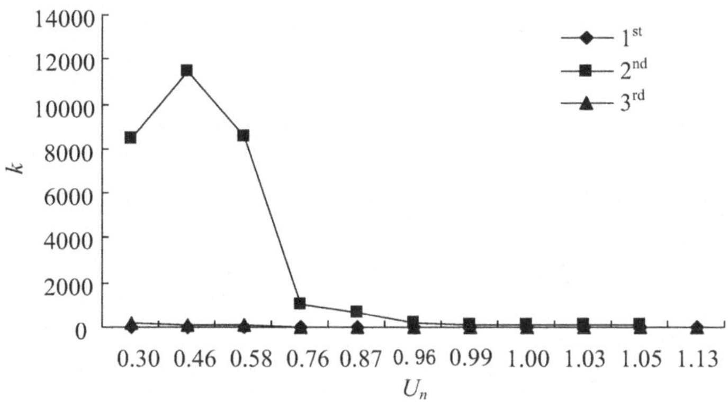

Fig. 15.96 Comparison of RMS values of fundamental, second and third harmonics at $I_{dc} = 1.9$ A

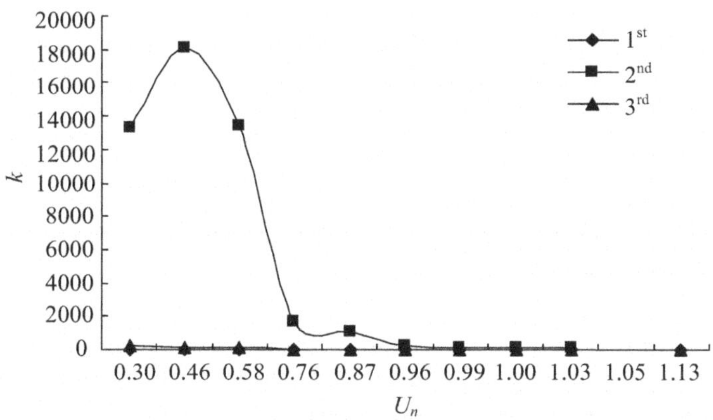

Fig. 15.97 Comparison of RMS values of fundamental, second and third harmonics at $I_{dc} = 3$ A

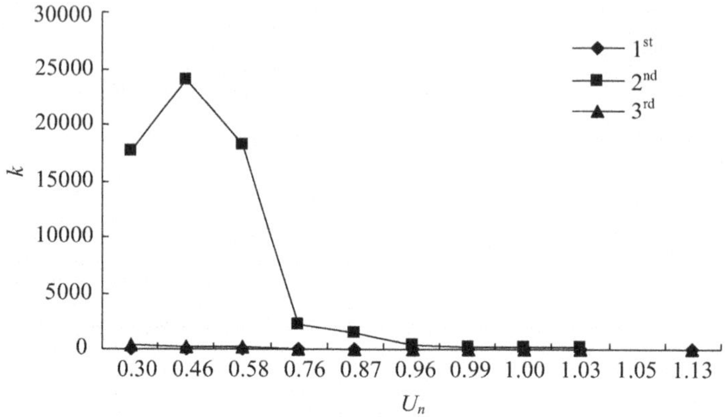

Fig. 15.98 Comparison of RMS values of fundamental, second and third harmonics at $I_{dc} = 4$ A

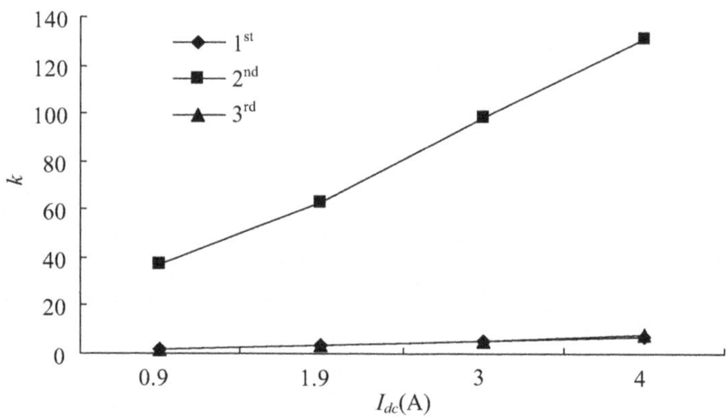

Fig. 15.99 Variation of 1st–3rd harmonic components with different DC bias conditions $U = U_n$

Harmonic analysis of exciting current is carried out in detail, which is of great significance to the power company to take measures to monitor and suppress the DC.

15.5.3 Measurement of Sound Level Under DC Bias Condition

The sound level of transformer is measured on three points which are 0.3 m away from the transformer body, and the height is 1/3 the height of the transformer body. The distribution of measuring points is shown in Fig. 15.100.

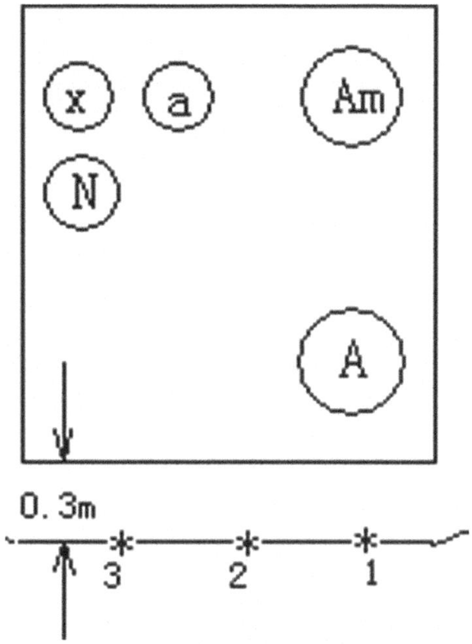

Fig. 15.100 Arrangement of measuring points for sound-level measurement

The variation of sound with DC under different working flux densities of transformer is shown in Fig. 15.101.

Figure 15.101 shows the sound level increases with the increase of DC, and to further understand the noise level of the large power transformer under DC bias, the histogram regarding the increment of sound level under different DC bias conditions relative to sound level when I_{dc} = 0 A is calculated at each flux density, as shown in Fig. 15.102. It can be seen that the sound increment decreases with the increase of DC although the sound level shows an increasing trend.

The rated flux density of the transformer is 1.73T, and the no-load current of the transformer is 0.83 A. The linear curves to fit sound level under different DC bias conditions are shown in Fig. 15.103, of which Fig. 15.103a, b corresponds to B = 1.73T and B = 1.92T, respectively.

As B = 1.73T and $0 \leq I_{dc} \leq 100\% \, I_o$, sound level can be given as follows:

$$Y = 8.11X + 73.7 \tag{15.8}$$

As B = 1.73T and $100\% \, I_o \leq I_{dc} \leq 480\% \, I_o$, sound level can be given as follows:

$$Y = 2.55X + 78.9 \tag{15.9}$$

As B = 1.92T and $0 \leq I_{dc} \leq 480\% \, I_o$, sound level can be given as follows:

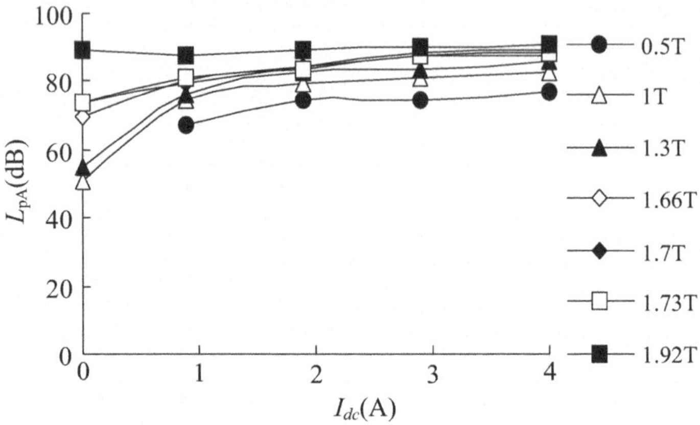

Fig. 15.101 Sound level under different DC bias conditions

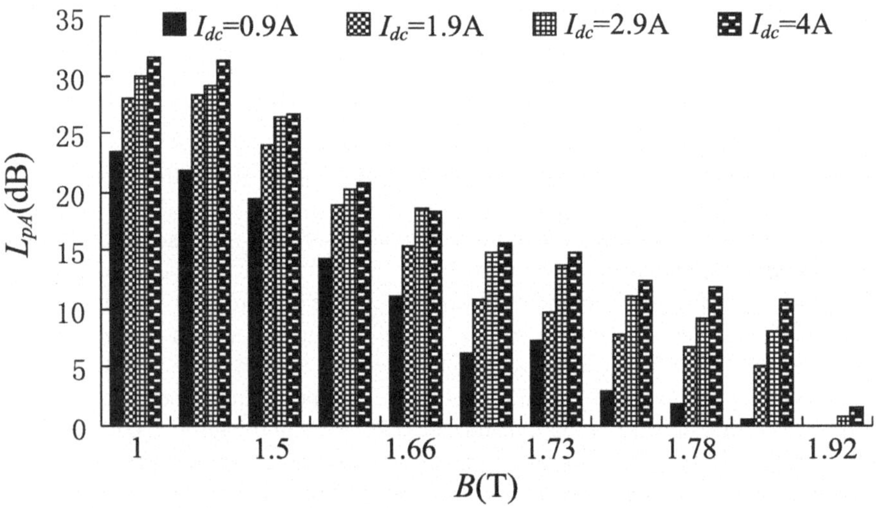

Fig. 15.102 Increment of sound level under DC bias conditions

$$Y = 0.5X + 88.3 \tag{15.10}$$

where Y is sound level in dB(A) and X is DC-biased current in A.

The actual maximum value of measured sound level under DC bias condition is about 90 dB(A), which is 17 dB(A) higher than that measured in rated condition with I_{dc} = 0 A. The sound level can be predicted by Eqs. (15.8) and (15.9) for the transformer subjected to DC bias in practical operation.

In conclusion, the sound level of single-phase three-limb transformer increases with the increment of DC under no-load condition. When the DC reaches some

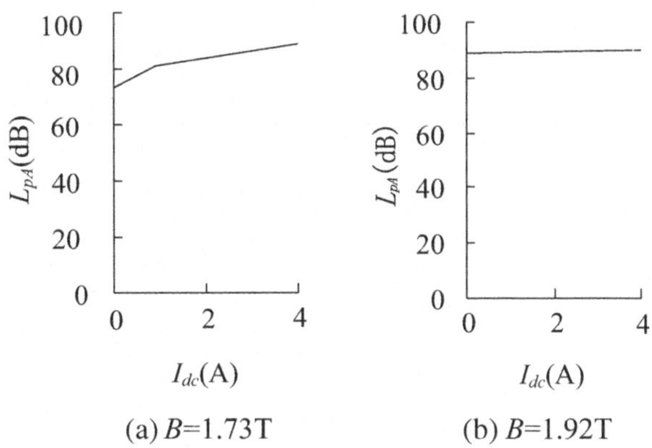

Fig. 15.103 Simulated acoustic level under different DC bias conditions

value high enough, the sound level becomes saturated. Furthermore, the maximum sound level under DC bias condition can be one of the criteria for the ability of the transformer to withstand DC bias.

15.6 On the Ability to Withstand DC Bias for Power Transformers

In general, the working flux density for a transformer depends on the technical specifications of the product, and the ability to withstand direct current is related to the working flux density of the transformer directly. The ΔB can be obtained from the $\Delta \Phi$ as shown in Fig. 15.17 based on the cross-sectional area of the core, thus based on DC flux density ΔB under each working condition. A curve reflecting variation of flux density (ΔB) under DC bias condition (I_{dc} = 3.2 A, i.e., H_{dc} = 126 A/m) with the AC flux density can be plotted as shown in Fig. 15.104.

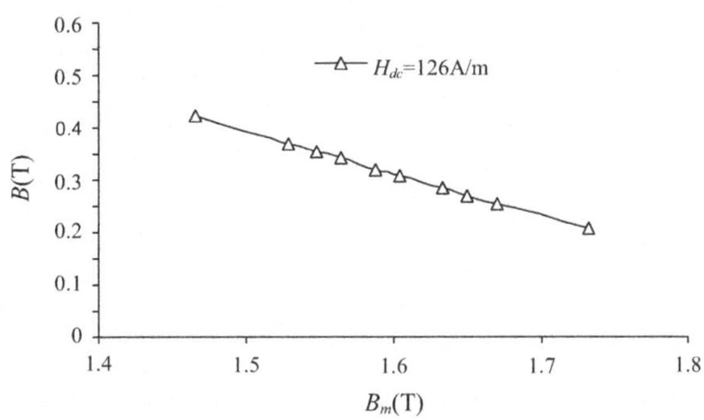

Fig. 15.104 Variation of DC flux density (ΔB) with B_m under DC bias condition as I_{dc} = 3.2 A

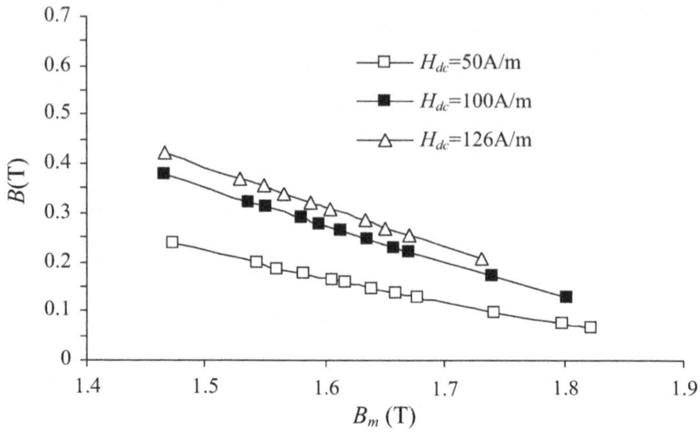

Fig. 15.105 Variation of DC flux density (ΔB) with B_m under DC bias condition as I_{dc} = 1.26 A, I_{dc} = 2.53 A, and I_{dc} = 3.2 A

Similarly, curves corresponding to different DC bias conditions (I_{dc} = 1.26 A, H_{dc} = 50 A/m and I_{dc} = 2.53 A, H_{dc} = 100 A/m) can be obtained, as shown in Fig. 15.105.

Furthermore, more curves corresponding to H_{dc} = 200, 300, 350, 400 A/m can be obtained by applying larger DC. Figure 15.106 shows the curves in the range from B_m = 1.5T to B_m = 1.8T.

The relationship between the magnitude of alternating flux density B_m, the DC flux density ΔB, and the DC biasing magnetic field strength H_{dc} (or I_{dc}) can be seen

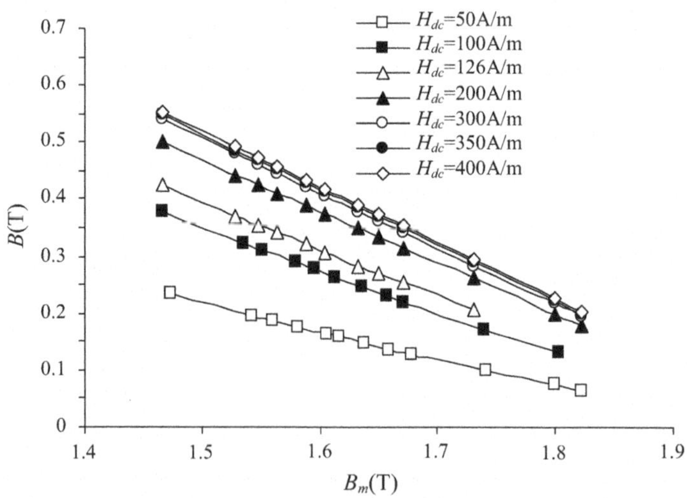

Fig. 15.106 Variation of flux density (ΔB) with B_m under DC bias condition

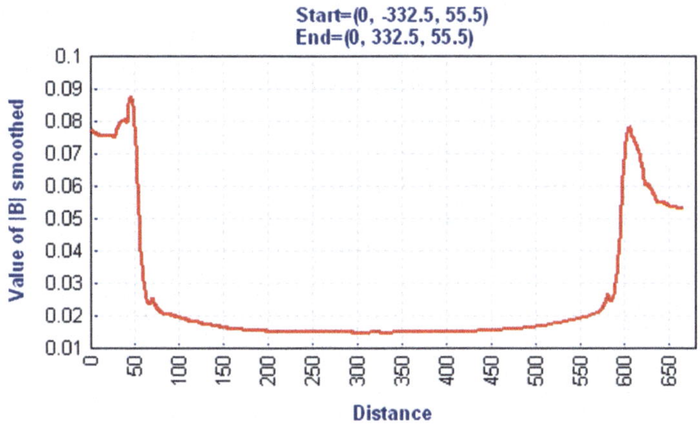

Fig. 15.107 Flux densities (along the line 0, 332.5, 55.5, 0, −332.5, 55.5) in the tie plate of transformer model under DC bias condition (H_{dc} = 320 A/m)

from the curves in Figs. 15.104, 15.105, and 15.106. Based on the curves, the DC biasing magnetic field strength H_{dc} (or direct current I_{dc}) can be obtained if B_m and ΔB are known.

The transformer model is solved and analyzed under DC bias condition with H_{dc} = 320 A/m and B_m = 1.7T. The flux density distribution in the tie plate is shown in Fig. 15.107.

The maximum flux densities in the tie plate under a different DC biasing magnetic field strength H_{dc} (or I_{dc}) are obtained from numerical calculation and shown in Table 15.14.

The tie plate can work reliably at B_m = 0.0878T under normal cooling conditions. If the flux density allowable for tie plate is taken as a criterion, the allowable ΔB for the transformer under DC bias conditions with different alternating flux densities applied can be obtained, and then an allowable boundary can be obtained by connecting these points, as shown in Fig. 15.108. If the temperature rise of the tie plate is taken as the criterion, the boundary of thermal criterion can be obtained too.

The transformer cannot operate in the area beyond the threshold curve for a long time, which can be used to evaluate the transformer's ability to withstand DC bias, based on the relationship of "$\Delta B - B_m - H_{dc}$ (I_{dc})". If the rated working flux density of the transformer is known, the corresponding ΔB can be obtained by the

Table 15.14 Maximum flux densities in the tie plate of transformer model under DC bias condition with different H_{dc} applied (or I_{dc})

H_{dc} (A/m)	0	50	100	126	250	320
Tie plate B_m (T)	0.0058	0.0102	0.0153	0.0176	0.0538	0.0878

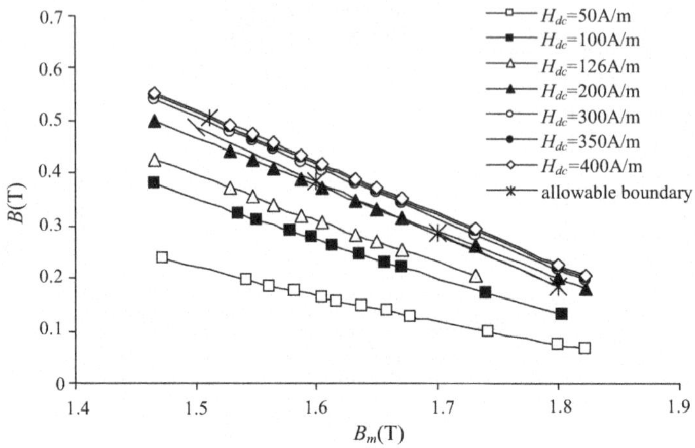

Fig. 15.108 Allowable threshold for transformer to withstand DC bias

above threshold curve, and ΔB is associated with the corresponding H_{dc}; therefore, the relationship of $B_m - H_{dc}$ can be obtained.

Apparently, at the same DC biasing magnetic field strength, the smaller the allowable DC, the higher the working flux density and vice versa. At the same rated alternating flux density, the greater the DC magnetic field strength, the greater the corresponding DC. Hence, the DC is directly proportional to the DC biasing magnetic field strength and inversely proportional to the working alternating flux density, which depends primarily on ΔB. For transformers operating with specific alternating flux density, ΔB will limit the magnitude of the DC. The DC that a transformer can withstand can be preliminarily determined by the number of coil turns and the length of the magnetic circuit if the allowable maximum H_{dc} of the transformer is known.

For the transformer with specific material, structure, and flux density, the H_{dc} can be determined by the performance of the transformer. In addition to the temperature rise of the tie plate, sound level and loss of transformer body or the harmonic increment of excitation current under DC bias condition can be regarded as a criterion of limiting DC applied on the transformer. After the DC is determined by one criterion, the other technical index including but not limited to what was mentioned above should be used for the recalculation of the DC. After correcting and confirming, the DC can be finally determined.

Based on the experimental results of model and product, it can be concluded that for a single-phase transformer the H_{dc} can reach 200 A/m if the rated flux density is 1.7, while the H_{dc} can reach 150 A/m if the rated flux density is 1.74T. It is noticed that the transformer uses magnetic shielding to prevent overheating and the transformer core is made of 30RGH120 with mitered joint gaps. The DC that the transformer can withstand is calculated by Eq. (15.1).

15.7 Summary and Outlook

The main characteristics of electromagnetic field problems of large transformers under DC bias condition include hybrid AC/DC excitation, unknown exciting current, strong nonlinearity of materials and time asymmetry of field quantity, thermal field, vibration, and noise coupling, which can be regarded as a comprehensively complex problem. In this chapter, based on the typical transformer models for DC bias test, large-scale finite element analysis (e.g., the number of tetrahedron elements: more than 700,000; matrix nonzero entries: over 8 million) is carried out to investigate the electromagnetic behaviors of the transformer. A large number of experimental studies on a transformer model and practical product are also completed. The main works are as follows:

- Based on DC bias test on a transformer model, the average Φ–I and B–H curves of transformer cores, corresponding to different DC bias levels, are systematically established to reflect DC biasing characteristics of a transformer; meanwhile, the transformer loss ($B_m - W$) curves corresponding to different DC bias levels are also obtained. The differences between the magnetic properties of silicon steel materials (single sheet) and cores (lamination) under no-load condition are investigated. The results show that the actual Φ–I curve of transformer core should be used to determine the exciting current under DC bias condition. The application of curves mentioned and described above makes it possible to calculate DC biasing magnetic field of large-scale power transformers.
- The program to determine the exciting current is developed and verified by comparing the simulated results with those measured. The program can be used to calculate the exciting current of a transformer under different DC bias conditions based on the provided B–H curve of the transformer core.
- Simulations and tests are carried out on the 50 kVA single-phase transformer model to compute the magnetic field and loss under no-load DC bias as well as on-load DC bias conditions. When I_{dc} = 3.2 A, the calculated results show that the no-load loss increases by 9.15% and the load loss increases by 5.72%. The test results show that the no-load loss increases by 11.6% and the load loss increases by 2.5%.
  It can be seen that the influence of DC bias on no-load operation is greater than that on load. The deviation between the calculated loss and the measured one of the single-phase transformer model under no-load condition is less than 5%, which can meet the engineering requirements of accuracy.
- The harmonic analysis of the exciting current of the 500 kV autotransformer shows that when no DC is injected, there are mainly odd harmonics increasing with the increment of applied voltage. When DC is injected into the transformer, there are even harmonics in addition to odd harmonics, and all harmonic components increase with the increment of the applied voltage. When the applied voltage is equal to or greater than the rated voltage, the second and fourth harmonics increase significantly. By quantitative analysis, it can be seen that THD, I_2/I_1, and I_3/I_1 reach their respective maximum values from U = 80%

U_n to $U = 100\%$ U_n. When $I_{dc} = 0.9$ A (the ratio of direct current to RMS values of no-load exciting current is 108.4%), THD reaches up to 175%, and the second harmonic I_2 and the third harmonic I_3 reach up to 100%. It can be found that the increment of the second harmonic component is much larger than that of the fundamental and third harmonic, based on the comparison of harmonic component under different DC bias conditions with those without DC applied. When $I_{dc} = 4$ A at rated voltage, the second harmonic increases by 131%.

The harmonic analysis of exciting current shows that with the increase of DC, all harmonic components increase. Based on the ratio of high-order harmonic component to the fundamental component under various DC bias conditions, it can be seen that the increment of second, fourth, and sixth harmonics is larger than that for the third, fifth, and seventh harmonics.

- It can be seen from the experimental results of the 500 kV autotransformer that when DC up to 4 A (482% I_{0RMS}/256% I_{0PEAK}) is injected into the transformer, the no-load loss is 1.065 times of that without DC bias. The sound level is about 90 dB(A), which is 17 dB(A) higher than that measured underrated condition. That indicates the transformer can withstand DC up to 4 A.
- The relation curve between DC flux density and AC working flux density under different DC biasing magnetic field strengths is established based on the DC flux density (ΔB), and a method to calculate the DC that the transformer with a certain material, structure, and rated flux density can withstand is proposed. Based on the $\Delta B - B_m - H_{dc}$ (I_{dc}) curve, the performance parameters of transformer, such as the temperature rise of the tie plate, sound level and loss of transformer body, and the allowable harmonic increment of excitation current under DC bias, can be used to determine the allowable H_{dc} for the transformer, and then the corresponding DC that the transformer can withstand can be obtained by using the equation $I_{dc} = H_{dc} * L/N$.

On the basis of the above principles, if a single-phase transformer uses magnetic shielding to prevent overheating and the core is made of 30RGH120 with mitered joint gaps, it is proposed that the H_{dc} is 200 A/m when the rated flux density is 1.7T, and the H_{dc} is 150 A/m when the rated flux density is 1.74T.

The DC bias phenomenon is an abnormal operation state of power transformers, and it involves complicated electromagnetic field problems. The works completed in this chapter provide only a concise engineering idea and a route map. If an accurate solution to the DC-biased problem is required, several aspects below need to be concerned:

- The magnetic and loss characteristics of a transformer core under non-sinusoidal excitations (such as DC-biased excitation and harmonic excitation), including *B–H* magnetization curve, loss curve, excitation power curve, and hysteresis loops, should be comprehensively analyzed and used for the investigation of DC-biased power transformers. A database of material characteristics under non-sinusoidal excitation should be established to lay a foundation for analysis of electromagnetic properties of transformer cores in engineering.

- The magnetic hysteresis loops and other factors should be considered to accurately analyze and obtain the magnetization characteristic curve of a transformer core under DC bias condition, so as to replace the simplified characteristics curve adopted herein. A mature mathematical model of hysteresis should be explored and established to calculate the exciting current in the transformer under DC bias condition, combined with the characteristics of magnetic and electric circuits.
- On the basis of the work mentioned above, effective method to compute loss of transformer core under bias condition should be presented to analyze various characteristics of the DC-biased transformer core. The loss distribution in transformer components under DC bias condition should be calculated by using the finite element method in combination with the loss characteristics of material corresponding to each component under bias condition. Furthermore, the local overheating of transformer components under DC bias condition should be of concern and investigated.
- The DC bias test on transformer product should be carried out actively as long as experimental conditions permit.

Acknowledgements This work was supported by China Southern Power Grid Co., Ltd. (CSG), the National Natural Science Foundation of China (NSFC) (no.50677016), and the Hebei Natural Science Foundation (no.E2006000772). In this chapter, the numerical simulations were completed by Mansheng Guo and Junjie Zhang, and the experiments were completed by Zhongxiang Li, Lanrong Liu, Deliang Zhang, and other colleagues. The author thanks the relevant leaders of Baobian Electric, including Dongsheng Liu, Qifan Hu, Zhongji Gao, and Cuiping Jiao, for their strong support.

Appendix: Magnetic Property Data Under DC Bias Conditions

In addition to showing the corresponding curves, some important data is listed in tabular form for ease of access.

See Tables 15.15, 15.16, 15.17, and 15.18.

Table 15.15 No-load *B–H* curve data of DC-biased single-phase transformer model

B_+/T	H_+/(A/m)	B_-/T	H_-/(A/m)
0.0	0.0	−0.1079	−3.6911
0.1079	3.6911	−0.1801	−5.2681
0.1801	5.2681	−0.2521	−6.2716
0.2512	6.2716	−0.3187	−7.1328
0.3187	7.1328	−0.3943	−8.3338
0.3943	8.3338	−0.4620	−8.9383
0.4620	8.9383	−0.5706	−11.2770
0.5706	11.2770	−0.7116	−14.0840
0.7116	14.0840	−0.8872	−20.6698
0.8872	20.6698	−1.0644	−38.3532
1.0644	38.3532	−1.2406	−72.7324
1.2406	72.7324	−1.4088	−124.4521
1.4088	124.4521	−1.4807	−158.3319
1.4807	158.3319	−1.5449	−195.5279
1.5449	195.5279	−1.5535	−198.9136
1.5535	198.9136	−1.5883	−221.9739
1.5883	221.9739	−1.6025	−239.3632
1.6025	239.3632	−1.6223	−251.4054
1.6223	251.4054	−1.6396	−267.5692
1.6396	267.5692	−1.6617	−289.4368
1.6617	289.4368	−1.6757	−305.2442
1.6757	305.2442	−1.7456	−403.6049
1.7456	403.6049	−1.8100	−538.5521
1.8100	538.5521	−1.8384	−624.0049
1.8384	624.0049	−1.8657	−727.6194
1.8657	727.6194	−1.8903	−842.8227
1.8903	842.8227	−1.9163	−1007.6816
1.9163	1007.6816	−1.9394	−1193.5334
1.9394	1193.5334	−1.9603	−1391.1329
1.9603	1391.1329	−1.9804	−1687.1664
1.9804	1687.1664	−1.9945	−2028.3045
1.9945	2028.3045		

Table 15.16 Data sheet of magnetization curve of DC-biased single-phase transformer model when I_{dc} = 3.2 A

Φ (mWb)	I (A)	Φ (mWb)	I (A)
−0.1692	3.2	−10.9164	1.895
10.9164	4.950	−17.9880	1.275
17.9880	6.066	−25.3338	0.746
25.3338	7.275	−32.4052	0.353
32.4052	8.464	−39.2227	0.099
39.2227	9.605	−46.7458	−0.039
46.7458	10.786	−56.7633	−0.109
56.7633	12.082	−70.4987	−0.163
70.4987	12.600	−87.5127	−0.208
87.5127	15.050	−104.9878	−0.244
104.9878	16.689	−121.6631	−0.327
121.6631	17.900	−139.4383	−0.633
139.4383	19.300	−145.9496	−0.978
145.9496	20.118	−152.1998	−1.499
152.1998	20.900	−154.0689	−1.680
154.0689	21.194	−155.7183	−1.878
155.7183	21.569	−158.0739	−2.190
158.0739	22.000	−159.7061	−2.423
159.7061	22.400	−162.5095	−2.880
162.5095	22.950	−164.1845	−3.150
164.1845	23.450	−166.2543	−3.560
166.2543	23.969	−172.2788	−5.070
172.2788	25.950		

Table 15.17 Data sheet of different bias current losses of DC-biased single-phase transformer model

I_{dc} = 1.26 A		I_{dc} = 2.53 A		I_{dc} = 3.2 A	
B(T)	P(W/kg)	B(T)	P(W/kg)	B(T)	P(W/kg)
0.1146	0.013	0.1191	0.020	0.1137	0.022
0.1889	0.034	0.1910	0.047	0.1873	0.053
0.2647	0.061	0.2665	0.083	0.2638	0.093
0.3330	0.091	0.3360	0.121	0.3374	0.139
0.4132	0.125	0.4116	0.167	0.4084	0.187
0.4784	0.156	0.4838	0.208	0.4867	0.240
0.5953	0.211	0.5896	0.277	0.5910	0.312
0.7334	0.284	0.7359	0.368	0.7340	0.408
0.9208	0.387	0.9126	0.484	0.9111	0.529
1.1015	0.507	1.0928	0.614	1.0931	0.655
1.2820	0.650	1.2735	0.752	1.2667	0.798
1.4613	0.834	1.4494	0.923	1.4517	0.973
1.5269	0.922	1.5199	1.014	1.5195	1.044
1.5986	1.017	1.5916	1.103	1.5846	1.132

(continued)

Table 15.17 (continued)

I_{dc} = 1.26 A		I_{dc} = 2.53 A		I_{dc} = 3.2 A	
B(T)	*P*(W/kg)	*B*(T)	*P*(W/kg)	*B*(T)	*P*(W/kg)
1.6173	1.052	1.6073	1.126	1.6040	1.155
1.6388	1.100	1.6368	1.153	1.6212	1.182
1.6634	1.122	1.6534	1.193	1.6457	1.217
1.6747	1.155	1.6724	1.218	1.6627	1.251
1.6974	1.197	1.6948	1.252	1.6919	1.286
1.7185	1.246	1.7169	1.289	1.7094	1.313
1.7387	1.289	1.7314	1.336	1.7309	1.353
1.8047	1.464	1.8033	1.479	1.7936	1.485

Table 15.18 Data sheet of non-bias current losses of DC-biased single-phase transformer model

B (T)	*P* (W/kg)	*B* (T)	*P* (W/kg)
0.0000	0.0000	1.6459	1.0454
0.1118	0.0054	1.6634	1.0790
0.1866	0.0142	1.6812	1.1209
0.2603	0.0264	1.6991	1.1653
0.3303	0.0424	1.7219	1.1995
0.4086	0.0630	1.7365	1.2489
0.4788	0.0852	1.8089	1.4266
0.5912	0.1280	1.8757	1.6459
0.7374	0.1938	1.9051	1.7612
0.9194	0.3002	1.9334	1.8905
1.1030	0.4306	1.9588	2.0172
1.2856	0.5892	1.9858	2.1555
1.4599	0.7844	2.0097	2.2969
1.5344	0.8785	2.0314	2.4398
1.6009	0.9769	2.0523	2.5742
1.6099	1.0063	2.0668	2.7164

References

1. P. R. Price, "Geomagnetically induced current effects on transformers," *IEEE Trans. on Power delivery,* vol.17, no.4, pp. 1002–1008,2002.
2. C. Li, "Influence of DC Monopole Operation on AC Transformer," *East China Power*, 2005, 1 (33): 36–39.
3. G. Mei, B. Xu, X. Wang, Z. Deng, and C. Liu, "Influence of DC Transmission on Transformers in AC System," *Guangdong Electric Power*, 2006,1 (1): 1–7.
4. D. Kuai, D. Wan, and Y. Zou,"Influence of DC Bias on Transformer," *China Power*, 2004, 37 (8): 41–43.
5. P. Picher, L. Bolduc, A. Dutil, and V. Q. Pham, "Study of the acceptable DC current limit in core-form power transformers," *IEEE Trans. on Power Delivery*, vo.12, no.1, pp. 257–263, 1997.

6. L. Liu and X. Xie, "Analysis of Abnormal Noise of 500 kV Main Transformer," *High Voltage Technology*. 2005, 31 (4): 85–87.
7. D. Kuai, D. Wan, and Y. Zou,"Analysis and Handling of the Impact of Geomagnetically induced Current Upon Electric Network Equipment in DC Transmission," *Power System Automation*. 2005,29 (2): 81–82.
8. X. Wang, and B. Xu, "On the Problem of Transformer Bias," *Transformer*, 1992,29 (8): 11–14.
9. Z. Ma, "Simulation of Transformer DC Magnetic Biasing," *Guangdong Electric Power*. 2004,17 (2): 5–9.
10. L. Zeng, "Impact of HVDC Ground Electrode Current on the Adjacent Power Transformers," *Electric Power Construction*, 2004,12 (25): 22–24.
11. L. Zhong, P. Lu, Z. Qiu, H. Cai,"The Influence of Current of DC Earthing Electrode on Directly Grounded Transformer," *High Voltage Technology*, 2003, 8 (29): 12–13.
12. C. Shang, "Measure to Decrease the Neutral Current of the AC Transformer in HVDC Ground-Return System", *High Voltage Technology*, 2004,30 (11): 52–54.
13. Y. Wang, and S. Liu, "The Measurement of AC Hysteresis Loop with DC Magnetic Biasing," *Magnetic Materials and Devices*, 2001,3 (32): 47–50.
14. Y. Yao, "Research on Direct-current Magnetic Bias of Large Power Transformer," A Dissertation Submitted to Shenyang University of Technology for the Doctoral Degree, 2000.11.
15. X. Li, "Study on the Transformers under DC Bias," A Dissertation Submitted to Wuhan University for the Doctoral Degree, 2006.3.
16. N.Takasu, T.Oshi, F.Miyawaki, S.Saito, and Y.Fujiwar, "An experimental analysis of DC excitation of transformers by geomagnetically induced currents," *IEEE Trans. on Power Delivery*, vol.9, no.2, pp. 1173–1179, 1994.
17. E.F.Fuchs, Y.You, and D.J.Roesler, "Modeling and simulation, and their validation of three-phase transformer with three legs under DC bias," *IEEE Trans. on Power Delivery*, vol.14, no.2, pp. 443–449, 1999.
18. L.Bolduc, M.Granger, G.Pare, J.Saintonge, and L.Brophy, "Development of DC current-blocking device for transformer neutrals," *IEEE Trans. on Power delivery*,vol.20, no.1, pp. 163–168, 2005.
19. T.Yoshida, M.Nakano, D.Miyagi, K.Fujiwara, and N.Takahashi, "Development of measuring equipment of DC-biased magnetic properties using open type single sheet tester,", *IEEE Trans. on Magn.*, vol..42, no10, pp. 2846–2848, 2006.
20. M. Guo, G. Mei, et al. "B–H Curve Based on Core and Asymmetric Magnetizing Current in DC-Biased Transformers," *Transactions of China Electrotechnical Society*, 2009,24 (5): 46–51.
21. N.Takahashi, T.Sakura and Z.Cheng, "Non-linear analysis of eddy current and hysteresis losses of 3-D stray-field loss model (Problem 21)," *IEEE Trans. on Magn.*, vol.37, no.5, pp. 3672–3675, 2001.
22. M. Guo, G. Mei,et al. "Simulation Calculation and Test Research on Harmonic Wave in Power Transformer under DC Bias," *Transformer*, 2010,47 (12): 28–31.
23. M. Guo, Z.Cheng, et al. "Three Dimensional Magnetic Field Analysis of Single-Phase Three-Limb Power Transformer under Direct Current Bias," Transformer, 2007,44 (4): 31–35.
24. Z.Cheng, M. Guo, and L. Li, "Eddy Current Loss Analysis and Validation in Electrical Engineering," (supported by National Natural Science Foundation of China), Higher Education Press, 2001.
25. K. Preis, O. Biro, G. Buchgraber, and I.Ticar, "Thermal-electromagnetic coupling in the finite-element simulation of power transformers," *IEEE Trans. on Magn.*, vol.42, no.4, pp. 999–1002, 2006.
26. O. Biro and K. Preis, "An efficient time domain method for nonlinear periodic eddy current problems," *IEEE Trans. on Magn.*, vol.42, no.4, 2006, pp. 695–698.

27. Z. Cheng, R. Hao, N. Takahashi, Q. Hu and C. Fan, "Engineering-oriented benchmarking of Problem 21 family and experimental verification," *IEEE Trans. on Magn.*, vol. 40, no.2, pp. 1394–1397, 2004.
28. Z. Cheng, N. Takahashi, S. Yang, T. Asano, Q. Hu, S. Gao, X. Ren, H. Yang, L. Liu and L. Gou, "Loss spectrum and electromagnetic Behavior of Problem 21 family," *IEEE Trans. on Magn.*, vol.42, no.4, pp. 1467–1470, 2006.
29. Z. Cheng, N. Takahashi, S. Yang, C. Fan, et al, "Eddy Current and loss analysis of multi-steel configuration and validation," *IEEE Trans. on Magn.*, vol. 43, no.4, 2007, pp. 1737–1740.
30. M. Guo, G. Mei, X. Zhang, et al, "Calculation of Losses in Single-Phase Three-Limb Power Transformer under DC-Biasing," *Transactions of China Electrotechnical Society*, 2010,25 (7): 67–71.

Chapter 16
Modeling and Validation of Thermal-Fluid Field of Transformer Winding Based on a Product-Level Heating and Cooling Model

Weige Wu and Gang Liu

Abstract The temperature rise and hotspot of transformer winding is an important index to judge the reliability of transformer and is one of the key issues concerned by the transformer manufacturers and researchers. At present, there are many simulation studies on the temperature rise and hotspots of transformer windings, but they are often limited to the algorithm study or software application and lack of experimental validation, especially the tests on a product-level platform. To this end, our laboratory has built a product-level platform for testing the winding temperature rise and hotspot in Baobian Electric. The heating and cooling model of the transformer winding's platform is mainly composed of the air core non-inductive coil, active-part insulation, heat-insulating tank, panel-type radiator, oil pump, fan, oil conduit and thermoelectric couple. Based on this model, the temperature rise and hotspot experiments under various operating conditions are carried out, and the experimental data are compared with the results of the ANSYS FLUENT. The simulation results show that the accuracy of the simulation results can be further improved by considering the influence of spacers, strips and other factors when modeling so as to better meet the engineering application requirements.

Keywords Thermal-fluid field · Heating and cooling model · Transformer winding · Modeling and simulation

W. Wu (✉)
Institute of Power Transmission and Transformation Technology, Baobian Electric, Baoding, China
e-mail: wuweige@btw.cn

G. Liu
Department of Electrical Engineering, North China Electric Power University, Baoding, China

Z. Cheng et al. (eds.), *Modeling and Application of Electromagnetic and Thermal Field in Electrical Engineering*, https://doi.org/10.1007/978-981-15-0173-9_16

16.1 Introduction

When transformer is operating, the losses are generated in the core, windings and structural components. These losses are dissipated as heat in the surrounding medium, causing the temperature rise in the transformer. The heat of transformer is transmitted to the cooling medium by means of heat conduction, convection and radiation, all of which follow their inherent laws of thermal field. In oil-immersed transformer, the insulating oil is also used as the cooling medium, which transferring the heat by oil circulation to reduce the temperature rise of each part. For example, the heat inside the winding is transferred to the surface of the winding by heat conduction; the heat on the surface of the winding is dispersed to the transformer oil by convection; then the heat in the transformer oil is also transferred to the tank wall or cooling device by convection; finally, the heat is transmitted to the surrounding air by convection or heat radiation from the tank wall or cooling device. Oil-immersed transformers are usually cooled by ONAN (Oil Natural Air Natural), ONAF (Oil Natural Air Forced), OFAF (Oil Forced Air Forced), ODAF (Oil Directed Air Forced) and OFWF (Oil Forced Water Forced). Generally, the cooling type of a transformer is determined by the working environment, capacity, shape and volume, etc.

Temperature rise is an important parameter to measure the quality of transformer design and operation reliability. The limits of the top oil temperature, average winding temperature rise and winding hotspot temperature rise in operation are stipulated in IEC 60076 standards, and the temperature rise test is defined as a transformer's type test. With the increasing capacity of a single transformer, the heating and cooling problems become more and more prominent, especially the hotspot temperature of the winding, which is the key factor restricting the safe operation and service life of a transformer. It is stipulated in IEC 60076-7:2005 *Power transformers—Part 7: Loading guide for oil-immersed power transformers* that the ageing rate of conventional oil-immersed transformers is doubled for every 6 K increase in hotspot temperature in the range of 98–140 °C. With the multi-types of cooling system, the complexity of temperature distribution and computation is increasing. Therefore, for the purpose of safe and stable operation, it is of great significance to carry out simulation and experimental research on the winding temperature distribution of large transformers [1].

In order to investigate the relationship between the oil flow and temperature distributions, the effects of winding temperature, various flow conditions, geometric structures and other factors on oil temperature, researchers all over the world have done a great deal of work [2–8], but there are few studies on experiments and simulation validations on the product-level modeling platforms. As we all know, it is unrealistic for a manufacturer to repeatedly test and study the temperature rise of a commercial product. In order to further study the temperature rise of large transformer windings, according to the testing capabilities of the laboratory,

Baobian Electric established a research model for heating and cooling of the oil-immersed transformers in stages, where the demands of instant testing and trial-and-error can be achieved.

16.2 Test Model

In an operating transformer, the hotspot temperature of the winding is closely related to the resistance loss, leakage flux distribution, oil flow distribution and heat dissipation system. That is why it is impossible to measure the temperature rise contributed by each factor in the finished commercial product. For this reason, the laboratory tries to eliminate the unnecessary interference factors when designing the experimental model of winding for the heating and cooling. The heating and cooling model of the transformer winding is mainly composed of the air core non-inductive coil, active-part insulation, heat-insulating tank, panel-type radiator, oil pump, fan, oil conduit and thermoelectric couple. In order to realize the technical route of the model from simplicity to complexity, the experimental model has the following characteristics:

(1) Oil pump and fan can be freely combined to realize three cooling modes: ONAN/ONAF/OFAF;
(2) Panel-type radiator can be moved in the vertical direction, i.e. the heating centre can be adjusted up and down;
(3) Bottom and side blowing of the fan are allowed;
(4) Oil washers are detachable.

The instruments of the experimental system consists of voltage regulator, precision power analyser, liquid turbine flowmeter, temperature recorder and other components, wherein the voltage regulator is for changing the heating power of the winding by adjusting the voltage; the precision power analyser is for reading the numerical values, such as voltage, current and power of the winding; the liquid turbine flowmeter is used to measure the oil flow of the transformer in the connecting pipe, and then the oil velocity in the connecting pipe can be calculated, and the temperature recorder is used for reading the temperature value of each measuring point. The schematic diagram and product-level platform of the experiment model are shown in Figs. 16.1 and 16.2, respectively.

The winding is composed of the paper-covered flat copper conductors, end rings, an inner cylinder, an outer cylinder, oil washers, spacers and strips, etc. It is a continuous type with two parallel conductors outgoing at the terminal [9]. Fifteen turns of copper conductor are radiated from each disc that is 30 flat copper conductors in total. The two parallel conductors are welded together in the last disc, so that the two conductors are wound in the same way with the currents flowing in opposite directions; the magnetic fields generated cancel each other out, eliminating the eddy current losses of the conductors, and the loss of the winding can be considered as resistance loss only (the power factor of the winding tested by the

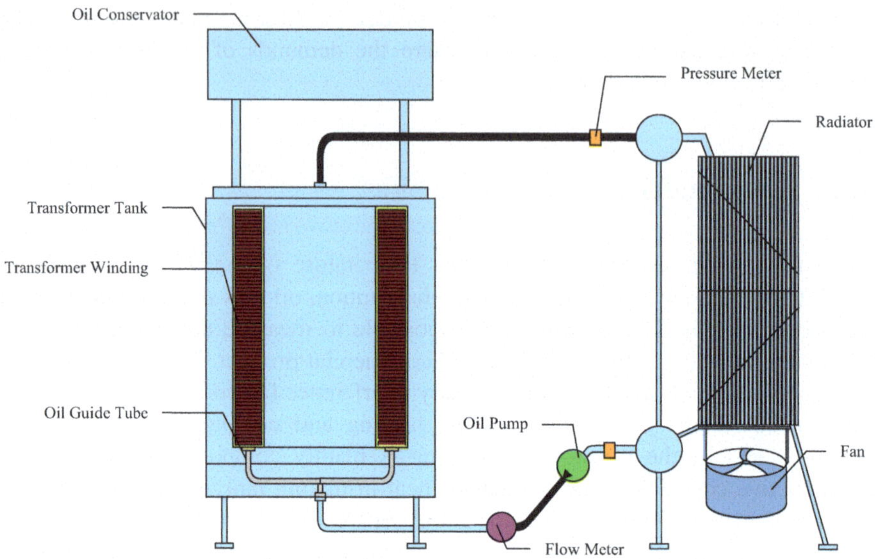

Fig. 16.1 Schematic diagram of test model

Fig. 16.2 On-site heating and cooling experiment

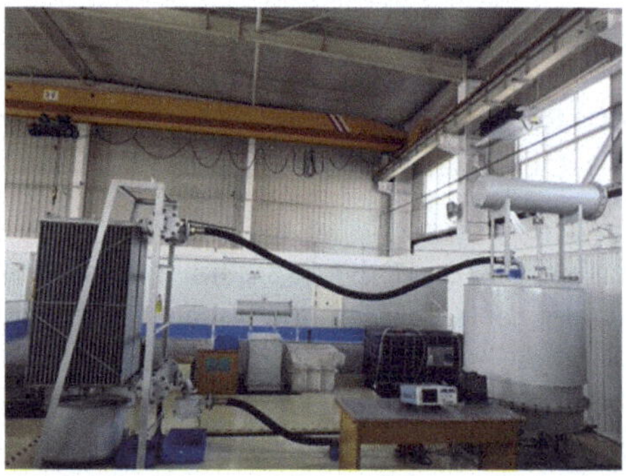

precision power tester YOKOGAWA WT3000 from Japan is 0.99982), so that the loss densities under different currents can be accurately obtained. The winding is divided into 8 passes in the height direction, each of the first 3 Passes contains 7 discs, and each of the last 5 Passes contains 9 discs, totalling 66 discs. In the winding, the thickness of the radial duct is 6 mm, and the widths of the inner and outer axial duct are 8 and 10 mm, respectively, as shown in Fig. 16.3, while an actual model of winding is shown in Fig. 16.4.

The model tank adopts the bell type, the upper part of the tank wall is formed in a two-tier structure, and the middle part of the tank wall is filled with the thermal insulation material, such as glass fibre, asbestos, rock wool and the like, and the heat-insulating material is tightly sealed to prevent the heat-insulating material from

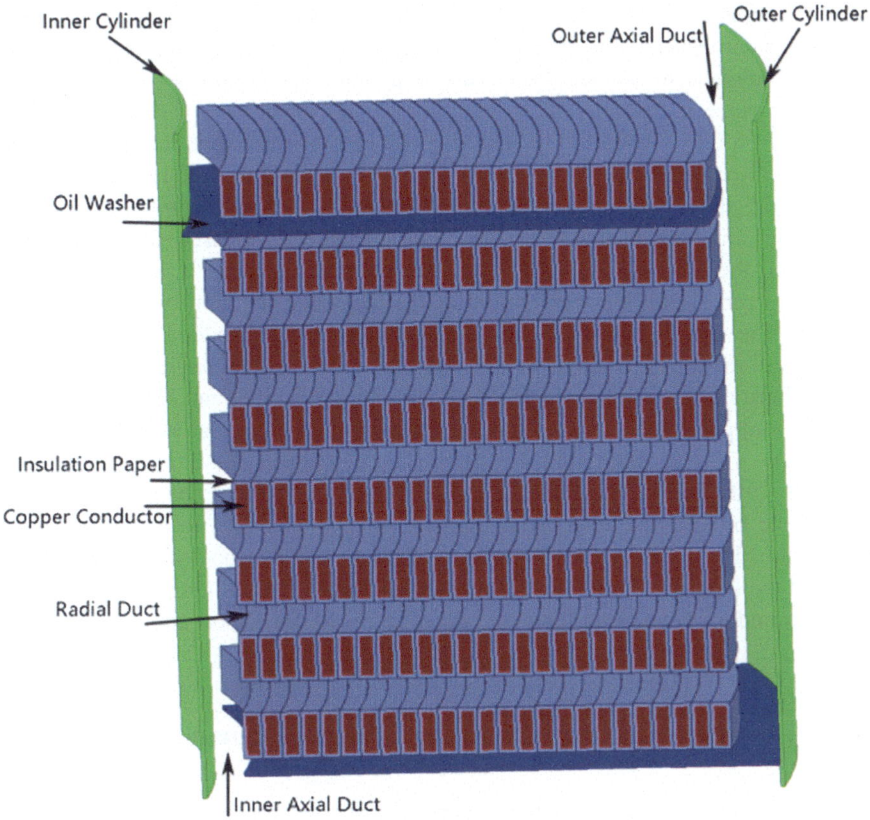

Fig. 16.3 Schematic diagram of inner structure of winding

Fig. 16.4 Experimental winding

absorbing moisture, thereby causing an increase in thermal conductivity and losing its heat-insulating function after a long time of use. The purpose of setting the insulation layer is to use the tank wall as a boundary condition of the thermal insulated surface in the simulation. The bottom part of the tank is designed with a sealed oil conduit so that all the transformer oil can be introduced into the winding.

The thermoelectric couple is commonly used as the temperature measurement element with the advantages of simple structure, easy manufacturing, wide measuring range and high accuracy and can be used to directly measure the temperature and convert the temperature signal into a thermoelectromotive force signal, which may be converted into the temperature of the measured medium by a temperature recorder. The copper-constantan thermoelectric couple embedded in this experimental platform is used to measure the temperature of the flat copper conductor and oil. The thermoelectric couples, placed on the 12th, 20th, 30th and 38th discs (in the order from upper end to lower end of the winding), measure the axial temperature distribution of the winding, and the thermoelectric couples, placed on the 1st, 4th, 7th, 10th, 13th, 16th, 19th, 22nd, 25th, 28th and 30th copper conductors of each disc (in the order from inner axial duct to outer axial duct), measure the radial temperature distribution of the winding. Further, two thermoelectric couples are placed at the inlet and outlet of the tank for measuring the oil temperatures, which are the boundary conditions for the simulation of the forced oil circulation.

16.3 Experiment Instruments and the Performance

The main instruments used in the experiment are the precision power analyser, liquid turbine flowmeter, temperature recorder, LCR tester, power supply, oil pump and so on. The experiment instruments required, as well as the specifications, are listed in Table 16.1.

16.4 Measurement Methodology

The experimental circuit of the test model is shown in Fig. 16.5, where FU is a three-phase fuse, K1 is a three-phase circuit breaker and K2 is a single-phase circuit breaker.

Five different input powers (55, 48, 40, 32, 24 kW) are applied to the test model under the experimental conditions of the OFAF, and five different flow rates (42, 35, 25, 18, 10 m^3/h) are tested under each input power, totalling 25 experimental operating conditions. The experiment procedures under different conditions are the same. For example, when the input power is 55 kW and the flow rate is 18 m^3/h, the procedures are summarized as follows: [10].

Table 16.1 Experiment instruments and specifications

Equipment	Specifications	Performance index
Precision power analyser (Japan)	YOKOGAWA WT3000	Current and voltage reading accuracy: ±0.02%; range accuracy: ±0.04%; power accuracy: ±0.06%; frequency range: DC, 0.1 Hz–1 MHz
Liquid turbine flowmeter (China)	LWGYS-80	Measurement range: 10–100 m^3/h; instrument accuracy: 0.5%; level of protection: IP65
Temperature recorder (China)	TP1000	Intrinsic display error of measurement: ±0.2%F S; 24-bit AD converter; display range: −999.99 to 1999.99; measurement resolution: 1/60,000; real-time inter-record gap: 1–999 s
LCR tester (Japan)	NF ZM2353	Intrinsic accuracy of measurement: 0.1%; display resolution: 4–1/2 digits (19,999); measurement signal level: 10mV-5 Vrms; measuring speed: 25 ms (fastest)
Power supply (China)	SOYI-33200	Rated capacity: 200 kVA; input voltage: 380 V (line voltage); output voltage: 260–280 V for single-phase gear; output frequency: 50 Hz (fixed)/60 Hz (fixed)/ 40–70 Hz (continuously adjustable)
Transformer oil pump (China)	$6BF_645$-4.5/ 1.1 V	Power: 1.1 kW; current: 3.8 A; frequency: 50 Hz; flow: 45 m^3/h; Pumping lift 4.5 m; rotating speed: 900 r/min
Panel-type radiator (China)	PC1800-22/ 460	Centre distance: 1800 mm; number of panels: 22; panel width: 460 mm; geometric area: 39.44 m^2; radiator weight: 327.8 kg
Fan (China)	CFZ-5Q6	Impeller diameter: 500 mm; air volume 6500 m^3/h; rated power: 0.37 kW; voltage: 380 V; rated current: 1.33 A; frequency: 50 Hz; total pressure: 120 Pa

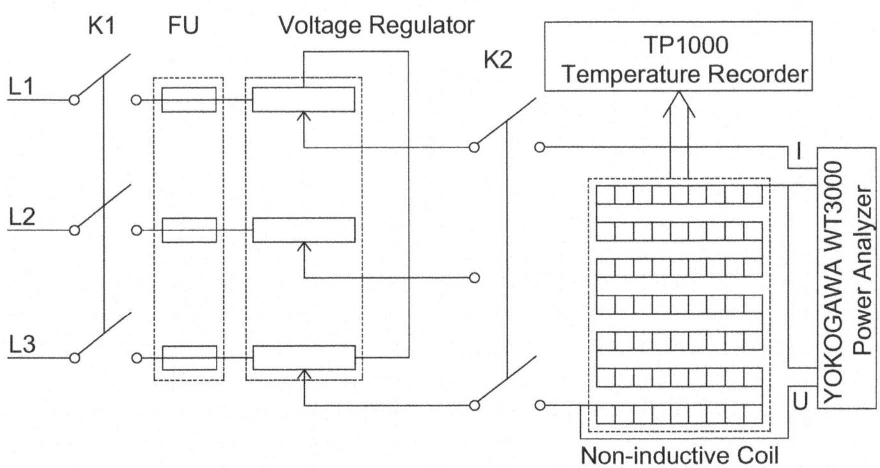

Fig. 16.5 Schematic diagram of experimental circuit

(1) Check to ensure that there is no oil leakage at each interface of the model, the circuit is connected correctly, and the thermoelectric couple is firmly installed;
(2) Close the circuit breaker K1 of the three-phase AC power supply and adjust the output voltage knob of the regulator power SOYI-33200 to the minimum. Turn on power analyser WT3000 and temperature recorder TP1000;
(3) The single-phase circuit breaker K2 is turned on and the output power of voltage regulator SOYI-33200, adjusted, to heat the winding; the oil temperature rising time can be shortened by increasing the input power and reducing the cooling of the transformer at the beginning of the experiment (using only the oil pump, for which the fan should be stopped);
(4) The voltage regulator is lowered immediately after the temperature rise (e.g. top oil temperature rise) at the monitoring site reaches 70% of the expected temperature rise, and the input power should be adjusted to the value required for the experimental condition (55 kW), where such value can be read from the power analyser WT3000; then the valve between the radiator and the oil tank is slowly adjusted in such a way that the value read from the turbine flowmeter is the one required by the experimental condition (18 m^3/h); the input power and flow rate should remain unchanged in the subsequent steps;
(5) The temperature rise is considered stable when the top oil temperature rising rate is less than 1 K/h and lasts for 3 h; the temperature values in the last hour at each measuring point of the winding, inlet and outlet should be imported into a memory device by the temperature recorder TP1000;
(6) The single-phase circuit breaker K2 is disconnected and the hot DC resistance of the winding, tested by LCR tester ZM2353, is used to calculate the average winding temperature;
(7) The power analyser WT3000, temperature recorder TP1000, oil pump and fan are turned off, and the three-phase circuit breaker K1 is disconnected.

16.5 Numerical Modeling and Simulation of Thermal-Fluid Field in Transformer Winding

As popular commercial software of the computational fluid dynamics (CFD), the ANSYS FLUENT, which is based on finite volume method (FVM), can be used to solve the fluid and heat transfer related problems. The ANSYS FLUENT can use 2-D plane, 2-D axisymmetric and 3-D models to solve the steady and unsteady flow problems, laminar and turbulent flow problems, solid and fluid coupled heat transfer problems. In the design of a power transformer, the ANSYS FLUENT can be used to calculate the temperature distribution and hotspot location of the winding disc, flow velocity and oil temperature distribution. It can analyse the influence of the parameters' change, such as the size of oil duct and different structures, on the temperature rise, so as to optimize the transformer's thermal design.

Due to the symmetry of the experiment model, a 2-D axisymmetric model is set up, and the steady-state field is solved by the ANSYS FLUENT [11]. For 2-D axisymmetric coordinate, the mass conservation and the momentum conservation laws in fluid region are expressed by Eqs. (16.1), (16.2) and (16.3), respectively.

$$\frac{\partial(\rho u)}{\partial z} + \frac{1}{r}\frac{\partial(\rho r v)}{\partial r} = 0 \tag{16.1}$$

$$\frac{\partial(\rho u u)}{\partial z} + \frac{\partial(\rho v u)}{\partial r} = F_z - \frac{\partial P}{\partial z} + \frac{\partial}{\partial z}\left(\mu\frac{\partial u}{\partial z}\right) + \frac{1}{r}\frac{\partial}{\partial r}\left(\mu r\frac{\partial u}{\partial r}\right) \tag{16.2}$$

$$\frac{\partial(\rho u v)}{\partial z} + \frac{\partial(\rho v v)}{\partial r} = F_\mathrm{r} - \frac{\partial P}{\partial r} + \frac{\partial}{\partial z}\left(\mu\frac{\partial v}{\partial z}\right) + \frac{1}{r}\frac{\partial}{\partial r}\left(\mu r\frac{\partial v}{\partial r}\right) \tag{16.3}$$

where z is the axial coordinate; r is the radial coordinate; u is the axial fluid velocity; v is the radial fluid velocity; ρ is the fluid density; F_z is the body force of the volume element in the axial direction; F_r is the body force of the volume element in the radial direction; P is the volume element pressure; and μ is the hydrodynamic viscosity. Because the temperature rise of the transformer oil and winding is a coupling problem of the fluid and thermal field, the energy equation of the thermal field is expressed by Eq. (16.4) at the 2-D axisymmetric coordinate.

$$\frac{\partial(\rho u c_\mathrm{p} T)}{\partial z} + \frac{\partial(\rho v c_\mathrm{p} T)}{\partial r} = S_\mathrm{E} + \frac{\partial}{\partial z}\left(\lambda\frac{\partial T}{\partial z}\right) + \frac{1}{r}\frac{\partial}{\partial r}\left(\lambda r\frac{\partial T}{\partial r}\right) \tag{16.4}$$

where λ is the fluid's heat transfer coefficient, T is the temperature, c_p is the specific heat capacity and S_E is the heat source. The governing equation used in the solid region is the heat conduction equation [12], which can be expressed by Eq. (16.5).

$$\frac{\partial}{\partial z}\left(k\frac{\partial T}{\partial z}\right) + \frac{1}{r}\frac{\partial}{\partial r}\left(kr\frac{\partial T}{\partial r}\right) + S_\mathrm{E} = 0 \tag{16.5}$$

where k is the solid thermal transfer coefficient.

Considering the axisymmetry of the winding and oil tank, and assuming that the distribution of the winding temperature field does not change in gradient along its circumferential direction, the axisymmetric model is adopted. The fine modeling of the flat copper conductor and turn insulation is built, and the oil washer is built in the middle position of the oil duct according to the actual model [13, 14]. The 2-D axisymmetric geometric model of the winding and the local enlarged drawing of the Passes 1 and 2 are shown in Fig. 16.6.

The entire winding is divided into 8 Passes (Pass 1–Pass 8), of which each Pass in the first 3 Passes (Pass 1–Pass 3) contains 7 discs, and each Pass in the last 5 Passes (Pass 4–Pass 8) contains 9 discs, totalling 66 discs; the rightmost disc is disc No. 1, and the leftmost disc is disc No. 66. Each disc is composed of 30 flat copper conductors in radial direction, of which the conductor near the inner axial duct is

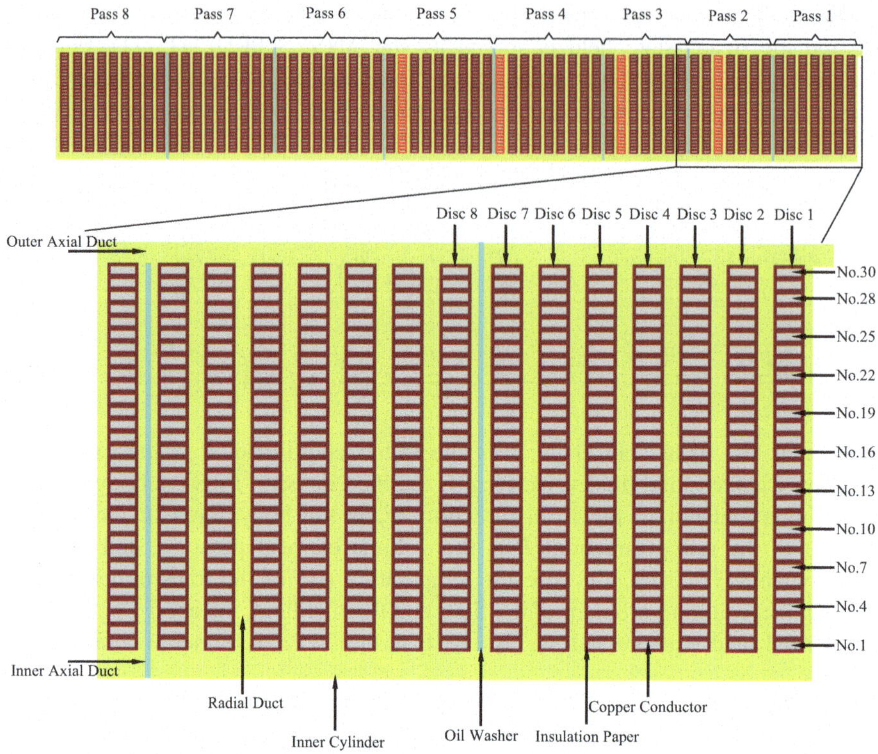

Fig. 16.6 2-D axisymmetric model of the winding

conductor No. 1, while the one near the outer axial duct is conductor No. 30; the size of each flat copper conductor is 3 mm × 10 mm, and the turn insulation thickness is 1.45 mm. The thickness of oil duct between adjacent discs is 6 mm, and the widths of vertical oil duct in internal and external axis are 8 mm and 10 mm, respectively. The thermoelectric couples are placed on the discs of No. 12, No. 20, No. 30 and No. 38 for measuring the axial temperature distribution, and for each disc, the thermoelectric couples are placed on the copper conductors of No. 1, No. 4, No. 7, No. 10, No. 13, No. 16, No. 19, No. 22, No. 25, No. 28 and No. 30 for measuring the radial temperature distribution.

The quality of mesh is directly related to the computational efficiency and results of ANSYS FLUENT. The area near the boundary layer between the oil and winding has great gradient change in velocity and temperature, and it is the focus of the mesh generation. The geometric model is meshed by the ANSYS MESHING in a quadrilateral-dominated manner as a whole. The inflation layer is divided into three layers at the boundary of the oil and winding. The first layer is 0.2 mm and the thickness increment rate is 1.2. Focus on checking the grid whose Skewness value is less than 0.8, and ensure the grid can be correctly calculated after importing the ANSYS FLUENT.

The natural circulation effect of the transformer oil under the gravity is considered in the simulation of the forced circulation. There are two methods to study the natural oil circulation in ANSYS FLUENT, one is the defined density method and the other is the Boussinesq model method. For the Boussinesq model, it is assumed that the oil density is constant and only changes in the body force of the momentum equation along the direction of gravity; thus, for the oil, whose density changes little change with temperature, it has a better convergence for the oil. The physical property parameters of oil are input according to Eqs. (16.6)–(16.9).

$$k_{\text{oil}} = 0.1509 - 7.101 \times 10^{-5} T_{\text{oil}} \left(\text{W m}^{-1}\ \text{K}^{-1}\right) \tag{16.6}$$

$$\rho_{\text{oil}} = 1098.72 - 0.712 T_{\text{oil}} \left(\text{kg m}^{-3}\right) \tag{16.7}$$

$$\mu_{\text{oil}} = 0.08467 - 4.0 \times 10^{-4} T_{\text{oil}} + 5.0 \times 10^{-7} T_{\text{oil}}^2\ (\text{Pa s}) \tag{16.8}$$

$$c_{\text{oil}} = 807.163 + 3.58 T_{\text{oil}} \left(\text{W kg}^{-1}\ \text{K}^{-1}\right) \tag{16.9}$$

The heat transfer in the winding is mainly carried out on the radial oil duct, which is composed of the two adjacent discs and spacers, similar to the structure between two panels, and can be treated according to the fluid and heat transfer problems between parallel panels. Because the velocity of oil flow at the inlet is low and the Reynolds number is small, the laminar flow model is adopted in the simulation, and the pressure-based coupling solver is selected to accelerate the convergence [15].

The inlet can be a velocity or a mass flow boundary condition, either of which is acceptable, due to the simulation model is regarded as incompressible flow (fluid density is independent of pressure and only related to temperature), and there is no difference in choosing the inlet of velocity or mass flow. The inlet velocity can be calculated from the values read from the turbine flowmeter, and the inlet oil temperature can be read from the temperature recorder. For the outlet, the pressure outlet boundary condition is adopted. For the same reason that the oil is incompressible, the relative pressure drop value of the whole fluid field is constant, and the total pressure set at the pressure outlet has only reference values. All wall surfaces of the fluid are set to 'Wall', and considering that the heat preservation effect of the tank wall surface is better and the heat exchange at the inlet and outlet is far greater than the heat dissipation of the wall surface, they are simplified to the adiabatic surface. As a heating element, the winding is provided as an internal heat source. Since the winding of the experiment model is composed of the non-inductive coils, the loss is only the winding's ohmic loss, and the unit heat source is the total power of the input winding divided by the volume of all the discs.

16.6 Results and Discussions

Five input powers (55, 48, 40, 32, 24 kW) were applied to the experiment model under the forced oil circulation conditions, and five different flow rates (42, 35, 25, 18, 10 m^3/h) were applied to the experiment model under each input power, a total of 25 forced oil circulation tests were carried out. The results of the test and simulation under the condition of input power 55 kW and flow rate 18 m^3/h are analysed and compared in the following.

The results of simulation and experimental test are given in Table 16.2. The graphical method is used to analyse and research in the following contents in order to find out the distribution conveniently.

The overall oil velocity distribution of the winding and the local distributions of the oil flow in Passes 1, 2 and 3 are shown in Fig. 16.7. As can be seen from Fig. 16.7, the velocity distribution of the oil in the whole winding is non-uniform. The velocity of the inner axial duct is greater than that of the outer axial duct, which is related to the different widths of the inner (8 mm) and outer (10 mm) axial ducts. For each Pass (e.g. Pass 1), the velocity of the 4 radial ducts near the upper part of the Pass is significantly greater than that of the other radial ducts.

In order to observe the oil flow in the oil passage, the velocity vectors of the oil flow on both sides of the oil washer between Pass 1 and Pass 2 are given in Figs. 16.8 and 16.9, which are the oil flow velocity vectors near the outer and inner axial duct, respectively. It can be seen from Figs. 16.8 and 16.9 that when the oil flow along the outer axial duct is blocked by the oil washer, the second radial duct near the lower part of the oil washer has the greatest velocity. When the oil flows into the upper Passes through the oil washer of the inner axial duct, it flows mainly in the inner axial duct and produces a backflow at the end of the radial duct. However, the flow velocity of the oil entering the radial duct is very small, even forming a dead oil zone.

The overall disc temperature distribution and the local distributions of the disc temperature in Passes 1, 2 and 3 are shown in Fig. 16.10. It is learned, from Fig. 16.10, that in each Pass, the temperature of disc decreases along the axial height of the winding, with the hottest disc located at the bottom of the Pass. The position of the hottest temperature in the radial direction is related to the oil flow direction, in that, such position is close to the outer axial duct when the oil flows from the inner axial duct to the outer axial duct and is close to the inner axial duct when the oil flows from the outer to the inner axial duct. The simulation results show that the hottest spot of the whole winding is located in the flat copper conductor No. 22 of the disc No. 7, wherein the calculated value is 96.32 °C. The average temperature simulation value of the whole winding is 82.81 °C with an error of 1.91% (the average temperature of the whole winding is 84.42 °C calculated from the thermal resistance measured by LCR tester).

Table 16.2 Temperature comparison between the measured and calculated values

Measured position	DISC 12			DISC 20			DISC 30			DISC 38		
	Measured (°C)	Calculated (°C)	Error (%)	Measured (°C)	Calculated (°C)	Error (%)	Measured (°C)	Calculated (°C)	Error (%)	Measured (°C)	Calculated (°C)	Error (%)
1	89.45	82.35	−7.94	79.72	73.85	−7.36	86.76	86.19	−0.66	74.64	70.87	−5.05
4	93.52	86.68	−7.31	83.81	80.27	−4.22	91.13	90.81	−0.35	80.84	78.48	−2.92
7	93.43	86.73	−7.17	85.92	82.24	−4.28	91.87	90.48	−1.51	83.42	80.60	−3.38
10	92.72	86.26	−6.97	87.27	83.37	−4.47	90.68	89.82	−0.95	84.81	81.90	−3.43
13	90.95	85.83	−5.63	87.80	84.37	−3.91	90.16	89.25	−1.01	85.45	82.94	−2.94
16	90.46	85.24	−5.77	88.45	85.10	−3.79	89.70	88.37	−1.48	86.09	83.72	−2.75
19	90.03	84.75	−5.86	88.90	85.82	−3.46	89.95	87.33	−2.91	86.87	84.44	−2.80
22	90.22	83.97	−6.93	89.80	86.48	−3.70	89.46	86.07	−3.79	87.57	85.16	−2.75
25	89.14	82.75	−7.17	88.88	86.94	−2.18	87.30	84.36	−3.37	86.97	85.62	−1.55
28	85.75	80.24	−6.43	84.39	85.85	1.73	84.38	80.88	−4.15	84.78	85.50	0.85
30	81.27	74.85	−7.90	83.72	82.26	−1.74	78.22	74.06	−5.32	78.65	80.15	1.91

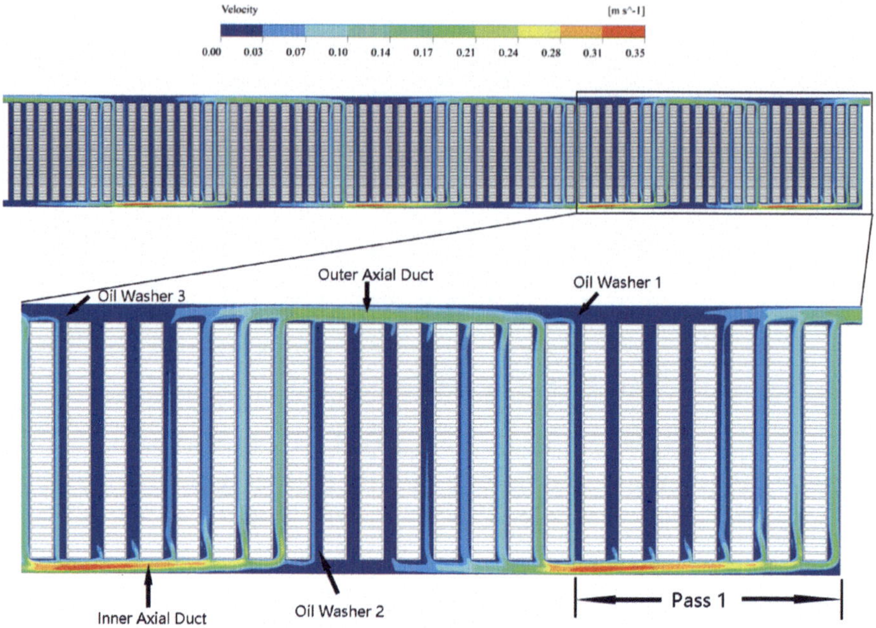

Fig. 16.7 Oil flow distribution of the winding

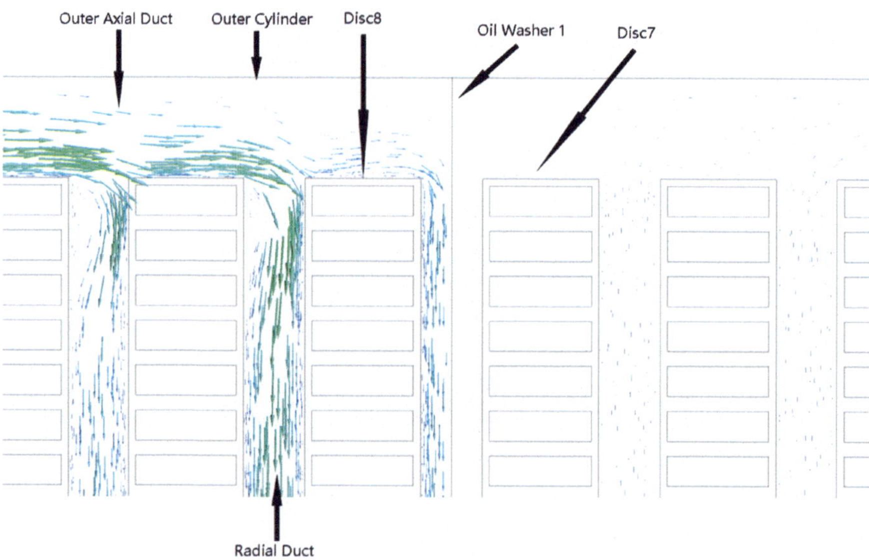

Fig. 16.8 The oil flow velocity vector on both sides of the oil washer near the outer axial duct

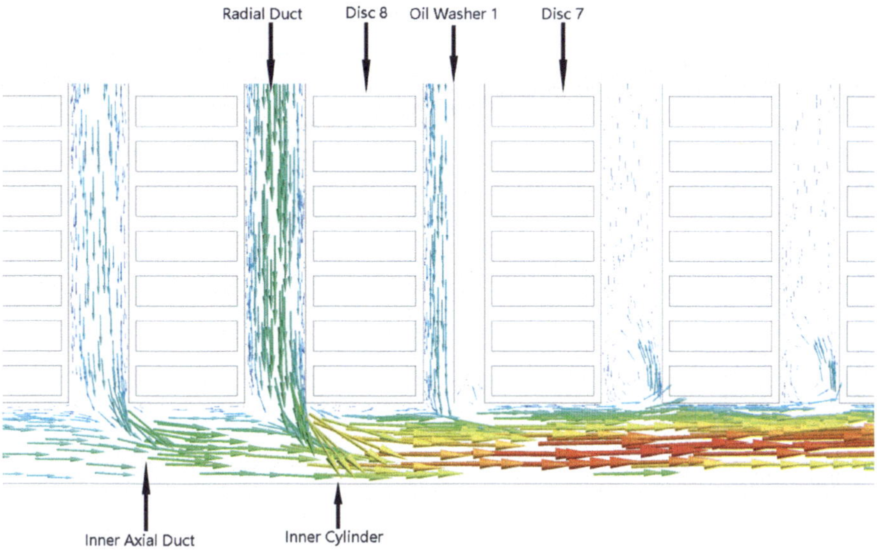

Fig. 16.9 The oil flow velocity vector on both sides of the oil washer near the inner axial duct

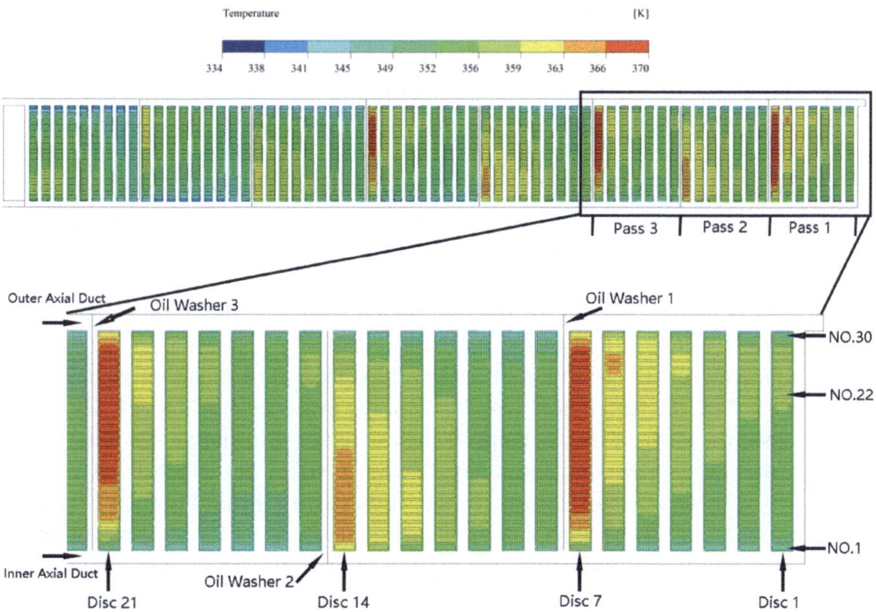

Fig. 16.10 Disc temperature distribution

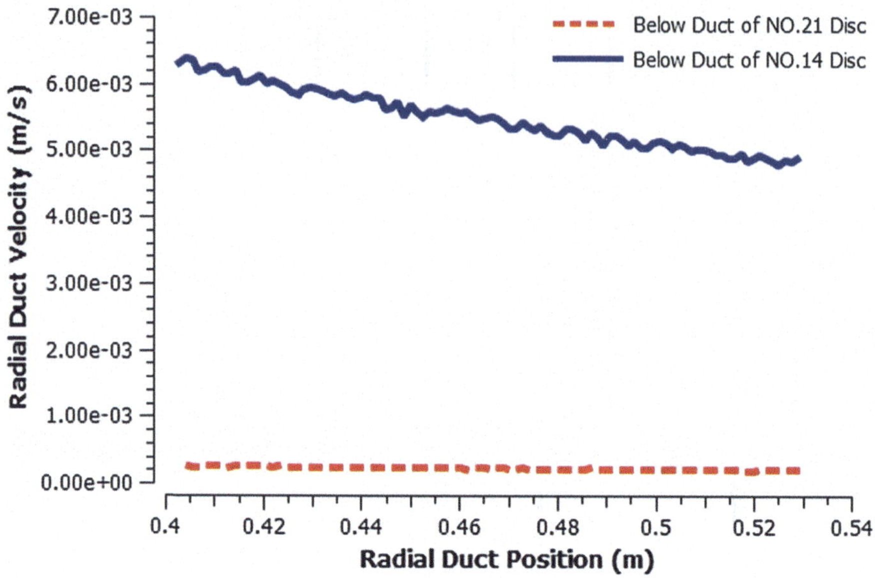

Fig. 16.11 Comparison of flow velocity of lower oil duct

It can also be seen from Fig. 16.10 that the disc temperature of disc No. 14 (the disc at the bottom of Pass 2) is lower than that of disc No. 21 (the disc at the bottom of Pass 3), which is related to the oil velocity on both sides of the disc. For this reason, the comparison of oil flow velocities in oil ducts on both sides of discs No. 14 and No. 21 is given in Figs. 16.11 and 16.12. As illustrated in Fig. 16.11, the oil velocity in the lower duct of disc No. 21 is lower than that in the lower duct of No. 14 and approaches zero, forming a dead oil zone. Also, the oil velocity in the upper duct of disc No. 21 is lower than that in the upper duct of No. 14, as illustrated in Fig. 16.12. This demonstrates that the convection dissipation effect of disc No. 14 is better than that of No. 21, which results in the temperature of the former being lower than that of the latter [16].

The temperature comparison between the measured and simulation values of the discs No. 12, No. 20, No. 30 and No. 38 is shown in Figs. 16.13, 16.14, 16.15 and 16.16, respectively. Each disc contains 11 temperature measurement points, which placed in the flat copper conductors No. 1, No. 4, No. 7, No. 10, No. 13, No. 16, No. 19, No. 22, No. 25, No. 28 and No. 30, respectively. From Figs. 16.13, 16.14, 16.15 and 16.16, it can be seen that the temperature curves of the measured points at the 4 discs are low at both ends and high in the middle. The lowest and highest temperature of discs No. 12 and No. 30 appears on the flat copper conductors No. 30 (at the outer axial duct) and No. 4 (near the inner axial duct), respectively. The lowest and highest temperature of discs No. 20 and No. 38 appears on the flat copper conductors No. 1 (at the inner axial duct) and No. 28 (near the outer axial duct), respectively. This is because the oil on both sides of the discs No. 12 and

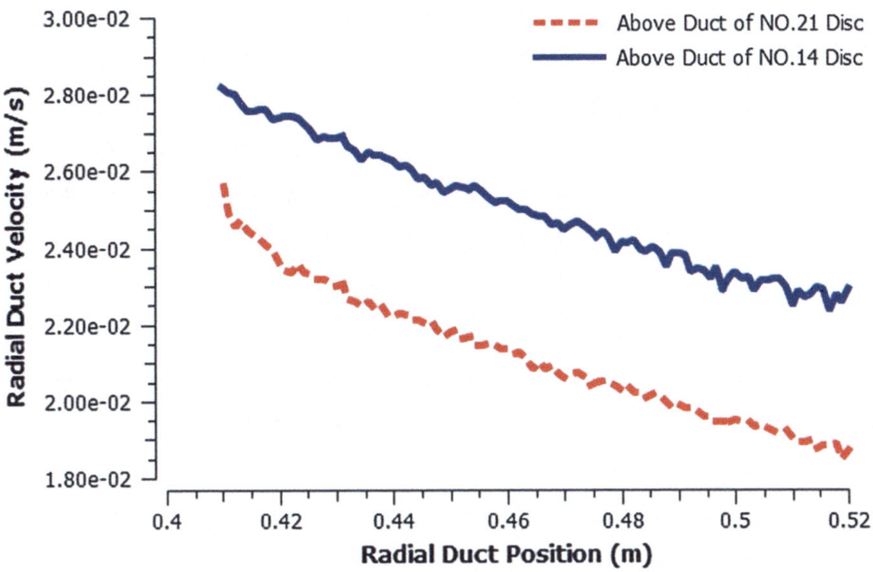

Fig. 16.12 Comparison of flow velocity of upper oil duct

Fig. 16.13 Comparison of measured and simulated temperatures of disc No. 12

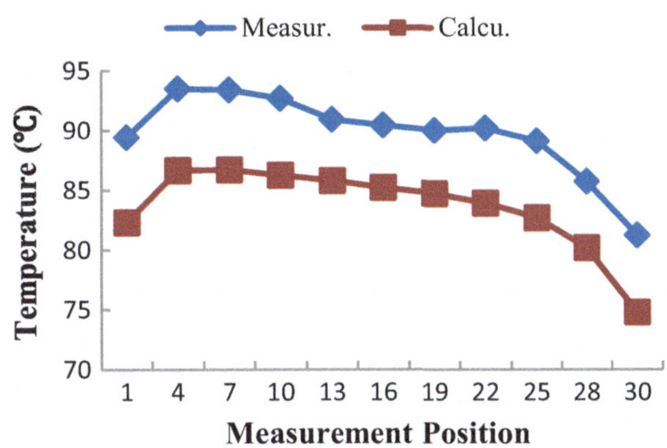

No. 30 flows from the outer axial duct to the inner axial duct, and the closer the oil flows to the inner axial duct, the greater its velocity is, while an opposite phenomenon of the oil flow in discs No. 20 and No. 38 could be observed.

The relative errors between the measured and simulated values of the discs No. 12, No. 20, No. 30 and No. 38 are shown in Fig. 16.17. It can be seen from these figures that most of the temperature simulation values are less than the measured ones. The main reasons are as follows:

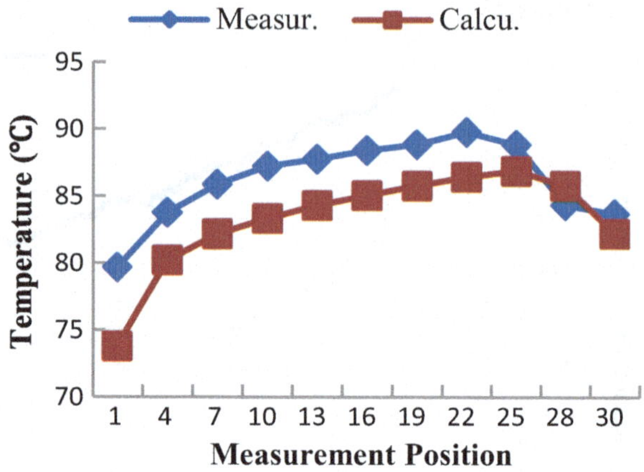

Fig. 16.14 Comparison of measured and simulated temperatures of disc No. 20

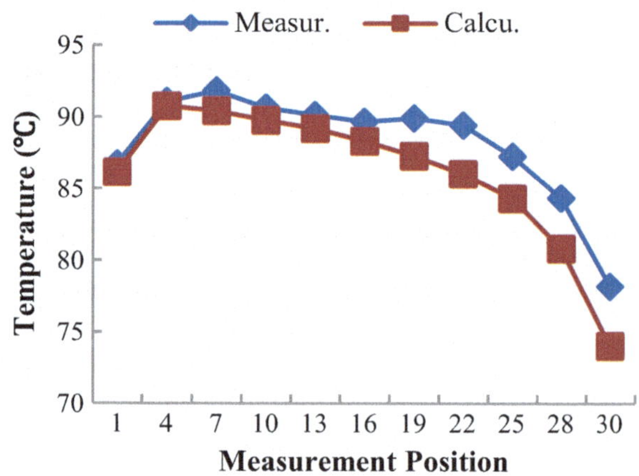

Fig. 16.15 Comparison of measured and simulated temperatures of disc No. 30

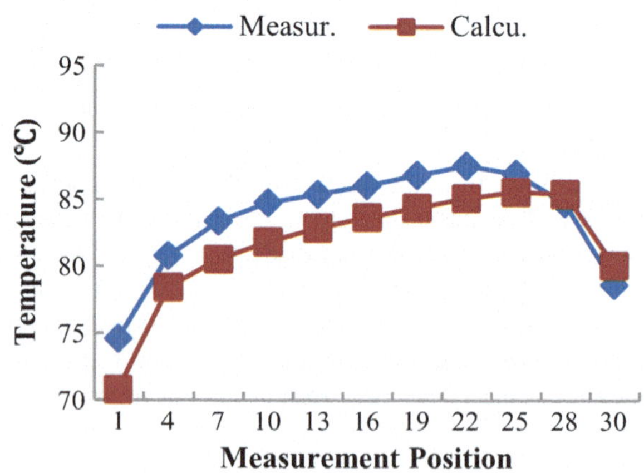

Fig. 16.16 Comparison of measured and simulated temperatures of disc No. 38

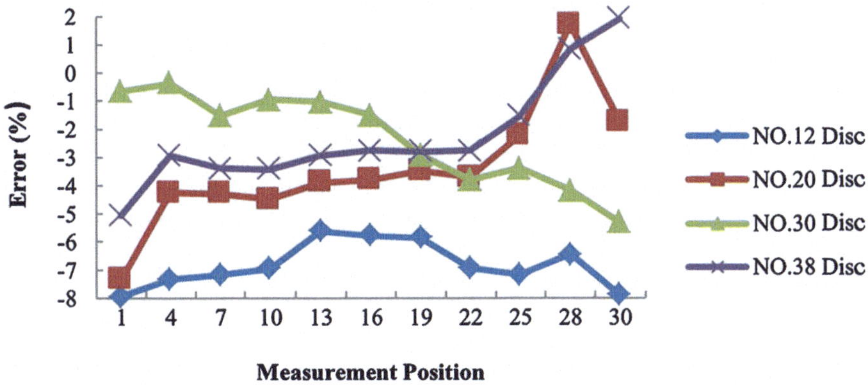

Fig. 16.17 Comparison of measured and simulated temperatures of discs No. 12, No. 20, No. 30 and No. 38

(1) the cover effect of the spacers and strips is not considered fully in the 2-D axisymmetric simulation [17];
(2) the flow velocity of transformer oil is affected when the coil conductor transits from one disc to another;
(3) the actual size of the insulating element used in the oil duct between discs and the inner/outer axial ducts of the winding after processing is negative compared with the size of 6, 8 and 10 mm used in the simulation.

By comparing the measured values with the simulated values under the OFAF cooling condition, as shown in Figs. 16.13, 16.14, 16.15, 16.16 and 16.17, it is shown that the simulation results obtained by determining the calculation parameters through repeated model tests are in better agreement with the measurement results, and the relative error of the 2-D simulation can be reduced to less than 7.94%.

Other tests and 2-D simulations under various types of cooling and heat dissipation conditions are also carried out by the team using the test model herein. The influence of related factors on simulation results are analysed one by one. The research shows that if the influence of the cover effect and negative deviation of the spacer are fully considered, the relative error of temperature can be reduced to less than 5%, and then the requirements of the engineering application can be fully met. In the subsequent study, the 3-D simulation of the winding thermal field and thermal-fluid coupled field based on the product-level model are planned to be carried out.

16.7 Summary

This chapter introduces the composition and experiment procedure of the temperature rise test platform with the product-level transformer's winding, which can be used for winding temperature rise and hotspot tests under different heat dissipation conditions. The simulation and test results of the fluid and thermal fields of the winding model are analysed and compared in details. The results show that the model can be used to validate the effectiveness of the fluid-thermal simulation method and computational software.

Acknowledgements The authors wish to express gratitude to Prof. Yinhan Zhao for his design and guidance of model-based experiment. The authors wish to thank Zhenbin Du, director of the Institute of Power Transmission and Transformation Technology, for his review of the manuscript and for his many important suggestions. The authors would also like to thank Guisheng Han, Jie Li, Lanron Liu and Meilin Lu for their strong support for this work.

References

1. Y. Xie,Power Transformer Manual, China Machine Press, Beijing, 2014. ISBN 978-7-111-46903-2.
2. D. Susa, M. Lehtonen, and H. Nordman, "Dynamic Thermal Modeling of Distribution Transformers," IEEE Transactions on Power Delivery, 2005,20(3), 1919–1929.
3. N. El Wakil, N.C. Chereches, and J. Padet. "Numerical study of heat transfer and fluid flow in a power transformer," International Journal of Thermal Sciences, 2006,45, 615–626.
4. D. Susa, M. Lehtonen, "Dynamic Thermal Modeling of Power Transformers: Further Development—Part I," IEEE Transactions on Power Delivery, 2006,21(4), 1961–1970.
5. M. A. Taghikhani, A. Gholami. "Temperature distribution in ONAN power transformer windings with finite element method," Euro. Trans. Electr. Power, 2009, 19,718–730.
6. J. Gastelurrutia, J. C. Ramos, Gorka S. Larraona, et al. "Numerical modelling of natural convection of oil inside distribution transformers," Applied Thermal Engineering, 2011,31,493–505.
7. Miguel E. Rosillo, Carlos A. Herrera, and G. Jaramillo, "Advanced Thermal Modeling and Experimental Performance of Oil Distribution Transformers," IEEE Transactions on Power Delivery, 2012,27(4), 1710–1717.
8. J. Zhang, and X. Li, "Oil Cooling for Disk-Type Transformer Windings—Part II: Parametric Studies of Design Parameters," IEEE Transactions on Power Delivery, 2006,21(3), 1326–1332.
9. J. Zhao, "Transformer Manufacturing Process," China Electric Power Press, Beijing, 2009. ISBN 978-7-5083-8328-6.
10. Q. Hu, "Transformer Test Technology," China Electric Power Press, Beijing, 2009. ISBN 978-7-5083-9503-6.
11. F. Wang, "Computational Fluid Dynamics Analysis - Principles and Applications of CFD Software," Tsinghua University Press, Beijing, 2004. ISBN 978-7-302-09503-3.
12. S. Yang, and W. Tao, "Heat Transfer". Higher Education Press, Beijing, 2006. ISBN 7-04-018918-6.
13. J.Zhang, X.Li, "Oil Cooling for Disk-Type Transformer Windings-Part 1: Theory and Model Development," IEEE Transactions on Power Delivery, 2006,21(3),1318–1325.

14. J. Zhang, X. Li, "Michael Vance. Experiments and modeling of heat transfer in oil transformer winding with zigzag cooling ducts," Applied Thermal Engineering, 2008,28, 36–48.
15. F. Torriano, M. Chaaban, and P. Picher, "Numerical study of parameters affecting the temperature distribution in a disc-type transformer winding," Applied Thermal Engineering, 2010,30, 2034–2044.
16. A. Skillen, A. Revell, H. Iacovides, and W. Wu, "Numerical prediction of local hot-spot phenomena in transformer windings," Applied Thermal Engineering, 2012, 36, 96–105.
17. F. Torriano, P. Picher, and M. Chaaban, "Numerical investigation of 3D flow and thermal effects in a disc-type transformer winding," Applied Thermal Engineering, 2012,40, 121–131.

The manufacturer's authorised representative in the EU is Springer Nature Customer Service Centre GmbH, Europaplatz 3, 69115 Heidelberg, Germany. If you have any concerns regarding our products, please contact ProductSafety@springernature.com

Printed and bound by CPI Group (UK) Ltd, Croydon, CR0 4YY
25/09/2026
02222884-0018